D. Demus, J. Goodby, G. W. Gray,
H.-W. Spiess, V. Vill

Handbook
of Liquid Crystals

WILEY-VCH

Handbook of Liquid Crystals

D. Demus, J. Goodby,
G. W. Gray,
H.-W. Spiess, V. Vill

Vol. 1:
Fundamentals

Vol. 2 A:
**Low Molecular
Weight Liquid
Crystals I**

Vol. 2 B:
**Low Molecular
Weight Liquid
Crystals II**

Vol. 3:
**High Molecular
Weight Liquid
Crystals**

Further title of interest:

J. L. Serrano:
Metallomesogens

ISBN 3-527-29296-9

D. Demus, J. Goodby, G. W. Gray,
H.-W. Spiess, V. Vill

Handbook of Liquid Crystals

Vol. 2 B:
Low Molecular
Weight Liquid
Crystals II

 **WILEY-VCH**

Weinheim • New York • Chichester
Brisbane • Singapore • Toronto

Prof. Dietrich Demus
Veilchenweg 23
06118 Halle
Germany

Prof. John W. Goodby
School of Chemistry
University of Hull
Hull, HU6 7RX
U. K.

Prof. George W. Gray
Merck Ltd.
Liquid Crystals
Merck House
Poole BH15 1TD
U.K.

Prof. Hans-Wolfgang Spiess
Max-Planck-Institut für
Polymerforschung
Ackermannweg 10
55128 Mainz
Germany

Dr. Volkmar Vill
Institut für Organische Chemie
Universität Hamburg
Martin-Luther-King-Platz 6
20146 Hamburg
Germany

This book was carefully produced. Nevertheless, authors, editors and publisher do not warrant the information contained therein to be free of errors. Readers are advised to keep in mind that statements, data, illustrations, procedural details or other items may inadvertently be inaccurate.

Library of Congress Card No. applied for.

A catalogue record for this book is available from the British Library.

Deutsche Bibliothek Cataloguing-in-Publication Data:

Handbook of liquid crystals / D. Demus ... – Weinheim ; New York
; Chichester ; Brisbane ; Singapore ; Toronto : Wiley-VCH
  ISBN 3-527-29502-X
  Vol. 2B. Low molecular weight liquid crystals. – 2. – (1998)
  ISBN 3-527-29491-0

The Editors

D. Demus
studied chemistry at the Martin-Luther-University, Halle, Germany, where he was also awarded his Ph. D. In 1981 he became Professor, and in 1991 Deputy Vice-Chancellor of Halle University. From 1992–1994 he worked as a Special Technical Advisor for the Chisso Petrochemical Corporation in Japan. Throughout the period 1984–1991 he was a member of the International Planning and Steering Commitee of the International Liquid Crystal Conferences, and was a non-executive director of the International Liquid Crystal Society. Since 1994 he is active as an Scientific Consultant in Halle. He has published over 310 scientific papers and 7 books and he holds 170 patients.

J. W. Goodby
studied for his Ph. D. in chemistry under the guidance of G. W. Gray at the University of Hull, UK. After his post-doctoral research he became supervisor of the Liquid Crystal Device Materials Research Group at AT&T Bell Laboratories. In 1988 he returned to the UK to become the Thorn-EMI/STC Reader in Industrial Chemistry and in 1990 he was appointed Professor of Organic Chemistry and Head of the Liquid Crystal Group at the University of Hull. In 1996 he was the first winner of the G. W. Gray Medal of the British Liquid Crystal Society.

G. W. Gray
studied chemistry at the University of Glasgow, UK, and received his Ph. D. from the University of London before moving to the University of Hull. His contributions have been recognised by many awards and distinctions, including the Leverhulme Gold Medal of the Royal Society (1987), Commander of the Most Excellent Order of the British Empire (1991), and Gold Medallist and Kyoto Prize Laureate in Advanced Technology (1995). His work on structure/property relationships has had far reaching influences on the understanding of liquid crystals and on their commercial applications in the field of electro-optical displays. In 1990 he became Research Coordinator for Merck (UK) Ltd, the company which, as BDH Ltd, did so much to commercialise and market the electro-optic materials which he invented at Hull University. He is now active as a Consultant, as Editor of the journal "Liquid Crystals" and as author/editor for a number of texts on Liquid Crystals.

H. W. Spiess
studied chemistry at the University of Frankfurt/Main, Germany, and obtained his Ph. D. in physical chemistry for work on transition metal complexes in the group of H. Hartmann. After professorships at the University of Mainz, Münster and Bayreuth he was appointed Director of the newly founded Max-Planck-Institute for Polymer Research in Mainz in 1984. His main research focuses on the structure and dynamics of synthetic polymers and liquid crystalline polymers by advanced NMR and other spectroscopic techniques.

V. Vill
studied chemistry and physics at the University of Münster, Germany, and acquired his Ph. D. in carbohydrate chemistry in the gorup of J. Thiem in 1990. He is currently appointed at the University of Hamburg, where he focuses his research on the synthesis of chiral liquid crystals from carbohydrates and the phase behavior of glycolipids. He is the founder of the LiqCryst database and author of the Landolt-Börnstein series *Liquid Crystals.*

List of Contributors

Volume 2B, Low Molecular Weight Crystals II

Boden, N.; Movaghar, B. (**IX**)
Centre for Self-Organising Molecular
Systems (SOMS)
Dept. of Chemistry
University of Leeds
Leeds LS9 9JT
U.K.

Bushby, R. J. (**VII**)
University of Leeds
Leeds LS2 9JT
U.K.

Cammidge, A. N. (**VII**)
University of East Anglia
Norwich NR4 7TJ
U.K.

Chandrasekhar, S. (**VIII**)
Centre for Liquid Crystal Research
P. O. Box 1329
Jalahalli
Bangalore – 560 013
India

Diele, S.; Göring, P. (**XIII**)
Martin-Luther-Universität
Halle-Wittenberg
FB Chemie
Inst. f. Physikal. Chemie
06099 Halle
Germany

Fukuda, A.; Miyachi, K. (**VI**:3)
Shinshu University
Faculty of Textile Science and Technology
Dept. of Kansei Engineering
Tokida 3-15-1, Ueda-shi
Nagano-ken 386
Japan

Giroud, A.-M. (**XIV**)
Chimie de Coordination
Unité de Recherche Associée au CNRS N.
1194
CEA Grenoble-DRFMC/SCIB
17, rue des Martyrs
38054 Grenoble Cedex
France

Imrie, C. T. (**X**)
University of Aberdeen
Dept. of Chemistry
Meston Walk,
Old Aberdeen AB24 3UE
U.K.

Kato, T. (**XVII**)
University of Tokyo
Institute of Industrial Science
7-22-1 Roppongi, Minato-ku
Tokyo 106
Japan

Kelly, S. M. (**VI**:1)
The University of Hull
Liquid Crystals & Advanced Organic
Materials Research Group
Dept. of Chemistry
Hull HU6 7RX
U.K .

Lagerwall, S. T. (**VI**:2)
Chalmers University of Technology
Physics Department
Liquid Crystal Group
41296 Göteborg
Sweden

Luckhurst, G. R. (**X**)
University of Southampton
Dept. of Chemistry
Highfield
Southampton SO17 1BJ
U.K.

Lydon, J. E. (**XVIII**)
University of Leeds
Dept. of Biochemistry
& Molecular Biology
Leeds LS2 9JT
U.K.

Malthête, J. (**XII**)
Institut Curie
Section de Recherche
UMR – CNRS 168
11, Rue Pierre et Marie Curie
75231 Paris Cedex 05
France

Nguyen, H. T.; Destrade, C. (**XII**)
Centre de Recherche Paul Pascal
Avenue A. Schweitzer
33600 Pessac
France

Praefcke, K.; Singer, D. (**XVI**)
TU Berlin
IOC
Straße des 17. Juni 124
10623 Berlin
Germany

Sadashiva, B. K. (**XV**)
Raman Research Institute
CV Raman Avenue
Bangalore – 560 080
India

Weissflog, W. (**XI**)
Max-Planck-Gesellschaft
Arbeitsgruppe Flüssigkristalline Systeme
Martin-Luther-Universität
Mählpforte 1
06108 Halle
Germany

Outline

Volume 1

Volume 2 A

Volume 2B

Volume 3

Contents

Volume 2 A

Volume 2 B

**Chapter VIII: Discotic Liquid Crystals: Their Structures
and Physical Properties** 749
S. Chandrasekhar

Chapter XI: Laterally Substituted and Swallow-Tailed Liquid Crystals 835
Wolfgang Weissflog

Chapter XII: Plasmids and Polycatenar Mesogens 865
Huu-Tinh Nguyen, Christian Destrade, and Jacques Malthête

Chapter XIII: Thermotropic Cubic Phases 887
Siegmar Diele and Petra Göring

Chapter XIV: Metal-containing Liquid Crystals 901
Anne Marie Giroud-Godquin

Chapter XV: Biaxial Nematic Liquid Crystals 933
B. K. Sadashiva

Chapter XVI: Charge-Transfer Systems 945
K. Praefcke and D. Singer

General Introduction

Liquid crystals are now well established in basic research as well as in development for applications and commercial use. Because they represent a state intermediate between ordinary liquids and three-dimensional solids, the investigation of their physical properties is very complex and makes use of many different tools and techniques. Liquid crystals play an important role in materials science, they are model materials for the organic chemist in order to investigate the connection between chemical structure and physical properties, and they provide insight into certain phenomena of biological systems. Since their main application is in displays, some knowledge of the particulars of display technology is necessary for a complete understanding of the matter.

In 1980 VCH published the *Handbook of Liquid Crystals*, written by H. Kelker and R. Hatz, with a contribution by C. Schumann, which had a total of about 900 pages. Even in 1980 it was no easy task for this small number of authors to put together the *Handbook*, which comprised so many specialities; the *Handbook* took about 12 years to complete. In the meantime the amount of information about liquid crystals has grown nearly exponentially. This is reflected in the number of known liquid-crystalline compounds: in 1974 about 5000 (D. Demus, H. Demus, H. Zaschke, *Flüssige Kristalle in Tabellen*) and in 1997 about 70000 (V. Vill, electronic data base LIQCRYST). According to a recent estimate by V. Vill, the cur-

rent number of publications is about 65000 papers and patents. This development shows that, for a single author or a small group of authors, it may be impossible to produce a representative review of all the topics that are relevant to liquid crystals – on the one hand because of the necessarily high degree of specialization, and on the other because of the factor of time.

Owing to the regrettable early decease of H. Kelker and the poor health of R. Hatz, neither of the former main authors was able to continue their work and to participate in a new edition of the *Handbook*. Therefore, it was decided to appoint five new editors to be responsible for the structure of the book and for the selection of specialized authors for the individual chapters. We are now happy to be able to present the result of the work of more than 80 experienced authors from the international scientific community.

The idea behind the structure of the *Handbook* is to provide in Volume 1 a basic overview of the fundamentals of the science and applications of the entire field of liquid crystals. This volume should be suitable as an introduction to liquid crystals for the non-specialist, as well as a source of current knowledge about the state-of-the-art for the specialist. It contains chapters about the historical development, theory, synthesis and chemical structure, physical properties, characterization methods, and applications of all kinds of liquid crystals. Two subsequent volumes provide more specialized information.

The two volumes on *Low Molecular Weight Liquid Crystals* are divided into parts dealing with calamitic liquid crystals (containing chapters about phase structures, nematics, cholesterics, and smectics), discotic liquid crystals, and non-conventional liquid crystals.

The last volume is devoted to polymeric liquid crystals (with chapters about main-chain and side-group thermotropic liquid crystal polymers), amphiphilic liquid crystals, and natural polymers with liquid-crystalline properties.

The various chapters of the *Handbook* have been written by single authors, sometimes with one or more coauthors. This provides the advantage that most of the chapters can be read alone, without necessarily having read the preceding chapters. On the other hand, despite great efforts on the part of the editors, the chapters are different in style, and some overlap of several chapters could not be avoided. This sometimes results in the discussion of the same topic from quite different viewpoints by authors who use quite different methods in their research.

The editors express their gratitude to the authors for their efforts to produce, in a relatively short time, overviews of the topics, limited in the number of pages, but representative in the selection of the material and up to date in the cited references.

The editors are indebted to the editorial and production staff of WILEY-VCH for their constantly good and fruitful cooperation, beginning with the idea of producing a completely new edition of the *Handbook of Liquid Crystals* continuing with support for the editors in collecting the manuscripts of so many authors, and finally in transforming a large number of individual chapters into well-presented volumes.

In particular we thank Dr. P. Gregory, Dr. U. Anton, and Dr. J. Ritterbusch of the Materials Science Editorial Department of WILEY-VCH for their advice and support in overcoming all difficulties arising in the partnership between the authors, the editors, and the publishers.

The Editors

Part 2:
Discotic Liquid Crystals

Chapter VI
Chiral Smectic Liquid Crystals

1   Synthesis of Chiral Smectic Liquid Crystals

Stephen M. Kelly

1.1   Introduction

Although it was known from seminal investigations early in this century [1] that smectic phases can be characterized by a one dimensional density wave, the exact layer structure was not known. Most smectic phases were characterized as smectic A, B or C (SmA, SmB and SmC) until the 1960s, depending on their optical characteristics under polarized light [2–4]. At this time it was known that in the SmA phase the time-averaged director orientation is perpendicular to the layer plane. Therefore, uniformly aligned domains appear optically uniaxial. The structure of the SmB phase was unknown and this term was used to describe a range of orthogonal and tilted smectic phases [2–4], whose number and nature were determined much later [5, 6] The structure within the layers of the SmC phase, where the molecules are tilted at an azimuthal angle, θ, to the layer normal [7–9] giving rise to the observed optical biaxiality [8], had

just been determined [7–9]. Liquid Crystals (LCs) incorporating an optically active center and exhibiting chiral smectic phases including the chiral smectic C phase (SmC*) had been synthesized along with nonoptically active LCs from the very beginning of LC research [10–13]. They were the result of investigations of the relationship between molecular structure and mesomorphic behavior and transition temperatures [10–13]. The first SmC* compounds synthesized intentionally and recognized as such [14] were prepared in order to study its structure. These were optically active versions [14] of laterally substituted Schiff's bases (see Structure 1) [7]. The lateral substituent had been introduced in order to reduce the high melting point (T_m) of the original Schiff's bases [15, 16], which probably also possess an SmC phase. Theoretical studies had indicated that the molecular orientation (i.e. the director) in an SmC* phase should spiral around an axis normal to the layer plane giving rise to a helical structure [17]. If the pitch, p, the distance for a rotation of the di-

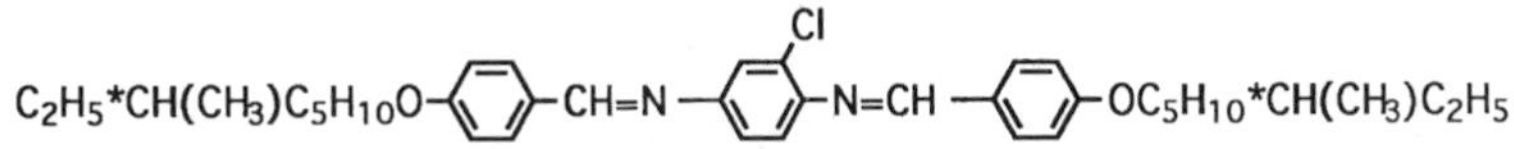

$$C_2H_5{}^*CH(CH_3)C_5H_{10}O-\!\!\bigcirc\!\!-CH=N-\!\!\bigcirc\!\!-N=CH-\!\!\bigcirc\!\!-OC_5H_{10}{}^*CH(CH_3)C_2H_5$$

Cr-SmC* 29°C; SmC*-N* 94°C; N*-I 146.5°C [14]

Structure 1

Table 1. Comparison of the transition temperatures and spontaneous polarization measured at TSmC* − 10 °C of (S)-4-decyloxybenzylidene-, (S)-α-methyl 4-decyloxybenzylidene-, (S)-2-methylbutyl α-chloro 4-decyloxy-benzylidene-, (S)-2-methylbutyl α-cyano 4-decyloxybenzylidene- and (S)-2-methylbutyl 4-decyloxy-2-hydroxy-benzylidene-4-amino-cinnamate [21, 35] and (S)-1-methylpropyl 4-decyloxybenzylidene-4-amino-cinnamate [37, 38].

$$C_{10}H_{21}O \text{—} \bigcirc \text{—} \underset{X}{} \text{—} N \text{—} \bigcirc \text{—} \underset{Y \quad O}{} O(CH_2)_n \text{*}CH(CH_3)C_2H_5$$

n	X	Y	Cr–SmC* (°C)	H–SmC* (°C)	SmC*–SmA* (°C)	SmA*–I (°C)	P_s (nC cm^{-2})	
1	H	H	76	(63)	92	117	3	DOBAMBC
1	H	CH$_3$	45	–	68	99	–	
1	H	Cl	51	–	(45)	71	–	
1	H	CN	92	–	(70)	104	–	
1	OH	— H	88	–	114	145	≈3	HDOBAMBC
0	H	H	82	–	91	106	10	DOBA-1-MPC

Values in parentheses represent monotropic transition temperatures.

rector through 360°, is of the same magnitude as the wavelength of visible light, this can lead to the selective reflection of circularly polarized light in the visible part of the electromagnetic spectrum. θ was found to be temperature dependent and normally increases with decreasing temperature [9]. The other tilted smectic (SmI* and SmF*) or crystal (J*, G*, H* and K*) phases or orthogonal smectic (hexatic SmB) or crystal (B and E) phases exhibit varying degrees of higher order within and between the layers [5, 6]. The ordered chiral smectic phases can exhibit the same form chirality as the SmC* phase (see Sec. 2 and 3 of this chapter) [18].

The local C_2 symmetry in tilted chiral smectic layers leads to a polar inequivalence within the layers due to the chiral and polar nature of the individual molecules despite a considerable degree of molecular reorientation about the long axis (see Sec. 2 and 3 of this chapter) [19–22]. The consequence of this time-dependent alignment of the dipoles along the C_2 axis is a spontaneous polarization (P_s) parallel to the layer planes, which is averaged out macroscopically to

zero by the helical superstructure. However, a permanent polarization arises in an homogeneous monodomain, if the helix is unwound by an electric field or suppressed mechanically. Based on these symmetry arguments the ferroelectric properties of chiral smectic phases with a helical structure such as the SmC* phase were predicted and then found [21, 22] for the Schiff base 4-decyl-oxybenzylidene-4-amino-2-methylbutyl-cinnamate (DOBAMBC), see Table 1. The first eight members of this series had already been prepared [12]. This was before the structure and optical characteristics of the SmC* phase had been established.

Soon after the initial discovery of ferroelectricity in chiral smectic LCs it was predicted that, if the helix of an SmC* phase were suppressed by surface forces in very thin layers between two glass electrodes, then this would pin the molecules in their positions and allow switching between two energetically equivalent polarization directions, thereby giving rise to an electro-optic memory effect [22]. This is the basis of the electro-optic display device called the surface stabilized ferroelectric liquid crystal device (SSFLCD) [23] utilizing ferro-electric LCs (see Sec. 2 and 3 of this chapter). In this device the helix is suppressed by surface forces, if the cell gap is considerably lower than the pitch of the SmC* helix. Therefore, long-pitch SmC* materials are used. Compounds exhibiting the SmC* phase are chosen due to the low rotational viscosity (γ) of this phase compared to that of the more ordered tilted smectic phases. Chiral LCs with an SmA* phase are also receiving increased attention for potential applications utilizing the electroclinic effect, where the molecular tilt is coupled directly to the applied field [24].

1.2   Long Pitch Chiral Smectic Liquid Crystals or Dopants

Although several attempts were made soon after the initial discovery of ferroelectricity in DOBAMBC to reduce T_m and increase P_s in order to facilitate the study of this new phenomenon at temperatures closer to room temperature, the vast majority of syntheses of chiral smectic LCs were undertaken later in order to provide stable SmC* materials with a defined spectrum of physical properties for SmC* mixtures designed for electro-optic display devices. Although the first eutectic mixtures created for prototype SSFLCDs consisted of purely optically active SmC* components [25], it was soon found that γ and response time (τ) for mixtures consisting of purely nonracemic SmC* components are generally too large for practical applications. It was also necessary to mix SmC* compounds with the same sign of P_s, but opposing p, so that the P_s values are additive and the p of the resultant mixture would be large. Both of these parameters depend on the polar character of the substituents at the chiral center and the position of the chiral center in the terminal chain (see Sec. 2 and 3 of this chapter) [5, 6, 18]. Current commercially available SmC* mixtures designed for SSFLCDs consist of a non-optically active base SmC mixture with a low value for γ, a high $TSmC*$ and low T_m doped with at least one optically active (chiral) dopant [26–29].

The chiral dopant should induce the desired value of P_s and a long p in the base SmC mixture, in order to avoid the necessity of pitch compensation, without depressing $TSmC*$ or increasing γ, or the birefringence (Δn) of the mixture excessively. The dopant can also affect θ, which should be approximately 22° in order to produce a good optical contrast. All the mixture components must be chemically, thermally and photo- and electrochemically stable. The resultant induced P_s of the mixture must also not be too large in order to avoid charge effects, but a large intrinsic value of P_s for the chiral dopant allows smaller amounts to be used. Since many chiral dopants designed for LCD applications do not themselves possess a SmC* phase, parameters such as the P_s or τ often have to be determined in a standard SmC mixture. This is often of more relevance to display device applications than the value for the pure dopant, if it does possess an SmC* phase, as the P_s is matrix, concentration and temperature dependent. The P_s in a mixture often reaches a maximum at a certain concentration and then remains constant or even decreases as more dopant is added [26–29]. In contrast to the expectation that a higher P_s necessarily means a shorter τ an increase of P_s which is due to an increase in θ, leads to longer τ. If both τ and P_s increase (or decrease) upon changing the side chain within an homologous series of dopants, then this is probably a change of θ of the mixture induced by the dopant. If τ decreases and P_s increases, this is probably due to a change of P_o (where $P_s = P_o \sin\theta$), which is a constant characteristic of the dopant, especially for low dopant concentrations, where changes of viscosity are small [30].

Both microscopic and steric models of the SmC phase postulate a zigzag shape for the terminal aliphatic chains of essentially symmetric compounds for SmC formation [20, 31]. The microscopic model assumes additionally that antiparallel permanent dipoles at an angle to the molecular axis promote SmC behavior due to an induced torque [20]. It has been shown that while such dipole moments are not always absolutely essential for SmC formation, they are to be found in most SmC components and still mostly lead to higher SmC transition temperatures than those of the corresponding compounds without these dipoles [32, 33]. This is not valid for compounds incorporating isolated, nonconjugated dipoles next to aliphatic rings [34]. There are also indications that alternately a cis/trans, linearly-extended conformation of the terminal chains is preferred in the SmC phase [34].

1.2.1   Schiff's bases

To facilitate the study of ferroelectricity in chiral smectics lateral substituents were introduced into the phenyl ring next to the cinnamate group of DOBAMBC and other members of the series. This was in order to reduce the degree of rotation at the chiral center and to increase the lateral dipole moment in the core of the molecule in an attempt to increase P_s and at the same time lower the high T_m of the original Schiff's bases, see Table 1, [35]. Although the lateral substituents resulted in large decreases in T_m, TSmC* and the clearing point (T_c), except for the nitriles, where T_m increases, and the suppression of ordered smectic phases,

no significant effect on P_s could be determined. This is most probably due to the relatively large distance between the substituents Y and the chiral center in the (S)-2-methylbutyl moiety.

The introduction of a hydroxy group in a lateral position next to the central bond results not only in an improved stability, but the transition temperatures and the SmC* temperature range are also increased significantly, see Tables 1 and 2 [36–40]. This is caused by the formation of a pseudo-six-membered-ring at the core of the molecule due to hydrogen bonding between the hydroxy group and the nitrogen atom of the central linkage. This gives rise to a higher degree of planarity of the aromatic rings, without leading to a broadening of the molecular rotation volume. However, the large dipole (2.58 D) associated with the hydroxy group in a lateral position does not lead to a significant increase in P_s. This implies that a nonchiral dipole moment in the middle of the core of the molecule is of significantly less importance for the P_s value than the dipole moment at the chiral center.

Related Schiff's bases incorporating commercially available optically active secondary alcohols such as (R)-2-butanol and (R)-2-pentanol instead of the primary alcohol (S)-2-methylbutanol exhibit a broad SmC* phase at high temperatures [37, 38], see Table 1. A consequence of the closer proximity ($n = 0$) of the methyl group to the core in 4-decyloxybenzylidene-4-amino-1-methylbutylcinnamate (DOBA-1-MBC) is a higher degree of coupling between the dipole moment at the chiral center and that of the core as well as a lower degree of free rotation. This results in a relatively high value for the P_s and slightly lower transition temperatures than those of DOBAMBC ($n = 1$) [21].

To further increase the chemical and photochemical stability and reduce T_m more to-

Table 2. Comparison of the transition temperatures, tilt angle and spontaneous polarization measured at 25 °C for (S)-4-(2-methylbutyl)-2-hydroxyresorcylidene 4-octylaniline, (S)-4-(4-methylhexyl)-2-hydroxyresorcylidene 4-octylaniline and (S)-4-(6-methyloctyl)-2-hydroxyresorcylidene 4-octylaniline [39, 40].

$$C_2H_5(CH_3{}^*)\,CH(CH_2)_n\,O-\!\!\!\!\bigcirc\!\!\!\!\underset{OH}{}\!\!\!\!N-\!\!\!\!\bigcirc\!\!\!\!-C_8H_{17}$$

n	Cr–SmC* (°C)	Sm2–SmC* (°C)	SmC*–SmA* (°C)	SmA*–I (°C)	θ (degrees)	P_s (nC cm^{-2})	P_s/θ (nC cm^{-2} rad^{-1})	
1	36	–	50	57	8	3	22.0	MBRA8
3	–	≈20	77	–	≈40	2	2.9	MHxRA8
5	20	–	86	–	≈5	≈0.1	0.2	MORA8

wards room temperature to yield SmC* substances more suitable for use in SSFLCDs, new Schiff's bases without the cinnamate group were synthesized [39, 40]. The effect of chain length and proximity of the point of branching to the molecular core, which are general for SmC* LCs, is shown for these Schiff's bases in Table 2. The TSmC* is highest for the compounds with the longest chain and the chiral center furthest from the molecular core (n = 3). The steric effect of the methyl group at the point of branching decreases for positions further from the core due to the greater number of chain conformations possible. The P_s also decreases due to the smaller energy difference between the various positions of the dipole moment and the lower degree of coupling with the dipoles of the core. This is shown more clearly when differences in θ are taken into consideration. The low value for T_m and wide SmC* temperature range of MHxRA 8 allowed studies of the ferroelectric properties of these LCs to be carried out at room temperature.

1.2.2   Aromatic Esters with Alkyl Branched Alkyl Chains

To prepare stable SmC* LCs with comparable P_s values to those determined for Schiff's bases, but with greater resistance towards hydrolysis, higher resistivity and better alignment properties a wide variety of aromatic esters with alkyl-branched and, above all, methyl-branched terminal chains were synthesized, see Table 3 [41–51]. Several ester series with a nonracemic chiral center had already been prepared for thermochromic applications [44]. Several homologues also exhibited SmC* as well as N* phases at high temperatures. The optically active part of the molecules was usually derived form the same alcohols such as (S)-2-methylbutanol, (R)-2-butanol and (R)-2-pentanol as for the Schiff's bases, see Scheme 1. Similar dependencies of the transition temperatures, P_s and γ on the nature of the methyl-branched chain were found. Additionally it was found that the value of P_s increased with the number of carbon atoms in the chain after the point of branching, probably due to the dampening effect of the longer chain [6, 41–44]. Large substituents like an ethyl group at the point of branching reduce the transition temperatures substantially more than a methyl group. The absolute transition temperatures and the presence of other smectic phases or the nematic phase depends on the direction of the ester linkage in the molecular core, the presence and nature of polar groups between the core and the terminal chain and

Table 3. General structures for typical smectic C* liquid crystals with nonracemic, methyl-branched terminal chains derived from primary and secondary alcohols.

Structure	Ref.
$\left\{ \begin{array}{l} C_nH_{2n+1}\,O \\ C_nH_{2n+1}\,CO_2 \end{array} \right\}$—⬡—⬡—$COO\text{-}(CH_2)_p\text{-}\overset{CH_3}{\underset{*}{CH}}\text{-}C_mH_{2m+1}$	[25] [41–43]
$C_nH_{2n+1}\,O$—⬡—$\left\{ \begin{array}{l} CO_2 \\ O_2C \end{array} \right\}$—⬡—$\left\{ \begin{array}{l} CO_2 \\ O \end{array} \right\}$—$(CH_2)_p\text{-}\overset{CH_3}{\underset{*}{CH}}\text{-}C_mH_{2m+1}$	[25] [41–43]
$C_nH_{2n+1}\,CO_2$—⬡—CO_2—⬡—$COO\text{-}(CH_2)_p\text{-}\overset{CH_3}{\underset{*}{CH}}\text{-}C_m\,H_{2m+1}$	[25, 41]
$\left\{ \begin{array}{l} C_nH_{2n+1}\,O \\ C_nH_{2n+1} \end{array} \right\}$—⬡—⬡—$\left\{ \begin{array}{l} CO_2 \\ O_2C \end{array} \right\}$—⬡—$O(CH_2)_p\text{-}\overset{CH_3}{\underset{*}{CH}}\text{-}C_mH_{2m+1}$	[25, 41] [44]
C_nH_{2n+1}—⬡—$\cdots C_2H_4$—⬡—$\left\{ \begin{array}{l} OCH_2 \\ O_2C \end{array} \right\}$—⬡—$O(CH_2)_p\text{-}\overset{CH_3}{\underset{*}{CH}}\text{-}C_mH_{2m+1}$	[45]
$\left\{ \begin{array}{l} C_nH_{2n+1}\,O \\ C_nH_{2n+1} \end{array} \right\}$—⬡(N)—⬡—$\left\{ \begin{array}{l} CO_2 \\ O \end{array} \right\}$—$(CH_2)_p\text{-}\overset{CH_3}{\underset{*}{CH}}\text{-}C_m\,H_{2m+1}$	[52–54]
$\left\{ \begin{array}{l} C_nH_{2n+1}\,O \\ C_nH_{2n+1} \end{array} \right\}$—⬡—⬡(N)—⬡—$\left\{ \begin{array}{l} CO_2 \\ O \end{array} \right\}$—$(CH_2)_p\text{-}\overset{CH_3}{\underset{*}{CH}}\text{-}C_mH_{2m+1}$	[52–54]
$\left\{ \begin{array}{l} C_nH_{2n+1}\,O \\ C_nH_{2n+1} \end{array} \right\}$—⬡—⬡—$CO_2$—⬡—$\overset{O}{C}(CH_2)_p\text{-}\overset{CH_3}{\underset{*}{CH}}\text{-}C_mH_{2m+1}$	[50]

the number of rings in the core [6, 41]. Aliphatic rings in the core such as 1, 4-disubstituted cyclohexane [45], bicyclo(2, 2, 2)octane [46], dioxane rings [47] or dioxaborinane rings [48] lead to the elimination of the SmC* phase in two-ring compounds and to lower $TSmC*$ in three-ring compounds compared to the corresponding fully aromatic analogs. However, γ and/or Δn are lower [45]. This results in lower τ or allows thicker cells to be used. The introduction of a *cis* carbon–carbon double bond in the middle or in a terminal position of the alkoxy chains of two-ring and three-ring esters leads to lower transition temperatures in general and sometimes to broader SmC* phases close to room temperature [49].

The additive effects of dipole moments near the chiral center and at the chiral center itself are demonstrated by a reference to Tables 4 and 5. The introduction of a large dipole moment in the form of a carbonyl function between the aromatic core and the chiral center to yield a number of alkanoyl-substituted (keto-) esters led to higher P_s values than those of the corresponding ethers and esters, see Table 4 [50]. The presence of a second optically active group in the second terminal chain can also lead to additive effects and a still higher P_s, although γ can be expected to be considerable [50]. A strong dependence of the observed P_s values on the number of carbon atoms in both terminal chains and the presence and

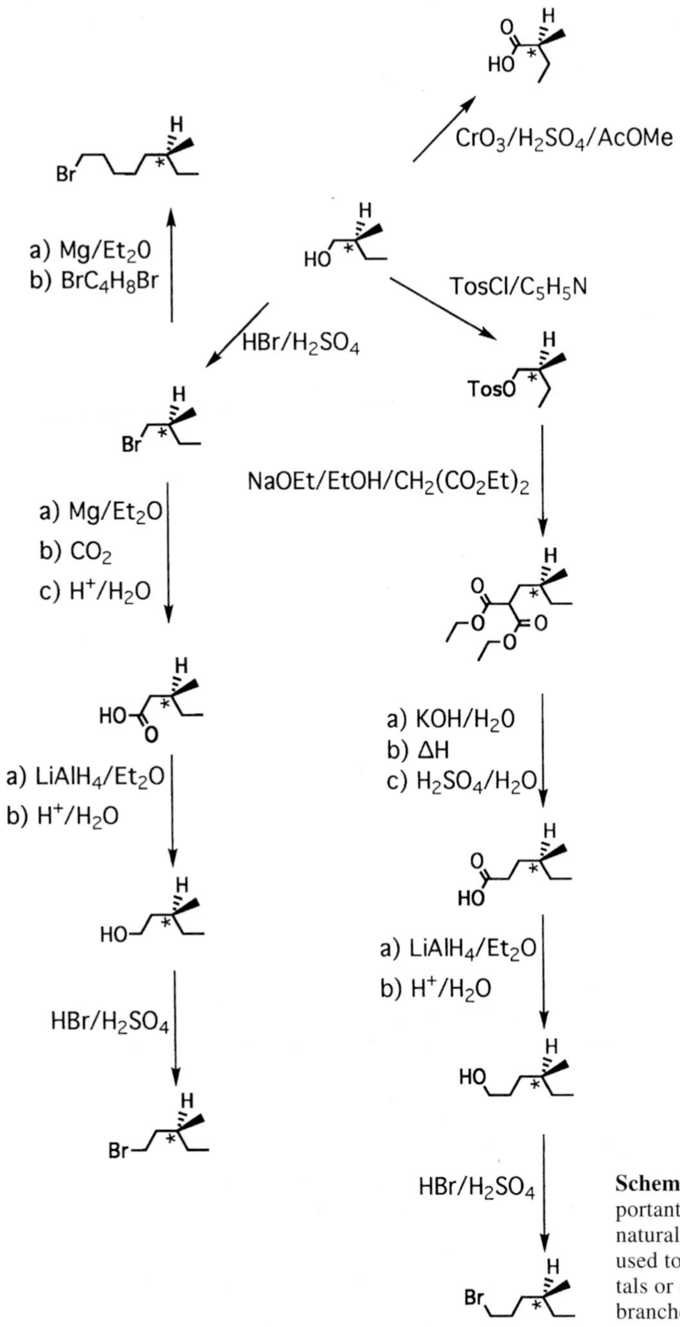

Scheme 1. Typical reaction pathways to important reaction intermediates derived from naturally occurring (S)-2-methylbutanol and used to synthesize chiral smectic liquid crystals or dopants with nonracemic, methyl-branched terminal chains.

position of various lateral substituents was also observed. Dipole moments incorporated in the form of lateral substituents in the aromatic core next to the terminal chain, but further away from the chiral center can also give rise to a higher P_s, see Table 5 [28, 51]. However, the effect is smaller for similar dipole moments and the increase is al-

Table 4. Comparison of the spontaneous polarization measured at TSmC*–10 °C of (S)-4-([2-octyl]oxy)phenyl 4-octyloxy-1,1′-biphenyl-4′-yl-carboxylate (X = O), (S)-2-octyl 4-(4-octyloxy-1,1′-biphenyl-4′-yl-carboxy)ben-zoate (X = COO) and (S)-4-(2-methyloctanoyl)phenyl 4-octyloxy-1,1′-biphenyl-4′-yl-carboxylate (X = CO) [50].

$$C_8H_{17}O - \bigcirc - \bigcirc - CO_2 - \bigcirc - X-*CH(CH_3)\,C_6H_{13}$$

X	μ (D)	P_s (nC cm^{-2})	P_0 (nC cm^{-2})
-O-	0.68	38	130
-COO-	1.23	70	195
-CO-	1.99	112	313

Table 5. Comparison of the transition temperatures, tilt angle and spontaneous polarization measured at TSmC*–10°C of 4-octyloxy-1,1′-biphenyl-4′-yl 3-substituted-4-([(S)-2-octyl]oxy)benzoates [51].

$$C_6H_{13}C*H(CH_3)\,O - \bigcirc\overset{X}{} - CO_2 - \bigcirc - \bigcirc - OC_8H_{17}$$

X	μ (D)	Cr–SmC* (°C)	SmC*–SmA*/N* (°C)	SmA*–I (°C)	N*–I (°C)	θ (degrees)	P_s (nC cm^{-2})	P_0 (nC cm^{-2})
H	0	78	99	–	122	45	45	64
F	-1.47	65	97	–	100	36	78	134
Cl	-1.59	73	95	–	97	40	123	191
Br	-1.57	43	87	–	89	41	131	202
CN	-4.05	68	90	105	–	25	160	386

most certainly also due to steric effects as the energy difference between the most stable, time-dependent conformations of the chiral center is increased. This is seen as p decreases and γ increases with the size of the lateral substituent [28, 51]. However, the small p requires pitch compensation for practical application in SSFLCDs. This contributes to the higher γ and τ values caused by the presence of the lateral substituents.

1.2.3 Aromatic Heterocycles with Alkyl-Branched Alkyl Chains

A further development was the introduction of an optically active center into the terminal chains of two- and three-ring SmC com-

pounds with a phenyl-pyrimidine, pyrazine or pyridazine group in the molecular core to produce the corresponding methyl-branched SmC* materials [52–55]. Some of these new compounds exhibited a broad SmC* phase, sometimes close to room temperature (Structure 2). They were characterized by low γ values. The corresponding alkenyloxy-substituted phenylpyrimidines with a cis-carbon–carbon double bond or a carbon–carbon double bond in a terminal position often possessed an SmC* phase closer to room temperature [56].

$$C_8H_{17}O - \bigcirc - \overset{N}{\underset{N}{\bigcirc}} - C_5H_{10}*CH(CH_3)C_2H_5$$

Cr-SmC*: 23°C; SmC*-N*: 34°C; N*-I: 47°C [52]

Structure 2

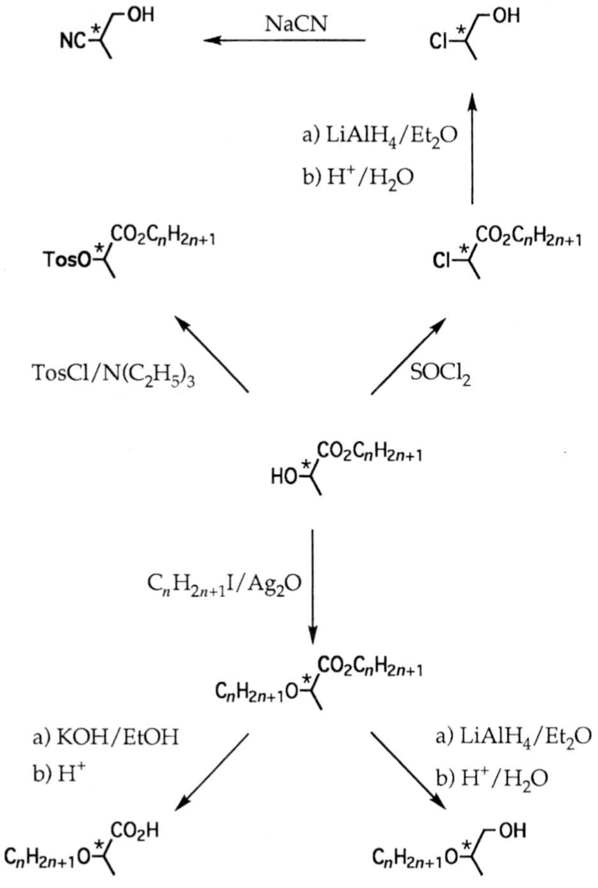

Scheme 2. Typical reaction pathways to important reaction intermediates derived from naturally occurring alkyl L-(+)-lactates (alkyl (S)-2-hydroxypropanoates) and used to synthesize chiral smectic liquid crystals or dopants with nonracemic terminal chains with polar substituents.

1.2.4   Esters and Ethers in the Terminal Chain

Large P_s values can be achieved by increasing the polarization of the branching methyl group at the chiral center by the introduction of the electron withdrawing oxygen atom. Simple esters and ethers were usually prepared from naturally occurring alkyl L-(+)-lactates (alkyl (S)-2-hydroxypropa-noates), see Scheme 2 [57–60] or readily accessible, synthetic alkyl (R)-3-hydroxybutanoates [61]. More sophisticated syntheses of two- and three-ring phenylpyrimidines with an alkoxy group and a methyl group at the chiral center in the terminal chain (Structure 3) involved resolution of a prochiral ester into the corresponding optically active alcohol and ester by an enzyme catalyst [62]. Lactate ethers [60] exhibit higher P_s

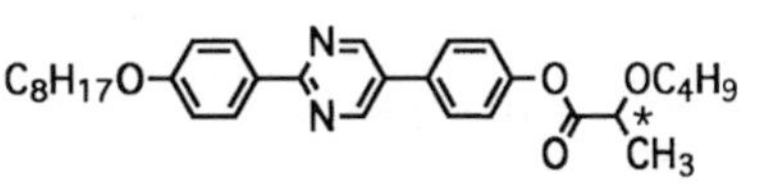

Cr-SmC*: 88°C; SmC*-I: 148°C [58]

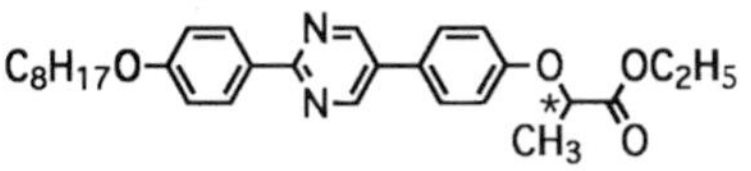

Structure 3

Cr-SmC*: 51°C; S_X-SmC*: 46°C; SmC*-N: 69°C; N*-I: 91.5°C [60]

Table 6. Comparison of the transition temperatures and spontaneous polarization of 4′-heptyloxy-1, 1′-biphenyl-4-yl (S)-2-chloro-3-methylbutanoate, (2S, 3R)-2-chloro-3-methylpentanoate and (S)-2-chloro-4-methylpentanoate [64, 65].

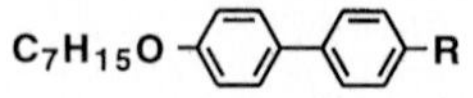

$$C_7H_{15}O - \text{〈〉} - \text{〈〉} - R$$

R	Cr–SmC*/SmA* (°C)	S_3*–SmC*/SmA* (°C)	SmC*–SmA* (°C)	SmA*–I* (°C)	P_s (nC cm^{-2})
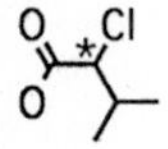 [structure: O, Cl, *]	76	–	(71)	80	−98
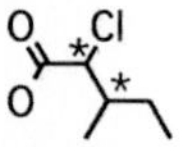 [structure: O, Cl, *, *]	48	(36)	56	66	−210
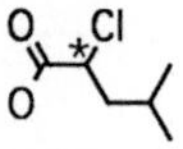 [structure: O, Cl, *, *]	71	–	(65)	74	+58

Values in parentheses represent monotropic transition temperatures.

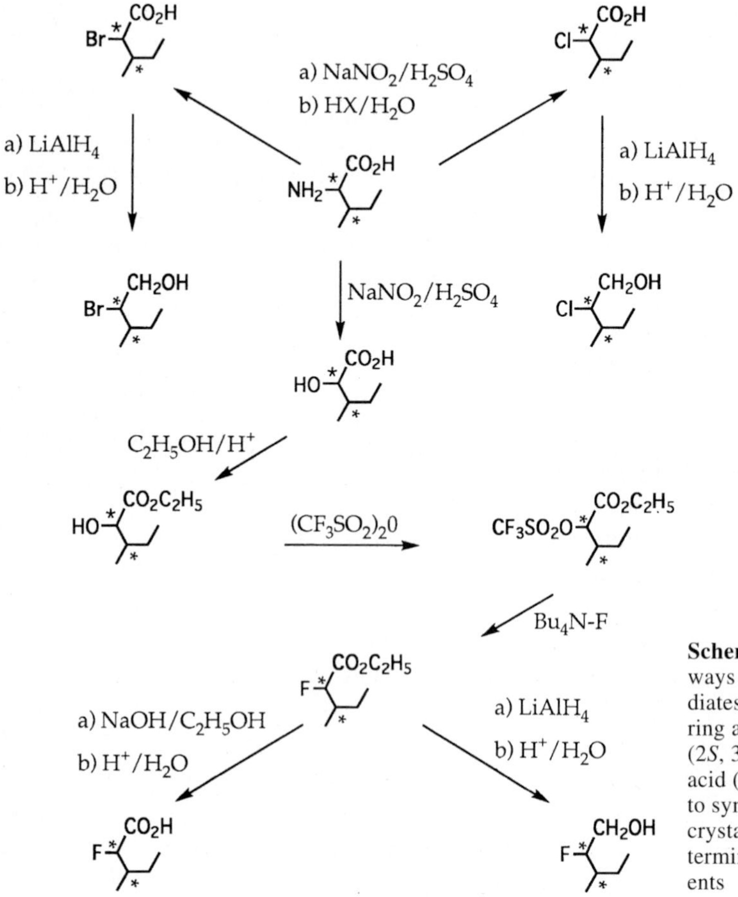

Scheme 3. Typical reaction pathways to important reaction intermediates derived from naturally occurring amino acids such as (2R, 3S) or (2S, 3R)-2-amino-3-methylpentanoic acid (D- or L-alloleucine) and used to synthesize chiral smectic liquid crystals or dopants with nonracemic terminal chains with polar substituents

Table 7. Comparison of the spontaneous polarization (extrapolated to 100%), response times (10 V_{pp}/µm square wave, time to maximum current), pitch and tilt angle measured at 25°C for mixtures of 7 wt% 4-octyloxy-1,1'-biphenyl-4'-yl (R)-2-fluorohexanoate and 7 wt% *trans*- 4-(4-octyloxyphenyl)cyclohexyl (R)-2-fluorohexanoate and a host mixture [69].

X	P_s (nC cm^{-2})	τ (µs)	p (µm)	θ (degrees)
	56	950	0.75	24.5
	35	700	1.52	22

and higher γ values than comparable lactate ester mixtures with similar θ and TSmC* values. This results in similar τ for similar mixtures containing one or the other chiral dopant. The high TSmC* temperatures for the pure lactate esters [58] are not reflected in high TSmC* values for mixtures containing them. This is typical of the nonlinear and nonideal behavior of chiral dopants in SmC* mixtures.

1.2.5   Halogens at the Chiral Center

If the dipole at the chiral center is increased by replacing the methyl branching group by a more polar moiety such as a halogen atom, then P_0 increases due to the larger dipole moment perpendicular to the long axis of the molecule, see Table 6. Most ferroelectric LCs incorporating halogen atoms at the optically active center are esters or ethers derived either from naturally occurring lactic acid or amino acids, see Schemes 2 and 3 [63–67]. There is still a considerable degree of freedom of rotation at the chiral center and the observed value for P_s is still considerably lower than that

expected if the dipole moment at the chiral center is taken into account. For a large value of the P_s it is important that the dipoles in the vicinity of the optically active center point in the same direction, see Table 6. The dipoles due to the two chiral centers in the (2S, 3R)-2-chloro-3-methyl-pentanoyloxy-substituted esters are additive and lead to the highest P_s values [63–67]. The pure esters exhibit high negative P_s values and enantiotropic SmC* phases. The values extrapolated to 100% from SmC* mixtures are often considerably lower. The small van der Waals radius (1.47 Å) of fluorine and relatively large dipole moment (1.41 D) of the C–F bond lead to SmC* dopants with a relatively high P_s, but with a lower γ and higher TSmC* than the corresponding chloro- or bromo-substituted compounds [68, 69].

1.2.6   Cyclohexyl α-Fluorohexanoates

Substitution of the aliphatic *trans*-1, 4-disubstituted cyclohexane ring for the aromatic benzene ring leads to a lower electron density at the chiral center and to a higher degree of conformational mobility of the dipole moment, but P_s is only reduced by about 40%, see Table 7 [30, 69]. However, the cyclohexane ring gives rise to lower values for Δn, γ and a substantially longer pitch, as expected.

The data collated in Table 8 allow a valid comparison of the relative effects of either a fluorine or a chlorine atom attached directly to the optically active center of the chiral dopant [69]. The transition temperatures and P_s of the mixture containing the

α-fluoroester are all higher than those observed for the corresponding mixture containing an equal amount of the otherwise identical α-chloroester. This could be due to the stronger electronegativity of the fluorine atom. Additionally, γ of the α-fluoroester must also be lower than that of the α-chloroester as shown by the significantly lower τ, which is only partially due to the higher value of P_0. This is a result of the smaller size of the fluorine atom and the shorter carbon–fluorine bond compared to that of chlorine.

Despite the presence of the cyclohexane ring, which promotes SmB behavior, several optically active α-fluoroesters incorporating the aliphatic cyclohexane ring and a number of different cores possess an enantiotropic SmC* phase at elevated temperatures [69]. However, the tendency to form the SmC* phase is much reduced and a number of dipole moments in various parts of the molecular core are required, see Table 9. This is consistent with the microscopic, dipole–dipole theory of the SmC phase [20]. Only the combination of the phenylpyrimidine core and an alkoxy chain in the ester gives rise to the SmC* phase. Two-ring compounds incorporating a cyclohexane ring do not generally exhibit an SmC phase.

Table 8. Comparison of the transition temperatures, spontaneous polarization and response times (10 $V_{pp}/\mu m$ square wave, time to maximum current) measured at $25°C$ for mixtures of 7 wt% of the *trans*-4-(2′,3′-difluoro-4′-decyloxy-1,1′-biphenyl-4-yl)cyclohexyl (*R*)-2-fluorohexanoate and *trans*-4-(2′, 3′-Difluoro-4′-decyloxy-1, 1′-biphenyl-4-yl)cyclohexyl (*R*)-2-chlorohexanoate and a host mixture (SmC–SmA: 76°C; SmA–N: 81°C; N–I: 103°C) [30].

X	P_s nC cm^{-2}	τ (μs)	SmC*–SmA (°C)	SmA–N* (°C)	N*–I (°C)
F	6.2	415	75.5	88.3	104.3
Cl	4.5	630	74.5	86.3	102.3

Table 9. Comparison of the transition temperatures for the *trans*-4-(4′-decyl-1,1′-biphenyl-4-yl)cyclohexyl (*R*)-2-fluorohexanoate, *trans*-4-(4-[5-decylpyrimidin-2-yl]phenyl)cyclohexyl (*R*)-2-fluorohexanoate and *trans*-4-[4-(5-nonyloxypyrimidin-2-yl)phenyl]cyclohexyl (*R*)-2-fluorohexanoate (SmC*–SmA: 76°C; SmA–N: 81°C; N–I: 103°C) [30].

R	X	C–SmB*/SmC*/I (°C)	SmB*–SmC*/SmA* (°C)	SmC*–SmA* (°C)	SmA*–I (°C)
$C_{10}H_{21}$	CH	145	–	–	–
$C_{10}H_{21}$	N	48	116	–	145
$C_9H_{19}O$	N	60	90	108	165

Table 10. Comparison of the spontaneous polarization, response times (10 V_{pp}/µm square wave, time to maximum current) and pitch measured at 25 °C for mixtures of 7 wt% of the ether *trans*-4-[4-decyloxyphenyl]cyclohexyl (*R*)-2-fluorohexanoate and the ester *trans*-4-[4-decanoyloxyphenyl]cyclohexyl (*R*)-2-fluorohexanoate and a host mixture (SmC–SmA: 76 °C; SmA–N: 81 °C; N–I: 103 °C) [30].

X	P_s (nC cm^{-2})	τ (µs)	SmC*–SmA* (°C)	SmA*–N* (°C)	N*–I (°C)
CH$_2$	3.8	460	64.5	82.6	97.8
CO	5.7	400	68.0	81.5	97.1

Table 11. Comparison of the transition temperatures, spontaneous polarization and response times (10 V_{pp}/µm square wave, time to maximum current) measured at 25 °C for mixtures of 7 wt% of the diester *trans*-4-[4-(2, 3-difluoro-4-decyloxybenzoyloxy)phenyl]cyclohexyl (*R*)-2-fluorohexanoate and 7 wt% of the ester *trans*-4-(2',3'-difluoro-4'-decyloxy-1,1'-biphenyl-4-yl)cyclohexyl (*R*)-2-fluorohexanoate and a host mixture (SmC–SmA: 76 °C; SmA–N: 81 °C; N–I: 103 °C) [30].

X	P_s (nC cm^{-2})	τ (µs)	SmC*–SmA* (°C)	SmA*–N* (°C)	N*–I (°C)
COO	5.4	650	76.7	83.5	103.9
–	6.2	415	75.5	88.3	104.3

The data collated in Table 10 allow the effect of an ester group in a terminal position in the core of a chiral dopant to be determined. The ether [30] and the diester [69] differ only in the presence of a carbonyl group (CO) instead of a methylene unit (CH$_2$), that is, the chain lengths are the same. TSmC* for the mixture containing the ester (X = CO) is higher than that incorporating an equal amount of the ether (X = CH$_2$). The P_s is higher for the ester than for the ether, indicating a greater value for P_0 as shown by the shorter τ.

The effect of an ester group in a central position of a chiral dopant can be elucidated from the data in Table 11 [69]. TSmC* for the mixture containing the α-fluoro diester with a second ester group in a central position is higher than that incorporating an equal amount of the α-fluoro (mono-) ester. This is not surprising as the α-fluoro di-ester exhibits an enantiotropic SmC* phase at elevated temperatures in the pure state. The P_s is higher for the ester than for the diester. This is probably due to a change of P_0 as well as of γ, because of presence of the second carboxy (ester) group. Even if P_0 were constant and θ were fully responsible for the increase of P_s, γ of the mono ester would still be substantially lower.

The P_s values for related fluoro-alkyl- or alkoxy-substituted compounds are lower due to the lower dipole moment at the chiral center [70–75]. The presence of the trifluoromethyl group at the chiral center in the side-chain of three-ring compounds leads to high P_s values and an SmC* phase for the pure substances [76]. Although a trifluoromethyl group at the chiral center in the central linkage between two aromatic rings in the core of a similar three-ring phenylpyrimidine leads to very low transition temperatures and the absence of a SmC* phase for the pure substance and to a low $TSmC*$ for mixtures containing these chiral dopants the values of P_s, τ and θ are otherwise attractive [77].

1.2.7   Cyano Groups at the Chiral Center

The large dipole moment and steric bulk of the cyano group at the chiral center gives rise to a larger energy difference between the most probable time-averaged conformations of the chiral center and therefore to a high P_s value, although low transition temperatures are also observed [78–81]. This is a consequence of the large steric effect of

Cr-SmC*: 102 °C; SmC*-SmA: 145 °C; SmA-I: 155 °C [79]

Structure 4

the cyano group close to the molecular core (Structure 4) leading to low transition temperatures for the pure compounds. Therefore, enantiotropic LC phases are only observed for three-ring compounds.

1.2.8   Optically Active Oxiranes and Thiiranes

Another way to maximize P_s is to limit the degree of free rotation at the chiral center and increase the energy difference between the two major conformations of the dipole at the chiral center. This can be achieved by incorporating the chiral center in a cyclic system such as in an oxirane [27, 82, 83]. The molecular dipole associated with an epoxide moiety is oriented normal to the tilt plane. The rigidity of the epoxide ring minimizes the conformational averaging out of the dipole moment and should lead to a high value for the P_s. Oxiranes are prepared by Sharpless epoxidation of allyl alcohols,

Table 12. Transition temperatures and spontaneous polarization of mixtures of 7.37 wt% of 4-[(R)-1-fluoro-(2S, 3S)-epoxynonyl]phenyl 4-decyloxybenzoate (X = H; Y = F) or 7.58 wt% of 4-[(S)-1-fluoro-(2S, 3S)-epoxynonyl]phenyl 4-decyloxybenzoate (X = F; Y = H) and a host compound (Cr–SmC: 35 °C; SmC–SmA: 70.5 °C; SmA–I: 72 °C) [83].

X	Y	Cr–SmC* (°C)	SmC*–SmA* (°C)	SmA*–I (°C)	P_s (nC cm^{-2})
H	F	53	66	70	+2
F	H	25	61	66	−13

Table 13. Helical twisting power (HTP) (measured just above SmA*–N*), rotational viscosity, response time and spontaneous polarization measured at 25 °C for mixtures of 10 mol% of 5-([cis-(2S, 3S)-epoxy]octanoyloxy)phenyl-2-octyloxyprimidine or 5-([trans-(2R, 3S)-epoxy]octanoyloxy)phenyl-2-octyloxyprimidine and a host mixture (C–SmC: 10 °C; SmC–SmA: 84.5 °C; SmA–N: 93.5 °C; N–I: 105 °C) [27].

Configuration	HTP μm^{-1}	τ μs	γ (mPa)	P_s (nC cm^{-2})
(2S, 3S) cis	6.2	38	240	77
(2R, 3S) trans	0.8	422	36	2.3

which can be easily attached to a phenol, or oxidized to the corresponding acid and attached to the phenol in a normal esterification reaction [27, 82].

If the dipoles associated with the fluorine atoms and the epoxy moiety in the two diastereomeric oxiranes shown in Table 12 [83] point in the same direction, then a high P_s value is observed. If they are opposed, then they cancel each other out to some extent, and a low P_s value is the result. Converting the ether linkage to an ester group

does not lead to a significant change in the observed P_s value, see Table 13 [27]. However, changing the configuration at the second carbon atom of the epoxy group (i.e. changing from trans to cis) results in an enhanced stereo coupling between the chiral moiety and the lateral molecular dipole. Thus, significantly higher P_s values are observed for the cis-substituted oxiranes [27]. The cis configuration at the epoxy moiety leads automatically to a nonlinear (i.e. fully extended) conformation of the chain. This leads to a substantially higher value for γ. However, γ is increased to a lesser extent than the P_s is increased and this results in significantly shorter τ for the cis-substituted esters.

Replacement of the oxygen atom in the epoxy moiety by a sulfur atom to produce the analogous thiiranes [84] results not only in lower transition temperatures, but also significantly lower values for P_s, if the optically active unit is situated on the periphery of the molecule. The opposite is observed with respect to P_s, if the chiral entity is situated between two mesogenic units such as six-membered rings. A Walden inversion during the reaction produces an inversion of configuration at both asymmet-

Table 14. Transition temperatures, response time (10 V$_{pp}$/µm square wave, time to maximum current), tilt angle and spontaneous polarization measured at 25°C in a 2 µm cell with polyimide alignment containing mixtures of 2 mol% of the (2S, 4R)-trans- (α-[4′-pentyl-1,1′-biphenyl-4-yl]-γ-pentyl)lactone or (2R, 4R)-cis- (α-[4′-pentyl-1,1′-biphenyl-4-yl]-γ-pentyl)lactone and a host mixture (Cr–SmC: <25 °C; SmC–SmA: 51 °C; SmA–N: 63 °C; N–I: 69 °C) [85].

Configuration	SmC*–SmA* (°C)	SmA*–N* (°C)	N*–I (°C)	τ (µs)	θ (degrees)	P_s (nC cm^{-2})
2S, 4R (trans)	52	61	68	655	17	<0.5
2R, 4R (cis)	52	61	68	100	21	+5.3

Table 15. Transition temperatures, response time (10 $V_{pp}/\mu m$ square wave, time to maximum current), tilt angle and spontaneous polarization measured at 25 °C in a 2 μm cell with polyimide alignment containing mixtures of 2 mol% of the (2R, 5R)-trans-(α-[4′-hexyloxy-1,1′-biphenyl-4-ylcarboxy]-δ-hexyl)lactone or (2S, 5R)-cis-(α-[4′-Hexyloxy-1,1′-biphenyl-4-ylcarboxy]-δ-hexyl)lactone and a host mixture (SmC–SmA: 51 °C; SmA–N: 61 °C; N–I: 68 °C) [86].

Configuration	Cr–SmC* (°C)	SmC*–SmA* (°C)	SmA*–N* (°C)	N*–I (°C)	τ (μs)	θ (degrees)	P_s (nC cm^{-2})
2R, 5R (trans)	2.5	53	60	67	195	21.0	-2.4
2S, 5R (cis)	2.6	53	59	67	126	22.3	+4.2

ric carbon atoms resulting in an inverse absolute configuration. The thiiranes are significantly less thermally stable than the corresponding oxiranes.

C-SmC*: 111°C; SmC*-I: 128°C [86]

Structure 5

1.2.9   Optically Active γ-Lactones

In the optically active, 5-membered γ-lactone ring there are two asymmetric (chiral) carbon atoms and a large dipole moment due to the carboxy function. The dipole moments associated with the carbonyl and ether components of the ester linkage are in the same plane, point in the same direction and are arranged perpendicular to the long axis of the molecule [85]. The degree of free rotation of the chiral centers and the dipole moments are greatly reduced. Therefore a large value for P_s can be expected and this is indeed observed to be the case, see Table 14. It is clear from the data in the table that the values for P_s, θ and τ, but not the transition temperatures, are strongly dependent on the configuration of the lactone ring (i.e. cis or trans). Both configurational isomers are generated as a result of the same reaction pathway [85].

1.2.10   Optically Active δ-Lactones

A similar geometry is also observed for the related 6-membered δ-lactone ring. Therefore similarly large values for P_s are to be expected and this is observed to be the case, see Table 15 [86]. Large differences in P_s are observed for the cis and trans configurations (Structure 5), although the value for the trans-isomer is greater than that of the cis-isomer, where the opposite is the case for the γ-lactones. As the concentration of dopant is so low little difference is seen in the transition temperatures of the mixtures. [86]

1.2.11   Miscellaneous Optically Active Heterocycles

Dopants incorporating a series of optically active rings such as 1,3-dioxolan-4-ones

[87], 1,3-dioxolan-2-ones [88], 2-oxeta-nones [89, 90], oxazolidines [27] and pro-lines [27] have also been prepared. These dopants generally induce lower P_s values and/or higher τ and γ and/or lower transition temperatures in hosts than do the oxiranes, γ-lactones and δ-lactones discussed above.

1.3   Short Pitch Chiral Smectic Liquid Crystals or Dopants

A major problem inherent to bistable FLCDs such as SSFLCDs and SBFLCDs [91] is the d.c.-bias, which results from the P_s if an area of the display remains in one of the degenerate states for any length of time [92]. This may result in a build up of space charge in the FLC or in the alignment or insulating layers. These so called ghost images can be alleviated, but not complete-ly eliminated, by sophisticated driving schemes and electrode designs. This is also valid for the inherent lack of grey scale in FLCDs based on bistable effects required for full color displays. An alternative tech-nology utilizes the deformed helix ferro-electric effect DHFLCD [93] on an active matrix substrate driven via simple capaci-tive coupling in a symmetric, dipolar driv-ing mode [94]. This eliminates the build up of space charges and of ghost images. Ac-tive addressing of each pixel is required due to the absence of bistability and of multi-plexability. The DHFLCD effect is based on the distortion of a helix consisting of SmC* layers with p significantly shorter than the wavelength of visible light. The birefrin-gence, Δn, is averaged out over the helical structure and can be modulated by an electric field. A good contrast necessitates a wide switching angle, 2θ. This requires optically active dopants with a significant-ly different spectrum of properties to those

Table 16. The pitch and spontaneous polarization measured at 25 °C of mixtures of 5 wt% of the bis(S)-1-methylalkyl 4,4″-terphenyl dicarboxylates in 4-hex-yloxyphenyl 4-octyloxybenzoate (Cr–SmC: 55 °C; SmC–N: 66 °C; N–I: 89.5 °C) [95].

n	p (μm)	P_s (nC cm^{-2})
1	15.7	4.4
2	17.5	9.4
6	6.9	8.8
8	5.0	3.1

for SSFLCDs such as very high P_s and very short p.

1.3.1   Optically Active Terphenyl Diesters

The ability to incorporate two or more chi-ral centers close to the rigid core of the mole-cule render 4, 4″-terphenyl dicarboxylates of interest for the above application. The usual optically active primary and secon-dary alcohols with various polar substitu-ents at the chiral center have been used to prepare a multitude of diesters [95]. The usual trends found for mono-substituted es-ters were also found for the diesters, al-though most of them do not exhibit enantio-tropic phases due to the combined effect of the two branching groups at the chiral cen-ters. However, an advantageous combina-tion of high P_s, short p, high solubility, good stability and low γ was found for the bis(R or S)-2-alkyl 4, 4″-terphenyl dicarboxy-lates, see Table 16. The P_s rises as the length of the chain after the chiral center is in-creased, reaches a maximum and then de-creases as also observed for SmC* dopants incorporating only one chiral center. A sim-

Table 17. Comparison of the tilt angle, helical twisting power (HTP) and spontaneous polarization (extrapolated to 100%) measured at $TSmC*-15\,°C$ of a mixture of 7 wt% of the dioxanes in a host mixture [96].

n	θ (degrees)	HTP (μm^{-1})	P_s ($nC\ cm^{-2}$)
1	29	9.1	27
2	30	–	59

ilar trend is also observed for p. Thus the optimal combination of P_s and short p is seen for intermediate chain lengths ($n = 8$). This diester also exhibits a high virtual $TSmC*$ which results in an increase in $TSmC*$ of the mixture as the concentration of the dopant is increased [95]. The corresponding biphenyl-dicarboxylates lead to low transition temperatures and surprisingly long response times.

1.3.2   Optically Active Methyl-Substituted Dioxanes

The optically active 2,5-disubstituted-4-methyl-dioxane ring contains three chiral centers and is synthetically accessible [96]. A tetracyclic compound incorporating two such rings possesses a high P_s value, which is approximately twice as large as that of corresponding monodioxanes, see Table 17. The lower value for the tricyclic dioxane infers that the optically active parts are not completely decoupled from each other. The high P_s, high helical twisting power (HTP) and relatively large θ render these compounds highly attractive as dopants for DHFLCDs. However, the low solubility and high γ of the tetracyclic dioxanes prohibit their use in more than small quantities.

1.4   Antiferroelectric Liquid Crystals

The first antiferroelectric phases were found for chiral and racemic 2-octyl N, N′-terephthalidene-bis(4-aminocinnamate) and were classified as smectic O (SmC(alt)) [97]. Both the chiral and achiral smectic phase possess a herringbone structure (Structure 6), where the molecules in alternating layers are tilted at an angle $+\theta$ or $-\theta$ to the layer normal [98]. The helical structure of the chiral phase results in a zero macroscopic P_s as the very high P_s of alternating ferroelectric layers point in different directions and cancel each other out. In an electric field the helix is unwound and a normal ferroelectric SmC* phase is produced. If the field is reversed then the FLC is switched into the second degenerate state as usual. On removal of the field the molecules relax to the intermediate position parallel to the rubbing direction. This behavior is characteristic of the phase denominated as the antiferroelectric smectic C (SmC$_A^*$) phase, which has now been shown to be the same as the SmO phase [99–101]. The tristable switching in the SmC$_A^*$ phase [102, 103] is the basis of antiferroelectric liquid crystal display devices (AFLCDs) [104]. The use of an alternating bias prohibits the appearance of ghost or memory effects. The sharp threshold voltage and hysteresis allow a high degree of multiplexing. Several subphases have also been identified, at least one of which exhibits a ferrielectric structure [105–111]. Several of these compounds with a high P_s and HTP also exhibit the recently discovered twisted grain boundary (TGB) phases [105].

Several of the SmC_A^* compounds collated in Table 18 were prepared during systematic studies of LCs exhibiting a high degree of molecular and form chirality [106–110], while others were the product of attempts to synthesize optimized materials for SSFLCDs [51] and later for AFLCDs [111–117]. The synthesis of SmC_A^* LCs is in its infancy and most of the compounds in Table 18 show a strong similarity. The chiral center is close to the aromatic core and the aliphatic chain after the chiral center is long. The necessity of a rigid aromatic core close to the chiral center for SmC_A^* formation is illustrated by the comparison of the transition temperatures of cinnamates and

$C_6H_{13}*CH(CH_3)O_2CCH=CH-\langle \rangle-CH=N-\langle \rangle-N=CH-\langle \rangle-CH=CHCO_2*CH(CH_3)C_6H_{13}$ (with Cl substituent)

C-SmO*: 95°C; SmO*-SmQ*: 130°C; SmQ*-I: 133°C [97]

Structure 6

Table 18. Structures of typical antiferroelectric smectic SmC_A^* compounds.

Structure	Ref.
$\{C_nH_{2n+1}O,\ C_nH_{2n+1}\}-\langle\rangle-\langle\rangle-CO_2-\langle\rangle-CO_2*CH(CH_3)C_6H_{13}$	[51, 111]
$\{C_nH_{2n+1}CO_2,\ C_nH_{2n+1}O_2C\}-\langle\rangle-\langle\rangle-CO_2-\langle\rangle-CO_2*CH(CH_3)C_6H_{13}$	[111]
$\{C_nH_{2n+1}O,\ C_nH_{2n+1}\}-\langle\rangle-\langle\rangle-CO_2-\langle\rangle-CO_2*CH(CF_3)C_6H_{13}$	[112, 113]
$\{C_nH_{2n+1}CO_2,\ C_nH_{2n+1}O_2C\}-\langle\rangle-\langle\rangle-CO_2-\langle\rangle-CO_2*CH(CF_3)C_6H_{13}$	[103]
$C_nH_{2n+1}-\langle\rangle-\langle\rangle-CO_2-\langle\rangle-CO_2*CH(CF_3)C_4H_8OC_mH_{2m+1}$	[103]
$C_nH_{2n+1}-\langle\rangle-\langle\rangle-CO_2-\langle\rangle(F)-CO_2*CH(CF_3)C_4H_8OC_mH_{2m+1}$	[103]
$C_nH_{2n+1}O-\langle\rangle-\langle\rangle-CO_2-\langle\rangle-CO*CH(CH_3)C_6H_{13}$	[114]
$C_nH_{2n+1}O-\langle\rangle-CO_2-\langle\rangle-\langle\rangle-CO_2*CH(C_mH_{2m+1})C_6H_{13}$	[108]
$C_nH_{2n+1}-\langle N,N\rangle-\langle\rangle-CO_2-\langle\rangle-CO_2*CH(CF_3)C_6H_{13}$	[115]
$C_nH_{2n+1}O-\langle\rangle-CO_2-\langle\rangle-\langle\rangle-CO_2*CH(C_mH_{2m+1})C_6H_{13}$	[106, 117]
$C_nH_{2n+1}O-\langle\rangle\equiv CO_2-\langle\rangle-\langle\rangle-CO_2*CH(C_mH_{2m+1})C_6H_{13}$	[109, 117]
$C_nH_{2n+1}O-\langle\rangle\equiv CO_2-\langle\rangle-\langle\rangle-CO_2CH(C_3H_7)_2$	[109, 118]

the corresponding phenylethyl-carboxy-lates, where the conjugation has been broken by effective hydrogenation of the carbon–carbon double bond [116]. However, it is noteworthy that the phenylethyl-carboxy-lates still exhibit an SmC_A^* phase, although at much lower temperatures. It is postulated that the SmC_A phase of nonoptically active, swallow-tail compounds, which cannot be caused by chirality, is the result of the bent or zigzag shape of dimers formed by the steric interaction between the swallow tail of one molecule and the normal alkyl chain of a second molecule in an adjacent layer, although dipole–dipole interactions probably also make a contribution [118]. Similar steric interactions are probably responsible for the SmC_A phase exhibited by some tetracyclic dimers [119].

Acknowledgements. The author is grateful to Professor J. Fünfschilling and Professor J. W. Goodby for many helpful and instructive discussions.

1.5 References

[1]   G. Friedel, *Ann. Phys.* **1922**, *18*, 273–474.
[2]   H. Sackmann, D. Demus, *Mol. Cryst. Liq. Cryst.* **1966**, *2*, 81–102.
[3]   H. Arnold, *Mol. Cryst. Liq. Cryst.* **1966**, *2*, 63–70.
[4]   H. Sackmann, *Mol. Cryst. Liq. Cryst.* **1989**, *5*, 43–55.
[5]   G. W. Gray, J. W. Goodby, *Smectic Liquid Crystals: Textures and Structures*, Blackie Press, Glasgow and London, **1984**.
[6]   J. W. Goodby in *Ferroelectric Liquid Crystals* (Ed.: G. W. Taylor), Gordon and Breach, Philadelphia, **1991**, pp. 99–241.
[7]   S. L. Arora, J. L. Fergason, A. Saupe, *Mol. Cryst. Liq. Cryst.* **1970**, *10*, 243–257.
[8]   T. R. Taylor, J. L. Fergason, S. L. Arora, *Phys. Rev. Lett.* **1970**, *24*, 359–362.
[9]   T. R. Taylor, S. L. Arora and J. L. Fergason, *Phys. Rev. Lett.* **1970**, *25*, 722–726.
[10] F. M. Jäger, *Rec. Trav. Chim. Pays Bas* **1906**, *25*, 334–351.
[11] W. Urbach, J. Billard, *C. R. Hebd. Séan. Acad. Sci.* **1972**, *274B*, 1287–1290.
[12] G. W. Gray, *Mol. Cryst. Liq. Cryst.* **1969**, *7*, 127–151.
[13] M. Leclercq, J. Billard, J. Jacques, *Mol. Cryst. Liq. Cryst.* **1969**, *8*, 367–387.
[14] W. Helfrich, C. S. Oh, *Mol. Cryst. Liq. Cryst.* **1971**, *14*, 289–292.
[15] G. W. Gray, *J. Chem. Soc.* **1955**, 4359–4368.
[16] S. L. Arora, T. R. Taylor, J. L. Fergason, A. Saupe, *J. Am. Chem. Soc.* **1969**, *91*, 3671–3673.
[17] A. Saupe, *Mol. Cryst. Liq. Cryst.* **1969**, *7*, 59–74.
[18] J. W. Goodby, *J. Mater. Chem.* **1991**, *1*, 307–318.
[19] P. G. de Gennes, *The Physics of Liquid Crystals*, Oxford University Press, Oxford, **1974**, p. 321.
[20] W. L. McMillan, *Phys. Rev.* **1973**, *A-8*, 1921–1929.
[21] R. B. Meyer, L. Liebert, L. Strzelecki and P. Keller, *J. Phys. (Paris) Lett.* **1975**, *36*, L69–L71.
[22] R. B. Meyer, *Mol. Cryst. Liq. Cryst.* **1977**, *40*, 33–48.
[23] N. A. Clark, S. T. Lagerwall, *Appl. Phys. Lett.* **1980**, *36*, 899–901.
[24] S. Garoff, R. B. Meyer, *Phys. Rev Lett.* **1977**, *38*, 848–851.
[25] J. W. Goodby, T. M. Leslie in *Liquid Crystals and Ordered Fluids*, Vol. 4, (Eds.: A. C. Griffin, J. F. Johnson), Plenum, New York, **1984**, p. 1–32.
[26] T. Geelhaar, *Ferroelectrics* **1988**, *85*, 329–349.
[27] H. R. Dübal, C. Escher, D. Günther, W. Hemmerling, Y. Inoguchi, I. Müller, M. Murikami, D. Ohlendorf, R. Wingen, *Jpn. J. Appl. Phys.* **1988**, *27*, L2241–L2244.
[28] F. Leenhouts, J. Fünfschilling, R. Buchecker, S. M. Kelly, *Liq. Cryst.* **1989**, *5*, 1179–1186.
[29] U. Finkenzeller, A. E. Pausch, E. Poetsch, J. Suermann, *Kontakte* **1993**, *2*, 3–14.
[30] S. M. Kelly, R. Buchecker, J. Fünfschilling, *J. Mater. Chem.* **1994**, *4*, 1689–1697.
[31] A. Wulf, *Phys. Rev. A* **1975**, *11*, 365–375.
[32] J. W. Goodby, G. W. Gray, D. G. McDonnell, *Mol. Cryst. Liq. Cryst. Lett.* **1977**, *34*, 183–188.
[33] J. W. Goodby, G. W. Gray, *Ann. Phys.* **1978**, *3*, 123–130.
[34] S. M. Kelly, J. Fünfschilling, *J. Mater. Chem.* **1994**, *4*, 1689–1697.
[35] P. Keller, L. Liebert, L. Strzelecki, *J. Phys. (Paris) Coll.* **1976**, *37*, C3–C27.
[36] B. I. Ostrovski, A. Z. Rabinovitch, A. S. Sonin, E. L. Sorkin, *Ferroelectrics* **1980**, *24*, 309–312.
[37] T. Sakurai, K. Sakamoto, M. Hanma, K. Yoshino, M. Ozaki, *Ferroelectrics* **1984**, *58*, 21–32.
[38] K. Yoshino, T. Sakurai in *Ferroelectric Liquid Crystals*, (Ed.: G. W. Taylor), Gordon and Breach, Philadelphia, **1991**, pp. 317–361.

[39] A. Hallsby, M. Nilsson, B. Otterholm, *Mol. Cryst. Liq. Cryst. Lett.* **1982**, *82*, 61–68.

[40] B. Otterholm, M. Nilsson, S. T. Lagerwall, K. Skarp, *Liq. Cryst.* **1987**, *2*, 757–768.

[41] J. W. Goodby, T. M. Leslie, *Mol. Cryst. Liq. Cryst.* **1984**, *110*, 175–203.

[42] P. Keller, *Ferroelectrics* **1984**, *58*, 3–7.

[43] G. Decobert, J. C. Dubois, *Mol. Cryst. Liq. Cryst.* **1984**, *114*, 237–247.

[44] G. W. Gray, D. G. McDonnell, *Mol. Cryst. Liq. Cryst*, **1976**, *37*, 189–211.

[45] S. M. Kelly, R. Buchecker, *Helv. Chim. Acta*, **1988**, *271*, 451–460.

[46] R. Dabrowski, J. Dziaduszek, J. Szulc, K. Czuprynski, B. Sosnowska, *Mol. Cryst. Liq. Cryst.* **1991**, *209*, 201–211.

[47] Y. Haramoto, H. Kamogawa, *Mol. Cryst. Liq. Cryst.* **1990**, *182*, 195–200.

[48] H. Matsubara, S. Takahashi, K. Seto, H. Imazaki, *Mol. Cryst. Liq. Cryst.* **1990**, *180*, 337–342.

[49] S. M. Kelly, R. Buchecker, M. Schadt, *Liq. Cryst.* **1988**, *3*, 1115–1113; and 1125–1128.

[50] A. Yoshizawa, I. Nishiyama, M. Fukumasa, T. Hirai, M. Yamane, *Jpn. J. Appl. Phys. Lett.* **1989**, *28*, L1269–L1270.

[51] K. Furakawa, K. Terashima, M. Ichihashi, S. Saitoh, K. Miyazawa and T. Inukai, *Ferroelectrics*, **1988**, 85, 451–459.

[52] H. Nonoguchi, H. Suenaga, K. Oba, M. Kaguchi, T. Harada in *Preprints of the 12th Jpn. Liq. Cryst. Meeting*, Japan, **1985**, 1Z07, 54–55.

[53] H. Suenaga, K. Oba, M. Kaguchi T. Harada, S. Shimoda, in *Preprints of the 12th Jpn. Liq. Cryst. Meeting,* Japan, **1985**, 3FO1, 152–153; 2F13, 106–107.

[54] M. Murakami, S. Miyake, T. Masumi, T. Ando, A. Fukami, *Jpn. Display '86* **1986**, 344–347.

[55] S. Matsumoto, A. Murayama, H. Hatoh, Y. Kinoshita, H. Hirai, M. Ishikawa, S. Kamagami, *SID 88 Digest*, **1988**, 41–44.

[56] S. M. Kelly, A. Villiger, *Liq. Cryst.* **1988**, *3*, 1173–1182.

[57] K. Yoshino, M. Osaki, K. Nakao, H. Taniguchi, N. Yamasaki, K. Satoh, *Liq. Cryst.* **1989**, *5*, 1203–1211.

[58] S. Sugita, S. Toda, T. Yamashita, T. Teraji, *Bull. Chem. Soc. Jpn.* **1993**, *66*, 568–572.

[59] E. Chin, J. W. Goodby, J. S. Patel, *Mol. Cryst. Liq. Cryst.* **1988**, *157*, 163–191.

[60] S. Sugita, S. Toda, T. Yoshiyasu, T. Teraji, *Mol. Cryst. Liq. Cryst.* **1993**, *237*, 399–406.

[61] J. Nakauchi, Y. Kageyama, S. Hayashi, K. Sakashita, *Bull. Chem. Soc. Jpn.* **1989**, *62*, 1685–1691.

[62] C. Sekine, T. Tani, Y. Ueda, K. Fujisawa, T. Higasii, I. Kurimoto, S. Toda, N. Takano, Y. Fujimoto, M. Minai, *Ferroelectrics* **1993**, *148*, 203–212.

[63] P. Keller, S. Jugé, L. Liebért, and L. Strzelecki, *Compt. Rend. Acad. Sci.* **1976**, *282*, C639–C641.

[64] T. Sakurai, N. Mikami, R. Higurchi, M. Honma, K. Yoshino, *Ferroelectrics* **1988**, *85*, 469–478.

[65] C. Bahr, G. Heppke, *Mol. Cryst. Liq. Cryst.* **1987**, *148*, 29–43.

[66] G. Heppke, D. Lötsch, N. K. Sharma, D. Demus, S. Diele, K. Jahn, M. Neundorf, *Liq. Cryst.* **1994**, *241*, 275–288.

[67] S. Arakawa, K. Nito, J. Seto, *Mol. Cryst. Liq. Cryst.* **1991**, *204*, 15–25.

[68] J. Bömelburg, G. Heppke, A. Ranft, *Z. Naturforsch.* 1989, **44b**, 1127–1131.

[69] R. Buchecker, S. M. Kelly, J. Fünfschilling, *Liq. Cryst.* **1990**, *8*, 217–227.

[70] S. Nakamura, H. Nohira, *Mol. Cryst. Liq. Cryst.* **1990**, *185*, 199–207.

[71] H. Nohira, S. Nakamura, M. Kamei, *Mol. Cryst. Liq. Cryst.* **1990**, *180*, 379–388.

[72] D. M. Walba, H. A. Razavi, N. A. Clark, D. S. Parma, *J. Am. Chem. Soc.* **1988**, *110*, 8686–8691.

[73] M. D. Wand, R. Vohra, D. M. Walba, N. A. Clark, R. Shao, *Mol. Cryst. Liq. Cryst.* **1991**, *202*, 183–192.

[74] K. Yoshino, M. Ozaki, H. Taniguchi, M. Ito, K. Satoh, N. Yamasaki, T. Kitazume, *Jpn. J. Appl. Phys.* **1987**, *26*, L77–L78.

[75] Y. Suzuki, T. Hagiwara, I. Kawamura, N. Okamura, T. Kitazume, M. Kakimoto, Y. Imai, Y. Ouchi, H. Takezoe, A. Fukuda, *Liq. Cryst.* **1989**, *6*, 167–174.

[76] A. Sakaigawa, Y. Tashiro, Y. Aoki, H. Nohira, *Mol. Cryst. Liq. Cryst.* **1991**, *206*, 147–157.

[77] Y. Aoki, H. Nohira, *Chem. Lett.* **1993**, L113–L116.

[78] L. A. Beresnev, L. M. Blinov, *Ferroelectrics* **1981**, *33*, 129–138.

[79] L. K. M. Chan, G. W. Gray, D. Lacey, R. M. Scrowston, I. G. Shenouda, K. J. Toyne, *Mol. Cryst. Liq. Cryst.* **1989**, *172*, 125–146.

[80] D. M. Walba, K. F. Eidman, R. C, Haltiwanger, *J. Org. Chem.* **1989**, *54*, 4939–4943.

[81] C. J. Booth, J. W. Goodby, J. P. Hardy, O. C. Lettington, K. J. Toyne, *Liq. Cryst.* **1994**, *16*, 925–940.

[82] D. M. Walba, R. T. Vohra, N. A. Clark, M. A. Handschy, J. Xue, D. S. Parma, S. T. Lagerwall, K. Skarp, *J. Am. Chem. Soc.* **1986**, *108*, 724–725.

[83] D. M. Walba, N. A. Clark, *Ferroelectrics* **1988**, *84*, 65–72.

[84] G. Scherowsky, J. Gay, *Liq. Cryst.* **1989**, *5*, 1253–1258.

[85] K. Sakaguchi, T. Kitamura, *Ferroelectrics* **1991**, *114*, 265–272.

[86] K. Sakashita, M. Shindo, J. Nakauchi, M. Uematsu, Y. Kageyama, S. Hayashi, T. Ikemoto, *Mol. Cryst. Liq. Cryst.* **1991**, *199*, 119–127.

[87] G. Scherowsky, M. Sefkow, *Mol. Cryst. Liq.* *Cryst.* **1991**, *202*, 207–216.
[88] G. Scherowsky, J. Gay, M. Gunaratne, *Liq. Cryst.* **1992**, *11*, 745–752.
[89] G. Scherowsky, M. Sefkow, *Liq. Cryst.* **1992**, *12*, 355–362.
[90] S. Takehara, M. Osawa, K. Nakayama, T. Kusumoto, K. Sato, A. Nakayama, T. Hiyama, *Ferroelectrics* **1993**, *148*, 195–202.
[91] J. Fünfschilling, M. Schadt, *SID Digest* **1990**, 106–109.
[92] Y. Inaba, K. Katagiri, H. Inoue, J. Kanabe, S. Yoshihara, S. Iijama, *Ferroelectrics* **1988**, *85*, 255–264.
[93] L. A. Beresnev, V. G. Chigrinov, D. I. Dergachev, E. P. Poshidaev, J. Fünfschilling, M. Schadt, *Liq. Cryst.* **1989**, *5*, 1171–1177.
[94] M. Schadt, J. Fünfschilling, *SID IDW'94* Hamamatsu City, Japan, **1994**, 67–70.
[95] M. Loseva, N. Chernova, A. Rabinovitch, E. Poshidaev, J. Narkevitch, O. Petrashevitch, E. Kazachkov, N. Korotkova, M. Schadt, R. Buchecker, *Ferroelectrics* **1991**, *114*, 357–377.
[96] R. Buchecker, J. Fünfschilling, M. Schadt, *Mol. Cryst. Liq. Cryst.* **1992**, *213*, 259–267.
[97] A. M. Levelut, C. Germain, P. Keller, L. Liebert, J. Billard, *J. Phys. (Paris)*, **1983**, *44*, 623–629.
[98] Y. Galerne, L. Liebert, *Phys. Rev. Lett.* **1990**, *64*, 906–909; and **1991**, *66*, 2891–2894.
[99] P. Cladis, H. Brand, *Liq. Cryst.* **1993**, *14*, 1327–1349.
[100] G. Heppke, P. Kleineberg, D. Lötsch, S. Mery, R. Shashidar, *Mol. Cryst. Liq. Cryst.* **1993**, *231*, 257–262.
[101] Y. Takanishi, M. Johno, T. Yui, H. Takezoe, A. Fukuda, *Jpn. J. Appl. Phys.* **1993**, *32*, 4605–4610.
[102] A. D. L. Chandani, T. Hagiwara, Y. Suzuki, Y. Ouchi, H. Takezoe, A. Fukuda, *Jpn. J. Appl. Phys.*, **1988**, *27*, L729–L732.
[103] A. Fukuda, *J. Mater. Chem.* **1994**, *4*, 997–1016.
[104] Y. Yamada, N. Yamamoto, K. Mori, K. Nakamura, T. Hagiwara, Y. Suzuki, I. Kawamura, H.

Orihara, Y. Ishibashi, *Jpn. J. Appl. Phys.* **1990**, *29*, 1757–1764.
[105] J. W. Goodby, M. A. Waugh, S. M. Stein, E. Chin, R. Pindak, J. S. Patel, *Nature*, **1989**, *337*, 449–452.
[106] I. Nishiyama, J. W. Goodby, *J. Mater. Chem.* **1993**, *3*, 149–159.
[107] I. Nishiyama, E. Chin, J. W. Goodby, *J. Mater. Chem.* **1993**, *3*, 161–168.
[108] J. W. Goodby, J. S. Patel, E. Chin, *J. Mater. Chem.* **1992**, *2*, 197–207.
[109] J. W. Goodby, I. Nishiyama, A. J. Slaney, C. J. Booth, K. J. Toyne, *Liq. Cryst.* **1993**, *14*, 37–66.
[110] J. W. Goodby, A. J. Slaney, C. J. Booth, I. Nishiyama, J. D. Vuijk, P. Styring, K. J. Toyne, *Mol. Cryst. Liq. Cryst*, **1994**, *243*, 231–298.
[111] N. Okabe, Y. Suzuki, I. Kawamura, T. Isozaki, H. Takezoe, A. Fukuda, *Jpn. J. Appl. Phys.* **1992**, *31*, L793–L796.
[112] Y. Suzuki, T. Hagiwara, I. Kawamura, N. Okamura, T. Kitazume, M. Kakimoto, Y. Imai, Y. Ouchi, H. Takezoe, A. Fukuda, *Liq. Cryst.* **1989**, *6*, 167–174.
[113] Y. Yamada, K. Mori, N. Yamamoto, H. Hayashi, K. Nakamura, M. Yamawaki, H. Orihara, Y. Ishibashi, *Jpn. J. Appl. Phys.* **1989**, *28*, L1606–L1608.
[114] I. Nishiyama, A. Yoshisawa, M. Fukumasa, T. Hira, *Jpn. J. Appl. Phys.* **1989**, *28*, L2248–L2250.
[115] S. Inui, S. Kawano, M. Saitoh, H. Iwane, Y. Takanishi, K. Hiraoka, Y. Ouchi, H. Takezoe, A. Fukuda, *Jpn. J. Appl. Phys.* **1990**, *29*, L987–L990.
[116] S. Inui, T. Susuki, N. Iimura and H. Iwane, *Mol. Cryst. Liq. Cryst.* **1994**, *239*, 1–9.
[117] C. J. Booth, D. A. Dunmur, J. W. Goodby, J. S. Kang, K. J. Toyne, *J. Mater. Chem.* **1994**, *4*, 747–759.
[118] I. Nishiyama, J. W. Goodby, *J. Mater. Chem.* **1992**, *2*, 1015–1023.
[119] J. Watanabe, M. Hayashi, *Makromolecules*, **1989**, *22*, 4083–4088

2 Ferroelectric Liquid Crystals

Sven T. Lagerwall

2.1 Introduction

Ferroelectric liquid crystals are a novel state of matter, a very recent addition to the science of ferroelectrics which, in itself, is of relatively recent date. The phenomenon which was later called ferroelectricity was discovered in the solid state (on Rochelle salt) in 1920 by Joseph Valasek, then a PhD student at the University of Minnesota. His first paper on the subject [1] had the title *Piezo-Electric and Allied Phenomena in Rochelle Salt*. This was at the time when solid state physics was not a fashionable subject and it took several decades until the importance of the discovery was recognized. Valasek had then left the field. Later, however, the development of this branch of physics contributed considerably to our understanding of the electrical properties of matter, of polar materials in particular and of phase transitions and solid state physics in general. In fact, the science of ferroelectrics is today an intensely active field of research. Even though its technical and commercial importance is substantial, many breakthrough applications may still lie ahead of us. The relative importance of liquid crystals within this broader area is also constantly growing. This is illustrated in Fig. 1, showing how the proportion of the new materials, which are liquid-crystalline, has steadily increased since the 1980s.

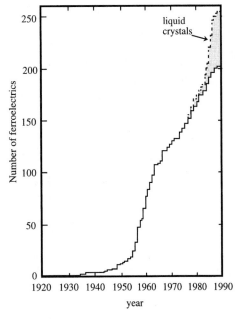

Figure 1. Number of known ferroelectrics. Solid line: solid state ferroelectrics, where each pure compound is counted as one. Dashed line: total number, including liquid crystal ferroelectrics, for which a group of homologs is counted as one. From about 1984, the proportion of liquid crystals has steadily grown which has been even more pronounced after 1990. (After Deguchi [2] as cited by Fousek [3]).

The general level of knowledge of ferro-electricity, even among physicists, is far lower than in the older and more classical subjects like ferromagnetism. It might therefore be worthwhile to discuss briefly the most important and characteristic features of solid ferroelectrics and polar materials, before turning to liquid crystals. This will facilitate the understanding and allow us to appreciate the striking similarities as well as distinctive differences in how polar phenomena appear in solids and how they appear in liquid crystals. One of the aims, of course, is also to make a bridge to existing knowledge. Those not aware of this important knowledge are apt to coin new words and concepts, which are bound to be in contradiction to already established concepts or even contradictory to themselves.

When dealing with ferroelectric liquid crystals, we use the same conceptual framework already developed for solid polar materials. An important part of this is the Landau formalism describing phase transitions (still not incorporated in any textbook on thermodynamics), based on symmetry considerations. It is important to gain some familiarity with the peculiarities of this formalism before applying it to ferroelectric liquid crystals. In this way it will be possible to recognize cause and effect more easily than if both subject matters were introduced simultaneously.

Concepts like piezoelectric, pyroelectric, ferroelectric, ferrielectric, antiferroelectric, paraelectric, electrostrictive, and several more, relate to distinct phenomena and are themselves interrelated. They are bound to appear in the description of liquid crystals and liquid crystal polymers, as they do in normal polymers and crystalline solids. Presently, great confusion is created by the uncritical use of these terms. For example, in the latest edition of the Encyclopedia Britannica [4] it is stated that pyroelectric-

ity was discovered in quartz in 1824. This is remarkable, because quartz is not pyroelectric at all and cannot be for symmetry reasons. To clarify such issues (and the confusion is no less in the area of liquid crystals), we will have to introduce some simple symmetry considerations that generally apply to all kinds of matter. In fact, symmetry considerations will be the basic guidelines and will play a probably more important role here than in any other area of liquid crystals. Chirality is a special property of dissymmetry with an equally special place in these considerations. It certainly plays a fundamental role for ferroelectric liquid crystals at least so far. Therefore we will have to check how exactly the appearance of polar properties in liquid crystals is related to chiral properties, and if chirality is dispensable, at least in principle. Finally, flexoelectricity is also a polar effect, and we will have to ask ourselves if this is included in the other polar effects or, if not, if there is an interrelation.

Can liquids in which the constituents are dipoles be ferroelectric? For instance, if we could make a colloidal solution of small particles of the ferroelectric $BaTiO_3$, would this liquid be ferroelectric? The answer is no, it would not. It is true that such a liquid would have a very high value of dielectric susceptibility and we might call it superparaelectric in analogy with the designation often used for a colloidal solution of ferromagnetic particles, which likewise does not show any collective behavior. An isotropic liquid cannot have polarization in any direction, because every possible rotation is a symmetry operation and this of course is independent of whether the liquid lacks a center of inversion, is chiral, or not. Hence we have at least to diminish the symmetry and go to anisotropic liquids, that is, to liquid crystals, in order to examine an eventual appearance of pyroelectricity or ferroelectricity. To search for ferroelectricity in an isotropic liquid would be futile, because a ferroelectric liquid cannot be isotropic. In order to have a bulk polarization, a medium must have a direction, the polarity of which cannot be reversed by any symmetry operation of the medium. On the other hand, an isotropic liquid consisting of dipoles may show a polarization during flow, because a shear diminishes the symmetry and will partially order the dipoles, thus breaking the randomness. This order will be polar if the liquid is chiral. However, we would not consider such a liquid ferroelectric or pyroelectric – no more than we would consider a liquid showing flow birefringence to be a birefringent liquid. It is clear that there may be lots of interesting polar effects yet to be explored in flowing liquids, particularly in fluids of biological significance (which are very often chiral). Nevertheless, these effects should not be called "ferroelectric". They should not even be called piezoelectric, even if setting up shear flow in a liquid certainly bears some resemblance to setting up shear strain in a crystal.

Are there magnetic liquids? Yes, there are. We do not mean the just-mentioned "ferrofluids", which are not true magnetic liquids, because the magnetic properties are due to the suspended solid particles (of about 10 nm size). As we know, ferromagnetic materials become paramagnetic at the Curie temperature and this is far below the melting point of the solid. However, in some cases it has been possible (with extreme precautions) to supercool the liquid phase below the Curie temperature. This liquid has magnetic properties, though it is not below its own Curie temperature (the liquid behaves as if there is now a different Curie temperature), but it would be wrong to call the liquid ferromagnetic. The second example is the equally extreme case of the quantum liquid He-3 in the A1 phase. Just like

the electrons in a superconductor, the He-3 nuclei are fermions and have to create paired states to undergo Bose–Einstein condensation. However, unlike the electron case, the pairs are created locally, and the axis between two He-3 then corresponds to a local director. Thus He-3 A is a kind of nematic-like liquid crystal, and because of the associated magnetic moment, He-3 A is undeniably a magnetic liquid, but again, it is not called *ferro*magnetic. Thus, in the science of magnetism a little more care is normally taken with terminology, and a more sound and contradiction-free terminology has been developed: a material, solid or liquid, may be designated magnetic, and then what kind of magnetic order is present may be further specified. When it comes to polar phenomena, on the other hand, there is a tendency to call everything "ferroelectric", a usage that leads to tremendous confusion and ambiguity. It would be very fortunate if in future we could reintroduce the more general concepts of "electric materials" and "electric liquids" in analogy with magnetic materials and magnetic liquids. Then, in every specific case, it would be necessary to specify which electric order (paraelectric, dielectric, etc.) is present, just as in the magnetic case (paramagnetic, diamagnetic, etc.).

Coherent and contradiction-free terminology is certainly important, because vagueness and ambiguity are obstacles for clear thinking and comprehension. In the area of liquid crystals, the domain of ferroelectric and antiferroelectric liquid crystals probably suffers from the greatest problems in this respect, because in the implementation of ideas, concepts, and general knowledge from solid state physics, which have been of such outstanding importance in the development of liquid crystal research, the part of ferroelectrics and other polar materials has generally not been very well represented.

Presumed ferroelectric effects in liquid crystals were reported by Williams at RCA in Princeton, U. S. A., as early as 1963, and thus at the very beginning of the modern era of liquid crystal research [5]. By subjecting nematics to rather high dc fields, he provoked domain patterns that resembled those found in solid ferroelectrics. The ferroelectric interpretation seemed to be strengthened by subsequent observations of hysteresis loops by Kapustin and Vistin and by Williams and Heilmeier [7]. However, these patterns turned out to be related to electrohydrodynamic instabilities, which are well understood today (see, for instance, [8], Sec. 2.4.3 or [9], Sec. 2.4.2), and it is also well known that certain loops (similar to ferroelectric hysteresis) may be obtained from a nonlinear lossy material (see [10], Sec. 2.4.2). As we know today, nematics do not show ferroelectric or even polar properties. In order to find such properties we have to lower the symmetry until we come to the tilted smectics, and further lowering their symmetry by making them chiral. The prime example of such a liquid crystal phase is the smectic C*.

In principle, the fascinating properties of the smectic C* phase could have been detected long before their discovery in 1974. Such materials were synthesized by Vorländer [11] and his group in Halle before the first World War. The first one seems to have been an amyloxy terephthal cinnamate with a smectic C* phase from 133 °C to 247 °C and a smectic A* phase from 247 °C to 307 °C, made in 1909 [11] at a time far earlier than the first description of the smectic C phase as such [12] in 1933. At that time it was not, and could not possibly have been realized as ferroelectric. The concept did not even exist.

In a classic review from 1969, Saupe (at Kent State University) discusses a hypothetical ferroelectric liquid crystal for the first time [13]. While a nematic does not have polar order, such order, he points out, could possibly be found in the smectic state. The ferroelectric smectic, according to Saupe, is an orthogonal nonchiral smectic in which all molecular dipoles are pointing along the layer normal in one single direction (a longitudinal ferroelectric smectic). He also discussed a possible antiferroelectric arrangement. Among the other numerous topics discussed in this paper (suggesting even the first blue phase structure), Saupe investigates the similarities between nematics and smectics C and introduces the twisted smectic structure as the analog of a cholesteric. In the same year, Gray in Hull [14] synthesized such materials (actually the first members of the DOBAMBC series), but only reported on an orthogonal smectic phase; no attention was paid to smectic polymorphism in those days. Actually, in the year before, Leclerq et al. [15] (in Paris) had reported on a material having two distinct chiral and strongly optically active phases, which they interpreted as two distinct nematic phases. They were thus very close to discovering the helicoidal smectic. (Their material had a first order N*–C* transition.) A helicoidal smectic was then reported for the first time by Helfrich and Oh, at RCA in 1971, who described the first smectic liquid crystal ("spiraling" or "conical" smectic) identified as optically active [16]. Like the substances in the above-mentioned examples, this one belongs to the category that is the topic of this chapter, but who could have expected them to have special polar properties?

While the smectic C* phase was gradually becoming recognized, the question of ferroelectricity was again brought up by McMillan at the University of Illinois, Urbana, U. S. A. In 1973 he presented a microscopic molecular theory of the smectic C phase [17] based on dipole–dipole interactions, which predicted three different polar phases. McMillan's model molecules have a central dipole and two outboard dipoles perpendicular to the long axis. Either all three can line up or only the outboard ones with the central dipoles random, or the central dipoles can line up with the outboard dipoles random. The transition from the A phase to these polar phases is thought to take place through different rotational transitions where the rotational freedom is lost or frozen out due to the dipole–dipole interaction. The net polarization in the condensed phases lies in the smectic plane and gives rise to a two-dimensional ferroelectric. Actually, whether the order is ferroelectric or antiferroelectric depends on the sign of the interplanar interaction, which cannot be predicted. McMillan's dipolar theory, which does not involve chirality at all, never really applied to liquid crystals and was rapidly superseded by the ideas of Meyer in the following year.

In fact, the discovery and introduction of practically all polar effects in liquid crystals go back to the ideas of Meyer, at that time working at Harvard. In 1969 he published an epoch-making paper entitled *Piezoelectric Effects in Liquid Crystals* [18]. It must be said that the new phenomena described in that paper are beautifully analogous to piezoelectric effects in solids. Nevertheless they are of a different nature. Therefore de Gennes instead proposed the name flexoelectric [19], in order to avoid misunderstanding. Seven years later, together with his student Garoff, Meyer presented a new, original effect which he called the piezoelectric effect in smectic A liquid crystals [20]. The analogies between this effect and the piezoelectric effect in solids are here perhaps even more striking, as we will see. However, it is not the same thing and, after much consideration by the authors, the new phenomenon was finally published under

the name electroclinic effect [21], a term which has since been generally adopted. In the following twenty years, there were numerous publications, in which different workers reported measuring a piezoelectric effect in liquid crystals (normally without stating what it meant and why they used this term). Obviously, they meant neither the flexoelectric nor the electroclinic effect, because the meanings of these are by now well established. Therefore the question arises as to whether a third effect exists in liquid crystals, which would finally qualify for the name piezoelectric. Obviously, this state of affairs is not very encouraging. A critical review of the terminology is therefore necessary in this area and should contribute to clarifying the concepts.

Ferroelectric liquid crystals have been a field of research for about twenty years, and have certainly been in the forefront of liquid crystal research, with an increasing number of researchers involved. The first state-of-the-art applications have recently appeared. This account concentrates on the basic physics, but also treats in considerable detail the topics of highest relevance for applications. Literature references have been given, as far as possible, for topics that, for space reasons, could not be treated [8 - 10, 22 - 58]. A big help for finding access to previous work is the bibliography of [52], which extends to 1989, as well as the series of conference proceedings published by *Ferroelectrics* [54–58], covering a great deal of the work from 1987 to 1995. New ferroelectric liquid crystal materials are continually included in the Liqcryst-Database [59] set up by Vill at the University of Hamburg.

2.2   Polar Materials and Effects

2.2.1   Polar and Nonpolar Dielectrics

A molecule that has an electric dipole moment in the absence of an external electric field is called a polar (or dipolar) molecule. Such a molecule will tend to orient itself in an electric field. In a material consisting of polar molecules, the induced polarization P due to the average molecular reorientation is typically 10–100 times larger than the contribution from the electronic polarization present in all materials. In contrast, a nonpolar molecule has its distributions of positive and negative charges centered at the same point. A characteristic of materials consisting of nonpolar molecules is that the polarization P induced by a field E is small and independent of temperature, whereas in the first-mentioned case, P is a function $P(T)$ with an easily observable temperature dependence. Hence the same goes for the dielectric constant ε and the susceptibility χ.

If we write the relations between dielectric displacement D, induced polarization P, and applied electric field E, assuming that P is linear in E

$$D = \varepsilon_0 E + P = \varepsilon_r \varepsilon_0 E = \varepsilon E \qquad (1)$$

and we get

$$P = (\varepsilon_r - 1)\varepsilon_0 E = \chi \varepsilon_0 E \qquad (2)$$

with ε_r as the relative, ε as the total dielectric constant or permittivity, and ε_0 the permittivity of free space, where $\varepsilon_0 = 8.85 \times 10^{-12}$ C/V m. The dielectric susceptibility χ is then, like ε_r, a dimensionless number, and lies in the range $0-10$ for most materials, although it may attain values higher than 10^4 for certain ferroelectric substances. We will equally use this term, the susceptibility, for its dimensional form $\varepsilon_0 \chi = \partial P / \partial E$.

As examples of nonpolar molecules we may take H_2, O_2, CO_2, CS_2, CH_4, and CCl_4, and as well-known polar molecules CO (0.10), NH_3 (1.47), C_6H_5OH (1.70), H_2O (1.85), $C_6H_5NO_2$ (4.23), where we have stated the value of the dipole moment in parentheses, expressed in Debye (D), a unit commonly used for molecules. One Debye equals 10^{-18} cgs units and, expressed in SI units, $1\,D = 3.3 \times 10^{-30}$ C m. For a comparison, let us consider a dipole consisting of two elementary charges $\pm e$ (i.e., with e the charge of the electron, 1.6×10^{-19} C) at a typical atomic distance of 1 Å or 0.1 nm from each other. This gives a dipole moment $p = q\,d = 1.6 \times 0^{-19} \times 10^{-10} = 1.6 \times 0^{-29}$ C m equal to 4.85 D. Let us assume that we had a gas consisting of molecules with this dipole moment and that we had a field sufficiently strong to align the dipoles with the field. With a density $N = 3 \times 10^{25}$ molecules/m^3 this would correspond to a polarization of $P = 5 \times 10^{-4}$ C m$^{-2} = 50$ nC cm^{-2}. However, this is a completely unrealistic assumption because the orientational effect of the field is counteracted by the thermal motion. Therefore the distribution of dipolar orientation is given by a Boltzmann factor $e^{-U/kT}$, where $U = -p \cdot E$ is the potential energy of the dipole in the electric field. Integrating over all angles for p relative to E gives the polarization P as a function of E according to the Langevin function L [which expresses the average of cos (p,E)]

$$P = NpL\left(\frac{pE}{kT}\right) \qquad (3)$$

shown in Fig. 2. For small values of the argument, $L[pE/(kT)] \approx pE/(3kT)$, and thus

$$P = \frac{Np^2 E}{3kT} \qquad (4)$$

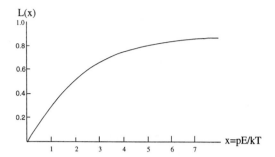

Figure 2. The Langevin function $L(x) = \coth x - 1/x$.

corresponding to the linear part around the origin. However, even at a field $E = 10^7$ V m^{-1}, corresponding to dielectric breakdown, the value of $pE/(kT)$ is only about 0.03 at room temperature, giving a resulting polarization of 1% of the saturation value. A similar result would be true for the liquid phase of the polar molecules. In liquid crystal phases, it will generally be even harder to polarize the medium in an external field. In the very special polar liquid crystals, on the other hand, the reverse is true: for quite moderate applied fields it is possible to align all dipoles, corresponding to polarization values in the range of 5– 500 nC cm^{-2} (depending on the substance).

According to Eq. (4), the polarization, at constant field, grows when we lower the temperature. By forming $\partial P/\partial E$, we may write for the susceptibility

$$\chi = \frac{C}{T} \qquad (5)$$

The fact that the susceptibility has a $1/T$ dependence, called the Curie law, is characteristic for gases and liquids, but may also be found in solids. Generally speaking, it indicates that the local dipoles are noninteracting.

For another comparison, consider a crystal of rock salt, NaCl. It has a χ value of 4.8. If we apply the quite high but still realistic field of 10^6 V m^{-1} (1 V/μm) we will, ac-

cording to Eq. (2), induce a polarization of 4.25 nC cm^{-2} (or 42.5 μC m^{-2}; the conversion between these units commonly used for liquid crystals is 1 nC cm^{-2} = 10 μC m^{-2}). The mechanism is now the separation of ionic charges and thus quite different from our previous case. It turns out that the displacement of the ions for this polarization is about 10^{-2} nm (0.1 Å), i.e., it represents only a small distortion of the lattice, less than 2%. Small displacements in a lattice may thus have quite strong polar effects. This may be illustrated by the solid ferroelectric barium titanate, which exhibits a spontaneous polarization of 0.2 C m^{-2} = 20 000 nC cm^{-2}. Responsible for this are ionic displacements of about 10^{-3} nm, corresponding to less than half a percent of the length of the unit cell. Similar small lattice distortions caused by external pressure induce considerable polar effects in piezoelectric crystals.

When atoms or molecules condense to liquids and solids the total charge is zero. In addition, in most cases the centers of gravity for positive and negative charges coincide. The matter itself is therefore nonpolar. For instance, the polar water molecules arrange themselves on freezing to a unit cell with zero dipole moment. Thus ice crystals are nonpolar. However, as already indicated, polar crystals exist. (In contrast, elementary particles have charge, but no dipole moment is permitted by symmetry.) These are then macroscopic dipoles and are said to be pyroelectric materials, of which ferroelectrics are a subclass. Pyroelectric materials thus have a macroscopic polarization in the absence of any applied electric field. Piezoelectric materials can also be polarized in the absence of an electric field (if under strain) and are therefore sometimes also considered as polar materials, though the usage is not general nor consistent. Clearly their polarization is not as "spontaneous" in the pyroelectric case.

Finally, there seems to be a consensus about the concepts of polar and nonpolar liquids. Water is a polar liquid and mixes readily with other polar liquids, i.e., liquids consisting of polar molecules, like alcohol, at least as long as the sizes of the molecules are not too different, whereas it is insoluble in nonpolar liquids like benzene. If in liquid form, constituent polar molecules interact strongly with other polar molecules and, in particular, are easily oriented in external fields. We will also use this criterion for a liquid crystal. That is, we will call a liquid crystal polar if it contains local dipoles that are easily oriented in an applied electric field.

2.2.2   The Nonpolarity of Liquid Crystals in General

The vast majority of molecules that build up liquid crystal phases are polar or even strongly polar. As an example we may take the cyanobiphenyl compound 8CB (Merck Ltd) with the formula

$$C_8H_{17} \!-\!\!\bigcirc\!\!-\!\!\bigcirc\!\!- CN$$

which has an isotropic–nematic transition at 40.1 °C and a nematic to smectic A transition at 33.3 °C. Whereas the molecules are strongly polar, the nematic and smectic A phases built up of these molecules are nonpolar. This means that the unit vector n, i.e., the director, describing the local direction of axial symmetry, is not a polar direction but rather one with the property of an optic axis. This means that n and $-n$ describe the same state, such that all physical properties of the phase are invariant under the sign reversal of n

$$n \rightarrow -n, \text{ symmetry operation} \qquad (6)$$

Repeatedly this invariance has been described as equivalent to the absence of fer-
roelectricity in the nematic phase, whereas it only expresses the much weaker condition that n is not a polar direction.

How can we understand the nonpolarity expressed by Eq. (6), which up to now seems to be one of the most general and important features of liquid crystals? Let us simplify the 8CB molecules to cylindrical rods with a strong dipole parallel to the molecular long axis. In the isotropic phase these dipoles could not build up a macroscopic dipole moment, because such a moment would be incompatible with the spherical symmetry of the isotropic phase. But in the anisotropic nematic phase such a moment would be conceivable along n. However, this would set up a very strong external electrostatic field and such a polar nematic would tend to diminish the electrostatic energy by adjusting the director configuration $n(r)$ to some complex configuration for which the total polarization and the external field would be cancelled. Such an adjustment could be done continuously because of the three-dimensional fluidity of the phase. It would lead to appreciably elastic deformations, but these deformations occur easily in liquid crystals and would carry small weight compared to the energy of the polar effects which, as we have seen, are very strong. Thus it is hard to imagine a polar nematic for energetic reasons. Moreover, the dipolar cancellation can, and therefore will, take place more efficiently on a local scale. This is illustrated in Fig. 3, where the local dipolar fields are shown for two molecules. Each dipole will want to be in the energetic minimum position in the total field of all its neighbors. If two molecules are end on, their dipoles tend to be parallel, but in a lateral position the dipoles will be antiparallel, as shown to the right. Now, on the average, there are many more neighbors in a lateral position, which is a result of the shape of the molecules, and hence the latter

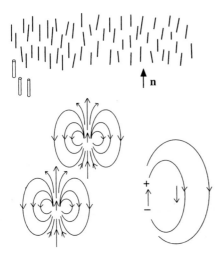

Figure 3. Why liquid crystals are in general nonpolar. Around any molecule a lateral neighbor molecule has the tendency to align with its dipole in an antiparallel fashion. This means that $n \rightarrow -n$ is a symmetry operation for liquid crystals.

configuration will prevail. We will therefore have an antiparallel correlation on a local scale. This leads to the fundamental property of liquid crystals of being invariant under sign reversal of the director, i.e., to Eq. (6). We may note that the argument given above is even more conclusive for smectic phases, which are thus characterized by the same nonpolar, "nematic" order.

The oblong (or more generally anisotropic) shape of the molecules, which both sterically and by the anisotropy of their van der Waals interactions leads to the liquid-crystalline order, at the same time makes this order nonpolar. It has nothing to do with the strength or weakness of the dipolar interaction. Whatever strength such an interaction may have, it cannot be assumed that it would favor a parallel situation giving a bulk polarization to a liquid, nor even to a solid.

Undeniably there is also an entropic contribution to this order, because a polar orientation has a lower entropy than an apolar orientation for the same degree of parallel order. Thus, if we turn all dipoles into the parallel position, we diminish the entropy and increase the free energy. However, the electrostatic contribution seems to be the important one. This is confirmed by computer simulations [60], which show that the antiparallel correlation is quite pronounced, even in the isotropic phase of typical mesogenic molecules.

Although the nematic phase is nonpolar, there are very interesting and important polar effects in this phase, in a sense analogous to piezoelectric effects in solid crystals. This was recognized by Meyer [18] in 1969. These so-called flexoelectric effects are discussed in Sec. 2.4 of this Chapter. Meyer also recognized [61] in 1974 that all chiral tilted smectics would be truly polar and the first example of this kind, the helielectric smectic C*, was presented [62] in 1975. Out of Meyer's discovery grew the whole research area of ferroelectric and antiferroelectric liquid crystals, which is today a major part of liquid crystal physics and chemistry.

2.2.3 Behavior of Dielectrics in Electric Fields: Classification of Polar Materials

All dielectrics become polarized if we put them in an electric field. The polarization is linear in the field, $P \sim E$, which means that it changes sign if we reverse the sign of the field. When the field is reduced to zero, the polarization vanishes. For very strong fields we will observe saturation effects (and eventually dielectric breakdown of the material). The typical behavior of a normal (nonpolar) dielectric is shown at the top of Fig. 4. In a piezoelectric, an appropriate strains s will have a similar influence to the electric field for a normal dielectric. The effect is likewise linear around the origin and shows saturation at high strains. Converse-

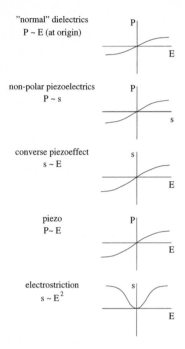

Figure 4. Response of nonpolar dielectrics (which do not contain orientable local dipoles) to an applied electric field. Piezoelectric materials react both with polarization and distortion and, because of the way these are related, they are also polarized by a distortion in the absence of an electric field.

ly, we can induce a strain $s \sim E$ by applying an external field. This is the converse piezoelectric effect, which is used to produce ultrasound, whereas the direct effect is used to transform a mechanical strain into an electric signal, for instance, in gramophone pick-ups. The linearity in $P \sim s$ and $s \sim E$ means that there is linearity between P and E, that is, a nonpolar piezoelectric behaves dielectrically just like a normal dielectric. In addition, all materials show the electrostrictive effect. This is normally a very small field-induced strain, which originates from the fact that the equilibrium distance between atoms and the distribution of dipoles are to some degree affected by an applied field. It corresponds to the small induced polarization and thus small dielectric constant

typical for ordinary materials. The electrostrictive effect therefore has an entirely different character; the strain itself is not related to any polarization so no converse effect exists (i.e., an electric field cannot be generated in ordinary materials simply by applying mechanical pressure). The electrostrictive strain is always superposed on, but can be distinguished from, a piezoelectric strain by the fact that it is quadratic and not linear in the field, thus $s \sim E^2$. The linear and quadratic dependence, resepectively, of the piezoelectric and electrostrictive strain, of course applies near the origin for small fields. In the more general case, the effect is called piezoelectric if it is an odd function of E, and electrostrictive if it is an even function [63]. The piezoelectric effect can only be present in noncentrosymmetric materials. If we apply a field to a material with a center of symmetry, the resultant strain must be independent of the field direction, hence

$$s^e = \alpha E^2 + \alpha' E^4 + \ldots \qquad (7)$$

whereas the reversibility of the piezoeffect will only admit odd powers

$$s^p = \beta E + \beta' E^3 + \ldots \qquad (8)$$

Sometimes care has to be taken not to confuse the two effects. If, for some reason, the sample has been subjected to a static field E, and then a small ac signal E_{ac} is applied, the electrostrictive strain will be

$$s^e = a(E_0 + E_{ac})^2 \approx aE_0^2 + 2aE_0E_{ac} \qquad (9)$$

giving a linear response with the same frequency as the applied field. An aligning field E_0 may, for instance, be applied to a liquid crystal polymer, for which the electrostrictive coefficient is often particularly large and the signal coming from this so-called biased electrostriction may easily be mistaken for piezoelectricity. Therefore, in the search for piezoelectricity, which would indicate the lack of a center of symmetry in a material, the direct piezoeffect should be measured, whenever possible, and not the converse effect.

In Fig. 5 we have in the same way illustrated how polar materials may behave in response to an external electric field E. The $P-E$ trace is the fingerprint of the category that we are dealing with and is also characteristic of the technological potential. At the top there is the normal dielectric response with P increasing linearly up to a saturation value at high fields. In principle this behavior is the same whether we have local dipoles or not, except that with local dipoles the saturation value will be high and strongly temperature-dependent. If the molecules lack permanent dipoles, the induced local polarization is always along the field and temperature-independent. Dipolar molecules may, on the other hand, align spontaneously at a certain temperature (Curie temperature) to a state of homogeneous polarization. On approaching such a temperature the susceptibility $(\partial P/\partial E)$ takes on very high values and is strongly temperature-dependent. We will describe this state – which is unpolarized in the absence of a field, but with a high and strongly temperature-dependent value of the dielectric susceptibility – as paraelectric, independent of whether there actually is a transition to an ordered state or not. It should be noted from the previous figure that a piezoelectric material has the same shape of the $P-E$ curve as a normal dielectric, but it often shows paraelectric behavior with a large and even diverging susceptibility.

Next, the contrasting, very strongly nonlinear response of a ferroelectric is shown in Fig. 5. The two stable states $(+P, -P)$ at zero field are the most characteristic feature of this hysteresis curve, which also illustrates the threshold (coercive force) that the external field has to overcome in order to

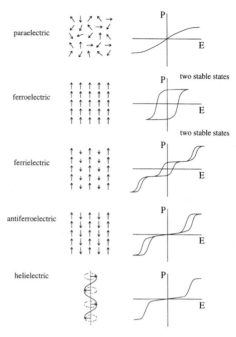

Figure 5. Response of polar dielectrics (containing local permanent dipoles) to an applied electric field; from top to bottom: paraelectric, ferroelectric, ferrielectric, antiferroelectric, and helielectric (helical antiferroelectric). A pyroelectric in the strict sense hardly responds to a field at all. A paraelectric, antiferroelectric, or helielectric phase shows normal, i.e., linear dielectric behavior and has only one stable, i.e., equilibrium, state for $E=0$. A ferroelectric as well as a ferrielectric (a subclass of ferroelectric) phase shows the peculiarity of two stable states. These states are polarized in opposite directions $(\pm P)$ in the absence of an applied field $(E=0)$. The property in a material of having two stable states is called bistability. A single substance may exhibit several of these phases, and temperature changes will provoke observable phase transitions between phases with different polar characteristics.

flip over from one state to the other. In the solid state this behavior may be represented by $BaTiO_3$.

The response of an antiferroelectric is shown two diagrams below. The initial macroscopic polarization is zero, just as in a normal dielectric and the $P-E$ relation is linear at the beginning until, at a certain threshold, one lattice polarization flips over to the direction of the other (the external field is supposed to be applied along one of the sublattice polar directions). This is the field-induced transition to the so-called ferroelectric state of the antiferroelectric. (Not ferroelectric phase, as often written – the phase is of course antiferroelectric. Transitions between different thermodynamic phases by definition only occur on a change of temperature, pressure, or composition, that is, the variables of a phase diagram.) The very characteristic double-hysteresis loop reveals the existence of two sublattices as opposed to a random distribution of dipoles in a paraelectric. From the solid state we may take $NaNbO_3$ as representative of this behavior. It is not without interest to note that the structure of $NaNbO_3$ is isomorphous with $BaTiO_3$. The fact that the latter is ferroelectric whereas the former is antiferroelectric gives a hint of the subtleties that determine the character of polar order in a lattice.

At the limit where the hysteresis loops shrink to thin lines, as in the diagram at the bottom, we get the response from a material where the dipoles are ordered in a helical fashion. Thus this state is ordered but has no macroscopic polarization and therefore belongs to the category of antiferroelectrics. It is called helical antiferroelectric or heliectric for short. If an electric field is applied perpendicular to the helical axis, the helix will be deformed as dipoles with a direction almost along the field start to line up, and the response $P-E$ is linear. As in the normal antiferroelectric case, the induced P value will be relatively modest until we approach a certain value of E at which complete unwinding of the helix takes place rather rapidly. Although the helielectric is a very special case, it shares the two characteristics of normal antiferroelectrics: to have an ordered distribution of dipoles (in

contrast to random) and a threshold where the linear response becomes strongly nonlinear (see Fig. 5). In the solid state this behavior is found in $NaNO_2$.

In the middle diagram of Fig. 5 we have also traced the $P-E$ response for the modification of an antiferroelectric, which we get in the case where the two sublattices have a different polarization size. This phase is designated ferrielectric. Because we have, in this case, a macroscopic polarization, ferrielectrics are a subclass of ferroelectrics. It also has two stable states as it should, although the spontaneous macroscopic polarization is only a fraction of that which can be induced. If the polarization values of the sublattices are P_1 and $P_2<P_1$, the bistable states have the values $\pm(P_1-P_2)$, whereas the saturation polarization after sublattice reversal is $\pm(P_1+P_2)$.

The presence of two stable states indicates that a ferroelectric material basically has memory properties, just like a ferromagnetic material. Any of these two states can prevail when the acting external field has been removed ($E=0$). In the virgin state, the ferroelectric tries to minimize its own external field by the creation of macroscopic domains of opposite polarization, which is an important characteristic in identifying the state as ferroelectric. In all the other materials mentioned, including pyroelectrics and antiferroelectrics, we have no spontaneous polarization domains, i.e., polarization up and polarization down domains appearing in the field-free state. The fact that it has not yet been possible to develop solid state ferroelectric memory devices is related to the very involved interactions with charge carriers and to the fact that switching between the two states is normally coupled to a change in the lattice distortion, which may eventually cause a total breakdown of the lattice (crystal fatigue). If the same or analogous memory property could be found in a liquid, this might be an attractive way to rule out at least the second problem. This is one of the facts that make ferroelectric liquid crystals very promising. As a result of these inherent problems, solid state ferroelectrics have, strangely enough, not been used for their ferroelectric properties, but rather for their superior pyroelectric (heat detectors), piezoelectric, and dielectric (very high ε) properties. In addition, they are the dominating electrooptic materials as Pockels modulators as well as for nonlinear optics (NLO), e.g., for second-harmonic generation.

2.2.4   Developments in the Understanding of Polar Effects

The denominations dielectric, paralectric, and ferroelectric were of course created as analogs to the names of the previously recognized magnetic materials. In the first treatise on magnetism, *De Magnete* (1600), Gilbert (W. Gilbert, De Magnete, Engl. transl. Dover Publications, New York 1958) describes how iron loses its magnetic (like magnesian stone) properties when it is made red-hot. He also introduced the word "electric" (like amber) for bodies that seem to be charged (but differently than the magnetic ones) or may even produce sparks. It was Faraday who divided the magnetic materials into the groups diamagnetic, paramagnetic, and ferromagnetic and he also (1839) introduced the word dielectric for materials through which electric forces are acting (which could get polarized). Confirming Gilbert's observation, Faraday (M. Faraday, Experimental Researches in Electricity, Vol. 1–3, Taylor & Francis, London 1839–1855) also found that iron loses its ferromagnetic properties at high temperature (770 °C) and becomes paramagnetic. This temperature is today called the Curie point, or Curie temperature, for iron.

In his thesis (1895) Pierre Curie [64] investigated the magnetic properties of a number of different materials. He found that the magnetic susceptibility is independent of temperature for diamagnetic materials. For paramagnetic materials (like oxygen) he found a relation of the type of Eq. (5). A different relation

$$\chi = \frac{C}{T - T_c} \tag{10}$$

was found for such bodies that showed ferromagnetism, on approaching the ferromagnetic state from above. That is, the susceptibility does not diverge at $T = 0$ (in which case one would not observe a divergence), but at a finite temperature T_c, which is the temperature below which the body is ferromagnetic. When Curie's friend and colleague Paul Langevin in 1905 deduced the formulae corresponding to Eqs. (3), (4), and (5) it was the first microscopic physical theory relating to macroscopic phenomena [65], in this case to the measured magnetization M and magnetic susceptibility $\chi = \partial M / \partial H$. Thus Langevin found for C, the Curie constant, the value $N\mu^2/3k$ where μ is the magnetic moment of the atom. The electric case reflected in our Eqs. (3), (4), and (5) with p instead of μ, is of course modeled in exact analogy to the magnetic case. One year later, in 1906, Weiss [66], at that time professor at the ETH in Zürich, attacked the problem of ferromagnetism by a phenomenological theory similar to that presented in 1873 by van der Waals in his thesis on the condensation of a gas to a liquid [J. D. van der Waals, On the Continuity of Gas and Liquid State, Engl. transl. in Physical Memoirs, London 1890].

Weiss introduced the notion of a "molecular field", the average field on a certain magnetic moment caused by its magnetized environment, and set this field proportional to the external field H. In doing this, he could account for the spontaneous magnetization of ferromagnets, and he was able to deduce a relation for the susceptibility corresponding to Eq. (10), which since then has been called the Curie–Weiss law. The divergence of χ at the Curie temperature T_c corresponds to the divergence of the compressibility of a fluid at the critical point. The approach by Weiss has been very successful and a longstanding model for descriptions of condensed matter phenomena which go under the name of "mean field theories".

The word dielectric is analogous to diamagnetic. Physically, however, dielectric does not correspond to diamagnetic in any other sense than, that the effects are very general, whereas paraelectric and paramagnetic effects can only be found where we have orientable electric and magnetic dipoles. In fact, there is no electric effect corresponding to diamagnetism, the origins of which (in the classical description) are induced atomic and therefore lossless currents ("superconductivity on an atomic scale") and which therefore give a small magnetization against the applied field, corresponding to a negative susceptibility. As, in contrast, the general dielectric effect and the more special paraelectric effect do not differ in sign, the reason to make a distinction between them is not generally felt as very strong, except when we want to express the fact that in the dielectric effect (in the narrow sense) there is no temperature dependence, because the dipole does not exist prior to the application of the field, but is only created by the field, just as in the diamagnetic case. Dielectric and diamagnetic effects are therefore also always present in paraelectric and paramagnetic materials, but can on the other hand usually be neglected there.

While electric phenomena connected to matter seem to have been observed, like the magnetic ones, even in ancient times, their investigation and comprehension is of a more recent date. Curiosity about these phenomena was raised in about 1700, when Dutchmen in foreign trade brought tourmaline from Ceylon to Europe. They observed how these rodlike crystals acquired a force to attract light objects, like small pieces of paper, when heated. The first scientific description [67] was given in 1756 by the German Hoeck (Aepinus) working in Berlin and then mainly in Russian Saint Petersburg. Brewster in Scotland, who made a systematic study on a number of minerals [67], coined the name pyroelectric in 1824. However, it was necessary to wait until the present century to really get an understanding of the phenomenon, and the historical name (pyro etymologically related to fire) is a very unfortunate one. It is true that a change ΔT in temperature causes the appearance of a polarization P, and pyroelectricity is often defined this way. Still, it is not really the heart of the matter. Brewster's colleague in Glasgow, Lord Kelvin, who worked out (1878–93) the basic thermodynamics of the pyroelectric effect and its converse, the electrocaloric effect (heat dissipation on applying an electric field) suggested [69] that pyroelectric materials have a permanent electric polarization. This is the basic thing: a pyroelectric has a spontaneous macroscopic polarization, i.e., it is an electric dipole in the absence of an external field. The pyroelectric effect, however, important it may be for applications, is only a secondary effect, just a manifestation of the temperature dependence of this polarization. The charged poles in a pyroelectric cannot normally be observed, because they are neutralized in a rather short time by the ubiquitous free charges which are attracted by the poles. (As there are no free magnetic charges, such a screening does not occur in a magnet, which makes magnetic phenomena simpler in many respects than the corresponding electric phenomena). However, the magnitude (and in rare cases even the direction) of the polarization is generally a function of temperature, $P = P(T)$. Hence a change ΔT in temperature causes an additional polarization ΔP which is not compensated by the charges present and can be observed (normally in a matter of minutes) until the redistribution of charges to equilibrium has been completed. This is also one of the standard methods to measure polarization: the pyroelectric current is integrated through an external circuit after firing a heat pulse from an infrared laser. This gives $\partial P/\partial T$, which is determined as a function of T. From this, $P(T)$ is deduced.

Charles Friedel, the father of George Friedel, had studied pyroelectric phenomena in crystals since 1860 and in the 1870s published a number of papers on the topic with Jacques Curie, who was his assistant in the mineralogy laboratory at Sorbonne (and later went to Montpellier as a lecturer in mineralogy). When the brothers Jacques and Pierre Curie discovered piezoelectricity in 1880, their conjecture [70] was at first that when creating surface charges by compressing the crystal they had invented an alternative method for making a pyroelectric, because they believed that there was a correspondence between thermal and mechanical deformation. This was objected to by Hankel, professor of physics in Leipzig, Germany, who pointed out that the two effects must be fundamentally different in nature and proposed the name piezoelectricity [71], which was immediately adopted by the Curie brothers. Basically we can say that temperature, being a scalar, can never change the symmetry. Therefore the polarity must already exist and be intrinsic in a pyroelectric crystal. A strain, on the other hand, being a tensor, can alter the symmetry and thereby provoke the polarity.

The further theory of pyro- and piezoelectricity was later essentially worked out by Voigt (the man who also coined the word "tensor") and summarized in his monumental *Lehrbuch der Kristallphysik (Treatise of Crystal Physics)*, which appeared in 1910 [W. Voigt, Lehrbuch der Kristallphysik, Teubner, Leipzig 1910]. Voigt, who had been a student of Franz Neumann in Königsberg and then professor of theoretical physics in Göttingen from 1883, was able to show in detail how these polar phenomena were related to crystal symmetry. In particular he showed that among the 32 crystal classes, 20 have low enough symmetry to admit piezoelectricity and, within these 20 classes, 10 have the still lower symmetry to be polar, i.e., pyroelectric. We can therefore divide dielectric materials into hierarchical categories, and this is illustrated in Fig. 6.

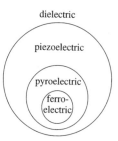

Figure 6. The hierarchy of dielectric materials. All are of course dielectrics in a broad sense. To distinguish between them we limit the sense, and then a dielectric without special properties is simply called a dielectric; if it has piezoelectric properties it is called a piezoelectric, if it further has pyroelectric but not ferroelectric properties it is called a pyroelectric, etc. A ferroelectric is always pyroelectric and piezoelectric, a pyroelectric always piezoelectric, but the reverse is not true. Knowing the crystal symmetry we can decide whether a material is piezoelectric or pyroelectric, but not whether it is ferroelectric. A pyroelectric must possess a so-called polar axis (which admits no inversion). If in addition this axis can be reversed by the application of an electric field, i.e., if the polarization can be reversed by the reversal of an applied field, the material is called ferroelectric. Hence a ferroelectric must have two stable states in which it can be permanently polarized.

In this figure, the higher, more narrow category has the special properties of the lower ones, but not vice versa. Thus, for instance, a pyroelectric is always also piezoelectric. We have also inserted in the figure the very important subclass of pyroelectrics called ferroelectrics, to which we will return in detail below.

For 35 years piezoelectric materials were rather a laboratory curiosity, but in 1916, during the first World War, Langevin started his epoch-making investigations in the harbour of Toulon in southern France, to range submarines and to develop navigation and ranging systems based on his technique of producing and detecting ultrasound by means of piezoelectric (quartz) crystals. He thereby became the founder of the whole industry of sonar and ultrasonics with very important military as well as civil applications. At the end of the war, Cady, in the United States, initiated the applications of piezoelectric crystals for oscillators and frequency stabilizers, which subsequently revolutionized the whole area of radio broadcasting, and set a new standard for the measurement of time prior to the arrival of atomic clocks (up until 1925, when quartz oscillators took over; the frequency of the electronic master clocks was still set by tuning forks).

In the meantime Debye, who the year before had been appointed professor of theoretical physics at the University of Zürich, in 1912 introduced the idea of "polar molecules", i.e., molecules with a permanent electric dipole moment (at that time a hypothesis) and worked out a theory for the macroscopic polarization in analogy with Langevin's theory of paramagnetic substances. He found, however, that the interactions in condensed matter could lead to a permanent dielectric polarization, corresponding to a susceptibility tending to infinity for a certain temperature, which he

proposed to be the analog to the Curie temperature for a ferromagnet. In the same year, Schrödinger in Vienna, in his habilitation thesis [72] applied the model to solids and concluded that all solids ought to become "ferroelectric" at sufficiently low temperatures. Thereby he coined the word ferroelectric even before such a similar state was to be found eight years later. (The state in Schrödinger's anticipation does not quite cover the concept in the sense it is used today.)

After the first World War, the strategic importance of piezoelectric materials was fully recognized and research was intensified on Rochelle salt, whose piezocoefficients had been found to be much higher than those of quartz. It became the thesis subject for Valasek of the University of Minnesota. Valasek recognized the electric hysteresis in this material as perfectly analogous to magnetic hysteresis [1] and also used the word Curie point for the transition at 24 °C. Thereby the phenomenon of ferroelectricity was discovered (although it was not quite recognized in the beginning). After some time, it was found that ferroelectric materials were not only extremely interesting as such, but also had the most powerful piezoelectric and pyroelectric properties. Important research projects were started in Leningrad headed by Kurchatov who founded the Russian school and wrote the first treatise on ferroelectricity (or "seignette electricity" as it was previously often called after the French name on Rochelle salt) [73]. In Zürich, Busch and Scherrer (the latter initially studied both in Königsberg and with Voigt in Göttingen) discovered [74] the important ferroelectric KDP (Curiepoint 123 K) in 1935. When Landau in 1937 presented his very general framework based on symmetry changes in phase transitions [75], it would gradually come to dominate the whole area of phase transitions and collective phenomena. Mueller at ETH, independently, worked out a similar theory [76] and applied it successfully to KDP. Finally, wartime research efforts during the second World War led to the discovery in 1945 of the very important perovskite class starting with barium titanate ($BaTiO_3$, with the conveniently located Curie point of 120 °C) by Wul and Goldman at the Lebedev Institute in Moscow [77, 78] and, probably independently, in 1946 by von Hippel and his group at MIT in the U. S. A. [79]. In the same year, Ginzburg applied the Landau theory for the first time to the ferroelectric case (of type KDP) [80]. Today, still no unified microscopic theory is available for these different ferroelectrics. This is a good argument for using phenomenological theories, which we will do in the following. These theories also have the advantage that they can be applied to ferroelectric liquid crystals.

Finally, a comment has to be made about the polar materials called *electrets*. The word itself was introduced by Heaviside at the end of the last century [81] to designate a permanently polarized dielectric, in analogy with the word magnet. It has, however, hardly been used in that general sense, but has instead acquired a special meaning. In current terminology, an electret is a dielectric that produces a permanent or rather quasi-permanent external field, which results from the ordering of molecular dipoles or stable, uncompensated surface or space charges [82]. The first electrets were prepared by Eguchi [83, 84] at practically the same time as the discovery of ferroelectrics by Valasek. Normally electrets are made by orienting dipoles, or separating, or injecting charges by a high electric field (for instance, in a corona discharge) acting on materials like natural waxes or resins or, in modern times, organic polymers, and then cooling down the material in the presence of the field such that the charges, or dipoles, stay trapped (thermoelectrets). A powerful modern method is to inject charge in the material directly by an electron or ion beam (charge implantation). Amorphous selenium and other inorganic materials can also be used as electrets in which charges are separated by illumination in the presence of the field (photoelectrets). Common to all electrets is the fact that the polarization or space charge is a non-equilibrium condition. These materials are therefore in a "glassy", metastable state and their polarization in principle slowly decays with time, more rapidly at higher temperatures. However, for practical purposes their long-term stability can be very high and their industrial importance is enormous both for electroacoustic devices (electrets are used in almost all small earphones, microphones, and loudspeakers), sensors and transducers (ultrasound and touch devices). The typical electroacoustic electret is a thin foil of polyimide, polyethylene, or teflon (polytetrafluoroethylene), 5–50 μm thick, with one side metallized and charged to about 20 nC cm^{-2} (corresponding to 10 V/μm) and in extreme cases up to 500 nC cm^{-2}. Although liquid-crystalline electrets are presently a subject of some research activity, especially in connection with NLO applications, electrets will not be considered further in this chapter. The development of ferroelectric materials and concepts has been reviewed by Fousek [3] and, with particular emphasis on historical details, by Busch [85] and Känzig [86].

2.2.5   The Simplest Descriptions of a Ferroelectric

As mentioned above, Debye concluded that the interaction between local dipoles could lead to a permanent dielectric polarization. The mechanism can be described as follows:

As long as the polarization is small, it is directly proportional to the applied field according to Eq. (4). However, as P becomes large, it begins to contribute to the field at the site of the dipoles. If we consider one of them, it will be in a field E superposed by an average field created by the other dipoles, which will be proportional to P. If we call the proportionality constant λ, we can replace E in Eq. (4) by $(E + \lambda P)$ to give

$$P = \frac{Np^2 (E + \lambda P)}{3kT} \qquad (11)$$

Solving for P gives

$$P = \frac{Np^2}{3k} \cdot \frac{E}{T - \lambda Np^2 / 3k} \qquad (12)$$

and for the susceptibility

$$\varepsilon_0 \chi = \partial P / \partial E = \frac{Np^2 / 3k}{T - \lambda Np^2 / 3k} \qquad (13)$$

We see that the susceptibility diverges and becomes infinite when T approaches the value

$$T = T_c = \frac{\lambda Np^2}{3k} \qquad (14)$$

which means that at this temperature we can have a nonzero polarization P even if the external field E turns to zero. T_c is therefore the Curie temperature at which the dielectric goes from the paraelectric to the ferroelectric state. In Eq. (5) we introduced the Curie constant C equal to $Np^2 / 3k$. With this inserted in Eq. (11), the susceptibility can be written as

$$\varepsilon_0 \chi = \frac{C}{T - T_c} \qquad (15)$$

This corresponds to the Curie–Weiss law for magnetic phenomena and has the same name for polar materials. In fact, Eqs. (11) to (15) are just the mean field description by Weiss, applied to polar materials.

It may be pointed out that, although Eq. (11) correctly connects the ferroelectric state to an instability, the numerical value required for the phenomenological coefficient λ turns out (just as in the Weiss theory of ferromagnetism) to be quite unrealistic. Most polar materials should also be expected to be ferroelectric according to this model, in disagreement with experience. On the contrary, the collective interaction of an assembly of dipoles may not mean that the lowest state is one where all dipoles are directed in the same manner. The general problem is still unsolved, but in 1946 Luttinger and Tisza [87] were able to show that an assembly of dipoles forming a simple cubic structure has a lower energy in antiparallel fashion (antiferroelectric order) than in one where all dipoles are in the same direction. As a matter of fact, antiferroelectrics are much more common in nature than ferroelectrics! The paraelectric–ferroelectric instability is thus considerably more complex than indicated by the ideas behind Eq. (11).

We may also notice that the instability described by this formalism corresponds to the very general concept of positive feedback, describing how an instability appears in a system. For instance, the diverging susceptibility describing the transition from paraelectric to ferroelectric at the instability point T_c corresponds to the diverging gain when an electronic amplifier makes the transition to a self-sustaining oscillator at a certain feedback coefficient β (the fraction βV of the output voltage is fed back to the input voltage V_i). These analogous cases are compared in Fig. 7. It may be pointed out in advance that in liquid crystals there is no such instability connected with polar interactions.

The Landau description, although in essence a mean field theory, is a phenomenological description of far greater generality

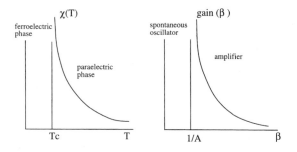

Figure 7. Phase transition from paraelectric to spontaneously polarized state in a ferroelectric, compared to the state transition in an amplifier when the gain $V/V_i = \dfrac{A}{1-\beta A}$ diverges for $\beta = 1/A$. A is the amplification at zero feedback.

and penetration than our discussion so far. Even from its simplest versions, deductions of high interest can be made. We begin by introducing an order parameter, and in this case we naturally choose the polarization P, describing the order in the low temperature phase ($P\neq0$) and the absence of order in the paraelectric phase ($P=0$). The order parameter is a characteristic of the transition and should conform to the symmetry of the high temperature phase as well as to that of the low temperature phase. We expand the free energy in powers of P

$$G(T,P) = G_0(T) + \frac{1}{2}a(T)P^2$$
$$+ \frac{1}{4}bP^4 + \frac{1}{6}cP^6 + \dots \qquad (16)$$

Often, the high temperature phase has a center of symmetry (for instance, the cubic paraelectric phase of $BaTiO_3$). If the high temperature phase is piezoelectric (as in the case of KH_2PO_4, abbreviated KDP), the free energy nevertheless cannot depend on the sign of strain and polarization. It is also true that the two states of opposite polarization in the low temperature ferroelectric phase are energetically equivalent. G is therefore invariant under polarization reversal ($P\rightarrow-P$) and only even powers can appear in the Landau expansion. It may be noted that no such invariance is valid for a pyroelectric material, and that a corresponding Landau expansion cannot be made. Sometimes expressions of this type (Eq. 16)

are called Taylor series expansions, which is rather misleading because the coefficients bear no relation to Taylor coefficients. In fact (and we will see examples of this later) the expression is of a far more general nature and, in addition to powers of several order parameters and their conjugates and coupling terms, may contain scalar invariants of their components as well as their derivatives of different order. Generally speaking, the terms included in the Landau expansion have to be invariant under all symmetry operations of the high temperature phase. (Then they are also automatically invariant under those of the low temperature phase, whose symmetry group is here assumed to be a sub-group of that of the high temperature phase.) As for the coefficients, we have indicated a temperature dependence of a, which we will express with the simplest straightforward choice

$$a(T) = \alpha(T - T_c), \quad \alpha > 0 \qquad (17)$$

It is further assumed that b and c in Eq. (16) are small and both positive in the first instance. The term $G_0(T)$ is often included in G on the left side. Equation (17) means that the coefficient a changes sign at the temperature T_c. For $T>T_c$, a is positive and the free energy (now with G_0 included) will be a simple parabolic function at the origin, according to Eq. (16), thus with a minimum for $P=0$. For $T<T_c$, the free energy will have a maximum instead (see Fig. 8 a). When P grows larger, however, the P^4 term

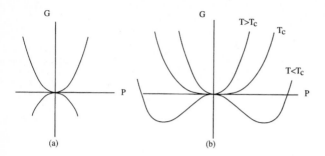

Figure 8. The free energy as a function of the order parameter according to the Landau expansion Eq. (16) with Eq. (17). For $T > T_c$, G has a minimum for the order parameter of zero ($P=0$, corresponding to the paraelectric phase); for $T < T_c$, there are two opposite states of nonzero order parameter in thermodynamic equilibrium. A second order phase transition takes place at $T = T_c$.

takes over and will yield two symmetrically situated minima, corresponding to a nonzero polarization which can take the opposite sign. At T_c, the P^2 term vanishes and G becomes $\sim P^4$, giving a curve which is very flat at the origin (see Fig. 8b).

The equilibrium value of P is obtained by minimizing the free energy with respect to this variable, thus

$$\frac{\partial G}{\partial P} = \alpha(T - T_c)P + bP^3 + cP^5 = 0 \qquad (18)$$

which gives, if we disregard the fifth power,

$$P[\alpha(T - T_c) + bP^2] = 0 \qquad (19)$$

For $T > T_c$, the only real solution is $P=0$, describing the paraelectric phase. For $T < T_c$, we get

$$P = \sqrt{\frac{\alpha}{b}}(T_c - T)^{1/2} \qquad (20)$$

describing how the spontaneous polarization increases smoothly from zero in a parabolic fashion when we decrease the temperature below the Curie point T_c. This is characteristic for a second-order phase transition and qualitatively conforms well with real behavior in most ferroelectric materials. If b was negative instead and c positive, we would have had to include the original P^6 term and would have found a (weakly) first-order transition with P making a (small) jump to zero at the Curie point, but we will not pursue this matter further.

If we apply an electric field, we have to include the term $-PE$ in the Landau free energy. The field E is here the thermodynamic intensive conjugate variable to P. In the presence of a field, we have of course a nonzero polarization even in the paraelectric phase for $T > T_c$. However, as P is small, we only keep the P^2 term from Eq. (14) and write

$$G = \frac{1}{2}\alpha(T - T_c)P^2 - PE \quad \text{for} \quad T > T_c \quad (21)$$

Minimizing the free energy now gives

$$\frac{dG}{dP} = \alpha(T - T_c)P - E = 0 \qquad (22)$$

i.e., a polarization proportional to the applied field according to

$$P = \frac{E}{\alpha(T - T_c)} \qquad (23)$$

and a corresponding susceptibility

$$\varepsilon_0 \chi = \frac{\partial P}{\partial E} = \frac{1}{\alpha(T - T_c)} \qquad (24)$$

i.e., the Curie–Weiss law describing the divergence of χ when we approach the Curie temperature from above. Let us compare this to the situation in the ordered (ferroelectric) phase ($T < T_c$). As P is now not small, we have to include the P^4 term in G and therefore write

$$G = \frac{1}{2}\alpha(T - T_c)P^2 + \frac{1}{4}bP^4 - PE \qquad (25)$$

for $T < T_c$

Minimizing with respect to P gives

$$\frac{dG}{\partial P} = \alpha(T - T_c)P + bP^3 - E = 0 \quad (26)$$

and, differentiating with respect to E (remembering that $\varepsilon_0\chi = \partial P/\partial E$)

$$\alpha(T - T_c)\varepsilon_0\chi + 3bP^2\varepsilon_0\chi = 1 \quad (27)$$

If we insert $P^2 = (\alpha/b)(T_c - T)$, according to Eq. (20), into this expression we get

$$\varepsilon_0\chi = \frac{1}{2\alpha(T_c - T)} \quad (28)$$

Our results are summarized in Fig. 9. Analytically, we can express them as

$$P \sim |T - T_c|^\beta \quad (29)$$

$$\chi \sim |T - T_c|^{-\gamma} \quad (30)$$

$$\chi_- = \frac{1}{2}\chi_+ \quad (31)$$

with the so-called critical exponents $\beta = 1/2$ and $\gamma = 1$, for the order parameter and susceptibility characteristic of the Landau description or, more generally, of mean field models. On approaching the Curie point from below, the susceptibility diverges just as on approaching from above, with the difference that $\chi(T < T_c) = \frac{1}{2}\chi(T > T_c)$. This is a general mean field result: the susceptibil-

ity in the condensed phase is half of that in the disordered phase. The reason for their difference is evident: the effect of an external field on the polarization must be smaller in the condensed phase, due to the existing internal field (molecular field) in that phase, which has a stabilizing effect on the polarization. We may also write this as

$$\frac{C_+}{C_-} = \frac{\chi_+}{\chi_-} = 2 \quad (32)$$

because it also means that the Curie constant above and below the transition differs by a factor of two. Graphically this is most easily seen if we plot the inverse of the susceptibility as a function of the temperature, as in Fig. 10. In passing it may be mentioned that no such critical behavior (no divergence in χ) would be observed if the low temperature phase has antiferroelectric order, corresponding to the fact that such a phase has zero macroscopic polarization.

Finally, let us look a little closer at the behavior in an external field both above and below the Curie point. We will rewrite the free energy from Eq. (25) in a slightly more expanded form

$$(33)$$

$$G = G_0 + \frac{1}{2}aP^2 + \frac{1}{4}bP^4 + \frac{1}{6}cP^6 + \dots - PE$$

Minimizing the free energy with respect to P at constant external field E

$$\left(\frac{\partial G}{\partial P}\right)_E = 0 \quad (34)$$

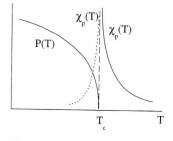

Figure 9. The growth of the order parameter $P(T)$, in this case representing ferroelectric polarization below the Curie point T_c, together with the corresponding susceptibility $\chi = \frac{1}{\varepsilon_0}\frac{\partial P}{\partial E}$ above and below the phase transition.

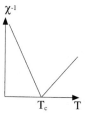

Figure 10. The inverse dielectric susceptibility following a Curie law with different Curie constants $(C_+/C_- = 2)$ above and below the Curie point T_c.

yields

$$E = aP + bP^3 + cP^5 + \ldots \tag{35}$$

Sometimes this is written (including ε_0 in χ)

$$E = \frac{1}{\chi} P + \xi P^3 + \zeta P^5 \tag{36}$$

where the values of ξ and ζ are given as measure of the dielectric nonlinearity. Let us again neglect the highest power and write

$$E = a(T - T_c)P + bP^3 \tag{37}$$

When $T > T_c$ we may also discard the P^3 term and see that

$$E \sim P \text{ (paraelectric state)} \tag{38}$$

whereas for $T = T_c$ we get a purely cubic relationship

$$E \sim P^3 \tag{39}$$

A graphic illustration of Eq. (37) is given in Fig. 11. The $P-E$ curves are analogous to van der Waals isotherms. For the isotherm at $T = T_c$, instead of Eq. (39) we write

$$P \sim E^{1/\delta} \tag{40}$$

introducing a third critical exponent δ, with

the value $\delta = 3$ for the critical isotherm of the order parameter. This value, $\delta = 3$, is also a characteristic of mean field theories.

For $T < T_c$, we clearly recognize the shape of a hysteresis loop (emphasized in Fig. 11) connected with the presence of an instability. Whereas P follows the field in a normal fashion when we reverse the sign of E at $T > T_c$, and much more dramatically at $T = T_c$, P does not change sign at first below the Curie point. But P is now not a single-valued function of E, and below the value $-E_c$, the decrease in P no longer follows the upper curve but continues, after a jump, along the lower curve. We also see that below T_c there are two stable states for $E = 0$.

In Fig. 12 the free energy as a function of polarization according to Eq. (33) is sketched (the P^6 term can here be neglected) for three different values of the external field. The stable states for P correspond to the minima of G. In a ferroelectric sample these are seen in the spontaneous domains. For a sufficiently high field, E_2 (corresponding to E_c in Fig. 11), the oppositely directed polarization state becomes metastable and the polarization is reversed.

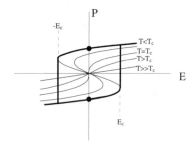

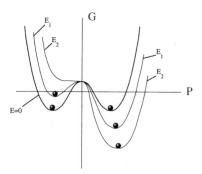

Figure 11. Graphic representation of $E = aP + bP^3$. This strongly nonlinear relationship $P(E)$ describes the ferroelectric hysteresis (emphasized) appearing for $T < T_c$. Two stable states of opposite polarization exist at $E = 0$. A field superior to the coercive field E_c has to be applied in order to flip one polarization state into the other.

Figure 12. The free energy as a function of polarization for different values of the external field. The symmetrical curve for $E = 0$ for which the two polarization states $\pm P_0$ are stable has been emphasized and $E_2 > E_1$ corresponds to the coercive field E_c which destabilizes the $-P_0$ state.

2.2.6 Improper Ferroelectrics

So far it has been assumed that the polarization is the variable that determines the free energy, but it may happen that the transition is connected to, i.e. occurs in a completely different variable, let us call it q, to which a polar order can be coupled. A polarization P appearing this way may affect the transition weakly, but it can only have a secondary influence on the free energy. We will therefore call q and P, respectively, the primary and secondary order parameters, and the simplest Landau expansion in those two parameters can be written

$$G = \frac{1}{2} a(T) q^2 + \frac{1}{4} b q^4 + \frac{1}{2} \chi_0^{-1} P^2 - \lambda q P \tag{41}$$

We note, already from the presence of the coupling term $-\lambda q P$, that the Landau expansion is not a "power series", cf. the previous comments made with reference to the expression (16). If q and P have the same symmetry, i.e., if they transform in the same way under the symmetry operations, then a bilinear term of the form $-\lambda q P$ must be admitted since it behaves like the other terms of even power. The minus sign means that the free energy is lowered due to the coupling between q and P. The coupling constant λ is supposed to be temperature-independent. The fact that q appears to the fourth power, but P now only to the second, reflects their relative importance. Instead of using c as the coefficient in front of the P^2 term, it is convenient to introduce the symbol χ_0^{-1}. We assume that χ_0^{-1}, in contrast to the coefficient $a(T)$ in front of q^2, does not depend on temperature, which also reflects the fact that P is not as a transition parameter. In fact, it can often be shown that $\chi_0^{-1} \sim T$, which in practice means that it is temperature independent compared to a coefficient varying like $T - T_c$, where T_c lies in the region of experimental interest. As will be discussed in detail later, χ_0 takes the form of a special kind of susceptibility.

The equilibrium values of q and P are obtained by minimizing G with respect to these variables. Let us begin with P. From

$$\frac{\partial G}{\partial P} = \chi_0^{-1} P - \lambda q = 0 \tag{42}$$

we see that P will be proportional to the primary order parameter

$$P = \lambda \chi_0 q \tag{43}$$

Continuing with

$$\frac{\partial G}{\partial q} = \alpha (T - T_c) q + b q^3 - \lambda P = 0 \tag{44}$$

and inserting P from Eq. (43), we get

$$[\alpha (T - T_c) - \lambda^2 \chi_0] q + b q^3 = 0 \tag{45}$$

or

$$\alpha (T - T_c') q + b q^3 = 0 \tag{46}$$

with

$$T_c' = T_c + \frac{\lambda^2 \chi_0}{\alpha} \tag{47}$$

Equation (46) is the same relation as Eq. (19), hence q in the ordered phase will grow in the same way as the primary order parameter P in our previous consideration, i.e., according to Eq. (20), or

$$q(T) \sim (T_c' - T)^{1/2} \tag{48}$$

Equation (43) shows that this is also true for P as the secondary order parameter. However, we have a new Curie temperature given by Eq. (47), shifted upwards by an amount $\Delta T = T_c' - T_c$

$$\Delta T = \frac{\lambda^2 \chi_0}{\alpha} \tag{49}$$

Thus the coupling between the order parameters will always raise the transition temperature, i.e., stabilize the condensed phase. As we can see, the shift grows quadratically with the coupling constant λ.

Let us now look at the dielectric susceptibility in the high temperature phase $(T > T'_c)$. Both the order parameters q and P are zero in the field-free state, but when an electric field is applied it will induce a polarization and, by virtue of the coupling between P and q, also a nonzero value of q. In the high temperature phase we will thus have a kind of reverse effect brought about by the same coupling mechanism: whereas in the low temperature phase a non-zero q induces a finite P, here a non-zero P induces a finite q.

The values of P and q above T'_c are small, so we can limit ourselves to the quadratic terms for G and write

$$G = \frac{1}{2} a q^2 + \frac{1}{2} \chi_0^{-1} P^2 - \lambda q P - P E \quad (50)$$

We then find

$$\frac{\partial G}{\partial q} = a q - \lambda P = 0 \quad (51)$$

giving

$$q = \left(\frac{\lambda}{a} \right) P \quad \text{for} \quad T > T'_c \quad (52)$$

Because P is proportional to E, in the high temperature phase we induce a nonzero value of the order characteristic of the low temperature phase by applying a field, and this order is proportional to the field. In liquid crystals, the electroclinic effect is an example of this mechanism. This is not a normal pretransitional effect, because it is mediated by the kind of order that shows no translational instability and its susceptibility $\varepsilon_0 \chi = (\partial P / \partial E)$. Sometimes it will also be convenient to introduce a corresponding susceptibility for q, $\chi_q = \partial q / \partial E$. If we minimize G with respect to P in Eq. (50)

$$\frac{\partial G}{\partial P} = \chi_0^{-1} P - \lambda q - E = 0 \quad (53)$$

and insert q from Eq. (52), we find

$$P \left(1 - \frac{\lambda^2 \chi_0}{a} \right) = \chi_0 E \quad (54)$$

Therefore

$$\varepsilon_0 \chi = \frac{\chi_0 (T - T_c)}{T - T_c - \lambda^2 \chi_0 / \alpha} \quad (55)$$

or

$$\varepsilon_0 \chi = \chi_0 \frac{T - T_c}{T - T'_c} \quad (56)$$

Finally, with $T'_c - T_c = \Delta T = \lambda^2 \chi_0 / \alpha$, we obtain the expression

$$\varepsilon_0 \chi = \chi_0 + \frac{\lambda^2 \chi_0}{T - T_c} \quad (57)$$

If the coupling constant λ is sufficiently large, we see that χ will satisfy the Curie–Weiss law. Then the dielectric susceptibility will indeed show a divergent behavior, although P is not the transition parameter. If λ is small, the susceptibility will instead be practically independent of the temperature except in the vicinity of T_c, where we might observe some slight tendency to a divergent behavior.

In conclusion, the case where q and P have the same symmetry in many respects resembles the case where P is the primary order parameter. When there is a strong coupling between the two order parameters we might not see too much of a difference.

In the case where q and P do not have the same symmetry properties, the term $-\lambda q P$ is not an invariant and cannot appear in the Landau expression. Therefore it must be replaced by some other invariant like $-q^2 P$ (depending on the actual symmetries) or similar terms like $q^2 P_i P_j$. The coupling will then take on a different character. In the case where we have two order parameters q_1 and q_2, which couple to P, we may have a term $-q_1 q_2 P$, and so on. In this last case, Blinc and Žekŝ found ([30], pp. 45, 58) that P will not be proportional to q_1 and q_2, but to their squares, which means that P will grow linearly below the transition point T_0

$$P \sim (T_0 - T) \qquad (58)$$

instead of parabolically [Eq. (20)], as in normal ferroelectrics. Furthermore, the susceptibility is found to be practically independent of temperature above T_0, as well as below it, and makes a jump to a lower value when we pass below T_0. The nondivergence of the susceptibility is due to the fact that the fluctuations in q_1 and q_2 are not coupled to those of P in the first order in the high temperature phase [30], hence χ does not show a critical behavior.

Ferroelectrics in which the polarization P is not the primary order parameter are called improper ferroelectrics. The concept was introduced by Dvořak [88] in 1974. As we have seen in the last example, improper ferroelectrics may behave very differently from proper ferroelectrics. The differences are related to the nature of the coupling. If we write the coupling term $-\lambda q^n P$, our first example showed that for $n = 1$, the behavior may, however, be very similar to that of proper ferroelectrics. This class of improper ferroelectrics for which $n = 1$ is called pseudoproper ferroelectrics. Ferroelectric liquid crystals belong to this class.

2.2.7   The Piezoelectric Phase

The paraelectric phase has higher symmetry than the ferroelectric phase. However, it might have sufficiently low symmetry to allow piezoelectricity. In the case where the paraelectric phase is piezoelectric, it will become strained when electrically polarized. In addition to the term $-PE$ there will appear a term $-s\sigma$ in the free energy, where s designates the strain and σ the applied stress. The stress σ is a "field" corresponding to the field E and in a thermodynamic sense the intensive conjugate variable to the extensive variable s, just like E is to P (or H to M, or T to S, etc.). If we suppose that the elastic energy is a quadratic function of the strain, we can write (for materials in general)

$$G = \frac{1}{2\chi_0} s^2 - s\sigma \qquad (59)$$

so that we deduce the equilibrium strain from

$$\frac{\partial G}{\partial s} = \frac{1}{\chi_0} s - \sigma = 0 \qquad (60)$$

or

$$s = \chi_0 \sigma \qquad (61)$$

Thus the strain is a linear function of the applied stress (Hooke's law). The Landau expansion for a paraelectric phase with piezoelectric properties will now contain this new term in addition to a coupling term $-\lambda s P$, of the same form as in our previous case

$$G = \frac{1}{2} a P^2 + \frac{1}{2\chi_0} s^2 - \lambda s P - s\sigma - PE \qquad (62)$$

Now, $\partial G/\partial P = 0$ and $\partial G/\partial s = 0$ give the equilibrium values of P and s, according to

$$aP - \lambda s - E = 0 \qquad (63)$$

or

$$E = aP - \lambda s \qquad (64)$$

and

$$\frac{1}{\chi_0} s - \lambda P - \sigma = 0 \qquad (65)$$

or

$$\sigma = \frac{1}{\chi_0} s - \lambda P \qquad (66)$$

If we arrange the experimental conditions such that we apply an electric field at the same time as we mechanically hinder the crystal's deformation (this is called the "clamped crystal" condition), then $s = 0$, and Eq. (64) gives

$$E = aP \tag{67}$$

Hence

$$1 = a \frac{\partial P}{\partial E} \tag{68}$$

whence

$$\varepsilon_0 \chi = \frac{1}{a} = \frac{1}{\alpha(T - T_c)} \tag{69}$$

Thus the Curie law is fulfilled. In this situation a stress will develop to oppose the effect of the field and its induced polarization. Its value is, from Eq. (66)

$$\sigma = -\lambda P \tag{70}$$

If instead we have the "free crystal" condition, there can be no stress and, with $\sigma = 0$ in Eq. (65), we get

$$s = \lambda \chi_0 P \tag{71}$$

The insertion of s in Eq. (64) now gives

$$E = aP - \lambda^2 \chi_0 P \tag{72}$$

or, differentiating with respect to E

$$1 = (a - \lambda^2 \chi_0) \frac{\partial P}{\partial E} \tag{73}$$

Thus

$$\varepsilon_0 \chi = \frac{1}{a - \lambda^2 \chi_0} = \frac{1}{\alpha(T - T_c) - \lambda^2 \chi_0} \tag{74}$$

or

$$\varepsilon_0 \chi = \frac{1}{\alpha(T - T_c')} \tag{75}$$

with a new Curie point

$$T_c' = T_c + \frac{\lambda^2 \chi_0}{\alpha} > T_c \tag{76}$$

Thus we see again that the coupling raises the transition temperature. As an example, the paraelectric–ferroelectric transition point in a "free" crystal KDP lies 4 °C higher than for a clamped crystal. The reason that the Curie point is lower in the clamped crystal is that the mechanical clamping eliminates the piezocoupling.

Finally, we want to see how the piezo-coefficients behave at the transition. These coefficients d_{ijk} are tensors of rank three defined by

$$P_i = d_{ijk}\sigma_{jk} \tag{77}$$

but as the tensorial properties do not interest us for the moment we continue to use a scalar description and write

$$d = \frac{\partial P}{\partial \sigma} \tag{78}$$

Due to the appearance of the two conjugate pairs $s\sigma$ and PE in the free energy expression (Eq. 62), a Maxwell relation

$$\left(\frac{\partial P}{\partial \sigma}\right)_E = \left(\frac{\partial s}{\partial E}\right)_\sigma \tag{79}$$

exists, which allows a very simple calculation of d. Rewriting Eqs. (64) and (66) to eliminate P in order to form $d = (\partial s/\partial E)_\sigma$ yields

$$\lambda E - a\sigma = \left(\frac{a}{\chi_0} - \lambda^2\right)s \tag{80}$$

and

$$\lambda = \left(\frac{a}{\chi_0} - \lambda^2\right)\left(\frac{\partial s}{\partial E}\right)_\sigma \tag{81}$$

This gives

$$d = \frac{\lambda \chi_0}{a - \lambda^2 \chi_0} \tag{82}$$

If we compare this with Eqs. (74) and (75), it is evident that it can be reshaped to

$$d(T) = \frac{\lambda \chi_0}{\alpha(T - T_c')} \tag{83}$$

Hence the piezocoefficient diverges just like the dielectric susceptibility at the para-electric–ferroelectric transition. This divergence has other consequences: because of the piezoelectric coupling between polarization and elastic deformation, it will influence the elastic properties of the crystal. If we look at the coefficient relating the strain to an applied stress, it can be measured under two different conditions, either keeping the electric field constant or keeping the polarization constant. In the latter case the crystal behaves normally, it stays "hard" when we approach the transition. In the former case the elastic coefficient diverges just like d, i.e., the crystal gets anomalously soft when we approach the transition.

The formalism developed in this section was first used to describe the piezoelectric effect. It was later employed to treat the electroclinic effect in orthogonal chiral smectics. As will be seen in Sec. 2.5.7–2.5.8 there are very close analogies, but also some characteristic differences between the two effects.

2.3   The Necessary Conditions for Macroscopic Polarization in a Material

2.3.1   The Neumann and Curie Principles

When we are dealing with the question of whether a material can be spontaneously polarized or not, or whether some external action can make it polarized, there are two principles of great generality which are extremely useful, the Neumann principle and the Curie principle. Good discussions of these principles are found in a number of books, for instance [24, 36, 89]. The first of

these principles, from 1833 and named after Franz Neumann, who founded what must be said to be the first school of theoretical physics in history (in Königsberg), says that any physical property of a medium must be invariant under the symmetry operations of the point group of the medium.

The second, from 1894, after Pierre Curie, says that a medium subjected to an external action changes its point symmetry so as to preserve only the symmetry elements common with those of the influencing action. Symbolically we may write the Neumann principle as

$$K \subseteq P \tag{84}$$

i.e., the crystal group either coincides with P, the property group, or is a subgroup of P, and the Curie principle as

$$\tilde{K} = K \cap E \tag{85}$$

i.e., the point symmetry K changes to $\tilde{K}$, the highest common subgroup of K (the crystal group) and E (the external-influence symmetry group). This means that, in particular, if

$$K \subset E \tag{86}$$

then

$$\tilde{K} = K \tag{87}$$

i.e., if K is a subgroup of E, then the crystal symmetry is not influenced by the external action.

The Neumann and Curie principles have long been the dominating symmetry principles in condensed-matter physics. Both can be formulated in a number of different ways. For instance, the Neumann principle may be stated as follows: 'the symmetry elements of an intrinsic property must include the symmetry elements of the medium'. This formulation stresses that every physical property may and often does have higher symmetry, but never less than the medium.

A well-known example of this is that cubic crystals are optically isotropic, which means that the dielectric permittivity has spherical symmetry in a cubic crystal. Another example is that the thermal expansion coefficient of a cubic crystal is independent of direction. In fact, if it were not, the crystal would lose its cubic symmetry if it were heated. Thus, as far as thermal expansion is concerned, a cubic crystal "looks isotropic" just as it does optically. Since, according to Neumann's principle, the physical properties of a crystal may be of higher symmetry than the crystal, we will generally find that they range from the symmetry of the crystal to the symmetry of an isotropic body. A more general example of higher symmetry in properties is that such physical properties characterized by polar second rank tensors must be centrosymmetric, whether the crystal has a center of symmetry or not, cf. Fig. 27. For, if a second rank tensor T connects the two vectors p and q according to

$$p_i = T_{ij} q_j \qquad (88)$$

and we reverse the directions of p and q, the signs of all the components p_i and q_j will change. The equation will then still be satisfied by the same T_{ij} as before.

How can we know that the Neumann principle is always valid? Consider, for the sake of argument, that the crystal group K had a symmetry operation that was not contained in the property group P. Then, under the action of this operation, the crystal would on the one hand coincide with itself, and on the other, change its physical properties. This inherent contradiction proves the validity of the principle. This principle is often used in two ways, although it works strictly in only one direction. Evidently, it can be used to find out if a certain property is permitted in a medium, the symmetry of which is known. However, it may and has also been used (with caution) as an aid in the proper crys-

tallographic classification of bodies from the knowledge of their physical properties.

2.3.2 Neumann's Principle Applied to Liquid Crystals

The general application of the Neumann principle in condensed matter is normally much more formalized (for instance, involving rotation matrices) than we will have use for. Few textbooks perform such demonstrations on an elementary level, but there are exceptions, for example, the excellent treatise by Nussbaum and Phillips [90]. For a liquid crystal, we illustrate the simplest way of using the symmetry operations of the medium in Fig. 13. (The same discussion, in more detail, is given in [46]). We choose the z-direction along the director as shown in Fig. 13a and assume that there is a nonzero polarization $P = (P_x, P_y, P_z)$ in a nematic or smectic A. Rotation by 180° around the y-axis transforms P_x and P_z into $-P_x$ and $-P_z$, and hence both of these components must be zero, because this rotation is a symmetry operation of the medium. Next, we rotate by 90° around the z-axis, which transforms the remaining P_y into P_x. If therefore P_y were nonzero, we would see that the symmetry operations of the medium are not symmetry operations of the property, in violation of Neumann's principle. Hence $P_y = 0$ and P will vanish.

In Fig. 13·b we have illustrated the C phase, normally occurring at a second-order phase transition A→C at a certain temperature T_c, below which the molecules start to tilt, and with an order parameter than can be written as $\Psi = \theta e^{i\varphi}$, where θ is the tilt angle with respect to the normal and φ is the azimuthal angle indicating the direction of tilt in the layer plane. The variable φ is a phase variable (the transition is helium-like) with huge fluctuations; the director in the C phase

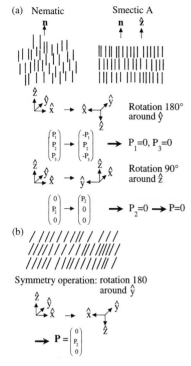

(a) Nematic   Smectic A

Rotation 180°
around $\hat{y}$

$\begin{pmatrix} P_1 \\ P_2 \\ P_3 \end{pmatrix} \rightarrow \begin{pmatrix} -P_1 \\ P_2 \\ -P_3 \end{pmatrix}$ → $P_1=0, P_3=0$

Rotation 90°
around $\hat{z}$

$\begin{pmatrix} 0 \\ P_2 \\ 0 \end{pmatrix} \rightarrow \begin{pmatrix} P_2 \\ 0 \\ 0 \end{pmatrix}$ → $P_2=0 \rightarrow P=0$

(b)

Symmetry operation: rotation 180
around $\hat{y}$

→ $P = \begin{pmatrix} 0 \\ P_2 \\ 0 \end{pmatrix}$

Figure 13. Neumann's principle applied to (*a*) the nematic and smectic A, (*b*) the smectic C and smectic C* phases.

therefore has a large freedom to move on a cone with half apex angle θ around the layer normal. In the figure, the tilt is chosen to be in the xz-plane. Hence, a 180° rotation around the y-axis is still a symmetry operation so P can have a component in the y-direction, whereas a 90° rotation around the z-axis is no longer a symmetry operation. However, if the xz-plane is a mirror plane, then P_y must be zero, because a nonzero P_y goes to $-P_y$ on reflection. On the other hand, if the medium lacks reflection symmetry, i.e., is chiral, such a component must be admitted. The important concepts chiral and chirality were introduced by Lord Kelvin, who derived them from the Greek word for hand. After having contemplated such symmetry–dissymmetry questions for at least a decade, he used them for the first time in his Baltimore lectures in the fall of 1884, pub-

lished twenty years later [91]. He there stated, "I call any geometrical figure or group of points chiral and say it has chirality if its image in a plane mirror ideally realized, can not be brought to coincide with itself". These concepts are used, and should be used, in exactly the same manner today.

Thus, in order to let P_y survive, we can put "small propellers" on the molecules, i.e., make them chiral. We could for instance synthesize molecules with one or more asymmetric carbon atoms, but it would, in principle, also be sufficient to dissolve some chiral dopant molecule in the nonchiral smectic C in order to remove the mirror plane. This can, symbolically, be written

$$C + * \rightarrow C* \qquad (89)$$

The smectic is now chiral and may be denoted C* (which is independent of whether there is any observable helix or not).

Whether the constituent molecules are chiral or whether we make the medium chiral by dissolving chiral molecules to a certain concentration $c > 0$, in both cases the only symmetry element left would be the twofold rotation axis along the y-direction, and a polarization along that direction is thus allowed. The symmetry of the medium is now C_2, which is lower than the symmetry of the property. The polarization P, like the electric field, is a polar vector, hence with the symmetry C_∞ (or ∞m in the crystallographic notation).

The fact that P, if admitted, must be 90 degrees out of phase with the director, could have been said from the very beginning. The fundamental invariance condition Eq. (6) means that if there is a polarization P in the medium, it cannot have a component along n, because the symmetry operation (6) would reverse the sign of that component, and thus P must be perpendicular to n

$$P \perp n \qquad (90)$$

It was the Harvard physicist Meyer who in 1974 first recognized that the symmetry properties of a chiral tilted smectic would allow a spontaneous polarization directed perpendicular to the tilt plane [61]. In collaboration with French chemists, he synthesized and studied the first such materials [62]. These were the first polar liquid crystals recognized and as such something strikingly new. As mentioned before, substances showing a smectic C* phase had been synthesized accidentally several times before by other groups, but their very special polar character had never been surmised. Meyer called these liquid crystals ferroelectric. In his review from 1977 [43] he also discussed the possible name antiferroelectric, but came to the conclusion that ferroelectric was more appropriate.

We will call the polarization P permitted by the symmetry argument just discussed, the spontaneous polarization. Leaving out its vectorial property, it is often indicated as P_s. The spontaneous polarization is so far a local property, not a macroscopic one.

2.3.3   The Surface-Stabilized State

According to the symmetry argument already given, the spontaneous polarization P is sterically connected with the molecule, lying along $\pm z \times n$. (The plus or minus sign is a characteristic of the material.) However, the prerequisite for the existence of P was that the molecules are chiral, and the chiral interactions between them lead to an incommensurate helical superstructure, as shown in Fig. 14. Actually there are two interactions involved in the helix, one chiral leading to a spontaneous twist and one nonchiral (steric) leading to a spontaneous bend deformation in the director field (see Sec. 2.4). The phase angle φ describes a right- or left-handed helix with a period of

Figure 14. The helical configuration of the sterically coupled variables $n - P$ in the unperturbed helical C* state. The phase angle φ changes at a constant rate if we move in a vertical direction. The variables n, P, and φ are all functions of z, the vertical coordinate. The twist is exaggerated about ten times relative to the densest possible twist occurring in reality.

typically several micrometers, whereas the smectic layer periodic length is several nanometers. The helical arrangement makes the macroscopic polarization zero for the bulk sample at the same time as satisfying $\nabla \cdot P = 0$ everywhere. It is therefore an alternative to the domain formation in a solid ferroelectric. However, there is no coercive force in such a structure, and if we apply an electric field perpendicular to the helix axis we will wind up the helix and turn more and more of the local polarization into the field direction (see Fig. 15). The $P-E$ response is dielectric – in fact it has the shape of the curve shown at the bottom of Fig. 5, corresponding to an antiferroelectric with an infinitely thin hysteresis loop.

This behavior is also known in the solid state, e.g., in the chiral structure $NaNO_2$, and the order is called helicoidal antiferroelectric. A shorter useful name is helielectric. The helielectric smectic C* has zero macroscopic polarization (like an antiferroelectric), no hysteresis, no threshold, and no bistability. However, by an artifice it can be turned into a structure with very different properties. This is illustrated in Fig. 16. If the smectic layers are made perpendicular to the confining glassplates, there is no boundary condition compatible with the

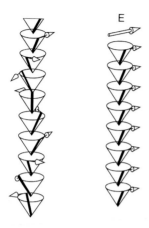

E

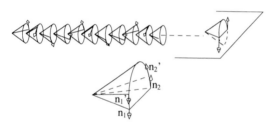

Figure 15. The helical configuration of the director–polarization couple is unwound by a sufficiently strong electric field E. The increasing field induces a macroscopic polarization (which is thus not spontaneous) and finally polarizes the medium to saturation (all dipolar contributions lined up parallel to the field), as shown to the right.

Figure 16. Elastic unwinding, by the surfaces, of the helical twist in the bookshelf geometry of smectic C*. The helical bulk state is incompatible with the surface conditions and therefore can never appear in sufficiently thin cells. The surface acts as an external symmetry breaking agent, reducing the degeneracy of the bulk to only two selected states. In the most attractive case these are symmetric, which also leads to a symmetric bistability of the device. The two memorized director states, n_1 and n_2 (n_1' and n_2' in the case of a pretilt different from zero) represent polarization states of opposite (or nearly opposite) direction.

helical arrangement. Let us assume that we can prepare the surface in such a way that the boundary condition imposed on the molecules it such that they have to be parallel to the surface, but without specified di-

rection. As we make the distance between the surfaces smaller and smaller, the conflict between the helical order and the surface order will finally elastically unwind the helix via the surface forces, and below a certain sample thickness the helix cannot appear: the nonhelical configuration has now a lower energy. The physical problem has one characteristic length, which is the helical period z. We can therefore expect that for a sample thickness $d < z$, i.e., of the order of one micrometer, the only allowed director positions will be where the surface cuts the smectic cone, because here both the intrinsic conical constraint and the constraint of the surface are simultaneously satisfied. There are two such positions, symmetrical around the cone axis and corresponding to polarization up and down, respectively. Energetically these states are equivalent, which leads to a symmetric bistability. Indeed, when very thin samples of $d \approx 1\ \mu m$ are made with the appropriate boundary conditions, spontaneous ferroelectric domains all of a sudden appear in the absence of any applied field (Fig. 17). This second step was realized five years later than Meyer's first paper [93]. By applying an external field we can now get one set of domains to grow at the cost of the other and reverse the whole process on reversing the field. There are two stable states and a symmetric bistability; the response has the characteristic form of a ferroelectric hysteresis loop, as represented by the second curve of Fig. 5. We might therefore call this structure a surface-stabilized ferroelectric liquid crystal (SSFLC). The surface stabilization brings the C* phase out of its natural crystallographic state and transfers macroscopic polarization to the bulk.

We may note that the polarization in the helical state of Fig. 14 is denoted P and not P_s. This is because the local polarization in this state does not correspond to P_s, but has

Figure 17. Spontaneous ferroelectric domains appearing in the surface-stabilized state.

a second contribution due to the flexoelectric effect. This contribution, which will be discussed in detail later, is of the same order of magnitude as P_s. There are several different methods to measure P_s, most of them involving the application of an electric field to saturate the polarization between the two extreme states, as represented to the right of Fig. 15, and the corresponding state with the field reversed. In such a state there is no flexoelectric contribution, hence P corresponds to P_s except for a contribution due to the electroclinic effect. This is an induced polarization, P_i, connected to an extra tilt $\delta\theta$ of the director caused by the field. The electroclinic tilt is negligible compared to θ for $T \ll T_c$, thus P_i is normally quite small, but

may be observable at high fields near $T = T_c$, and may then slightly change the shape of the observed polarization curve, as illustrated in Fig. 18.

The helical smectic C* state has the point symmetry D_∞ ($\infty 22$), illustrated in Fig. 19, which does not permit a polar vector. It is therefore neither pyroelectric nor ferroelectric. Nor can it, of course, be piezoelectric, which is also easily realized after a glance at Fig. 14: if we apply a pressure or tension vertically, i.e. across the smectic layers (only in this direction can the liquid crystal sustain a strain), we may influence the pitch of the helix but no macroscopic po-

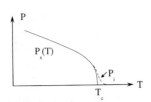

Figure 18. Polarization as a function of temperature, as often observed in a measurement. P corresponds to the spontaneous polarization P_s except in a small region around T_c where an induced polarization due to the electroclinic effect may influence the measurements.

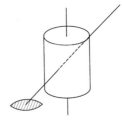

Figure 19. The point symmetry of the helical smectic C* state is D_∞ ($\infty 22$) illustrated by a twisted cylinder. The principal rotation axis is along the smectic layer normal and there are an infinite number of twofold rotation axes perpendicular to this axis, one of them illustrated to the right. The symmetry does not allow pyroelectricity.

larization can appear. On the other hand, if we do the same to the surface-stabilized structure (see Fig. 20), we change the tilt θ and thereby the local polarization P, hence in conformity with the statement according to the scheme of Fig. 6 that if a structure is ferroelectric it also, by necessity, has to be piezoelectric. That the SSFLC structure with symmetry C_2 is also pyroelectric, as it must be, is also evident, because θ and thereby P is a function of the temperature T. As there is now a macroscopic polarization in this state, we have a strong coupling with the external field leading to a high-speed response according to

$$\tau^{-1} \sim \frac{P \cdot E}{\gamma} \qquad (91)$$

where γ is a characteristic viscosity for the motion around the cone. The difference 2θ between the two optic axis directions (see Fig. 16) leads to an electrooptic effect of high contrast. The high available contrast and the very high (microsecond level) speed in both directions at moderate applied voltages, together with the inherent memory, make this a very attractive electrooptic effect for a variety of applications.

While the basic idea behind the work of [93] was to get rid of the helical space mod-

ulation of the tilt direction (i.e., of the infinity of polar axes) and achieve bistability by fitting the director to satisfy the cone condition and the surface condition at the same time, it was soon recognized in subsequent papers [94, 95] that the bistable switching demonstrated in [93] and the observed threshold behavior, the symmetric bistability, and the shape of the hysteresis curve (see Fig. 21) all came very close to the properties of materials previously classified as ferroelectrics, whereas these properties are absent in the helical smectic C* state. Therefore the concept surface-stabilized ferroelectric liquid crystal (SSFLC) was coined and used for the first time in print three years later in [96] from 1983. This was also first paper in which acceptable monodomain samples were demonstrated, awakening the interest in the display industry and pointing to the future potential of the surface-stabilized smectic C* state. With the developing skill in alignment techniques the first simple device prototypes could also be demonstrated [96a] even before room temperature mixtures were available. Finally, the study

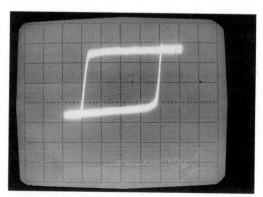

Figure 21. Oscilloscope picture showing photodiode response (change in optical transmittivity of the cell) to triangular pulses of opposite polarity (4 V peak-to-peak), demonstrating threshold, saturation, memory, and symmetry bistability; cf. Fig. 11. The material is DOBAMBC at 60 °C. Horizontal scale 1.5 V/div (from [95]).

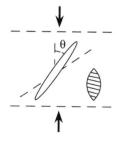

Figure 20. The surface-stabilized smectic C* has C_2 symmetry. It has a single polar axis. A pressure applied across the layers increases the tilt angle and thereby the polarization ($P \sim \theta$) along the twofold axis which is perpendicular to the paper plane.

of the very characteristic defect structures which showed up in monodomain samples led to the discovery of the chevron local layer structure, which will be discussed in detail in Sec. 2.8.

As already pointed out, one condition for achieving the fast bistable switching and all the additional characteristic properties is to get rid of the "antiferroelectric helix" (to use the expression employed in the 1980 paper [93]). Therefore, achieving the same properties by adding appropriate chiral dopants to the smectic C* could be imagined, in order to untwist the helical structure but, as the same time, keeping a residual polarization. This is illustrated by the structure to the left in Fig. 22. However, such a hypothetical bulk structure is not stable in the chiral case. It would transform to a twisted state where the twist does not take place from layer to layer, but in the layers, in order to cancel the macroscopic polarization. For this reason, surface-stabilization requires not only that $d < z$, but that the sample thickness is sufficiently small to prevent the ap-

pearance of this different twist state, by forcing it to appear on such a small length scale that its elastic deformation energy is higher than the energy of the untwisted state. The first analysis of these questions was made by Handschy and Clark [98].

2.3.4   Chirality and its Consequences

Chirality is a symmetry concept. We note from Kelvin's definition that being chiral is a quality and that this quality is perfectly general, regardless of the nature of the object (except that it must be three-dimensional, if we speak about chirality in three dimensions as Kelvin did). Thus we may speak of a "chiral molecule" and of a "chiral phase" or medium made up of chiral molecules or of chiral or nonchiral molecules ordered in a chiral fashion. Examples of nonchiral molecules ordered in a chiral fashion in the crystal state are α-quartz and sodium chlorate where, respectively, SiO_2 and $NaClO_3$ molecules are arranged in a helical order. We thus find right- and left-crystals of both substances. The reason for this is simply that the energy is lower in the helical state. Other examples are sulfur and selenium, which are found to form helical structures, thus ordering to a chiral structure although the atoms themselves are nonchiral. (Sometimes the designation "structure chirality" or "phase chirality" is used to distinguish this phenomenon where nonchiral objects build up a chiral structure.) In these examples we see the important one-wayness of chirality: a phase, i.e., an ensemble of molecules or atoms, may be chiral whether the constituents are chiral or not, but a phase built from chiral objects is always chiral, i.e. it can never possess reflection symmetry. When we here talk about chiral objects, we do not consider the trivial case which might

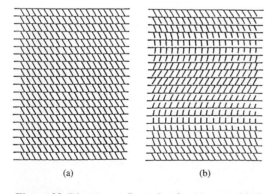

(a)   (b)

Figure 22. Director configuration for (a) a nonchiral smectic C and (b) a chiral smectic C* in its natural (helicoidal) structure (after [97]). The first could also be imagined as the bulk structure for a chiral smectic C* with infinite pitch, which would then correspond to a macroscopic polarization P pointing towards (or from) the reader. However, such a structure is not stable.

be taken as an exception, when there are exactly as many right-handed as left-handed objects, which are each others mirror images (racemic mixture). We may also remember that when talking about a chiral phase, we consider its static properties for which all molecules are in their time-averaged configurations (organic molecules would otherwise have very few symmetry elements). This corresponds to considering atoms to be situated on regular points in a lattice, without considering individual displacements due to thermal motion. In the case where only the building blocks are chiral, the optical activity is relatively weak (as in liquid solutions), whereas if there is a helical superstructure, this activity can be large as in quartz, $NaClO_3$, or immense as in liquid crystals, because the effect sensitively depends on d/λ, the ratio of the stereospecific length to the wavelength of light, which is appreciable in the latter cases. When the building blocks have higher symmetry than the structure, as in $NaClO_3$, there is, as we just mentioned, an inherent tendency to arrange them to a structure without a mirror plane and a center of symmetry. However, if we take a pure left-$NaClO_3$ crystal, dissolve it in water and let it crystallize from this solution, there is an equal chance to build up both mirror image arrangements. Thus from a solution of a left-crystal $NaClO_3$, right- and left-crystals grow with equal probability. This is a local breaking of a non-chiral symmetry to chiral, although the symmetry is preserved on a global scale.

A considerable number of such cases exist in solid crystals, although they are, relatively speaking, not very frequent. In principle, chiral "domains" or "crystallites" could be expected to be created in the same way in liquid crystals from nonchiral mesogens; see the discussion in Sec. 2.5.6.

A chiral phase normally shows optical activity. On the other hand, if we observe optical activity we know for sure that the phase is chiral. Whereas chirality is a property that does not have a length scale, optical activity is a global, bulk property and it can only be measured on a large ensemble of molecules. It is therefore less correct to speak about "optically active molecules". Optical activity and chirality are of course strongly related, because the optical activity is one of the consequences of chirality. We might, for instance, have a chiral substance that does not show any optical activity, because different parts in the molecule give rise to antagonistic contributions. If we now mix this substance with a nonchiral substance, the result may very well be that we find an easily measurable optical activity, and we may even see immediately, in the case of liquid crystals, that the phase gets twisted, thus a clear manifestation of chirality. Descriptions of such a case as "the chiral molecule does not show any chirality in its pure state" are not uncommon, but should be avoided, apart from their lacking logic, because they confuse cause and effect. Optical activity (or more precisely optical rotatory dispersion) is a quantitatively measurable property, like the helical twisting power, and both are therefore very useful to quantify the manifestations of chirality. However, expressions like "more or less chiral" are just as devoid of sense as "more or less cubic". A structure can have cubic symmetry, or else it does not, but there can be nothing in between. Likewise it is not the chirality that "goes down by a factor of five on raising the temperature" (to cite one of many similar statements), but rather the twisting power or a similar quantitative measure. Nevertheless, this common misuse of the word "chiral" points to a real problem.

Now, let us go back to the example of the chiral smectic C, written C* whenever we want to emphasize the chirality. The symbolic relation (Eq. 89) means that if we have added a concentration $c \neq 0$ of chiral dopant to our smectic layer then a local polarization P is permitted perpendicular to the tilt plane. If $c = 0$, no polarization is permitted. This corresponds to the qualitative character of chirality. Lack of reflection symmetry is thus a prerequisite for polarity. The argument does not say anything about the size of the effect and could not, because it is a symmetry argument. We have to take recourse to other arguments in order to say anything about the size of the effect. For instance, if we only add one dopant molecule, only the rotational motion of this molecule will be biased if we disregard its weak interaction with its neighboring molecules (see further discussion in Sec. 2.5.2), and a fraction of its lateral dipole moment will show up in P. If we increase the concentration, it is highly probable that P will be proportional to c, at least for small values of the dopant concentration. This is also borne out by experiment. This case is very similar to the case of dissolving noncentrosymmetric molecules in a liquid solution. Such a solution shows an optical activity that is, at least for low concentrations c, proportional to c. The sense of rotation depends on the solvent and is thus not just specific of the noncentrosymmetric molecule. We can expect the same thing to happen in the smectic C* case: the sense of optical rotation, the sense of P, and the sense of the helical twist will all depend on both the solute and the solvent, and cannot (so far) be predicted from first principles but have to be measured, at least until sufficient empirical data has been collected. We will define the polarization as positive if it makes a right-handed system with the layer normal and the director, and hence lies in the same direction as $z \times n$ (see Fig. 23). A priori, there are two classes of chiral smectics, those with $P > 0$ and those with $P < 0$. The majority of synthesized materials belongs to the second class, which

Figure 23. Spatial relation between layer normal, director, and local polarization for the two possible classes (+/–) of chiral smectics.

means that their opposite enantiomers would belong to the first class. A right-handed helix would also correspond to a left-handed and vice versa between the enantiomers. However, there is no a priori relation between the sense of P and the sense of the helix in a certain compound. The recognition of chiral effects and their importance is growing and it is not improbable that chiral effects will bring new and important perspectives to liquid crystal research, as well as to physics and chemistry in general in the coming decades. Thus there is a great need to be able to quantify chirality and chiral effects in different ways. This might be done by introducing convenient order parameters for different chiral manifestations, but the problem is exceedingly complex because, while chirality itself does not have a length scale, the effects of chirality range from intramolecular to bulk.

2.3.5   The Curie Principle and Piezoelectricity

If we want to investigate the conditions for an elastic stress to induce a macroscopic polarization, we have to turn to the Curie principle, which in a sense is a generalization of Neumann's principle. It cannot be proven in the same way as that principle; in fact it is often the violations (or maybe seeming violations) that are the most interesting to study. Among these cases are the phase transitions (spontaneous symmetry breakings) where the temperature is the "external force" in Curie's language. As mentioned before, Curie's principle can be stated in different ways and the earliest formulation by Curie himself is that "when a cause produces an effect, the symmetry elements of the cause must be present in the effect". This means that the produced effect (the induced property) may have higher symmetry, but never less symmetry than the cause.

In particular in this formulation, the principle has to be used with care. An example of this is related by Weyl in his book *Symmetry* [99]. Weyl here tells about the intellectual shock the young Mach received when he learned about the result of the Ørsted experiment (see Fig. 24 taken from the book). The magnetic needle is deflected in a certain sense, clockwise or anti-clockwise when a current is sent in a certain direction through the conducting wire, and yet everything seems to be completely symmetric (magnet, current) with respect to the plane containing the needle and the conductor. If this plane is a mirror plane, the needle cannot swing out in any direction. The solution to this paradox is that this plane is not a mirror plane, because of the symmetry properties of the magnetic field. The problem is that while we easily recognize the reflection properties of geometric objects, we do not know a priori the corresponding properties for abstract things, e.g., physical quantities. In fact, the magnetic field has reflection properties such that it is rather well illustrated by Magritte's well-known surrealist painting where a man is looking into a mirror and sees his back. A lesson to be learned from this is that we cannot rely on appearances when we judge the symmetry of various fields and physical properties in general. Pierre Curie was the first person to study these symmetries in a systematic way and, in order to describe them, he introduced the seven limiting point symmetry groups (also called infinite or continuous point groups), which he added to the 32 crystallographic groups. With this combination he could classify the symmetry of all possible media and all possible physical properties, illustrating the continuous symmetries with drawings related to simple geometric objects as, e.g., in Fig. 25. Obviously, these continuous symmetries have a special relevance for liquid crystals, liquid crystal polymers, and liquids in general, being continuous media without a lattice.

Let us apply a stress to a general medium and ask under what conditions it could cause an electric displacement. The basic problem we then have to sort out is the proper description of the symmetry of stress. Clearly a stress cannot be described by a polar vector – there must be at least two. A simple illustration of this is given in Fig. 26. We see that in two dimensions a homogeneous tensile stress as well as a pure shear stress has two perpendicular mirror planes, one twofold rotation axis, and one center of symmetry (center of inversion). In three dimensions we have analogously three mirror planes, three twofold axes, and the center of symmetry, which we may enumerate as m, m, m, 2,2, 2, Z (see Fig. 27). These are the symmetry elements of the point group mmm (or D_{2h}). This is the orthorombic point

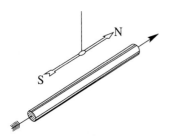

Figure 24. The symmetry paradox of the Ørsted experiment (after [99]).

552   2 Ferroelectric Liquid Crystals

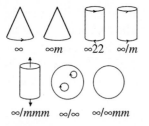

∞   ∞m   ∞22   ∞/m

∞/mmm   ∞/∞   ∞/∞mm

Figure 25. Pierre Curie's seven continuous point groups illustrated by geometric "objects". Among them we might distinguish ∞m representing a polar vector (like the electric field), and ∞/m representing an axial vector (like the magnetic field). Three of these symmetries, ∞, ∞22, and ∞/∞, can appear in a right-handed as well as in a left-handed form (enantiomorphic). Several equivalent ways of expressing the symmetry are in common use, for instance, the one in Fig. 19. Pyroelectricity (i.e., a macroscopic polarization) is only permitted by the first two groups (∞ and ∞m). They represent a chiral and a non-chiral version of a longitudinal ferroelectric smectic. If the continuous medium in question can sustain the mechanical strain, piezoelectricity would be allowed by the first three groups (∞, ∞m, and ∞22). The third of these (also written D∞) represents the symmetry of the cholesteric (N*) phase as well as the smectic A* phase and the helicoidal smectic C* phase, and the effects are, respectively, inverse to the flexoelectrooptic effect, the electroclinic effect and the deformed helix mode – all one-way effects. If we polymerize these phases (i.e. crosslink them to soft solids), the effects would become truly piezoelectric (two-way effects).

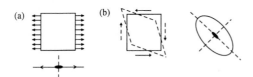

Figure 26. The symmetry of stress: (a) pure tensile stress, (b) pure shear stress.

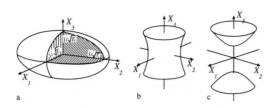

a   b   c

Figure 27. Characteristic surfaces of second-rank polar tensors: (a) ellipsoid, (b) hyperboloid of one sheet, (c) hyperboloid of two sheets. (After reference 36.)

group, which can be illustrated by a matchbox or a brick.

Now apply the stress in the most general way, i.e., such that its three axes of two-fold symmetry and its three planes of reflection symmetry do not have the same directions as the axes and planes in the unstrained medium. Curie's principle then says that the stressed medium will retain none of its symmetry axes and planes.

However, if the unstressed medium has a center of symmetry Z, it will retain that center because the stress also has a center of symmetry. Therefore, in a medium with inversion symmetry, no effect representable by an arrow can be induced by the stress, and hence no electric displacement or polarization, whatever stress is applied. In other words, a medium with inversion symmetry cannot be piezoelectric.

We could also have reasoned formally in the following way: If a property P (polarization in this case) should appear in the medium K as a result of the external action E (stress σ in this case), then P must be compatible with the symmetry of the strained medium, according to Neumann's principle

$$P \supseteq \tilde{K} \qquad (92)$$

but, according to Curie's principle

$$\tilde{K} = K \cap \sigma \qquad (93)$$

If we now insert ∞m for P and mmm for σ, we get from Eqs. (92) and (93)

$$\infty m \supseteq K \cap mmm \qquad (94)$$

The symmetry elements to the right (mmm) contain a center of inversion, but those to the left (∞m) do not. Therefore, if Eq. (94) should be satisfied (i.e., the Neumann and Curie principles together), K must not have a center of inversion, i.e.,

$$\bar{1} \notin K \qquad (95)$$

Out of the 32 crystallographic groups, 11 have symmetry elements including a center of inversion. Piezoelectricity should therefore be expected to appear in the other 21, but it appears only in 20. The exception is 432 ($\equiv$O), the octahedral group. However, this exception is sufficient to give the very important insight that the Neumann and Curie principles, and in fact all symmetry principles, only give necessary (never sufficient) conditions for a certain phenomenon to appear. The principles can only be used in the affirmative when they prevent things from happening.

2.3.6   Hermann's Theorem

It is often said that group 432 is "too symmetric" to allow piezoelectricity, in spite of the fact that it lacks a center of inversion. It is instructive to see how this comes about. In 1934 Neumann's principle was complemented by a powerful theorem proven by Hermann (1898–1961), an outstanding theoretical physicist with a passionate interest for symmetry, whose name is today mostly connected with the Hermann–Mauguin crystallographic notation, internationally adopted since 1930. In the special issue on liquid crystals by *Zeitschrift für Kristallographie* in 1931 he also derived the 18 symmetrically different possible states for liquid crystals, which could exist between three-dimensional crystals and isotropic liquids [100]. His theorem from 1934 states [101] that if there is a rotation axis C_n (of order n), then every tensor of rank $r<n$ is isotropic in a plane perpendicular to C_n, as shown in Fig. 28. For cubic crystals, this means that second rank tensors like the thermal expansion coefficient α_{ij}, the electrical conductivity σ_{ij}, or the dielectric constant ε_{ij}, will be isotropic perpendicular to all four space diagonals that have threefold symme-

try. This requires all these properties to have spherical symmetry, as already mentioned above for the optical wave surface and for the thermal expansion. It also means that crystals belonging to the trigonal, tetragonal, or hexagonal systems are all optically uniaxial (most liquid crystal cases can be included in these categories). The indicatrix is thus an ellipsoid of revolution, as illustrated in Fig. 29, with the optic axis along the threefold, fourfold, or sixfold crystallographic axis, respectively. If we have a fourfold axis C_4, then the piezocoefficients d_{ijk} will be isotropic perpendicular to C_4. Finally, if we have a sixfold axis C_6, the elastic constants c_{ijkl} will be isotropic perpendicular to C_6. Thus in a smectic B the layer can also be considered isotropic with respect to the elastic constants.

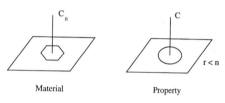

Material   Property

Figure 28. The crystal symmetry compared to the property symmetry around an *n*-fold rotation axis.

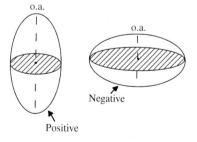

Positive   Negative

Figure 29. Media with a threefold, fourfold, or sixfold crystallographic axis must be optically uniaxial. Thus crystals belonging to the trigonal, tetragonal, or hexagonal system, including smectic B, must be uniaxial. The picture shows the indicatrix with its optic axis in the case of positive as well as negative birefringence.

There are now three cubic classes that do not have a center of symmetry. We should therefore expect these three – 23 (T), $\bar{4}3m$ (T_d), and 432 (O) – to be piezoelectric, but only the first two are. The latter has three fourfold rotation axes perpendicular to each other, making the piezoelectric tensor d_{ijk} isotropic in three dimensions. The piezo-effect would not then depend on the sign of the stress, which is only possible for all $d_{ijk} \equiv 0$.

It is clear that similar additional symmetries would occur frequently if we went to liquids, which are generally more symmetric than crystals.

2.3.7   The Importance of Additional Symmetries

A property that may be admitted by noncentrosymmetry may very well be ruled out by one of the other symmetry operations of the medium. As an example we will finally consider whether some of the properties discussed so far would be allowed in the cholesteric liquid crystals, which lack a center of inversion. A cholesteric is simply a chiral version of a nematic, abbreviated N*, characterized by the same local order but with a helical superstructure, which automatically appears if the molecules are chiral or if a chiral dopant is added (see Fig. 30). Could such a cholesteric phase be spontaneously polarized? If there were a polarization P, it would have to be perpendicular to n, because of the condition of Eq. (6), and thus along the helical axis direction m. However, the helical N* phase has an infinity of twofold rotation axes perpendicular to m and the symmetry operation represented by any of these would invert P. Hence $P=0$. A weaker requirement would be to ask for piezoelectricity. (Due to the helical configuration, the liquid has in fact some small

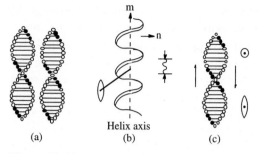

Figure 30. The symmetry of the cholesteric phase (a, b) and the effect of a shear (c).

elasticity for compression along m.) Nevertheless, as illustrated in the same figure, a compression does not change the symmetry, and hence a polarization $P \neq 0$ cannot appear due to a compression.

Finally, it is often stated that a medium that lacks a center of inversion can be used for second-harmonic generation (SHG). This is because the polarizability would have a different value in one direction compared with that in its antiparallel direction. Could this be the case for the cholesteric phase, i.e., could it have a nonlinear optical susceptibility $\chi^{(2)} \neq 0$? The answer must be no, as it cannot be in the n-direction, and the m-direction is again ruled out by the two-fold axis (one of the infinitely many would suffice). For the unwound (non-helical) N*, the two-fold axis along m works in the same way. The cholesteric example shows that it is not sufficient for a medium to lack a center of inversion in order to have SHG properties. Questions like these can thus be answered by very simple symmetry arguments, when we check how additional symmetries may compensate for lack of inversion symmetry. In contrast to the cholesteric phase, the unwound smectic C* phase does have a direction along which a second harmonic can be generated. This is of course the C_2 axis direction in the SSFLC geometry. Even if initially small, the existence of this effect was soon confirmed [102, 103] and studied in considerable detail by several groups [104]. The SSFLC structure is the only liquid (if it can be considered as such!) with SHG properties. No other examples – they can only be looked for among anisotropic liquids – are known, the externally poled electret waxes left aside (materials out of thermal equilibrium in which the polar axis is not an intrinsic property). Walba et al. were the first to synthesize C* molecules with powerful donor and acceptor groups active perpendicular to the director (thus along the C_2 axis), which is a necessity in order to achieve $\chi^{(2)}$-values of interest for practical applications [105, 106]. However, there is in fact really no point in having a surface-stabilized ferroelectric liquid crystals as an SHG material. A pyroelectric material would be much better, in which the polar axis is fixed in space and nonswitchable, but this cannot be achieved in the framework of liquids. The first such materials were recently presented [107], made by using an SSFLC as a starting material and crosslinking it to a nonliquid-crystalline polymer. It is thus a soft solid.

As we have seen, most liquid crystals have too high a symmetry to be macroscopically polar if they obey the $n \rightarrow -n$ invariance (which all "civilized" liquid crystals do, that is, all liquid crystal phases that are currently studied and well understood). The highest symmetry allowed is C_2 (monoclinic), which may be achieved in materials which are "liquid-like" at most in two dimensions. Even then external surfaces are required. Generally speaking, a polar liquid crystal tends to use its liquid translational degrees of freedom so as to macroscopically cancel its external field, i.e., achieve some kind of antiferroelectric order. For more "liquid-like" liquids, piezo-, pyro-, ferro-, and antiferroelectricity are a fortiori ruled out as bulk properties. These phenomena

would, however, be possible in crosslinked polymers (soft solids). A simple example may illustrate this. If we shear the cholesteric structure shown in Fig. 30c, Curie's principle tells us that a polarization is now permitted along the twofold axis shown in the figure. If the liquid crystal will not yield, this would constitute a piezoelectric effect. Hence liquid-crystalline polymers are expected to show a piezoelectric effect both in the N* phase, the A* phase, and the C* phase (as well as in chiral phases with lower symmetries). However, as already pointed out, these materials are not really liquid-like in any direction, thus not liquid crystals but soft solids (like rubber).

It may finally be pointed out that the symmetry operation $n \rightarrow -n$ describes a bulk property of liquid crystals. At surfaces, no such symmetry is valid. Therefore SHG signals can be detected from nematic surface states and have, indeed, been used to probe the order and directionality of such surface states [108]. As a general rule, a surface is always more or less polar, which certainly contributes to the complexity of alignment conditions at the interface between a surface and a liquid crystal with the special polar properties of the materials treated in this chapter. In this case, therefore, we have in principle a surface state with a polar property along n as well as perpendicular to n.

2.4 The Flexoelectric Polarization

2.4.1 Deformations from the Ground State of a Nematic

Let us consider a (nonchiral) nematic and define the ground state as one where the director n is pointing in the same direction everywhere. Any kind of local deviation from this direction is a deformation that involves a certain amount of elastic energy $\int_V G\,dV$, where G is the elastic energy density and the integration is over the volume of the liquid crystal. As $n(r)$ varies in space, G depends on the details of the vector field $n(r)$. Now, a vector field, with all its local variations, is known if we know its divergence, $\nabla \cdot n$, and curl, $\nabla \times n$, everywhere (in addition to how it behaves at the boundaries).

It was an advance in the theoretical description of liquid crystals (the continuum theory) when in 1928 Oseen, who had introduced the unit vector n, which de Gennes later gave the name "director" [38], showed [109] that the elastic energy density G in the bulk (i.e., discarding surface effects) can be written in the diagonal form

$$G = \frac{1}{2}\,K_{11}(\nabla \cdot n)^2 + \frac{1}{2}\,K_{22}\big[n \cdot (\nabla \times n)\big]^2$$
$$+ \frac{1}{2}\,K_{33}\big[n \times (\nabla \times n)\big]^2 \qquad (96)$$

This relation expresses the fact that the elastic energy is a quadratic form in three curvature deformations or strains, now called splay, twist, and bend, which we can treat as independent. They are sketched in Fig. 31. The splay is described by a scalar (a pure divergence), $\nabla \cdot n$, the twist is described by a pseudoscalar (it changes sign on reflection in a plane parallel to the twist axis or when we go from a right-handed to a left-handed reference frame), which is the component of $\nabla \times n$ along the director, $|\nabla \times n|_{\parallel}$, whereas the bend is described by a vector with the component of $\nabla \times n$ perpendicular to the director, $|\nabla \times n|_{\perp}$.

Oseen was also the first to realize the importance of the $n \rightarrow -n$ invariance [110], which he used to derive Eq. (96). This expression was rederived thirty years later by Frank in a very influential paper [111],

Figure 31. The three elementary deformations splay, twist, and bend. None of them possesses a center of symmetry.

which led to a revival in the international interest in liquid crystals. The denominations splay, twist, and bend stem from Frank. The Oseen–Frank constants K_{11}, K_{22}, and K_{33} are components of a fourth rank tensor, just like the ordinary (first order) elastic constants.

The question we will ask and answer in this section is whether these deformations will polarize the nematic. The reason for such a conjecture is, of course, the analogy with piezoelectric phenomena: as we have seen in Sec. 2.3.5 an elastic deformation may polarize a solid. Do we find the same phenomenon in the liquid crystal?

We recall from the discussion in Sec. 2.3.5 that a necessary condition for the appearance of a polarization was that the medium lacks a center of symmetry. The reason for this was that since at equilibrium the stress as well as the strain will be centrosymmetric, the piece of matter cannot develop charges of opposite sign at opposite ends of a line through its center if it has a center of symmetry, in accordance with Curie's principle. On the other hand, inspection of Fig. 31 immediately reveals that none of the three strains splay, twist, and bend has a center of symmetry. Hence Curie's principle allows a local polarization to appear as a result of such local deformations in the director field, even if the medium itself has a center of symmetry. We see from Fig. 31 that the splay deformation violates the $n \rightarrow -n$ invariance, whereas twist and bend do not. Therefore a polarization may appear along n in the case of splay, but has to be perpendicular to n in the case of twist and bend. In fact, as a result of a general local deformation in a nematic, a local polarization density P will appear in the bulk, given by

$$P = e_s n (\nabla \cdot n) + e_b n \times (\nabla \times n) \qquad (97)$$

This polarization is called flexoelectric and the phenomenon itself the flexoelectric effect.

2.4.2 The Flexoelectric Coefficients

Equation (97) consists of two parts, one of which is nonzero if we have a nonzero splay, the other of which is nonzero for a nonzero bend. The coefficients e_s and e_b for splay and bend are called flexoelectric coefficients and can take a plus or minus sign (a molecular property). In the case where e_s and e_b are positive, the polarization vector is geometrically related to the deformation in the way illustrated in Fig. 32. For a splay deformation, P is along the director and in the direction of splay. For a bend, P is perpendicular to the director and has the direction of the arrow if we draw a bow in the same shape as the bend deformation.

An important feature of Eq. (97) is that it does not contain any twist term, although the Curie principle would allow a polarization caused by a twist. However, as P would have to be perpendicular to n, it should then lie along the twist axis. Now, it is easy to see (see Fig. 33) that there is always a twofold rotational symmetry axis along the director in the middle of a twist. Thus any P along the twist axis would be reversed by a symmetry operation and must therefore be zero. In other words, the symmetry of twist does not permit it to be related to a polar

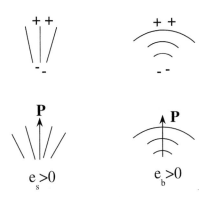

Figure 32. Geometrical relation between the local polarization density and the director deformation in the case of positive values for the flexoelectric coefficients. For negative values the direction of the induced dipole should be reversed. The size and sign of e_s and e_b (alternatively called e_1 and e_3) are a molecular property.

Figure 33. The symmetry of twist. If we locally twist the adjacent directors in a nematic on either side of a reference point, there is always a twofold symmetry axis along the director of the reference point. Therefore a twist deformation cannot lead to the separation of charges. Thus a nematic has only two nonzero flexoelectric coefficients.

vector (see our earlier discussion of additional symmetries in Secs. 2.3.6 and 2.3.7). Hence a twist deformation cannot lead to a separation of charges. This property singles out the twist from the other deformations, not only in a topological sense, and explains why twisted states are so common. The well-known fact that a nematic which is noninvariant under inversion is unstable against twist and normally adopts a helical (cholesteric) structure, would not have been the same if the twist had been connected with a nonzero polarization density in the medium. This also means that a "twisted nematic" as used in displays is not polarized by the twist.

2.4.3   The Molecular Picture

Recognition of the flexoelectric effect is due to Meyer [18], who in 1969 derived an expression equivalent to Eq. (97). He also had a helpful molecular picture of the effect, as illustrated in Fig. 34a. To the left in this picture, the unstrained nematic structure is shown with a horizontal director in the case of wedge-shaped and crescent-shaped molecules. To the right the same molecules are shown adjusted in their distribution corresponding to a splay and a bend distortion, respectively. In either case the distortion is

coupled to the appearance of a nonzero polarization density. As we can see, this is a steric (packing) effect due to the asymmetry of the molecular shape. The inverse effect is shown in Fig. 34b. When an electric field is applied it induces a distortion. In this figure we have illustrated a hypothetical compound splay–bend deformation and also used a slightly more general shape for the model molecules. It is clear that the observable effects will depend on the molecular shape as well as the size and distribution of the dipoles in the molecules. The inverse flexoelectric effect simply offers another mechanism for polarizing the medium, since the distortion will imply a polarization parallel to E. It is therefore a special kind of dielectric mode involving orientational polarization. In this respect, the inverse effect bears a certain resemblance to the field-induced lining up of dipoles in a paraelectric material, as the dipoles are already present in the molecules in both cases (see Fig. 35). Flexoelectric polarization leads to a separate contribution to the general term $G=-E \cdot P$ in the free energy. With Eq. (97) this contribution takes the form

$$G_f = -e_s E \cdot n (\nabla \cdot n) - e_b E \cdot [n \times (\nabla \times n)] \tag{98}$$

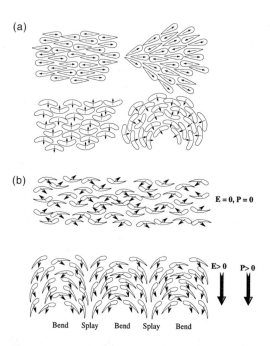

(a)

(b)

E = 0, P = 0

E > 0   P > 0

Bend   Splay   Bend   Splay   Bend

Figure 34. (a) The flexoelectric effect. A polarization is coupled to a distortion (from [18]). (b) The inverse flexoelectric effect. An applied field induces a distortion (from [112]). With our sign conventions, (a) corresponds to $e_s>0$, $e_b<0$, and (b) to $e_s<0$, $e_b<0$.

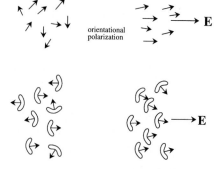

orientational polarization

E

E

Figure 35. Orientational polarization in an applied field is generally coupled to a director distortion in the liquid-crystalline state. The drawn direction of the dipoles corresponds to the case $e_b<0$.

2.4.4   Analogies and Contrasts to the Piezoelectric Effect

The flexoelectric effect is the liquid crystal analogy to the piezoelectric effect in solids. To see this we only have to make the connection between the translational variable in solids and the angular variable in liquid crystals. For both effects there is a corresponding inverse effect. However, the differences are notable. The piezoeffect is related to an asymmetry of the medium (no inversion center). The flexoelectric effect is due to the asymmetry of the molecules, regardless of the symmetry of the medium. We have seen how the elementary deformations in the director field destroy the center of symmetry in the liquid crystal. Therefore the liquid may itself possess a center of symmetry. In other words, the situation is just the opposite to the one in solids!

The fact that the flexoelectric effect is not related to the symmetry of the medium, in particular means that it is not related to chirality. It means that the effect is common to all liquid crystals. Therefore expressions like, for instance, a "flexoelectric nematic", which are encountered now and then, are nonsensical.

2.4.5   The Importance of Rational Sign Conventions

The flexoelectric effect is still a "submarine" phenomenon, which has so far mostly been of academic interest. One of the reasons is its complexity when we go beyond nematics, but another reason is that, even in nematics, the size and sign of the flexocoefficients are largely unknown. A condition contributing to this is that Eq. (97) can be written in a number of ways, not only in terms of the vector operators, but also corresponding to different sign conventions (which are mostly not stated). This leads to great confusion and difficulties in extracting correct values from published data (see the discussion in [112]). We may write the flexoelectric polarization (Eq. 97) in the form

$$P = e_s S + e_b B \tag{99}$$

As far as the director gradients are concerned we have chosen to use the form

$$B = n \times (\nabla \times n) \tag{100}$$

in the bend polarization density, corresponding to the most common form

$$G_b = \frac{1}{2} K_{33} [n \times (\nabla \times n)]^2 \tag{101}$$

in the Oseen–Frank expression for the bend elastic energy. Often this last expression is, however, written in the equivalent form

$$G_b = \frac{1}{2} K_{33} [(n \cdot \nabla)n]^2 \tag{102}$$

or

$$G_b = \frac{1}{2} K_{33} (n \cdot \nabla n)^2 \tag{102a}$$

for instance, found with Oseen [109] and Nehring and Saupe [113], which then corresponds to a change of sign in B

$$B = -e_b (n \cdot \nabla)n \tag{103}$$

This is due to the equality

$$n \times (\nabla \times n) = -(n \cdot \nabla)n \tag{104}$$

It should be noted that Eq. (104) is not a general equality for vectors, but is only valid for a unit vector. In that case we may use the expansion rule for a triple vector product

$$n \times \nabla \times n = \nabla(n \cdot n) - (n \cdot \nabla)n \tag{105}$$

which is equal to $-(n \cdot \nabla)n$, since $n \cdot n = 1$.

If we use Eq. (103), the expression for the flexoelectric polarization (Eq. 97) changes to

$$P = e_s n (\nabla \cdot n) - e_b (n \cdot \nabla) n \qquad (106)$$

On the other hand, the signs in the expression could also change due to a different geometric convention adopted for the direction of P relative to the deformation. Thus Meyer in his first paper uses the opposite convention (Fig. 34) to ours (Fig. 32) for the positive direction of the P arrow, which leads to the form

$$P = e_s n (\nabla \cdot n) + e_b (\nabla \times n) \times n \qquad (107)$$

or

$$P = e_s n (\nabla \cdot n) - e_b n \times (\nabla \times n) \qquad (107\,a)$$

although in later work he changes this to conform with Fig. 32 as well as Eq. (97). This geometric convention for the direction of the P arrow was first proposed by Schmidt et al. [114]. The different sign conventions are discussed at length in [112]. It is important that a unique sign convention be universally adopted for the flexoelectric effect. Our proposal is to write P in the form given by Eq. (97), which corresponds to the commonly used form for bend in the elastic energy, combined with the geometric convention as expressed by Fig. 32, which is most natural and easy to remember. This will be applied in the following.

2.4.6   The Flexoelectrooptic Effect

The periodic deformation in Fig. 34 b cannot be observed in a nonchiral nematic because it does not allow for a space-filling splay–bend structure. Instead, such a pattern would require a periodic defect structure. However, we can continually generate such a space-filling structure without defects in a cholesteric by rotating the director everywhere in a plane containing the helix axis.

In one of his early classic papers from 1969, Bouligand [115] showed that if a cut is made in a cholesteric structure at an oblique angle to the helix axis, an arc pattern of the kind in Fig. 34b will be observed as the projection of the director field onto the cut plane, as illustrated in Fig. 36. We will call this oblique plane the Bouligand plane or the Bouligand cut. Evidently, if we turn the director around an axis perpendicular to the helix axis until it is aligned along the Bouligand plane, we will have exactly

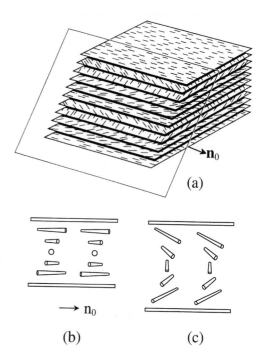

Figure 36. (a) Oblique cut through a cholesteric structure showing the arc pattern produced by the director projection onto the cut plane (Bouligand plane). In (b) the same right-handed twist is seen looking perpendicular to the twist axis, and in (c) looking perpendicular to the cut plane. Here the splay–bend pattern becomes evident, even if the directors are not lying in this plane (after Bouligand [116]). If an electric field is applied along n_0, which corresponds to the n direction at the top and bottom in (b), the directors will swing out around n_0 into a Bouligand plane corresponding to the value of the field.

the pattern of Fig. 36, and therefore a polarization along the plane. This consideration will facilitate the understanding of the flexoelectrooptic effect. This new linear dielectric mode was reported [117] by Patel and Meyer in 1987. It consists of a tilt ϕ of the optic axis in a short-pitch cholesteric when an electric field is applied perpendicular to the helix axis, as illustrated in Fig. 37. The optic axis is perpendicular to the director and coincides with the helix axis in the field-free state. Under an applied field, in its polarized and distorted state, the cholesteric turns slightly biaxial.

The physical reason for the field-induced tilt is that the director fluctuations, which in the field-free state are symmetrical relative to the plane perpendicular to the helix axis, become biased in the presence of a field, because for tilt in one sense, keeping E fixed, the appearing splay–bend-mediated polarization lowers the energy by $-E \cdot \delta P$, whereas a tilt in the opposite sense raises the energy by $+E \cdot \delta P$.

The mechanism is clear from a comparison of Fig. 36b, c. If we assume that the twist between top and bottom is 180°, the director at these planes is n_0, as indicated in Fig. 36b. Viewed perpendicular to the Bouligand plane, the cholesteric structure corresponds to "one arc" in the splay–bend pattern, but the directors do not yet lie in this plane. However, if they all turn around n_0 as the rotation axis, also indicated in Fig. 36a, they will eventually be in the Bouligand plane and the medium will thereby have acquired a polarization (see the lower part of Fig. 37) along n_0. Hence, if we apply a field along n_0, increasing its value continuously from zero, the optic axis will swing out continuously, as shown in Fig. 38, and the Bouligand plane (perpendicular to the axis) will be more and more inclined. At the same time, the polarization P increases according to Eq. (97) with the growing amplitude to the splay–bend distortion.

The flexoelectrooptic effect is a field-sensitive electrooptic effect (it follows the sign of the field), which is fast (typically $10-100$ µs response time) with two outstanding characteristics. First, the induced tilt ϕ has an extremely large region of linearity, i.e., up to 30° for materials with dielectric anisotropy $\Delta\varepsilon \approx 0$. Second, the induced tilt is almost temperature-independent. This is illustrated in Fig. 39 for the Merck cholesteric mixture TI 827, which has a temperature-independent pitch but not designed or optimized for the flexoelectrooptic effect in other respects.

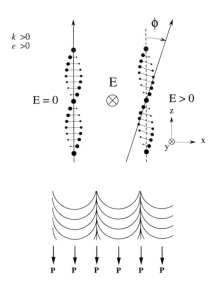

Figure 37. An electric field E applied perpendicular to the helix axis of a cholesteric will turn the director an angle $\phi(E)$ and thereby the optic axis by the same amount. The director tilt is coupled to the periodic splay–bend director pattern shown below, which is generated in all cuts perpendicular to the new optic axis. In this inverse flexoelectric effect, splay and bend will cooperate if e_s and e_b have the same sign. The relation between E and ϕ is shown for a positive helical wave vector k (right-handed helix) and a positive average flexoelectric coefficient $e = \frac{1}{2}(e_s + e_b)$. When the sign of E is reversed, the optic axis tilts in the opposite direction ($\phi \rightarrow -\phi$).

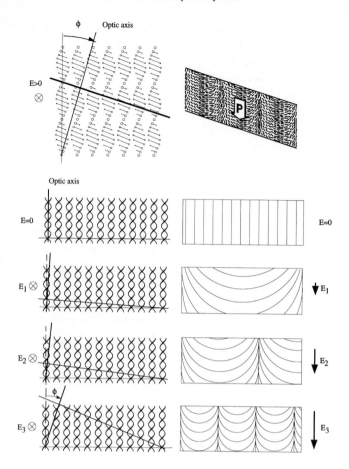

Figure 38. Field-induced tilt (*left*) and corresponding splay – bend distortion when looking at the Bouligand plane along the field direction (*right*). The same director pattern will be found in any cut made perpendicular to the tilted optic axis, whereas the director is homogeneous in any cut perpendicular to the nontilted axis.

It can be shown [117, 118] that the induced tilt is linear in E according to

$$\phi = \frac{eE}{Kk} \quad (108)$$

where

$$e \equiv \frac{1}{2}(e_s + e_b) \quad (109)$$

and

$$K \equiv \frac{1}{2}(K_{11} + K_{33}) \quad (110)$$

and k is the cholesteric wave vector. The response time is given by

$$\tau = \frac{\gamma}{Kk^2} \quad (111)$$

where γ is the characteristic viscosity. We note, in advance of discussing the same feature in the electroclinic effect, cf. Sec. 2.5.8, that the response time does not depend on the value of the applied electric field. The temperature independence is seen immediately from Eq. (108), because both e and K should be proportional to S^2, the square of the scalar nematic order parameter. Therefore a mixture with temperature-independent pitch (not too difficult to blend) will have a $\phi(E)$ that is independent of temperature.

Since the original work of Patel and Meyer [117], the effect has been further investigated [119 – 122], in the last period [112, 118, 123 – 126] with a special empha-

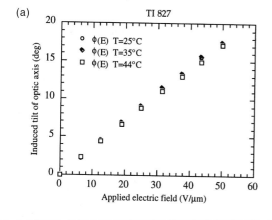

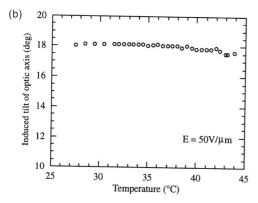

Figure 39. The flexoelectrooptic effect measured on the Merck mixture TI 827. (a) Induced tilt as a function of E. (b) Induced tilt as a function of temperature.

sis on ruling out the normal dielectric coupling and increasing the linear range, which is now larger than for any other electrooptic effect in liquid crystals. With the available cholesteric materials, the applied field is quite high but can be estimated to get lowered by at least a factor of ten if dedicated synthetic efforts are made for molecules with convenient shapes and dipoles.

The flexoelectrooptic effect belongs to a category of effects that are linear in the electric field, and which all have rather similar characteristics. These include the electroclinic effect, the deformed helical mode in the C* phase, and the linear effect in anti-

ferroelectrics. They all belong to the category of "in-plane switching" with a continuous grayscale and still have a great potential to be exploited in large scale applications.

2.4.7 Why Can a Cholesteric Phase not be Biaxial?

The nonchiral nematic is optically a positive uniaxial medium. A cholesteric is a nematic with twist. The local structure of a cholesteric is believed to be the same as that of the nematic except that it lacks reflection symmetry. This means that the director and therefore the local extraordinary optic axis is rotating around the helix axis making the cholesteric a negative uniaxial medium with the optic axis coinciding with the twist axis. The question has been asked as to why the nematic with twist could not be biaxial, and attempts have been made to measure a slight biaxiality of the cholesteric phase. In other words, why could the twist not be realized in such a way that the long molecular axis is inclined to the twist axis? Why does it have to be perpendicular?

Section 2.4.6 sheds some light on this question. As we have seen, as soon as we make the phase biaxial by tilting the director the same angle out of the plane everywhere, we polarize the medium. Every fluctuation δn out of the plane perpendicular to the helix axis is thus coupled to a fluctuation δP raising the free energy, $\sim(\delta P)^2$. The energy of the state is thus minimized for $\phi=0$, which is the ground state. In other words, the N* state cannot be biaxial for energetic reasons, just as the N state cannot be polar for the same reason (see the discussion in Sec. 2.2.2). In other words, the cholesteric phase is uniaxial because this is the only nonpolar state.

2.4.8   Flexoelectric Effects in Smectic A Phases

In smectics such deformations which do not violate the condition of constant layer thickness readily occur. In a smectic A (or A*) this means that $\nabla \times n = 0$ and the only deformation is thus pure splay, in the simplest approximation. We will restrict ourselves to this case, as illustrated in Fig. 40. Special forms are spherical or cylindrical domains, for instance, the cylindrical structure to the right in the figure, for which a biologically important example is the myelin sheath which is a kind of "coaxial cable" around a nerve fiber. Let us consider such a cylindrial domain.

The director field in the xy plane is conveniently described in polar coordinates ρ and θ as

$$n = (n_\rho, n_\theta) = (1, 0) \tag{112}$$

from which we get the divergence

$$\nabla \cdot n = \frac{1}{\rho}\frac{\partial}{\partial \rho}(\rho n_\rho) + \frac{1}{\rho}\frac{\partial n_\theta}{\partial \theta} = \frac{1}{\rho}\frac{\partial}{\partial \rho}\rho = \frac{1}{\rho} \tag{113}$$

This gives us the flexoelectric polarization density

$$P = e_s\, n (\nabla \cdot n) = e_s\, n\left(\frac{1}{\rho}\right) \tag{114}$$

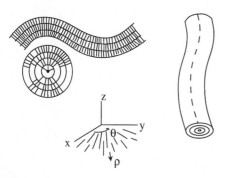

Figure 40. The easy director deformation in a smectic A is pure splay which preserves the layer spacing everywhere.

The polarization density thus falls off as $1/\rho$. In the center of the domain we have a disclination of strength one. Because the disclination has a finite core we do not have to worry about P growing infinite. It is interesting to take the divergence of this polarization field. We get

$$\nabla \cdot P = \frac{1}{\rho}\frac{\partial}{\partial \rho}(\rho P_\rho) = 0 \tag{115}$$

Thus the polarization field is divergence free in this geometry, which makes it particularly simple, because a nonzero $\nabla \cdot P$ would produce space charges and long range coulomb interactions. Hence, a myelin sheath only has charges on its surface from the flexoelectric effect. It is further a well-known experience when working with both smectic A and smectic C samples, under the condition that the director prefers to be parallel to the boundaries, and that cylindrical domains are quite abundant. In contrast, spherical domains do generate space charge because, as is easily checked, $\nabla \cdot P$ is nonzero, falling off as $\sim 1/r^2$.

2.4.9   Flexoelectric Effects in Smectic C Phases

Flexoelectric phenomena in smectic C phases are considerably more complicated than in smectic A phases, even in the case of preserved layer thickness. Thus under the assumption of incompressible layers, there are not less than nine independent flexoelectric contributions. This was first shown in [127]. If we add such deformations which do not preserve the layer spacing, there are a total of 14 flexocoefficients [128]. If we add chirality, as in smectic C*, we thus have, in principle, 15 different sources of polarization to take into account.

In a smectic C, contrary to a smectic A, certain bends and twists are permitted in the director field without violating the condition of constant smectic layer spacing. Such deformations we will call "soft" in contrast to "hard" distortions which provoke a local change in layer thickness and therefore require much higher energy. As in the case of the N* phase, soft distortions may be spontaneous in the C* phase. Thus the chiral smectic helix contains both a twist and a bend (see Fig. 41) and must therefore be equal to a flexoelectric polarization, in contrast to the cholesteric helix which only contains a twist. This may even raise the question of whether the spontaneous polarization in a helical smectic C* is of flexoelectric origin. It is not. The difference is fundamental and could be illustrated by the following example: if starting with a homogeneously aligned smectic C you twist the layers mechanically (this could be done as if preparing a twisted nematic) so as to produce a helical director structure identical to that in a smectic C*, you can never produce a spontaneous polarization. What you do achieve is a flexoelectric polarization of a

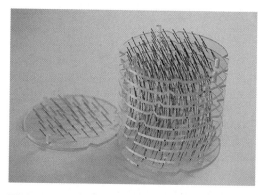

Figure 41. Model of the director configuration in a helical smectic C*. To the left is shown a single layer. When such layers are successively added to each other, with the tilt direction shifted by the same amount every time, we obtain a space-filling twist–bend structure with a bend direction rotating continuously from layer to layer. This bend is coupled, by the flexoelectric effect, to an equally rotating dipole.

certain strength and sign, which is always perpendicular to the local tilt plane, just as the spontaneous C* polarization would be. However, the flexoelectric polarization is strictly fixed to the deformation itself and cannot, unlike the spontaneous one, be switched around by an external electric field. On the other hand, it is clear from the example that the two effects interfere: the flexoelectric contribution, which is independent of whether the medium is chiral or nonchiral, will partly cancel or reinforce the spontaneous polarization in a smectic C* in all situations where the director configuration is nonhomogeneous in space. This is then true in particular for all chiral smectic samples where the helix is not unwound, and this might influence the electrooptic switching behavior.

The simplest way to get further insight into the flexoelectric polarization just described is to use the "nematic description" [46] of the C* phase. We then look, for a moment, at the smectic as if it were essentially a nematic. This means that we forget about the layers, after having noticed that the layers permit us to define a second vector (in addition to n), which we cannot do in a nematic. The second vector represents the layers and is the layer normal, z (or k), which differs from the n direction by the tilt angle. The implicit understanding that z is constant in space represents the condition of undeformed smectic layers.

We use the Oseen–Frank elastic energy expression [Eq. (96)] for a nematic medium as a starting point. Now, according to our assumption, the medium is chiral, and an ever so slight chiral addition to a nematic by symmetry transforms the twist term according to [111]

$$[n \cdot (\nabla \times n)]^2 \rightarrow [n \cdot (\nabla \times n) + q]^2 \quad (116)$$

The ground state now corresponds to a twisted structure with a nonzero value of

$n \cdot (\nabla \times n)$ given by a wave vector q, the sign of which indicates the handedness. Equation (116) is written for a right-handed twist for which $n \cdot (\nabla \times n) = -q$. Note that the reflection symmetry is lost but the invariance condition [Eq. (6)] is still obeyed.

Chirality thus here introduces a new scalar quantity, a length characteristic of the medium. If the medium is also conjectured to be polar, it might be asked if it is possible, in a similar way, to introduce a true vector (n and z are not true vectors, since there is one symmetry operation that changes the sign of both). A look at the first term of Eq. (96) clearly shows that this would not allow such a thing. In fact, it is not possible to add or subtract any true scalar or vector in the splay term without violating the invariance of $(\nabla \cdot n)^2$ under the operation $n \rightarrow -n$. Thus no ground state can exist with a spontaneous splay. Is there a way to introduce a vector in the bend term? There is. The bend transforms, obeying Eq. (6), as follows

$$[n \times (\nabla \times n)]^2 \rightarrow [n \times (\nabla \times n) - B]^2 \quad (117)$$

where B is a vector. Only a vector B parallel to $n \times (\nabla \times n)$ can be introduced in the bend term. For undeformed smectic layers, $n \times (\nabla \times n)$ lies in he $|z \times n|$ direction, and B can be written

$$B = \beta z \times n \quad (118)$$

where β is a scalar that must be zero in the nonchiral case. If β is nonzero, the medium is characterized by the local vector B and the reflection symmetry is lost. The form of the bend expression in the presence of a local polarization then corresponds to a constant spontaneous bend in the local frame of the director. The converse of this is the flexoelectric effect. Note that Eq. (117) conforms to Eq. (100).

From the above reasoning we see two things: First, that this description permits the smectic C* to be polar and requires the polarization vector to be perpendicular to the tilt plane, a result that we achieved before. Second, that the chiral and polar medium will be characterized by both a spontaneous twist and a spontaneous bend. The smectic C* is, in fact, such a medium where we have a space-filling director structure with uniform twist and bend.

This nematic description has been very helpful in the past and permitted rapid solutions to a number of important problems [129, 130]. We will return to it in Sec. 7.1.

If we insert the new twist term according to Eq. (116) into Eq. (96), the free energy attains its minimum value for $n \cdot \nabla \times n = -q$ if splay and bend are absent (the cholesteric ground state). It means that we have a spontaneous twist in the ground state. This is connected to the fact that we now have a linear term in the free energy. For, if we expand the square in the twist term we could write the free energy as

$$G_{N^*} = G_N + \frac{1}{2} K_{22} q^2 + K_{22} q n \cdot (\nabla \times n) \quad (119)$$

The last term is not reflection invariant and secures the lowest energy for right-handedness, which is therefore the lowest energy state.

In the same way, if we expand the bend term (Eq. 117) we get the linear term $-K_{33} B [n \times (\nabla \times n)]$ in the free energy. With both these linear terms present, the liquid crystal thus has both spontaneous twist and spontaneous bend.

The general problem of tracing the deformations coupled to polarization in smectic C phases is complex, as has already been stated. We refer to [127] for the derivation in the incompressible case and to the discussion in [128] for the general case. In the following we only want to illustrate the results in the simplest terms possible. This is done in Fig. 42, where we describe the deformations with regard to the reference system, k,

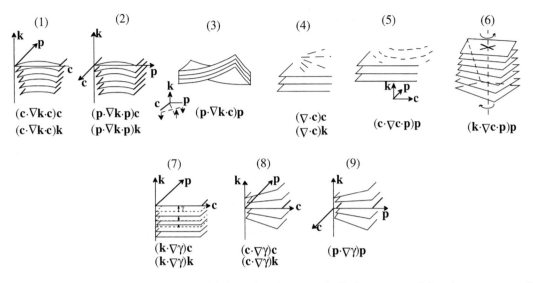

Figure 42. The six soft and the three hard deformations in a smectic C; these are coupled to the appearance of local polarization. Below each deformation is stated the covariant form of the independent vector field corresponding to the deformation. Five of the distortions (1,2,4,7,8) create a dipole density along the c (tilt) direction as well as along the k direction. The four other distortions (3,5,6,9) create a dipole density along the p direction, which corresponds to the direction of spontaneous polarization in the case of a chiral medium (C*). By the inverse flexoelectric effect, the distortions will be provoked by electric fields along certain directions. In more detail they can be described as follows: (1) is a layer bend (splay in k) with the bending axis along p, (2) is the corresponding layer bend with bending axis along c. Both deformations will be provoked by any electric field in the tilt plane (c, k). (3) is a saddle-splay deformation of the layers, with the two bending axes making 45° angles to c and p. This deformation will be caused by a field component along p. (4) is a splay in the c director (corresponding to a bend in the P field if the smectic is chiral). It is generated by any field component in the tilt plane. (5) is a bend in the c director (in the chiral case coupled to a splay in P) generated by a p-component in the field. (6) is a twist in the c director. It generates a dipole in the p direction (and is itself thus generated by a field in this direction). As this distortion is the spontaneous distortion in the helicoidal smectic C*, it means that we have a flexoelectric polarization along the same direction as the spontaneous polarization in such a material. Whereas these six deformations do not involve compression, the last three are connected with changes in the interlayer distance, described by the variable $\gamma = \partial u/\partial z$, where u is the layer displacement along the layer normal (z). Thus (7) is a layer compression or dilatation which varies along k and thereby induces a bend in the n director, connected with a dipole density, in the tilt plane. (8) is a layer splay inducing a splay-bend deformation in the n field, and thereby a dipole density, in the tilt plane. (7) and (8) are thus generated by field components in that plane. Finally, (9) is a layer splay perpendicular to the tilt plane, which induces a twist–bend in the n field along p. The bend component then means a dipole along p. Distortions (1), (2), and (4) are coupled among themselves as are (3), (5), and (6). The latter three, and in principle also (9), occur spontaneously in chiral materials.

c, p, where k is the local layer normal (along the direction of the wave vector for a helical smectic C*), c is the local tilt direction, i.e., n tilts out in this direction (hence k, c is the tilt plane), and p is the direction of $k \times c$ (corresponding to the direction along which the spontaneous polarization has to be in the chiral case). It should be stressed that k, c, and p are all considered unit vectors in this scheme, thus c does not give any indication of the magnitude of the tilt, only its direction.

All in all there are 14 independent flexoelectric coefficients, 10 of which describe five distinct deformations generating dipolar densities in the tilt plane. By the converse effect, these deformations are generated by electric field components lying in the tilt plane. A field component perpendicular to the tilt plane will generate the other four deformations (3, 5, 6, 9), which, consequently, themselves generate dipole densities along the p direction. In the figure, the deformations have been divided into three categories: those with gradients in k (1, 2, 3), those with gradients in c (4, 5, 6), and those with gradients in $\gamma \equiv \dfrac{\partial u}{\partial z}$, i.e., gradients in layer spacing (7, 8, 9) hard deformations. However, it is illuminating to make other distinctions. Thus deformations (1), (2), (4), (7), and (8) give quadratic terms in the free energy and are not influenced by whether the medium is chiral or not.

As we have repeatedly stated, the flexoelectric effect is not related to chirality. However, a certain deformation which gives rise to a flexoelectric polarization, whether the material is chiral or not, may turn out to be spontaneous in the chiral case, like the C* helix. Thus the four deformations (3), (5), (6) and (9) turn out to give linear terms (like Eq. 119) in the free energy in the chiral case. Therefore we may expect them, at least in principle, to occur spontaneously. They all give contributions to the flexoelectric polarization in the direction along p. At the incompressible limit, deformation (9) may be omitted. The remaining three deformations (3), (5), and (6) are coupled to each other and have to be considered as intrinsic in the smectic C* state. Of these, the only space-filling structure is (6) (the same as illustrated in Fig. 41) and will therefore give rise to the dominating flexoelectric effect. Of the other two, deformation (3) in the chiral case means an inherent spontaneous tendency to twist the flat layers. This deformation actually only preserves a constant layer

thickness very locally, but not in a macroscopic sample, and will therefore be suppressed. In cyclindrical domains it will tend to align the c director 45° off the cylinder axis [127]. Finally, deformation (5) amounts to a spontaneous bend in the c director (corresponding to a splay in P). This means that P has the tendency to point inwards or outwards at the edges of the smectic layers. It means, on the other hand, that the effects are transformed to a surface integral so that the term does not contribute to the volume energy. Thus the general contributions to the flexoelectric effect are exceedingly complex and, at the present state of knowledge, hard to evaluate in their relative importance, except that the dominating contribution by far is the twisted (helicoidal) deformation. The consequences of the inverse flexoelectric effect are even harder to predict, especially in the dynamic case, without any quantitative knowledge of the flexocoefficients. When strong electric fields are applied to a smectic C (or C*) any of the nine deformations of Fig. 12 will in principle be generated, if not prohibited by strong boundary conditions. This consideration applies equally well to antiferroelectric liquid crystals as to ferroelectric liquid crystals. The fact that sophisticated displays work well in both cases seems to indicate that the threshold fields for these detrimental deformations are sufficiently high.

2.5 The SmA*–SmC* Transition and the Helical C* State

2.5.1 The Smectic C Order Parameter

By changing the temperature in a system we may provoke a phase transition. The thermodynamic phase that is stable below the transition generally has a different symmetry (normally lower) than the phase that is stable above. The exceptions to this are rare, for instance, the liquid–gas transition where both phases have the same symmetry. In a number of cases of transitions between different modifications in solids there is no rational relation between the two symmetries. In such cases it may be difficult to construct an order parameter. In the majority of cases, however, the transition implies the loss of certain symmetry elements, which are thus not present any longer in the low temperature (condensed or ordered) phase. The symmetry group of the ordered phase is then a subgroup of the symmetry group of the disordered phase. In such cases we can always construct an order parameter, and in this sense we may say, in a somewhat simplifying manner, that for every phase transition there is an order parameter. Thus, in liquid crystals, the transitions isotropic $\leftrightarrow$ nematic $\leftrightarrow$ smectic A $\leftrightarrow$ smectic C $\leftrightarrow$ smectic F, etc. are all described by their different specific order parameters. (There may be secondary order parameters, in addition.)

The order parameter thus characterizes the transition, and the Landau free energy expansion in this order parameter, and in eventual secondary order parameters coupled to the first, has to be invariant under the symmetry operations of the disordered phase, at the same time as the order parameter itself should describe the order in the condensed phase as closely as possible. In addition to having a magnitude (zero for $T > T_c$, nonzero for $T < T_c$), it should have the same symmetry as that phase. Further requirements of a good order parameter are that it should correctly predict the order of the transition, and that it should be as simple as possible. As an example, the tensorial property of the nematic order parameter

$$Q_{ij} = S\left(n_i n_j - \frac{1}{3}\delta_{ij} \right) \qquad (120)$$

correctly predicts that the isotropic–nematic transition is first order. Very often, however, only its scalar part S is used as a reduced order parameter, in order to facilitate discussions.

In the case of the smectic A–smectic C transition, the tilt θ is such a reduced order parameter. The symmetry is such that positive or negative tilt describes identical states in the C phase (see Fig. 43), hence the free energy can only depend on even powers in θ and we may write, in analogy with the discussion in Sec. 2.2.5

$$G = \frac{1}{2}\alpha(T - T_c)\theta^2 + \frac{1}{4}b\theta^4 + \frac{1}{4}c\theta^6 \qquad (121)$$

For the time being, this simple expansion will be sufficient to assist in our discussion. A first-order SmA–SmC transition occurs if $b<0$. The equilibrium value of the tilt is the one that minimizes the free energy. For $b>0$ and $T-T_c \approx 0$, θ will be small and the θ^6 term can be dropped. Putting $\frac{\partial G}{\partial \theta} = 0$ then gives

$$\alpha(T - T_c)\theta + b\theta^3 = 0 \qquad (122)$$

with two solutions

$$\theta = 0 \qquad (123)$$

which corresponds to the SmA phase (a check shows that here G attains a maximum value for $T < T_c$, but a minimum value for

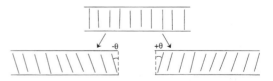

Figure 43. Simple diagram of the SmA–SmC transition. Positive and negative tilt describe identical states.

$T > T_c$), and

$$\theta = \left(\frac{\alpha}{b}\right)^{1/2} (T_c - T)^{1/2} \qquad (124)$$

corresponding to the SmC phase (the extremal value of G is a minimum for $T < T_c$). The transition is second order and the simple parabolic function of $\Delta T = T_c - T$ at least qualitatively describes the temperature behavior of the tilt for materials having an SmA–SmC transition. However, the director has one more degree of freedom and we have to specify the tilt direction. In Fig. 43 we have chosen the director to tilt in the plane of the paper.

By symmetry, infinitely many such planes can be chosen in the same way, and evidently we have a case of continuous infinite degeneracy in the sense that if all the molecules tilt in the same direction, given by the azimuthal angle φ, the chosen value will not affect the free energy. The complete order parameter thus has to have two components, reflecting both the magnitude of the tilt θ and its direction φ in space, and can conveniently be written in complex form

$$\Psi = \theta e^{i\varphi} \qquad (125)$$

With a complex scalar order parameter, the SmA–SmC transition is expected [131] to belong to the 3D XY universality class, which does not have the critical exponent β equal to 1/2. Nevertheless, experiments show [132] that it is surprisingly well described by mean field theory, although with an unusually large sixth order term.

As all tilt directions are equivalent, the free energy can only depend on the absolute value squared of the order parameter. Actually, the first intuitive argument would say that there could be a linear term in the absolute value. However, physical descriptions avoid linear terms in absolute values, since these mean that the derivatives (with

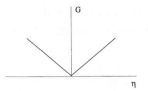

Figure 44. Functions of the kind $G \sim |\eta|$, i.e., linear dependence in an absolute value, are with few exceptions not allowed in physics, because they imply discontinuities in the first derivative, which are mostly incompatible with the physical requirements.

physical significance) would have to be discountinous (see Fig. 44). Therefore, we assume that the free energy can only depend on

$$|\Psi|^2 = \Psi^* \Psi = \theta e^{-i\varphi} \times \theta e^{i\varphi} = \theta^2 \qquad (126)$$

The Landau expansion then has to be in powers of $|\Psi|^2 = \Psi^* \Psi$ and we see that the free energy is independent of φ, as it has to be. This is just equivalent to our first expansion. Generally speaking, G is invariant under any transformation

$$\Psi \rightarrow \Psi e^{i\psi} \qquad (127)$$

This is an example of gauge invariance and the azimuthal angle is a gauge variable. The complex order parameter Ψ is shown schematically in Fig. 45 illustrating the conical degeneracy characteristic of the SmC

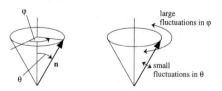

Figure 45. Illustrations of the two-component order parameter $\Psi = \theta e^{i\varphi}$ describing the SmA–SmC transition. The thermodynamic variable θ and the gauge variable φ are completely different in their fluctuation behavior. The fluctuations in φ may attain very large values and, being controlled by an elastic constant scaling as θ^2, may actually become larger than 2π at a small but finite value of θ.

phase. The gauge variable φ is fundamentally different in character from θ. The latter is a "hard" variable with relatively small fluctuations around its thermodynamically determined value (its changes are connected to compression or dilation of the smectic layer, thus requiring a considerable elastic energy), whereas the phase φ has no thermodynamically predetermined value at all. Only gradients $\nabla\varphi$ in this variable have relevance in the energy. The result is that we find large thermal fluctuations in φ around the cone, of long wavelengths relative to the molecular scale, involving large volumes and giving rise to strong light scattering, just as in a nematic.

The thermally excited cone motion, sometimes called the spin mode (this is very similar to the spin wave motion in ferromagnets), or the Goldstone mode, is characteristic of the nonchiral SmC phase as well as the chiral SmC* phase, but is of special interest in the latter because in the chiral case it couples to an external electric field and can therefore be excited in a controlled way. This Goldstone mode is of course the one that is used for the switching mechanism in surface-stabilized ferroelectric liquid crystal devices. The tilt mode, often, especially in the SmA phase, called the soft mode (although "hard" to excite in comparison with the cone mode, it may soften at a transition), is very different in character, and it is convenient to separate the two motions as essentially independent of each other. Again, this mode is present in the nonchiral SmA phase but cannot be detected there by dielectric methods, because a coupling to an electric field requires the phase to be chiral. In the SmA* phase this mode appears as the electroclinic effect.

2.5.2 The SmA*–SmC* Transition

If the medium is chiral the tilt can actually be induced in the orthogonal phase by the application of an electric field in a direction perpendicular to the optic axis, as shown in Fig. 46. The tilt is around an axis that is in the direction of the electric field. This is the electroclinic effect, presented in 1977 by Garoff and Meyer [21]. Soft mode fluctuations occur in the SmA phase just as in the SmA* phase, but only in the SmA* phase can we excite the tilt by an electric field. This is because a tilt fluctuation in the medium with chiral symmetry is coupled to a local polarization fluctuation, resulting from the transverse dipoles being ever so slightly lined up when the tilt disturbs the cylindrical symmetry of the molecular rotation. Superficially, it might be tempting to think that chiral and nonchiral smectic A (SmA* and SmA) would be very similar or even have the same symmetry, because they are both orthogonal smectics with the same organization of the molecules in the layers and with the same cylindrical symmetry around the layer normal. Nothing could be more wrong. As we have seen (in one striking example – several more could be given),

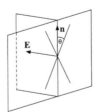

Figure 46. If an electric field E is applied perpendicular to the optic axis in a smectic A* (or any chiral orthogonal smectic), the optic axis will swing out in a plane perpendicular to the plane E, n defined by the optic axis and the field (electroclinic or soft-mode effect).

their physics is very different, due to the fact that they have different symmetry. Reflection is not a symmetry operation in the SmA* phase, and the electroclinic effect does not exist in the SmA phase just because the SmA phase has this symmetry operation.

We may also apply the Curie principle to this phenomenon. The SmA phase has $D_{\infty h}$ (or ∞/m) symmetry, with one C_∞ axis along the director (optic axis), infinitely many C_2 axes perpendicular to this axis, and in addition one horizontal and infinitely many vertical mirror planes. The mirror planes are absent in the SmA* phase (D_∞ or $\infty 22$). If we apply an electric field E (of symmetry C_∞) perpendicular to the C_∞ axis, the only common symmetry element left is one C_2 axis along E, which permits a tilt around C_2. In Fig. 46 we also see that, in particular, the plane (E, n) is a mirror plane in the nonchiral SmA phase, and consequently neither of the two tilting directors shown in the figure are allowed. In the SmA* phase (E, n) is not a mirror plane, hence one of these tilt directions will be preferred. (Which one cannot be predicted as this is a material property.)

An interesting aspect of this phenomenon is that it means that an electric field acting on the chiral medium actually exerts a torque on the medium. This is further discussed in Sec. 2.5.8. The torque is of course inherent in the chirality. The axial symmetry character is provided by the medium.

The electroclinic effect is a new form of dielectric response in a liquid crystal. If we increase the electric field E, the induced polarization P will increase according to the first curve of Fig. 5. With a small field, P is proportional to E, and then P saturates. As the coupling between tilt and polarization is also linear at small values of tilt, we get a linear relationship between θ and E. With $\theta = e^*E$, the proportionality factor e^* is called the electroclinic coefficient. It has an intrinsic chiral quality. (We have used an

asterisk here to emphasize this and also to clearly distinguish it from the flexoelectric e-coefficients.) In nonchiral systems, e^* is identically zero. With present materials, θ is quite small ($\leq 15°$), but this may change with new dedicated materials.

The electroclinic effect is also appropriately called the soft-mode effect, because the tilt deformation is a soft mode in the SmA phase, the restoring torque of which softens when we approach the SmA*–SmC* transition, at which the deformation starts to "freeze in" to a spontaneous tilt. When we have such a spontaneous tilt as we have in the SmC* phase, we will also have a spontaneous polarization because of the rotational bias. It is important to note that the molecular rotation is biased in the nonchiral SmC phase as well as in the SmC* phase, but in a different manner. We may start with the "unit cell" of cylindrical symmetry of the SmA (or SmA*) phase and the corresponding cell of monoclinic symmetry of the SmC (or SmC*) phase, as in Fig. 47. We put a molecule in each one and ask whether the molecular motion (which, like the shape of the molecule, is much more complicated than here indicated) is directionally biased in its rotation around the optic axis (only roughly represented by the core part of the molecule). One simple way of representing this bias is to draw the surface segments corresponding to some interval in the rotation angle. Let us imagine that the molecule has a dipole perpendicular to its average rotation axis. We then find that the probability of finding the dipole in any particular angular segment is same in all directions in the SmA (SmA*) phase, as it has to be, because the optic axis is a C_∞ axis.

In the nonchiral C phase, the rotation cannot have this circular symmetry, but it must be symmetrical with respect to the tilt plane, as this is a mirror plane. This means that the rotational bias has a quadrupolar symmetry,

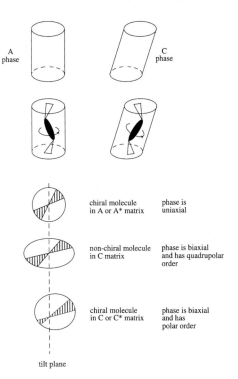

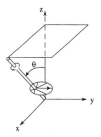

Figure 48. Origin of the spontaneous polarization in a smectic C* phase. The illustration shows the directional bias for a lateral dipole in the rigid rod model of a molecule, with the directionality diagram for the short axis when the molecule is rotating about its long axis (from Blinov and Beresnev [45]).

Figure 47. The origin of the spontaneous polarization. The probability for a lateral dipole (averaged in time and space) to be in a unit angular segment is compared for the orthogonal, nonchiral tilted and chiral tilted phases. The common statement that the origin of polarization is a "hindered rotation" is quite misleading because, hindered or not, a rotation will only result in nonzero polarization for a specific (polar) directional bias.

which we can describe by a quadrupolar order parameter. If we now drop the mirror plane by making the phase chiral we will, in addition to this quadrupolar dissymmetry, have a polar bias giving a different probability for the dipole to be directed at one side out of the tilt plane rather than in the opposite direction. This will give a nonzero polarization density in a direction perpendicular to the tilt plane (see also the illustration in Fig. 48). Which direction cannot be predicted a priori but it is a molecular property. It is fairly obvious that the bias will increase as we increase the tilt angle, and

hence the polarization will grow when we lower the temperature below the SmA*–SmC* transition. This transition occurs when the tilt θ becomes spontaneous and is only weakly influenced by whether the phase is chiral or not. However, chirality couples θ to P (which is a dramatic effect in a different sense) and brings P in as a secondary order parameter. There might also be a coupling between P and the quadrupolar order parameter, which will, however, be ignored at this moment.

2.5.3 The Smectic C* Order Parameters

If we keep the tilt angle θ in Fig. 48 fixed (just by keeping the temperature constant), we can move the molecule around the tilt cone, and we see that the rotational bias stays sterically fixed to the molecule and the tilt plane. The direction of the resulting P lies in the direction $z \times n$. If we change the azimuthal angle by 180°, this corresponds to a tilt $-\theta$ (along the $-x$ direction), and results in a change of P direction from y to $-y$. This means that we could tentatively write the relation between the secondary and primary order parameters as

$$P \sim \theta \tag{128}$$

or more generally that P is an odd function of θ

$$P = a\theta + b\theta^3 + \ldots \tag{129}$$

As $\sin\theta$ is an odd function, an expression that has this form is

$$P = P_0 z \times n = (P_0 \sin\theta)\boldsymbol{\theta} \tag{130}$$

where we have introduced P_0 (with its sign, see Fig. 23 in Sec. 2.3.4), the magnitude of spontaneous polarization per radian of tilt. With $\boldsymbol{\theta} \equiv z \times n$ we may write the tilt as a vector $\boldsymbol{\Theta} = \theta\boldsymbol{\theta}$ to compare its symmetry with that of P. $\boldsymbol{\Theta}$ is then an axial vector, while P is a polar. As we have no inversion symmetry in the chiral phase, this difference does not cause any problem, and if P and $\boldsymbol{\Theta}$ transform in the same way we may make a Landau expansion according to the improper ferroelectric scheme of Sec. 2.2.6. If we take the scalar part of Eq. (130), it takes the form

$$P = P_0 \sin\theta \tag{131}$$

This expression is often used but does not have the correct symmetry because it is not invariant under the C_2 symmetry operation. For, if we change θ to $\theta + \pi$, $\sin\theta$ changes to $-\sin\theta$ although P cannot change, because this is the same state (see Fig. 49). Therefore Eq. (131) should be replaced by

$$P = \frac{1}{2} P_0 \sin 2\theta \tag{132}$$

which is a C_2 invariant expression. In practice the expressions do not differ very much; P just increases somewhat slower at large tilts according to Eq. (132). For small values of tilt the expressions do of course give the same results, i.e., $P = P_0\theta$. Experimentally, this simple linear relation between P and θ often holds quite well, as shown in

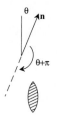

Figure 49. Under the action of the twofold symmetry axis in the smectic C* state the tilt changes from θ to $\theta + \pi$.

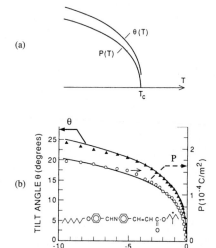

Figure 50. (a) Schematic temperature dependence of the tilt angle θ and polarization P at a second-order SmA*–SmC* transition. (b) Tilt angle versus temperature for DOBAMBC below the SmA*–SmC* transition at 95 °C (from Dumrongrattana and Huang [133]).

Fig. 50. However, there are exceptions, for instance, cases where P is a linear function of temperature rather than parabolic.

Although the tilt angle has certain limitations, we will continue to use θ as the order parameter for the time being. Finally, however, we will construct a different order parameter (Sec. 2.5.11) to describe the tilting transition.

2.5.4 The Helical Smectic C* State

We did not change anything in the nonchiral order parameter (Eq. 125) when we proceeded to the chiral case, except that in addition we introduced a secondary order parameter P which does not exist in the nonchiral case. Now the normal state of the smectic C* is the helical C* state, i.e., one in which the tilt has a twist from layer to layer. At constant θ, φ is therefore a function of z (see Fig. 51), with $\varphi(z) = qz$, q being the wave vector of the helix

$$q = \frac{2\pi}{Z} \tag{133}$$

where the helix is right-handed for $q > 0$ and Z is the helical periodicity. For this case there is an obvious generalization of the nonchiral order parameter (Eq. 125) to

$$\Psi_q = \theta e^{i\varphi(z)}$$

with

$$\varphi(z) = qz \tag{134}$$

where we have introduced a new length q^{-1} as an effect of chirality. Ψ_q is now chiral by construction but does not make any change in the Landau expansion since $\Psi_q^* \Psi_q$ is still equal to θ^2, and we will still have the SmA*–SmC* transition as second order.

We can also write Ψ_q as

$$\Psi_q = \theta(\cos qz + i \sin qz) \tag{135}$$

and as a vector

$$\Psi_q = (\theta \cos qz, \theta \sin qz, 0) \tag{136}$$

or

$$\Psi_q = \theta c \tag{136a}$$

where we have introduced the "c director", the projection of the director n on the smectic layer plane (see Fig. 51b). Note that the c director in this description (which is the common one) is not a unit vector, as it is in the movable reference frame k, c, p used, in contrast to the space-fixed reference frame x, y, z, in Sec. 2.4.9 to describe the flexoelectric deformations.

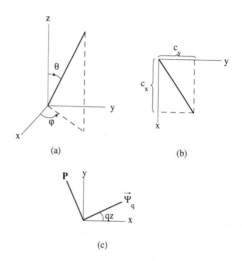

Figure 51. (a) In the helical smectic C* state the constant tilt θ changes its azimuthal direction φ such that φ is a linear function in the coordinate z along the layer normal. (b) The c director is a two component vector (c_x, c_y) of magnitude $\sin \theta$, which is the projection of n on the smectic layer plane. (c) P makes a right angle with the tilt direction. The P direction here corresponds to a material with $P_0 > 0$.

As the polarization vector is phase-shifted by 90° relative to the tilt vector Ψ_q, we see from Eq. (136) that it can be written (see Fig. 51c) as

$$P = (-P \sin qz, P \cos qz, 0) \tag{137}$$

It is interesting to form the divergence of this vector. We find

$$\nabla \cdot P = \frac{\partial P_x}{\partial x} + \frac{\partial P_y}{\partial y} + \frac{\partial P_z}{\partial z} = 0 \tag{138}$$

This is an important result as it means that the helicoidal C* state is a divergence-free vector structure $P(r)$ in space. Hence we have no appearance of space charges anywhere. Thus not only does the helix cancel the macroscopic polarization and thereby any external coulomb fields, although we have a local polarization P everywhere, but the fact that $\nabla \cdot P = 0$ also secures that there are no long range coulomb interactions in the material due to space charge. The helical state is therefore "low cost" and "natural" from several points of view. A homogeneous state of the director, as can ideally be realized by surface stabilization, of course also has $\boldsymbol{\nabla}\cdot\boldsymbol{P}=0$, although it has an external coulomb field and therefore is not stable by itself. However, as we will see later, the condition $\boldsymbol{\nabla}\cdot\boldsymbol{P}=0$ cannot be maintained everywhere if we have a local layer structure of chevron shape (see Sec. 2.8.3).

For the sake of completeness, but also for distinction, let us finally make a comment on an entirely different kind of chiral order parameter. Let us imagine that the substance we are dealing with is a mixture of two enantiomers (R) and (S). We can then define a scalar quantity

$$\kappa = \frac{R-S}{R+S} \tag{139}$$

where R and S stand for the relative concentration of (R) and (S) enantiomers. This quantity can in a sense be regarded as an order parameter. $\kappa=1$ for "all R", -1 for "all S", and zero for a racemic mixture. Thus in general, $0\le|\kappa|\le1$. This quantity is nothing less than what organic chemists introduced long ago, but in a completely different context. It is sometimes called "optical purity", but is today rather called enantiomeric excess. It is an extremely useful quantity in its right place, but it is not relevant as an order parameter because it does not depend on the temperature T, only on the concentration, which is trivial in our context. It does not help us in introducing chirality and we will not have use for it.

2.5.5   The Flexoelectric Contribution in the Helical State

We note from Fig. 51 that the projection of the director on the smectic layer plane is

$n\sin\theta=\sin\theta$. If we write the components of $\boldsymbol{n}$ in the x,y,z system, we therefore have

$$n_x = \sin\theta\,\cos\varphi$$
$$n_y = \sin\theta\,\sin\varphi$$
$$n_z = \cos\theta \tag{140}$$

If we insert $\varphi(z)=qz$ for the helical state this gives,

$$n_x = \sin\theta\,\cos qz$$
$$n_y = \sin\theta\,\sin qz$$
$$n_z = \cos\theta \tag{141}$$

This is a twist–bend structure, as we have repeatedly stated, and is therefore connected to a flexoelectric polarization of size (see Eq. 97)

$$P_f = e_b\boldsymbol{n}\times(\boldsymbol{\nabla}\times\boldsymbol{n}) \tag{142}$$

We will now calculate this contribution. Forming the curl as

$$\boldsymbol{\nabla}\times\boldsymbol{n} = \boldsymbol{x}\left(\frac{\partial n_z}{\partial y}-\frac{\partial n_y}{\partial z}\right) \tag{143}$$
$$+\boldsymbol{y}\left(\frac{\partial n_x}{\partial z}-\frac{\partial n_z}{\partial x}\right)+\boldsymbol{z}\left(\frac{\partial n_y}{\partial x}-\frac{\partial n_x}{\partial y}\right)$$

we get

$$\boldsymbol{\nabla}\times\boldsymbol{n} = \boldsymbol{x}\sin\theta(q\cos qz)$$
$$-\boldsymbol{y}\sin\theta(q\sin qz)+0$$
$$= -q\sin\theta(\cos qz,\sin qz,0) \tag{144}$$

and

$$\boldsymbol{n}\times(\boldsymbol{\nabla}\times\boldsymbol{n})$$

$$= -q\sin\theta\begin{vmatrix}\boldsymbol{x}\cdot & \boldsymbol{y} & \boldsymbol{z}\\ \sin\theta\cos qz & \sin\theta\sin qz & \cos\theta\\ \cos qz & \sin qz & 0\end{vmatrix}$$

$$= q\sin\theta\cos\theta(\sin qz,-\cos qz,0) \tag{145}$$

This is a vector in the layer plane which is antiparallel to the P_s vector for a smectic C* material with $P_0>0$. Its magnitude is

$$|\boldsymbol{n}\times(\boldsymbol{\nabla}\times\boldsymbol{n})| = q\sin\theta\cos\theta = \frac{1}{2}q\sin2\theta \tag{146}$$

Therefore we can write the flexoelectric polarization due to the helical deformation as

$$P_f = -e_b \cdot \frac{1}{2} q \sin 2\theta \qquad (147)$$

Note that this polarization grows with the tilt according to the same function as the spontaneous polarization P_s in Eq. (132). We may then write the total polarization density $P = P_s + P_f$ as

$$P = \frac{1}{2}(P_0 - e_b q)\sin 2\theta \qquad (148)$$

or, for small angles

$$P = (P_0 - e_b q)\theta \qquad (148\,a)$$

If P_0 and e_b have different signs, the two contributions will cooperate, otherwise they tend to cancel. (This is not changed if we go from a certain chiral molecule to its enantiomer, as q and P_0 will change sign simultaneously.) In the absence of a smectic helix, $q=0$, and the flexoelectric contribution will vanish.

The sign of P_0 has been determined for a large number of compounds, but sign (and size) determinations of flexoelectric coefficients are almost entirely lacking. Such measurements are highly important and should be encouraged as much as possible. However, recent measurements by Hoffmann and Kuczyński [134] show that the flexoelectric polarization is of the same magnitude as the spontaneous one.

Equation (148) may also be a starting point for a new question. In the absence of chirality, thus with $P_0=0$, could the flexoelectric effect lead to a helical smectic state in a nonchiral medium? This will be the topic of our next section.

2.5.6   Nonchiral Helielectrics and Antiferroelectrics

As we have repeatedly stressed, flexoelectricity is a phenomenon that is a priori independent of chirality. But we have also seen that some flexoelectric deformations do have a tendency to occur spontaneously in a chiral medium. All except the helical C* state are, however, suppressed, because they are not space-filling. A flexoelectric deformation may of course also occur spontaneously in the nonchiral case, namely, under exactly the same conditions where the deformation is space-filling and does not give rise to defect structures. In other words, in creating the twist–bend structure which is characteristic of a helielectric. Imagine, for instance, that we have mesogens which have a pronounced bow shape and, in addition, some lateral dipole. Sterically they would prefer a helicoidal structure, as depicted in Fig. 52, which would minimize the elastic

Figure 52. Space-filling twist–bend structure of strongly bow-shaped achiral molecules equally split into right-handed and left-handed helical domains. (Here a right-handed domain is indicated.) A bow-shaped molecule of this kind is illustrated in Fig. 53. The kind of domain depicted in this figure would correspond to the normal helical antiferroelectric organization in a smectic C*. As a result of the two-dimensional fluidity of these smectics, the cancellation can, however, be expected to occur on the smallest possible space scale.

energy, because the spontaneous bend B would cancel the bend term (Eq. 117) in the Oseen–Frank expression. The resulting local polarization from Eq. (148) with $P_0=0$ may still not be too costly because the external field is cancelled and $\nabla \cdot P = 0$. This would have an interesting result: because the starting material is nonchiral, we would observe a spontaneous breaking of the nonchiral symmetry leading to equal regions with left-handed and right-handed chirality. It would correspond to the well-known examples of SiO_2 and $NaClO_3$ discussed in Sec. 2.3.4.

It seems that finally this kind of phenomenon may have been observed in liquid crystals [135]. However, in contrast to the cases of SiO_2 and $NaClO_3$, the helical structures are probably possible on a molecular level as well as on a supermolecular level. Thus we may expect domains of nonchiral molecules in different conformations, right and left-handed, which behave as if they belong to different enantiomeric forms. The possibility that a space-filling flexoelectric deformation will be spontaneous for certain molecular shapes and thereby create a chiral structure out of nonchiral molecules will be much enhanced if the deformation can take place in the layer rather than in the interlayer twist–bend structure of Fig. 52, and may then lead to antiferroelectric (rather than helielectric) order similar to that in antiferroelectric liquid crystals made of chiral compounds. The polarization may very

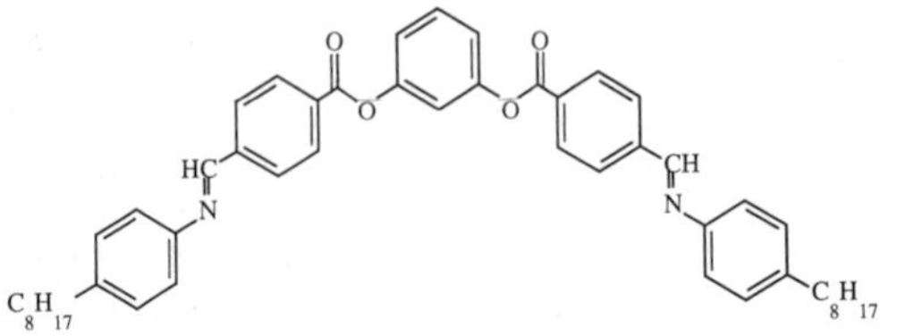

Figure 53. Bow-shaped nonchiral molecule which may create chiral C* domains (from [135]).

well be switchable, because it is not connected to a supramolecular director deformation. In other words, the deformation represented to the lower right of Fig. 32 now applies to the single molecule and this can be flipped around by the electric field.

Further investigations of these new materials are important and will shed light on a number of problems related to polarity and chirality. Whether we will find mechanisms for complete spatial separation reminiscent of the classic discovery by Pasteur, who could separate crystallites of tartaric acid into right-handed and left-handed pieces, is too early to speculate about, but the possibilities, as always in liquid crystalline systems, are quite rich.

2.5.7 Simple Landau Expansions

In order to sharpen the observance for inconsistencies in a theory, it is sometimes instructive to present one which does not work. For instance, let us see if we can describe the transition to a polar nematic phase by a Landau expansion. It might also be thought of as describing the transition isotropic-polar smectic A or nematic-polar smectic A. (In either interpretation it will suffer from serious deficiencies.) We write the free energy as

$$G = \frac{1}{2} aS^2 + \frac{1}{2} bP^2 - cSP^2 \qquad (149)$$

where S is the (reduced) nematic order parameter and P the polarization. The same abbreviation $a = \alpha (T - T_0)$ and sign conventions are used as before. The form of the coupling term is motivated by the fact that S and P have different symmetries, quadrupolar and polar, respectively. (It is also easy to persuade oneself that a term $\sim SP$ would lead to absurdities.) G now has to be minimized with respect to both order parameters: putting $\partial G/\partial P$ and $\partial G/\partial S$ equal to zero gives

$$bP - 2c\,SP = 0 \tag{150}$$

$$aS - c\,P^2 = 0 \tag{151}$$

from which we deduce

$$S = \frac{b}{2c} \tag{152}$$

$$P^2 = \frac{a}{c}\,S = \frac{\alpha b}{2c^2}\,(T - T_0) \tag{153}$$

Whether it is already not clear physically how a non-zero S-value would couple to a non-zero polarization, we see that Eq. (153) describes a $P(T)$ increasing below the transition point only if $b < 0$, at the same time as S is essentially independent of temperature.

In order to give $S > 0$, we have then to require that $c < 0$ which means that the coupling raises the energy. This is confirmed by checking that the solution of Eqs. (152) and (153) does not correspond to a minimum in the free energy ($\partial^2 G/\partial P^2 = 0$; $\partial^2 G/\partial S^2 = a$, < 0 for $T < T_0$). Other similar expansions along the same lines would give equally or even more absurd results. We will now contrast this against what happens at the transition $A^* \rightarrow C^*$.

Let us take our starting point in a general polar material, disregarding which are the sources of polarization, writing the free energy in an external field along P simply as

$$G = \frac{1}{2\chi_0\,\varepsilon_0}\,P^2 - PE \tag{154}$$

Why we have chosen to write the coefficient in front of the P^2 term in this form will be evident in a moment. $\partial G/\partial P = 0$ gives

$$\frac{1}{\chi_0\,\varepsilon_0}\,P - E = 0 \tag{155}$$

or

$$P = \chi_0\,\varepsilon_0\,E \tag{156}$$

where $\chi_0\,\varepsilon_0$ is the dielectric susceptibility, by definition. For the chiral smectic phase we now write the free energy with the tilt θ as the primary order parameter

$$G = \frac{1}{2}\,a\theta^2 + \frac{1}{4}\,b\theta^4 + \frac{1}{2\chi_0\,\varepsilon_0}\,P^2$$
$$-c^*\,P\theta - PE \tag{157}$$

where the $\theta - P$ coupling coefficient is written c^* to emphasize that it is chiral in its nature, while all other coefficients are non-chiral. We have disregarded any coupling with the nematic order parameter S. This is totally justified. Although it is true that S increases to some extent when θ increases, this is certainly of no significance around the $A^* - C^*$ transition. At this stage we also neglect the complications of an eventual helical structure. This is admissible because the helix is not a necessary condition for a chiral tilted smectic. We are first interested in the spontaneous polarization, i.e. the case $E = 0$. The variable P means just "polarization", i.e. the total polarization, even if a number of sources may contribute – we cannot have different P variables for different "sorts" of polarization. We may however ask: what is the origin of the P^2 term and what does it signify? Answer: take the coupling away, i.e. put the coupling coefficient $c^* = 0$. Then minimizing G with respect to P just gives

$$\frac{\partial G}{\partial P} = \frac{1}{\chi_0\,\varepsilon_0}\,P = 0 \tag{158}$$

i.e.

$$P = 0 \tag{159}$$

In other words, the third term in the expansion counteracts polar order. It secures that P vanishes, unless there is a coupling between θ and P. Its origin is entropic and has nothing to do with dielectric effects. The third term is just not "electric" and therefore

χ_0 is often referred to as a "generalized" susceptibility. What this P^2 term means is that other factors being equal, polar order increases the order, hence decreases the entropy and increases the free energy: a non-zero polarization is always connected with an entropic cost in energy – whence the plus sign in Eq. (157) – and therefore will not occur spontaneously if there is no other contribution that diminishes the energy even more for $P \neq 0$. We might therefore expect that the result depends on the strength of the chiral coupling constant c*. With c* $\neq 0$ Eq. (157) gives in the field-free case ($E = 0$)

$$\frac{\partial G}{\partial P} = \frac{1}{\chi_0 \varepsilon_0} P - c*\theta = 0 \tag{160}$$

or

$$P = \chi_0 \varepsilon_0 c*\theta \tag{161}$$

Perhaps to our surprise, this means that however weak the coupling, any non-zero coupling constant c* actually leads to a non-zero value of spontaneous polarization P. This polarization, at least for small values of θ, is proportional to θ

$$P \sim \theta \tag{162}$$

i.e. the polarization increases with increasing tilt, and the proportionality factor $\chi_0 \varepsilon_0 c*$ only vanishes for c* strictly equal to zero. This means that however weak the coupling, it dominates the counteracting entropy factor. This is also consistent with our discussion in Sec. 2.3.2 of this Chapter: the coupling is via chirality which brings the decisive change in symmetry. We could symbolize the strength c* of the coupling constant by the concentration of chiral dopant which we add to a non-chiral smectic C host. Any non-zero concentration, however small, is sufficient to change the symmetry.

If we now allow a non-zero external field along P, Eq. (157) in the same manner

gives

$$\frac{\partial G}{\partial P} = \frac{1}{\chi_0 \varepsilon_0} P - c*\theta - E = 0 \tag{163}$$

Hence

$$P = \chi_0 \varepsilon_0 c*\theta + \chi_0 \varepsilon_0 E \tag{164}$$

Now we see that the total polarization has two origins. On one hand a polarization is induced by an external field, the proportionality factor being the susceptibility $\chi_0 \varepsilon_0$, and on the other, there is a tilt induced polarization $\chi_0 \varepsilon_0 c*\theta$ independent of any external field E. If we would have applied the external field along just any direction, the E in Eq. (164) would correspond to the component perpendicular to the tilt plane. Therefore we could also write Eq. (164) as

$$P = \chi_\perp \varepsilon_0 (c*\theta + E) \tag{165}$$

expressing the two contributions to P and emphasizing that the susceptibility is the one perpendicular to the tilt plane. Without changing anything else, we can make c* $= 0$ by replacing the chiral substance by its racemate. Eq. (165) then reduces to

$$P = \chi_\perp \varepsilon_0 E \tag{166}$$

Hence our $\chi_\perp \varepsilon_0$, or $\chi_0 \varepsilon_0$, is the suceptibility of the racemate perpendicular to the tilt plane.

Let us now continue with the case $E = 0$, that is with the spontaneous polarization below T_c. Introducing the structure coefficient s* we can write the relation Eq. (161) as

$$P = s*\theta \tag{167}$$

The structure coefficient

$$s* = \chi_0 \varepsilon_0 c* \tag{168}$$

is the dielectric susceptibility of the racemate multiplied by the chiral coupling coefficient and thus another related chiral parameter, which we can expect to be essentially temperature independent. We can expect it to grow very slowly towards lower temperatures, like $\sim 1/kT$, originating in the entropic character of the P^2 term ($\chi_0^{-1} \sim kT$) in Eq. (157).

The linear relation Eq. (167) and the temperature independence of s* for reasonable temperature ranges have been experimentally confirmed in dielectric measurements, especially by Bahr and Heppke [188]. The quotient $P/\theta = s^*$ is free of divergences and a characteristic for each molecular species. Its meaning is the dipole moment per unit volume for unit tilt angle. In rare cases of strong conformational changes in the molecule it may behave anomalously and even change sign. It can be looked upon as a kind of susceptibility like χ and we will understand all these and similar susceptibilities more strictly as response functions in the limit of small tilts or small applied fields, for instance $\chi_0 \varepsilon_0 = (\partial P/\partial E)_{E=0}$.

The relation Eq. (167) may look trivial but hides something fundamental that we should just not let go unnoticed. The polar order expressed by P is a result of a rotational bias expressed by θ. It is counteracted by thermal disorder, represented by the term $\frac{1}{2}(\chi_0 \varepsilon_0)^{-1} P^2$ in the Landau expansion which is proportional to kT. Thus in Eq. (167) $s^* \sim 1/T$. But this is nothing else than the old Langevin balance, just the same in principle as between a magnetic field H wanting to align magnetic dipoles and thermal motion wanting to destroy the alignment, leading to a magnetization

$$M \sim \frac{H}{T} \tag{169}$$

or between an electric field E wanting to align electric dipoles and thermal motion, leading to a polarization

$$P \sim \frac{E}{T} \tag{170}$$

We already discussed this in Sec. 2.2 of this Chapter (cf. Eqs. 4 and 5). Now, with $\chi_0 \sim 1/kT$ we can write Eq. (167) in the form

$$P \sim \frac{\theta}{T} \tag{171}$$

The tilt here acts like a (biasing) field corresponding to H and E in Eq. (169) and (170). This is thus nothing else than the Curie law, only unusual in the sense that the field variable is a tilt. As we already noticed in Sec. 2 of this Chapter, this law indicates that the local dipoles are non-interacting. The collective behavior in the medium is in θ, not in P.

So far, except for discussing the physical character of the terms in the Landau expansion, we have mainly derived a very simple relation (Eqs. 167 and 168) between P and θ in the case that both appear spontaneously, that is, in the C* phase. But what is their relation if P is a polarization induced by an electric field? When we consider the A* phase, i.e. $T > T_c$, P and θ can be nonzero only if $E \neq 0$, thus the $-PE$ term is now important. And now we have a situation which in a sense is opposite to the one below T_c: in the C* phase we have a non-zero tilt which gives rise to a non-zero polarization; here we have a non-zero polarization causing a non-zero tilt. How do we get the equilibrium value of θ for a certain constant value of E? We here recall the expression Eq. (157).

$$G = \frac{1}{2} a\theta^2 + \frac{1}{4} b\theta^4 + \frac{1}{2\chi_0 \varepsilon_0} P^2$$
$$-c^* P\theta - PE \tag{157}$$

and want to proceed by forming $\partial G/\partial \theta$ and putting it equal to zero. But the sensitive reader might react to this: we cannot vary θ while keeping both E and P constant. In $(\partial G/\partial \theta)_{P,E}$ we can keep E constant but we cannot vary θ while keeping P constant. A variation of θ involves a variation of P, or does it not? The answer to this question reveals the very special character of the Landau expansion. This is an expansion around equilibrium. The variables θ and P appearing in the expression are pre-thermodynamic variables in the sense that only the minimization of G will give the relation between those variables in equilibrium. It is quite possible to imagine that we tilt the director without changing P, but then we disturb the statistics which costs energy. Conversely we may imagine that the rotational bias changes in such a way that P increases (we decrease the rotational entropy) without changing θ, but then we are not in equilibrium. In order to go back to equilibrium without changing θ we would have to apply an electric field counteracting the θ change (which costs energy), and so on. Hence, the variables in the Landau expansion have to be considered independent, and only when we have determined the equilibrium values by $\partial G/\partial P = 0$, $\partial G/\partial \theta = 0$, we may look upon them as dependent, interrelated variables. We will now take that step and see which physical results can be derived from the expansion. Putting the two derivatives equal to zero gives, respectively,

$$\frac{P}{\chi_0 \varepsilon_0} - c^*\theta - E = 0 \qquad (172)$$

and

$$a\theta + b\theta_3 - c^*P = 0 \qquad (173)$$

Writing the first one (as previously Eq. 164)

$$P = \chi_0 \varepsilon_0 c^*\theta + \chi_0 \varepsilon_0 E \qquad (174)$$

and inserting P in the second, we find

$$a\theta + b\theta^3 - \chi_0 \varepsilon_0 c^{*2}\theta - \chi_0 \varepsilon_0 c^*E = 0 \quad (175)$$

For $E = 0$ the first solution $\theta = 0$ corresponds to the A* phase. In the second solution

$$a - \chi_0 \varepsilon_0 c^{*2} + b\theta^2 = 0 \qquad (176)$$

we rewrite the first two terms according to

$$\alpha(T - T_0) - \chi_0 \varepsilon_0 c^{*2}$$
$$= \alpha\left(T - T_0 - \frac{\chi_0 \varepsilon_0 c^{*2}}{\alpha} \right) = \alpha(T - T_c) \quad (177)$$

with

$$T_c = T_0 + \frac{\chi_0 \varepsilon_0 c^{*2}}{\alpha} \qquad (178)$$

Thereby Eq. (134) is reshaped into

$$\alpha(T - T_c) + b\theta^2 = 0 \qquad (179)$$

giving the wellknown parabolic increase of θ below T_c. The reader will now remember from Sec. 2.5.1 that in the Landau expansion Eq. (157) the coefficient $a(T)$ is equal to $\alpha(T - T_c)$ where T_0 is the temperature at which the tilt goes to zero in the nonchiral system (the racemate). Therefore we see from Eq. (178) that the phase transition temperature has been raised to a slightly higher value than the transition temperature T_0 for the racemate, cf. Fig. 54. This increase in transition temperature ΔT corresponds to the additional energy needed to break the coupling between polarization and tilt in the chiral case. Conversely, a spontaneous tilt

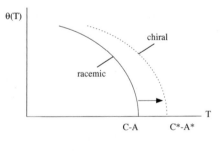

Figure 54. The chiral coupling between θ and P shifts the tilting transition to a higher temperature than the C→A transition. The shift is proportional to the coupling constant squared. Doping a non-chiral SmC host we should therefore expect a shift ΔT proportional to the square of the concentration of the dopant.

appears earlier on cooling a chiral material than the corresponding nonchiral material. We note that the offset in the transition point

$$\Delta T = \frac{\chi_0 \, \varepsilon_0 c^{*2}}{\alpha} \qquad (180)$$

is proportional to the chiral coupling coefficient squared and inversely proportional to the nonchiral thermodynamic coefficient α, the latter fact being reasonable because α is proportional to the restoring torque density $(-\alpha\theta)$ trying to establish the $\theta = 0$ value of the orthogonal A* state, thereby diminishing the rotational bias.

We have, so far, considered the non-helical C* case. With the helix present we would have found the transition C* → A* pushed even slightly further upwards in temperature because now, in addition, we have to bring up energy to unwind the helix at the transition. We would then have found another contribution to ΔT being proportional to q^2, where q is the value of the helical smectic C* wave vector at the transition. In any case, ΔT still turns out to be quite small (like all chiral perturbations), in practice often less than a degree.

With the same reshaping of $a - \chi_0 \varepsilon_0 c^{*2}$ to $\alpha(T - T_0)$, which we just used in Eq. (179), our previous Eq. (175) will look like

$$\alpha(T - T_c)\theta + b\,\theta^3 - \chi_0\varepsilon_0 c^* E \qquad (181)$$

With this we now specifically turn to the case $T > T_c$. As the tilt induced in the A* phase by an external electric field is quite small we first discard the θ^3 term and get

$$\alpha(T - T_c)\theta = \chi_0\varepsilon_0 c^* E \qquad (182)$$

This gives the induced tilt angle

$$\theta = \frac{\chi_0\,\varepsilon_0 c^*}{\alpha(T - T_c)} E \qquad (183)$$

It is thus linear in the electric field in this approximation, but strongly temperature

dependent, diverging as we approach T_c from above. In the A* phase the relation between θ and E is almost as simple as that between P and θ in the C* phase, while the relation between P and θ is not as simple. Equation (183) describes the electroclinic effect. We may write it

$$\theta = e^*(T)E \qquad (184)$$

where e^* is the electroclinic coefficient

$$e^* = \frac{\chi_0\,\varepsilon_0 c^*}{\alpha(T - T_c)} = \frac{s^*}{\alpha(T - T_c)} \qquad (185)$$

showing the close relationship between the chiral material parameters e^*, c^* and s^*. Specifically we see that the structure coefficient s^* is the non-diverging part of e^*.

2.5.8 The Electroclinic Effect

If we assume (which is not quite true, but empirically fairly well justified) that the same simple relation $P = s^*\theta$ is valid also for $T > T_c$, the combination

$$P = s^*\theta, \quad \theta = e^* E \qquad (186)$$

is equivalent to

$$P = s^* e^* E \qquad (187)$$

With e^* taken from Eq. (185) we see that

$$P = \frac{s^{*2}}{\alpha(T - T_c)} E \qquad (188)$$

i.e., that part of the polarization which corresponds to the induced tilt, diverges in the same way as the primary order parameter. (There is of course also a non-diverging part of P). While $\theta \sim s^*$ in Eq. (184), $P \sim s^{*2}$, which has to be the case as P, unlike θ, cannot depend on the sign of s^*. We note, furthermore, that as e^* and s^* only differ in the temperature factor they always have the same sign. This means that the electroclinically induced tilt or polarization will never counteract a spontaneous tilt or polarization, except in the sense that in a switching process – because the electroclinic effect is so fast – the electroclinic contribution may temporarily get out of phase from the ferroelectric contribution.

In order to get the correct relation between P and θ for $T > T_c$ we have to start from the generally valid Eq. (174) which we write as

$$P = s^*\theta + \left(\frac{s^*}{c^*}\right)E \tag{189}$$

If we assume the linear relationship $\theta = e^* E$ to hold – which is not true near T_c – then

$$P = s^*\theta + \frac{s^*}{c^* e^*}\theta \tag{190}$$

and inserting e^* from Eq. (185)

$$\frac{P}{\theta} = \chi_0 \varepsilon_0 c^* + \frac{\alpha}{c^*}(T - T_c) \tag{191}$$

The first term slowly decreases with increasing temperature ($\chi_0 \sim 1/T$) whereas the second slowly increases. This explains the empirically found result that P/θ is essentially the same ($=s^*$) in the C* phase as well as in the A* phase. The ratio is thus practically the same whether P and θ are spontaneous or induced.

Within the approximation of small tilt, Eq. (182), we now insert the value for θ from Eq. (183) in the expression Eq. (189) for P. This gives

$$P = \chi_0 \varepsilon_0 c^* \frac{\chi_0 \varepsilon_0 c^* E}{\alpha(T - T_c)} + \chi_0 \varepsilon_0 E \tag{192}$$

or

$$\frac{P}{E} = \chi_0 \varepsilon_0 + \frac{s^{*2}}{\alpha(T - T_c)} \tag{193}$$

The second part is the electroclinic contribution to the dielectric susceptibility corre-

sponding to Eq. (188). The first is the "background" part corresponding to $s^* = 0$, ie.e to a racemic or non-chiral substance.

The electroclinic effect is linear only if we do not approach too closely to T_c. At the transition, $T = T_c$, the first term in Eq. (181) vanishes and the relation between θ and E becomes

$$b\theta^3 = s^* E \tag{194}$$

or

$$\theta = \left(\frac{s^*}{b}\right)^{1/3} E^{1/3} \tag{195}$$

This non-linear behavior with saturation is well known experimentally. It means that the electroclinic coefficient becomes field-dependent and falls off rather strongly ($\sim E^{-2/3}$) at high fields. It also means the same for the electroclinically induced polarization which acquires relatively high values at low applied fields but then shows saturation. It explains the shape of the measured $P(T)$ curves near T_c as illustrated in Fig. 18 (dashed line).

For $T \neq T_c$ but near T_c we have to keep all the terms in Eq. (181). We may write this expression in the form

$$A\theta + B\theta^3 = E \tag{196}$$

with

$$A = \frac{\alpha}{s^*}(T - T_c) \tag{197}$$

and

$$B = \frac{b}{s^*} \tag{198}$$

Generally it can be said that Eq. (196) very well describes the experimental results. A particularly careful study has been made by Kimura, Sako and Hayakawa, confirming both the linear relationship Eq. (197) and the fact that B, and therefore b as well as s*, are independent of temperature [189].

The electroclinic effect is intrinsically very fast. The torque Γ^θ acting on the director, and giving rise to a change in tilt, is obtained by taking the functional derivative of the free energy with respect to the tilt,

$$-\Gamma^\theta = \frac{\partial G}{\partial \theta} \qquad (199)$$

Equation (157) combined with Eq. (164) then gives

$$
\begin{aligned}
-\Gamma^\theta &= a\theta + b\theta^2 - c*P \\
&= \alpha(T - T_c)\theta + b\theta^3 - s*E \qquad (200) \\
&= a\theta - \chi_0 \varepsilon_0 c*^2 \theta + b\theta^3 - \chi_0 \varepsilon_0 c* E
\end{aligned}
$$

As evident from the middle one of these equations, we confirm, by Eq. (194) that this torque vanishes when we approach $T = T_c$, where θ shows a diverging tendency and we can expect a critical slowing down in the response.

The viscous torque Γ^υ counteracting any change in θ is, by definition,

$$\Gamma^\upsilon = -\gamma_\theta \frac{\partial \theta}{\partial t} \qquad (201)$$

where γ_θ is the electroclinic or soft mode viscosity. In dynamic equilibrium the total torque $\Gamma = \Gamma^\theta + \Gamma^\upsilon$ is zero. This gives

$$\alpha(T - T_c)\theta + b\theta^3 + s*E + \gamma_\theta \frac{\partial \theta}{\partial t} = 0 \qquad (202)$$

This equation describes the dynamics of the electroclinic effect in a material where we have a lower-lying phase $(T < T_c)$ characterized by a spontaneous tilt of the optic axis. If we divide all terms by s*, it can be written

$$A\theta + B\theta^3 + \left(\frac{\gamma_\theta}{s*}\right)\dot{\theta} = E \qquad (203)$$

This is the dynamic equation corresponding to Eq. (196). For small induced tilts, however, we skip the θ^3 term in Eq. (202). If we further put $E = 0$, we will see how the optic axis relaxes back to its state along the layer normal from a beginning state of nonzero tilt. Equation (202) in this case simply reads

$$\theta + \frac{\gamma_\theta}{\alpha(T - T_c)} \frac{\partial \theta}{\partial t} = 0 \qquad (204)$$

which is directly integrated to

$$\theta(t) = \theta_0 e^{-t/\tau} \qquad (205)$$

with the characteristic time constant

$$\tau = \frac{\gamma_0}{\alpha(T - T_c)} \qquad (206)$$

If instead a constant field E_0 is applied at $t = 0$, the growth of θ will be given by

$$\theta + \tau \frac{\partial \theta}{\partial t} = \frac{s*}{\alpha(T - T_c)} E_0 \qquad (207)$$

which is integrated to

$$\theta(t) = \theta_0 (1 - e^{-t/\tau}) \qquad (208)$$

with the saturation value of θ equal to

$$\theta_0 = \frac{s*}{\alpha(T - T_c)} E_0 \qquad (209)$$

which of course also corresponds to Eq. (183). If we finally apply a square wave, where we reverse the polarity from $-E_0$ to $+E_0$ at $t = 0$, the optic axis will change direction by an amount $2\theta_0$ according to

$$\theta(t) = \theta_0 (1 - 2e^{-t/\tau}) \qquad (210)$$

corresponding to the initial and final values $\theta(0) = -\theta_0$, $\theta(\infty) = +\theta_0$.

We note that the response time given by Eq. (206) is independent of the applied electric field. In comparison, therefore, the response time due to dielectric, ferroelectric and electroclinic torque is characterized by different powers of E, as

$$\tau_{\text{diel}} \sim E^{-2} \qquad (211)$$

$$\tau_\varphi \sim E^{-1} \qquad (212)$$

$$\tau_\theta \sim E^0 \qquad (213)$$

The reason for this apparent peculiarity in the electroclinic effect is of course that if the field E is increased by a certain factor, the saturation value of the tilt is increased by the same factor. The angular velocity in the switching is doubled if we double the field, but the angle for the full swing is also doubled. Thus, electrooptically speaking, the speed of a certain transmission change in a modulator certainly does increase with increased field. But one mystery remains to be solved. How come, that the characteristic time τ in Eq. (205) describing the relaxation back to equilibrium, is the same as the "response time" τ in Eq. (208) describing the change of θ under the influence of the electric field? Should not the relaxation back be much slower? This is what we are used to in liquid crystals: if a field is applied to a nematic, the director responds very quickly, but it relaxes back slowly when we take the field off. The answer is: the electric field exerts no torque at all on the director in the electroclinic action. Its effect is to shift the equilibrium direction in space of the optic axis. If the field is high the offset is large and the angular velocity in the director motion towards the new equilibrium state will be high. Equation (207) describes this motion and tells that the rate of change of θ is proportional to the angular difference between the initial and final director state. In fact Eqs. (204) and (207) are the same equation and can be written

$$-\frac{\partial \theta}{\partial t} = \frac{\theta - \theta_0}{\tau} \qquad (214)$$

The difference is only that the final state θ_0 in Eq. (204) is zero, whereas in Eq. (207) θ_0 is the value given by the electric field according to Eq. (209). Note that in Eq. (214) the electric field does not appear at all. The electric field E does, however, exert a torque on the whole medium during the electroclinic action. This torque, by conservation of an-gular momentum, is opposite in sense to the torque exerted by the "thermodynamic force" on the director.

As already mentioned in Sec. 1 of this Chapter, the electroclinic effect was first announced as a "piezoelectric" effect [20], but shortly thereafter renamed. The similarities are interesting but the differences sufficient to characterize the electroclinic effect as a new dielectric effect. The variable θ in which the "distortion" takes place is here an angular variable, cf. Fig. 55. If E is reversed, P and θ are reversed. But there is no converse effect. (One cannot choose $\pm\theta$ in the distortion variable to induce a P of a given sign.) The electroclinic tilt is connected with a mechanical distortion, a shrinking of the smectic layer thickness d. In the simplest model, assuming tilting rigid rods instead of molecules, thus exaggerating the effect, the thickness change is

$$\delta d = d\,(1 - \cos\theta) \sim \theta^2 \qquad (215)$$

hence $\delta d \sim E^2$, not $\sim E$ and we have an electrostrictive, not a piezoelectric distortion. This is also consistent with the fact that no

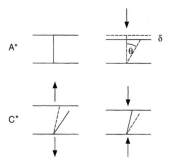

Figure 55. Analogy between the piezoelectric and the electroclinic effect. A translational distortion in the former corresponds to an angular distortion in the latter. If the sign of the applied field E is reversed, the sign of P and θ is also reversed. But there is no converse effect in the electroclinic case. The mechanical deformation is electrostrictive, i.e. proportional to E^2. In the smectic C* case there is also a component proportional to E.

dilatation is possible in the A* phase, thus the effect is inherently unsymmetric. (Note that this is not so in the C* phase; with a finite θ dilatation as well as compression is allowed.) The electrostriction may be a problem in the practical use of the electroclinic effect, because the field-induced layer shrinking, at least sometimes, leads to the appearance of layer kink defects, which may be observed as striations in the texture [190]. However, very little is known today of how the electrostrictive properties vary between different materials.

The electroclinic effect is the fastest of the useful electro-optic effects so far in liquid crystals, with a speed allowing MHz switching rates in repetitive operation. Its electro-optical performance is discussed and compared with other modes in the review of reference [191]. A number of device applications are discussed in reference [140]. The effect appears in all chiral orthogonal smectics but has been studied in a very restricted class of materials. It suffers so far, except from the fairly strong temperature sensitivity, in particular from a limitation in the value of induced tilt angle (12 to 15 degrees). This means that only a very small part of the typical V-shaped transmission curve can be utilized for a modulator using this kind of material, cf Fig. 56. In this respect short pitch C* or N* materials are much better (exploiting the deformed helical and flexoelectro-optic mode, respectively) with a sweep in either direction of more than 22 degrees in the first and more than 30 degrees in the second case. However, new A* materials may very soon change the scene. The ideal material for the electroclinic effect would be a so-called de Vries compound [191a] in which the molecules have a large but unbiased tilt in the A phase (adding up just as in the short pitch C* or antiferroelectric case, to an optic axis along the layer normal) but a biased tilt in the C* phase.

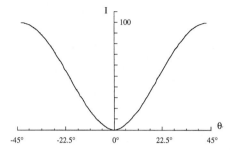

Figure 56. Modulation of the transmitted light intensity I for all electro-optic effects describing a tilt of the optic axis in the plane parallel to the cell plates (in-plane switching), for the case that the initial axis direction is along one of the two crossed polarizer directions. The curve is drawn for a cell thickness corresponding to $\lambda/2$ plate condition, for which a tilt sweep of 45 degrees gives full modulation (from zero to 100 percent transmitted light). If this condition is not met, the degree of modulation will be reduced.

We will return to such materials in Sec. 2.8.10 because they are also of eminent interest for high resolution displays. Up to now they have been somewhat hypothetical but seem recently to have been confirmed [192]. Such de Vries materials would be ideal even if they have to be somewhat slower (the torque counteracting tilt fluctuations would not be so strong for such compounds) because first of all the electroclinically induced tilt would be very large, namely up to the full value of unbiased individual tilt, and this for an applied field that can be expected to be very low. Furthermore the tilt biasing does not involve any electrostrictive effect or, at least this would be minimal. Finally the temperature sensitivity can be expected to be very low as it is in the absence of collective behavior, essentially T^{-1} instead of $(T-T_c)^{-1}$.

2.5.9   The Deformed Helix Mode in Short Pitch Materials

An interesting electro-optic effect utilizing the helical SmC* state of a short pitch material was presented in 1980 by Ostrovskij, Rabinovich and Chigrinov [193]. In this case the twisting power was so high that the pitch came below the wavelength of light. In such a densely twisted medium the light wave averages out the twist to feel an optic axis directed along the helical axis, i.e. in the direction of the layer normal z (Fig. 57). When an electric field is applied perpendicular to this axis it starts to partly untwist the helix and in this way perturb the spatial direction of the average. The result is a linear

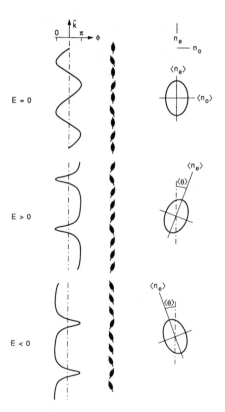

Figure 57. Partial untwisting of the helical axis in short-pitch materials (pitch in the range of 0.1 μm to 0.5 μm) has the effect of turning the averaged optic axis away from the layer normal as shown by the direction of the space averaged indicatrix. The twist also has the effect of averaging out part of the inherent large birefringence n_e-n_o to a much smaller value $\langle n_e \rangle - \langle n_o \rangle$ (averaged over a length > pitch) (from Beresnev et al. [194]).

effect very similar to that of the electroclinic effect and the flexoelectro-optic effect: the angular deviation $\langle \theta \rangle$ of the effective optic axis is proportional to the applied field. Like the other two mentioned effects this deformed-helix mode (DHM) has no memory but a continuous grey scale. Though not as rapid as the electroclinic effect it has at least two distinctive advantages. First, $\langle \theta \rangle$ can attain much higher values than the soft-mode tilt. Second, the apparent birefringence $\langle \Delta n \rangle$ is quite low, allowing a larger cell thickness to be used and still matching the $\lambda/2$ condition.

The short-pitch materials have shown another highly interesting property. When they are used in the SSFLC mode, i.e. when a field sufficient for complete unwinding is applied and then reversed, in order to actuate the switching between the two extreme cone positions, a new mode of surface-stabilization is found, in the limit $Z \gg d$, where Z is the pitch or helical periodicity and d is the cell gap thickness. It may at first seem surprising that surface stabilization can be achieved not only for $d \leq Z$ but also for $d \gg Z$, in fact the smaller the value of Z the better. In the limit $Z \ll d$ it is a pinning phenomenon which makes it distinctly different from the first case of surface-stabilization. In other words, whereas for $Z > d$ the non-helical state is the equilibrium state, for $Z \ll d$ it is a metastable state. The helix thus tends to rewind but, for topological reasons (the back-transition is defect-mediated) it may stay for a considerable time. This means that short-pitch materials are very interesting alternatives to pitch-compensated materials for display applications. Their demonstrated inconvenience so far has been the difficulty to align then well, generally giving somewhat scattering textures. One further advantage of this effect is that the response time is only weakly temperature dependent. A general discussion of this and other features is found in chapter VI of reference [41]. The DHM has been developed further by the Roche and Philips groups and applied in active matrix displays with grey scale, cf references [195] and [196].

2.5.10 The Landau Expansion for the Helical C* State

As we repeatedly pointed out, the Landau expansion Eq. (157) used so far

$$G = \frac{1}{2} a\theta^2 + \frac{1}{4} b\theta^4 + \frac{1}{2\chi_0 \varepsilon_0} P^2$$
$$-c*P\theta - PE \qquad (157)$$

refers to the non-helical C* state. Even if the non-helical state is somewhat special – the general case is one with a helix – and if it is true that for many purposes we may even regard the presence of the helix as a relatively small perturbation, it is now time to take the helix into account. The first thing we have to do then is to add a so-called Lifshitz invariant. This invariant is a scalar

$$L = n_x \frac{\partial n_y}{\partial z} - n_y \frac{\partial n_x}{\partial z} \qquad (216)$$

permitted by the local C_2 symmetry which describes the fact that the structure is modulated in space along the z-direction. If in Fig. 49 we let the x direction be along the C_2 axis, we see that the C_2 symmetry operation involves $y \rightarrow -y$, $z \rightarrow -z$, $n_y \rightarrow -n_y$ and $n_z \rightarrow -n_z$, of which the first and last do not interfere and the middle two leave Eq. (216) invariant. We remember from Fig. 29 that the cholesteric structure has infinitely many C_2 axes perpendicular to the helix axis. The expression Eq. (216) is therefore also an invariant for the N* phase. If q is the cholesteric wave vector, we write the director as

$$n = (\cos\phi, \sin\phi) = (\cos qz, \sin qz) \qquad (217)$$

Inserting this in Eq. (216) we find that the Lifshitz invariant (which has a composition rule slightly reminiscent of angular momentum, cf. $L_x = x p_y - y p_x$) in the cholesteric case has the value equal to q, the wave vector. In fact we can gain some familiarity with this invariant by starting from an expression we know quite well, the Oseen-Frank expression for the elastic free energy Eq. (96). Because of its symmetry, this expression cannot describe the cholesteric state of a nematic which lacks reflection symmetry and where the twisted state represents the lowest energy. Now, if there is a constant twist with wave vector q, the value of $n \cdot \nabla \times n$ in the K_{22} term equals $-q$. The expression Eq. (96) therefore has to be "renormalized" to

$$G = \frac{1}{2} K_{11} (\nabla \cdot n)^2 + \frac{1}{2} K_{22} (n \cdot \nabla \times n + q)^2$$
$$+ \frac{1}{2} K_{33} (n \times \nabla \times n)^2 \qquad (218)$$

which, if there is no splay and bend present, becomes zero for $n \cdot \nabla \times n = -q$. If we expand the square in the K_{22} term we can write the energy in the form

$$G_{N*} = G_N + K_{22} qn \cdot \nabla \times n + \frac{1}{2} K_{22} q^2 \qquad (219)$$

where G_N corresponds to the energy expression Eq. (96) for a non-chiral nematic. Two new terms have appeared in the energy as a result of dropping the reflection symmetry. The first is a linear term which is the Lifshitz term, the second is an energy term quadratic in the wave vector. In fact, a simple check shows that

$$n \cdot \nabla \times n = n_x \left(\frac{\partial n_z}{\partial y} - \frac{\partial n_y}{\partial z} \right) + n_y \left(\frac{\partial n_x}{\partial z} - \frac{\partial n_z}{\partial x} \right)$$
$$+ n_z \left(\frac{\partial n_y}{\partial x} - \frac{\partial n_x}{\partial y} \right)$$
$$= -\left(n_x \frac{\partial n_y}{\partial z} - n_y \frac{\partial n_x}{\partial z} \right) = -L \qquad (220)$$

We could therefore alternatively write

$$G_{N*} = G_N$$
$$- K_{22}\, q \left(n_x \frac{\partial n_y}{\partial z} - n_y \frac{\partial n_x}{\partial z} \right) + \frac{1}{2} K_{22}\, q^2 \tag{221}$$

or use

$$L = -\boldsymbol{n} \cdot \boldsymbol{\nabla} \times \boldsymbol{n} \tag{222}$$

instead of Eq. (216). We may finally note that if we go from a right-handed to a left-handed reference frame then the Lifshitz invariant (like angular momentum) changes sign because $\boldsymbol{\nabla} \times \boldsymbol{n}$ does. We can also see this in Eq. (216) because in addition to the C_2 operation the inversion involves that $x \rightarrow -x$ and $n_x \rightarrow -n_x$.

In the helical smectic C* case, again with $\varphi(z) = q z$, the director components were stated in Eq. (141)

$$n_x = \sin\theta \, \cos\phi$$
$$n_y = \sin\theta \, \sin\phi$$
$$n_z = \cos\theta \tag{141}$$

Inserting this in Eq. (216) we find that the Lifshitz invariant equals

$$L = q \sin^2\theta \tag{223}$$

which is also often written

$$L = \frac{\partial \varphi}{\partial z} \sin^2\theta \tag{224}$$

for the more general case that the twist is not homogeneous. We may also note, that in the smectic C* case we can equally well write Eq. (216) in the form

$$L = c_x \frac{\partial c_y}{\partial z} - c_y \frac{\partial c_x}{\partial z} \tag{225}$$

i.e. using the two-dimensional c director, as the n_z component does not appear in L.

Before we add the Lifshitz invariant to our Landau expansion Eq. (157) another consequence of the helix may be pointed out. It is the fact that, as we have already seen in Sec. 2.5.5, with a helical deformation we also have a flexoelectric contribution to P. This is opposite in sign to P_s for $e_b > 0$, cf. Eq. (148), and proportional to the wave vector q. These two contributions must now be separated in the Landau expansion because we have to regard q as a new independent variable. In that case we not only have a coupling term $-c^* P\theta$ but also one corresponding to the flexoelectric contribution which we could write $eqP\theta$. With this convention, Eq. (157) would be extended, for the helical C* case, to

$$G = \frac{1}{2} a\theta^2 + \frac{1}{4} b\theta^4 + \frac{1}{2\chi_0\,\varepsilon_0}\, P^2$$
$$- c^* P\theta + eq P\theta - \lambda^* q\theta^2$$
$$+ \frac{1}{2} K_{22}\, q^2 + \frac{1}{2} K_{33}\, q^2\,\theta^2 \tag{226}$$

where the $-PE$ term has been skipped since we are here only interested in the field-free case. For the Lifshitz invariant we have used Eq. (223) in the limit of small θ, but chosen the minus sign, corresponding to $\lambda^* > 0$ in Eq. (219). In accordance with Eq. (219) we have also added a term proportional to q^2. The origin of the last term, proportional to $q^2\theta^2$ is the fact that the smectic helix is a combined twist-bend deformation and that its energy therefore has to depend on the tilt angle as well as on q. By symmetry this dependence has to be $\sim\theta^2$ (or, actually proportional to the square of $\sin 2\theta$). But it is illustrative to examine these last terms in a less ad hoc way. Thus, if we go back to our "nematic" description, introduced in Sec. 2.4.9, we can write

$$G_{C*} = \frac{1}{2} K_{22}\,(\boldsymbol{n} \cdot \boldsymbol{\nabla} \times \boldsymbol{n} + q)^2 \tag{227}$$
$$+ \frac{1}{2} K_{33}\,(\boldsymbol{n} \times \boldsymbol{\nabla} \times \boldsymbol{n} - \boldsymbol{B})^2$$
$$= G_C + \frac{1}{2} K_{22}\, q^2 + \frac{1}{2} K_{33}\, B^2$$
$$+ K_{22}\, q\,\boldsymbol{n} \cdot \boldsymbol{\nabla} \times \boldsymbol{n} - K_{33}\,\boldsymbol{B} \cdot \boldsymbol{n} \times \boldsymbol{\nabla} \times \boldsymbol{n}$$

where G_C corresponds to the expression (96) without splay term. As the spontaneous bend B is equal to $\frac{1}{2} q \sin 2\theta \approx q\theta$ for small tilts, cf. Eq. (147), the third term on the right side is equal to the last term in Eq. (226). From Eq. (145) we further know that $n \times \nabla \times n = -q \sin\theta \cos\theta \; z \times n$ for the helical C* case, hence the last term in Eq. (227) can be written $K_{33} B q\theta$ for small tilts. With $P = e_b B$ for the flexoelectric contribution we write this term $e q P \theta$, with the abbreviation $K_{33}/e_b = e$. This term was already included in Eq. (226). Before continuing with the Landau expansion we note that with $n \cdot \nabla \times n = -q$ and $n \times \nabla \times n = B$ inserted in Eq. (227), the free energy in the helical C* state can be written in the form

$$G_{C*} = G_C - (K_{22} + K_{33}\theta^2)\, q^2 \qquad (228)$$

With the new terms in Eq. (226), minimizing with respect to P, θ and q gives

$$\frac{\partial G}{\partial P} = \frac{1}{\chi_0 \varepsilon_0} P - c^*\theta + e q\theta = 0 \qquad (229)$$

$$\begin{aligned}\frac{\partial G}{\partial \theta} &= a\theta + b\theta^3 - c^* P + e q P \\ &\quad - 2\lambda^* q\theta + K_{33}\,\theta q^2 = 0 \end{aligned} \qquad (230)$$

$$\begin{aligned}\frac{\partial G}{\partial q} &= e P\theta - \lambda^*\theta^2 + K_{22}\, q \\ &\quad + K_{33}\, q\theta^2 = 0\end{aligned} \qquad (231)$$

The first of these yields

$$P = \chi_0 \varepsilon_0 (c^* - e q)\,\theta \qquad (232)$$

Comparing this with Eq. (148) which, for small tilts we can write

$$P = (P_0 - q\, e_b)\,\theta \qquad (233)$$

we find that P_0 corresponds to $\chi_0 \varepsilon_0 c^*$, equal to s^* in Eq. (167), and that e_b corresponds to $\chi_0 \varepsilon_0 e$, where e is the coefficient in the Landau expansion. We will now insert P from Eq. (232) into Eq. (231) which gives

$$(K_{33}q - \chi_0\varepsilon_0 e^2 q)\theta^2 + K_{22}q$$
$$= (\lambda^* - \chi_0\varepsilon_0 c^*)\,\theta^2 \qquad (234)$$

or

$$K_f q^2\theta^2 + K_{22}q = \lambda_f^*\theta^2 \qquad (235)$$

We have here introduced the renormalized coefficients

$$K_f = K_{33} - \chi_0\varepsilon_0 e^2, \quad \lambda_f^* = \lambda^* - \chi_0\varepsilon_0 e c^* \quad (236)$$

which contain the flexoelectric contributions. The wave vector is then obtained as

$$q = \frac{\lambda_f^* \,\theta^2}{K_{22} + K_f\,\theta^2} \qquad (237)$$

For a non-chiral material, $\lambda_{f*} = 0$ (both λ^* and c^* are chiral coefficients in Eq. (236), and hence $q = 0$. The pitch of the smectic helix Z equals $2\pi/q$, thus

$$Z = \frac{K_{22} + K_f\,\theta^2}{\lambda_f^*\,\theta^2} \qquad (238)$$

indicating that the pitch should diverge at the transition C* → A*, cf Fig. 58. Experimentally, the situation is still not very clear regarding $Z(T)$. While some measurements do indicate that a pure divergence exists if the layers are parallel to the cell plates, the majority of measurements give a result that the pitch goes through a maximum value

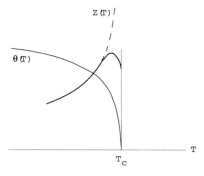

Figure 58. The measured helical pitch Z at the transition C* → A*. The pitch should diverge according to Eq. (238).

slightly below the transition and attains a finite value at $T=T_c$. If we write the pitch dependence

$$Z = Z_0 + \zeta \theta^{-2} \qquad (239)$$

there should be a saturation value $Z_0 = 2\pi K_f / \lambda_f^*$ at low temperatures. Z_0 as well as $\zeta = 2\pi K_{22}/\lambda_f^*$ shows the balance between the non-chiral elastic coefficients (K) wanting to increase the pitch and the chiral coefficients (λ^*) wanting to twist the medium harder.

When we finally insert the value of P from Eq. (232) into Eq. (230) we get

$$a\theta + b\theta^3 - \chi_0 \varepsilon_0 c^{*2} \theta$$
$$+ 2\chi_0 \varepsilon_0 e c^* q\theta - \chi_0 \varepsilon_0 e^2 q^2 \theta$$
$$- 2\lambda^* q\theta + K_{33}\theta q^2 = 0 \qquad (240)$$

After skipping the solution $\theta = 0$, we are left with

$$- \chi_0 \varepsilon_0 c^{*2} - 2(\lambda^* - \chi_0 \varepsilon_0 e c^*) q \qquad (241)$$
$$+ (K - \chi_0 \varepsilon_0 e^2) q^2 + b\theta^2$$
$$a - \chi_0 \varepsilon_0 c^{*2} - 2\lambda_f^* q + K_f q^2 + b\theta^2 = 0$$

Together with Eq. (237) this leads to a complicated expression for $\theta(T)$, which we will not pursue further here. However, in the approximation that we take the low temperature limit for the wave vector

$$q_0 = \frac{\lambda_f^*}{K_f} \qquad (242)$$

(corresponding to Z_0) we see that Eq. (241) simplifies to

$$a - \chi_0\varepsilon_0 c^{*2} - K_f q_0^2 + b\theta^2 \qquad (243)$$

which is reshaped in the usual way to

$$\alpha(T - T_c) + b\theta^2 \qquad (244)$$

with a phase transition shift $T_c - T_0$ equal to

$$\Delta T = \frac{\chi_0 \varepsilon_0 c^{*2} + K_f q_0^2}{\alpha} \qquad (245)$$

While ΔT is still expected to be generally small, it is certainly true that the contribution from unwinding the helix may raise the transition temperature more than the contribution from the chiral coupling, at least in the case of hardtwisted media like short-pitch smectics C* where the pitch is smaller than the wavelength of light.

2.5.11 The Pikin-Indenbom Order Parameter

A group theoretical symmetry analysis by Indenbom and his collaborators in Moscow [197, 198] led, around 1977, to the introduction of a very attractive order parameter for the A*–C* transition which usually is referred to as the Pikin-Indenbom order parameter. It is attractive because it presents, in the simplest form possible, the correct symmetry and a very lucid connection to the secondary order parameter P. It was adopted by the Ljubljana group around Blinc for the description of both static and dynamic properties of the C* phase [199–204]. The basic formalism has been described in particular detail by Pikin [40] and by Pikin and Osipov [41].

In the smectic C* phase, the layer normal z and the director n represent the only two natural directions which we can define. That is, z and n are the only natural vectors in the medium. Using these we now want to construct an order parameter which is a vector, is chiral, has C_2 symmetry, leads to a second order A*–C* transition and is proportional to θ for small values of θ. We can do this by combining their scalar and vector products, multiplying one with the other,

$$N = (n \cdot z)(n \times z) \qquad (246)$$

We then obtain a vector in which the first factor is invariant, while the second changes sign on inversion. N thus has the two first properties. Performing the multiplication we find

$$N = n_z(n_y, -n_x, 0) \qquad (247)$$

which only has two components, hence is written

$$N = (n_z n_y, -n_z n_x) \qquad (248)$$

The $n_z n_y$ and $n_z n_x$ combinations do not change on inversion but the x and y directions do, thus we see that N changes sign on inversion.

The angle between n and z is θ. Therefore, taking the absolute value in Eq. (246)

$$N = \cos\theta \sin\theta = \frac{1}{2}\sin 2\theta \qquad (249)$$

We then recognize that this order parameter automatically reflects the C_2 symmetry, cf. the discussions around Eqs. (131), (132) and Figure 49, where we had to introduce this symmetry in an ad hoc way. Clearly $N \approx \theta$ is valid for small tilts according to Eq. (249). Finally, we can check that N and $-N$ describe the same state from Eq. (248), by virtue of the fact that n and $-n$ describe the same state. Thus, if we make a Landau expansion in N, only even powers can appear and the transition will be one of second order.

Before going on to this expansion we should, however, make the connection to the secondary order parameter P. Looking onto the smectic layer plane in Figure 59 we have (for a positive material, $P_0 > 0$) a relation between the P direction and the tilt direction such that P advances in phase by 90 degrees in the positive φ direction. If we express P as a complex quantity $P_x + iP_y$, we can relate P directly to the $\dot{c}$ director expressed in complex form, $c = c_x + ic_y$, by

$$P_x + iP_y \sim -i(c_x + ic_y) \qquad (250)$$

Let us now introduce the tilt vector ξ as the complex quantity $n_z c = (n_z c_x + in_z c_y) =$

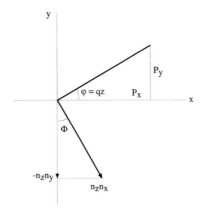

Figure 59. The tilt vector $n_z c$ lags P in phase by 90 degrees (for $P_0 > 0$), corresponding to a multiplication by $-i$ in the complex plane.

$(n_z n_x + in_z n_y)$ and rewrite Eq. (250) as

$$P_x + iP_y$$
$$\sim -i(n_z n_x + in_z n_y) = (n_z n_y - in_z n_x) \qquad (251)$$

In vector notation the tilt vector is $\xi = (\xi_1, \xi_2) = (n_z n_x, n_z n_y)$ and Eq. (251) in vector form reads

$$(P_x, P_y) \sim (\xi_2, -\xi_1) \qquad (252)$$

This is the relation between the secondary order parameter $P = (P_x, P_y)$ and the primary order parameter $N = (n_z n_y, -n_z n_x)$, when the latter is expressed in the tilt vector $\xi = (\xi_1, \xi_2)$.

We may now proceed to ask what invariants we may form in N and P, and their combinations, to include in the Landau expansion. N^2 and N^4 give $\frac{1}{2}a(\xi_1^2 + \xi_2^2)$ and $\frac{1}{4}b$ $(\xi_1^2 + \xi_2^2)^2$ corresponding to our earlier θ^2 and θ^4 terms, P^2 gives in the same way the $\frac{1}{2\chi_0\varepsilon_0}(P_x^2 + P_y^2)$ term. But because P and N transform in the same way (cf. Eq. (252) under the symmetry operations, their scalar product $P \cdot N$ will also be an invariant,

$$P \cdot N = P_x \xi_2 - P_y \xi_1 \qquad (253)$$

This corresponds to our previous $P\theta$ and gives a term $-c^*(P_x\xi_2 - P_y\xi_1)$ in the Landau expansion. In contrast to $P\theta$ the form $P_x\xi_2 - P_y\xi_1$ directly shows the chiral character. Thus, this invariant changes sign when we change from a right- to a left-handed reference frame, which means that the optical antipode of a certain C* compound will have a polarization of opposite sign. Further, we have the Lifshitz invariant

$$\xi_1 \frac{\partial \xi_2}{\partial z} - \xi_2 \frac{\partial \xi_1}{\partial z} \tag{254}$$

corresponding to Eq. (225) and, because both ξ and P behave like the c director, also one in P,

$$P_x \frac{\partial P_y}{\partial z} - P_y \frac{\partial P_x}{\partial z} \tag{255}$$

Both have the same chiral character as the form in Eq. (253). Only one of these can be chosen, because they are linearly dependent, according to Eq. (252). This results in a Landau expansion

$$G = \frac{1}{2}\, a(\xi_1^2 + \xi_2^2) + \frac{1}{4}\, b(\xi_1^2 + \xi_2^2)$$
$$+ \frac{1}{2\chi}(P_x^2 + P_y^2) + c^*(P_x\,\xi_2 - P_y\,\xi_1)$$
$$- \lambda^*\!\left(\xi_1 \frac{\partial \xi_2}{\partial z} - \xi_2 \frac{\partial \xi_1}{\partial z}\right)$$
$$+ e\!\left(P_x \frac{\partial \xi_1}{\partial z} + P_y \frac{\partial \xi_2}{\partial z}\right)$$
$$+ \frac{1}{2}\, K_{33}\!\left[\left(\frac{\partial \xi_1}{\partial z}\right)^2 + \left(\frac{\partial \xi_2}{\partial z}\right)^2\right] \tag{256}$$

In this expansion the last two terms describe the flexoelectric (previously $eqP\theta$) energy and the elastic energy due to the helix. The wave vector q in these terms is hidden in the derivatives $\partial\xi_1/\partial z$ and $\partial\xi_2/\partial z$.

The expression Eq. (256) is the "classical" Landau expansion in the formulation of Pikin and Indenbom. We now have to insert

convenient expressions for ξ_1, ξ_2, P_x and P_y. This can be done in several ways if we only correctly express the spatial relationship of ξ and P. For a positive material ($P_0 > 0$), P is always 90 degrees ahead. If the director tilts out in the x direction, this will cause a polarization along y, symbolically expressed by

$$\xi = (\theta, 0), \quad P = (0, P) \tag{257}$$

and similarly, with tilt along y

$$\xi = (0, \theta), \quad P = (-P, 0) \tag{258}$$

Generally, with $\varphi = qz$, and small tilts,

$$\xi_1 = n_z n_x = \cos\theta \sin\theta \cos\varphi \approx \theta \cos qz$$
$$\xi_2 = n_z n_y = \cos\theta \sin\theta \sin\varphi \approx \theta \sin qz \tag{259}$$

corresponds to

$$P_x = -P \sin qz$$
$$P_y = P \cos qz \tag{260}$$

Inserting these in Eq. (256) immediately gives

$$G = \frac{1}{2}\, a\theta^2 + \frac{1}{4}\, b\theta^4 + \frac{1}{2\chi_0\varepsilon_0}\, P^2$$
$$- c^* P\theta + eq P\theta - \lambda^* q\theta^2$$
$$+ \frac{1}{2}\, K_{33} q^2 \theta^2 \tag{261}$$

This is the same as Eq. (226) except that here the term $\frac{1}{2}K_{22}q^2$ is lacking. While minimizing with respect to P gives the same equilibrium value as before, $\partial G/\partial q = 0$ gives

$$eP\theta - \lambda^*\theta^2 + K_{33}q\theta^2 = 0 \tag{262}$$

Since $P \sim \theta$, all terms will be proportional to θ^2 and the θ-dependence simply cancels out. This means that the wave vector, according to this model, will simply be $q = \lambda_f^*/K_f$, equal to the low temperature value of Eq. (242). Thus the pitch Z is temperature independent, equal to Z_0. This is of course in serious disagreement with experiment and the reason for this deficiency is the failure to recognize the twist component of the helix. We can learn from this that a good order parameter will not by itself take care of all the physics. But our previous Landau expansion is likewise far from perfect, even if it gives a more reasonable behavior for the pitch. We should therefore stop for a moment to contemplate how capable the theory is, so far, to reproduce the experimental facts. As we already discussed, the pitch has a diverging behavior as we approach T_c (this is a priori natural, as no helix is allowed in the A* phase) but seems to change its behavior (within about 1 K from the transition) to decrease to a finite value at the transition. Another important fact is that the ratio $P/\theta = s^*$ experimentally is constant, as predicted by theory, up to about 1 K from the transition after which it decreases to a lower value [133]. A third fact, which we had no space to develop, is that the dielectric response, according to the Landau expansion used so far, should exhibit a cusp at the transition, which does not correspond to experimental facts.

What is then the deficiency in the theory? Did we make an unjustified simplification when taking $\sin\theta \approx \theta$ in Eq. (259)? Probably not since, as the problem seems to lie in the very vicinity of T_c, where θ is small, it would not do any good to go to the more accurate expressions valid for large values of θ. Then how do we go further? Should we extend our expansion to include θ^6 beyond θ^4, or P^4 beyond P^2? No, rather the solution ought to come from physical reasoning instead of formal procedures. This step was taken by Žekš in 1984 [201]. The argument is as follows. Our Landau expansion so far has failed to take into account a quite important non-chiral contribution. At the tilting transition rotational bias appears, whether the material is chiral or not, which is quadrupolar in character, cf.

Fig. 47. But if there is a strong quadrupolar order it will certainly contribute to increasing the polar order, once we remove the reflection symmetry. This leads to a coupling term in P and θ which is biquadratic, $\sim P^2\theta^2$. If we now only consider those terms in the free energy which contain P and q, we can write

$$G = \frac{1}{2\chi_0\,\varepsilon_0}P^2 - c^*\,P\theta + eqP\theta$$
$$- \frac{1}{2}\,\Omega P^2\,\theta^2 + \frac{1}{4}\,\eta P^4 - \lambda^*q\theta^2$$
$$+ \frac{1}{2}\,K_{33}\,q^2\,\theta^2 - v^*q\theta^4 \qquad (263)$$

The biquadratic coupling term is $-\frac{1}{2}\Omega(P_x\xi_2 - P_y\xi_1)^2$ using the Pikin-Indenbom order parameter. The $v^*q\theta^4$ term is the higher invariant $-v^*(\xi_1^2 + \xi_2^2)\left(\xi_1\frac{\partial\xi_2}{\partial z} - \xi_2\frac{\partial\xi_1}{\partial z}\right)$ which gives a coupling between tilt and twist. The P^4 term has been added to stabilize the system. $\partial G/\partial P = 0$ gives

$$\qquad\qquad\qquad\qquad\qquad\qquad (264)$$
$$\frac{1}{\chi_0\,\varepsilon_0}\,P - c^*\,\theta + eq\theta - \Omega P\theta^2 + \eta P^3 = 0$$

For small θ we can skip θ^2 and P^3 and get the same result as before, $P = \chi_0\varepsilon_0c^*\theta$ or $P/\theta = s^*$. For large P we get

$$P^3 - \frac{\Omega}{\eta}\,P\theta^2 = 0 \qquad (265)$$

or

$$P = \theta\sqrt{\frac{\Omega}{\eta}} \qquad (266)$$

Thus the larger constant value of P/θ is $\sqrt{\Omega/\eta}$, which drops to s^* near T_c where the rotational bias is not so efficient, as we expect for small tilts. Finally, $\partial G/\partial q$ gives

$$eP\theta - \lambda^*\theta^2 + K_{33}\,q\theta^2 - v^*\theta^4 = 0 \qquad (267)$$

from which follows

$$q = \frac{\lambda^*}{K_{33}} - \frac{eP}{K_{33}\,\theta} + \frac{v^*\theta^2}{K_{33}} \qquad (268)$$

If θ is not too small, P/θ in the second term is constant and q increases with increasing θ, i.e. Z decreases on lowering the temperature. Near T_c, q increases again with increasing T since P/θ decreases. An interesting feature is that the flexoelectric coefficient e has to be positive in order to yield a maximum in the $Z(T)$ curve. It is therefore conceivable that certain materials show a divergence, while others do not. In general the coupled equations have to be solved numerically but give, at least qualitatively, the correct temperature dependence for the different parameters.

We would like to close this section with some remarks, in particular concerning the different order parameters. First of all we note a surprising thing about the very smart Pikin-Indenbom order parameter. It has been very important in the basic symmetry discussion, but it has never really been used as such in practice. Perhaps because it is too sophisticated in its simplicity. Everybody instead uses the tilt vector ξ as order parameter. But $\xi = (\xi_1, \xi_2)$ is non-chiral, hence does not have the correct symmetry, whereas $N = (\xi_2, -\xi_1)$ is chiral by construction. Thus the tilt vector ξ should not be confused with the chiral vector $(n_z n_y, -n_z n_x)$. Having said this, however, it is clear that ξ is easiest to handle, if we only combine it with $(-P_2, P_1)$. Note further that ξ is not the projection of the director on the smectic plane, $\xi \neq \mathbf{c} = (c_x, c_y) = (n_x, n_y)$. Although ξ and $\mathbf{c}$ are parallel they are not identical: $\xi = n_z \mathbf{c}$. The θ dependence of ξ and $\mathbf{c}$ is different but for small tilts the difference between the vectors is not big. After having defined ξ, the first thing we did in using it in Eq. (256) was in fact to go to the small tilt limit. Thus in practice we have not used anything else than

the $\mathbf{c}$ director as order parameter. Which means that we have in essence, all the time, in different shapes, used the two component complex order parameter from the beginning of Sec. 2.5, cf. Eq. (125)

$$\Psi = \theta e^{i\varphi} = \theta e^{iqz} = \theta\cos qz + i\sin qz \quad (269)$$

As primary and secondary order parameters we have thus used two two-component vectors lying in the smectic plane, perpendicular to the layer normal. The number of degrees of freedom, i.e. the number of components, is evidently more important than subtle differences in symmetry, if the order parameters are used with sound physical arguments.

2.6 Electrooptics in the Surface-Stabilized State

2.6.1 The Linear Electrooptic Effect

The basic geometry of a surface-stabilized ferroelectric liquid crystal is illustrated in Fig. 60. The director $\mathbf{n}$ (local average of the

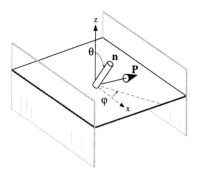

Figure 60. Local symmetry of the smectic C* phase. The direction of P is shown for a positive material ($P>0$); by definition, one in which z, $\mathbf{n}$, and P make a right-handed system. If electrodes are applied as shown, P will follow the direction of the electric field generated between them.

molecular axes) may be identified with the optic axis (we disregard the small biaxiality) and is tilted at the angle θ away from the layer normal (z). In the figure the director is momentarily tilted in the xz plane with a tilt angle θ, which is fixed for a fixed temperature. Its only freedom of motion is therefore described by the azimuthal variable φ indicated in the xy plane. If the phase is chiral, no reflections are symmetry operations and therefore a local polarization P is permitted in the (positive or negative) y direction. If electroded glass plates are mounted as in the figure, such that the layer normal z runs between them in the middle, we see that this most peculiar geometry with P perpendicular to the optic axis is ideal for a display: by applying the electric field across the thin slice, we can move the optic axis back and forth between two positions in the plane of the slice. This so-called in-plane switching cannot be achieved in a solid ferroelectric (nor in a nematic) without the use of a much more complicated electrode configuration. A typical cell structure is shown in Fig. 61. It illustrates the fact that we have two equivalent stable states; this is the reason for the symmetric bistability. The coupling is linear in the electric field and described by a torque

$$\Gamma = P \times E \tag{270}$$

driving n (which is sterically coupled to P) both ways on reversal of the field direction and, as we will see below, leading to a microsecond response speed according to

$$\tau^{-1} \sim \frac{P \cdot E}{\gamma} \tag{271}$$

dominating over any quadratic (dielectric) terms at low fields.

In Fig. 62 this linear electrooptic effect is compared with the quadratic effect controlling the state of a twisted nematic device.

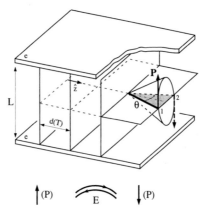

Figure 61. Originally assumed FLC cell structure. The field E is applied between the electroded glass plates e. The smectic layers are perpendicular to these glass plates and have a layer thickness $d(T)$ of the order of 5 nm, which is temperature-dependent. The thickness of the smectic slab is $1-2$ μm and equal to L, the distance between the plates. The shaded area illustrates the achievable switching angle of the optic axis. The symbols below illustrate that the polarization vector can be actively switched in both directions ($1 \leftrightarrow 2$). Strangely enough, this simple geometry, which corresponds to the original concepts of [93–95], has only been realized in recent times, thanks to a devoted synthetic effort by chemists over the last 15 years. Before these materials are coming into use (not yet commercial) the layer structure is much more complicated. These problems are discussed in Sec. 2.8.

With a torque $\sim E^2$ the latter effect is insensitive to the direction of E and is driven electrically only in the direction from twisted to untwisted state, whereas it has to relax back elastically in the reverse direction. The difference is also striking optically. The important feature of the in-plane switching is that there is very little azimuthal variation in brightness and color, because the optic axis is always reasonably parallel to the cell plane. If the optic axis pointed out of this plane (see Fig. 63), which it often does in the field-on state of a twisted nematic, a considerable variation with viewing angle will occur, including reversal of contrast when the display is viewed along angular direc-

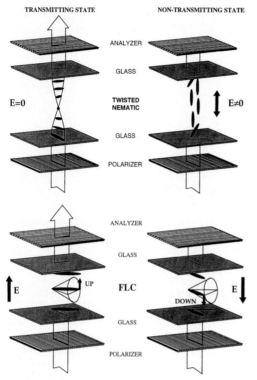

TRANSMITTING STATE   NON-TRANSMITTING STATE

ANALYZER

GLASS

E=0   TWISTED
NEMATIC   E≠0

GLASS

POLARIZER

ANALYZER

GLASS

E   UP   FLC
DOWN   E

GLASS

POLARIZER

Figure 62. Comparison between twisted nematic and FLC switching geometry.

tions nearly coinciding with the optic axis. The basic FLC optical effect is also illustrated in Fig. 64.

In-plane switching has another advantage than the superior viewing angle, which may be equally important but has so far passed fairly unnoticed. This lies in its color neutrality. In the top part of Fig. 62, it is seen that when intermediate values of the electric field are applied to a twisted nematic in order to continuously vary the transmission, the tilt of the optic axis and thus the birefringence are changed at the same time. Thus the hue is influenced at the same time as the grey level in a twisted nematic, which makes the color rendition less perfect in a display where the colors are generated either by red–green–blue filter triads or by sequential color illumination. In-plane switching, on the other hand, in principle allows perfect separation of color and grey shades.

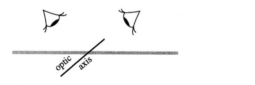

optic axis

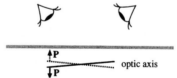

P

P   optic axis

Figure 63. The presentation of the first Canon monochrome A4 size panel in 1988 not only demonstrated the high resolution capability based on speed and bistability, it was also a definite breakthrough in LCD optics, demonstrating the essentially hemispherical viewing angle connected with in-plane switching.

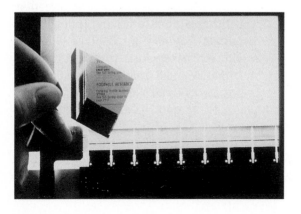

Figure 64. The basic FLC optic effect is illustrated by the 1988 Canon monochrome prototype in which the front polarizer sheet has been removed; this also makes the drivers along the bottom visible. The optical contrast between the two polarization states is given by the small hand-held polarizer to the lower left.

2.6.2 The Quadratic Torque

When we apply an electric field across an FLC cell there will always be a dielectric torque acting on the director, in addition to the ferroelectric torque. This is of course the same torque as is present in all liquid crystals and in particular in nematics, but its effect will be slightly different here than it is in a nematic. This is because in the smectic C phase the tilt angle θ is a "hard" variable, as earlier pointed out. It is rigid in the sense that it is hardly affected at all by an electric field. This is certainly true for a nonchiral smectic C and also for the chiral C*, except in the immediate vicinity of $T_{C^*A^*}$, where we may not neglect the electroclinic effect. Hence we will consider the tilt angle θ uninfluenced by the electric field, and this then only controls the phase variable φ of Fig. 60.

In order to appreciate the difference, it is illustrative to start with the effect in a nematic. We will thus study the torque on the director due to an electric field E that is not collinear with n, as illustrated in Fig. 65. If we introduce the dielectric anisotropy

$$\Delta\varepsilon = \varepsilon_{\parallel} - \varepsilon_{\perp} \qquad (272)$$

we may write the dielectric displacement $D = \varepsilon_0 \varepsilon E$ caused by the field in the form of the two contributions

$$D = \varepsilon_0 \varepsilon_{\parallel} E_{\parallel} + \varepsilon_0 \varepsilon_{\perp} E_{\perp} \qquad (273)$$

Subtracting $\varepsilon_0 \varepsilon_{\perp} E_{\parallel}$ from the first and adding it to the second term gives

$$D = \varepsilon_0 \Delta\varepsilon E_{\parallel} + \varepsilon_0 \varepsilon_{\perp} E$$
$$= \varepsilon_0 \Delta\varepsilon (n \cdot E) n + \varepsilon_0 \varepsilon_{\perp} E \qquad (274)$$

Figure 65. The dielectric torque will diminish the angle α between the director n and the electric field for $\Delta\varepsilon > 0$.

This gives a contribution to the free energy (which is to be minimized for the case of a fixed applied voltage)

$$-\int D \cdot dE \qquad (275)$$
$$= -\varepsilon_0 \Delta\varepsilon \int (n \cdot E) n \cdot dE - \varepsilon_0 \varepsilon_{\perp} \int E \cdot dE$$
$$= -\frac{1}{2} \varepsilon_0 \Delta\varepsilon (n \cdot E)^2 - \frac{1}{2} \varepsilon_0 \varepsilon_{\perp} E^2$$

We discard the second term as it does not depend on the orientation of n and write the relevant free energy contribution

$$\qquad (276)$$
$$G = -\frac{1}{2} \varepsilon_0 \Delta\varepsilon (n \cdot E)^2 = -\frac{1}{2} \varepsilon_0 \Delta\varepsilon E^2 \cos^2\alpha$$

where α is the angle between n and E. With positive dielectric anisotropy, $\Delta\varepsilon > 0$, we see that this energy is minimized when n is parallel to E. If $\Delta\varepsilon < 0$, it is minimized for $n \perp E$. If the situation is not one of minimum energy, the electric field will exercise a torque Γ^ε on the molecule and the molecule will respond with a motion, counteracted by viscosity, towards the equilibrium state, which corresponds to $\alpha = 0$ for $\Delta\varepsilon > 0$.

The dieelectric torque Γ^ε is obtained by taking the functional derivative of G with respect to α

$$\Gamma^\varepsilon = -\frac{\partial G}{\partial \alpha} \qquad (277)$$

which yields

$$\Gamma^\varepsilon = -\frac{1}{2} \varepsilon_0 \Delta\varepsilon E^2 \sin 2\alpha \qquad (278)$$

The viscous torque Γ^v is always counteracting the motion, and can be written

$$\Gamma^v = -\gamma_1 \frac{\partial \alpha}{\partial t} \qquad (279)$$

where γ_1 is the so-called twist viscosity counteracting changes in α. In dynamic equilibrium the total torque is zero. From this we finally obtain the dynamic equation for the local director reorientation in an externally applied field E

$$\frac{1}{2}\varepsilon_0 \Delta\varepsilon E^2 \sin 2\alpha + \gamma_1 \frac{d\alpha}{dt} = 0 \qquad (280)$$

Equation (280) describes the dielectric response to an electric field. It is easily integrated, especially if we look at the case of very small deviations of the director from the field direction (α), for which we have $\sin 2\alpha \approx 2\alpha$ and

$$\varepsilon_0 \Delta\varepsilon_0 E^2 \alpha + \gamma_1 \frac{\partial\alpha}{\partial t} = 0 \qquad (281)$$

In this case, α will approach zero simply as

$$\alpha = \alpha_0 \exp\left(\frac{-t}{\tau}\right) \qquad (282)$$

where α_0 is the initial angular deviation in Fig. 65 and τ is given by

$$\tau = \frac{\gamma_1}{\Delta\varepsilon\varepsilon_0 E^2} \qquad (283)$$

The characteristic response time is thus inversely proportional to the square of the applied field.

What we have derived so far is valid for a nematic, where there are no particular restrictions to the molecular motion. In smectic media there will be such restrictions, and therefore Eqs. (278) and (280) will be modified in a decisive way.

In any liquid crystal the induced polarization P_i caused by an applied electric field is obtained from $D = \varepsilon_0 \varepsilon E = \varepsilon_0 E + P_i$ as $P_i = D - \varepsilon_0 E$ or, from Eq. (274)

$$P_i = \varepsilon_0 \Delta\varepsilon(n \cdot E)n + \varepsilon_0 \varepsilon_\perp E - \varepsilon_0 E$$
$$= (\varepsilon_\perp - 1)\varepsilon_0 E + \varepsilon_0 \Delta\varepsilon(n \cdot E)n \qquad (284)$$

The first term represents a component parallel to the electric field and does not therefore give any contribution to the torque $P \times E$. From the second we get

$$\Gamma^\varepsilon = P_i \times E = \Delta\varepsilon\varepsilon_0 (n \cdot E)(n \times E) \qquad (285)$$

In the nematic case we would have, with $n \cdot E = E \cos\alpha$ and $n \times E = -E \sin\alpha$

$$\Gamma^\varepsilon = -\Delta\varepsilon\varepsilon_0 E^2 \sin\alpha \cos\alpha \qquad (286)$$

where α is the angle between n and E, which is equivalent to Eq. (278). In the smectic case the rigidity of the tilt angle θ prevents any other change than in the phase variable φ. Consequently, we have to take the z component of the torque working on φ and causing the conical motion of the director. From Fig. 51 and Eq. (140) we find, with the external field E applied in the y direction, i.e., $E = (0, E, 0)$

$$n \cdot E = E \sin\theta \sin\varphi \qquad (287)$$

and

$$n \times E = E(n_z, 0, -n_x)$$
$$= E(\cos\theta, 0, -\sin\theta \cos\varphi) \qquad (288)$$

from which Eq. (285) takes the shape

$$\Gamma^\varepsilon = \Delta\varepsilon\varepsilon_0 E^2 (\sin\theta \cos\theta, 0,$$
$$-\sin^2\theta \sin\varphi \cos\varphi) \qquad (289)$$

The z component of this torque is thus

$$\Gamma^\varepsilon_z = -\Delta\varepsilon\varepsilon_0 E^2 \sin^2\theta \sin\varphi \cos\varphi \qquad (290)$$

Compared to the nematic case we get a formally similar expression, except for the appearance of the $\sin^2\theta$ factor, but the involved angle φ has a completely different meaning than α. The azimuthal angle φ is not the angle between the director and the field, but in the SmC* case the angle between the spontaneous polarization and the field, as seen when we imagine the director to move out in direction φ in Fig. 60.

2.6.3 Switching Dynamics

We are now able to write down the equation of motion for the director moving on the smectic cone and being subjected to ferroelectric and dielectric torques, as well as the elastically transmitted torques $K\nabla^2\varphi$, which tend to make $\boldsymbol{n}$ spatially uniform. Disregarding the inertial term $J\ddot{\varphi}$ (of negligible importance) and the elastic torques from the surfaces (a far more serious omission), we can write the dynamic director equation

$$\gamma_\varphi \frac{\partial\varphi}{\partial t} = -PE\sin\varphi$$
$$-\Delta\varepsilon\,\varepsilon_0\,E^2\sin^2\theta\,\sin\varphi\,\cos\varphi + K\,\nabla^2\varphi \quad (291)$$

For $E=0$ this reduces to

$$\nabla^2\varphi = \frac{\gamma_\varphi}{K}\frac{\partial\varphi}{\partial t} \quad (292)$$

which has the form of the equation for heat conduction or diffusion and means that spatial gradients in φ will damp out smoothly with a rate determining quantity K/γ_φ corresponding to the heat conductivity or diffusion coefficient, respectively.

With regard to the signs, we note that (see Fig. 60) φ is the angle between the tilt plane (or $\boldsymbol{c}$ director) and $\boldsymbol{x}$, and therefore also between $\boldsymbol{P}$ and $\boldsymbol{E}$. The torque $\boldsymbol{P}\times\boldsymbol{E}=-PE\sin\varphi$ is consequently in the negative z direction for $\varphi>0$ (see Fig. 66). Equation (291) describes the overdamped approach towards equilibrium given by the field E, together with the material parameters P, θ, γ_φ, K, and $\Delta\varepsilon$. Let us divide all terms by PE

$$\frac{\gamma_\varphi}{PE}\frac{\partial\varphi}{\partial t} = -\sin\varphi - \frac{\Delta\varepsilon\,\varepsilon_0\,E}{P}$$
$$\cdot\sin^2\theta\,\sin\varphi\,\cos\varphi + \frac{K}{PE}\,\nabla^2\varphi \quad (293)$$

Introducing a characteristic time $\tau\equiv\dfrac{\gamma_\varphi}{PE}$,

a characteristic length $\xi=\sqrt{\dfrac{K}{PE}}$, and the di-

mensionless quantity $\chi=\dfrac{P}{\varepsilon_0\,\Delta\varepsilon\,E}$, we can

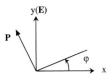

Figure 66. If the xz plane is parallel to the electrodes and the electric field is applied in the positive y direction, the ferroelectric torque will turn the tilt plane (given by φ) in the negative φ direction.

write Eq. (293) as

$$\tau\frac{\partial\varphi}{\partial t} = -\sin\varphi$$
$$-\frac{1}{\chi}\sin^2\theta\,\sin\varphi\,\cos\varphi + \xi^2\,\nabla^2\varphi \quad (294)$$

Evidently we can write Eq. (294) in an invariant form using the variables $t'=t/\tau$ and $r'=r/\xi$. For, with $\dfrac{\partial}{\partial t}=\dfrac{1}{\tau}\dfrac{\partial}{\partial t'}$ and $\dfrac{\partial}{\partial x}=\dfrac{1}{\xi}\dfrac{\partial}{\partial x'}$, etc., Eq. 294 takes the dimensionless shape

$$\frac{\partial\varphi}{\partial t'} = -\sin\varphi(r',t') - \frac{1}{\chi}\sin^2\theta\,\sin\varphi(r',t')$$
$$\cdot\cos\varphi(r',t') + \nabla^2\varphi(r',t') \quad (295)$$

The new variables are scaled with

$$\xi = \sqrt{\frac{K}{PE}} \quad (296)$$

and

$$\tau = \frac{\gamma_\varphi}{PE} \quad (297)$$

which set the basic length and time scales for the space-time behavior of $\varphi(r, t)$. In addition, the behavior is governed by the dimensionless parameter

$$\chi = \frac{P}{\varepsilon_0\,\Delta\varepsilon E} \quad (298)$$

The characteristic time τ determines the dynamic response towards the equilibrium state when we apply an electric field. If we put in the reasonable values

$\gamma_\varphi = 50$ cP $(=0.05$ N s/m$^2)$ (see Fig. 67), $P = 10$ nC/cm^2, and $E = 10$ V/μm, we get a τ value of about 5 μs. The characteristic length ξ expresses the balance between elastic and electric torques. An external field will align P along the field direction except in a thin layer of the order of ξ, where a de-

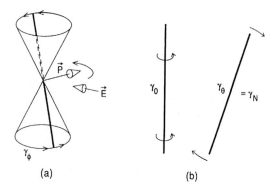

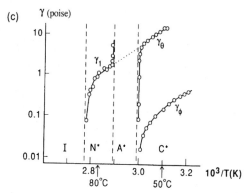

Figure 67. (a, b) An external electric field E perpendicular to the smectic layer normal is rotating the local polarization P as indicated in (a), forcing P to end up parallel to E. The corresponding reorientation of the optic axis takes place through the variable φ around a conical surface, i.e., the tilt cone, with no change in tilt θ. In contrast to a pure θ motion (corresponding to viscosity γ_θ, or the equivalent twist viscosity γ_1 in the nematic case), the φ reorientation has a significant component around the long molecular axis, which leads to low effective viscosity. (c) Arrhenius plot of the viscosity γ_1 in the nematic phase, as well as soft mode and Goldstone mode viscosities γ_θ and γ_φ in the smectic C* phase of a two-component mixture (from Pozhidayev et al. [148]).

viating boundary condition will be able to resist the torque of the field. As this layer thickness falls off according to $\xi \sim E^{-1/2}$, it will be much smaller than the sample thickness d for sufficiently high fields. In fact, it also has to be much smaller than the wavelength of light in order for the sample to be optically homogeneous. Only then will Eq. (297) also apply for the optical response time. It is easy to check that this is the case. With the P and E values already used and with $K = 5 \times 10^{-12}$ N, we find from Eq. (296) that ξ is less than a nanometer.

Finally, the parameter ζ expresses the balance between the ferroelectric and the dielectric torque. As it appears in the dielectric term of Eq. (295) in the combination $\sin^2\theta/\chi$, which is $\approx 0.15/\chi$ for $\theta \approx 22°$, χ would have to be less than about 0.5 in order for us to have to keep this term. With the same P and E values as before, this requires a $|\Delta\varepsilon|$ value as high as ≥ 2.5. We will therefore skip this term for the moment, but return to this question later.

2.6.4   The Scaling Law for the Cone Mode Viscosity

Figure 67 gives an indication that, for the motion on the cone, the effective viscosity ought to depend on the tilt angle. Evidently, for small values of tilt it approaches the limiting case, where all the motion is around the long inertial axis, which must have the lowest possible viscosity. We also note from the experimental data of that figure that the cone mode viscosity, γ_φ, is even lower than the nematic viscosity at a higher temperature, which at first may seem puzzling. We will look at the viscosity in a more general way in Sec. 2.6.9, but here only try to illuminate this question. We will treat the viscosity in the simplest way possible, regarding it for a moment as a scalar.

To a first approximation, any electric torque will excite a motion around the θ-preserving cone, i.e., around the z axis of Fig. 60. The essential component of the viscous torque will therefore be Γ_z, which is related to $d\varphi/dt \ (\equiv \dot{\varphi})$. We can thus write for the cone motion

$$-\Gamma_z = \gamma_\varphi \dot{\varphi} \tag{299}$$

For a general rotational motion of the director n, the relation between the viscous torque and the viscosity has the form

$$-\boldsymbol{\Gamma} = \gamma \boldsymbol{n} \times \dot{\boldsymbol{n}} \tag{300}$$

Taking the z component we get

$$-\Gamma_z = \gamma |\boldsymbol{n} \times \dot{\boldsymbol{n}}|_z = \gamma |\boldsymbol{c} \times \dot{\boldsymbol{c}}| \tag{301}$$

expressed by the two-dimensional c-director, which is the projection of n onto the plane of the smectic layer. Their relation means that $c = n \sin\theta = \sin\theta$ (see Fig. 51). However

$$|\boldsymbol{c} \times \dot{\boldsymbol{c}}| = c^2 \dot{\varphi} \tag{302}$$

Hence

$$-\Gamma_z = \gamma \sin^2\theta \dot{\varphi} \tag{303}$$

Comparing Eqs. (299) and (303) we obtain the important scaling law with respect to the tilt angle for the viscosity γ_φ describing the motion around the cone (Goldstone mode)

$$\gamma_\varphi = \gamma \sin^2\theta \tag{304}$$

The index φ refers to the φ variable, and the index-free γ corresponds to the nematic twist viscosity. Equation (304) means that the cone mode viscosity is lower than the standard nematic viscosity ($\gamma_\varphi < \gamma_1$). It also seems to indicate that the viscosity γ_φ tends to zero at the transition $T \to T_{CA}$ ($\theta \to 0$). This, of course, cannot be strictly true, but we will wait till Sec. 2.6.9 to see what actually happens.

2.6.5 Simple Solutions of the Director Equation of Motion

If we disregard both the dielectric and elastic terms in Eq. (291), the dynamics is described by the simple equation

$$\gamma_\varphi \frac{\partial\varphi}{\partial t} + PE\sin\varphi = 0 \tag{305}$$

For small deviations from the equilibrium state, $\sin\varphi \approx \varphi$ and Eq. (305) is integrated directly to

$$\varphi = \varphi_0 \, e^{-t/\tau} \tag{306}$$

where $\tau = \dfrac{\gamma_\varphi}{PE}$ and φ_0 is the angle between E and P at time $t = 0$. The response time τ is inversely proportional to the field, instead of being inversely proportional to the square of the field, as in the dielectric case. Generally speaking, this means that the FLC electrooptic effect will be faster than normal dielectric effects at low voltages, and also that these dielectric effects, at least in principle, will ultimately become faster when we go to very high applied fields. The unique advantage of FLC switching is, however, that it is not only very fast at moderate fields, but that it switches equally fast in both directions; a feature that cannot be achieved if the coupling to the field is quadratic.

If φ is not small in Eq. (305), we change the variable to half the angle ($\sin\varphi = 2\sin\dfrac{\varphi}{2}\cos\dfrac{\varphi}{2}$) and easily find the analytical solution

$$\tan\frac{\varphi}{2} = \tan\frac{\varphi_0}{2} \, e^{-t/\tau} \tag{307}$$

with the same $\tau = \dfrac{\gamma_\varphi}{PE}$. This can alternatively be expressed as

$$\varphi(t) = 2\arctan\left(\tan\frac{\varphi_0}{2} \, e^{-t/\tau}\right) \tag{308}$$

2.6.6   Electrooptic Measurements

Equations (307) and (308) are a solution to Eq. (305) if the electric field is constant with time. The expression is therefore also useful when the field is a square wave. Such a field is commonly used for polarization reversal measurements when switching between $+P_s$ and $-P_s$ states. In order to switch from $-P_s$ to $+P_s$ we have to supply the charge $2P_s$ to every unit area, i.e., $2P_sA$ over the sample, if its area is A. The reversal of the vector P_s thus gives rise to an electric current pulse $i(t) = \dfrac{dQ}{dt}$ through the curcuit in which the sample cell is connected. This current is given by the time derivative of the charge due to P_s on the electrodes of area A. When P_s makes an angle φ with E, the polarization charge on the electrodes is $Q = P_sA \cos \varphi$. The electric current contribution due to the reversal of P_s is thus given by

$$i(t) = P_s A \frac{d}{dt}[\cos\varphi(t)] = -P_s A \frac{d\varphi}{dt}\sin\varphi(t) \qquad (309)$$

Substituting the time derivative of φ from Eq. (305), the polarization reversal current is directly obtained as

$$i(t) = \frac{P_s A}{\tau}\sin^2\varphi(t) \qquad (310)$$

When φ passes $90°$ during the polarization reversal, $i(t)$ in Eq. (310) goes through a maximum. The shape of the function $i(t)$ is that of a peak. This is shown in Fig. 68 where Eqs. (308) and (310) have been plotted for the case $\varphi_0 = 179°$.

By measuring the area under the polarization reversal peak, we obtain the total charge transferred and can thus determine the value of the spontaneous polarization P_s. The time for the appearance of the current peak then allows a determination of the viscosity γ_φ. The curves in Fig. 68 have been traced for the case where P makes an angle of $1°$ relative to the surface layer normal, corresponding to the same pretilt of the director out of the surface. This pretilt also has to be included as a parameter determining the electrooptic properties. On the other hand, the integration of Eq. (305) giving the current peak of Fig. 68 can only give a crude estimation of the electrooptic parameters, because the equation does not contain any term describing the surface elastic torques. The value of the method lies therefore in the

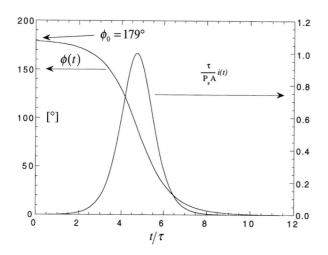

Figure 68. The polarization reversal $\varphi(t)$ according to Eq. (304) for a half-period of an applied square wave and the resulting current pulse $i(t)$. The current peak appears roughly at $t = 5\tau$ and the reversal is complete after twice that time. The latter part of the φ curve corresponds to the simple exponential decay of Eq. (32). The optical transmission is directly related to $\varphi(t)$. If the state at $t = 0$ is one of extinction, the transmission will increase to its maximum value, but in a steeper way than $|\partial\varphi/\partial t|$ (from Hermann [136]).

fact that it is simple and rapid. The same can be said about the alternative formalism of [137], which has proven useful for routine estimations of electrooptic parameters, at least for comparative purposes. In [137] a hypothetical ad hoc surface torque of the form $K \cos \varphi$ has been added to the ferroelectric torque in Eq. (305). With only these two torques present, the equation can still be easily integrated in a closed form. However, the chosen form of the elastic torque means that it cannot describe the properties of a bistable cell.

Numerous other approaches have been proposed in order to describe the switching dynamics, for instance by neglecting the surface elastic torques, but including the dielectric torque using the correct representation of the dielectric tensor [138]. Common to most of them is the fact that they describe the shape of the polarization reversal current and the electrooptic transmission curve reasonably well. They are therefore easy to fit to experimental data, but may give values to the electrooptic parameters that may deviate within an order of magnitude. (An important exception to this is the polarization value P_s, which is essentially insensitive to the model used because it is achieved by integration of the current peak.) Hence no model put forward so far accounts generally for the switching dynamics, and there is a high degree of arbitrariness in the assumptions behind all of them.

Experiments in which the parameters in the equation of motion are measured may be carried out with an electrooptic experimental setup, usually based on a polarizing microscope. In the setup, the sample is placed between crossed polarizers and an electric field $E(t)$ is applied across the cell. The optical and electric responses of the sample to the field are then measured.

The light intensities $I(t)$ through the crossed polarizers and the sample are mea-

sured by a photodiode after the analyzer. The electric current $i(t)$ through the sample is obtained by measuring the voltage $u(t)$ across a resistance R connected in series with the liquid crystal cell (see Fig. 69). All three signals $E(t)$, $I(t)$, and $i(t)$ are fed into an oscilloscope so that the switching dynamics of the liquid crystal cell can be monitored. The oscilloscope is connected to a computer so that the waveforms can be stored and analyzed in detail. From the optical transmission, the tilt angle and the optical response time can be measured. From the electric response the spontaneous (and induced) polarization can be measured, as well as the response time, as discussed earlier. If we also include the SmA* phase (the electroclinic effect), the coefficients of the Landau expansion of the free energy density may be derived. Estimations of the viscosities for the φ- and θ-motions may also be obtained from the electrooptic measurements.

In Fig. 69 we see how the quasi-bookshelf (QBS) smectic constitutes a retarder with a switchable optic axis. In the smectic A* phase this axis can be switched continously, but with small amplitude. The same setup is also used to study the DHM effect in short pitch chiral smectics and the flexoelectrooptic effect in the cholesteric phase, in both of which we achieve much higher tilts. Inserted between crossed polarizers, the retarder transmits the intensity

$$I = I_0 \sin^2 2\Psi \sin^2 \frac{\pi d \Delta n}{\lambda} \qquad (311)$$

where Ψ is the angle between the polarizer direction and the projection of the optic axis in the sample plane, Δn is the birefringence, λ is the wavelength of light in a vacuum, and d is the thickness of the sample. The function $I(\Psi)$ is shown in Fig. 70. A very important case is the lambda-half condition given by $d \Delta n = \lambda/2$. A cell with this

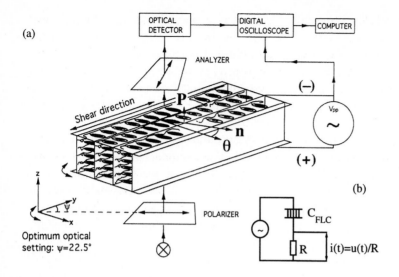

(a)

Figure 69. (a) The experimental setup for electrooptic measurements. The sample is placed between crossed polarizers. (b) The RC-circuit used in the switching experiments. The current response from the sample is measured across the resistor R (from Hermann [136]).

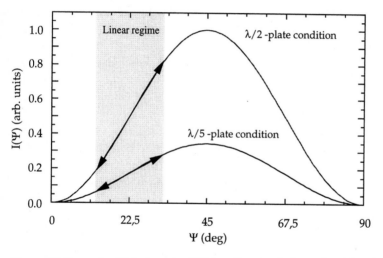

Figure 70. Transmitted light intensity $I(\Psi)$ for a linear optical retarder with switchable optic axis placed between crossed polarizers. Ψ is the angle between the polarizer direction and the switchable optic axis, and maximum transmitted light intensity is achieved for $\Psi = 45°$ combined with a sample thickness fulfilling the $\lambda/2$ condition. A smectic C* in QBS geometry and tilt angle $\theta = 22.5°$ could ideally be switched from $\Psi = 0$ to $\Psi = 45°$ in a discontinuous way. DHM and flexoelectrooptic materials could be switched in a continuous way. $I(\Psi)$ is proportional to Ψ in a region around $\Psi = 22.5°$, which is the linear regime (from Rudquist [139]).

value of $d\Delta n$ will permit maximum transmission and flip the plane of polarization by an angle 2Ψ. Since the position of the optic axis, i.e., the angle Ψ, may be altered by means of the electric field, we have electrooptic modulation. As seen from Eq. (311) and from Fig. 70, the minimum and maximum transmission is achieved for $\Psi = 0$ and $\Psi = 45°$, respectively. If we adjust the polarizer to correspond to the first state for one sign of E across an SSFLC sample, reversal of sign would switch the polarization to give the second state, if $2\theta = 45°$. For natural grey-scale materials (electroclinic, DHM,

flexoelectrooptic, antiferroelectric), the electrooptic response to a field-controlled optic
axis cell between crossed polarizers is
modelled in Fig. 71. Here $\Psi = \Psi_0 + \Phi(E)$,
where Ψ_0 is the position of the optic axis for
$E=0$ and $\Phi(E)$ is the field-induced tilt of
the optic axis.

The tilt angle in the SmC* phase can be
measured by the application of a square
wave of very low frequency (0.2–1 Hz), so
that the switching can be observed directly
in the microscope. By turning the angular
turntable of the microscope, the sample is
set in such an angular position that one of
the extreme tilt angle positions, say $-\theta_0$,
coincides with the polarizer (or analyzer),
giving $I=0$. When the liquid crystal is then

switched to $+\theta_0$, I becomes $\neq 0$. The sample is now rotated so as to make $I=0$ again,
corresponding to $+\theta_0$ coinciding with the
polarizer (or analyzer). It follows that the
sample has been rotated through $2\theta_0$,
whence the tilt angle is immediately obtained.

The induced tilt angle in the SmA* phase
is usually small, and the above method becomes difficult to implement. Instead, the
tilt can be obtained [140] by measuring the
peak-to-peak value ΔI_{pp} in the optical response, from the relation

$$\theta_{ind} = \frac{1}{4} \arcsin \frac{\Delta I_{pp}}{I_0} \qquad (312)$$

which is valid for small tilt angles.

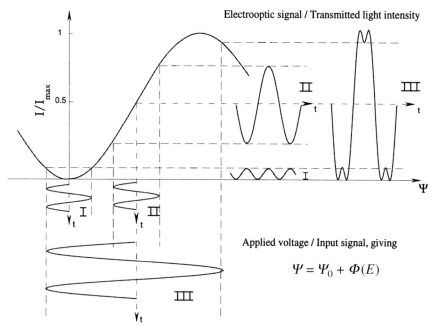

Figure 71. Electrooptic modulation due to the field-controlled position of the optic axis between crossed polarizers. I) The optic axis is swinging around $\Psi_0 = 0$ giving an optical signal with double frequency compared to
the input signal. II) The linear regime. When the optic axis is swinging around $\Psi_0 = 22.5°$, the variation of the
transmitted intensity is proportional to Φ. For a liquid crystal $\lambda/2$ cell with quasi-bookshelf (QBS) structure, it
follows from Eq. (311) that the intensity I varies linearly with Φ if Φ is small ($I/I_0 \approx 1/2 + 2\Phi(E)$), which, for
instance, is the case if the electroclinic effect in the SmA* phase is exploited. As electroclinic tilt angles are still
quite small, certain other materials, though slower, have to be used for high modulation depths. III) Characteristic response for $\Phi(E)$ exceeding $\pm 22.5°$. (from Rudquist [139]).

On applying a square wave electric field in the SmC* phase the electric response has the shape of an exponential decay, on which the current peak $i(t)$ (see Fig. 68) is superposed. The exponential decay comes from the capacitive discharge of the liquid crystall cell itself, since it constitutes a parallel-plate capacitor with a dielectric medium in between. The electrooptic response $i(t)$ on application of a triangular wave electric field to a bookshelf-oriented liquid crystal sample in the SmC* and SmA* phases is shown as it appears on the oscilloscope screen in Fig. 72, for the case of crossed polarizers and the angular setting $\Psi_0 = 22.5°$. Applying such a triangular wave in the SmC* phase gives a response with the shape of a square wave on which the polarization reversal current peak is superposed. The square wave background is due to the liquid crystal cell being a capacitor; an RC-circuit as shown in Fig. 69b delivers the time derivative of the input. The time derivative of a triangular wave is a square wave, since a triangular wave consists of a constant slope which periodically changes sign. The optical response $I(t)$ also almost exhibits a square wave shape when applying a triangular wave, due to the threshold of the two stable states of up and down polarization corresponding to the two extreme angular positions $-\theta$ and $+\theta$ of the optic axis (the

director) in the plane of the cell. The change in the optical transmission coincides in time with the polarization reversal current peak, which occurs just after the electric field changes sign.

In the SmC* phase, the optical transmission change in the course of the switching is influenced by the fact that the molecules move out of the plane of the glass plates, since they move on a cone. They thus make some angle $\zeta(t)$ with the plane of the glass plates at time t. The refractive index seen by the extraordinary ray $n_c(\zeta)$ depends on this angle. The birefringence thus changes during switching according to $\Delta n[\zeta(t)] = n_c[\zeta(t)] - n_0$. After completion of the switching, the molecules are again in the plane of the glass plates and $\zeta = 0$, so that the birefringence is the same for the two extreme positions $-\theta$ and $+\theta$ of the director.

When raising the temperature to that of the SmA* phase, the polarization reversal current peak gradually diminishes. In the SmA* phase near the transition to SmC*, a current peak can still be observed, originating from the induced polarization due to the electroclinic effect. On application of a triangular wave, the change of the optical response can most clearly be observed: it changes from a squarewave-like curve to a triangular curve, which is in phase with the applied electric field. This very clearly illustrates the change from the nonlinear cone switching of the SmC* phase to the linear switching of the SmA* phase due to the electroclinic effect.

2.6.7   Optical Anisotropy and Biaxiality

The smectic A phase is optically positive uniaxial. The director and the optic axis are in the direction of highest refractive index, $n_\parallel - n_\perp = \Delta n > 0$. At the tilting transition

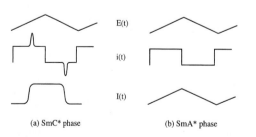

(a) SmC* phase   (b) SmA* phase

Figure 72. The electrooptic response when applying a triangular wave over the cell in the SmC* and the SmA* phases. The ohmic contribution due to ionic conduction has been disregarded.

SmA → SmC, the phase becomes positive biaxial, with $n_\parallel \to n_3$ and $n_\perp$ splitting in two, i.e., $n_\perp \to n_1$, n_2. The ε ellipsoid is prolate and

$$n_3 > n_2 > n_1 \tag{313}$$

or

$$\Delta n > 0, \quad \partial n > 0 \tag{314}$$

where we use, with some hesitation, the abbreviations

$$\Delta n = n_3 - n_1 \tag{315}$$

$$\partial n = n_2 - n_1 \tag{316}$$

The hesitation is due to the fact that the refractive index is not a tensor property and is only used for the (very important) representation of the index ellipsoid called the optical indicatrix, which is related to the dielectric tensor by the connections

$$\varepsilon_1 = n_1^2, \quad \varepsilon_2 = n_2^2, \quad \varepsilon_3 = n_3^2 \tag{317}$$

between the principal axis values of the refractive index and the principal components of the dielectric tensor.

For orthorhombic, monoclinic, and triclinic symmetry the indicatrix is a triaxial ellipsoid. The orthorhombic system has three orthogonal twofold rotation axes. This means that the indicatrix, representing the optical properties, must have the same symmetry axes (Neumann principle). Therefore the three principal axes of the indicatrix coincide with the three crystallographic axes and are fixed in space, whatever the wavelength. This is not so for the monoclinic symmetry represented by the smectic C. Because the symmetry element of the structure must always be present in the property (again the Neumann principle), the crystallographic C_2 axis perpendicular to the tilt plane is now the twofold axis of the indicatrix and the ε tensor, but no other axes are fixed. This means that there is ambiguity in

the direction of ε_3 along the director, (which means ambiguity in the director as a concept) as well as in the direction ε_1 perpendicular to both the "director" (ε_3) and the tilt axis (ε_2). Only the latter is fixed in space, and therefore only ε_2 can be regarded as fundamental in our choice of the principal axes of the dielectric tensor (ε_1, ε_2, ε_3). Now, it is well known that the triaxial character, i.e., the difference between ε_2 and ε_1 (or n_2 and n_1), $\partial \varepsilon = \varepsilon_2 - \varepsilon_1$, is very small at optical frequencies.

The quantity $\partial \varepsilon$ is now universally called the biaxiality, although this name would have been more appropriate for the degree of splitting of the two optic axes as a result of the tilt. Thus optically speaking we may still roughly consider the smectic C phase as uniaxial, with an optic axis (director) tilting out a certain angle θ from the layer normal. However, as has recently been pointed out by Giesselmann et al. [141], due to the ambiguity of the ε_1 and ε_3 directions, the optical tilt angle must be expected to be a function of the wavelength of the probing light. As they were able to measure, the blue follows the tilt of the core, whereas the red has a mixture of core and tail contributions. The optical tilt is thus higher for blue than for red light, and the extinction position depends on the color. While this phenomenon of dispersion in the optic axis is not large enough to create problems in display applications, it is highly interesting in itself and may be complemented by the following general observation regarding dispersion. At low frequencies, for instance, 10^5 Hz, the dielectric anisotropy $\Delta \varepsilon = \varepsilon_3 - \varepsilon_1$ is often negative, corresponding to a negatively biaxial material. As it is positively biaxial at optical frequencies ($\sim 10^{15}$ Hz), the optical indicatrix must change shape from prolate to oblate in the frequency region in between, which means, in particular, that at some frequency the ε tensor must become isotropic.

2.6.8   The Effects of Dielectric Biaxiality

At lower frequencies we can no longer treat the smectic C phase as uniaxial. Therefore our treatment of dielectric effects in Sec. 2.6.2 was an oversimplification, valid only at low values of tilt. In general, we now have to distinguish between the anisotropy, $\Delta\varepsilon = \varepsilon_3 - \varepsilon_1$, as well as the biaxiality, $\partial\varepsilon = \varepsilon_2 - \varepsilon_1$, both of which can attain important nonzero values typically down to -2 for $\Delta\varepsilon$ and up to $+3$ or even higher for $\partial\varepsilon$. The latter parameter has acquired special importance in recent years, due to a special addressing method for FLC displays, which combines the effects of ferroelectric and dielectric torques. As for the background, it was successively discovered in a series of investigations by the Boulder group between 1984 and 1987 that the smectic layers are generally tilted and, moreover, form a so-called chevron structure, according to Fig. 73, rather than a bookshelf structure. The reason for the chevron formation is the effort made by the smectic to fill up the space given by the cell, in spite of the layer thickness shrinking as the temperature is lowered through the SmA – SmC transition and further down into the SmC phase. The chevron geometry (to which we will return at length in Sec. 2.8) reduces the effective switching angle, which no longer corresponds to the optimum of $2\theta = 45°$, thereby reducing the brightness – contrast ratio (we have in the figure assumed that the memorized states have zero pretilt, i.e., that they are lying in the surface). This can, to some extent, be remedied by a different surface coating requiring a high pretilt.

Another way of ameliorating the optical properties is to take advantage not only of the ferroelectric torque ($\sim E$) exerted on the molecules, but also of the dielectric torque ($\sim E^2$). This torque always tries to turn the highest value of the permittivity along the direction of the electric field. The tilted smectic has the three principal ε values in the directions shown in the figure, ε_1 along the chosen tilt direction, ε_2 along the local polarization, and ε_3 along the director. If $\partial\varepsilon$ is large and positive, it is seen that the dielectric torque exerted by the field will actually lift up the director away from the surface-stabilized state along the cone surface to an optically more favorable state, thus increasing the effective switching angle. The necessary electric field for this action can be provided by the data pulses acting continuously as AC signals on the columns. This addressing method thus uses the ferroelectric torque in the switching pulse to force the director from one side to the other between the surface-stabilized states (this could not be done by the dielectric torque, because it is insensitive to the sign of E), after which high frequency AC pulses will keep the director dynamically in the corresponding extreme cone state. This AC enhanced contrast mode can ameliorate the achievable contrast in cases where the memorized director positions give an insufficient switching angle, which is the case in the so-called C2 chevron structures (see Sec. 8.8). It requires specially engineered materials with a high value of $\partial\varepsilon$. Its main drawback is the requirement of a relatively high voltage (to increase the dielectric torque relative to the ferroelectric torque), which also increases the power consumption of the device. The AC contrast enhancement is often called "AC stabilization" or "HF (high frequency) stabilization". A possible inconvenience of this usage is that it may lead to a misunderstanding and confusion with surface-stabilization, which it does not replace. The dielectric contrast enhancement works on a surface-stabilized structure. There is no conflict in the concepts; the mechanisms rather work together.

(a)   (b)

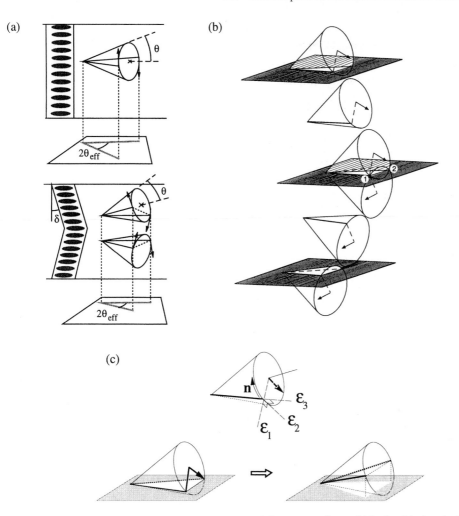

(c)

Figure 73. The simple bookshelf structure with essentially zero pretilt would lead to ideal optical conditions for materials with 2θ equal to $45°$. In reality the smectic layers adopt a much less favorable chevron structure. This decreases the effective switching angle and leads to memorized P states that are not in the direction of the field (a, b). A convenient multiplexing waveform scheme together with a properly chosen value of the material's biaxiality $\partial\varepsilon$ may enhance the field-on contrast relative to the memorized (surface-stabilized) value (c), by utilizing the dielectric torque from the pulses continuously applied on the columns (after [142, 143]).

In Eq. (291) we derived an expression for the director equation of motion with dielectric and ferroelectric torques included. If the origin of the dielectric torque is in the dielectric biaxiality, the equation (with the elastic term skipped) will be the closely analogous one

$$\gamma_\varphi \frac{\partial\varphi}{\partial t} = -PE \sin\varphi - \varepsilon_0 \, \partial\varepsilon \, E^2 \sin\varphi \cos\varphi \tag{318}$$

In the present case, where the layers are tilted by the angle δ relative to E, as in Fig. 73a, the effective applied field along the layer will be $E \cos\delta$, thus slightly modifying the equation to

$$\gamma_\varphi \frac{\partial\varphi}{\partial t} = -PE \cos\delta \sin\varphi$$
$$- \varepsilon_0 \, \partial\varepsilon E^2 \cos^2\delta \sin\varphi \cos\varphi \tag{319}$$

In addition to the characteristic time τ and length ξ, we had earlier introduced the dimensionless parameter χ describing the balance between ferroelectric and dielectric torques. Its character appears even more clearly than in Eq. (298) if we write it as

$$\chi = \frac{PE}{\varepsilon_0\, \partial\varepsilon\, E^2} \tag{320}$$

If we again replace E by $E\cos\delta$, our new χ will be

$$\chi = \frac{P}{\varepsilon_0\, \partial\varepsilon\, E \cos\delta} \tag{321}$$

From χ and $\tau = \dfrac{\gamma_\varphi}{PE\cos\delta}$ we can form a characteristic time

$$\tau_0 = \frac{\tau}{\chi} = \gamma_\varphi \frac{\varepsilon_0\, \partial\varepsilon}{P^2} \tag{322}$$

and from χ and Ed ($\cos\delta$ factors canceling in Ed), a characteristic voltage

$$V_0 = \chi Ed = \frac{Pd}{\varepsilon_0\, \partial\varepsilon \cos\delta} \tag{323}$$

The values of the parameters τ_0 and V_0 are decisive in situations where the ferroelectric and dielectric torques are of the same order of magnitude, and they play an important role in the electrooptics of FLC displays addressed in a family of modes having the characteristics that there is a minimum in the switching time at a certain voltage (see Fig. 74a), because the ferroelectric and dielectric torques are nearly balancing. This means that the switching stops both on reducing the voltage (due to insufficient ferroelectric torque) and on increasing it (due to the rapidly increasing dielectric torque, which blocks the motion, i.e., it increases the delay time for the switching to take off). The existence of a minimum in the switching time when dielectric torques become important was discovered by Xue et al. [138] in 1987, but seems to have appeared in novel addressing schemes some years later in the British national FLC collaboration within the JOERS/Alvey program. It turns out [145] that the minimum switching time

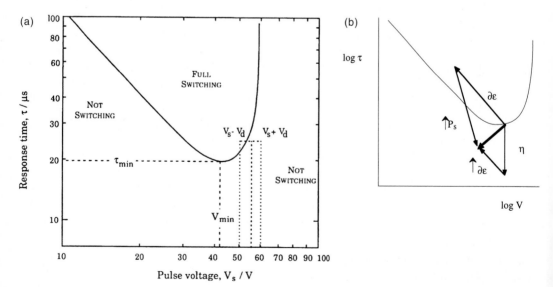

Figure 74. (a) Typical pulse switching characteristic for a material with parameters giving ferroelectric and dielectric torques of comparable size; (b) Possible routes for engineering materials to minimize both τ_{min} and V_{min} (from [144]).

depends on γ_φ, $\partial\varepsilon$, and P according to Eq. (322), and that this minimum occurs for a voltage that depends on $\partial\varepsilon$ and P according to Eq. (323). At present, considerable effort is given to reduce both τ_0 and V_0. Figure 74b indicates how this could be done: either (1) by lowering the viscosity γ_φ whilst increasing the biaxiality $\partial\varepsilon$, or (2) by increasing both $\partial\varepsilon$ and P by a relatively large amount.

2.6.9   The Viscosity of the Rotational Modes in the Smectic C Phase

In Sec. 2.6.4 we derived the scaling law for the cone mode viscosity with respect to the tilt angle θ. In this section we want to penetrate a little deeper into the understanding of the viscosities relevant to the electrooptic switching dynamics. Let us therefore first review the previous result from a new perspective. With $\theta = \text{const}$, an electric torque will induce a cone motion around the z axis (see Fig. 75a). We can describe this motion in different ways. If we choose to use the angular velocity $\dot\varphi$ of the c director, that is, with respect to the z axis, then we have to relate it to the torque component Γ_z^v

$$-\Gamma_z^v = \gamma_\varphi \dot\varphi \qquad (324)$$

In Fig. 75b (γ_φ is denoted as η) we have instead illustrated the general relation, according to the definition of viscosity, between the angular velocity $\dot\alpha$ of the director $\boldsymbol{n}$ and the counteracting viscous torque $\boldsymbol{\Gamma}^v$

$$-\boldsymbol{\Gamma}^v = \gamma\dot\alpha \qquad (325)$$

As $\boldsymbol{\alpha}$ is the coordinate describing the motion of the head of the $\boldsymbol{n}$ arrow (on a unit sphere), $\dot{\boldsymbol{\alpha}} = \boldsymbol{n}\times\dot{\boldsymbol{n}}$, and the relation can be written

$$-\boldsymbol{\Gamma}^v = \gamma\boldsymbol{n} \times \dot{\boldsymbol{n}} \qquad (326)$$

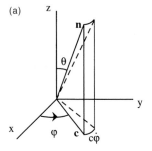

(a)

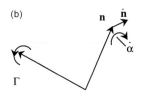

(b)

Figure 75. (a) The variable φ describes the motion of the c director around the layer normal. (b) The relation between the rate of change of the director $\dot{\boldsymbol{n}}$, the angular velocity $\dot{\boldsymbol{\alpha}}$, and the (always counteracting) viscous torque $\boldsymbol{\Gamma}$.

We have used this before in Eq. (300). Several things should now be pointed out. The difference between γ_φ and γ is, in principle, one of pure geometry. Nevertheless, it is not an artifact and the scaling law of γ_φ with respect to θ has a real significance. In a nematic we cannot physically distinguish between the θ motion and the φ motion on the unit sphere. Therefore γ in Eq. (325) is the nematic viscosity. In a smectic C, on the other hand, we have to distinguish between these motions, because the director is moving under the constraint of constant θ. As is evident from Fig. 75b, however, this motion is not only counteracted by a torque Γ_z; instead Γ must have nonzero x and y components as well.

We might also add that in general we have to distinguish between the θ and φ motions as soon as we go to a smectic phase, either SmA or SmC. For instance, we have a soft mode viscosity γ_θ in these phases, which has no counterpart in a nematic.

Let us now continue from Eq. (326). With

$$n = \begin{pmatrix} \sin\theta\cos\varphi \\ \sin\theta\sin\varphi \\ \cos\theta \end{pmatrix} \qquad (327)$$

we have

$$\dot{n} = \begin{pmatrix} -\sin\theta\sin\varphi\dot{\varphi} \\ \sin\theta\cos\varphi\dot{\varphi} \\ 0 \end{pmatrix} \qquad (328)$$

hence

$$(329)$$
$$-\Gamma^v = \gamma\,\dot{\varphi} \begin{pmatrix} -\sin\theta\cos^2\varphi \\ -\sin\theta\sin\varphi\cos\varphi \\ \sin^2\theta\cos^2\varphi + \sin^2\theta\sin^2\varphi \end{pmatrix}$$

Therefore the z component is

$$-\Gamma_z^v = \gamma\,\dot{\varphi}\,\sin^2\theta \qquad (330)$$

A comparison of Eq. (330) with Eq. (324) then gives

$$\gamma_\varphi = \gamma\,\sin^2\theta \qquad (331)$$

which is the same as Eq. (304) (γ is a nematic viscosity). We may note at this point that Eq. (331) is invariant under $\theta \to \theta + \pi$, as it has to be (corresponding to $n \to -n$). We have still made no progress beyond Eq. (304) in the sense that we have stuck with the unphysical result $\gamma_\varphi \to 0$ for $\theta \to 0$. The nonzero torque components Γ_x and Γ_y in Eq. (329) mean that the cone motion cannot take place without torques exerting a tilting action on the director. With θ constant, these have to be taken up by the layers and in turn counteracted by external torques to the sample.

We also note that since $\theta = const$ in Eq. (327), we could just as well have worked with the two-dimensional c director from the beginning

$$c = (c_x, c_y) = \begin{pmatrix} \sin\theta\cos\varphi \\ \sin\theta\sin\varphi \end{pmatrix} \qquad (332)$$

to obtain the result of Eq. (329).

One of the drastic oversimplifications made so far has been in treating the viscosity as a scalar, whereas in reality it is a tensor of rank 4. In the hydrodynamics, the stresses σ_{ij} are components of a second rank tensor, related to the velocity strains $v_{k,l}$

$$\sigma_{ij} = \eta\,\frac{\partial v_k}{\partial x_l} \qquad (333)$$

in which η is thus a tensor of rank 4 just like the elastic constants c_{ijkl} for solid materials, for which Hooke's law is written

$$\sigma_{ij} = c_{ijkl}\varepsilon_{kl} \qquad (334)$$

As is well known, however, the number $3^4 = 81$ of the possibly independent components is here reduced by symmetry, according to

$$c_{ijkl} = c_{jikl} = c_{jilk} \qquad (335)$$

giving only 36 independent components. They therefore permit a reduced representation using only two indices

$$c_{ijkl} = c_{\mu\nu}, \quad \sigma_\mu = c_{\mu\nu}\varepsilon_\nu \qquad (336)$$

which has the advantage that we can write them down in a two-dimensional array on paper, but has the disadvantage that $c_{\mu\nu}$ no longer transforms like a tensor. The same is valid for the Oseen–Frank elastic constants and the flexoelectric coefficients in liquid crystals, which are also tensors of rank 4, but are always written in a reduced representation, K_{ij} and e_{ij}, respectively, where K_{ij} and e_{ij} cannot be treated as tensors. The viscosity tensor in liquid crystals is a particular example of a fourth rank tensor. The viscous torque Γ is supposed to be linear in the time derivative of the director and in the velocity gradients, with the viscosities as proportionality constants. This gives five independent viscosities in uniaxial (12 in biaxial) nematics, the same number in smectics A, but 20 independent components in the smectic C phase [146]. With nine independent elastic constants and as many flex-oelectric coefficients, this phase is certain-ly extraordinary in its complexity. Never-theless, the viscous torque can be divided in the rotational torque and the shearing torque due to macroscopic flow. When we study the electrooptic switching of SSFLC structures, we deal with pure rotations for which mac-roscopic flow is thus assumed to be absent. Even if this is only an approximation, be-cause the cone motion leads to velocity gra-dients and backflow, it will give us an im-portant and most valuable description be-cause of the fact that the rotational viscosity has a uniquely simple tensor representation.

If we go back to the Eq. (325) defining the viscosity, both the torque and the angu-lar velocity are represented by axial vectors or pseudovectors. This is because they ac-tually have no directional property at all (as polar vectors have), but are instead connect-ed to a surface in space within which a ro-tation can only be related to a (perpendicu-lar) direction by a mere convention like the direction for the advance of a screw, "right-hand rule", etc. In fact they are tensors of second rank which are antisymmetric, which means that

$$\Gamma_{ij} = -\Gamma_{ji} \quad \text{and} \quad \dot{\alpha}_{ij} = -\dot{\alpha}_{ji} \qquad (337)$$

with the consequence that they have (in the case of three-dimensional space) only three independent components and therefore can be written as vectors. As γ in Eq. (325) con-nects two second rank tensors, it is itself a fourth rank tensor like other viscosities, but due to the vector representation of Γ and α, the rotational viscosity can be given the very simple representation of a second rank ten-sor. Thus in cases where we exclude trans-lational motions, we can write γ as

$$\gamma = \begin{pmatrix} \gamma_1 & 0 & 0 \\ 0 & \gamma_2 & 0 \\ 0 & 0 & \gamma_3 \end{pmatrix} \qquad (338)$$

Unfortunately, no similar simple represen-tation exists for the tensor components re-lated to macroscopic flow. We will now fi-nally treat γ not as a scalar, but as a tensor in this most simple way.

In Eq. (338) γ is already written in the "molecular" frame of reference in which it is diagonalized, as illustrated in Fig. 76, with the principal axes 1, 2, and 3. Because the C_2 axis perpendicular to the tilt plane has to appear in the property (Neumann princi-ple – we use it here even for a dynamic pa-rameter), this has to be the direction for γ_2. As for the other directions, there are no com-pelling arguments, but a natural choice for a second principal axis is along the director. We take this as the 3 direction. The remain-ing axis 1 is then in the tilt plane, perpen-dicular to n. The 3 axis is special because, along n, the rotation is supposed to be char-acterized by a very low viscosity. In other words, we assume that the eigenvalue γ_3 is very small, i.e., $\gamma_3 \ll \gamma_1, \gamma_2$.

Starting from the lab frame x, y, z, γ is diagonalized by a similarity transformation

$$\hat{\gamma} = T^{-1} \gamma T \qquad (339)$$

where T is the rotation matrix

$$T = \begin{pmatrix} \cos\theta & 0 & -\sin\theta \\ 0 & 1 & 0 \\ \sin\theta & 0 & \cos\theta \end{pmatrix} \qquad (340)$$

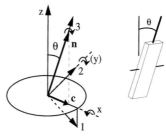

Figure 76. The principal axes 1, 2, and 3 of the vis-cosity tensor. In the right part of the figure is illustrat-ed how the distribution function around the director by necessity becomes biaxial, as soon as we have a nonzero tilt θ.

corresponding to the clockwise tilt θ (in the coordinate transformation the rotation angle θ is therefore counted as negative) around the y (2) axis, and T^{-1} is the inverse or reciprocal matrix to T; in this case (because rotation matrices are orthogonal) T is simply equal to its transpose matrix $\tilde{T}$ (i.e., $\tilde{T}_{ij} = T_{ij}$). Hence we obtain the viscosity components in the lab frame as

$$\gamma = T \hat{\gamma} T^{-1} \tag{341}$$

which gives

$$\gamma = \begin{pmatrix} \cos\theta & 0 & -\sin\theta \\ 0 & 1 & 0 \\ \sin\theta & 0 & \cos\theta \end{pmatrix} \begin{pmatrix} \gamma_1 & 0 & 0 \\ 0 & \gamma_2 & 0 \\ 0 & 0 & \gamma_3 \end{pmatrix} \begin{pmatrix} \cos\theta & 0 & \sin\theta \\ 0 & 1 & 0 \\ -\sin\theta & 0 & \cos\theta \end{pmatrix}$$

$$= \begin{pmatrix} \gamma_1 \cos^2\theta + \gamma_3 \sin^2\theta & 0 & (\gamma_1 - \gamma_3)\sin\theta\cos\theta \\ 0 & \gamma_2 & 0 \\ (\gamma_1 - \gamma_3)\sin\theta\cos\theta & 0 & \gamma_1 \sin^2\theta + \gamma_3 \cos^2\theta \end{pmatrix} \tag{342}$$

Assuming now the motion to be with fixed θ, the angular velocity $\dot{\alpha}$ is

$$\dot{\alpha} = (0, 0, \dot{\varphi}) \tag{343}$$

and Eq. (325) takes the form

$$-\Gamma^v = \gamma \begin{pmatrix} 0 \\ 0 \\ \dot{\varphi} \end{pmatrix} = \begin{pmatrix} \gamma_1 \sin\theta\cos\theta\,\dot{\varphi} \\ 0 \\ (\gamma_1 \sin^2\theta + \gamma_3 \cos^2\theta)\,\dot{\varphi} \end{pmatrix} \tag{344}$$

the z component of which is

$$-\Gamma_z^v = (\gamma_1 \sin^2\theta + \gamma_3 \cos^2\theta)\,\dot{\varphi} \tag{345}$$

Comparison with Eq. (324) now gives

$$\gamma_\varphi(\theta) = \gamma_1 \sin^2\theta + \gamma_3 \cos^2\theta \tag{346}$$

This relation illustrates first of all that the φ motion is a rotation that occurs simultaneously around the 1 axis and the 3 axis, and it reflects how the rotational velocities add vectorially, such that when θ gets smaller and smaller the effective rotation takes place increasingly around the long axis of the molecule. For $\theta=0$ we have

$$\gamma_\varphi(0) = \gamma_3 \tag{347}$$

whereas for $\theta = \pi/2$ we would find (formally) that

$$\gamma_\varphi\left(\frac{\pi}{2}\right) = \gamma_1 \tag{348}$$

corresponding to the rotational motion in a two-dimensional nematic. There is another reason than Eq. (347) not be neglect γ_3: as illustrated to the right of Fig. 76, a full swing of the director around the z axis in fact involves a simultaneous full rotation around

___

the 3 axis. Hence these motions are actually intrinsically coupled, and conceptually belong together.

If, on the other hand, we had neglected the small γ_3 in Eq. (342), the viscosity tensor would have taken the form

$$\gamma = \begin{pmatrix} \gamma_1 \cos^2\theta & 0 & \gamma_1 \sin\theta\cos\theta \\ 0 & \gamma_2 & 0 \\ \gamma_1 \sin\theta\cos\theta & 0 & \gamma_1 \sin^2\theta \end{pmatrix} \tag{349}$$

which would have given the same incomplete equation (304) for γ_φ as before.

The first experimental determination of the cone mode viscosity was made by Kuczyński [147], based on a simple and elegant analysis of the director response to an AC field. Pyroelectric methods have been used by the Russian school [136, 148], for instance, in the measurements illustrated in Fig. 67. Later measurements were normally performed [149] by standard electrooptic methods, as outlined earlier in this chapter. The methodology is well described by Escher et al. [150], who also, like Carlsson and Žekš [151], give a valuable contribution to the discussion about the nature of the viscosity.

The measured viscosities are not the principal components γ_1, γ_2, γ_3 in the molecular frame, but γ_φ and γ_2, which we can refer to the lab frame, as well as γ_3, for which we discuss the measurement method in Sec. 2.7. As γ_2 refers to a motion described by the tilt angle θ, we will from now on denote it γ_θ and often call it the soft mode viscosity (sometimes therefore denoted γ_s). It corresponds more or less, like γ_1, to a nematic viscosity (see Fig. 67 for experimental support of this fact). The cone mode viscosity γ_φ, which is the smallest viscosity for any electrooptically active mode, will correspondingly often be called the Goldstone mode viscosity (sometimes denoted γ_G). As γ_1 and γ_θ must be of the same order of magnitude, Eq. (304) makes us expect $\gamma_\varphi \ll \gamma_\theta$. The difference is often an order of magnitude. The value γ_3, finally, corresponding to the rotation about the long molecular axis, is normally orders of magnitude smaller than the other ones. The value should be expected to be essentially the same as in a nematic, where this viscosity is mostly denoted $\gamma_{||}$ or γ_1. We will here prefer the latter, thus in the following we will write γ_1 instead of γ_3. The value for twist motion in a nematic is mostly called γ_1 but also γ_t or, as opposed to $\gamma_{||}$, $\gamma_\perp$. As γ_1 and γ_2 cannot be distinguished in the nematic, the twist viscosity corresponds both to γ_1 and γ_2 (γ_θ) in the smectic C. The meanings of γ_θ and γ_1 are illustrated in Fig. 77.

If we symbolize the director by a solid object, as we have to the right of Fig. 76, we see that the coordinates θ, φ, and ψ, the latter describing the rotation around the long axis, thus corresponding to γ_1, are three independent angular coordinates which are re-

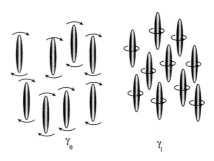

Figure 77. Rotations that correspond to viscosities γ_θ and γ_1, as drawn for the uniaxial SmA (SmA*) phase.

quired for specifying the orientation of a rigid body. Except for the sign conventions (which are unimportant here), they thus correspond to the three Eulerian angles. More specifically, φ is the precession angle, θ the nutation angle, and ψ is the angle of eigenrotation. The three viscosities γ_φ, γ_θ, and γ_1 determine the rotational dynamics of the liquid crystal. Only γ_φ can easily be determined by a standard electrooptic measurement. All of them can, however, be determined by dielectric spectroscopy. We will turn to their measurement in Sec. 2.7.

2.7 Dielectric Spectroscopy: To Find the $\hat{\gamma}$ and $\hat{\varepsilon}$ Tensor Components

2.7.1 Viscosities of Rotational Modes

Using dielectric relaxation spectroscopy, it is possible to determine the values of the rotational viscosity tensor corresponding to the three Euler angles in the chiral smectic C and A phases. These viscosity coefficients (γ_θ, γ_φ, γ_1) are active in the tilt fluctuations (the soft mode), the phase fluctuations (the Goldstone mode), and the molecular reorientation around the long axis of the molecules, respectively. It is found, as expected, that these coefficients obey the inequality $\gamma_\theta > \gamma_\varphi \gg \gamma_1$. While the temperature dependence of γ_θ and γ_φ in the SmC* phase can be described by a normal Arrhenius law as long as the tilt variation is small, the temperature dependence of γ_θ in the SmA* phase can only be modeled by an empirical relation of the form $\gamma_\theta = A e^{E_a/kT} + \gamma_\theta^0 (T - T_C)^{-\nu}$. At the smectic A* to SmC* transition, the soft mode viscosity in the smectic A* phase is found to be larger than the Goldstone mode viscosity in the SmC* phase by one order of magnitude. The corresponding activation energies can also be determined from the temperature dependence of the three viscosity coefficients in the smectic A* and SmC* phases. The viscosity coefficient connected with the molecular reorientation around the long axis of the molecules is not involved in electrooptic effects, but is of significant interest for our understanding of the delicate relation between the molecular structure and molecular dynamics.

2.7.2   The Viscosity of the Collective Modes

Different modes in the dipolar fluctuations characterize the dielectric of chiral smectics. As the fluctuations in the primary order parameter, for instance represented by the tilt vector

$$\xi_1 = n_z n_x = \frac{1}{2} \sin 2\theta \cos \varphi$$

$$\xi_2 = n_z n_y = \frac{1}{2} \sin 2\theta \sin \varphi \qquad (350)$$

are coupled to P, some excitations will appear as collective modes. Dipolar fluctuations which do not couple to ξ give rise to non-collective modes. From the tempera-

ture and frequency dependence of these contributions, we can calculate the viscosities related to different motions.

In the standard description of the dielectric properties of the chiral tilted smectics worked out by Carlsson et al. [152], four independent modes are predicted. In the smectic C* the collective excitations are the soft mode and the Goldstone mode. In the SmA* phase the only collective relaxation is the soft mode. Two high frequency modes are connected to noncollective fluctuations of the polarization predicted by the theory. These two modes become a single noncollective mode in the smectic A* phase. There is no consensus [153] as yet as to whether these polarization modes really exist. Investigations of the temperature dependence of the relaxation frequency for the rotation around the long axis show that it is a single Cole–Cole relaxation on both sides of the phase transition between smectic A* and smectic C* [154]. The distribution parameter α of the Cole–Cole function is temperature-dependent and increases linearly ($\alpha = a_T T + b_T$) with temperature. The proportionality constant a_T increases abruptly at the smectic A* to SmC* transition. This fact points to the complexity of the relaxations in the smectic C* phase.

The dielectric contribution of each of these modes has been worked out using the extended Landau free energy expansion of Sec. 2.5.11. The static electric response of each mode is obtained by minimizing the free energy in the presence of an electric field. The relaxation frequency of the fluctuations in the order parameter is obtained by means of the Landau–Khalatnikov equations, which control the order parameter dynamics.

The kinetic coefficients Γ_1 and Γ_2 of these equations have the dimensions of inverse viscosity. The soft mode and Goldstone mode are assumed to have the same kinetic coefficient, Γ_1, which, however, does not mean that they have the same measured value of viscosity.

The viscosity of the soft mode in the smectic A* phase can be written, according to [152]

$$\gamma_\theta = \frac{\varepsilon_0 \chi_0^2 C^{*2}}{2\pi} \frac{1}{\varepsilon_s f_s} \qquad (351)$$

where C* is the coupling coefficient between tilt and polarization, ε_s is the dielectric contribution from the soft mode, and f_s is the relaxation frequency of the soft mode in the smectic A* phase.

In order to calculate the viscosity, a measure of the coefficient C* is needed. It can be extracted from the tilt angle dependence on the applied field, from Sec. 2.5.7

$$\theta = \frac{\chi_0 \varepsilon_0 C^*}{\alpha(T - T_C)} E \qquad (352)$$

The corresponding expression for the soft mode viscosity in the SmC* phase is not equally simple, and reads

$$\gamma_\theta = \frac{1}{4\pi \varepsilon_0} \frac{1}{\varepsilon_s f_s} \left(\frac{b_3}{b_7}\right)^2 \qquad (353)$$

where b_3/b_7 is the following combination of parameters from the expansion (263)

$$\left(\frac{b_3}{b_7}\right) = \frac{C^* - e q + 2\Omega P \theta}{\dfrac{1}{\chi_0 \varepsilon_0} - \Omega \theta^3 + 3\eta P^2} \qquad (354)$$

This ratio can be obtained by fitting the experimentally measured values of ε_s versus temperature. In the frame of this Landau theory, the soft mode dielectric constant is

$$\varepsilon_s = \frac{\left(\dfrac{b_3}{b_7}\right)^2}{4\varepsilon_0 \alpha(T - T_C) + \varepsilon_0 \chi_0} \qquad (355)$$

where the term $\varepsilon_0 \chi_0$ is a cut-off parameter.

Assuming the ratio b_3/b_7 is independent of temperature, γ_θ can be calculated in the smectic C* phase. This approximation is valid deeper in the smectic C* phase, where P, θ, and q have their equilibrium values.

The Goldstone mode viscosity can be calculated [152] according to the expression

$$\gamma_\theta = \frac{1}{4\pi \varepsilon_0} \frac{1}{\varepsilon_G f_G} \left(\frac{P}{\theta}\right)^2 \qquad (356)$$

where P is the polarization, θ is the tilt angle, and ε_G and f_G are the dielectric constant and relaxation frequency of the Goldstone mode. The polarization and the tilt angle need to be measured in order to calculate the viscosity.

Figure 78 shows a plot of the soft mode and Goldstone mode rotational viscosities measured on either side of the phase transition between the smectic A* and SmC*. It can be seen that, except in the vicinity of the phase transition, the viscosity γ_θ seems to connect fairly well between the two phases. The activation energies of these two processes are, however, different. This result may be compared to results obtained by Pozhidayev et al. [148], referred to in Fig. 67. They performed measurements of γ_θ beginning in the chiral nematic phase of a liquid crystal mixture with corresponding measurements in the SmC* phase, and have shown the viscosity values on an Arrhenius plot for the N* and SmC* phases. Despite missing data of γ_θ in the smectic A* phase they extrapolate the N* values of γ_θ down to the smectic C* phase and get a reasonably smooth fit. Their measurements also show that γ_θ is larger than γ_φ, and this is universally the case.

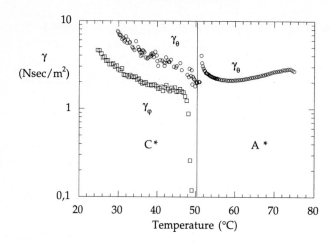

Figure 78. The soft mode and Goldstone mode rotational viscosities as a function of temperature. The material is mixture KU-100 synthesized at Seoul University, Korea (courtessy of Prof. Kim Yong Bai) (from Buivydas [155]).

2.7.3 The Viscosity of the Noncollective Modes

The viscosity of the noncollective modes can be estimated, as shown by Gouda [154] and Buivydas [155], using the original Debye theory of dielectric response. Assuming that the Stokes law is valid, the viscosity for the long axis rotation around the short axis scales as l^{-3}, where l is the length of the long axis. For this transverse rotation we have

$$\gamma_t = \frac{kT}{8\pi^2 f_t l^3} \tag{357}$$

where f_t is the relaxation frequency of the molecular rotation around the short axis and l is the length of the molecule, as shown in Fig. 79. The viscosity related to the molecular rotation around the long axis has the corresponding expression

$$\gamma_l = \frac{kT}{8\pi^2 f_l d^3} \tag{358}$$

where f_l is the relaxation frequency and d is the cross diameter of the molecule, cf. Fig. 79.

The substance used in these measurements is a three-ring compound synthesized by Nippon Mining Inc., which we may call LC1. It has the structural formula

C4H9-O-⬡-COO-⬡⬡-C-CH-C6H13

O
‖
-C-ĊH-C6H13
|
CH3

Cl

and the phase sequence I 105.0 SmA* 68.1 SmC* 8.0 Cr.

The relaxation frequencies f_l (of the order of 3 MHz) and f_l (of the order of 110 MHz) are now measured as a function of temperature, the former in quasi-homeotropic alignment (layers parallel to the cell plates), the latter in QBS geometry. With a cross section d taken from a simple molecular model to be about 4.5 Å, the corresponding function $\gamma_l(T)$ is obtained from (358).

Now, the length and width of a molecule used in Eqs. (357) and (358) are not self-ev-

Figure 79. The molecular dimensions d and l. The depicted length/width ratio corresponds to the substance LC1 (see later).

ident values. Therefore the Landau description may be used to calculate the viscosity related to the molecular rotation around the long axis. In the smectic C* phase, the viscosity γ_1 is then expressed by

$$\gamma_1 = \frac{1}{2\pi\varepsilon_0}\left(\frac{1}{\varepsilon_1 f_1}\right)\left(\frac{P}{\theta}\right)^2 \qquad (359)$$

In order to determine the same viscosity in the smectic A* phase, the normalization factor $(P/\theta)^2$ should be preserved, otherwise the viscosity will show a jump at T_C [154]. Thus the corresponding equation will read

$$\gamma_1 = \frac{1}{2\pi\varepsilon_0}\left(\frac{1}{\varepsilon_1 f_1}\right)\left(\frac{P}{\theta}\right)^2_{T_C} \qquad (360)$$

where $(P/\theta)_{T_C}$ is the limiting value of the P/θ ratio measured at the transition.

The viscosity γ_1 calculated by the Landau theory and the γ_1 evaluated by using the Debye formulas are shown for comparison in Fig. 80.

We see that they agree fairly well in the limit of small tilt, thus particularly in the A* phase. The non-Arrhenius increase at higher tilts is here an artifact inherent in the

Landau method. In order to calculate $\gamma_1(T)$ we also need a value for the molecular length l. This can be obtained by the requirement that we can have only one single viscosity in the isotropic phase. Combining (357) and (358) gives

$$\frac{\gamma_t}{\gamma_1} = \frac{f_1 d^3}{f_t l^3} \qquad (361)$$

On putting $\gamma_t = \gamma_1$ we find

$$l = \sqrt[3]{\frac{f_1}{f_t}} \cdot d \qquad (362)$$

From this l is obtained as 15 Å. When we check the molecular length by searching the lowest energy conformation of LC1 in a simple computer program, we see that this value actually corresponds to the length of the core. We could of course also have slightly adjusted the input value of d such that the Debye and Landau viscosity values coincide exactly in the A* phase. This is found to occur for $d = 4.7$ Å, corresponding to $l = 16$ Å.

The results from all viscosity measurements on the substrate LC1 are summarized in Fig. 81.

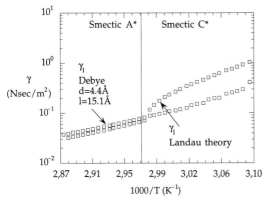

Figure 80. Viscosities of noncollective motions evaluated by two different methods. The substance is LC1 (from Buivydas [155]).

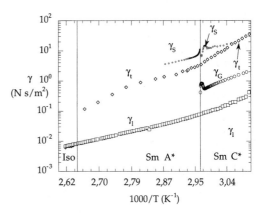

Figure 81. The three components of the viscosity tensor measured by dielectric relaxation spectroscopy. The substance is LC1 (from Buivydas [155]).

The noncollective rotational viscosities are denoted by γ_l (longitudinal) and γ_t (transverse). The collective motions have the corresponding designations γ_s (for soft mode) and γ_G (for Goldstone mode). Equivalently, γ_θ and γ_φ could also have been used. It is remarkable that the Goldstone mode viscosity is almost as low as the viscosity around the long axis of the molecule. As pointed out earlier, this is explained by the fact that a large component of the angular velocity is around this axis. The soft mode viscosity lies much higher and, as the results show, is essentially the same as the viscosity for the corresponding individual molecular rotation.

2.7.4　The Viscosity γ_φ from Electrooptic Measurements

In the smectic C* phase, the rotational viscosity γ_φ can be estimated by observing the polarization reversal or the electrooptic properties of the cell, as described in Sec. 2.7.6. The estimation may, for instance, be based on the approximation mentioned there, using the elastic torque [137]

$$P_S\, E \sin\varphi - K \cos\varphi + \gamma_\varphi\, \dot{\varphi} = 0 \qquad (363)$$

Solving this equation and fitting the observed polarization reversal process to the solution, the viscosity can be estimated and compared to the one measured by means of dielectric relaxation spectroscopy. This method is fast, but less accurate.

The polarization reversal measurement is a large signal method requiring full switching of the liquid crystal, and therefore cannot be expected to coincide with the much more precise results from the low signal dielectric relaxation spectroscopy measurements. Basically, a γ_φ value from the polarization reversal technique involves spurious contributions of elastic effects due to the

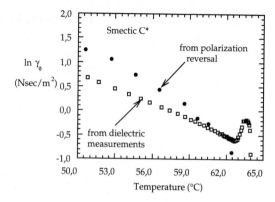

Figure 82. Comparative study of the γ_φ rotational viscosity measured by the polarization reversal technique based on Eq. (363) and dielectric relaxation spectroscopy. The material is LC1 (from Buivydas [155]).

surfaces of the sample. However, the values from the polarization reversal method do not differ by more than typically a factor of two, and the difference tends to be smaller at higher temperatures (see Fig. 82).

2.7.5　The Dielectric Permittivity Tensor

Phases having biaxial symmetry (tilted smectic phases) exhibit dielectric biaxiality in particular. At frequencies of 1 MHz and below, the biaxiality becomes important and critically influences the electrooptic switching behavior of the SmC* phase. It is therefore important to be able to measure the biaxiality at these frequencies. Being a symmetrical second rank tensor, the dielectric permittivity can always be diagonalized in a proper frame and described by three components along the principal directions. The three principal values can then be expressed by a single subscript and can be determined by three independent measurements performed at three different orientations of the director relative to the measuring electric field. In practice, this may not be that straightforward, for instance, in the case of a thick sample (measured on a scale of the pitch) of a smectic C* when the helix is not quenched by the surfaces. In such a case the local frame of the ε tensor is rotating helically through the sample.

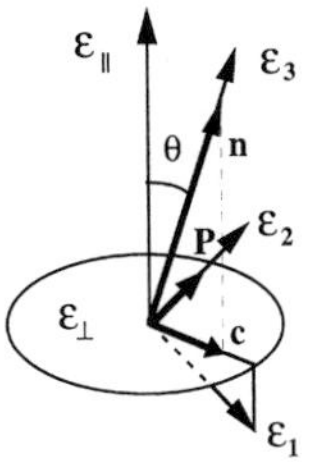

2.7.6   The Case of Helical Smectic C* Compounds

In the molecular frame (Fig. 83) we write the tensor ε as

$$\hat{\varepsilon} = \begin{pmatrix} \varepsilon_1 & 0 & 0 \\ 0 & \varepsilon_2 & 0 \\ 0 & 0 & \varepsilon_3 \end{pmatrix} \tag{364}$$

As for γ in Sec. 2.6.9, we may consider $\hat{\varepsilon}$ being brought to diagonal form (starting from the laboratory frame) by a similarity transformation

$$\hat{\varepsilon}(1, 2, 3) = T^{-1}\varepsilon(x, y, z)\,T \tag{365}$$

where T is a rotation matrix and T^{-1} is the inverse or reciprocal matrix to T, in this case identical to its transpose.

Figure 83. The definitions of the molecular axes and components of the dielectric tensor.

In Fig. 83 we may imagine a laboratory frame with the z axis along the layer normal (thus along $\varepsilon_\parallel$ in the figure), such that we have turned the director n a certain angle θ around the y direction (along ε_2 in the figure). The x direction is along c.

In this laboratory frame xyz, the permittivity tensor is then given by

$$\varepsilon(\theta) = T_\theta\,\hat{\varepsilon}\,T_\theta^{-1} \tag{366}$$

with the rotation matrix T_θ

$$T_\theta = \begin{pmatrix} \cos\theta & 0 & -\sin\theta \\ 0 & 1 & 0 \\ \sin\theta & 0 & \cos\theta \end{pmatrix} \tag{367}$$

Performing the multiplication – the same as in Eq. (342) – we find

$$\varepsilon(\theta) = \begin{pmatrix} \varepsilon_1\cos^2\theta + \varepsilon_3\sin^2\theta & 0 & (\varepsilon_1 - \varepsilon_3)\sin\theta\cos\theta \\ 0 & \varepsilon_2 & 0 \\ (\varepsilon_1 - \varepsilon_3)\sin\theta\cos\theta & 0 & \varepsilon_1\sin^2\theta + \varepsilon_3\cos^2\theta \end{pmatrix} \tag{368}$$

which can be written in the form

$$\varepsilon(\theta) = \begin{pmatrix} \varepsilon_\perp & 0 & -\Delta\varepsilon\sin\theta\cos\theta \\ 0 & \varepsilon_2 & 0 \\ -\Delta\varepsilon\sin\theta\cos\theta & 0 & \varepsilon_\parallel \end{pmatrix} \tag{369}$$

by introducing the dielectric anisotropy $\Delta\varepsilon = \varepsilon_3 - \varepsilon_1$ and the abbreviations

$$\varepsilon_\perp = \varepsilon_1\cos^2\theta + \varepsilon_3\cos^2\theta \tag{370}$$

and

$$\varepsilon_\parallel = \varepsilon_1\sin^2\theta + \varepsilon_3\cos^2\theta \tag{371}$$

If we now further turn the director by an angle φ around the z direction, the permittivity tensor will change to

$$\varepsilon(\theta, \varphi) = T_\varphi \varepsilon(\theta) T_\varphi^{-1} \tag{372}$$

As this rotation is performed counterclock-wise (positive direction) in the xy plane, the rotation matrix is now instead

$$T_\varphi = \begin{pmatrix} \cos\varphi & \sin\varphi & 0 \\ -\sin\varphi & \cos\varphi & 0 \\ 0 & 0 & 1 \end{pmatrix} \tag{373}$$

This gives

$$\varepsilon(\theta, \varphi) = T_\varphi T_\theta \, \hat{\varepsilon} \, T_\theta^{-1} T_\varphi^{-1} \tag{374}$$

or

$$\varepsilon(\theta,\varphi) = \begin{pmatrix} \varepsilon_1 \cos^2\varphi + \varepsilon_2 \sin^2\varphi & -(\varepsilon_1 - \varepsilon_2)\sin\varphi\cos\varphi & -\Delta\varepsilon\sin\theta\cos\theta\cos\varphi \\ -(\varepsilon_1 - \varepsilon_2)\sin\varphi\cos\varphi & \varepsilon_1\sin^2\varphi + \varepsilon_2\cos^2\varphi & \Delta\varepsilon\sin\theta\cos\theta\sin\varphi \\ -\Delta\varepsilon\sin\varphi\cos\theta\cos\varphi & \Delta\varepsilon\sin\theta\cos\theta\sin\varphi & \varepsilon_{\parallel} \end{pmatrix} \tag{375}$$

In the helical smectic C*, the tilt θ is a constant while the phase angle φ is a function of the position along the layer normal, $\varphi = \varphi(z)$. Averaging over one pitch, with

$$\langle \cos^2\varphi \rangle = \langle \sin^2\varphi \rangle = \frac{1}{2} \tag{376}$$

and

$$\langle \sin\varphi \rangle = \langle \cos\varphi \rangle = \langle \sin\varphi \cos\varphi \rangle = 0 \tag{377}$$

ε simplifies to

$$\langle \varepsilon(\theta) \rangle_\varphi = \begin{pmatrix} \langle \varepsilon_\perp \rangle & 0 & 0 \\ 0 & \langle \varepsilon_\perp \rangle & 0 \\ 0 & 0 & \langle \varepsilon_{\parallel} \rangle \end{pmatrix} \tag{378}$$

with the new abbreviations

$$\langle \varepsilon_\perp \rangle = \frac{1}{2} (\varepsilon_1 \cos^2\theta + \varepsilon_2 + \varepsilon_3 \sin^2\theta) \tag{379}$$

and

$$\langle \varepsilon_{\parallel} \rangle = \varepsilon_1 \sin^2\theta + \varepsilon_3 \cos^2\theta \tag{380}$$

Equation (378) reveals that the helical state of the smectic C* is effectively uniaxial, if the pitch is sufficiently short to allow the averaging procedure. The shorter and more perfect the helix, the more completely the local biaxiality is averaged out (both dielectric and optical). This is consistent with the empirical evidence, showing almost perfect uniaxiality for helical samples of short pitch materials.

Before we can use our derived relations to determine ε_1, ε_2 and ε_3, we have to perform one more refinement. The simple measurement geometry where smectic layers are parallel to the glass plates is easy to obtain in practice. On the other hand, it is very hard to turn them into being perfectly parallel to the plates; normally, the result is that they meet the surface with a slight inclination angle δ. To account for this case we have to see how the tensor $\langle \varepsilon(\theta) \rangle_\varphi$ transforms when we turn the layer normal by a certain angle δ around the x axis. We perform the rotation clockwise (in this case the direction does not matter), corresponding to

$$T_\delta = \begin{pmatrix} 1 & 0 & 0 \\ 0 & \cos\delta & -\sin\delta \\ 0 & \sin\delta & \cos\delta \end{pmatrix} \tag{381}$$

and giving

$$\langle \varepsilon(\theta,\delta)\rangle_\varphi = T_\delta \,\langle \varepsilon(\theta)\rangle_\varphi \, T_\delta^{-1} = \begin{pmatrix} \langle \varepsilon_\perp \rangle & 0 & 0 \\ 0 & \langle \varepsilon_\perp \rangle \cos^2\delta + \langle \varepsilon_\| \rangle \sin^2\delta & \langle \varepsilon_\perp \rangle - \langle \varepsilon_\| \rangle \sin\delta\cos\delta \\ 0 & \langle \varepsilon_\perp \rangle - \langle \varepsilon_\| \rangle \sin\delta\cos\delta & \langle \varepsilon_\perp \rangle \sin^2\delta + \langle \varepsilon_\| \rangle (\cos^2\delta) \end{pmatrix} \quad (382)$$

In the case where we assume a helix-free sample and fully addressed or memorized director states lying on extreme ends of the smectic cone, so that we might put φ equal to 0 or π in Eq. (375), would instead have given the tensor components

$$\varepsilon(\delta,\theta) = \begin{pmatrix} \varepsilon_1\cos^2\theta + \varepsilon_3\sin^2\theta & \Delta\varepsilon\sin\theta\cos\theta\sin\delta & -\Delta\varepsilon\sin\theta\cos\theta\sin\delta \\ \Delta\varepsilon\sin\theta\cos\theta\sin\delta & \varepsilon_2\cos^2\delta + \langle \varepsilon_\| \rangle\sin^2\delta & (\varepsilon_2 - \langle \varepsilon_\| \rangle)\sin\delta\cos\delta \\ -\Delta\varepsilon\sin\theta\cos\theta\cos\delta & (\varepsilon_2 - \langle \varepsilon_\| \rangle)\sin\delta\cos\delta & \varepsilon_2\sin^2\delta + \langle \varepsilon_\| \rangle\cos^2\delta \end{pmatrix} \quad (383)$$

2.7.7   Three Sample Geometries

With the expressions derived in the last section, in the first version worked out by Hoffmann et al. [156] and further developed by Gouda et al. [157], we can calculate ε_1, ε_2, and ε_3 from three independent measurements, using three different sample geometries.

Consider first that we have the smectic layers oriented perpendicular to the glass plates so that the applied field is perpendicular to the layer normal. In the Smectic A* phase we would call this planar alignment, and a dielectric measurement would immediately give the value $\varepsilon_p = \varepsilon_\perp$ in the SmA* phase. If we cool to the SmC* phase, the helix will appear in a direction parallel to the glass plates. By applying a bias field we unwind the helix and the signal field will now measure ε_{unw}, which is identical to ε_2

$$\varepsilon_{unw} = \varepsilon_2 \quad (384)$$

the permittivity value along the direction of the polarization P. If we take away the bias field the helix will eventually relax. The value of the permittivity measured in the presence of a helix is denoted by ε_{helix}. It is equal to ε_{yy} in the lab frame, i.e., to $\langle \varepsilon_\perp \rangle$ in

Eq. (378) or, by Eq. (379)

$$\varepsilon_{helix} = \frac{1}{2}(\varepsilon_1\cos^2\theta + \varepsilon_2 + \varepsilon_3\sin^2\theta) \quad (385)$$

The third geometry normally requires a second sample aligned with the layers along the glass plates. We may call this geometry homeotropic and in the SmA* phase we measure the value $\varepsilon_h = \varepsilon_\|$ along the layer normal. After cooling to the SmC* phase we may call the geometry quasi-homeotropic and denote the measured value ε_{hom}. The helix axis (z) is now parallel to the measuring field, which means that ε_{hom} is ε_{zz} in the lab frame, which is $\langle \varepsilon_\| \rangle$ in Eq. (378) or, by Eq. (380)

$$\varepsilon_{hom} = \varepsilon_1\sin^2\theta + \varepsilon_3\cos^2\theta \quad (386)$$

From Eqs. (384) to (386) the three principal values of the dielectric tensor can thus be calculated. After solving for ε_1, ε_2, and ε_3, the components of the diagonalized tensor can be expressed in the measured values as

$$\varepsilon_1 = \frac{1}{1 - 2\sin^2\theta}$$
$$\cdot \left[(2\varepsilon_{helix} - \varepsilon_{unw})\cos^2\theta - \varepsilon_{hom}\sin^2\theta \right] \quad (387)$$

$$\varepsilon_2 = \varepsilon_{unw} \quad (388)$$

$$\varepsilon_3 = \frac{1}{1-2\sin^2\theta}$$
$$\cdot \left[(2\varepsilon_{\text{helix}} - \varepsilon_{\text{unw}})\sin^2\theta - \varepsilon_{\text{hom}}\cos^2\theta \right] \tag{389}$$

The three measurements require preparation of two samples. By using a field-induced layer reorientation, first investigated by Jakli and Saupe [158], Markscheffel [158a] was able to perform all three measurements on a single sample, starting in the quasi-homeotropic state, and then, by a sufficiently strong field, turning the layers over to the quasi-bookshelf structure. Generally, one would start with the quasi-bookshelf geometry and then warm up the sample to the smectic A* phase and strongly shear. The advantage of this technique is a very homogeneous, homeotropic orientation. The quality of homeotropic orientation may even improve (without degrading the quality of planar orientation) if the glass plates are coated by a weak solution of tenside.

be

$$\varepsilon_{\text{helix}} = \langle\varepsilon_\perp\rangle\cos^2\delta + \langle\varepsilon_\parallel\rangle\sin^2\delta \tag{391}$$

while ε_{hom} stays unaffected, i.e., $\langle\varepsilon_\parallel\rangle$, given by Eq. (386), because of the absence of the layer tilt in this geometry. The value of δ is often known, growing roughly proportional to θ, and can in many cases be approximated as $\delta\approx0.8\,\theta$. The principal values of ε_{ij} can then be extracted from Eqs. (386), (390), and (391). It turns out that the values are not very sensitive to small errors in δ.

A special case of tilted layers is the chevron structure, for which the tilt makes a kink either symmetrically in the middle or unsymmetrically closer to one of the surfaces. The chevron structure means that two values $+\delta$ and $-\delta$ appear in the sample. As can be seen from Eqs. (389) and (390), the angle δ only appears as a quadratic dependence, hence the relations are also valid in the chevron case. Solving for the principal values in the chevron or tilted layers case, we obtain

$$\varepsilon_1 = \frac{1}{1-2\sin^2\theta}\left[\frac{\cos^2\theta}{\cos^2\delta}(2\varepsilon_{\text{helix}} - \varepsilon_{\text{unw}} - \varepsilon_{\text{hom}}\sin^2\delta) - \varepsilon_{\text{hom}}\sin^2\theta\right] \tag{392}$$

$$\varepsilon_2 = \frac{1}{\cos^2\delta}(\varepsilon_{\text{unw}} - \varepsilon_{\text{hom}}\sin^2\delta) \tag{393}$$

$$\varepsilon_3 = \frac{-1}{1-2\sin^2\theta}\left[\frac{\sin^2\theta}{\cos^2\delta}(2\varepsilon_{\text{helix}} - \varepsilon_{\text{unw}} - \varepsilon_{\text{hom}}\sin^2\delta) - \varepsilon_{\text{hom}}\cos^2\theta\right] \tag{394}$$

2.7.8   Tilted Smectic Layers

When the layers make the inclination δ with the normal to the glass plates, we see from Eq. (383) that $\varepsilon_{\text{unw}} (=\varepsilon_{yy})$ will be expressed by

$$(\varepsilon_{\text{unw}} = \varepsilon_2\cos^2\delta + \langle\varepsilon_\parallel\rangle\sin^2\delta \tag{390}$$

whereas $\varepsilon_{\text{helix}} = \varepsilon_{yy}$ in Eq. (382) turns out to

Typical measured variations of the permittivities ε_1, ε_2, and ε_3, as we pass the phase boundaries from isotropic to tilted smectic phase, are shown in Fig. 84. The values of ε_1 and ε_2 have a similar temperature variation. Therefore their difference, the biaxiality $\delta\varepsilon$, has a weak but noticeable temperature dependence. The difference between the values ε_1 and ε_3 is the dielectric anisotropy $\Delta\varepsilon$. This depends on temperature much

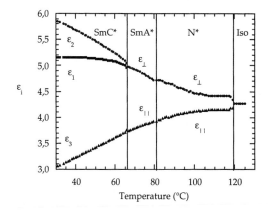

Figure 84. The principal values of the dielectric permittivity tensor at 100-kHz measured by the short pitch method. The material is the mixture SCE12 by BDH/Merck (from Buivydas [155]).

more strongly than the dielectric biaxiality. As a result of the fact that the relaxation frequency of ε_3 in the nematic phase lies below 100 kHz, the dielectric anisotropy is negative at that frequency. A negative value of $\Delta\varepsilon$ is a common feature, which may greatly affect the electrooptic switching properties at room temperature. In contrast, the relaxation frequency of ε_2 lies much above 100 kHz, whence the positive biaxiality $\delta\varepsilon$ (which may rather be considered as a low frequency value). This means that for information about $\Delta\varepsilon$ we have to go to relatively low frequencies and inversely to high frequencies for corresponding information about $\delta\varepsilon$.

2.7.9   Nonchiral Smectics C

When the cell has another director profile than the uniform helix, averaging of the angle φ over the pitch length cannot be performed in a straightforward way. It is necessary to know about the director profile through the cell before making simplifications. This profile is hard to get at but may be estimated by optical methods, as de-

scribed by Sambles et al. [159]. They found the profile to be uniform or nearly so within each surface-stabilized domain. However, other reports show that the surface-stabilized state is more accurately modeled by some triangular director profile [160].

When a uniform director profile is assumed, then the smectic director field $\boldsymbol{n}$ can be described by the three angles φ, θ, and δ. These angles are uniform throughout the sample and the dielectric tensor component ε_{yy} corresponds to the dielectric permittivity ε_p measured in the planar orientation. Under certain conditions this may be written

$$\varepsilon_p = \varepsilon_2 - \delta\varepsilon \, \frac{\sin^2 \delta}{\sin^2 \theta} \qquad (395)$$

where

$$\delta\varepsilon = \varepsilon_2 - \varepsilon_1 \qquad (396)$$

is the dielectric biaxiality.

When measuring in the quasi-homeotropic orientation, we will observe the component $\varepsilon_h = \varepsilon_{zz}$ and, assuming the layer tilt δ to be zero, ε_h takes the form

$$\varepsilon_h = \varepsilon_2 + \Delta\varepsilon \cos^2 \delta \qquad (397)$$

Again we have to look for an independent measurement of a third value, in order to cal-

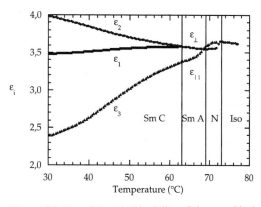

Figure 85. The dielectric biaxiality of the nonchiral smectic C mixture M4851 (Hoechst), measured by the long pitch method (from Buivydas [155]).

culate all three principal components of the dielectric permittivity tensor. At the second order transition SmA→SmC, the average value of ε is conserved, thus

$$\bar{\varepsilon} = \frac{2\varepsilon_\perp + \varepsilon_\parallel}{3} = \frac{\varepsilon_1 + \varepsilon_2 + \varepsilon_3}{3} \tag{398}$$

With knowledge of the tilt angle θ and the layer inclination δ, we then can calculate ε_1, ε_2, and ε_3 from Eqs. (395), (397) and (398).

2.7.10 Limitations in the Measurement Methods

Each of the models serving as basis for the dielectric biaxiality measurements has been derived using certain assumptions. The method described by Gouda et al. [157] assumes an undisturbed ferroelectric helix, and this assumption is best fulfilled in short pitch substances. Actually, the pitch/cell thickness ratio is the important parameter. For measurement convenience, the cells cannot be made very thick, hence the method works well for materials with a helical pitch of up to about 5 µm. This includes practically all cases of single smectic C* substances. The method cannot be applied to nonchiral materials.

The method described by Jones et al. [161] (long pitch method) was developed for materials with essentially infinite pitch. It may be used for long pitch materials in cells where the helical structure is unwound by the surface interactions and a uniform director profile is established. The weakness of the method is that it requires the director profile to be known. On the other hand, it works for the nonchiral smectic C. It also works for materials with $\theta = 45°$ (typically materials with N→SmC transitions, lacking the SmA phase), which is a singular case where the short pitch method fails.

The reason for this failure is that for a 45° tilt angle θ there is no difference in average dielectric anisotropy between the planar orientation and the quasi-homeotropic orientation, and thus the two equations in the system (Eqs. 385 and 386) are linearly dependent as far as that they give the same geometric relation between the direction of the measuring field and the direction of the principal axes ε_1 and ε_3 (ε_2 is not involved in ε_{hom}). Equations (385) and (386) in this case give

$$\varepsilon_{helix} = \frac{1}{2}(\varepsilon_1 + \varepsilon_2) + \varepsilon_2 \tag{399}$$

and

$$\varepsilon_{hom} = \frac{1}{2}(\varepsilon_1 + \varepsilon_3) \tag{400}$$

respectively, whence it is not possible to split ε_1 and ε_3, but only measure their sum. Using tilted layers ($\delta \neq 0$) or knowing the average value ε is not remedy for this, because only the combination $\varepsilon_1 + \varepsilon_3$ appears in all equations.

As a result of this the errors in the calculated values of ε_1 and ε_3 increase and diverge when θ approaches the value 45°. This is also evident from the denominator $(1 - 2 \sin^2\theta)$ in Eqs. (392) and (394). In Fig. 86 we show how the errors in ε_1 and ε_3 grow with the tilt angle under the assumption that the value ε_{helix} used in the calculation formula was measured with a 2% error and that the value of ε_{hom} was assumed to be exact.

The presence of this singularity limits the usability of the short pitch method to substances with tilt angle θ up to about 35–37°. In practice, however, this is not too serious, as materials with a tilt approaching 40° are rare.

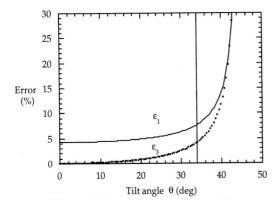

Figure 86. The dependence of the errors in the tensor components ε_1 and ε_3 calculated by the short pitch method when the smectic tilt angle θ approaches $45°$. The input value ε_{helix} is assumed to be measured with a 2% error (from Buivydas [155]).

2.8 FLC Device Structures and Local-Layer Geometry

2.8.1 The Application Potential of FLC

Ten years of industrial development of ferroelectric liquid crystal (FLC) devices have now passed. This may be compared with 25 years for active matrix displays. The potential of FLC was certainly recognized early enough following the demonstration of ferroelectric properties in 1980. In principle the FLC has the potential to do what no other liquid crystal technology could do: in addition to high-speed electrooptic shutters, spatial light modulators working in real time, and other similar hardware for optical processing and computing (all applications requiring very high intrinsic speed in repetitive operation), the FLC (and the closely related and similar AFLC) technology offers the possibility to make large-area, high-resolution screens without the need for tran-

sistors or other active elements, i.e., in passive matrix structures using only the liquid crystal as the switching element. Several kinds of such structures (Canon, Matsushita, Displaytech) have now been demonstrated which give very high quality rendition of color images, and one polymer FLC technology has been developed (Idemitsu) capable of monochrome as well as multi-color panels.

When the SSFLC structure was presented in 1980, some obvious difficulties were immediately pointed out: the liquid crystal had to be confined between two glass plates with only about 1.5 μm spacing or even less and, furthermore, in a rather awkward geometry ("bookshelf" structure), which meant that both the director alignment and the direction of smectic layers had to be carefully controlled. Luckily, a number of more serious problems were not recognized until much later. In addition, only a few molecular species had been synthesized at that time; in order to develop a viable FLC technology, a considerable effort in chemical synthesis would have had to be started and pursued for many years. (We may recall that the submicrosecond switching reported in the first paper had been observed on a single substance, HOBACPC, at an elevated temperature of 68 °C.) This chemical problem alone would have discouraged many laboratories from taking up the FLC track.

Nevertheless, in 1983 excellent contrast and homogeneity in the optical properties were demonstrated on laboratory samples of DOBAMBC, one of the other single substances available (thus not on a room temperature mixture) and sufficiently well aligned for that purpose by a shearing technique [96]. Shortly afterwards, two Japanese companies, Canon Inc. and Seiko Instruments and Electronics, were engaged in R&D. By 1986 about twenty Japanese and five European companies were engaged in

FLC research, supported by about ten chemical companies worldwide. Common to almost all these ventures was the fact that the FLC teams were very small. When really troublesome obstacles began to show up, one company after the other left the scene, favoring STN or TFT technology. It was clear that the obstacles were not only in manufacturing technology, but there were tremendous scientific difficulties of a very basic nature as well.

After having presented the first mature FLC prototypes in 1988 (black and white) and 1992 (color), Canon in Tokyo is now manufacturing the first 15 in. (37.5 cm) FLC panel with 1280×1024 pixels, where each pixel, of size $230\ \mu m \times 230\ \mu m$, can display 16 different colors because it is composed of four dots or subpixels. This screen (Fig. 87) has a remarkable performance and does not resemble anything else in its absolute freedom of flicker. At the other end of the scale, Displaytech Inc., U.S.A.,

is marketing FLC microdisplays, slightly larger than 5 mm by 5 mm in size with VGA resolution (640×480), where every dot is capable of 512 colors (see Fig. 88). Another remarkable display is Idemitsu Kosan's fully plastic monochrome screen in reflection, of A4 size (15 in) with 640×400 dots (see Fig. 89). This is a passively multiplexed (1:400) sheet of FLC polymer, thus entirely without glass plates and all in all 1 mm thick. Regarding these achievements, it may be time to look at the real SSFLC structures that are used in these devices.

2.8.2 Surface-Stabilized States

It is important to point out that for a given material there is a variety of different realizations of surface-stabilized states (states with nonzero macroscopic polarization), on the one hand, and also a variety of other states. Among the former, those with a non-

Figure 87. Canon 15 in (37.5 cm) color screen which went into production 1995. High quality rendition of artwork is extremely sensitive to flicker. The FLC freedom of flicker therefore gives an unusual means of presenting art, here illustrated by the turn of the century artist Wyspiański (self-portrait) and Bonnard (Dining Room with Garden).

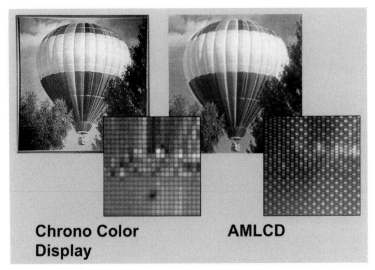

Figure 88. The first version of Displaytech's miniature display (ChronoColor), which went into production in early 1997, in comparison with a conventional active matrix display. This microdisplay utilizes sequential color to produce full color on each pixel, resulting in a brighter, crisper image than that of an AMLCD, which uses a triad of red, green, and blue pixels to form a color. In the insets, the pixels of the ChronoColor display and AMLCD are magnified to show the difference in fill factors. The color is what makes the image that you see, and it occupies 75% of the display on the left, but only 35% of the corresponding active matrix display. (Courtesy of Displaytech, Inc.)

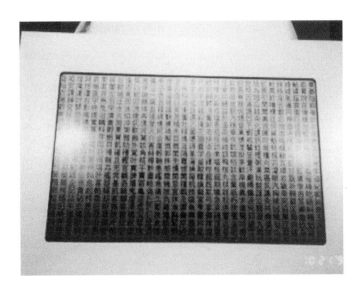

Figure 89. All-polymer FLC prototype presented by Idemitsu Kosan Co., Ltd., in 1995. This reflective monochrome display is 0.5 mm thick, has a 1:400 multiplex ratio, and a size of 20 cm×32 cm (400×640 pixels each 0.5 mm×0.5 mm), corresponding to a diagonal of 15 in (37.5 cm).

zero P along the applied field direction will have a strong nonlinear dielectric response in contrast to other states, which behave linearly until saturation effects dominate. Only the former may be bistable, i.e., show a stable memory ($P \neq 0$ in the absence of an applied field E). We will give two examples to illustrate this. The first is a case with supposedly strong anchoring of the director along a specified azimuthal direction φ, corresponding to parallel rubbing at top and bottom plate of the cell [127]. This case is illustrated in Fig. 90, where successive $n-P$ states are shown across the cell. Clearly we have two completely symmetrical arrangements, which have a certain effective polarization down and up, respectively, and which are, each of them, a stable state. They can be switched back and forth between each other by the application of convenient pulses, and have a symmetrical relationship similar to the two cases of Euler buckling of a bar in two dimensions under fixed boundary conditions. The cell is thus surface-stabilized and ferroelectric with a symmetrically bistable reponse. However, it is immediately clear that this configuration may not be the most desirable one, because we have a splayed P state with only a small part of P that can actually be switched by a short pulse, requiring a high voltage for a large optical effect. This is because the optical

contrast between the two memorized up and down states is very low. In order to enhance this contrast, we would have to apply a considerable AC holding voltage across the cell after application of the switching pulse, which seems to make this use of ferroelectric structure rather pointless. Clearly the structure is intrinsically inferior to the structure in Fig. 61 where the director profile is supposed to be homogeneous across the cell. However, it not only serves the purpose of illustrating that surface-stabilized structures can be of many kinds and that they might be characterized by their electrical (speed) as well as their optical efficiency, but also this structure actually turns out to be of high practical interest, as we will see in Sec. 2.8.6.

The second example is what we call a twisted smectic [162], and is illustrated in Fig. 91. Ideally a smectic C* with $\theta=45°$ should be used as a twisted smectic. In Fig. 91 the anchoring is again assumed to be strong (although this is not a necessity for

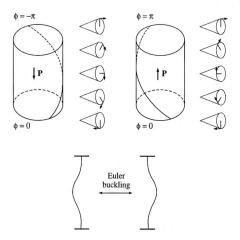

Figure 90. Surface-stabilized configuration with less than optimum efficiency, switchable between two symmetric states with low optical contrast. The surface pretilt angle has been chosen equal to the smectic tilt angle θ in this example. For a strong boundary condition with zero pretilt, a different extreme limiting condition with θ approaching zero at the boundary is also conceivable, without any essential difference in the performance of the cell.

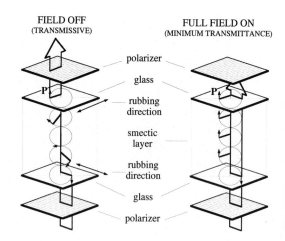

Figure 91. The twisted smectic preferably uses a 45° tilt material. It is a fast-switching device giving an electrically controlled, continuous grayscale. This is a smectic C* working in a dielectric mode. It should not be mistaken for a ferroelectric device (after Pertuis and Patel [162]).

the working principle), and the alignment directions are at right angles (with zero pre-tilt at the surface) at top and bottom, introducing a 90° twist across the cell. This also results in zero macroscopic polarization along the direction in which the electric field is applied. With no field applied, the light will be guided in a way similar to that in a twisted nematic, and the cell will transmit between crossed polarizers. On application of an electric field, the twisted structure will unwind, until the whole structure except for one boundary layer has a homogeneous arrangement of the director, giving zero transmittance. The response is linear at low electric field and we have a fast switching device with a continuous grey scale. It has to be pointed out that this, of course, is not a ferroelectric mode. Thus a smectic C* liquid crystal does not have a ferroelectric response per se, but it can be used in many different ways. As already pointed out, bulk structures of smectic C* tend to have an anti-ferroelectric-like ordering in one way or another. The point about surface stabilization is that it transfers macroscopic polarization to the bulk, giving it ferroelectric properties in certain geometries. These properties are best recognized by the appearance of spontaneous domains of up and down polarization, and by the fact that the response to an electric field is now strongly nonlinear.

Let us finally look back to Fig. 61 and consider the material parameters θ, P, and d, i.e., tilt angle, polarization, and smectic layer thickness. They all depend on temperature: $\theta(T)$, $P(T)$, and $d(T)$. The temperature variations of θ and P, if not desirable, at least turned out to be harmless. On the other hand, the much smaller temperature dependence of $d(T)$ turned out to be significantly harmful.

If the liquid crystal molecules behaved like rigid rods, the layer thickness would diminish with decreasing temperature according to $d_A\cos\theta$, as the director begins to tilt at the SmA $\rightarrow$ SmC transition, i.e., we would have a layer shrinkage

$$\Delta d(T) \sim \theta^2 \qquad (401)$$

in the smectic C phase. In reality, the shrinkage is less (i.e., the molecules do not behave like rigid rods), but still a phenomenon that is almost universal, i.e., present in almost all materials. If, in Fig. 61, we consider the translational periodicity d to be imprinted in the surface at the nematic–smectic A transition, and if we further assume that there is no slip of the layers along the surface, the only way for the material to adjust to shrinking layers without generating dislocations is the creation of a folded structure in one direction or the other on entering the SmC phase, as illustrated in Fig. 92. As the layer thickness d_C decreases with decreasing temperature, the chevron angle δ increases according to

$$\cos\delta = \frac{d_C}{d_A} \qquad (402)$$

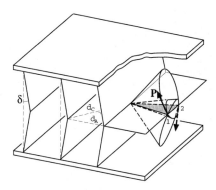

Figure 92. The shrinkage in smectic layer thickness due to the molecular tilt $\theta(T)$ in the SmC* phase results in a folding instablity of the layer structure ("chevrons"). Even if the fold can be made to go everywhere in the same direction (in the figure to the right) to avoid invasive zigzag defect structures, the switching angle is now less than 2θ, which lowers brightness and contrast.

where δ is always smaller than θ. Often δ and θ have a similar dependence, such that $\delta = \lambda \theta$, with $\lambda = 0.85 - 0.90$ in typical cases. The resulting "chevron" structure constitutes one of the most severe obstacles towards a viable FLC technology, but its recognition in 1987 by Clark and his collaborators by careful X-ray scattering experiments [163, 164] was one of the most important steps both in revealing the many possibilities of smectic local layer structures and in controlling the technical difficulties resulting from these structures.

2.8.3   FLC with Chevron Structures

The extremely characteristic zigzag defects had been observed for many years before the subtleties of the local layer structure were unveiled. Very often they appear as long "streets" or "lightning flashes" of thin zigzag lines, with the street going essentially along the smectic layers, whereas the lines themselves are at a small angle or almost perpendicular to the layers. In Fig. 93 the angle between the lines and the layer normal is roughly 5°. Often these thin lines are "short-circuited" by a broad wall, with a clear tendency to run parallel to the layers, as in Fig. 93a where it is almost straight and vertical, or slightly curved, as in Fig. 93b. It is evident that if the layers are not straight in the transverse section of the sample but folded, as in Fig. 92, then where we pass from an area with one fold direction to an area with the opposite direction there must be a defect region in between, and hence it is likely that the zigzags mediate such

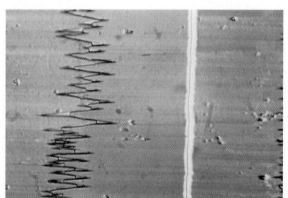

(a)

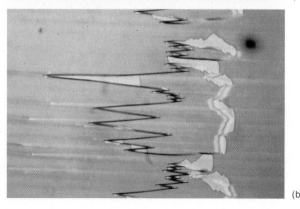

(b)

Figure 93. Appearance of zigzag walls in a smectic C structure with smectic layers initially (in the SmA phase) standing perpendicular to the confining glass plates, running in a vertical direction on the micrographs (thin lines to the left, broad walls to the right). (Courtesy of Monique Brunet, University of Montpellier, France).

changes. The question is how. This may be illustrated starting from Fig. 94. In (a) we show the typical aspect of a zigzag tip with γ being the small angle between the line and the layer normal (which is often the same as the rubbing direction). In (b) is shown the folded structure making chevrons of both directions, suggesting that at least two kinds of defect structure must be present. In the picture the fold is, as is most often the case, found in the middle of the sample – we will call this fold plane the chevron interface – although its depth may vary, as in (c). We will mainly restrict ourselves to the symmetric case (b) in the following discussion of the general aspects. As we will see, the chevron interface actually acts as a third surface, in fact the one with the most demanding and restrictive conditions on the local director and polarizations fields.

It is clear that the transition from one chevron to the other cannot take place abruptly in a layer as sketched in (d), as this would lead to a surface (2D) defect with no continuity of layer structure nor director field. Instead it is mediated by an unfolded part of width w (this region will be observed as a wall of thickness w) in (e), looked at from the top as when observing the sample

through a microscope. The wall thickness w is of the order of the sample thickness L. A more adequate representation of the three-dimensional structure is given in Fig. 95. The vertical part of the layer has the shape of a lozenge (in the general case, with unsymmetric chevrons it becomes a parallelogram). The sterically coupled fields $n(r)$ and $P(r)$ can now be mapped on this structure. Along the folds, $n(r)$ must simultaneously satisfy two cone conditions (i.e., to be on the cone belonging to a particular layer normal) indicated in the figure. This is always possible for $\delta \leq \theta$ and gives a limiting value for the chevron angle δ.

After the discovery of the chevron structures, a number of complexities in observed optical and electrooptical phenomena could be interpreted in these new terms. For a more thorough discussion in this matter, we refer to references [166–168]. Our emphasis will be on the important consequences for the physics due to the presence of chevrons, but even this requires dealing with at least some structural details. First of all, the chevron fold forces the director n to be in the interface, i.e., to be parallel to the plane of the sample in this region, regardless of how $n(r)$ may vary through the rest of the sample and independent of the surface conditions. Although the director n is continuous at the chevron interface, the local polarization field P cannot be, as shown in Fig. 96. It makes a jump in direction at the interface, nevertheless, in such a way that

$$\nabla \cdot P = 0 \qquad (403)$$

which means that the discontinuity is not connected with a build-up of local charge density. If we now apply an electric field between the electrodes, i.e., vertically as in (a), we see that there is an immediate torque because E and P are not collinear. Hence no nucleation is needed for switching from state 1 to state 2. Neither are fluctuations

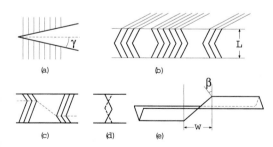

(a)   (b)

(c)   (d)   (e)

Figure 94. Thin zigzag lines run almost perpendicular to the smectic layers making a small angle γ with the layer normal as shown in (a). In (b) we look at the chevron folds in the plane of the sample, in (e) perpendicular to the plane of the sample. Asymmetric chevrons (c) cannot preserve anchoring conditions at both surfaces and are therefore less frequent.

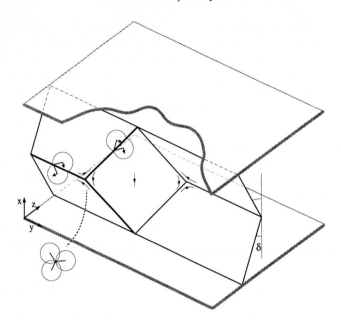

Figure 95. Section of a thin wall mediating the change in chevron direction. The layers in the chevron structure make the angle δ (the chevron angle) with the normal to the glass plates. Along a chevron fold where two surfaces meet, two cone conditions have to be satisfied simultaneously for n, which can be switched between two states. At the two points of the lozenge where three surfaces meet, three conditions have to be satisfied and n is then pinned, thus cannot be switched.

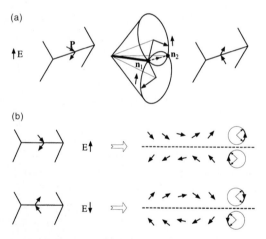

Figure 96. (a) At the chevron interface the local polarization P is discontinuous, making a jump in direction. (b) When switching from the down to the up state, P rotates everywhere anticlockwise above and clockwise below the chevron interface (time axis to the right). The director is locked in the chevron plane and can move between n_1 and n_2.

ish to a smaller value (midway between 1 and 2), which gives a contribution to the threshold for the process. Generally speaking, the bulk switching on either side of the chevron interface preceeds the switching in the interface. The latter contributes to the latching and thus to the bistability.

As seen from Fig. 96b, the switching process is unambiguous as regards the motion of n and P (sterically bound to n): on the upper side of the chevron, P rotates counterclockwise, on the lower side it rotates clockwise when we switch from 1 to 2; everything turns around in the reverse switching direction. This explains why there are no twist and antitwist domains like the ones observed in twisted nematics prior to the time when chiral dopants were added in order to promote a certain twist sense.

So far we have described the switching concentrating on the chevron interface, completely disregarding what could happen at the two bounding (electrode) surfaces. In fact, if the anchoring condition on the surfaces is very strong, switching between up and down states of polarization will only

needed for this. In other words, the switching process is not fluctuation-controlled. On the other hand, the chevron region requires an elastic deformation of the local cone in order to switch: the local tilt θ must dimin-

take place at the chevron interface. At high voltage this will more or less simultaneously take place in the whole sample. At low voltage it will be possible to observe the appearance of down domains as "holes" created in an up background, or vice versa, in the shape of so-called boat domains (see Fig. 105) in the chevron interface (easily localized to this plane by optical microscopy). The walls between up and down domains have the configuration of strength one (or 2π) disclinations in the P field.

It should be pointed out that the uniqueness of director rotation during the switching process is not a feature related to the chevron per se, but only to the fact that the chevron creates a certain P-tilt at the chevron interface. If the boundary conditions of the glass surfaces involved a similar P-tilt, this will have the same effect.

A glance at Fig. 97 reveals another important consequence of the chevron structure. As P is not along E (applied field) there will always be a torque $P \times E$ tending to straighten up the chevron to an almost upright direction. Especially in antiferroelectric liquid crystals, which are used with very high P values, this torque is sufficiently strong for almost any applied field, for instance, normal addressing pulses, to raise and keep the structure in a so-called quasi-bookshelf structure (QBS) under driving conditions. In ferroelectric liquid crystals,

presently with considerably lower P values, the same effect was previously employed to ameliorate contrast and threshold properties, by conditioning the chevron FLC to QBS FLC by the application of AC fields [169]. The effect on the switching threshold can be extracted from Fig. 96. When the chevron structure is straightened up, δ decreases and the two cones overlap more and more, leading to an increasing distance between 1 and 2, as well as further compression of the tilt angle θ in order to go between 1 and 2. The threshold thus increases, in agreement with the findings of the Philips (Eindhoven) group. On the other hand, this straightening up to QBS violates the conservation of smectic layer thickness d_C, which will lead to a breaking up of the layers in a direction perpendicular to the initial chevrons, thus causing a buckling out of the direction running perpendicular to the paper plane of Fig. 97.

2.8.4 Analog Grey Levels

As we just pointed out, in the chevron structure the polarization is no longer collinear with the external field. This can be used (for materials with a high value of P) to straighten up the chevron into a so-called quasi-bookshelf structure, combining some of the advantages from both types of structure. For instance, it can combine a high contrast with a continuous gray scale.

How to produce analog gray levels in an SSFLC display is perhaps not so evident, because the electrooptic effect offers two optical states, hence it is digital. Nevertheless, the shape of the hysteresis curve reveals that there must be small domains with a slightly varying threshold, in some analogy with the common ferromagnetic case. Normally, however, the flank of the curve is not sufficiently smeared out to be controlled and to

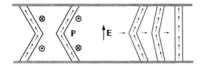

Figure 97. The fact that even after switching the polarization P is not entirely in the direction of the applied field will tend to raise the chevron structure into a more upright position, so decreasing the effective δ but breaking up the layers in a perpendicular direction. This gives a characteristic striped texture from the newly created, locked-in defect network.

accommodate more than a few levels. Curve I of Fig. 98 shows the transmission–voltage characteristics for a typical SSFLC cell with the layers in the chevron configuration [165]. The threshold voltage is fairly low, as well as the achievable transmission in the bright state, leading to a low brightness–contrast ratio. The position and sharpness of the threshold curve reflect the relatively large and constant chevron angle δ_0 in the sample. If a low frequency AC voltage of low amplitude (6–10 V) is applied, the smectic layers will be straightened up towards the vertical due to the $P-E$ coupling, so that the local polarization vector increases its component along the direction of the field. This field action, which requires a sufficiently high value of P, breaks the layer ordering in the plane of the sample and introduces new defect structures, which are seen invading the sample. The result is that the chevron angle δ is reduced, on average, and the threshold smeared out, as shown by curve II. Lower δ means a larger switching angle (and higher threshold), and thus higher transmission. Still higher transmission can be achieved by an additional treatment at a somewhat higher voltage (± 25 V), giving threshold curve III, corresponding to a new distribution around a lower δ-value and a microdomain texture on an even finer scale.

The actual switching threshold is a complicated quantity, not fully understood (no successful calculation has been presented so far), and usually expressed as a voltage–time area threshold for the switching pulse. For a given pulse length it is, however, reasonable that the amplitude threshold increases according to Fig. 98 when the average value of δ decreases. There are at least two reasons for this, as illustrated by Fig. 96. First, it is seen that the distance between the two positions n_1 and n_2 in the chevron kink level (which acts as a third,

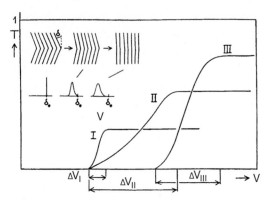

Figure 98. Amplitude-controlled gray scale in SSFLC. The chevron structure is transformed to a quasi-bookshelf (QBS) structure by external field treatment. In addition to giving gray shades, the QBS structure increases the brightness and the viewing angle. This method of producing gray levels was developed by the Philips (Eindhoven) group, who called it the "texture method".

internal surface), as well as the corresponding positions at the outer surfaces and in between, increase when δ decreases. It would therefore take a longer time to reach and pass the middle transitory state, after which the molecules would latch in their new position. In addition, it is seen that the local deformation of the cone i.e., a decrease of the tilt angle θ, which is necessary to actuate the transition from n_1 to n_2, increases when δ decreases. (A paradox feature of this deformation model is that it works as long as $\delta \neq 0$, whereas $\delta = 0$ gives no deformation at all – but also no chevron – at the chevron kink level.)

The smectic layer organization corresponding to curves II and III of Fig. 98 is generally characterized as a quasi-bookshelf (QBS) structure, denoting that the layers are essentially upright with only a small chevron angle. The QBS structure has a very large gray scale capacity. This might, however, possibly not be utilized to advantage in a passively driven display (as it can in the AFLC version). Its drawback in this respect is that the shape of the threshold curve is temperature-dependent, which leads to the requirement of a very well-controlled and constant temperature over the whole area of a large display. Furthermore, the QBS structure is a metastable state. Finally, the microdomain control of gray shades requires an additional sophistication in the electronic addressing: in order to achieve the same transmission level for a given applied amplitude, the inherent memory in the microdomains has to be deleted, which is done by a special blanking pulse. Using this pulse, the display is reset to the same starting condition before the writing pulse arrives. As a result of these features, it is not clear whether the microdomain method will be successfully applied in future FLC displays. A combination FLC–TFT might be required; if so it will be quite powerful.

2.8.5   Thin Walls and Thick Walls

The micrographs of Fig. 93 indicate that the thin wall normally runs with a certain angle γ to the smectic layer normal. What determines this angle γ, what determines the thickness of the wall, and what is the actual structure of it? To answer these questions we have to develop the reasoning started for Figs. 94 and 95. This is done in Fig. 99 where we look at the layer kink (the lozenge-shaped part in Fig. 95) from above. By only applying the condition of layer continuity, the following geometric relations are easily obtained, which describe the essential features of thin walls. Starting with Fig. 99a and remembering that the layer periodicity along the glass surface is d_A, we obtain a relation between γ and the layer kink angle β with the chevron angle δ as a parameter. With $d_C/d_A = \cos \delta$, this is written

ten

$$\frac{\cos(\beta - \gamma)}{\cos \gamma} = \cos \delta \tag{404}$$

For small angles we can solve either for β

$$\beta = \gamma + \sqrt{\gamma^2 + \delta^2} \tag{405}$$

or for γ

$$\gamma = \frac{1}{2} \beta - \frac{\delta^2}{2\beta} \tag{406}$$

This relation shows that γ increases monotonously with β. Thus the larger the smectic layer kink, the more the wall will run obliquely to the layer normal. The fact that there is a physical upper limit for β (see below) explains why we never observe zigzag walls with a large inclination, γ. For ease in computation, when we make estimations Eq. (406) can more conveniently be written

$$\gamma = \frac{(\beta + \delta)(\beta - \delta)}{2\beta} \tag{407}$$

Next we get the obvious relations between the width w of the wall and the width b of the lozenge or the length c of the cross section of the wall cut along the layer normal

$$w = b \cos(\beta - \gamma) = c \sin \gamma \tag{408}$$

as well as the width b of the lozenge from

$$c \tan \gamma = b \cos \delta \tag{409}$$

But this width b is also related to the kink angle and the sample thickness L (see Fig. 99c, d and Fig. 94e).

$$b \sin \beta = L \tan \delta \tag{410}$$

We may now use Eqs. (408), (410), and (404) to obtain the relation between the wall thickness and the sample thickness

$$w = L \frac{\cos \gamma \sin \delta}{\sin \beta} \tag{411}$$

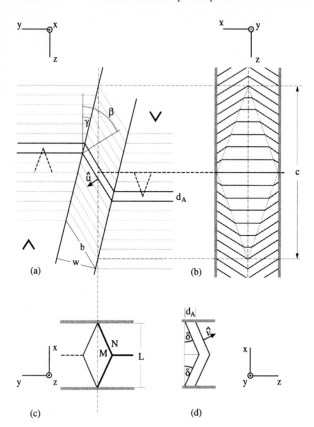

Figure 99. (a) Chevron structure like the one in Fig. 36, as seen from above. A thin wall of width w joins the region with a chevron bend in the negative z direction (left part of the figure) with another having the chevron bend in the positive z direction (right part of the figure). (b) Section through the wall cut along the z direction, (c) projection along the z direction, and (d) a cut parallel to the xz plane.

Finally, we can express the angle σ between the adjacent surfaces M and N in Fig. 40c by taking the scalar product of the corresponding layer normals $\boldsymbol{u}$ and $\boldsymbol{v}$ [marked in (a) and (d)]. This gives

$$\cos\sigma = \cos\beta \cos\delta \qquad (412)$$

Let us now extract the physical consequences contained in all these relations and start with the last one. As numerical examples we will use the data from Rieker and Clark [168] on the mixture W7–W82, for which the tilt angle θ saturates at about $21°$ at low temperature, and the corresponding saturation value for the chevron angle δ has been measured as $18°$. To begin with we see that Eq. (412) sets an upper limit to the layer kink β, because of the requirement of uniqueness of the director $\boldsymbol{n}$ at each chevron interface. Figure 96a illustrates the fact

that where the two inclined layers meet they can share the director only to a maximum layer inclination of up to twice the tilt angle θ. At this point the two cones touch, and beyond it no uniqueness in the director field can be maintained. Thus σ in Eq. (412) has a maximum value equal to 2θ, which gives a maximum value, in practice only somewhat lower than 2θ, for the allowable layer kink β. This, in turn, gives a limiting value for γ. With the data for W7–W82 we get $\sigma = 42°$, which gives (Eq. 412), $\beta \approx 39°$ and (Eq. 407) $\gamma \approx 15°$. The width of the wall in this case [Eq. (411)] is found to be roughly $0.5\,L$, i.e., half the sample thickness. This is thus the limiting case for an inclined zigzag in this material. Let us now assume that we observe another wall running at a $5°$ inclination to the layer normal. In this case Eq. (405) gives $\beta \approx 24°$ and Eq. (411) gives

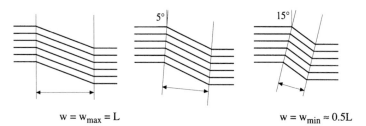

Figure 100. Thin walls with limiting cases, $\gamma=0$ and $\gamma=\gamma_{max}$ (in this case about $15°$). The width may vary by a factor of two, being of the order of the sample thickness L. The layer kink angle is at least equal to the chevron angle and at most slightly more than twice that value.

$w \approx 0.8\,L$. Finally, let us check a wall running exactly perpendicular to the layers ($\gamma=0$). In this case Eq. (404) shows β to be equal to δ, hence $\beta=18°$, and $w=L$ according to Eq. (411). This is the other limiting case with a wall of maximum width. Summarizing (see Fig. 100) we may say that a thin wall has its maximum width (equal to the sample thickness) when running perpendicular to the layers. The layer kink then has its minimum value, equal to δ, i.e., $\approx \theta$. Inclined walls are thinner, represent a higher energy density, and cease to exist when the layer kink approaches $\approx 2\,\theta$, which sets the maximum value of obliqueness to about $15-20°$.

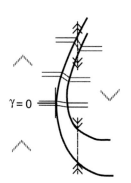

$\gamma = 0$

Figure 101. Thin wall curving until it finally runs parallel to the smectic layers, then being a broad wall. The character changes from $\lll\ggg$ to $\ggg\lll$ as soon as the wall reverses the inclination angle γ, with the angle counted as positive if it is in the same direction as the layer kink β; otherwise it is negative. That the sign of γ relative to the layer normal is irrelevant for the character can be seen if we mirror the picture along a vertical line: the mirror image must have the same character in every part of the wall.

Let us now imagine that the thin wall of Fig. 99, if we follow it downwards in the positive z direction, will curve so that γ changes sign, eventually to form a closed loop. Any cut along z will have the structure $\lll\ggg$ as long as $\gamma>0$, whereas this configuration changes to $\ggg\lll$ as soon as $\gamma<0$. This is illustrated in Fig. 101, where γ finally approaches $-90°$, which means that the wall runs parallel to the smectic layers. The thin wall has now turned to a thick wall, which thus mediates the transition $\ggg\lll$. With the nomenclature used so far, we can say that, while a thin wall in its course may change between $\lll\ggg$ and $\ggg\lll$, a thick wall can only have the structure $\ggg\lll$. Figure 102 explains why. In (a) we have assumed, for the sake of argument, the opposite configuration. The translation period at the boundary is d_A as before, and below it the layer can tilt in any direction to preserve this horizontal period while shrinking to layer thickness d_C. However, when we reach the lozenge, layer continuity would demand the layers to expand not only to $d>d_C$, but even to $d>d_A$. This means that this whole area would stand under great tension. Note that this is entirely different from the seemingly similar configuration in Fig. 99b. In that case the layers run obliquely to the cut, and hence the periodicity in this cut does not at all correspond to the layer thickness. In contrast, the layers in Fig. 102 run along the wall, and thus perpendicular to the cut. If we assume the $\ggg\lll$ struc-

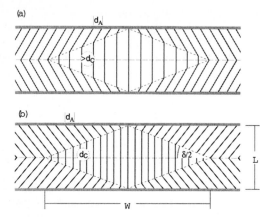

Figure 102. A thick wall in its relaxed state has a width W much larger than the sample thickness L. It can only have the structure >>><<< as shown in (b), not the one in (a). While the chevron angle δ changes with the smectic C layer thickness, the surface is assumed to maintain the pitch of the layers at the smectic A value, independent of the temperature.

ture, on the other hand, as in (b) we see that the thick wall can relax to an equilibrium configuration preserving the layer thickness everywhere. It is easy to see that this gives the equilibrium width W for the thick wall according to

$$\frac{L}{W} = \tan\frac{\delta}{2} \tag{413}$$

or, for small values of δ

$$W \approx \frac{2L}{\delta} \tag{414}$$

In a material with $\delta = 18°$ this would give an equilibrium thickness of 25 μm for a broad wall in a 4 μm thick sample. Thus a broad wall is much thicker than L, whereas a narrow wall is thinner than L. This certainly conforms very well with observation. Naturally, walls will also be observed that are not in equilibrium. If a thick wall is thicker than the equilibrium value, the smectic layers in the lozenge will be under tension, corresponding to a tilt $\theta < \theta_{eq}$. If it

is thinner, the layers will be under compression, corresponding to $\theta > \theta_{eq}$, and thus with a tilt that is larger than the equilibrium tilt angle. Because the layers are upright in the middle of a broad wall, the optical contrast between the two states is particularly high, and may be further enhanced if the width is smaller than the equilibrium value W corresponding to Eq. (413).

As is evident from Fig. 102b, a broad wall is thicker the smaller the chevron angle δ. This is also most clearly expressed by Eq. (414). In contrast, the width of a thin wall is not very much affected by δ. However, it is useful to take a closer look at its behavior with varying γ. For this, let us go back to Fig. 101. We have already seen that the character of the wall changes from <<<>>> to >>><<< when γ changes from >0 to <0, and also pointed out that a positive γ means that the wall's inclination relative to the layer normal in the region it traverses is in the same sense as the layer kink angle relative to that normal, while a negative γ means it is in the opposite sense. With this convention, all the relations (Eqs. 404 to 413) are valid with the appropriate sign inserted for γ. If we still take the material W7–W82 at a low temperature as an example, the Eqs. (405) and (413) will yield the value $\beta = 14°$ for the layer kink and the wall thickness $w = 1.3L$ when we insert $\gamma = -5°$. At $\gamma = -30°$, the layer kink has diminished to about $4°$ and the wall thickness increased to more than $3L$. On further increasing the inclination towards $-90°$, the angle β goes smoothly towards zero, while the width w goes towards its asymptotic value W given by Eq. (413), which in the present case gives $W \approx 6L$.

In Fig. 101 we have indicated how the kink flattens out continuously from its maximum value (at the maximum $\gamma > 0$) as the wall curves in the other direction, until the layers are perfectly flat when the wall runs parallel to the layers ($\gamma=-90°$). It has been customary to denote a wall in this limiting state a "thick wall", with the consequence that other walls, called "thin", can sometimes have the configuration <<<>>> and sometimes >>><<< in the cross section along the layer normal on either side of the wall. However, it would make much more sense to use the fact that the character of the wall changes when γ changes sign. Thus we propose the designation "thin" for $\gamma>0$ and "thick" for $\gamma<0$. This means that, unambiguously, a thin wall has a <<<>>> configuration only, and hence that any wall of configuration >>><<< should be called a thick wall. As a consequence, $w=L$ becomes the natural demarcation line between thick and thin walls: any wall with configuration <<<>>> has $w<L$, whereas any wall with configuration >>><<< has $w>L$.

The proposed criterion should not, and does not, depend on the chevron angle δ, and thus not on temperature. For example, in the same material W7–W82 at a higher temperature, such that $\delta=5°$, w still has the maximum width equal to L when $\gamma=0$ and attains its minimum width $\approx L/2$ when $\gamma\approx4°$ (γ_{max}). For $\gamma=-5°$ and $-30°$, $w\approx2.5\,L$ and $w\approx12\,L$, respectively.

This discussion may be briefly summarized by Figs. 103 and 104. In the first we have indicated a wall in the form of a closed loop, with the corresponding layer kinks and the chevron configuration. In particular, we have marked out that the inclination of the wall relative to the layer normal does not determine the sign of γ. The thin walls ultimately merging into a fine tip thus have the same character. Figure 99 gives an idea of what happens at the tip: when the walls

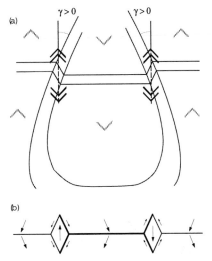

Figure 103. Inside a closed loop the chevron has a unique direction: (a) downwards or, (b) seen from below, towards the reader. As the lozenges connect two opposite tips in the chevron folds (b), we likewise have a uniquely determined kink direction in the wall. It is relative to this direction that we decide whether γ should be counted as positive (same sense) or negative (opposite sense). Thus note that $\gamma>0$ for both thin walls, even if they have geometrically opposite inclinations.

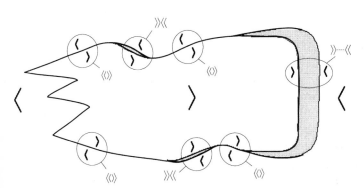

Figure 104. Zigzag wall making a closed loop. The smectic layers are assumed to be vertical in the picture. The chevron fold is always away from the zigzag inside and towards the thick wall. When the wall curves, the chevron character changes from <<<>>> for a thin wall ($w<L$) to >>><<< for a thick wall ($w>L$). The wall attains its maximum thickness (relaxed state) when it runs parallel to the layers.

merge, the kink becomes twice as large (2β), which requires β to be less then θ. For $\theta = 21°$, this gives a maximum value for γ of only 3° at the very tip. In fact, this narrowing of the tip angle can often be observed (see Fig. 93). In Fig. 104 finally, we have indicated how the character of the wall changes around a closed loop with zigzags when the wall curves in different directions. As for the chevron fold, inside the loop it can only be directed away from the zigzag and towards the thick wall.

2.8.6   C1 and C2 Chevrons

The presence of chevrons has numerous consequences for optics and electrooptics, above all for the transmission, contrast, bistability, and switching behavior. Some of these have already been hinted at earlier. The effects will be particularly drastic if we combine chevrons with polar boundary conditions that are very common, as most aligning materials favor one or the other direction for P (most often into and less frequently out of the liquid crystal). In contrast, the chevron interface represents a nonpolar boundary, by symmetry favoring neither the up nor the down state (see Fig. 96). These states are distinct and switchable as long as the chevron angle δ is less than the tilt angle θ of the director. If the limiting case ($\delta = \theta$) is reached, the two states merge into one and the chevron no longer contributes to the bistability.

In Fig. 103 we have indicated the P field distribution (one of many possible states) around two adjacent thin walls. The field is predominantly up in one of them and predominantly down in the other. Therefore alternate walls often serve as nucleation centers for domain switching taking place in the chevron plane, i.e., for the motion of walls between up and down domains. Such

domains often appear as boat-shaped "holes" of up state on a background of down state, or vice versa (see Fig. 105). On the other hand, the zigzags themselves do not normally move in an electric field, but are quite stationary. This is because zigzags normally separate regions with the same average P direction (up or down, as illustrated in Fig. 103b).

Also, the optical state (transmission, color) is very often practically the same on both sides of a zigzag wall, as in Fig. 93b. Indeed, if the director lies parallel to the surface (pretilt $\alpha = 0$) at the outer boundaries, the chevron looks exactly the same whether the layers fold to the right or to the left. However, if the boundary condition demands a certain pretilt $\alpha \neq 0$, as in Fig. 105, the two chevron structures are no longer identical. The director distribution across the cell now depends on whether the director at the boundary tilts in the same direction relative to the surface as does the cone axis, or whether the tilt is in the opposite direction. In the first case we say that the chevron has a C1 structure, in the second a C2 structure (see also Fig. 106). We may say that the C1 structure is "natural" in the sense that if the rubbing direction (r) is the same at both surfaces, so that the pretilt α is symmetrically inwards, the smectic layer has a natural tendency (already in the SmA phase) to fold accordingly. However, if less evident at first sight, the C2 structure is certainly possible, as demonstrated in Figs. 105 and 106.

The very important discovery of the two chevron structures and the subsequent analysis of them and their different substructures was made by the Canon team [170]. The presence of C1 and C2 as distinct structures has obvious consequences. Any wall now becomes a complete configuration consisting of one C1 and one C2 part. Its contrast changes with its running direction. A broad relaxed wall may be thick enough that its C1

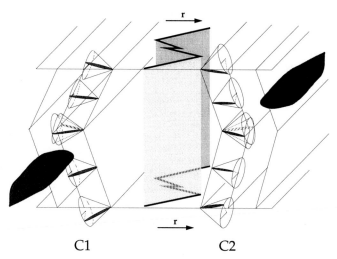

C1   C2

Figure 105. A zigzag street (lightning) running between two regions of opposite chevron fold for the case of nonzero pretilt α. This gives a different structure to the left (C1) and to the right (C2). The two regions then have different properties, like transmission, color contrast, switchability, etc., which get more pronounced the higher α is. Only when $\alpha=0$ are the C1 and C2 structures the same thing. For $\alpha \neq 0$ the director positions are different for C1 and C2, except in the chevron plane where $\boldsymbol{n}$ has to be horizontal. For the purpose of illustrating different possibilities, we have made α very small in the C1 case, whereas we have drawn the C2 case for medium α towards the upper, with high α towards the lower substrate. Thus the boundary conditions to the right and left to not correspond to the same surface treatment.

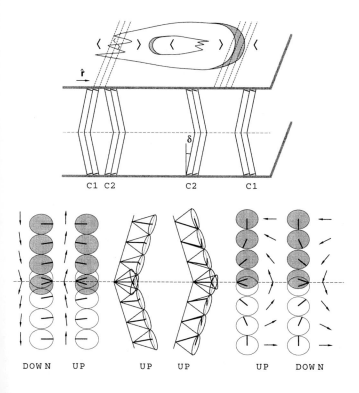

DOWN   UP   UP   UP   UP   DOWN

Figure 106. Closed zigzag loop with thin and thick walls and a smaller loop inside. Inside any loop the chevron always points towards the broad wall. With a polymer coating on the inner side of the glass plates the rubbing direction is indicated by the unit vector $\boldsymbol{r}$, giving the pretilt α to the inside. In the lower part of the figure, the C1 structure to the left and the C2 structure to the right of the thin wall are compared with their director and polarization fields in the case of switchable surfaces for C1 and nonswitchable surfaces for C2. The figure is drawn for a material with $P<0$, and for a case roughly corresponding to $\alpha + \delta = \theta$. Shaded cone surfaces belong to the upper chevron half.

and C2 parts, and even the small, almost pure bookshelf section in between, can be distinguished by different optical contrast. The regions representing opposite fold direction, separated by a wall, acquire different colors even if the chevron plane is in the middle, and so on. However, the optical differences are small for small pretilts of α.

If we take a nonzero pretilt into consideration, leading to C1 and C2 structures, thin and thick walls look as illustrated in Fig. 106. In the lower part of that figure we have also illustrated the quite important difference in the polarization and director fields across a C1 and across a C2 structure. It is hard to draw these figures to scale and yet demonstrate the characteristic features. In Fig. 106 the situation may roughly correspond to $\theta \approx 30°$, $\delta \approx 20°$, and $\alpha \approx 10°$, such that $\alpha + \delta \approx \theta$. We may note that the C2 structure is none other than the chevroned version of the surface-stabilized configuration already discussed for Fig. 90. The same C2 configuration $n - P$ would be found for any case with $\alpha + \delta = \theta$, for instance, with $\alpha = 3°$, $\delta = 15°$, and $\theta = 18°$, as long as $\delta < \theta$.

We mentioned that $\boldsymbol{\nabla} \cdot \boldsymbol{P} = 0$ at the chevron interface, which means that even if $\boldsymbol{P}$ changes direction abruptly, P_x (as well as any other component) is continuous across that surface (Fig. 96). As the boundary conditions at the substrates do not normally correspond to the director condition at the chevron, $\partial P_x / \partial x$ is nonzero between the chevron and the substrates, but is small enough to be ignored. That this is not true when we have polar boundary conditions, is shown in Fig. 107. Let us, for instance, assume that the polarization $\boldsymbol{P}$ prefers to be directed from the boundary into the liquid crystal. Whether we have a chevron or not, we will have a splay state $\boldsymbol{P}(x)$ corresponding to a splay – twist state in the director. With a chevron the splay is taking place in the upper or lower half of the cell when this is in its

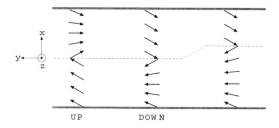

Figure 107. Polarization switching taking place in the chevron plane at fixed polar boundary conditions. In this case the condition is one of high pretilt α for $\boldsymbol{n}$ with the $\boldsymbol{P}$ vector pointing into the liquid crystal from the boundary.

up or down state, respectively. These half-splayed states, which are most often just called twist states, occur in both C1 and C2 structures and are then abbreviated C1T and C2T, in contrast to the uniform states C1U and C2U.

The twist states are normally bluish and cannot be brought to extinction, hence they give very low contrast in the two switchable states. Moreover, due to the different sign of $\boldsymbol{\nabla} \cdot \boldsymbol{P}$, there will be an unsymmetric charge distribution between adjacent areas switched to the up and the down state, tending to burn in any already written static picture. For high P_s materials, even the electrostatic energy has to be taken into account which, together with the elastic energy, may tend to shift the chevron plane to an unsymmetric state as illustrated to the right of Fig. 107, in order to relieve the high local energy density. Such a state is then completely unswitchable (monostable). It is evident that, whether we choose C1 or C2, a twisted state has to be avoided.

In their first analysis of C1 and C2 structures, Kanbe et al. [170] put forward the essential criteria for their stability. Both are allowed at low pretilt α. Whereas C1 can exist at high pretilts, the cone condition ($\boldsymbol{n}$ must be on the cone) cannot be fulfilled for C2 if α is larger than $(\theta - \delta)$, as illustrated in

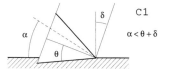

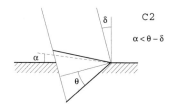

Figure 108. Stability conditions for chevrons of C1 and C2 type (after [170]).

Fig. 108. This means that for $\alpha \neq 0$ it cannot be fulfilled at the phase transition point SmA $\rightarrow$ SmC when θ and δ are both zero. Hence the chevron that is first created when the sample is cooled down to the SmC phase is always C1. However, C1 is not stable when the tilt angle θ increases. Therefore C2 appears together with C1 at lower temperatures and the sample is marred by zigzag defects. One of the reasons for this is that C1 has a higher splay–twist elastic deformation energy, which increases with increasing θ. Another reason is that the director is more parallel to the rubbing direction in the C2 case; it has to split more to fulfil the cone condiditon for C1. This effect will favor C2 under strong anchoring conditions. Finally, as pointed out in the cited Canon paper [170], while a gradual transition C1 $\rightarrow$ C2 to an almost chevron-free state might be observed on cooling, the once such-created C2 state tends to be quite stable if the temperature is subsequently raised; it is almost necessary to go back to the SmA* phase in order to recreate C1. Thus C2 is stable over a much broader range of temperature than C1. All this speaks for the C2 state, in spite of the considerably lower effective switching angle between its memorized states.

Several FLC projects have also been based on C2 (JOERS/Alvey, Thorn/CRL, Sharp). The fact that bistable switching only takes place through being latched by the chevron surface is compensated for by a number of advantages. First of all, the surface alignment is relatively simple, as the outer surfaces do not switch but only work together with the chevron. It is also much simpler to avoid twist states for C2 than for C1. This is easily seen from Fig. 106. Here the structure has been sketched for $\alpha \approx (\theta - \delta)$ such that the director is exactly along the rubbing direction and strongly anchored. As already pointed out, the nonchevron version of this is shown in Fig. 90. In practice it is found [171] that a medium pretilt α of about 5° is convenient for achieving this cone surface condition, whereas α values near zero give both C2U and C2T. For several materials, $(\theta - \delta)$ is of the order of about 5°; however, observations have been made [172] of chevron-free C2U structures for $\alpha = 5°$, although $(\theta - \delta)$ seems to be only about 1.5°, thus violating the stability condition for C2. Whether this is due to a change of δ under driving conditions (δ decreasing towards QBS case), is unclear.

A further advantage with the C2 structure is that zigzags induced by (light) mechanical shock, which causes a local transition to C1, easily heal out by themselves in the C2 structure [170], whereas the opposite is not true. Fortunately, C2 also has a substantially higher threshold to so-called boat–wake defects, which appear on the application of a very high voltage.

The main disadvantages with the C2 structure are obviously the low optical contrast in the true memorized state and the quite high voltages applied so far in the electronic addressing of the cell, when the so-called $\tau - V_{min}$ addressing modes are used. This is also combined with a high power consumption. Thus the switching pulse is

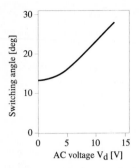

Figure 109. Effective switching angle in a C2 structure as a function of the amplitude V_d of the applied data pulses for the so-called Malvern-3 scheme. The switching pulse amplitude is 30 V. $V_d=0$ corresponds to the memorized states as latched by the chevron interface. The P_s value of the material is 4.4 nC/cm^2 (after 171).

used to accomplish latching in the chevron plane, whereas the bipolar AC data pulses are used to force the director as much as possible into the in-plane condition between the chevron surface and the outer surfaces (see the lower part of the right of Fig. 106). In this way the effective switching angle can be increased from a typical value of 14°, corresponding to the memorized state, to the more significant value of 24° (Fig. 109), when data pulses of 10 V amplitude are applied [171]. This higher switching angle thus corresponds to what we might call "forced memorized states".

2.8.7 The FLC Technology Developed by Canon

As indicated above, the first Canon prototype from 1988 had $\alpha \approx 0$ giving a hemispherical viewing angle, which was a complete novelty at that time. However, the contrast (~7:1) was only good enough for a monochrome display. Moreover, the C1 structure could not be made sufficiently stable at low temperatures. To solve this problem the team decided not to abandon C1

in favor of C2, but instead to pursue the much more difficult road to go to a very high pretilt. The idea was to squeeze out C2 completely from the material because its cone condition $\alpha < (\theta - \delta)$ can no longer be satisfied. Thus, by going to the rather extreme pretilt of 18° (a technological achievement in itself), Canon was able to produce zigzag-free structures, as the layers can only be folded to C1 chevrons. A considerable development of FLC materials also had to take place, among others, mixtures with $(\theta - \delta)$ values between 3 and 5°. In this way C2 could be successfully suppressed under static conditions, for instance, in a material with $\theta = 15°$ and $\delta = 10°$, which is not surprising, because α is not only larger than $(\theta - \delta)$ but is even larger than θ alone. However, it turned out [173] that C2 would appear to some extent under dynamic, i.e., driving conditions. This obviously cannot be explained by assuming that δ gets smaller under driving conditions. Thus, it seems that the stability criterion can hold at most under static conditions. This indicates that the underlying physics of the C1/C2 structure problem is extremely complex. In order to avoid C2 at any temperature, several parameters had to be optimized, such that both θ and δ showed only a small variation as functions of temperature.

The obvious advantage of working with C1 is that the switching angle is high, at least potentially, between the memorized states. This is particularly true if the cell switches not only at the chevron surface, but also at both outer boundaries. However, surface switching does not normally occur at polymer surfaces with a low pretilt (although it does at SiO-coated surfaces). This is another advantage of going to high pretilt. The situation may be illustrated as in Fig. 110. At high pretilt the surface state not only switches, but the switching angle may be just as large as permitted by the cone angle of the

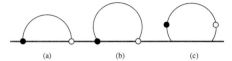

Figure 110. Switching angle under different conditions. (a) To the left QBS geometry is supposed. If the surface cannot be switched (for instance on polyimide), this will lead to a strongly twisted state. In the middle (b) a chevron geometry is assumed. If the surface is nonswitchable, it leads to a smaller twist than in (a), if it is switchable, the switching angle is less than in a switchable version of (a). With high pretilt (c) a polymer surface normally switches and the switching angle may be optimized, similar to the situation in (a).

material used. This cone angle is limited by the possible appearance of C2 at low temperatures and also by the fact that the twisted state C1T would appear at higher θ. Thus θ is limited, quite below the ideal performance value, in the C1 state as well as in the C2 state.

Whether the twisted state appears in the C1 case or not is, however, not so much determined by the tilt θ, but preeminently by the polar strength of the surface condition. It is therefore important to realize that this influence may be at least partly neutralized by a suitable cross-rubbing, i.e., obliquely to what is going to be the direction of the smectic layer normal. The idea is quite simple, but can give rise to a number of different interpretations, e.g., suppose that the lower surface demands the **P** vector to point into the liquid crystal. However, if **n** at this surface is tilted in a direction clockwise (seen from above) relative to the layer normal, then this **n** direction corresponds to the same direction of **P** (for $P<0$), namely, into the liquid crystal. Therefore rubbing in this direction will enhance the polar character of the surface [174]. Hence, if the surface is rubbed in the opposite angular direction, the polar character will be, at least partly, neu-

tralized. Canon, in fact, uses this kind of cross-rubbing in order to increase the symmetric bistability and avoid the twisted state [173, 175]. The rubbing directions deviate $\pm10°$ from the layer normal, roughly corresponding to the switching angle.

The basic geometry of rubbing, pretilt, etc., as applied to the actual Canon screen is illustrated in Fig. 111. While the contrast is far better in this version than in the 1988 version, the viewing angle – though still very good – has been compromised due to the fact that the local optic axis is pointing out of the flat surface by 18°. The one hundred million chevrons across the screen all have the same direction, making a completely defect-free surface. Even so, Fig. 112 reminds us that on a local scale the order has to be disturbed everywhere where spacers are localized.

As should be evident from this account, the development of FLC technology within the Canon group involved a lot of surprises. The group had to discover, and subsequently solve, a number of very complicated problems related to the polar nature of smectic C* materials. Among these problems were the appearance of ghost and shadow pictures (slow erasure of an already written image), and in addition to these was also the problem of mechanical fragility of the bookshelf or chevron layer configuration. The most complicated phenomenon, i.e., the strange electrodynamic flow (backflow net mass transport) that started to occur at high pretilt when electric fields were applied, could not even be superficially dealt with within the frame of this article. Already the materials problems were quite important. The hard to achieve high pretilt condition required several years of dedicated R&D in polymer materials. Finally, the group had to develop their own FLC materials and design manufacturing processes capable of handling large cells [48 in. (120 cm) diagonal]

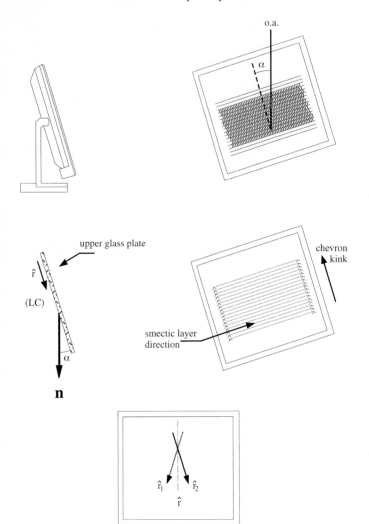

Figure 111. Geometric features in the Canon screen of Fig. 87 showing rubbing directions, smectic layer direction, chevron direction, and director pretilt α, equal to the optic axis tilt out of the screen surface.

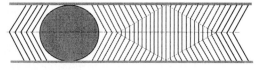

Figure 112. Every spacer represents a distortion, which by necessity traps a thick wall. In this picture the chevron kink is to the left everywhere except in the local neighborhood of the spacer (after [176]).

at a cell gap of 1.1 µm. Considering all this, the development and manufacturing of the present FLCD is a real tour de force performed by Kanbe and his team.

2.8.8   The Microdisplays of Displaytech

This is an interesting contrasting case, not only because it is the question of an extremely small display with completely different application areas, but also because the small size permits a different approach in almost every respect. In this case, the FLC layer is placed on a reflective backplane, which is a CMOS VLSI chip providing ± 2.5 V across each picture element [177]. This is thus a case of active driving. Therefore the question of the quality of the memorized states is unimportant, and we can safely use the C2 structure as the one with the simplest aligning technology avoiding any twisted state. (Displaytech uses rubbed nylon, which gives a low pretilt.) Because each pixel is fully driven and not multiplexed, we do not have the disadvantage of high voltage. We may recall the peculiarity of FLC switching in that the threshold is one of voltage – time area. Thus when the switching pulse time is replaced by the frame address time, which is now the relevant time for the applied voltage, the threshold in voltage becomes correspondingly low and well within the CMOS range. This is of course characteristic for all direct or "active" FLC driving. Furthermore, because each pixel is in its fully driven state, we have in-plane switching with its characteristic large viewing angle.

A transistor can in principle be used for implementing microdomain grey levels by charge control (a method pioneered by the Philips Eindhoven group), but in the Displaytech case it is only used to write one of two states. The gray levels are instead produced by time modulation, which is quite natural for direct drive. Video color pictures are produced at a frame rate of 76 Hz. Each frame is, however, subdivided into three pictures, one for each color, so that the actual frame rate is 228 Hz, under the sequential illumination from red, green, and blue LEDs. (The LEDs are GaN for blue and green, and AlInGaP for red.) During each of the 4.38 ms long color sequences (corresponding to 228 Hz), a gray level is now defined for every pixel by rapidly repeating the scanning of the FLC matrix a number of times, with a pixel being on or off for the fraction of time corresponding to the desired level. For instance, if the VLSI matrix is scanned $2^5 = 32$ times during illumination of one color, then five bits of gray per color are

achieved. This requires a scanning time (basic frame rate) of little less than 150 μs for a matrix consisting of 1000 lines. Displaytech has presented several versions of this miniature screen, for instance, one 7.7 by 7.7 mm with 256×256 picture elements and an aperture ratio of 90%, and some with higher resolution of up to 1280×1024. In the latter case the dimensions are 9.7 by 7.8 mm with individual pixel size of 7.5×7.5 μm. The display is viewed through a magnifying glass and may be headmounted. Obvious applications are virtual reality, high resolution viewers, but also projection displays which are good candidates for high definition television (HDTV) using FLC. The first version of this "Chronocolor" display had two bits per color giving 64 colors per pixel; the second has three bits giving 512 colors. The version now under development will give six bits per color corresponding to $(2^6)^3 = 262\,144$ colors per pixel, i.e., essentially full color. The first version is the one illustrated in Fig. 88.

2.8.9   Idemitsu's Polymer FLC

In 1985, around the same time as Canon started their FLC development project, another Japanese company, Idemitsu Kosan, started to synthesize materials with the goal of making a polymer version of FLC. There is a conflict in a liquid crystal polymer in the sense that a long polymer chain wants to be as disordered as possible, while the liquid crystal monomers want to align parallel to a local director. Nevertheless, it is possible to make both nematic and smectic structures and, in particular, chiral materials having an SmA* – SmC* transition. In one of Idemitsu's successful realizations, the main chain is a siloxane to which side group monomers (which would be normal FLC materials by themselves) are attached via spacing groups. Whereas a nematic polymer reacts too slowly to electric fields to be of any electrooptic interest, the internal cone motion can be activated in the polymer SmC* phase, with its characteristic low viscosity γ_φ giving a (relatively speaking) very fast switching. At high temperatures the switching time can go below a millisecond, i.e., the FLC polymer (FLCP) can be faster than a typical monomer nematic; however, the viscosity increases much more rapidly at low temperatures than in a nematic.

The FLCPs have many similarities to the monomer materials. By measuring the polarization reversal current [177] Idemitsu found that the P_s value is about the same (typically a little larger) as in the corresponding monomer. As the layer spacing decreases with decreasing tilt angle in the SmC* phase, this also results in chevron structures. In thick samples the smectic helix is present. In thin QBS cells FLCPs show good bistability, and so on. The switching mechanism is in principle a little more complicated in the aspect that the main chain to some extent takes part in the switching. A simplifying aspect is that the polymerization may stabilize the SmC* phase, which may then have a very broad temperature interval (something that in the monomer case has to be made by a multicomponent mixture). The polymer state also seems to stabilize a major chevron direction such that zigzag defects are not so important. Nor has a QBS cell in the polymer version the mechanical fragility that is so characteristic of monomer FLCs.

On the other hand, the differences are highly interesting. An FLCP display is flexible, light-weight, and only 0.5 mm thick. It is made by spreading the polymer onto an ITO-coated plastic substrate, which is then laminated in a continuous process to a second substrate (see Fig. 113) using rollers, one of which in its bending action shear-

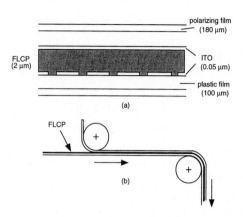

Figure 113. (a) Cross-sectional view of the all-plastic 0.5 mm thick Idemitsu FLCP display showing the SmC* polymer (black) between the substrates. (b) This shows how in its preparation the display is first laminated and then shear-aligned (after [178]).

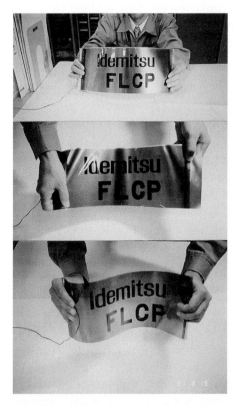

Figure 114. The FLCP display can be bent and shaped to a different curvature. The bottom picture also demonstrates the transmissive mode. (Courtesy of Idemitsu Kosan Co., Ltd.).

Figure 115. Two examples of simple color FLCP panels. The sizes are 20 cm×32 cm and 12 cm×60 cm, respectively. The color is produced by vertical color stripes giving eight colors per pixel. These 0.5 mm thin displays are mechanically durable and work at very low power thanks to the inherent memory. The picture can be changed twice a second. (Courtesy of Idemitsu Kosan Co., Ltd.).

aligns the smectic. The display thus needs neither spacers nor alignment layers, and can be made very large, in principle "by the meter". Its flexibility is demonstrated in Fig. 114. Evident future applications for such panels that can be bent and are extremely lightweight are, for instance, in electrically controllable motorcycle or welding goggles. However, as already demonstrated in Fig. 89, a switching time of 1 ms at room temperature is also sufficient for fairly sophisticated static displays. Although the switching is between a hundred and a thousand times slower than in a monomer FLC, 1 ms is sufficient to give this display an update rate of 2 Hz. Two examples of simple large size color panels are given in Fig. 115.

2.8.10 Material Problems in FLC Technology

FLC materials are complex and we are just at the beginning of exploring their physics and chemistry. As we have mentioned, they have nine independent elastic and 20 viscous coefficients, to which we may add that there are 15 different independent sources of local polarization (10 if we assume incompressible layers). One of them is related to chirality, whereas all the others are flexoelectric coefficients (thus equally relevant to the nonchiral SmC phase) describing local polarization resulting from the different deformations in the director field. With a local polarization vector and with complicated smectic layer structures, it is evident that we will also have an extremely complicated hydrodynamic coupling to an external electric field E. These hints may suffice to indicate the completely new level of complexity in these materials. However, already the fact that they are not three-dimensional liquids like nematics, but rather two-dimensional liquids with solid-like properties along one space dimension, makes them extremely fragile in their useful geometry – a pressure by the thumb is in practice sufficient to irreversible destroy the ordered bookshelf structure. On the other hand, the wealth of physical phenomena and the richness of potential effects to be exploited in future in these materials should be recognized.

The molecular engineering of chiral tilted smectics is correspondingly complicated with a large number of parameters to optimize. In Table 1 we have listed the most important (so far recognized), in comparison with the standard nematic parameters. Beside phase sequence and sufficiently large phase range (not minor problems), we have to be able to control all the parameters with which we have already made acquaintance. The elastic constants can be taken as an example illustrating the relative progress on the way to full control of the materials in the nematic and in the smectic case. Whereas in nematics we can balance and adjust the three K_{ij} values to suit any particular application, the relevant elastic constants have hardly been identified yet in smectics, and even less measured – there have been much more serious problems to take care of first, for instance, the matching temperature behavior of θ and δ (layer tilt) mentioned earlier in the description of the Canon C1 technology. The most serious problem was initially recognized alongside the growing understanding of the relation between zigzag defects and chevrons and is connected with the parameter $d(T)$ listed earlier, i.e.,

the smectic layer thickness as a function of temperature. And here is the challenge: to synthesize and mix materials to form a smectic C that keeps its layer thickness constant in spite of an increasing tilt angle θ below the SmA – SmC transition. This would revolutionize the further development of FLC technology by avoiding all complica-

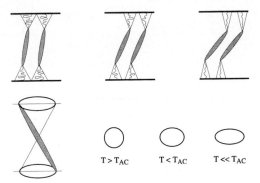

Figure 116. Two possible mechanisms for having a tilted smectic phase with a temperature-independent smectic layer thickness below the SmA – SmC transition. In (a) the end chains are first disordered but get more straight as the temperature is lowered; in (b) each individual molecule keeps its tilt constant relative to the layer normal, but the azimuthal direction of the tilt is unbiased in the SmA phase. At the SmA – SmC transition the average direction of the molecules begins to get biased.

Table 1. Parameters for nematics and smectics C*.

Material requirements	Nematics	Smectics C*
Phase range	(Cr)–N–(I)	(Cr)–SmC–(SmA) (SmC)–SmA–(N) (SmA)–N–(I)
Preferred phase sequence		N–SmA–SmC
Birefringence	Δn	Δn
Dielectric anisotropy	$\Delta\varepsilon$	$\Delta\varepsilon$
Dielectric biaxiality		$\partial\varepsilon$
Twist viscosity	γ_1	γ_φ
Elastic constants	K_{ij}	K_{ij}, B_{ij}
Tilt angle		θ
Spontaneous polarization		P
Smectic C* pitch		Z
Cholesteric pitch	(p)	p
Smectic layer thickness		$d(T)$

tions with chevrons and finally achieving the initially conceived structure of Fig. 61, with superior brightness, contrast, and viewing angle. That such materials could even be imagined to exist is illustrated in Fig. 116. In this very simplistic picture (which may or may not be realistic), let us assume that (a), as the central aromatic part (essentially representing the optic axis of the molecule, and therefore θ) tilts more and more away from the layer normal on lowering the temperature, the aliphatic end chains get less disordered and stiffen, making them effectively longer. Various degrees of interdigitation (decreasing with higher tilt) of molecules in the layer could also be imagined for materials in which the layer corresponds to more than the length of one

However, now and then, evidence has been reported indicating that some materials may behave like this. They would not only be of eminent interest for FLC applications, but just as much for use as electroclinic materials.

2.8.11   Nonchevron Structures

At the Japan Display 1989 Conference in Kyoto, two contributions independently reported on new materials which seemed to almost automatically give a defect-free alignment, avoiding the zigzag defect. The first one, a collaboration between Fujitsu Laboratories, Mitsui Toatsu Chemicals, and Tokyo Institute of Technology [179] presented mixtures based of the naphthalene structure

with $n=9$ and 10 and $m=2-6$. The second contribution [180] was from the central research laboratories of the company 3M in St. Paul, Minnesota, U. S. A., and showed the result of fluorination on a number of more conventional liquid crystal structures. The naphthalene compounds have since been used in research as well as in some Fujitsu FLC display prototypes. They are highly interesting, but suffer from a rather high viscosity.

molecule. A different mechanism (b) goes back to early ideas of the late crystallographer de Vries at Kent State University, who proposed that the molecules could already be tilted in the smectic A phase, but in an unbiased rotation making the phase uniaxial [191a]. At the SmA – SmC transition, a rotational bias appears, growing stronger on lowering the temperature, which is equivalent to a tilt of the optic axis. This tilt would then appear without being accompanied by a decrease of the layer thickness.

Today we know that the majority of liquid crystals undergoing an SmA – SmC transition do not behave according to the de Vries model. This would seem to require a very low degree of interaction between the constituents from layer to layer, or even within layers, i.e., a high degree of randomness, yet connected with smectic order.

In the 3M research the starting point was 4'-alkyloxyphenyl 4-alkyloxybenzoates

or corresponding thiocompounds

which were partially fluorinated to structures like

$C_n F_{2n+1} CH_2 O$ —⟨benzene ring⟩— C(=O) — O —⟨benzene ring⟩— O —

$C_n F_{2n+1} CH_2 O$ —⟨benzene ring⟩— C(=O) — O —⟨benzene ring⟩— O —

—O —⟨benzene ring⟩— C(=O) — O —⟨benzene ring⟩— O CH$_2$ C$_n$ F$_{2n+1}$

This partial fluorination has a number of interesting and even surprising effects [181]. First, it almost always totally suppresses the nematic phase and strongly enhances the smectic phases, in general both smectic A and smectic C. Moreover, it lowers the birefringence of the liquid crystal: Δn may go from typically 0.20 to 0.07 or 0.10. It also makes the smectic C* phase less hardtwisted: the helical pitch may increase from about 2 µm to 5–10 µm.

Another important effect is that it makes the smectic C phase occur at a shorter chain length than in the unfluorinated analogs. This means that fluorinated SmC* is faster used as FLC material; it has remarkably low viscosity. But first and foremost: these materials do not make chevrons. This means that the smectic C layer thickness must be essentially temperature-independent.

Partial fluorination leads to large terminal dipole effects. The fluorocarbon tail is also stiffer than the hydrocarbon tail. It is therefore relatively easy to understand the suppression of the nematic phase and strengthening of layered phases. However, it is less clear, as also in the naphthalene case, how to explain why the layer spacing should be temperature-independent. In both materials there seems to be a tendency for molecular interdigitation, and thus the smectic layer is thicker than the molecular

length. This may be important, but we still have no precise molecular interpretation of the resulting effect.

The 3M synthetic research in SmC* materials began in 1983 and has been pursued with a high degree of purposefulness. (It is thus not true that only Japanese companies invest in long-term research goals.) A number of pure compounds have been developed that exhibit almost continuous layer expansion on lowering the temperature in the smectic C phase, as well as in the smectic A phase. By mixing with other compounds, a practically temperature-independent spacing can be achieved. In addition to phenyl benzoate cores, 3M has largely used phenylpyrimidine cores, of which a general structure can be exemplified as [182]

$R'' — (C)_j — (O)_j —$⟨pyrimidine-phenyl⟩$— OCH_2(C_xF_{2x}O)_zC_yF_{2y+1}$

As an example of how different materials behave, we compare their smectic layer spacing as a function of temperature in Fig. 117. For the Chisso-1013 mixture the

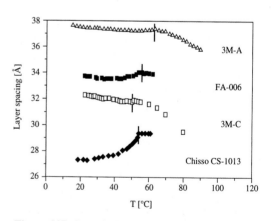

Figure 117. Smectic layer spacing for a number of different FLC materials as a function of temperature. The SmA–SmC transition is indicated by a vertical bar. Chisso CS-1013 is a mixture showing conventional behavior, FA-006 is a Fujitsu naphthalene-based mixture, 3M-C is the pure pyrimidine compound with formula, whereas 3M-A is a mixture (after [183]).

decrease in the layer thickness is pronounced as soon as we enter the smectic C
phase. The naphthalene mixture FA-006 has
a behavior that is qualitatively similar,
though much less dramatic. The fluorinated
compound 3M-C, of formula

$$C_8H_{17} - \text{(ring)} - \text{(ring)} - OCH_2CF_2 - OC_2F_4 - OC_4F_9$$

shows the opposite behavior, with a small,
but characteristic negative layer expansion
coefficient, and the mixture 3M-A is qualitatively the same with a weaker temperature dependence.

Although the 3M materials lack a nematic phase and thus have the phase sequence
I–SmA–SmC, they align fairly well on nylon 6/6 to a bookshelf structure with a small
pretilt at the surface. They switch symmetrically according to Fig. 51 and it has been
confirmed that the switching angle corresponds to the full cone angle. In principle,
therefore, with a θ optimized to about 22°
they should have memorized states corresponding to the ideal switching angle of 45°
(in comparison see Fig. 109, illustrating the
"forced" C2 switching states). Although it
must be a particularly hard task to develop
such materials with all other parameters optimized, it seems that they are now very
close to being used in FLC panels. Figure 118 shows a Canon prototype (not yet
commercialized) which evidently is the first
attempt to put chevron-free materials into a
display. The lower birefringence of these
materials has permitted the enlargement of
the cell gap from 1.1 μm to 2.0 μm, which
is an important advantage in manufacturing. The display has an XGA resolution
(1024×768) where every pixel is 300 μm
×300 μm and consists of three RGB stripes,
each of which is subdivided into four dots.
This gives 16 different states per color and
4096 different color shades per pixel. This

"digital full color" display is interesting because it demonstrates that a binary technology (based on just two states, i.e., without
any gray shade whatsoever) is capable of
quite satisfactory rendition of full color pictures. It is only a question of resolution, as
in newspapers and magazines.

The materials development is far from
finished in the case of chiral smectic materials. The observation of highest importance
for smectic C materials now relates to the
property of a negative layer expansion coefficient δ_d. In order to significantly increase the useful temperature range for high
performance FLC displays, especially the
lower end (for shipping by air a SmC*
phase stable down to –40 °C is desirable)
we have to increase the available pool of
molecular structures with this property of
negative γ_d. This is an important task for
synthetic chemists.

With the most critical problems now being understood or even solved, several
groups are attacking the problem of sufficient depth in the rendition of gray levels.
At least this could relatively soon be ap-
proached by going to a higher resolution, of
which FLC is eminently capable.

Another approach is active matrix driving, which uses charge control, and then can
suppress the use of a high resolution color
filter by use of color sequential back-
lighting. Other methods have been proposed
which use amplitude control in the same
kind of symmetric driving as known from
antiferroelectric displays [184], eventually
in combination with color sequential back-
lighting of high P_s QBS structures [185]. Finally, special addressing methods are being
worked out using a number of different
physical principles (for instance, resistance
electrodes). Some of the most interesting of
these are certainly of the kind represented
by the so-called binary group addressing
method [186, 187], combining a time-dith-

Figure 118. Canon's 1997 "digital full color" prototype is important in its demonstration that color pictures can be produced utilizing only two electrooptic states. The screen is here shown with two successive close-ups, the last of which with sufficient magnification to show the individual pixels. (Courtesy of B. Stebler.) The SSFLC structure in this panel is nonchevron and corresponds to Fig. 61.

ered gray scale with a substantial extension of the illumination period to nearly over the entire frame. Once the FLC devices are on the market, we may expect a diversity of techniques and a rapid development into different directions.

2.9   Is There a Future for Smectic Materials?

The driving force for academic research is not only the generation of new knowledge, but also the development of industrial applications, even if these applications may lie far in the future. In liquid crystals it is evident that the industrial aspects are dominated by displays. The fact that displays are the most severe bottleneck in the present-day information technology motivates the enormous investments in this industry.

The liquid crystal display market is today entirely dominated by nematic materials, mainly in form of twisted nematics for simple displays, supertwisted nematics, and, in particular, twisted nematics in combination with thin film transistor arrays for very sophisticated panels. During the last two decades, where practically everything has happened in the development of smectic display technologies, the nematic key technologies have constantly become better, and huge industrial investments have now been made which lock the development into a fixed direction for a very long time. Especially for those who extrapolate present trends, it is therefore doubtful whether there is any future for smectic materials at all, in spite of the enormous potential in the form of much higher speed, brightness, color neutrality, ideal viewing angle, memory, and so on, which these materials have already demonstrated at the same time as they have shown to be very difficult to handle. The situation has some similarity to the one in the semiconductor industry where gallium arsenide and similar materials, in spite of their much higher speed (in particular), have a hard time entering a marked completely dominated by classic and much simpler silicon technology, which by huge industrial investments is constantly becoming better and, in fact, always seems to be good enough for the present demands.

As always in history, the few pioneers, especially in Japan, have found a number of unexpected problems and difficulties in smectic materials that set back the development and which just have to be solved in order for the development to proceed. Regarding the serious nature and complexity of these problems, it is indeed remarkable what has been achieved, and the advent of the first FLC panels and also of AFLC prototypes in the last few years were important events. Moreover, the all-polymer FLC displays simply have no counterparts in the nematic domain. At the same time, it is clear that the chemistry that has been so instrumental for the perfection of the nematic technology has not yet, by far, reached the corresponding level of refinement for smectic materials. First of all this has to do with the simple fact that the materials development started much later for smectics, at the same time as it was much less intense. However, the task is also more complicated by the very rich polymorphism found in smectics, and it is harder to implement the requirements set by industry. For instance, it is generally harder to phase-stabilize any smectic such that it meets the requirements of storage and transport ($-40\,°C$ to $90\,°C$) and not only for the operation temperature interval, and this is still, to some extent, a limiting factor. On the other hand, there are yet many exciting discoveries to be made by synthetic chemists working with smectic materials, considering the richness in interactions when we go to the more complicated molecules that organize in smectic layers. There is no doubt that, in particular, Japanese industrial chemists have been very successful in this more speculative chemistry in the last decade, and their innovative activity will be very important for the future use of smectic materials.

Smectic materials have already demonstrated their potential use in very large [15–24 in. (37.5–60 cm)] flat panels with very high resolution, a potential that nematic materials do not have if not used in combination with thin film transistors. All smectic materials essentially use the same in-plane switching mechanisms, the cone mode or soft mode, based on the very fragile bookshelf geometry. This has long been considered a serious obstacle, but today well-known methods exist in spacer technology for making such displays mechanically rigid. With this and some similar problems eliminated, what we are probably going to see is not only powerful smectic displays working in passive matrix mode, but also a merging of the TFT technology and the smectic technology in panels of yet unachieved quality and sophistication.

2.10   References

[1] J. Valasek, *Phys. Rev.* **1920**, *15*, 537; **1920**, *17*, 475.
[2] K. Deguchi in Landolt-Börnstein, New Series, Vol III/28, Springer, Berlin **1990**.
[3] J. Fousek, *Ferroelectrics* **1991**, *113*, 3–20.
[4] *The New Encyclopedia Britannia*, Vol. 9, 15th ed., Micropedia, **1995**, p. 825.
[5] R. Williams, *J. Chem. Phys.* **1963**, *39*, 384.
[6] A. P. Kapustin, L. K. Vistin, *Kristallografiya* **1965**, *10*, 118.
[7] R. Williams, G. H. Heilmeier, *J. Chem. Phys.* **1966**, *44*, 638.
[8] P. G. de Gennes, J. Prost, *The Physics of Liquid Crystals*, Clarendon, Oxford **1993**.
[9] L. M. Blinov, V. G. Chigrinov, *Electrooptic Effects in Liquid Crystal Materials*. Springer, New York **1994**.
[10] M. E. Lines, A. M. Glass, *Principles and Applications of Ferroelectrics and Related Materials* Clarendon, Oxford **1977**.
[11] M. E. Huth, Dissertation, University of Halle **1909**; D. Vorländer, M. E. Huth, *Z. Phys. Chem. A* **1911**, *75*, 641.
[12] K. Hermann, *Trans. Faraday Soc.* **1933**, *29*, 972.
[13] A. Saupe, *Mol. Cryst. Liq. Cryst.* **1969**, *7*, 59–74.
[14] G. W. Gray, *Mol. Cryst. Liq. Cryst.* **1969**, *7*, 127.
[15] M. Leclercq, J. Billard, J. Jacques, *C. R. C* **1968**, *266*, 654.
[16] W. Helfrich, C. S. Oh, *Mol. Cryst. Liq. Cryst.* **1971**, *14*, 289–292.
[17] W. L. McMillan, *Phys. Rev. A* **1973**, *8*, 1921–1929.
[18] R. B. Meyer, *Phys. Rev. Lett.* **1969**, *22*, 918–921.
[19] P. G. de Gennes, *The Physics of Liquid Crystals*, Clarendon, Oxford **1974**, p 97.
[20] S. Garoff, R. B. Meyer in *6th Int. Liquid Crystal Conference*, Kent, OH, August 1976, Conference Abstracts, **1976**, Paper (B1–11).
[21] S. Garoff, R. B. Meyer, *Phys. Rev. Lett.* **1977**, *38*, 848–851; S. Garoff, R. B. Meyer, *Phys. Rev. A* **1979**, *19*, 338–347.
[22] W. G. Cady, *Piezoelectricity*, McGraw-Hill, New York **1946**; reprinted by Dover Publications, New York **1964**.
[23] C. Kittel, *Introduction to Solid State Physics*, Wiley, New York **1956** and later editions, Chap. 13.
[24] J. F. Nye, *Physical Properties of Crystals*, Oxford University Press, Oxford **1957**.
[25] A. J. Dekker, *Solid State Physics*, Prentice-Hall, Englewood Cliffs, NJ **1957**, Chap. 8.
[26] W. Känzig, *Solid State Physics Vol 4*, Academic, New York **1957**, pp. 1–197.
[27] F. Jona, G. Shirane, *Ferroelectric Crystals*, Pergamon, New York **1962**; reprinted by Dover Publications, New York **1993**.
[28] G. S. Zhdanov, *Crystal Physics*, Academic, New York **1965**.
[29] I. S. Zheludev, *Physics of Crystalline Dielectrics*, Plenum, New York **1971**.
[30] R. Blinc, B. Žekš, *Soft Modes in Ferroelectrics and Antiferroelectrics*, North-Holland, Amsterdam **1974**.
[31] S. B. Lang, *Sourcebook of Pyroelectricity*, Gordon and Breach, New York **1974**.
[32] T. Mitsui, I. Tatsuzaki, E. Nakamura, *An Introduction to the Physics of Ferroelectrics*, Gordon and Breach, New York **1976**.
[33] J. C. Burfoot, G. W. Taylor, *Polar Dielectrics and their Applications*, Macmillan, London **1979**.
[34] G. W. Taylor, J. J. Gagnepain, T. R. Meeker, T. Nakamura, L. A. Shuvalov (Eds.), *Piezoelectricity*, Gordon and Breach, New York **1981**.
[35] G. A. Smolenskii, V. A. Bokov, V. A. Isupov, N. N. Krainik, R. E. Pasynkov, A. I. Sokolov, *Ferroelectrics and Related Materials*, Gordon and Breach, New York **1984**.
[36] L. A. Shuvalov (Ed.), *Modern Crystallography*, Volume IV, Springer, Heidelberg **1988**.
[37] T. Ikeda, *Fundamental of Piezoelectricity*, Clarendon, Oxford **1990**.
[38] P. G. de Gennes, *The Physics of Liquid Crystals*, Clarendon, Oxford **1974**.
[39] G. W. Gray, J. W. G. Goodby, *Smectic Liquid Crystals*, Leonard Hill, Glasgow **1984**.
[40] S. A. Pikin, *Structural Transformations in Liquid Crystals*, Gordon and Breach, New York **1991**, Chap. V.
[41] J. W. Goodby, R-Blinc, N. A. Clark, S. T. Lagerwall, M. A. Osipov, S. A. Pikin, T. Sakurai, K. Yoshino, B. Žekš, *Ferroelectric Liquid Crystals, Principles, Properties and Applications*, Gordon and Breach, Philadelphia, PA **1991**.
[42] S. Chandrasekhar, *Liquid Crystals*, 2nd ed., Cambridge University Press, Cambridge **1992**, Chap. 5.
[43] R. B. Meyer, *Mol. Cryst. Liq. Cryst.* **1977**, *40*, 33–48.

[44] G. Durand, P. Martinot-Lagarde, *Ferroelectrics* **1980**, *24*, 89–97.

[45] L. M. Blinov, L. A. Beresnev, *Sov. Phys. Usp.* **1984**, *27*, 492–514.

[46] S. T. Lagerwall, I. Dahl, *Mol. Cryst. Liq. Cryst.* **1984**, *114*, 151–187.

[47] L. A. Beresnev, L. M. Blinov, M. A. Osipov, S. A. Pikin, *Mol. Cryst. Liq. Cryst.* **1988**, *158A* (Special Topics XXIX), 3–150.

[48] S. T. Lagerwall, N. A. Clark, J. Dijon, J. F. Clerc, *Ferroelectrics* **1989**, *94*, 3–62.

[49] J. Dijon in *Liquid Crystals, Applications and Uses*, Vol. 1 (Ed.: B. Bahadur), World Scientific, Singapore **1990**, pp. 305–360.

[50] R. Blinc, Condensed Matter News, **1991**, *1*, 17–23.

[51] S. T. Lagerwall, *J. Phys. Condensed Matter* **1996**, *8*, 9143–9166.

[52] H. P. Hinov, *Mol. Cryst. Liq. Cryst. Bull.* **1990**, *5*, 315–484. A bibliography for 1970–1989. (With authors in alphabetical order and a subject index.)

[53] *Ferroelectrics* **1984**, *58–59*.

[54] *Ferroelectrics* **1988**, *84–85*.

[55] *Ferroelectrics* **1991**, *113–114*.

[56] *Ferroelectrics* **1991**, *121–122*.

[57] *Ferroelectrics* **1993**, *147–149*.

[58] *Ferroelectrics* **1996**, *178–181*.

[59] V. Vill, Liqcryst-Database of Liquid Crystalline Compounds, Landolt-Börnstein, New Series, Group IV, Vol. 7a–d, Springer, Berlin, **1992–1994** (and continually updated).

[60] G. Krömer, D. Paschek, A. Geiger, *Ber. Bunsenges. Phys. Chem.* **1993**, *97*, 1188–1192.

[61] R. B. Meyer in *5th Int. Liquid. Crystal Conference*, Stockholm, June 1974.

[62] R. B. Meyer, L. Liebert, L. Strzelecki, P. Keller, *J. Phys. (Paris) Lett.* **1975**, *36*, L69–71.

[63] P. W. Forsbergh, Jr. *Handbook of Physics*, Volume XVII (Dielectrics) (Ed.: S. Flügge) **1956**, p. 265.

[64] P. Curie, *Ann. Chim. Phys.* **1895**, *5*, 289.

[65] P. Langevin, *J. Phys.* **1905**, *4*, 678.

[66] P. Weiss, *C. R. Acad. Sci.* **1906**, *143*, 1136.

[67] F. M. Aepinus, *Mem. Acad. R. Sci. Berlin* **1756**, *12*, 105–121.

[68] D. Brewster, *Edinbg. J. Sci.* **1824**, *1*, 208–218.

[69] W. Thomson, *Phil. Mag.* **1878**, *5*, 4; *C. R. Acad. Sci.* **1893**, *117*, 463.

[70] J. Curie, P. Curie, *C. R. Acad. Sci.* **1880**, *91*, 294.

[71] W. G. Hankel, *Abh. Math.-Phys. Kl. Saechs. Akad. Wiss.* **1881**, *12*, 457.

[72] E. Schrödinger, *Sitz. ber. Kaiserl. Akad. Wiss.* Bd. CXXI, Abt. IIa, **1912**.

[73] I. V. Kourtschatov, *Le Champs moleculaire dans les dielectriques.* (Le sel de Seignette), Hermann et Cie, Paris **1936**.

[74] G. Busch, P. Scherer, *Naturwiss.* **1935**, *23*, 737.

[75] L. Landau, *Phys. Z. Sov. Union* **1937**, *11*, 26.

[76] H. Mueller, *Phys. Rev.* **1935**, *47*, 175.

[77] B. Wul, I. M. Goldman, *C. R. Acad. Sci. URSS* **1945**, *46*, 139.

[78] B. Wul, *Nature* **1945**, *156*, 480.

[79] A. von Hippel, R. G. Breckenridge, F. G. Chesley, L. Tisza, *Ind. Eng. Chem.* **1946**, *38*, 1097.

[80] V. L. Ginzburg, *J. Phys. USSR* **1946**, *10*, 107.

[81] O. Heaviside in *Electrical Papers* Vol. 1. Macmillan, London **1892**, p. 488.

[82] B. Hilczer, J. Malecki, *Electrets*, Elsevier, Amsterdam **1986**.

[83] M. Eguchi, *Proc. Phys.-Math. Soc. Jpn.* **1920**, *2*, 169.

[84] M. Eguchi, *Phil. Mag.* **1925**, *49*, 178.

[85] G. Busch, *Condensed Matter News* **1991**, *1*, 20–29.

[86] W. Känzig, *Condensed Matter News* **1992**, *3*, 21–24.

[87] J. M. Luttinger, L. Tisza, *Phys. Rev.* **1946**, *70*, 954.

[88] V. Dvořak, *Ferroelectrics* **1974**, *7*, 1.

[89] A. Holden, The Nature of Solids, Columbia University Press, New York **1965**; reprinted by Dover Publications, New York **1992**.

[90] A. Nussbaum, R. A. Phillips, *Contemporary Optics for Scientists and Engineers*, Prentice-Hall, Englewood Cliffs NJ **1976**, p 379–381.

[91] W. Thomson (Lord Kelvin), *The Baltimore Lectures on Molecular Dynamics and the Wave Theory of Light*, C. J. Clay and Sons, London **1904**, p. 436 (see also p. 619).

[92] J. S. Patel, *Opt. Eng.* **1987**, *26*, 129–133.

[93] N. A. Clark, S. T. Lagerwall, *Appl. Phys. Lett.* **1980**, *36*, 899.

[94] N. A. Clark, S. T. Lagerwall in *Liquid Crystals of One and Two Dimensional Order* (Eds.: W. Helfrich, G. Heppke), Springer, Berlin **1980**, pp. 222–227.

[95] N. A. Clark, S. T. Lagerwall in *Recent Developments in Condensed Matter Physics* (Eds.: J. T. Devreese, L. F. Lemmens, V. E. Van Doren, J. Van Doren), Plenum, New York **1981**, pp. 309–319.

[96] N. A. Clark, M. A. Handsky, S. T. Lagerwall, *Mol. Cryst. Liq. Cryst.* **1983**, *94*, 213–234.

[96a] N. A. Clark, S. T. Lagerwall, J. Wahl, *Proceedings of the SID*, **1995**, *26*, 133.

[97] J. S. Patel, *Opt. Eng.* **1987**, *26*, 129–133.

[98] M. A. Handschy, N. A. Clark, *Ferroelectrics* **1984**, *59*, 69–116.

[99] H. Weyl, *Symmetry*, Princeton University Press, Princeton, NJ **1952**, p. 19.

[100] C. H. Hermann, *Z. Kristallographie* **1931**, *79*, 186–221.

[101] C. H. Hermann, *Z. Kristallographie* **1934**, *89*, 32–48.

[102] A. N. Vtyurin, V. P. Ermakov, B. L. Ostrovskii, V. F. Shabanov, *Phys. Status Solidi b* **1981**, *107*, 397.

[103] N. M. Shytkov, M. L. Barnik, L. A. Beresnev, L. M. Blinov, *Mol. Cryst. Liq. Cryst.* **1985**, *124*, 379.

[104] A. Taguchi, Y. Ouchi, H. Takezoe, A. Fukuda, *Jpn. J. Appl. Phys.* **1989**, *28*, L997.

[105] D. M. Walba, P. Keller, D. S. Parmar, N. A. Clark, M. D. Wand, *J. Am. Chem. Soc.* **1989**, *111*, 8273.

[106] D. M. Walba, M. B. Ros, T. Sierra, J. A. Rego, N. A. Clark, R. Shao, M. D. Wand, R. T. Vohra, K. E. Arnett, S. P. Velsco, *Ferroelectrics* **1991**, *121*, 247.

[107] A. Hult, F. Sahlén, M. Trollsås, S. T. Lagerwall, D. Herrmann, L. Komitov, P. Rudquist, B. Stebler, *Liq. Cryst.* **1996**, *20*, 23.

[108] H. Hsiung, Ph. D. Thesis, Physics Department, University of California, Berkeley **1985**.

[109] C. W. Oseen, *Arkiv Mat. Astron. Fysik* **1928**, *21A*, 1. This paper, like the earlier papers on the continuum theory, is written in German, a language that Oseen stoped using from 1933. References to Oseen's work are nowadays almost uniquely given to his article in English in *Trans. Faraday Soc.* **1933**, *29*, 883. While it is true that Eq. (1) is stated there, the article essentially contains the first comprehensive theory of the optical properties of cholesteric liquid crystals.

[110] C. W. Oseen, *Arkin Mat. Astron. Fysik* **1925**, *A19*, 1.

[111] F. C. Frank, *Discuss. Faraday Soc.* **1958**, *25*, 19.

[112] P. Rudquist, S. T. Lagerwall, *Liq. Cryst.* **1997**, *23*, 503.

[113] J. Nehring, A. Saupe, *J. Chem. Phys.* **1971**, *54*, 337.

[114] D. Schmidt, M. Schadt, W. Helfrich, *Z. Naturforsch* **1972**, *27a*, 277.

[115] Y. Bouligand, J. Physique (Paris) **1969**, *30*, C4–90.

[116] Y. Bouligand in *Mesomorphic Order in Polymers and Polymerization in Liquid Crystalline Media*, (Ed.: A. Blumstein), ACS Symposium series, Washington, DC **1978**, p. 74.

[117] J. S. Patel, R. B. Meyer, *Phys. Rev. Lett.* **1987**, *58*, 1538.

[118] P. Rudquist, T. Carlsson, L. Komitov, S. T. Lagerwall, *Liq. Cryst.* **1997**, *22*, 445.

[119] J.-S. Patel, S.-D. Lee, *J. Appl. Phys.* **1989**, *66*, 4.

[120] S.-D. Lee, J. S. Patel, R. B. Meyer, *J. Appl. Phys.* **1990**, *67*, 3.

[121] S.-D. Lee, J. S. Patel, *Phys. Rev. A* **1990**, *42*, 997.

[122] S.-D. Lee, J. S. Patel, R. B. Meyer, *Mol. Cryst. Liq. Cryst.* **1991**, *209*, 85.

[123] P. Rudquist, L. Komitov, S. T. Lagerwall, *Phys. Rev. E* **1994**, *50*, 4735.

[124] P. Rudquist, M. Buivydas, L. Komitov, S. T. Lagerwall, *J. Appl. Phys.* **1994**, *76*, 7778.

[125] P. Rudquist, L. Komitov, S. T. Lagerwall, *Liq. Cryst.* **1997**, in press.

[126] P. Rudquist, Thesis, Physics Department, Chalmers University of Technology **1997**.

[127] I. Dahl, S. T. Lagerwall, *Ferroelectrics* **1984**, *58*, 215.

[128] P. G. de Gennes, J. Prost, *The Physics of Liquid Crystals*, Clarendon, Oxford **1993**, p. 348.

[129] N. A. Clark, M. A. Handschy, S. T. Lagerwall, *Mol. Cryst. Liq. Cryst.* **1983**, *94*, 213.

[130] M. A. Handschy, N. A. Clark, *Ferroelectrics* **1984**, *59*, 69.

[131] P. G. de Gennes, *Mol. Cryst. Liq. Cryst.* **1973**, *15*, 49.

[132] C. C. Huang, J. M. Viner, *Phys. Rev. A* **1982**, *25*, 3385.

[133] S. Dumrongrattana, C. C. Huang, *Phys. Rev. Lett.* **1986**, *56*, 464.

[134] W. Kuczyński, J. Hoffmann, *25. Freiburger Arbeitstagung Flüssigkristalle*, **1996**, 34.

[135] T. Niori, T. Sekine, J. Watanabe, T. Furukawa, H. Takezoe, *J. Mater. Chem.* **1996**, *6*, 1231.

[136] D. S. Hermann, Ph. D. Thesis, Physics Department, Chalmers University of Technology, Göteborg **1997**.

[137] I. Dahl, S. T. Lagerwall, K. Skarp, *Phys. Rev. A* **1987**, *36*, 4380.

[138] J. Xue, M. A. Handschy, N. A. Clark, *Ferroelectrics* **1987**, *73*, 305.

[139] P. Rudquist, Ph. D. Thesis, Physics Department, Chalmers University of Technology, Göteborg **1997**.

[140] G. Andersson, I. Dahl, L. Komitov, S. T. Lagerwall, K. Skarp, B. Stebler, *J. Appl. Phys.* **1989**, *66*, 4983.

[141] F. Giesselmann, A. Langhoff, P. Zugenmaier in *26th Freiburger Arbeitstagung*, April 1997, **1997**, P13.

[142] H. Rieger, C. Escher, G. Illian, H. Jahn, A. Kattbeitzel, D. Ohlendorf, N. Rösch, T. Harada, A. Weippert, E. Lüder in *Conference Record of the SID May 6–10, 1991 Conference*, Anaheim, California, **1991**.

[143] J. C. Jones, E. P. Raynes, M. J. Towler, J. R. Sambles, *Mol. Cryst. Liq. Cryst.* **1991**, *199*, 277; J. C. Jones, M. J. Towler, E. P. Raynes, *Ferroelectrics* **1991**, *121*, 91; J. C. Jones, Ph. D. Thesis, University of Hull **1991**.

[144] J. C. Jones, M. J. Towler, J. R. Hughes, *Displays* **1993**, *14*, 86.

[145] M. J. Towler, J. C. Jones, E. P. Raynes, *Liq. Cryst.* **1992**, *11*, 365.

[146] T. Carlsson, F. M. Leslie, N. A. Clark, *Phys. Rev. E* **1995**, *51*, 4509.

[147] W. Kuczyński, *Ber. Bunsenges. Phys. Chem.* **1981**, *85*, 234.

[148] E. P. Pozhidaev, L. M. Blinov, A. Beresnev, V. V. Belyayev, *Mol Cryst. Liq. Cryst.* **1985**, *124*, 359.

[149] K. Skarp, I. Dahl, S. T. Lagerwall, B. Stebler, *Mol. Cryst. Liq. Cryst.* **1984**, *114*, 283.

[150] C. Escher, T. Geelhaar, E. Böhm, *Liq. Cryst.* **1988**, *3*, 469.

[151] T. Carlsson, B. Žekš, *Liq. Cryst.* **1989**, *5*, 359.

[152] T. Carlsson, B. Žekš, C. Filipiĉ, A. Levstik, *Phys. Rev. A* **1990**, *42*, 877.

[153] S. U. Valerien, F. Kremer, T. Geelhar, A. E. Wächter, *Phys. Rev. A* **1990**, *42*, 2482.

[154] F. M. Gouda, Ph. D. Thesis, Physics Department, Chalmers University of Technology **1992**.

[155] M. Buivydas, Ph. D. Thesis, Physics Department, Chalmers University of Technology **1997**.

[156] J. Hoffmann, W. Kuczyński, J. Malecki, J. Pavel, *Ferroelectrics* **1987**, *76*, 61.

[157] G. Gouda, W. Kuczyński, S. T. Lagerwall, M. Matuszczyk, T. Matuszczyk, K. Sharp, *Phys. Rev. A* **1992**, *46*, 951.

[158] A. Jákli, A. Saupe, *Appl. Phys. Lett.* **1992**, *60*, 2622.

[158a] S. Markscheffel, Ph. D. Thesis, Martin-Luther Universität Halle-Wittenberg **1996**.

[159] J. R. Sambles, S. Elston, M. G. Clark, *J. Mod. Opt.* **1989**, *36*, 1019.

[160] M. H. Anderson, J. C. Jones, E. P. Raynes, M. J. Towler, *J. Phys. D.* **1991**, *34*, 338.

[161] J. C. Jones, E. P. Raynes, M. J. Towler, J. R. Sambles, *MCLC* **1991**, *199*, 277.

[162] N. A. Clark, S. T. Lagerwall, U. S. Patent No. 4, 563, 059, Jan 7, **1986**. This structure has been investigated particularly by Patel; see V. Pertuis, J. S. Patel, *Ferroelectrics* **1993**, *149*, 193.

[163] T. P. Rieker, N. A. Clark, G. S. Smith, D. S. Parmar, E. B. Sirota, C. R. Safinya, *Phys. Rev. Lett.* **1987**, *59*, 2658.

[164] N. A. Clark, T. P. Rieker, *Phys. Rev. A* **1988**, *37*, 1053.

[165] W. J. A. M. Hartmann, *Ferroelectrics* **1991**, *122*, 1.

[166] N. A. Clark, T. P. Rieker, J. E. Maclennan, *Ferroelectrics* **1988**, *85*, 79.

[167] A. Fukuda, Y. Ouchi, H. Arai, H. Takano, K. Ishikawa, H. Takezoe, *Liq. Cryst.* **1989**, *5*, 1055.

[168] T. P. Rieker, N. A. Clark, *Phase Transitions in Liquid Crystals* (Eds.: S. Martellucci, A. N. Chester), Plenum, New York **1992**, p. 287.

[169] W. J. A. M. Hartmann, A. M. M. Luyckx-Smolders, *J. Appl. Phys.* **1990**, *67*, 1253.

[170] J. Kanbe, H. Inoue, A. Mizutome, Y. Hanyuu, K. Katagiri, S. Yoshihara, *Ferroelectrics* **1991**, *114*, 3.

[171] M. Koden, H. Katuse, N. Itoh, T. Kaneko, K. Tamai, H. Takeda, M. Kido, M. Matsaki, S. Miyoshi, T. Wada, *Ferroelectrics* **1993**, *149*, 183.

[172] N. Itoh, M. Koden, S. Miyoshi, T. Wada, T. Akahane, *Ferroelectrics* **1993**, *147*, 327.

[173] M. Terada, S. Yamada, K. Katagiri, S. Yoshihara, J. Kanbe, *Ferroelectrics* **1993**, *149*, 283.

[174] S. T. Lagerwall, J. Wahl, N. A. Clark, *IDRC 85 (San Diego), 1985*, 213.

[175] Y. Hanyu, K. Nakamura, Y. Hotta, S. Yoshihara, J. Kanbe, *1993 SID (Seattle) XXIV* **1993**, 364.

[176] J. Xue, Ph. D. Thesis, University of Colorado, Boulder, U. S. A. **1989**.

[177] K. Yuasa, S. Uchida, T. Sekiya, K. Hashimoto, K. Kawasaki, *Ferroelectrics* **1991**, *122*, 53.

[178] S. Hachiya, K. Tomoike, K. Yuasa, S. Togawa, T. Sekiya, K. Takahashi, K. Kawasaki, *J. SID* **1993**, *113*, 295.

[179] A. Mochizuki, T. Yoshihara, M. Iwasaki, M. Nakatsuka, Y. Takanishi, Y. Ouchi, H. Takezoe, A. Fukuda in *Proc. Japan Display '89* (Kyoto) **1989**, *Paper 3–4*, p. 32.

[180] E. P. Janulis, J. C. Novack, M. G. Tristani-Kendra, G. A. Papapolymerou, W. A. Huffman in *Proc. Japan Display '89* (Kyoto) **1989**, *Paper PD-3*, p. 3.

[181] E. P. Janulis, J. C. Novack, M. G. Tristani-Kendra, G. A. Papapolymerou, W. A. Huffman, *Ferroelectrics* **1988**, *85*, 375.

[182] K. A. Epstein, M. P. Keyes, M. D. Radcliffe, D. C. Snustad, U. S. Patent No. 5.417.883, May 23, **1995**.

[183] K. A. Epstein, M. D. Radcliffe, M. Brostrom, A. G. Rappaport, B. N. Thomas, N. A. Clark in *Abstract Book of FLC'93* (Tokyo) **1993**, P-46.

[184] S. J. Elston, *Displays* **1995**, *16*, 141.

[185] S. T. Lagerwall, *Liq. Cryst. Today* **1996**, *6*, 2.

[186] S. D. Bull, C. J. Morris, U. S. Patent No. 5.093.652, March 3, **1992**.

[187] B. J. Humphries, C. M. Waters, S. D. Bull, C. J. Morris, U. S. Patent No. 5.189.406, Feb. 23, **1993**.

[188] Ch. Bahr, G. Heppke, *Ber. Bunsenges. Phys. Chem.* **1987**, *91*, 925; *Liq.Cryst.* **1987**, *2*, 825; *Phys. Rev. A* **1988**, *37*, 3179.

[189] Y. Kimura, T. Sako, R. Hayakawa, *Ferroelectrics* **1993**, *147*, 315.

[190] J. Pavel, M. Glogarova, *Ferroelectrics* **1991**, *114*, 131.

[191] S. T. Lagerwall, M. Matuszczyk, P. Rodhe, L. Ödman, *The Electronic Effect,* in S. J. Elston, J. R. Sambles (eds) *The Optics of Thermotropic Liquid Crystals,* Taylor & Francis, London, **1997**.

[191a] A. de Vries, *Mol. Cryst. Liq. Cryst. Lett.* **1977**, *41*, 27; *49*, 143; *49*, 179.

[192] S. Inui, N. Iimura, T. Suzuki, H. Iwane, K. Miyachi, Y. Takanishi, A. Fukuda, *J. Mater. Chem.* **1996**, *6*, 671.

[193] B. I. Ostrovskij, A. Z. Rabinovichh, V. G. Chigrinov, L. Bata (Ed), *Advances in Liquid Crystal Research and Applications* **1980**, 426.

[194] L. A. Beresnev, L. M. Blinov, D. I. Dergachev, S. B. Kondratjev, *JETP Lett*, **1987**, 413.

[195] J. Fünfschilling, M. Schadt, *Proc. SID* **1990**, *31/2*, 119.

[196] A. G. H. Verhulst, G. Cnossen, *Ferroelectrics* **1996**, *179*, 141.

[197] V. L. Indenbom, S. A. Pikin, E. B. Loginov, *Sov. Phys. Crystallogr.* **1976**, *21*, 632.

[198] S. A. Pikin, V. L. Indenbom, *Sov. Phys. Usp.* **1978**, *21*, 487.

[199] R. Blinc, B. Žekš, *Phys. Rev. A* **1978**, *18*, 740.

[200] B. Žekš, A. Levstik, R. Blinc, *J. de Physique* **1979**, *40*, C3–409.

[201] B. Žekš, *Mol. Cryst. Liq. Cryst.* **1984**, *114*, 259.

[202] T. Carlsson, B. Žekš, C. Filipiĉ, A. Levstik, R. Blinc, *Phys. Rev. A* **1987**, *36*, 1484.

[203] T. Carlsson, B. Žekš, C. Filipiĉ, A. Levstik, R. Blinc, *Mol. Cryst. Liq. Cryst.* **1988**, *163*, 11.

[204] B. Žekš, C. Filipiĉ, T. Carlsson, *Physica Scripta* **1989**, *T25*, 362.

3 Antiferroelectric Liquid Crystals

Kouichi Miyachi and Atsuo Fukuda

3.1 Introduction

Chandani et al. [1] confirmed the existence of antiferroelectricity in liquid crystals by observing the electric-field-induced transition between the antiferroelectric smectic C_A^* (SmC$_A^*$) phase and the ferroelectric smectic C* (SmC*) phase. Since then, a large number of investigations have been performed and several review articles have been published [2–7]. In ordinary circumstances, it would not be easy for us to write another review. Fortunately, however, the last year or so has been full of excitement in the field of antiferroelectric liquid crystals (AFLCs). In the following sections we describe three unexpected findings: the origin of antiferroelectricity in liquid crystals, thresholdless antiferroelectricity (TLAF) and V-shaped switching, and AFLC materials with unusual chemical structures [8].

3.2 Origin of Antiferroelectricity in Liquid Crystals

3.2.1 Biased or Hindered Rotational Motion in SmC* Phases

Meyer and coworkers [9, 10] have described the mechanism for the emergence of ferroelectricity in liquid crystals. We begin with its review. In most of the ordinary liquid-crystalline phases (nematic, smectic A (SmA) and smectic C (SmC)), which appear successively on lowering the temperature from the isotropic phase, the symmetry is so high that free rotation about the molecular long axis and head–tail equivalence prevent the occurrence of ferroelectricity. In the chiral smectic C (SmC*) phase, however, the symmetry is low enough to allow the existence of chirality-induced improper ferroelectricity, which does not result from the Coulomb interaction between permanent dipoles. The molecules rotate thermally about their long axes, but this rotation is now biased or hindered so that a spontaneous polarization given by

$$P = P \frac{\hat{z} \times \hat{n}}{|\hat{z} \times \hat{n}|} \tag{1}$$

666   3 Antiferroelectric Liquid Crystals

emerges. Here $\hat{z}$ is a unit vector parallel to the smectic layer normal and $\hat{n}$ is the director; the sense of $\hat{z}$ is chosen in such a way that $\hat{z}$ coincides with $\hat{n}$ when the tilt angle θ becomes zero. The reversal of the tilting sense naturally results in a reversal of the polarization sense.

In this way, biased or hindered rotational motion about the molecular long axis plays an essential role in the emergence of the spontaneous polarization. However, only a few experiments have been aimed at the quantitative, microscopic examination of this motion [11,12]. Quite recently, Kim et al. [13] studied a prototype AFLC, MHPOBC, the structure of which is shown

in Figure 1, together with different aspects of the molecule and its dimensions. Using polarized infrared (IR) spectroscopy, they obtained microscopic proof, at least in its static aspect, of the biased or hindered rotational motion of the carbonyl (C=O) groups about the molecular long axis in the ferroelectric SmC* phase. A slightly unexpected conclusion derived by Kim et al. was that the most probable orientation of each individual C=O group may not be perpendicular to the tilt plane along the C_2 axis of the SmC* phase (the Y axis), although the spontaneous polarization appears perpendicular to the tilting plane along the two-fold axis because of the head–tail equivalence.

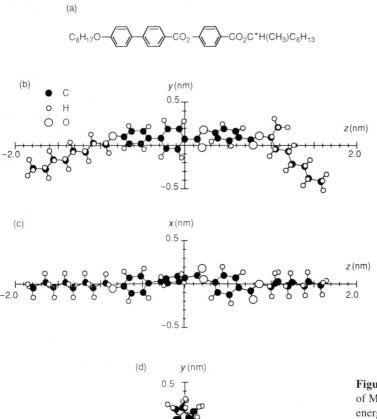

(a)

$C_8H_{17}O$-⬡-⬡-CO_2-⬡-$CO_2C^*H(CH_3)C_6H_{13}$

Figure 1. (a) Structural formula of MHPOBC and its minimum-energy molecular shape projected on (b) the yz plane, (c) the zx plane, and (d) the xy plane. The x, y and z axes are the eigen axes of the moment of inertia.

IR radiation is incident along the Y axis in Figure 2:

$$A(\omega') = \frac{1}{2} \int_0^{2\pi} f_{t+}(\psi) (\sin\omega' \sin\beta \sin\psi + \cos\omega' \cos\beta)^2 \, d\psi$$

$+$ [a corresponding term with $f_{b+}(\psi)$]   (4)

Here $\omega' = \omega - \theta$ is used for convenience. Some of the calculated results are shown in Figure 6. The absorbance shows maxima at $\omega' = \omega'_{max}$ and $\omega'_{max} + 180°$, and its polarizer-rotation angle dependence is symmetrical with respect to the line connecting the maxima through the origin. When the biased or hindered direction (the most probable orientation) in the molecular frame ψ_0 is

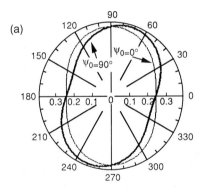

(a)

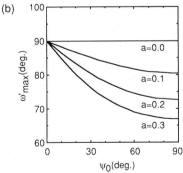

(b)

Figure 6. Calculated results in SmC* for the IR radiation incident along the tilt-plane normal by using the distribution function given in equation (2). (a) Normalized absorbance versus polarizer rotation angle; $a = 0.2$, $\omega' = \omega - \theta$, and $\omega = 0$ is the layer normal. (b) Rotation angle at which $A(\omega')$ becomes a maximum versus biased or hindered direction, $\omega'_{max}(\psi_0)$, for several values of a.

zero, we obtain $\omega'_{max} = 90°$, irrespective of the value of a, the degree of biasing or hindrance, so long as a is small. As ψ_0 increases, ω'_{max} rotates clockwise, as shown in Figure 6(a). Since the core C=O is distant from the chiral centre, mirror symmetry must exist and, in the present approximation, $\psi_0 = 90°$; i.e. the core C=O appears to lie on the tilt plane. Experimentally (Figure 3), $\omega'_{max} \approx 68°$ for the core C=O, and hence we obtain $a \approx 0.3$ or slightly smaller. Similarly (Figure 3) $\omega'_{max} \approx 83°$ for the chiral C=O, and hence we obtain $\psi_0 \approx 15°$ for the (R) enantiomer and 165° for the (S) enantiomer if the chiral C=O has the same degree of biasing or hindrance as the core C=O. Here, we have also used the fact that the (S) enantiomer of MHPOBC has a positive spontaneous polarization.

3.2.2   Biased or Hindered Rotational Motion in SmC$_A^*$ Phases

Let us consider the biased or hindered rotational motion of the core C=O in SmC$_A^*$. It is well established that the molecular orientation in SmC$_A^*$ in its unwound state is a herringbone structure. In each smectic layer, the director tilt with respect to the layer normal is uniform; in adjacent layers, the tilt is equal in magnitude, but opposite in sign (the azimuthal angles specifying the tilt directions differing from each other by 180°) [16, 17]. Because of this herringbone structure, the measuring geometry with the homogeneously aligned cells used for SmC* cannot be used to study the biased or hindered rotational motion of the chiral C=O in SmC$_A^*$. Instead, the measuring geometry obtained with free-standing films or homeotropically aligned cells, where IR radiation is incident along the Z axis in Figure 2, is used. If we average the two possible SmC* states, one tilted to the left and the other to the right, molecules are apparently configured equally for both SmC* and SmC$_A^*$; nevertheless, the results obtained experimentally offer a remarkable contrast.

Figure 7 summarizes the dependence of absorbance on the polarizer rotation angle in the SmC* and SmC$_A^*$ phases [13, 15]. Partial racemization not only simplifies the phase sequence, but also elongates the pitch of the helicoidal structure in SmC$_A^*$ so that its unwinding occurs at a relatively low electric field. However, a strong applied electric field was still needed because the anisotropy which causes unwinding is not very large. Although systematic investigations have not yet been performed and the data so far available are far from satisfactory, it is safe to conclude that the angular

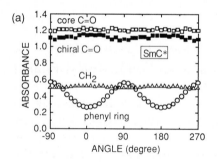

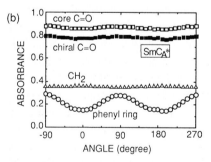

Figure 7. Absorbance versus the polarizer rotation angle in MHPOBC. (a) $A(\omega)$ in unwound SmC* obtained at 120 °C, with a 75 V mm^{-1} electric field, and using an approximately 8 µm thick, homeotropically aligned cell of (S)-MHPOBC. (b) $A(\omega)$ in unwound SmC$_A^*$ obtained at 85 °C, with an 800 Hz, ±1400 V mm^{-1} electric field, and using a 10 µm thick, homeotropically aligned cell of partially racemized ((R)-MHPOBC/(S)-MHPOBC = 84/16) MHPOBC.

dependence of the chiral C=O peak in SmC$_A^*$ is quite different from that in SmC*. The dependence in SmC$_A^*$ is very characteristic and conspicuously in phase with that of the phenyl peak, while it is out of phase in SmC*. The biased or hindered rotational motion of the chiral C=O in SmC$_A^*$ appears to be substantially different from that in SmC*.

In SmC$_A^*$, four equivalent configurations exist because the X and Y axes are the twofold symmetry axes; the chiral C=O rotational motion is biased or hindered not in a unique direction, but in four directions. To a first approximation, the hindered rotational motion is described by

$$f(\psi) = \frac{1}{4}[f_{t+}(\psi) + f_{b+}(\psi) + f_{t-}(\psi) + f_{b-}(\psi)] \quad (5)$$

where $f_{t+}(\psi)$, etc., are the same as in Eq. (3). In SmC*, either of the distribution functions describing the two possible states, one tilted to the left and the other to the right, given in Equation (2) results in the same normalized absorbance dependence on the polarizer rotation angle. Thus, for both SmC$_A^*$ and SmC* we obtain

$$A(\omega) = \frac{1}{4} \int_0^{2\pi} f_{t+}(\psi)$$
$$\cdot [\sin\omega (-\cos\beta \sin\theta + \sin\beta \sin\psi \cos\theta) + \cos\omega (\sin\beta \cos\psi)]^2 \, d\psi$$
$$+ [\text{three corresponding terms}] \quad (6)$$

where $\omega' = 0$ is the tilt plane normal [15].

Figure 8(a) illustrates some of the simulated results, which naturally depend on the degree of biasing or hindrance a, and the biased or hindered direction ψ_0. Figure 8(b) summarizes the degree of polarization versus the most probable orientation

$$D(\psi_0; a)$$
$$= \frac{2[A_{\omega=0°}(\psi_0; a) - A_{\omega=90°}(\psi_0; a)]}{A_{\omega=0°}(\psi_0; a) + A_{\omega=90°}(\psi_0; a)} \quad (7)$$

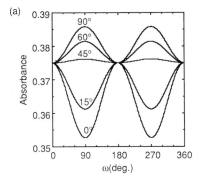

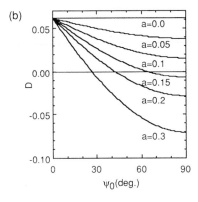

Figure 8. (a) Normalized absorbance versus the polarizer rotation angle calculated for SmC* and SmC$_A^*$ for radiation incident along the smectic layer normal by using the distribution function given in equations (3) and (5). The molecular tilt angle and the degree of hindrance are assumed to be $\theta = 25°$ and $a = 0.2$, respectively, and $\omega = 0$ is the tilt-plane normal. (b) Degree of polarization versus the most probable orientation for several values of a.

for various values of a. Note that positive and negative signs of $D(\psi_0; a)$ correspond, respectively, to the out-of-phase and in-phase cases with respect to the phenyl peak. As the molecules are tilting from the smectic layer normal along which the IR radiation is incident, the C=O peak shows the out-of-phase angular dependence even when the rotational motion is free. Nevertheless, the conspicuous out-of-phase and in-phase angular dependencies in SmC* and SmC$_A^*$, respectively, given in Figure 7, clearly indicate that ψ_0 in SmC$_A^*$ differs

from that in SmC*, provided that a does not change during the field-induced phase transition from SmC$_A^*$ to SmC*. Using the experimentally obtained degree of polarization, $D(\psi_0) = -0.03$, and assuming that the degree of biasing or hindrance is $a \approx 0.3$, we determine that $\psi_0 \approx 135°$ (45°) for the partially racemized (S) $((R))$ enantiomer in the SmC$_A^*$ phase. If $a \approx 0.2$, then $\psi_0 \approx 100°$ (80°). In the SmC* phase of (S)-MHPOBC at 120°C, however, the experimentally obtained value is $D(\psi_0) \approx 0.05$, and hence we obtained $\psi_0 \approx 175°$ (5°) for $a \approx 0.3$ and $\psi_0 \approx 165°$ (15°) for $a \approx 0.1$.

3.2.3   Spontaneous Polarization Parallel to the Tilt Plane

We can conclude, therefore, that the chiral C=O has a tendency to lie on the tilt plane in SmC$_A^*$, while it assumes a fairly upright position in SmC*. For several years antiferroelectricity has been modelled as arising from the pairing of transverse dipole moments at the smectic layer boundaries [17, 18]. Although the chiral C=O in question is not at the end of the molecule, the unexpected bent shape of MHPOBC in its crystal phase just below SmC$_A^*$, as revealed by X-ray crystallographic studies, has been considered to support the pairing model [18, 19]. The model based on the pairing of transverse dipole moments is physically (or, rather, chemically) intuitive in relation to an understanding of the stabilizaton of antiferroelectric SmC$_A^*$. However, because ψ_0 is as large as 45° or more, as concluded above, it is not reasonable to ascribe an essential role to the pairing alone, and we must look for other causes for the stabilization of SmC$_A^*$ [15].

The symmetry arguments that Meyer used to develop his speculation about ferroelectric SmC* may help us to clarify antiferroelectric SmC_A^*. There are two types of two-fold axis: one at the layer boundary is parallel to the X axes in the tilt plane ($C_{2X,b}$), and the other is located in the centre of the layer, perpendicularly ($C_{2Y,c}$) (Figure 9). In the pairing model, the $C_{2Y,c}$ axis is the only one taken into consideration explicitly; the spontaneous polarization along $C_{2Y,c}$, P_Y, alternates in sign from layer to layer [16–18, 20, 21]. However, Cladis and Brand [22, 23] insisted on the importance of the $C_{2X,b}$ axis in the tilt plane, and the above

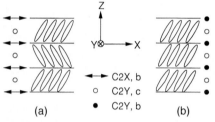

(a)   (b)

Figure 9. The two types of symmetry axes in unwound SmC_A^* and SmC^*. (a) Unwound SmC_A^* has two-fold axes, $C_{2X,b}$ and $C_{2Y,c}$, at the smectic layer boundary and at the smectic layer centre, respectively. (b) Unwound SmC^* has two-fold axes, $C_{Y,b}$ and $C_{2Y,c}$, at the smectic layer boundary and at the smectic layer centre, respectively.

conclusion about biased or hindered rotational motion of the chiral C=O supports their indication. The chiral C=O in question, although not in the core, is not at the end of the molecule either. Nevertheless, it is natural to consider that the in-plane spontaneous polarizations exist at smectic layer boundaries; they are parallel to the tilt plane along $C_{2X,b}$ in SmC_A^*, but are perpendicular to the tilt plane along $C_{2Y,b}$ in SmC^*. Their magnitude in SmC_A^* may not be equal to that in SmC^*. Note that $C_{2Y,b}$ exists, as well as $C_{2Y,c}$, even in SmC^*, as illustrated in Figure 10.

An unanswered question is whether or not the Coulomb interaction among the permanent dipoles plays an important role. If the interaction is considered to result in in-plane spontaneous polarizations parallel to the tilt plane P_X, we can designate this viewpoint as the P_X model. For the stabilization of ferroelectric SmC^*, the Coulomb interaction among the permanent dipoles has not been considered essential, as insisted upon initially by Meyer [10]. It is also possible that this interaction does not play any essential role in the stabilization of SmC_A^* [24]. A

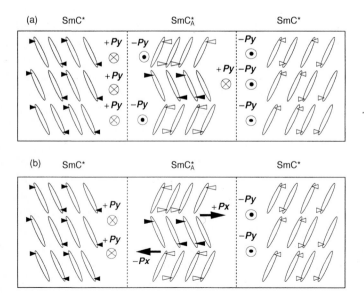

Figure 10. In-plane spontaneous polarizations in SmC_A^* and SmC^*. (a) Previous dipole pairing model. Only $C_{2Y,c}$, and hence the in-plane spontaneous polarization along the tilt-plane normal P_Y, are taken into consideration. (b) The present P_X model. Both $C_{2X,b}$ and $C_{2Y,b}$ are considered; hence in-plane spontaneous polarizations are considered to exist at the smectic layer boundaries, P_X in SmC_A^* and P_Y in SmC^*.

tempting feature of the P_X model is, however, the possibility of explaining the herringbone structure, where the tilt planes in adjacent smectic layers are parallel to each other; we could not devise any other simple explanation for the structure. When permanent dipoles of equal magnitude are distributed on a plane, they may align in the same direction in order to minimize the Coulomb interaction energy, resulting in the in-plane spontaneous polarization parallel to the tilt plane P_X. The sign of P_X alternates from boundary to boundary; this fact also contributes, through fluctuation forces, to minimizing the Coulomb interaction energy between P_X planes [25, 26].

Let us consider three dipoles which are located at the apices of a regular triangle with one side d, and oriented in the same direction and sense as illustrated in Figure 11. When another dipole approaches along a line perpendicular to the triangle and passing through its centre, it will align parallel to the three dipoles if $z < d/\sqrt{6}$, but antiparallel if $z > d/\sqrt{6}$, in order to minimize the Coulomb interaction energy. Although, in the pairing model, we have to invoke rather *ad hoc* assumptions (i.e. the local spontaneous optical resolution and the conformational chirality), the present P_X model can naturally explain the existence of SmC$_A$; the in-plane spontaneous polar-

ization parallel to the tilt plane P_X is independent of chirality [22, 23, 27] and emerges even in racemates and non-chiral swallow-tailed compounds with two terminal chains of equal length [24].

3.2.4   Obliquely Projecting Chiral Alkyl Chains in SmA Phases

The above discussion shows that it is most probable that in-plane spontaneous polarizations emerge at smectic layer boundaries and stabilize antiferroelectricity in the SmC$_A^*$ phase. In fact, Jin et al. [28] have quite recently shown unambiguously that the chiral chain projects obliquely from the core, even in SmA* and hence is considered as precessing around the core long axis; the angle between the chiral chain and core axes is more than the magic angle 54.7°. Such a bent molecular structure may allow the permanent dipoles in adjacent layers to interact adequately through Coulombic forces. By deuterating the chiral and achiral alkyl chains separately, Jin et al. observed the IR absorbance as a function of polarizer rotation angle for the CH$_2$ and phenyl ring stretching peaks in the SmA* phase of MHPOBC in a homogeneously aligned cell. Their results are summarized in Figure 12. Because of the selective deuteration, Jin et al. were able to obtain information about the achiral and chiral chains independently. The angular dependence of the CH$_2$ stretching peaks, both asymmetric (2931 cm^{-1}) and symmetric (2865 cm^{-1}), shows a marked contrast between the achiral and chiral chains. In the achiral chain it is out of phase with that of the phenyl ring stretching peak, while in the chiral chain it is in phase. Moreover, the degree of polarization is very small.

Jin et al. calculated the normalized absorbance by assuming an average transition di-

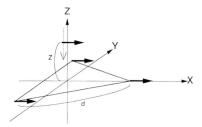

Figure 11. Illustration of three dipoles at the apices of a regular triangle with one side d and another dipole approaching on the Z axis (see the text for details).

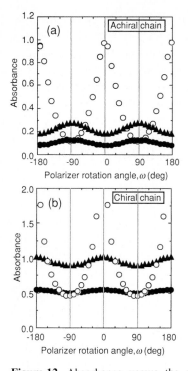

Figure 12. Absorbance versus the polarizer rotation angle in SmA* measured at 130°C using a homogeneously aligned cell. (a) 4.0 μm thick cell, (R)-MHPOBC-d_{13} (enantiomeric excess 30%); (b) 9.6 μm thick cell, racemic MHPOBC-d_{25}. (○) Phenyl ring stretching; (▲) CH$_2$ asymmetric stretching; (●) CH$_2$ symmetric stretching.

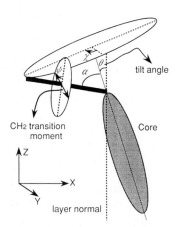

Figure 13. A model of the transition dipole moment for the CH$_2$ stretching vibrations, situated on the alkyl chain and freely rotating about the chain axis. The chain itself makes an angle α with the core axis, and precesses around it.

pole moment of the CH$_2$ stretching vibrations, which is perpendicular to the average chain axis and is rotating freely about it, and by neglecting the sample anisotropy. Several necessary angles are defined as depicted in Figure 13: α is the angle between the chain and core axes; χ is the azimuthal angle of the chain axis around the core axis; ψ is the angle of rotation of the transition dipole moment about the chain axis; and θ is the tilt angle in SmC* which is zero in SmA*. The relationship between the absorbance and the polarizer rotation angle is given by

$$A(\omega';\alpha) = \frac{1}{2\pi} \int_0^{2\pi} \int_0^{2\pi} f(\chi)$$
$$\cdot (\sin\omega' \cos\chi \sin\psi - \sin\omega' \cos\alpha \sin\psi$$
$$- \cos\omega' \sin\alpha \cos\psi)^2 \, d\chi \, d\psi \qquad (8)$$

Here the distribution function in SmA* is $f(\chi) = 1/(2\pi)$; the chain axis rotates freely about the core axis because of the uniaxial SmA* symmetry.

Figure 14 shows the calculated normalized absorbance in SmA*. At the magic angle $\alpha = \cos^{-1}(1/\sqrt{3}) = 54.7°$, the absorbance does not depend on the polarizer rotation angle $A(\omega;\alpha) = 1/\sqrt{3}$. The aforementioned in-phase and out-of-phase relationships corre-

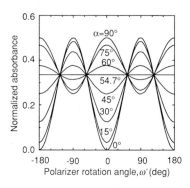

Figure 14. Normalized absorbance versus the polarizer rotation angle in SmA*, calculated for various values of α.

spond to $\alpha > 54.7°$ and $\alpha < 54.7°$, respectively. In this way, as shown unambiguously in Figure 12(b), the chiral chain in the SmA* phase projects obliquely, precessing around the core axis; the angle between the chain and core axes is more than 54.7°, but is probably close to the magic angle because of the fairly small observed degree of polarization. Figure 12(a) also clearly shows that, in this case, the other terminal chain in SmA* makes an angle of $\alpha < 54.7°$ with the core axis. By taking into account of the fact that not only the average transition moment, but also the average chain axis itself, is actually fluctuating, the appreciable degree of polarization observed may indicate that the angle is greatly different from the magic angle and rather close to zero.

Jin et al. also analysed how the chiral chain precession becomes biased or hindered in SmC*, and suggested that the biasing or hindering may occur to the tilt plane normal ($\chi_0 \approx 0°$). Although no experimental study of the biased or hindered precession in SmC$_A^*$ has yet been made, it is tempting to speculate that the biasing or hindrance may occur to the tilt plane ($\chi_0 \approx 90°$). As described in Sec. 3.2.1 and 3.2.2, the biased or hindered direction of the C=O group near to the chiral carbon atom in SmC* is quite different from that in SmC$_A^*$. Anyway, we can safely conclude that in a prototype AFLC, MHPOBC, the molecules have a bent structure so that the C=O groups near the chiral carbon atoms, i.e. the permanent dipoles, in adjacent smectic layers can adequately interact through Coulombic forces to stabilize the antiferroelectric SmC$_A^*$ (SmC$_A$), and that in-plane spontaneous polarizations are produced at layer boundaries along the two-fold $C_{2X,b}$ axes which are parallel to the molecular tilt plane. Figure 15 presents an illustrative P_X model based on bent shaped molecules.

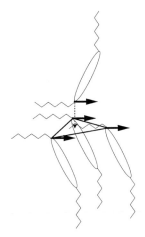

Figure 15. An illustrative P_X model based on bent shaped molecules.

3.3   Thresholdless Antiferroelectricity and V-Shaped Switching

3.3.1   Tristable Switching and the Pretransitional Effect

Ordinary antiferroelectricity in liquid crystals results in tristable switching, which is the electric-field-induced transition between antiferroelectric SmC$_A^*$ and ferroelectric SmC* and has the characteristic d.c. threshold and hysteresis shown in Figure 16 [29–35]. When a weak electric field is applied perpendicularly to the page, the dielectric anisotropy unwinds the helicoidal structure of SmC$_A^*$ and the tilt plane of the unwound SmC$_A^*$ becomes perpendicular to the page; all in-plane spontaneous polarizations, P_b, have the same magnitude, but point upwards and downwards alternately. After the phase transition into SmC*, the molecules are parallel to the page, tilting uniformly either to the right or to the left; all P_b have the same magnitude and point uniformly either downwards or upwards. During the pretransitional effect, the P_b stay

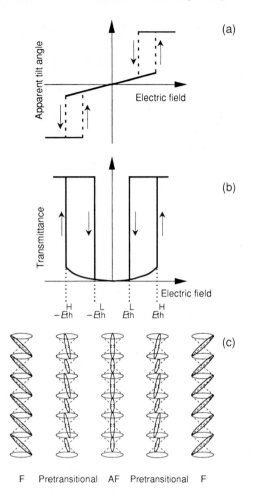

F   Pretransitional   AF   Pretransitional   F

Figure 16. Ordinary tristable switching; (a) apparent tilt angle versus applied electric field; (b) optical transmittance versus applied electric field; (c) molecular tilt directions during the pretransitional effect and in the antiferroelectric (AF) and ferroelectric (F) states.

perpendicular to the page, but their magnitude and sense may change alternately from boundary to boundary, so that a slight net spontaneous polarization, either downward or upward, may emerge. The molecular tilting changes slightly from the unwound SmC$_A^*$ directions towards either of the unwound SmC* directions, and the tilt planes in neighbouring layers are no longer parallel to each other; P_b at a boundary is neither par-

allel nor perpendicular to the tilt plane in either of the neighbouring smectic layers.

The unwinding process and the pretransitional effect may be virtually indistinguishable if the helicoidal pitch is short and the threshold field is low, or if the applied field changes very quickly. Because of the antiferroelectric unwound herringbone structure, however, the unwinding process in SmC$_A^*$ alone does not induce any apparent tilt angle; the small induced apparent tilt angle in SmC$_A^*$ clearly results from the pretransitional effect and increases almost linearly with the applied field from zero up to the threshold. The pretransitional effect causes a parabolic response in the optical transmittance versus applied electric field in a homogeneously aligned (smectic layer perpendicular to substrate) cell between crossed polarizers, one of which is set parallel to the smectic layer normal.

In order to see more clearly the temporal behaviour of electric-field-induced phase transitions between SmC$_A^*$ and SmC*, it is useful to observe the optical transmittance when the applied field is changed stepwise across the threshold, $E_{th,h}$ or $E_{th,l}$, the results for which are summarized in Figure 17 [36]. The antiferroelectric to ferroelectric (AF→F) switching (Figures 17(a) and 17(c)) critically depends on E_h, but not on E_l; it contains two components, fast and slow. The fast component increases at the expense of the slow component as E_h becomes higher. The slow component is seriously affected by E_h and the response time representing the transmittance change from 10% to 90% is approximately proportional to $(E_h - E_{th,h})^4$. The reverse switching (F→AF) (Figure 17b and d) critically depends on E_l, but not on E_h; it consists of the slow component only. The response time is approximately proportional to $(E_{th,l} - E_l)^7$. A hump is observed when E_l becomes sufficiently negative, i.e. when $E_l \leq -1.6$ V · μm^{-1}.

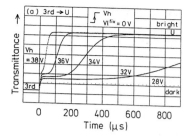

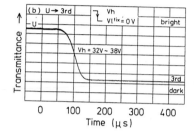

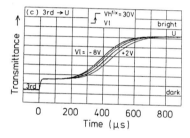

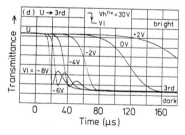

Figure 17. Optical transmittance change with crossed polarizers as a function of time after changing the electric field stepwise. Measurements were made using a 2.5 μm thick, homogeneously aligned cell of (R)-TFMNPOBC at 80°C by setting the polarizer in such directions that the antiferroelectric state becomes dark. The stepwise changes are (a) $E_l^{fix} = 0$ V μm^{-1} → $E_h = 11.2-15.2$ V μm^{-1}; (b) $E_h = 11.2-15.2$ V μm^{-1} → $E_l^{fix} = 0$ V μm^{-1}; (c) $E_l = -3.2-0.8$ V μm^{-1} → $E_n^{fix} = 12$ V μm^{-1}; (d) $E_n^{fix} = 12$ V μm^{-1} → $E_l = -3.2-0.8$ V μm^{-1}. U refers to uniform ferroelectric and 3rd to helicoidal antiferroelectric.

The observed ratio of the fast component to the total change in transmittance was confirmed to be almost consistent with the estimated ratio from the maximum induced apparent tilt angle during the pretransitional effect and the tilt angle after the transition to SmC*. Therefore, the fast component represents the pretransitional effect, which involves no macroscopic domain formation; when observing the effect through a microscope, the viewing field changes uniformly. The slow component represents a change due to the SmC$_A^*$–SmC* (AF→F) phase transition. Stroboscopic microscope observation has revealed that the transition always accompanies the movement of the domain (phase) boundaries of characteristic shape, as shown in Figure 18 [36, 37].

The hump observed in Figure 17(d) originates from the F→F switching (process) which occurs temporarily. Since the domain boundary propagates rather slowly, so that the middle part of the cell stays ferroelectric for a while (Figure 18), the director in this part may not switch directly from F to AF, but tends to rotate on the cone from F(+) to F(−), or vice versa, and then to return to AF, if E_l is negative. Therefore, the apparent tilt angle χ overshoots into the opposite side, crossing $\chi=0°$. The overshoot can actually be seen in the stroboscopic photomicrographs shown in Figure 18 (b and c), which were obtained by choosing the crossed polarizer directions in such a way that F(+) becomes dark. The bright (green) area in the middle is F(−) and the intermediate (blue) area appearing at the electrode edges (the right and left sides of the photomicrographs) is AF. The boundary of characteristic shape between the AF and F(−) regions moves towards the middle. The dark (black) area observed in Figure 18 (b and c) clearly indicates that the F(−) → F(+) process really does occur; although not shown in Figure 18, the F(+) → AF process follows

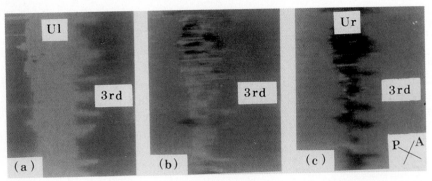

Figure 18. Stroboscopic photomicrographs recorded at (a) 0.145 V (253 ms), (b) 0.290 V (256 ms) and (c) 0.339 V (257 ms) by linearly changing the voltage from −48.4 V (0 ms) to 48.4 V (500 ms) in a 4.6 μm thick, homogeneous cell of (S)-MHPOBC at 80 °C. One of the ferroelectric states is chosen to be dark, overshooting to which is actually observed as clearly seen in (b) and (c). Ul and Ur refer to ferroelectric states tilted uniformly to the left F(−) and to the right F(+), respectively, and 3rd refers to helicoidal antiferroelectric state (A).

afterwards, and the whole area becomes blue.

These switching processes are also identified in the polarization reversal current measured by applying the triangular electric field (Figure 19). If the temporal rate of change of an applied electric field dE/dt is not large, two bell-shaped peaks may typically appear, which correspond to the two processes $F(−) \rightarrow AF$ and $AF \rightarrow F(+)$. As dE/dt increases, the $F(−) \rightarrow AF$ bell-shaped

peak shows a cusp; with further increase in dE/dt, the cusp becomes a conspicuous sharp peak, indicating the $F(−) \rightarrow F(+)$ process, and even the negative polarization reversal current, which results from the backward process $F(+) \rightarrow AF$, is observed (Figure 19). Note that the small cusp was observed in the initial measurement made by Chandani et al. [29], although its significance was not realized at that time.

In this way, it becomes clear that, phenomenologically, the pretransitional effect, as well as the phase transition between SmC_A^* (AF) and SmC^* (F), can be described by the rotation of the azimuthal angle ϕ; tilted molecules in alternate smectic layers rotate in the same direction, but those in the neighbouring smectic layers rotate in the opposite direction [38, 39]. Let us consider what happens during biased or hindered rotational motion about the molecular long axis. In Sec. 3.2.3 we concluded that the biased or hindered direction ψ_0 in SmC_A^* (ψ_0^{AF}) is substantially different from that in SmC^* (ψ_0^F); more generally speaking, ψ_0^{AF} and ψ_0^F may be regarded as representing the rotational states about the molecular long axes in SmC_A^* and SmC^*, respectively. From this

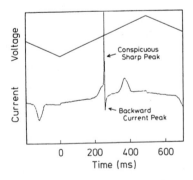

Figure 19. Polarization reversal current in a 4.6 μm thick cell of (S)-MHPOBC at 80 °C by applying a ±48.4 V triangular wave voltage of 1 Hz. Overshooting to the other ferroelectric state is observed as a conspicuous sharp peak near 0 V followed by a backward current peak.

point of view, the pretransitional effect is considered as a slight change in ψ_0 from ψ_0^{AF} towards ψ_0^{F}. Note the essential point that, not only the magnitude and direction of P_b at a boundary, but also the molecular tilt directions (azimuthal angle) in the neighbouring smectic layers of the boundary, are determined by the molecular rotational states. We have tried to observe this change in ψ_0 directly by means of polarized fourier transform IR (FTIR) spectroscopy, but have so far been unsuccessful. In the rotation of the azimuthal angle ϕ during the $F(-) \to F(+)$ process, or vice versa, the molecular rotational states have been observed to some extent by means of polarized dispersive IR spectroscopy in the submicrosecond regime [40].

3.3.2 Pretransitional Effect in Antiferroelectric Liquid Crystal Displays

Antiferroelectric liquid crystal displays (AFLCDs) have been relevant to flat panel displays since the discovery of AFLCs. Their characteristic features can be summarized as follows:

- In the SmC$_A^*$ phase stable at zero electric field, the optical axis is almost parallel to the rubbing direction. This favours the ease and stability of alignment of the molecules. Nakagawa et al. [41] have shown that the smectic layer normal in SmA* does not coincide exactly with the rubbing direction, deviating clockwise by several degrees for (S)-MHPOBC relative to the substrate.
- There is no permanent spontaneous polarization, but polarization emerges when needed. Hence the switching may not be hampered by the so-called 'ghost effect' or the unnecessarily strong, difficult to erase, memory effect.

- The switching is characterized by the distinctive d.c. threshold and hysteresis, which allow matrix addressing of large capacity displays, if use is made of a positive or negative bias field.
- The use of a symmetrical bias field, positive in one frame and negative in the next, also ensures an ideal viewing angle, much wider than that in SSFLDs, although the pretransitional effect that induces a small increase in the apparent tilt angle in SmC$_A^*$ may slightly decrease the contrast ratio.
- The quasi-bookshelf smectic layer structure is realized and the twisted state of the director alignment is inherently avoided. Hence the relatively high contrast ratio of 20–30 is easily obtained.
- The switching response time may become as short as 50 μs or better under acceptable driving conditions by optimizing the materials used.
- Stripe domains parallel to the smectic layer appear from the electrode edges quite regularly during switching, and the speed of movement of their boundaries depends critically on the field applied. This fact is utilized in controlling the grey scale, in combination with the distinctive d.c. threshold and hysteresis.
- Self-recovery from alignment damage caused by mechanical and thermal shocks is observed during operation. This results from the reversible layer switching between the bookshelf and the quasi-bookshelf structures, which occurs in the network of characteristic stripe defects parallel to the layer normal.

Because of these attractive characteristics, much attention has been paid to AFLCDs, and two prototype AFLCDs with passive matrix (PM) addressing have been developed and exhibited. One is a 6-inch, video rate, full-colour display by Nippondenso

Co., Ltd. [42], and the other is a 5.5-inch, VGA monochrome display by Citizen Watch Co., Ltd. [43], both of which have an extremely wide viewing angle with a relatively high contrast ratio of 30. What prevents AFLCDs from achieving much higher contrast ratios is the pretransitional effect, which manifests itself as a slight increase in the transmittance below the threshold (see Sec. 3.3.1). Ironically, during attempts to develop materials that can suppress the pretransitional effect, we have encountered materials which show a large pretransitional effect and, at the same time, a remarkable decrease in the threshold field strength. Hence we came to consider that enhancement of the pretransitional effect, in combination with the use of active matrix

(AM) addressing, is another way of endowing liquid crystals with attractive display characteristics [34,35].

Taking some compounds and their mixtures as examples, let us show in a concrete way how the threshold becomes lower as the pretransitional effect is enhanced. Table 2 [44, 45] shows that, in a series of compounds with the general molecular structure, an ether linkage and its position in the chiral end chain influence the threshold field $E_{th,h}$ in a subtle way. When the ether linkage is located at the chain terminus, $E_{th,h}$ becomes higher than that of the reference substance of equal chain length but having no ether linkage. As the linkage moves towards the core, $E_{th,h}$ becomes lower and, finally, SmC* (F) appears instead of SmC$_A^*$ (AF).

Table 2. The effect of an ether linkage position on the threshold field strength.

$C_{11}H_{23}O-\bigcirc\bigcirc-CO_2-\bigcirc-CO_2-\overset{*}{C}H(CF_3)C_mH_{2m}O_lC_nH_{2n+1}$ **1**

l	m	n	E_{th} (V μm^{-1})
1	4	3	–
1	5	2	3.4
1	6	1	15.7
0	0	8	6.5

Table 3. The effect on the contrast ratio of substituting the phenyl rings in **1** ($l=1$, $m=4$, $n=1$, **1a**) with fluorine.

$C_{11}H_{23}O-\overset{Y_3}{\bigcirc}\bigcirc-CO_2-\overset{Y_1}{\underset{Y_2}{\bigcirc}}-CO_2-\overset{*}{C}H(CF_3)C_4H_8OCH_3$ **1a**

Y_1	Y_2	Y_3	Contrast ratio [a]
H	H	H	36
H	F	H	27
F	H	H	12
H	H	F	31
H	F	F	28
F	H	F	10

[a] Measured at 20 °C below the SmC* (SmA*)–SmC$_A^*$ phase transition temperature by using a 1.7 μm thick cell and applying a 0.1 Hz, ±40 V triangular wave voltage.

Blending further reduces $E_{th,h}$ as illustrated in Figure 20 [35]. Furthermore, the pretransitional effect can be enhanced by suitably substituting phenyl rings with fluorine [35]. In order to evaluate the pretransitional effect we measured the contrast ratio T_F/T_{AF} just below $E_{th,h}$; the smaller the contrast ratio, the larger the pretransitional effect. Table 3 summarizes the results, indicating that substitution at Y_1 decreases the contrast ratio and hence enhances the pretransitional effect most remarkably.

In this way we can enhance the pretransitional effect and reduce the threshold $E_{th,h}$. An unanswered question is that of whether or not we can prepare a system where the threshold becomes zero, so that the pretransitional effect prevails and the AF–F phase transition occurs continuously. Surprisingly, it was not difficult to realize this property in a three-component mixture; the field-induced F–AF–F phase transition looks like a V-shaped switching (Figure 21) [34, 35]. Hereafter we will call this property 'thresholdless antiferroelectricity'. Although no detailed investigations have yet been made, the response of the system to an applied stepwise field is rather fast, as illustrated in Figure 22. When we observed the switching by means of polarizing optical microscopy, the visual field varied uniformly and continuously without showing any irregularities, indicating the boundary movement characteristic of tristable switching or the disclination lines caused by the helicoidal unwinding of a ferroelectric liquid crystal. The V-shaped switching is totally different from tristable switching and helicoidal unwinding. We still need to optimize the V-shaped switching characteristics by further developing suitable materials, but the results obtained with the materials available so far appear to indicate the possibility of realizing:

- a tilt angle of more than 35°;
- a field of less than 2 V μm^{-1} to complete the AF–F switching;
- light transmission that is almost linear with respect to the applied field, and free from hysteresis;
- a AF–F switching time of less than 50 μs;
- a contrast ratio as high as 90; and
- an extremely wide viewing angle of more than 60°.

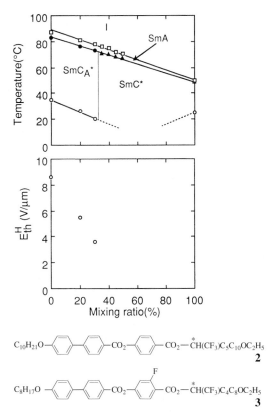

Figure 20. The phase diagram of a binary mixture of **2** and **3** and the threshold field versus mixing ratio. The phase boundary between the AF and F regions is parallel to the temperature axis and hence the stability of the AF phase becomes equal to that of the F phase in the binary mixture at this particular boundary mixing ratio. As the mixing ratio approaches the boundary, $E_{th,h}$ decreases, although it could not be confirmed that it diminished to zero.

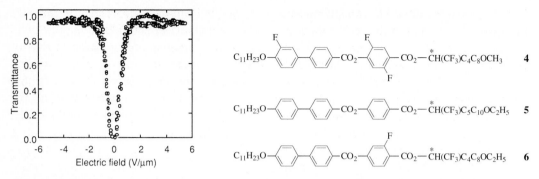

Figure 21. The V-shaped switching observed at 25 °C by applying a 0.179 Hz, ±7 Vpp triangular wave in a 1.4 µm thick, homogeneous cell of a mixture of **4**, **5** and **6** in the mixing ratio 40 : 40 : 20 (wt%), which has a tilt angle θ of 35.7°.

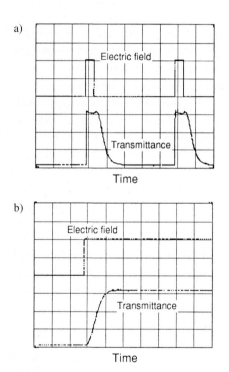

Figure 22. Temporal behaviour observed at 25 °C by applying a +4 V rectangular wave to the same cell as that in Figure 21: (a) 2 ms/division; (b) 50 µs/division.

A small cell, although very primitive, has actually been made, and demonstrates these attractive display characteristics.

3.3.3 Langevin-type Alignment in SmC$_R^*$ Phases

We will begin with speculation about thresholdless antiferroelectricity in terms of the molecular rotational motion which plays an essential role in the emergence of in-plane spontaneous polarizations in liquid crystals. We concluded in Sec. 3.2.3 that the biased or hindered direction ψ_0 in SmC$_A^*$ (ψ_0^{AF}) is substantially different from that in SmC* (ψ_0^F); more generally, ψ_0^{AF} and ψ_0^F may be regarded as representing the rotational states around the molecular long axes in SmC$_A^*$ and SmC*. We also concluded in Sec. 3.3.1 that the pretransitional effect is considered as a slight change in ψ_0 from ψ_0^{AF} towards ψ_0^F. Based on these conclusions, we speculate as follows. In the phase showing thresholdless antiferroelectricity, the biased or hindered rotational motions (more generally, the molecular rotational states in SmC$_A^*$ and SmC*) characterized by ψ_0^{AF} and ψ_0^F have nearly the same energies and, furthermore, the barrier between them diminishes, so that the states characterized by any ψ_0 between ψ_0^{AF} and ψ_0^F are equally thermally excited at zero electric field.

As noted in the previous section, this arbitrariness in ψ_0 makes the molecular tilting direction uncorrelated between adjacent layers. Figure 23 illustrates a simplified model of the phase SmC_R^* that shows thresholdless antiferroelectricity. The director tilting is uniform and has constant polar and azimuthal angles in a smectic layer, but its azimuthal angle varies randomly from layer to layer. In-plane spontaneous polarizations exist at the smectic layer boundaries, their orientation being random and their magnitude varying from boundary to boundary. Random tilting of the smectic layers ensures the disappearance of the net spontaneous polarization, but an applied electric field induces it according to a Langevin-type equation [46]. The difference from the ordinary Langevin equation is that we are dealing with very large effective dipole moments of variable magnitude because many molecules interact cooperatively and produce in-plane spontaneous polarizations, and their rotation is restricted in the two-dimensional space.

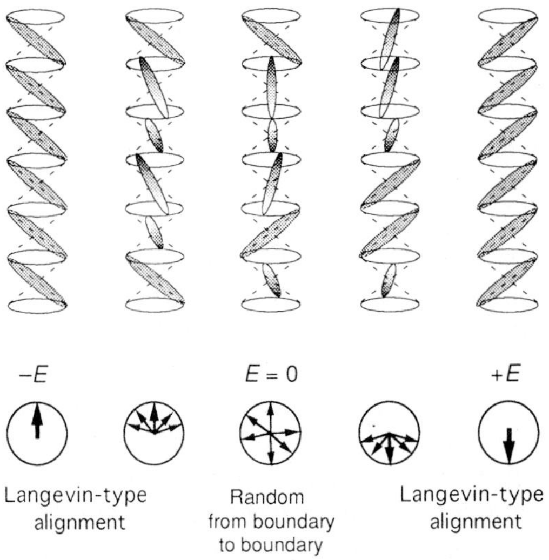

The light transmittance was obtained by assuming that the effective dipole moments have the same magnitude p_{eff} and follow the same distribution as the c director. The two-dimensional and three-dimensional Langevin alignments are given by

$$\langle\cos\phi\rangle = \frac{\int_0^\pi \exp(x\cos\phi)\cos\phi\,d\phi}{\int_0^\pi \exp(x\cos\phi)\,d\phi} \tag{9}$$

and

$$\coth x - \frac{1}{x} \tag{10}$$

respectively, where $x = p_{eff}E/(kT)$ (k is the Boltzmann constant and T is temperature) [46]. Figure 24 shows the numerically calculated results; the two-dimensional case aligns more steeply than the three-dimensional one. In a rough approximation, it is convenient to use an apparent tilt angle

$$\langle\cos\phi\rangle = \cos\theta_{ap} \tag{11}$$

We will further assume that the index of the ellipsoid is uniaxial with its eigen values, $n_\parallel$ and $n_\perp$, independent of the applied field.

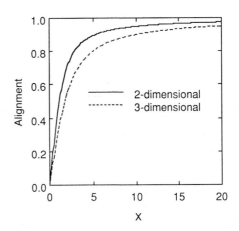

Figure 23. A simplified model of the phase that shows thresholdless antiferroelectricity.

Figure 24. The two- and three-dimensional Langevin alignments.

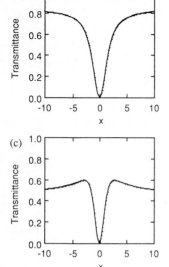

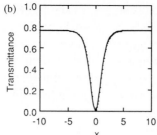

Figure 25. Calculated switching behaviour for cell thickness $d = 1.55$ μm, $n_\perp = 1.5$, wavelength of incident light $\lambda = 500$ nm:
(a) $n_\parallel = 1.65$, $\Delta n\, d/\lambda = 0.45$;
(b) $n_\parallel = 1.71$, $\Delta n\, d/\lambda = 0.63$;
(c) $n_\parallel = 1.75$, $\Delta n\, d/\lambda = 0.75$.

Some calculated results are shown in Figure 25. The switching behaviour in the final stage near the bright ferroelectric state naturally depends on a parameter $\Delta n \cdot d/\lambda$, where d is the cell thickness and λ is the wavelength of the incident light. In this approximation, when $\Delta n \cdot d/\lambda = 0.63$, the behaviour becomes ideal.

3.4 Antiferroelectric Liquid Crystal Materials

3.4.1 Ordinary Antiferroelectric Liquid Crystal Compounds

Because of their potential application in display devices, a large number of, probably almost 1000, AFLC and related compounds have been synthesized in Japan. Most of these have been reported in Japanese Patent Gazettes; some have been reported in scientific reports and papers, but many of these are written in Japanese, in particular, those in the Abstracts of Japanese Liquid Crystal Conferences. As access to these publications may not be easy for non-Japanese scientists and engineers in the liquid crystal community, and as these sources contain valuable information, we will try to summarize those AFLC compounds that have been reported mainly in Japan. Note that the materials characterized by the V-shaped switching due to thresholdless antiferroelectricity, which form one of the main topics of this discussion, are reported in these publications. Almost all the related Gazettes and Abstracts so far publicized are listed in [47] and [48]. Fortunately, the Abstracts contain titles and authors together with institutions written in English and chemical structural formulae used in the Gazettes and the Abstracts are international.

The phase sequence of a prototype AFLC, MHPOBC, has been known to be complex since the early stages of the investigation of antiferroelectricity in liquid crystals [1, 6, 49]. The origin and the structures of the subphases have been well documented [6, 50]. However, it is not easy to identify the subphases properly in newly synthesized compounds. Moreover, the existence of subphases as well as the structures may be critically influenced by the substrate interfaces. Both the subphases, SmC_α^*, SmC_γ^*, etc., and SmC_A^* itself are quite similar to SmC^*, and the differences in free energy between them must, therefore, be very small. As MHPOBC was first reported as a ferroelectric liquid crystal, the phase identifications given in the Gazettes and Abstracts are usually primitive and not always correct; they frequently disregard the details of the subphases, simply specifying them as ferroelectric or antiferroelectric. Therefore, when we refer to these publications, we must be aware of this ambiguity; nonetheless, these reference sources are very useful. Note that the most general subphase sequence currently known is [50–52]:

$$SmI_A^*/SmI^* - \boxed{SmC_A^*} - FI_L - SmC_\gamma^* - FI_H$$
$$- AF - Ferri/anti - \boxed{SmC^*} - SmC_\alpha^* - SmA^*$$

The chemical structural formulae of MHPOBC and another prototype AFLC, TFMHPOBC, are given in Figures 1 and 26, respectively. Most AFLC compounds so far synthesized have similar molecular structures to those of MHPOBC and TFMHPOBC. The structure can be considered to consist of three sub-structures: the chiral chain, the central core, and the nonchiral chain (Figure 26). In the chiral chain, it may be convenient to differentiate between the connector to the central core and the remaining part of the chiral chain; here-

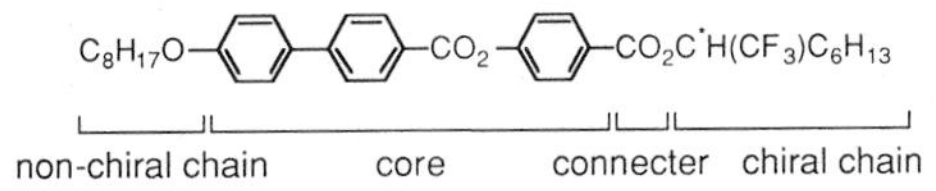

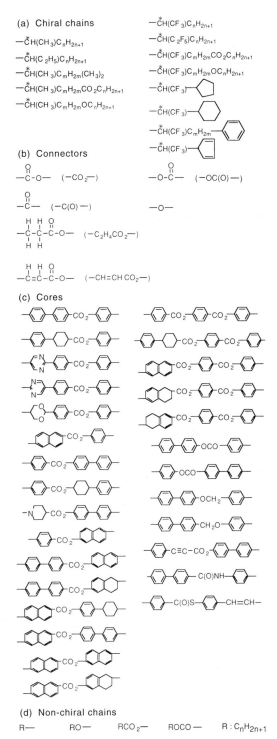

(a) Chiral chains

(b) Connectors

(c) Cores

(d) Non-chiral chains

R—   RO—   RCO$_2$—   ROCO—   R : C_nH_{2n+1}

Figure 27. Some typical sub-structures of AFLC compounds: (a) chiral chains; (b) connectors; (c) cores; (d) non-chiral chains.

$C_8H_{17}O$—⬡—⬡—CO_2—⬡—$CO_2\overset{*}{C}H(CF_3)C_6H_{13}$

⌞___non-chiral chain___⌟ ⌞__core__⌟ ⌞_connecter_⌟⌞_chiral chain_⌟

Figure 26. Structural formula of TFMHPOBC, illustrating four substructures: non-chiral chain, central core, connector, and chiral chain.

after we will designate these the 'connector' and the 'chiral chain', unless otherwise stated. The types of substructure so far studied (i.e. chiral chains, connectors, cores and non-chiral chains) are summarized in Figure 27. In this section, the chemical structural formulae are drawn in such a way that the chiral chains are located on the right-hand side.

Chiral chains play a critical role in determining the stability of the antiferroelectricity. The variety of chains is rather limited. The most common chiral chain is $-*CH(CF_3)C_nH_{2n+1}$, which stabilizes the antiferroelectricity much more than does $-*CH(CH_3)C_nH_{2n+1}$, as illustrated in Figure 28; the odd–even effect is clearly observed in $-*CH(CH_3)C_nH_{2n+1}$, while only the SmC_A^* phase occurs in $-*CH(CF_3)C_nH_{2n+1}$. Trifluoromethylalkoxyalkyl chains, $-*CH(CF_3)C_mH_{2m}OC_nH_{2n+1}$, give a low threshold field and fast response

(see Sec. 3.3.2). The pretransitional effect is very small in some compounds containing trifluoromethylcycloalkane chains, such as $-*CH(CF_3)-C_6H_{11}$ and $-*CH(CF_3)-C_5H_9$ (see e.g. **7** and **8**).

$$C_{10}H_{21}O-\langle\rangle-\langle\rangle-CO_2-\langle\rangle-CO_2-\overset{*}{C}H(CF_3)-\langle\rangle \qquad \textbf{7}$$

$$C_{10}H_{21}O-\langle\rangle-\langle\rangle-CO_2-\langle\rangle-CO_2-\overset{*}{C}H(CF_3)-\langle\rangle \qquad \textbf{8}$$

Monofluoro- and difluoromethylalkyl chains do not stabilize antiferroelectricity. Racemates do not exhibit antiferroelectricity, but they do frequently have the herringbone structure. Moreover, Nishiyama and coworkers [53, 54] have shown that chirality is not essential for the appearance of the herringbone structure; the swallow-tailed compounds (e.g. **9**),

$$C_9H_{19}O-\langle\rangle-C\equiv C-CO_2-\langle\rangle-\langle\rangle-CO_2-CH(C_3H_7)_2 \qquad \textbf{9}$$

which do not have an asymmetric carbon atom, have the herringbone structure, and

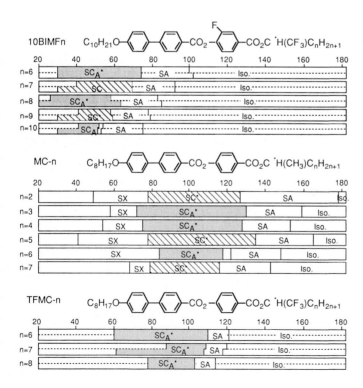

Figure 28. Odd–even effect with chiral end chains (Please note: In the figure the old nomenclature – S instead of Sm and Iso instead of I – is used).

the addition of a small amount of chiral dopant induces tristable switching.

Some connectors are listed in Figure 27(b). Almost all AFLC compounds have a carbonyloxy connector between the asymmetric carbon atom and the ring in the core, but some compounds [55] have a carbonyl or oxycarbonyl connector (Figure 29). Inui [56] studied the alkylcarbonyloxy connectors and found that $-C_2H_4COO-$ exhibits antiferroelectricity but $-CH_2COO-$ does not; although $-CH=CH-$ is better regarded as included in the core, cinnamates also show antiferroelectricity.

A characteristic feature of core structures is that there are three or more rings, at least two of which are not directly connected but are separated by a flexible group. Some examples are given in Figure 27(c). AFLC compounds containing only two rings are quite exceptional. Four-ring compounds usually show antiferroelectricity at temperatures above 100°C. The flexible connecting groups known so far are $-C(O)O-$, $-OC(O)-$, $-OCH_2-$, $-CH_2O-$, $C(O)S-$ and $-C(O)NH-$. Of these, $-COO-$ is the most commonly used. An increase in the number of flexible connecting groups may stabilize the antiferroelectricity, but makes the compound viscous. Rings used in the core are benzene, cyclohexane, tetralin, naphthalene, 1,3-dioxane, piperidine, pyrimidine, etc. Of these the benzene ring is the most commonly used. Six-membered rings may be connected directly, forming biphenyl, phenylpyrimidine, pyrimidine phenyl, dioxane phenyl, phenylcyclohexane, piperidine phenyl, etc. Halogen-substituted rings are also used, as illustrated in Figure 29. In this way, a variety of core structures can be used to obtain AFLC compounds.

$C_8H_{17}O$—◯—◯—CO_2—◯—$C(O)-\overset{*}{C}H(CH_3)C_6H_{13}$

$C_8H_{17}O$—◯—CO_2—◯—◯—$C(O)-\overset{*}{C}H(CH_3)C_6H_{13}$

$C_8H_{17}O$—◯—CO_2—◯—$C(O)-\overset{*}{C}H(C_2H_5)C_6H_{13}$

$C_{10}H_{21}O$—◯—◯—CO_2—◯—$OCO-\overset{*}{C}H(CH_3)C_3H_6CH(CH_3)_2$

$C_{10}H_{21}O$—◯—◯—CO_2—◯(F)—$OCO-\overset{*}{C}H(CH_3)C_3H_6CH(CH_3)_2$

$C_{10}H_{21}O$—◯(F)—◯—CO_2—◯(F)—$OCO-\overset{*}{C}H(CH_3)C_3H_6CH(CH_3)_2$

$C_{10}H_{21}O$—◯—CO_2—◯—◯—$OCO-\overset{*}{C}H(CH_3)C_3H_6CH(CH_3)_2$

$C_{10}H_{21}O$—◯(F)—CO_2—◯—◯(F)—$OCO-\overset{*}{C}H(CH_3)C_3H_6CH(CH_3)_2$

$C_{10}H_{21}O$—◯—◯—CO_2—◯—$CH_2CO_2-\overset{*}{C}H(CF_3)C_6H_{13}$ (FLC)

$C_{10}H_{21}O$—◯—◯—CO_2—◯—$C_2H_4CO_2-\overset{*}{C}H(CF_3)C_6H_{13}$

$C_{10}H_{21}O$—◯—◯—CO_2—[naphthalene]—$\overset{*}{C}H(CH_3)CO_2C_6H_{13}$

Figure 29. Some AFLC compounds containing a carbonyl, oxycarbonyl or alkylenecarbonyloxy connector.

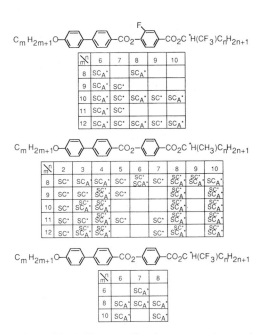

$C_mH_{2m+1}O$—◯—◯—CO_2—◯(F)—$CO_2C^*H(CF_3)C_nH_{2n+1}$

m \ n	6	7	8	9	10
8	SCA*		SCA*		
9	SCA*	SC*			
10	SCA*	SC*	SCA*	SC*	SCA*
11	SCA*	SC*			
12	SCA*	SC*	SCA*	SC*	SCA*

$C_mH_{2m+1}O$—◯—◯—CO_2—◯—$CO_2C^*H(CH_3)C_nH_{2n+1}$

m \ n	2	3	4	5	6	7	8	9	10
8	SC*	SCA*	SCA*	SC*	SC*/SCA*	SC*	SC*/SCA*	SC*/SCA*	SCA*
9	SC*	SC*	SC*/SCA*	SC*			SC*/SCA*		SC*/SCA*
10	SC*	SC*/SCA*	SC*/SCA*				SC*/SCA*		SC*/SCA*
11	SC*	SC*	SC*/SCA*	SC*			SC*	SC*/SCA*	SC*/SCA*
12	SC*	SC*/SCA*	SC*/SCA*				SC*	SC*/SCA*	SC*/SCA*

$C_mH_{2m+1}O$—◯—◯—CO_2—◯—$CO_2C^*H(CF_3)C_nH_{2n+1}$

m \ n	6	7	8
6		SCA*	
8	SCA*	SCA*	SCA*
10	SCA*		SCA*

Figure 30. Odd–even effect in compounds containing non-chiral and chiral end chains (Please note that the old nomenclature – S instead of Sm – is used).

Some non-chiral chains are listed in Figure 27(d). Alkoxy chains, RO−, are the most common. The alkanoyloxy chain, RCOO−, increases both the stability and the viscosity, while alkoxycarbonyl chains, ROCO−, decrease them. Simple alkyl chains, R−, may slightly decrease the stability and viscosity at the same time. The odd−even effect is less clear in non-chiral chains than in chiral chains, as illustrated in Figure 30.

3.4.2 Antiferroelectric Liquid Crystal Compounds with Unusual Chemical Structures

A homologous series of AFLC compounds, which have entirely different chemical structures from the ordinary ones characterized by having a carbonyl group in the connector, have been synthesized by Aoki and Nohira [57–59]. The chemical structures and the corresponding phase sequences are summarized in Figure 31. They identified the phases by checking miscibility with a TFMHPOBC-like compound as shown in Figure 32. The ordinary tristable switching, as illustrated in Figure 16(b), was observed optically, and is characterized by a d.c. threshold and hysteresis. Tristable switching was also observed electrically by detecting two current peaks that correspond to the appearance and disappearance of the spontaneous polarization. Aoki and Nohira further confirmed that the racemates have the SmC_A phase. According to the P_X model de-

$$C_8H_{17}O-\bigcirc-\bigcirc\stackrel{=N}{\underset{N}{\diagdown}}-\bigcirc-OC_2H_4\stackrel{*}{CH}(CF_3)-\bigcirc-OC_nH_{2n+1} \quad \textbf{10}$$

n	Phase transition temperatures on cooling step a)	Ps [nC/cm²] b)	θ [deg] b)
1		104	30
2		129	34
3		129	35
4		147	37
5		159	38
6		194	40

40 50 60 70 80 90 100 110

▨ Cr   ▦ SmcA*   ☐ Sm$_A$ ☐ I

a) cooling rate = 5K/min.   b) Tc-T = 5K.

Figure 31. Structural formula of AFLC compounds synthesized by Aoki and Nohira [57–59] and their phase sequences and physical properties (P_s, spontaneous polarization; θ, tilt angle).

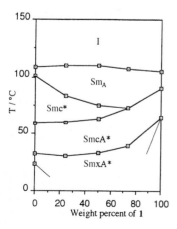

Figure 32. Miscibility test for **10** with $n = 1$ (see Figure 31) with a reference AFLC.

$$C_{10}H_{21}O-\bigcirc-\bigcirc-CO_2-\bigcirc-CO_2\stackrel{*}{CH}(CF_3)C_7H_{15} \quad \textbf{10a}$$

scribed in Secs. 3.2.3 and 3.2.4, the molecules are expected to be bent at the chiral carbon atom. It is an interesting problem for the future to study whether or not this is the case.

A second class of materials are pyrimidine host mixtures containing pyranose guests, some examples of which are given in Figure 33. Although the SmC$_A^*$ phase is not exhibited by the guest compounds themselves, Itoh et al. [60, 61]* have clearly confirmed the appearance of the SmC$_A^*$ in the mixture, the composition of which is specified in Figure 33, by observing the typical tristable switching. This finding is very important, suggesting as it does the possibility of developing practical guest–host type mixtures of AFLC materials.

Although detailed investigations have not yet been performed, some of the subphases associated with the SmC$_A^*$ phase seem to have been observed in the series of AFLC

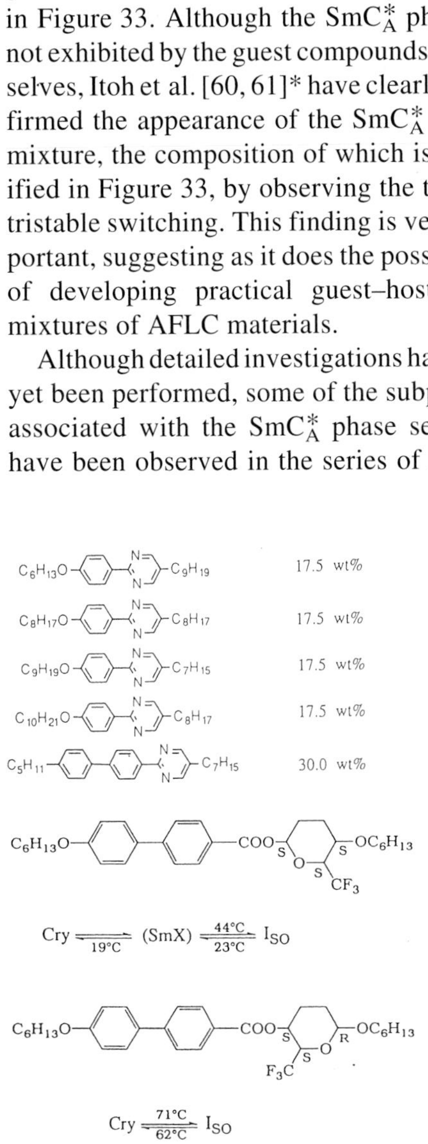

Cry $\xrightleftharpoons[19°C]{}$ (SmX) $\xrightleftharpoons[23°C]{44°C}$ I$_{SO}$

Cry $\xrightleftharpoons[62°C]{71°C}$ I$_{SO}$

Figure 33. Pyrimidine host mixtures containing pyranose guests.

compounds **11** synthesized by Wu et al. [62].

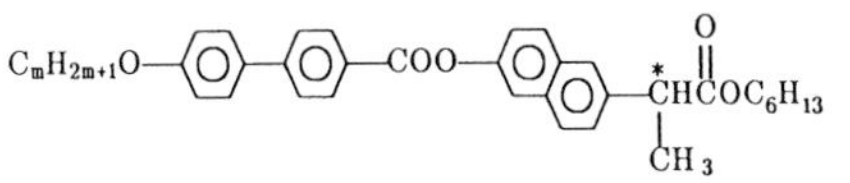

11

* **Note added in proof:** Quite recently, Itoh et al. reported that the phase in these pyrimidine host mixtures is not antiferroelectric but ferroelectric and that the typical tristable switching results from an ultrashort pitch of about 50 nm. (K. Itoh, Y. Takanishi, J. Yokoyama, K. Ishikawa, H. Takezoe and A. Fukuda, *Jpn. J. Appl. Phys.* **1997**, *36*, L784).

3.5 References

[1] A. D. L. Chandani, E. Gorecka, Y. Ouchi, H. Takezoe, A. Fukuda, *Jpn. J. Appl. Phys.* **1989**, *29*, L1265.
[2] J. W. Goodby, *J. Mater. Chem.* **1991**, *1*, 307.
[3] R. Blinc, *Condensed Matter News* **1991**, *1*, 17.
[4] P. E. Cladis, *Liq. Cryst. Today* **1992**, *2*, 7.
[5] A. Fukuda (Ed.), *The Future Liquid Crystal Display and its Materials*, CMC, Tokyo, 1992 (in Japanese): A. Fukuda, J. Li, Chap. 1, p. 1; Y. Yamada, Chap. 8, p. 102; K. Itoh, Chap. 13, p. 287; M. Johno, T. Yui, Chap. 14, p. 297; S. Inui, H. Iwane, Chap. 15, p. 308; and S. Nishiyama, Chap. 16, p. 318.
[6] A. Fukuda, Y. Takanishi, T. Isozaki, K. Ishikawa, H. Takezoe, *J. Mater. Chem.* **1994**, *4*, 997.
[7] I. Nishiyama, *Adv. Mater.* **1994**, *6*, 996.
[8] K. Miyachi, Ph. D. Thesis, Tokyo Institute of Technology, **1996**.
[9] R. B. Meyer, L. Liebert, L. Strzelecki, P. Keller, *J. Phys. (France)* **1975**, *36*, L69.
[10] R. B. Meyer, *Mol. Cryst. Liq. Cryst.* **1977**, *40*, 33.
[11] M. Luzar, V. Rutar, J. Seliger, R. Blinc, *Ferroelectrics* **1984**, *58*, 115.
[12] A. Yoshizawa, H. Kikuzaki, M. Fukumasa, *Liq. Cryst.* **1995**, *18*, 351.
[13] K. H. Kim, K. Ishikawa, H. Takezoe, A. Fukuda, *Phys. Rev. E* **1995**, *51*, 2166. (Erratum: *Phys. Rev. E* **1995**, *52*, 2120.)
[14] K. H. Kim, K. Miyachi, K. Ishikawa, H. Takezoe, A. Fukuda, *Jpn. J. Appl. Phys.* **1994**, *33*, 5850.
[15] K. Miyachi, J. Matsushima, Y. Takanishi, K. Ishikawa, H. Takezoe, A. Fukuda, *Phys. Rev. E* **1995**, *52*, R2153.

[16] A. D. L. Chandani, E. Gorecka, Y. Ouchi, H. Takezoe, A. Fukuda, *Jpn. J. Appl. Phys.* **1989**, *28*, L1265.

[17] A. Fukuda, Y. Takanishi, T. Isozaki, K. Ishikawa, H. Takezoe, *J. Mater. Chem.* **1994**, *4*, 997.

[18] Y. Takanishi, H. Hiraoka, V. K. Agrawal, H. Takezoe, A. Fukuda, M. Matsushita, *Jpn. J. Appl. Phys.* **1991**, *30*, 2023.

[19] K. Hori, K. Endo, *Bull. Chem. Soc. Jpn.* **1993**, *66*, 46.

[20] L. A. Beresnev, L. M. Blinov, M. A. Osipov, S. A. Pikin, *Mol. Cryst. Liq. Cryst.* **1988**, *158A*, 1.

[21] Y. Galerne, L. L. Liebert, *Phys. Rev. Lett.* **1991**, *66*, 2891.

[22] P. E. Cladis, H. R. Brand, *Liq. Cryst.* **1993**, *14*, 1327.

[23] H. R. Brand, P. E. Cladis, H. Pleiner, *Macromolecules* **1992**, *25*, 7223.

[24] I. Nishiyama, J. W. Goodby, *J. Mater. Chem.* **1992**, *2*, 1015.

[25] J. Prost, R. Bruinsma, *Ferroelectrics* **1993**, *148*, 25.

[26] R. Bruinsma, J. Prost, *J. Phys. II (France)* **1994**, *4*, 1209.

[27] Y. Galerne, L. Liebert, *Phys. Rev. Lett.* **1990**, *64*, 906.

[28] B. Jin, Z. H. Ling, Y. Takanishi, K. Ishikawa, H. Takezoe, A. Fukuda, M. Kakimoto, T. Kitazume, *Phys. Rev. E* **1996**, *53*, R4295.

[29] A. D. L. Chandani, T. Hagiwara, Y. Suzuki, Y. Ouchi, H. Takezoe, A. Fukuda, *Jpn. J. Appl. Phys.* **1988**, *27*, L729.

[30] A. D. L. Chandani, E. Gorecka, Y. Ouchi, H. Takezoe, A. Fukuda, *Jpn. J. Appl. Phys.* **1989**, *28*, L1265.

[31] M. Johno, A. D. L. Chandani, J. Lee, Y. Ouchi, H. Takezoe, A. Fukuda, K. Itoh, T. Kitazume, *Proc. SID* **1990**, *31*, 129.

[32] M. Yamawaki, Y. Yamada, N. Yamamoto, K. Mori, H. Hayashi, Y. Suzuki, Y. S. Negi, T. Hagiwara, I. Kawamura, H. Orihara, Y. Ishibashi in *Proceedings of the 9th International Display Research Conference (Japan Display '89, Kyoto)*, **1989**, p. 26.

[33] K. Miyachi, J. Matsushima, Y. Takanishi, K. Ishikawa, H. Takezoe, A. Fukuda, *Phys. Rev. E* **1995**, *52*, R2153.

[34] A. Fukuda in *Proceedings of the 15th International Display Research Conference (Asia Display '95, Hamamatsu)*, **1995**, p. S6-1.

[35] S. Inui, N. Iimura, T. Suzuki, H. Iwane, K. Miyachi, Y. Takanishi, A. Fukuda, *J. Mater. Chem.* **1996**, *6*, 671.

[36] M. Johno, K. Itoh, J. Lee, Y. Ouchi, H. Takezoe, A. Fukuda, T. Kitazume, *Jpn. J. Appl. Phys.* **1990**, *29*, L107.

[37] X. Y. Wang, P. L. Taylor, *Phys. Rev. Lett.* **1996**, *76*, 640.

[38] T. Akahane, A. Obinata, *Liq. Cryst.* **1993**, *15*, 883.

[39] H. Pauwels, A. de Meyere, J. Fornier, *Mol. Cryst. Liq. Cryst.* **1995**, *263*, 469.

[40] M. A. Czarnecki, N. Katayama, M. Satoh, T. Watanabe, Y. Ozaki, *J. Phys. Chem.* **1995**, *99*, 14101.

[41] K. Nakagawa, T. Shinomiya, M. Koden, K. Tsubota, T. Kuratate, Y. Ishii, F. Funada, M. Matsuura, K. Awane, *Ferroelectrics* **1988**, *85*, 39.

[42] N. Yamamoto, N. Koshoubu, K. Mori, K. Nakamura, Y. Yamada, *Ferroelectrics* **1993**, *149*, 295.

[43] E. Tajima, S. Kondon, Y. Suzuki, *Ferroelectrics* **1994**, *149*, 255.

[44] S. Inui, T. Suzuki, N. Iimura, H. Iwane, H. Nohira, *Ferroelectrics* **1993**, *148*, 79.

[45] M. Johno, T. Matsumoto, H. Mineta, T. Yui, *Abstracts II: Chemical Society of Japan 67th Spring Meeting*, Chemical Society of Japan, Tokyo, **1994**, Abstract 1 1B31 12.

[46] G. M. Barrow, *Physical Chemistry*, 5th edn, McGraw-Hill, New York, **1988**, p. 671.

[47] Japanese Patent Gazettes: H1-213390; H1-316339; H1-316367; H1-316372; H2-28128; H2-40625; H2-78648; H2-131450; H2-153322; H2-160748; H2-173724; H2-176723; H2-230117; H2-270862; H3-11041; H3-68686; H3-83951; H3-121416; H3-123759; H3-141323; H3-167518; H3-181445; H3-197444; H3-197445; H3-218337; H3-223263; H3-291270; H3-292388; H4-29978; H4-46158; H4-46159; H4-55495; H4-96363; H4-82862; H4-89454; H4-103560; H4-108764; H4-117345; H4-159254; H4-173779; H4-178353; H4-198155; H4-217942; H4-235950; H4-244047; H4-261162; H4-261163; H4-266853; H4-266855; H4-266876; H4-282344; H4-305554; H4-305555; H4-312552; H4-316562; H4-327576; H4-338361; H4-338362; H4-342546; H4-342547; H4-360850; H4-360854; H4-372929; H5-1041; H5-4945; H5-17407; H5-17408; H5-32675; H5-32973; H5-39246; H5-58959; H5-65484; H5-70392; H5-85988; H5-85989; H5-97821; H5-98258; H5-105644; H5-105877; H5-132449; H5-140046; H5-140119; H5-155880; H5-163208; H5-163248; H5-186401; H5-186402; H5-201929; H5-201983; H5-202028; H5-213881; H5-221928; H5-230032; H5-230033; H5-230035; H5-247026; H5-255193; H5-247026; H5-255193; H5-255196; H5-311171; H5-320125; H5-331107; H6-25098; H6-25099; H6-25100; H6-32764; H6-32765; H6-32770; H6-65154; H6-108051; H6-116208; H6-128236; H6-145111; H6-211749; H6-228056; H6-247898; H6-247901; H6-247902; H6-271852.

[48] *Abstracts of Japanese Liquid Crystal Conferences*: **1990**, *16*, 2K101, 2K102, 2K103, 2K104, 2K105, 2K106; **1991**, *17*, 2F302, 2F315, 3F313,

3F314, 3F316, 3F318, 3F321, 3F322, 3F324; **1992**, *18*, 2B408, 2B409, 2B410, 2B411, 2B412, 2B417, 3B416, 3B417, 3B419, 3B420, 3B421, 3B422, 3B423, 3B424, 3B425; **1993**, *19*, 1A11, 1A12, 1A14, 1B13, 2A03, 2A08, 2A10, 3D08; **1994**, *20*, 1F603, 1G303, 2F610, 2F614, 2F617, 3G201, 3G203, 3G205, 3G206, 3G208, 3G213, 3G215, 3G216, 3G218; **1995**, *21*, 2C01, 2C04, 2C05, 2C09, 2C10, 2C12, 2C13, 2C14, 2C18, 3C07, 3C08.

[49] M. Fukui, H. Orihara, Y. Yamada, N. Yamamoto, Y. Ishibashi, *Jpn. J. Appl. Phys.* **1989**, *28*, L849.

[50] K. Itoh, M. Kabe, K. Miyachi, Y. Takanishi, K. Ishikawa, H. Takezoe, A. Fukuda, *J. Mater. Chem.* **1996**, *7*, 407.

[51] H. Hatano, Y. Hanakai, H. Furue, H. Uehara, S. Saito, K. Muraoka, *Jpn. J. Appl. Phys.* **1994**, *33*, 5498.

[52] T. Sako, Y. Kimura, R. Hayakawa, N. Okabe, Y. Suzuki, *Jpn. J. Appl. Phys.* **1996**, *35*, L114.

[53] I. Nishiyama, J. W. Goodby, *J. Mater. Chem.* **1992**, *2*, 1015.

[54] Y. Ouchi, Y. Yoshioka, H. Ishii, K. Seki, M. Kitamura, R. Noyori, Y. Takanishi, I. Nishiyama, *J. Mater. Chem.* **1995**, *5*, 2297.

[55] K. Kondo in *Extended Abstracts of 52nd Research Meeting*, JSPS 142 Committee of Organic Materials for Information Science (Sub-Committee of Liquid Crystals) **1993**, 1.

[56] S. Inui, Ph. D. Thesis, Saitama University, **1995**, p. 1 (in Japanese).

[57] Y. Aoki, H. Nohira, *Liq. Cryst.* **1995**, *18*, 197.

[58] Y. Aoki, H. Nohira, *Liq. Cryst.* **1995**, *19*, 15.

[59] Y. Aoki, H. Nohira, *Ferroelectrics* **1996**, *178*, 213.

[60] K. Itoh, M. Takeda, M. Namekawa, S. Nayuki, Y. Murayama, T. Yamazaki, T. Kitazume, *Chem. Lett.* **1995**, 641.

[61] K. Itoh, M. Takeda, M. Namekawa, Y. Murayama, S. Watanabe, *J. Mater. Sci. Lett.* **1995**, *14*, 1002.

[62] S.-L. Wu, D.-G. Chen, S.-J. Chen, C.-Y. Wang, J.-T. Shy, G. Hsiue, *Mol. Cryst. Liq. Cryst.* **1995**, *264*, 39.

Chapter VII
Synthesis and Structural Features

Andrew N. Cammidge and Richard J. Bushby

1 General Structural Features

The discovery of discotic liquid crystals is generally dated to the work of Chandrasekhar on the hexaesters of benzene [1], although the concept that liquid crystal phases should be formed by disk-like as well as rod-like molecules had already been recognized by DeGennes and others [2]. Structural similarities also exist with the carbonaceous phases, which have been known for some time [3]. The least ordered (usually highest temperature) mesophase formed by disk-like molecules is the nematic (N_D) phase which is directly comparable to the nematic phase formed by calamitics. The molecules in such a phase have orientational order but no positional order. Examples of N_D phases remain rather few, and more common are those phases in which the molecules stack into columns, the columns then being organized on a two-dimensional lattice. In some respects these columnar phases are analogous to the smectic mesophases of calamitic liquid crystals. However, whereas the smectic phases can be thought of as two-dimensional liquids (they have positional order in one dimension and are disordered in the other two), the columnar phases can be thought of as one-dimensional liquids (they have positional order in two dimensions and are disordered in the third). Columnar phases show rich polymesomorphism and they are normally classified at three levels; according to the symmetry of the two-dimensional array, the orientation of the core with respect to the column axis, and (sometimes) the degree of order within the column. The latter criterion has been the subject of a variety of different interpretations. The main types [4] are illustrated in Figure 1 and they are discussed in more detail in Chapter 8 in this volume. Typically, if a compound forms both a D_r[1] phase (rectangular columnar) and a D_h phase (hexagonal columnar) the D_r phases are found at lower temperatures. It should be noted, however, that exceptions exist. Determination of columnar mesophase structure can often be difficult, since the differences in the X-ray diffraction results can be quite subtle.

The archetypal discogenic molecule has a rigid, planar aromatic core with three-, four- or six-fold rotational symmetry, and generally six or more flexible side chains, each more than five atoms. However, exceptions to all these 'rules' are now quite common. Systems are known with low symmetry, a non-planar, non-aromatic core, and with side chains as short as three atoms.

We survey here the different classes of discotic mesogens, the methods by which they have been synthesized, and the phase behaviour claimed for each system. Interest in the development of new routes to discotic mesogens has provided a welcome stimulus to a somewhat unfashionable area of organic

---

[1] In this contribution the nomenclature D for a columnar mesophase of disk-like molecules has been used, but the reader should appreciate that D_h, D_r, D_t etc. are the same as Col_h, Col_r, Col_t etc.

Nematic Phase N_D

Disordered Columnar Hexagonal Phase D_{hd}

Common variations in stacking within the columns

Ordered (o)   Disordered (d)   Tilted (t)

Common variations in stacking of the columns

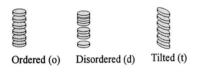

Hexagonal (h)   Rectangular (r)   Oblique (ob)

Figure 1. Schematic diagram illustrating the basis for the classification and nomenclature of the discotic mesophases. For the columnar phases the subscripts are often used in combination with each other. Hence D_{rd} refers to a (two dimensional) rectangular lattice of columns within which the discs are stacked in a disordered manner

synthesis, i.e. of polynuclear aromatic and heterocyclic systems. Quite a lot of what has been developed so far has exploited old, and in some cases very old, chemistry in which the polynuclear core is assembled through some 'one-off' condensation process. There is clearly a need to develop rational general synthetic strategies based on modern methods. Undoubtedly more of these will emerge in the next few years.

2 Aromatic Hydrocarbon Cores

2.1 Benzene

2.1.1 Esters and Amides

The hexasubstituted benzenes were the first discotic liquid crystals to be reported. Chandrasekhar and coworkers [1, 5] prepared a series of hexaesters of hexahydroxybenzene and found that the heptanoyl and octanoyl homologues were liquid crystalline. The synthetic route used, which has been adapted by others to prepare related derivatives, is shown in Scheme 1. Hexahydroxybenzene [6] is prepared from glyoxal and treated with an acid chloride to give the esters. Yields of the hexaester are improved if an argon atmosphere and/or a partial vacuum is used. Typically, no solvent is used for these reactions. The mesophase behaviour of a range of hexaesters is summarized in Table 1. All derivatives that have been characterized so far form an hexagonal mesophase, but the stability (temperature range) of the mesophase is extremely sensitive to

Scheme 1. Synthesis of the hexaesters of benzene. Reagents: (a) Na_2SO_3/O_2; (b) $SnCl_2/HCl$; (c) $RCOCl$ 14–87% [7].

Table 1. Transition temperatures for hexaesters (RCOO–) of benzene.

R	Phase transitions [a]	Refs.
C_6H_{13}	Cr 81 D_h 86 I	[1, 5, 8]
C_7H_{15}	Cr 82 D_h 84 I	[1, 5, 8]
OC_6H_{13}	Liquid	[7]
$CH_2OC_5H_{11}$	Cr 68 D_h 94 I	[7]
$C_2H_4OC_4H_9$	Cr 30 D_h 32 I	[7]
$C_3H_6OC_3H_7$	Cr 43 D_h 44 I	[7]
$C_5H_{10}OCH_3$	Cr 16 D_h 25 I	[7]
$CH_2OC_3H_7$	Cr 110 D_h 112 I	[7]
$CH_2OC_7H_{15}$	Cr 74 D_h 94 I	[7]
$CH_2SC_5H_{11}$	Cr 66 D_h 67 I (Cr 69 I)	[7, 9]
$C_2H_4SC_4H_9$	Cr 101 I	[9]
$C_3H_6SC_3H_7$	? 53 D 57 I	[9]
$C_5H_{10}SCH_3$	Cr 32 D 37 I	[9]
$CH(Me)C_6H_{13}$	Liquid	[10]
$CH_2CH(Me)C_5H_{11}$	Cr 96 I	[10]
$C_2H_4CH(Me)C_4H_9$	M 96 I	[10]
$C_3H_6CH(Me)C_3H_7$	M 102 I	[10]
$C_4H_8CH(Me)C_2H_5$	Cr 61 M 87 I	[10]
$C_5H_{10}CH(Me)CH_3$	Cr 85 M 87 I	[10]
$C_4H_8CH(Et)C_2H_5$	Cr 60 M_1 86 M_2 87 I	[10]
$C_5H_{10}CH(Me)C_2H_5$	Cr 52 M 75 I	[10]

[a] M, mesophase.

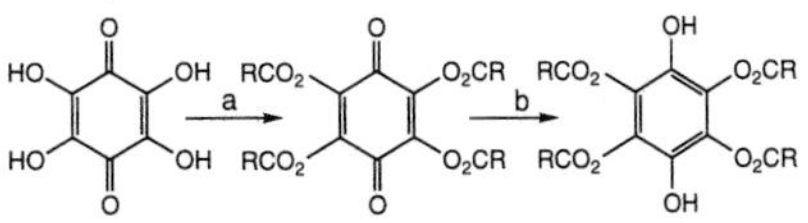

Scheme 2. Synthesis of the tetraesters of 1,4-benzo- and 1,4-hydro-quinone. Reagents: (a) RCOCl/pyridine reflux 30–60% [10]; (b) H⁺ 90–95% [11].

Table 2. Transition temperatures for the tetraesters (RCOO–) of 1,4-benzo- and 1,4-hydroquinone.

	R	Phase transitions	Refs.
Quinones	C_6H_{13}	Cr 55 D 57 I	[11]
	C_7H_{15}	Cr 68 D 70 I	[11]
Hydroquinones	C_6H_{13}	Cr 80 D 85 I	[13, 14]
	C_7H_{15}	Cr 79 D 82 I	[14]

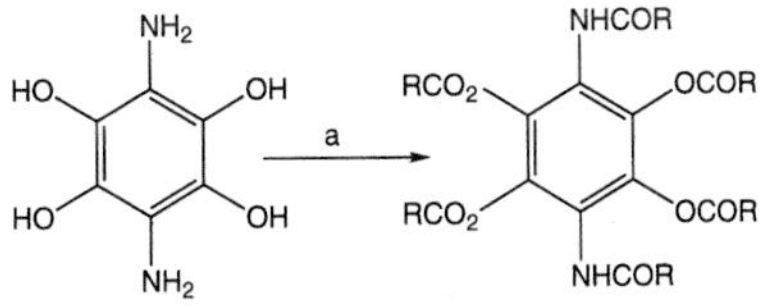

Scheme 3. Synthesis of *N,N'*-dialkanoyl-2,3,5,6-tetrakis(alkanoyloxy)-1,4-benzene diamines. Reagents: (a) RCOCl/pyridine [13].

heteroatoms in the side chain. The greatest stabilization is found for derivatives bearing a β-oxygen atom in the side chain. If the position of this atom is moved towards or away from the core, either the mesophase stability is drastically reduced or a mesophase is not observed at all [7]. Branched side chains stabilize the mesophase if the branch point is close to the middle, but destabilization is observed as it approaches the core [10].

Some partially (2,3,5,6-) acylated derivatives of hexahydroxybenzene have also been found to exhibit discotic liquid crystal phases, as have the related tetraesters of 2,3,5,6-tetrahydroxy-1,4-benzoquinone [11]. Their synthesis, shown in Scheme 2, is relatively straightforward. 2,3,5,6-Tetrahydroxy-1,4-benzoquinone [12] is acylated with an excess of the acid chloride in refluxing pyridine. Neutral work-up yields the quinone, whereas the hydroquinone is iso-

lated from an acidic work-up [13]. Both sets of compounds, however, have very narrow mesophase ranges (Table 2).

Another variation, employed by Matsunaga and coworkers, has been to replace some of the ester linkages of the benzene hexa-alkanoates with amide groups. The most heavily substituted members of this class of compounds, the *N,N'*-dialkanoyl-2,3,5,6-tetrakis(alkanoyloxy)-1,4-benzene diamines (**1**) [15], are prepared by total acylation of 3,6-diamino-1,2,4,5-tetrahydroxybenzene [16] (Scheme 3). The mesophase behaviour of this series is summarized in Table 3. It is immediately apparent that the introduction of the two amide groups stabiliz-

Table 3. Transition temperatures of N,N'-dialkanoyl-2,3,5,6-tetrakis(alkanoyloxy)-1,4-benzene diamines (**1**) (RCONH–/RCOO–).

R	Phase transitions [15]
C_4H_9	Cr 134 D_h 208 I
C_5H_{11}	Cr 71 D_h 209 I
C_6H_{13}	Cr 64 D_h 208 I
C_7H_{15}	Cr 75 D_h 205 I
C_8H_{17}	Cr 77 D_h 200 I
C_9H_{19}	Cr 77 D_h 199 I
$C_{10}H_{21}$	Cr 81 D_h 198 I
$C_{11}H_{23}$	Cr 89 D_h 197 I
$C_{13}H_{27}$	Cr 92 D_h 189 I
$C_{15}H_{31}$	Cr 94 D_h 190 I

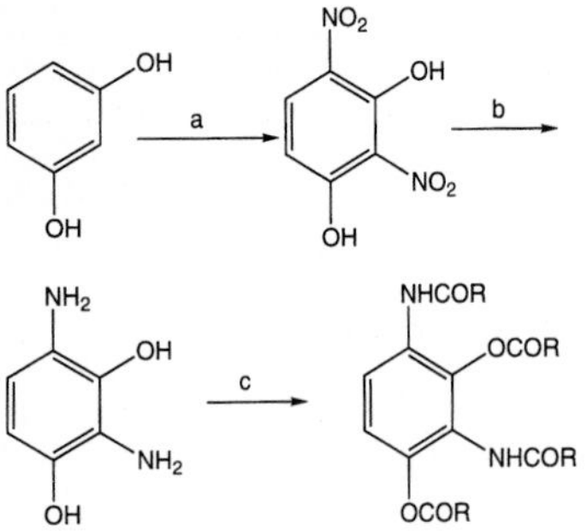

Scheme 4. Synthesis of benzene 1,3-diamide-2,4-diesters [18]. Reagents: (a) $NaNO_2$, $-2\,°C$ [17]; (b) $SnCl_2/HCl$. (c) RCOCl/pyridine.

Table 4. Transition temperatures for benzene 1,3-diamide-2,4-diesters (RCONH–/RCOO–) (see Scheme 4 for structure).

R	Phase transitions [18]
C_6H_{13}	Cr 89 D_h 124 I
C_7H_{15}	Cr 87 D_h 122 I
C_8H_{17}	Cr 90 D_h 121 I
C_9H_{19}	Cr 93 D_h 122 I
$C_{10}H_{21}$	Cr 95 D_h 120 I
$C_{11}H_{23}$	Cr 97 D_h 117 I
$C_{13}H_{27}$	Cr 101 D_h 114 I
$C_{15}H_{31}$	Cr 103 D_h 114 I

es the discotic mesophase substantially with respect to the parent hexaester. The melting points of the two sets of compounds are very similar for derivatives of the same chain length (e.g. for $C_7H_{15}CO-$, hexaester m.p.

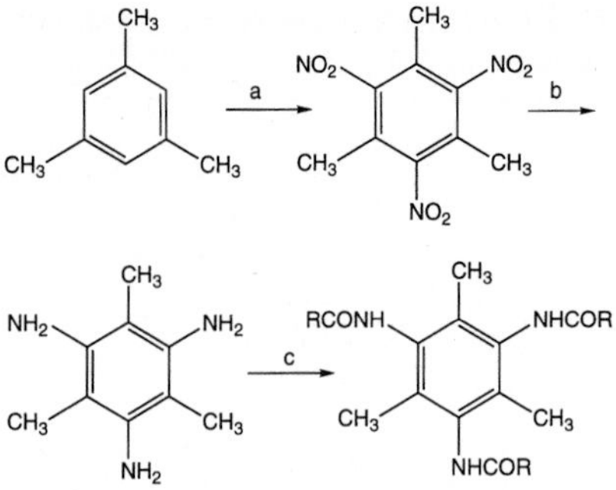

Scheme 5. Synthesis of 1,3,5-trimethylbenzene triamides [19]. Reagents: (a) HNO_3/H_2SO_4; (b) Sn/HCl; (c) RCOCl/pyridine.

Table 5. Transition temperatures for 1,3,5-trimethylbenzene triamides (RCONH–) (see Scheme 5 for structure).

R	Phase transitions [19]
C_3H_7	Cr 380 D_h 410 I
C_4H_9	Cr 315 D_h 380 I
C_5H_{11}	Cr 300 D_h 357 I
C_6H_{13}	Cr 257 D_h 357 I
C_7H_{15}	Cr 239 D_h 346 I
C_8H_{17}	Cr 200 D? 222 D_h 340 I
C_9H_{19}	Cr 189 D? 206 D_h 342 I
$C_{10}H_{21}$	Cr 185 D? 191 D_h 338 I
$C_{11}H_{23}$	Cr 183 D? 189 D_h 342 I
$C_{13}H_{27}$	Cr 118 D? 182 D_h 338 I
$C_{15}H_{31}$	Cr 120 D? 175 D_h 328 I

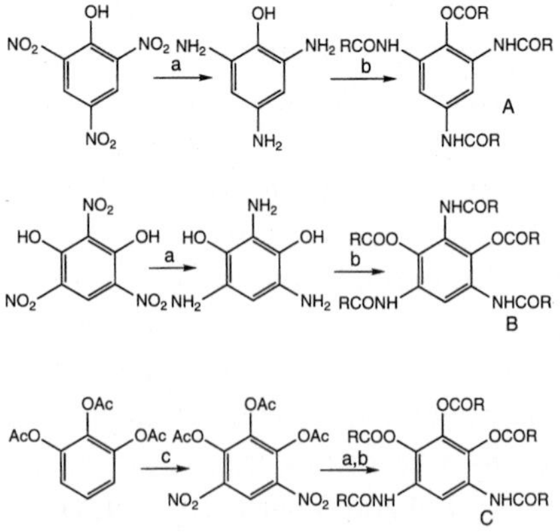

Scheme 6. Synthesis of mixed ester and amide derivatives of benzene [21]. Reagents: (a) Sn/HCl; (b) RCOCl/pyridine; (c) HNO_3 [20].

Table 6. Transition temperatures for mixed ester and amide derivatives of benzene (RCONH–/RCOO–) (see Scheme 6 for structures).

R	Phase transitions [21]		
	A	B	C
C_4H_9	Cr 144 D_{hd} 222 I	Cr 100 D_{hd} 241 I	Cr 113 D_{hd} 196 I
C_5H_{11}	Cr 133 D_{hd} 225 I	Cr 55 D_{hd} 238 I	Cr 67 D_{hd} 208 I
C_6H_{13}	Cr 129 D_{hd} 229 I	Cr 63 D_{hd} 241 I	Cr 88 D_{hd} 208 I
C_7H_{15}	Cr 123 D_{hd} 227 I	Cr 49 D_{hd} 242 I	Cr 77 D_{hd} 208 I
C_8H_{17}	Cr 109 D_{hd} 226 I	Cr 73 D_{hd} 227 I	Cr 93 D_{hd} 206 I
C_9H_{19}	Cr 111 D_{hd} 227 I	Cr 65 D_{hd} 225 I	Cr 92 D_{hd} 203 I
$C_{10}H_{21}$	–		Cr 96 D_{hd} 202 I
$C_{11}H_{23}$	Cr 107 D_{hd} 225 I	Cr 78 D_{hd} 224 I	Cr 101 D_{hd} 203 I
$C_{13}H_{27}$	Cr 112 D_{hd} 225 I	Cr 84 D_{hd} 224 I	Cr 103 D_{hd} 198 I
$C_{15}H_{31}$	Cr 113 D_{hd} 218 I	Cr 66 D_{hd} 225 I	Cr 103 D_{hd} 193 I

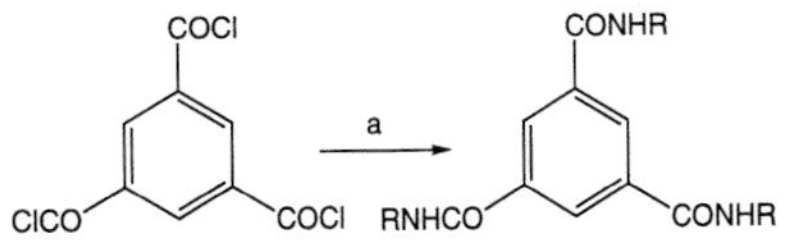

Scheme 7. Synthesis of N,N',N''-1,3,5-benzene tricarboxamides [22]. Reagents: (a) RNH$_2$/pyridine.

82 °C, bisamide m.p. 75 °C) and the meso-phase extension results primarily from an elevation of the clearing points.

The mesophase-stabilizing effect of the amide group allows discotic benzene derivatives to be prepared that carry fewer long chains. The syntheses of such materials are shown in Schemes 4–6, and their meso-phase properties are given in Tables 4–6.

The N,N',N''-1,3,5-benzene tricarbox-amides [22] also form mesophases, but it should be noted that their columnar structure has not been confirmed unambiguously. These compounds are simply synthesized from commercially available 1,3,5-benzene tricarbonyl chloride and alkyl-amines by refluxing in pyridine (Scheme 7). Their thermal behaviour is shown in Table 7. It should be pointed out that other 1,3,5-trisubstituted benzenes have been synthesized and shown to form monotropic mesophases [23].

Table 7. Transition temperatures for N,N',N''-1,3,5-benzene tricarboxamides (RCONH–).

R	Phase transitions [a] [22]
C_5H_{11}	Cr 119 M 206 I
C_6H_{13}	Cr 99 M 205 I
C_7H_{15}	Cr 116 M 208 I
C_8H_{17}	Cr 102 M 204 I
C_9H_{19}	Cr 65 M 215 I
$C_{10}H_{21}$	Cr 49 M 208 I
$C_{11}H_{23}$	Cr 72 M 216 I
$C_{12}H_{25}$	Cr 88 M 212 I
$C_{13}H_{27}$	Cr 81 M 216 I
$C_{14}H_{29}$	Cr 61 M 209 I
$C_{15}H_{31}$	Cr 88 M 214 I
$C_{16}H_{33}$	Cr 73 M 205 I
$C_{17}H_{35}$	Cr 87 M 211 I
$C_{18}H_{37}$	Cr 68 M 206 I

[a] M, mesophase.

2.1.2 Multiynes

A widely studied class of discotic benzene derivatives are the multiynes prepared by Praefcke and coworkers [24]. Like the structurally similar naphthalenes, these materials are prepared by palladium-catalysed coupling of polybromobenzenes with acetylene derivatives (Scheme 8). Their mesophase behaviour is given in Table 8. The hexakisalkynylbenzenes themselves exhib-

Table 8. Transition temperatures for benzene multiynes (see Scheme 8 for structures).

R	X	Phase transitions	Refs.
$C_5H_{11}C_6H_4$	$C_5H_{11}C_6H_4CC-$	Cr 170 N_D 185 I_{decomp}	[24]
$C_6H_{13}C_6H_4$	$C_6H_{13}C_6H_4CC-$	Cr 124 N_D 142 I	[24]
$C_7H_{15}C_6H_4$	$C_7H_{15}C_6H_4CC-$	Cr 98 N_D 131 I	[24]
$C_8H_{17}C_6H_4$	$C_8H_{17}C_6H_4CC-$	Cr 80 N_D 96 I	[25]
$C_7H_{15}OC_6H_4$	$C_7H_{15}OC_6H_4CC-$	Cr 109 N_D 193 I	[26]
$C_5H_{11}C_6H_4C_6H_4$	$C_5H_{11}C_6H_4C_6H_4CC-$	Cr 170 D >240 I_{decomp}	[27]
$C_6H_5C_6H_4$	CN	Cr 246 D >260 I_{decomp}	[27]
$C_6H_5C_6H_4$	$C_{16}H_{33}O-$	Cr 171 D ~230 N_D 250 I_{decomp}	[27]
$C_5H_{11}C_6H_4C_6H_4$	$C_{16}H_{33}O-$	Cr 106 D 285 I_{decomp}	[27]
C_5H_{11}	$C_5H_{11}CC-$	[I – 16 D]	[25]
C_8H_{17}	$C_8H_{17}CC-$	[I – 8 D]	[25]

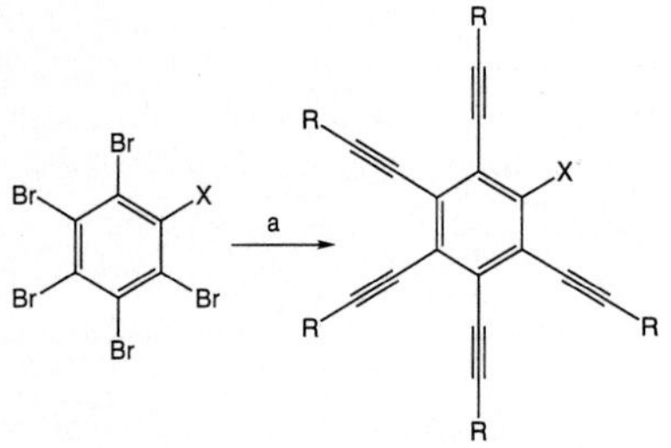

Scheme 8. Synthesis of benzene multiynes. Reagents: (a) RC≡CH/Pd[O].

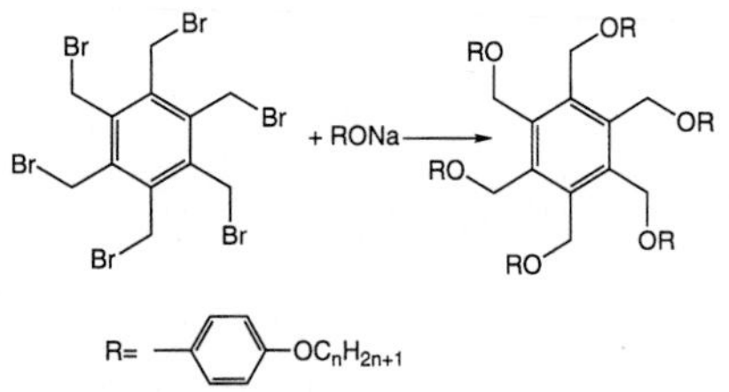

Scheme 9. Synthesis of hexakis[(alkoxyphenoxy)methyl]benzenes [28].

it only monotropic mesophases [25]. Formation of 'superdisk' structures such as benzene hexakisphenylacetylenes [24, 25] gives rise to materials that exhibit relatively stable discotic nematic phases. Enlarging the core even further (biphenyl analogues) leads to columnar discotic mesophases [27].

2.1.3 Hexakis(alkoxyphenoxy-methyl) Derivatives

One further class of discotic benzenes comprises the hexakis[(alkoxyphenoxy)methyl] derivatives [28]. These compounds are simply prepared from hexakis(bromomethyl)-benzene and an excess of p-alkoxyphenoxide (Scheme 9). The thermal behaviour of these materials is given in Table 9 [28], but it should be noted that the authors were un-

Table 9. Transition temperatures for hexakis[(alkoxyphenoxy)methyl]benzenes (ROCH$_2$–).

R	Phase transitions [28]
p-$C_5H_{11}OC_6H_4-$	Cr 68 D_{rd} 97 I
p-$C_6H_{13}OC_6H_4-$	Cr 76 D_{rd} 83 I
p-$C_7H_{15}OC_6H_4-$	Cr 67 D_{rd} 71 I

able to confirm unambiguously the presence of a discotic mesophase and point out that the 'mesophase' might be a special type of crystal [28].

2.1.4 Hexakis(alkylsulfone) Derivatives

Finally, a class of benzene derivatives that also form discotic mesophases is the hexa-

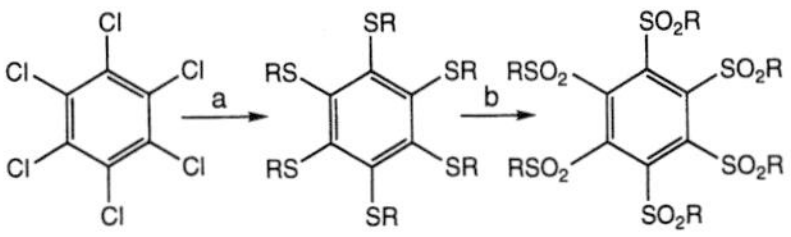

Scheme 10. Synthesis of hexakis(alkylsulfone) derivatives of benzene [29]. Reagents: (a) RSNa/HMPT 80–95%; (b) m-CPBA/CHCl$_3$ 60–70%.

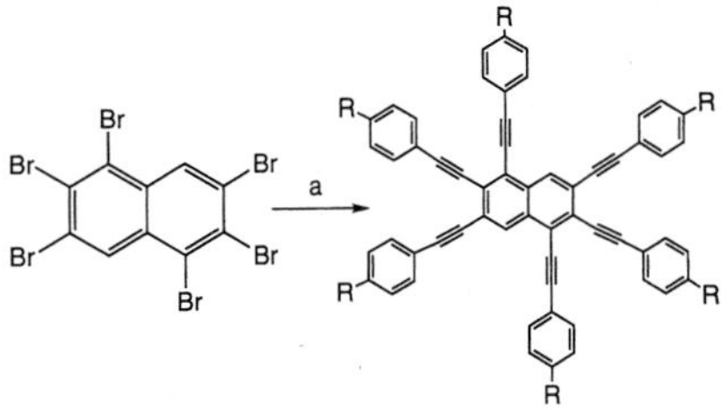

Scheme 11. Synthesis of naphthalene multiynes. Reagents: (a) RC$_6$H$_4$C≡CH/Pd[O].

Table 10. Transition temperatures for hexakis(alkylsulfone) derivatives of benzene (RSO$_2$–).

R	Phase transitions	Refs.
C$_6$H$_{13}$	I 164 Cr	[30]
C$_7$H$_{15}$	I 138 D$_h$ 120 Cr	[29, 30]
C$_8$H$_{17}$	I 134 D$_h$ 108 Cr	[30]
C$_9$H$_{19}$	I 130 D$_h$ 87 Cr	[29, 30]
C$_{10}$H$_{21}$	I 125 D$_h$ 77 Cr	[30]
C$_{11}$H$_{23}$	I 115 D$_h$ 65 Cr	[29, 30]
C$_{12}$H$_{25}$	I 103 D$_h$ 45 Cr	[30]
C$_{13}$H$_{27}$	I 90 D$_h$ 44 Cr	[30, 32]
C$_{14}$H$_{29}$	I 77 D$_h$ 42 Cr	[30, 31]
C$_{15}$H$_{31}$	I 63 D$_h$ 49 Cr	[30]
C$_{16}$H$_{33}$	I 69 Cr	[30]

Table 11. Transition temperatures for naphthalene multiynes (see Scheme 11 for structures).

R	Phase transitions	Refs.
C$_5$H$_{11}$	Cr 121 D$_r$ 156 D$_h$ ~260 I$_{decomp}$	[33]
C$_6$H$_{13}$	Cr 69 N$_D$ 98 D$_r$ 134 D$_h$ ~245 I$_{decomp}$	[33]
C$_7$H$_{15}$	Cr 49 N$_D$ 95 D$_r$ 159 D$_h$ ~230 I$_{decomp}$	[33]
C$_8$H$_{17}$	Cr 60 N$_D$ 112 D$_{ro}$ 137 D$_{ho}$ ~200 I	[33]
C$_9$H$_{19}$	Cr 68 N$_D$ 110 I	[33]
C$_5$H$_{11}$O	Cr 144 N$_D$ ~250 I$_{decomp}$	[33]
HC$_6$H$_4$	Cr 240 D >260 dI	[27]
C$_5$H$_{11}$C$_6$H$_4$	Cr 122 D >240 dI	[27]
C$_8$H$_{17}$C$_6$H$_4$	Cr 75 D >240 dI	[27]
C$_9$H$_{19}$C$_6$H$_4$	Cr 58 D ~180 N$_D$ ~230 I$_{decomp}$	[27]

kis(alkylsulfone) derivatives prepared by Praefcke and coworkers [29]. The synthesis (Scheme 10) is straightforward. Hexachlorobenzene is treated with an excess of sodium alkylthiolate, typically in HMPT, to give the hexakis(alkylthio)benzenes. Oxidation with m-CPBA affords the sulfones. The compounds appear to form relatively disordered hexagonal mesophases [30–32] (Table 10), which show significant supercooling.

2.2 Naphthalene

Two general series of naphthalene derivatives are known to give discotic liquid crystal phases, the multiynes [27, 33] and the p-alkoxyphenoxymethyl derivatives [28]. The multiynes are synthesized through a palladium-catalysed coupling of hexabromonaphthalene with aryl acetylenes [33] (Scheme 11) and their mesophase behaviour is summarized in Table 11. The multiynes exhibit a rich polymesomorphism and tend to form the relatively rare nematic discotic phase, which may be due to the 'super-disk' structure of the molecules [27, 33]. The most striking feature, however, is the reverse phase sequence for the lower homologues in the alkylphenylacetylene series. It should be noted that the related alkylacetylene derivatives do not give any liquid crystal phases.

The p-(alkoxyphenoxy)methylnaphthalenes are easily synthesized by a nucleophilic substitution on bromomethylnaphthalenes (prepared by bromination of polymethylnaphthalenes [34]) with p-alkoxyphenoxides [28] (Scheme 12). Their mesophase behaviour is summarized in Table 12 and it can

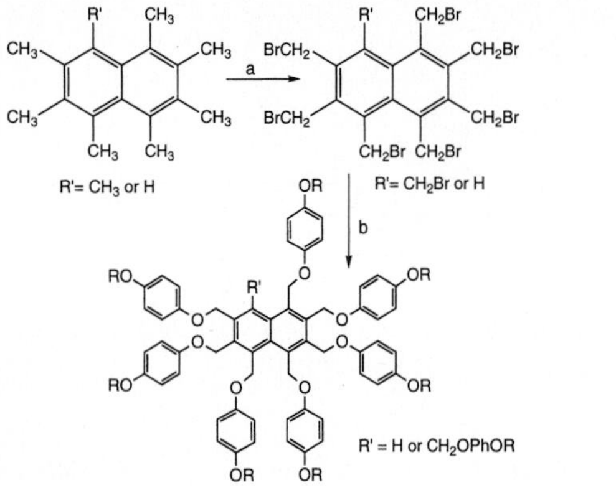

Scheme 12. Synthesis of (*p*-alkoxyphenoxymethyl) derivatives of naphthalene [28]. Reagents: (a) Br$_2$/hν; (b) NaOC$_6$H$_4$OR/DMF 80 °C, 2 h 20–63%.

Table 12. Transition temperatures for [(*p*-alkoxy-phenoxy)methyl]naphthalenes (see Scheme 12 for structures).

R	R′	Phase transitions [28]
C$_5$H$_{11}$	C$_5$H$_{11}$OPhOCH$_2$	Cr 162 D$_h$ 164 I
C$_6$H$_{13}$	C$_6$H$_{13}$OPhOCH$_2$	Cr 135 D$_h$ 167 I
C$_7$H$_{15}$	C$_7$H$_{15}$OPhOCH$_2$	Cr 105 D$_h$ 167 I
C$_8$H$_{17}$	C$_8$H$_{17}$OPhOCH$_2$	Cr 78 D$_h$ 167 I
C$_9$H$_{19}$	C$_9$H$_{19}$OPhOCH$_2$	Cr 62 D$_h$ 165 I
C$_6$H$_{13}$	H	Cr 65 D$_{ro}$ 147 D$_h$ 184 I
C$_7$H$_{15}$	H	Cr 79 D$_{ro}$ 92 D$'_{ro}$ 138 D$_h$ 182 I

be seen that the symmetrical (octa-substituted) derivatives give a single hexagonal mesophase. The unsymmetrical (hepta-substituted) materials, however, are polymesomorphic, with the higher homologue giving two distinct rectangular phases in addition to the high-temperature hexagonal phase.

2.3 Anthracene (Rufigallol)

The rufigallol esters occupy an important place in the history of discotic liquid crystals, as they were one of the earliest systems

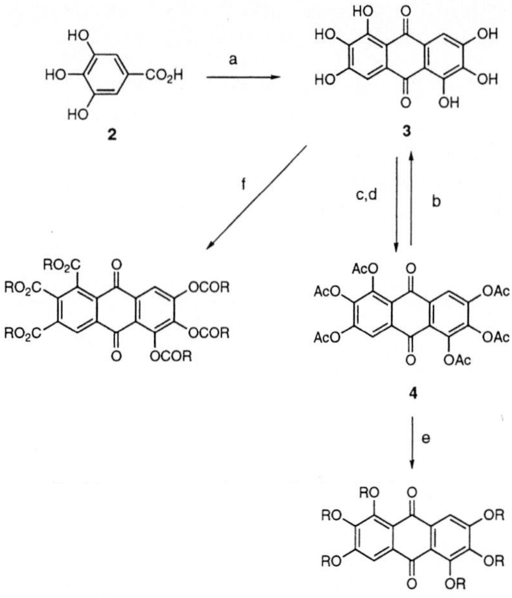

Scheme 13. Synthesis of Anthracene (Rufigallol) based discotic liquid crystals. Reagents: (a) 98% H$_2$SO$_4$/100 °C/2 h [36]; (b) Ac$_2$O/trace H$_2$SO$_4$/reflux/2 h [36]; (c) pyridine/boil [36]; (d) H$^+$ [36]; (e) RBr/Na$_2$CO$_3$/DMF/160 °C/16 h [38]; (f) RCOCl/pyridine – acetone/40 °C/1 h [37].

reported to form a columnar mesophase but to have a nucleus with only C_2 symmetry [35]. Rufigallol (**3**) is made in low yield through the acid-catalysed self-condensation of gallic acid (**2**) [36] and it is normally purified through its hexa-acetate (**4**). The esters are prepared from the hexaphenol (**3**) [35, 37], but the ethers are better made directly from the acetate (**4**) [38]. The syntheses are shown in Scheme 13 and the mesophase behaviour of the products is summarized in Table 13. All the ethers C$_4$ to C$_{13}$ [38] and esters C$_7$ to C$_{14}$ [37] show enantiotropic behaviour, and the esters C$_7$ to C$_9$ [35, 37] also show monotropic behaviour, a more ordered mesophase being formed when the normal columnar phase is cooled to 108 °C in the case of the heptanoate [37], 99 °C in the case of the octanoate [35] and 83 °C in the case of the nonanoate [37].

Table 13. Transition temperatures of anthracene (rufigallol) based discotic mesogens.

Substituent	Phase transitions[a]	Refs.
C_4H_9O	Cr 105 D_h 131 I	[38]
$C_5H_{11}O$	Cr 83 D_h 117 I	[38]
$C_6H_{13}O$	Cr 53 D_h 105 I	[38]
$C_7H_{15}O$	Cr 49 D_h 102 I	[38]
$C_8H_{17}O$	Cr 37 D_h 96 I	[38]
$C_9H_{19}O$	Cr 32 D_h 92 I	[38]
$C_{10}H_{21}O$	Cr 48 D_h 87 I	[38]
$C_{11}H_{23}O$	Cr 26 D_h 84 I	[38]
$C_{12}H_{25}O$	Cr 37 D_h 79 I	[38]
$C_{13}H_{27}O$	Cr 62 D_h 72 I	[38]
$C_3H_7CO_2$	Cr 216 I	[37]
$C_4H_9CO_2$	Cr 170 I	[37]
$C_5H_{11}CO_2$	Cr 152 I	[37]
$C_6H_{13}CO_2$	Cr 112 M[b] 134 I	[37]
$C_7H_{15}CO_2$	Cr 110 (108) M[c] 133 (128) I	[37 (35)]
$C_8H_{17}CO_2$	Cr 81 M[d] 128 I	[37]
$C_9H_{19}CO_2$	Cr 109 M 128 I	[37]
$C_{10}H_{21}CO_2$	Cr 92 126 I	[37]
$C_{11}H_{23}CO_2$	Cr 99 M 124 I	[37]
$C_{12}H_{25}CO_2$	Cr 96 M 120 I	[37]
$C_{13}H_{27}CO_2$	Cr 103 M 117 I	[37]

[a] M, mesophase.
[b] Gives a different mesophase on cooling to 108 °C.
[c] Gives a different mesophase on cooling to 99 °C.
[d] Gives a different mesophase on cooling to 83 °C.

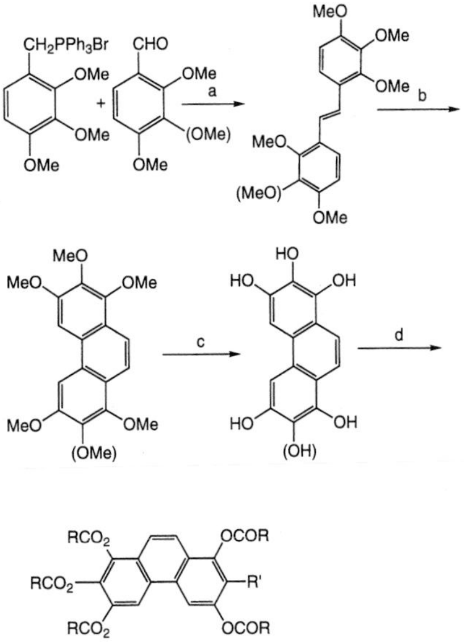

Scheme 14. Synthesis of phenanthrene based discogens [39]. Reagents: (a) LiOMe/MeOH; (b) $I_2/h\nu$; (c) BBr_3; (d) DCCI/DMAP/RCOOH.

2.4 Phenanthrene

Discotic phenanthrene derivatives have been prepared by Scherowsky and Chen [39]. Their synthesis is illustrated in Scheme 14. The phosphonium salt prepared by bromination of trimethoxybenzyl alcohol and reaction of the benzyl bromide with triphenyl phosphine was deprotonated and a Wittig reaction with 2,4-dimethoxy- or 2,3,4-trimethoxybenzaldehyde produced the stilbene. Photocyclization gives the phenanthrene, which can be demethylated with boron tribromide and acylated using the Steglich method.

The mesophase behaviour of discotic phenanthrenes is given in Table 14. It was

Table 14. Transition temperatures for phenanthrene based discotic mesogens (see Scheme 14 for structures).

R	R'	Phase transitions [39]
$C_7H_{15}OCH^*(CH_3)-$	H	D 51 I
$C_6H_{13}CH^*(CH_3)OCH_2-$	H	D 75 I
$C_7H_{15}OCH^*(CH_3)-$	RCO_2-	Cr 76 I
$C_6H_{13}CH^*(CH_3)OCH_2-$	RCO_2-	Cr -13 D 145 I

found that straight-chain esters of hexahydroxyphenanthrene gave only narrow, monotropic mesophases. Enantiotropic mesophases were formed for branched-chain penta- and hexaesters. A helical 'D_r' mesophase structure is proposed for these chiral derivatives, which show electro-optic switching.

2.5   Triphenylene

The hexasubstituted derivatives of triphenylene were among the first discotic mesogens to be discovered. The most commonly studied derivatives are the hexaethers and hexaesters of 2,3,6,7,10,11-hexahydroxytriphenylene, and their basic synthesis is illustrated in Scheme 15. 1,2-Dimethoxybenzene (veratrole) is oxidatively trimerized using chloranil in concentrated sulfuric acid to give hexamethoxytriphenylene [40]. The reaction takes about a week. Demethylation with boron tribromide [41] or hydrobromic acid [42] yields the hexaphenol,

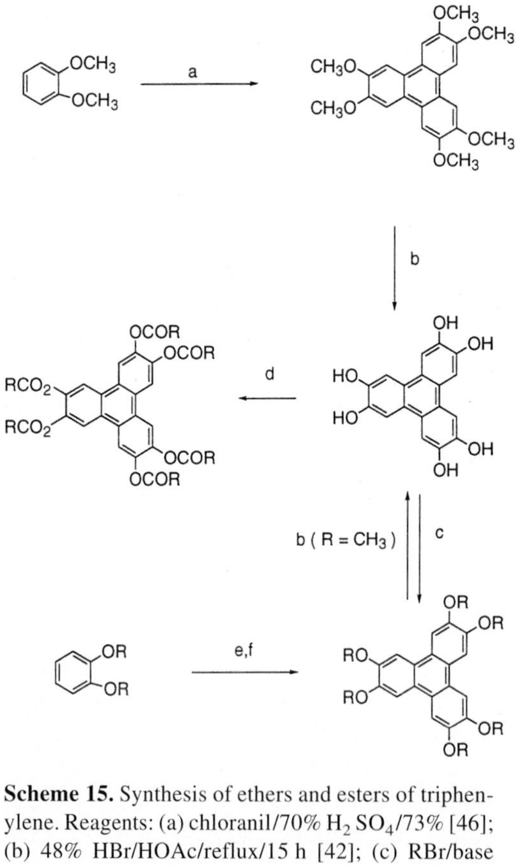

Scheme 15. Synthesis of ethers and esters of triphenylene. Reagents: (a) chloranil/70% $H_2 SO_4$/73% [46]; (b) 48% HBr/HOAc/reflux/15 h [42]; (c) RBr/base [49, 50]; (d) RCOCl/base [49, 50]; (e) $FeCl_3$/CH_2Cl_2 [45]; (f) MeOH/86% ($R=CH_3$) 73% ($R=C_6H_{11}$) two steps [45].

which is then alkylated or acylated to give the required mesogen. Alternatively, treatment of the hexamethoxy compound with trimethylsilyl iodide gives the hexatrimethylsilyloxy derivative, which can be alkylated or acylated directly [43]. Under the chloranil/concentrated sulfuric acid conditions, longer chain 1,2-dialkoxybenzene derivatives also undergo oxidative trimerization to give the required hexaethers directly, but the products are heavily contaminated with impurities [44] (mainly dealkylation products) which are difficult to remove so that, in general, the three-step procedure has been preferred. An improved synthesis of hexaalkoxytriphenylenes has been developed more recently in our laboratories [45], allowing dialkoxybenzenes to be converted directly to the corresponding triphenylene in a single high-yielding step using ferric chloride as the oxidizing agent. The reaction is typically carried out in dichloromethane solution, is complete in about an hour, and is worked up reductively with methanol. Alternative oxidizing agents used for the synthesis of related triphenylenes include ferric chloride/sulfuric acid [46] and electrosynthesis [47]. Indeed, it was in these electrochemical syntheses that the advantages of and the need for reductive work-up procedures was first recognized [48].

2.5.1   Ethers, Thioethers and Selenoethers

The mesophase behaviour of simple symmetrical hexaether derivatives of triphenylene is summarized in Table 15. All derivatives with side-chains longer than propyloxy form a single hexagonal mesophase, with both the transition temperatures and the mesophase range decreasing as the length of the side chain is increased. The corresponding thioethers and selenoethers are made

Table 15. Transition temperatures for ethers of triphenylene.

Substituent	Phase transitions	Refs.
C_3H_7O	Cr 177 I	[44, 45, 50]
C_4H_9O	Cr 89 D_{ho} 146 I	[44, 45, 50]
$C_5H_{11}O$	Cr 69 D_{ho} 122 I	[44, 50, 51]
$C_6H_{13}O$	Cr 68 D_{ho} 97 (100) I	[43, 44, 50 (45)]
$C_7H_{15}O$	Cr 69 D_{ho} 93 I	[43, 44, 50]
$C_8H_{17}O$	Cr 67 D_{ho} 86 I	[44, 45, 50]
$C_9H_{19}O$	Cr 57 D_{ho} 78 I	[44, 45, 50]
$C_{10}H_{21}O$	Cr 58 D_{ho} 69 I	[44, 50]
$C_{11}H_{23}O$	Cr 54 D_{ho} 66 I	[43, 44, 50]
$C_{13}H_{27}O$	D_{ho} 49 I	[44, 50]

from the reaction of hexabromotriphenylene [52] and the requisite nucleophile. As shown in Scheme 16, the hexabromide is readily obtained from triphenylene itself. Unfortunately, this is an expensive starting material and troublesome to make on any scale. Of the several routes available [57] perhaps the most convenient for a small-scale laboratory preparation is the ferric chloride oxidative cyclization of o-terphenyl [53], a relatively cheap starting material. Transition temperatures for the thioethers and selenoethers are summarized in Table 16.

Interest in the ether derivatives arises from the fact that they are photoconductive and that, when doped with a few mol per cent of an oxidant such as aluminium trichloride, they become semiconductors [59]. Oriented samples of the D_h phase can be produced, and it is found that the preferred direction of conduction is parallel to the director; the system behaves rather like a molecular wire, with the conducting stack of aromatic cores being surrounded by an insulating sheath of alkyl chains. The hexahexylthioether derivative gives a 'helical' ordered columnar phase at low temperatures [58, 60], which shows remarkably high charge-carrier mobility in experiments in

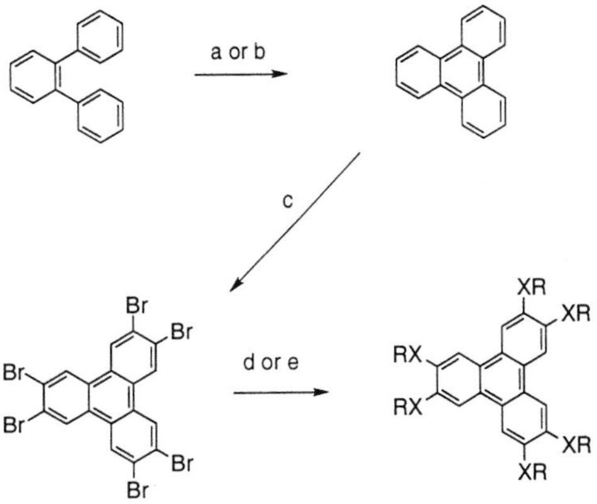

Scheme 16. Synthesis of thioethers and selenoethers of triphenylene. Reagents: (a) $FeCl_3$ [53]; (b) hv/I_2 [54]; (c) $Br_2/Fe/C_6H_5NO_2$/reflux/2 h, [52]; (d) RSNa/DMEU/100 °C/1 – 2 h, 40 – 55% [55]; (e), (i) RSeNa/HMPA/2 h/80 °C, (ii) RBr/15 h/60 °C, 38 – 55% [56].

Table 16. Transition temperatures for thioethers and selenoethers of triphenylene.

Substituent	Phase transitions	Refs.
$C_6H_{13}S$	Cr 62 D_{ho} 70 D_{hd} 93 I	[55, 58]
$C_8H_{17}S$	Cr 55 D_{hd} 87 I	[55, 58]
$C_{10}H_{21}S$	Cr 40 D_{hd} 71 I	[55, 58]
$C_6H_{13}Se$	Cr ~40 D ~72 I	[56]
$C_7H_{15}Se$	Cr ~40 D ~70 I	[56]

which the charge carriers are generated photochemically [61].

2.5.2   Esters

The mesophase behaviour of the discotic triphenylenehexakisalkanoates is summarized in Table 17. Simple, straight-chain derivatives tend to form a rectangular mesophase with the higher homologues also forming a hexagonal phase at higher temperatures. Melting and clearing temperatures do not alter significantly through the series. The effect of heteroatom substitution into the side chain depends on its nature [9] (oxygen or sulfur) and position [9, 64], and it is worth noting that mesophase behaviour is destroyed if a bromine atom is present in the side chain [65].

The hexabenzoate esters of triphenylene have received particular attention because they tend to form nematic phases. Their mesophase behaviour is summarized in Table 18. Increasing the chain length of the p-alkyl or p-alkoxy substituent in the benzene ring tends to reduce the transition temperatures slightly, but more pronounced effects are observed when lateral (ortho or meta) methyl substituents are introduced into the benzene rings. The number and position of these additional methyl groups influences both the transition temperatures and the phase sequence [68, 69]. The corresponding cyclohexane carboxylates (in which the benzene ring is formally reduced

to a trans-1,4-disubstituted cyclohexane) show wider mesophase ranges and they exhibit mesophase behaviour with shorter alkyl chain substituents. This is illustrated in Table 19 [51, 70].

2.5.3   Multiynes

As shown in Scheme 17, palladium catalysed cross-coupling of hexabromotriphenylene [52] and p-alkyl derivatives of phenylacetylene gives liquid crystalline hexaalkynyl derivatives of triphenylene [24]. These discogens, which have a core diameter of about 2.4 nm, give enantiotropic nematic phases only (7: $R = C_5H_{11}$, Cr 157 N_D 237 I; $R = C_7H_{15}$, Cr 122 N_D 176 I). Straight-chain alkylacetylene derivatives (like 7, but without the benzene rings) are not liquid crystalline [24].

Table 17. Transition temperatures of esters of triphenylene.

Substituent	Phase transitions	Refs.
$C_5H_{11}CO_2$	Cr 146 I	[44, 50]
$C_6H_{13}CO_2$	Cr 108 D_r 120 I	[9, 44, 50]
$C_7H_{15}CO_2$	Cr 64 D_r 130 I (Cr 66 D 126 I)	[44, 50 (51)]
$C_8H_{17}CO_2$	Cr 62 (66) D_r 125 (129) I	[(9), 44, 50]
$C_9H_{19}CO_2$	Cr 75 (65) D_r 126 I	[(9), 44, 50]
$C_{10}H_{21}CO_2$	Cr 67 D_r [a] 108 D_h 122 I	[44, 50]
$C_{11}H_{23}CO_2$	Cr 80 $D_?$ 93 D_r 111 D_h 122 I	[44, 49, 50]
$C_{12}H_{25}CO_2$	Cr 83 D_r [b] 99 D_h 118 I	[44, 50]
$C_{13}H_{27}CO_2$	Cr 87 D_r 96 D_h 111 I	[44, 50]
$C_6H_{13}C*H(CH_3)CH_2CO_2$	Cr 59 D 65 D_r 97 I	[62, 63]
$C_5H_{11}SCH_2CO_2$	Cr 75 D 138 I	[9]
$C_3H_7SC_2H_4CO_2$	Cr 86 D 198 I	[9]
$C_4H_9SC_2H_4CO_2$	Cr 57 D 202 I	[9]
$C_5H_{11}SC_2H_4CO_2$	D_r 203 I	[9]
$C_3H_7SC_3H_6CO_2$	Cr 98 I	[9]
$C_4H_9OC_2H_4CO_2$	Cr 94 I	[9]
$C_7H_{15}OCH_2CO_2$	Cr -14 D_h 227 I	[64]
$C_4H_9OCH(CH_3)CO_2$	Cr ? D 93 I	[63]
$C_6H_{13}OCH(CH_3)CO_2$	Cr ? D_h 156 I	[63]

[a] Gives another columnar phase on cooling to 56 °C.
[b] Gives another columnar phase on cooling to 81 °C.

Table 18. Transition temperatures for benzoate esters of triphenylene of the general type **5**.

5

X	A–D	Phase transitions	Refs.
C_7H_{15}	A–D=H	Cr 210 I (Cr 130 D 210 I)	[50, 66, (67)]
C_8H_{17}	A–D=H	Cr 208 N_D 210 I (Cr 179 N_D 192 I)	[50, 66 (67)]
		[Cr 125 D_1 179 D_2 222 I]	[51]
C_9H_{19}	A–D=H	Cr 175 D_r 183 N_D 192 I	[50, 66]
$C_{10}H_{21}$	A–D=H	Cr 185 D_r 189 I	[50, 66]
C_4H_9O	A–D=H	Cr 257 N_D 300 I	[50, 66]
$C_5H_{11}O$	A–D=H	Cr 224 N_D 298 I	[50, 66]
$C_6H_{13}O$	A–D=H	Cr 186 D_t 193 N_D 274 I	[50, 66]
$C_7H_{15}O$	A–D=H	Cr 168 N_D 253 I (Cr 142 N_D 248 I)	[50, 66 (68)]
$C_8H_{17}O$	A–D=H	Cr 152 D_r 168 N_D 244 I	[50, 66]
$C_9H_{19}O$	A–D=H	Cr 154 D_r 183 (180) N_D 227 (222) I	[50, 66, (67)]
		[Cr 147 D_1 182 D_2 230 I]	[51]
$C_{10}H_{21}O$	A–D=H	Cr 142 D_r 191 (155) N_D 212 (177) I	[50, 55 (67)]
$C_{11}H_{23}O$	A–D=H	Cr 145 D_r 179 N_D 185 I	[50, 66]
$C_{12}H_{25}O$	A–D=H	Cr 146 D_r 174 I	[50, 66]
$C_8H_{17}O$	A–D=F	<25 D 300 I	[51]
$C_{10}H_{21}O$	A=CH_3; B–D=H	Cr 109 N_D 164 I	[68]
$C_{10}H_{21}O$	A, B, D=H; C=CH_3	? 129 N_D 170 I	[68]
$C_6H_{13}O$	A, D=H; B, C=CH_3	Cr 150 D_{hd} 210 N_D 243 I	[69]
$C_8H_{17}O$	A, D=H; B, C=CH_3	Cr 170 D_{hd} 195 N_D 215 I	[69]
$C_{10}H_{21}O$	A, D=H; B, C=CH_3	Cr 157 D_{hd} 167 N_D 182 I	[69]
$C_{12}H_{25}O$	A, D=H; B, C=CH_3	D_{hd} 143 N_D 151 I	[69]
$C_6H_{13}CH(CH_3)O$	A, D=H; B, C=CH_3	Cr 125 D_h 156 N_D 183 I	[69]
$C_6H_{13}O$	A, D=CH_3; B, C=H	Cr 170 N_D 196 I	[69]
$C_8H_{17}O$	A, D=CH_3; B, C=H	Cr 155 N_D 170 I	[69]
$C_{10}H_{21}O$	A, D=CH_3; B, C=H	Cr 108 N_D 134 I	[69]
$C_{12}H_{25}O$	A, D=CH_3; B, C=H	Cr 88 N_D 99 I	[69]
$C_6H_{13}CH(CH_3)O$	A, D=CH_3; B, C=H	Cr 161 I	[69]

2.5.4 Unsymmetrically Substituted Derivatives

Unsymmetrically substituted derivatives of triphenylene have been prepared both by 'statistical' and 'rational' routes. An obvious 'statistical' approach to the problem is illustrated in Scheme 18. Oxidative trimerization of a mixture of two different 1,2-dialkoxy derivatives of benzene gives a mixture of all four possible triphenylene prod-

Table 19. Transition temperatures for cyclohexanoate esters of triphenylene of the type **6**.

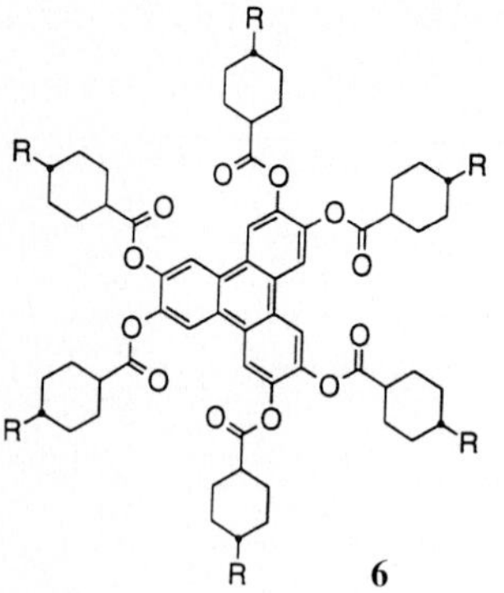

6

R	Phase transitions	Refs.
C_5H_{11}	Cr 158 D_{hd} 300 I	[70]
C_6H_{13}	Cr 146 D_{hd} 240 I	[70]
C_7H_{15}	Cr 150 D_{hd} 255 I	[70]
C_9H_{19}	Cr 108 D_{hd} 245 I (<25 D >345 I)	[70 (51)]

ucts [45, 50, 71]. If the two reactants are used in equal proportions the ratio of these products is almost exactly statistical, and often the mixture is very difficult to separate, although in the case of electro-oxidations it is possible to manipulate the product ratio to some extent by varying the reaction conditions [72]. In other cases it may be advantagous to use one reactant in excess, or favourable solubility properties may mean

that separation can be achieved fairly easily [45]. An alternative 'statistical' strategy that has been successfully exploited is that of partial alkylation of the hexa-acetate of triphenylene, as shown in Scheme 19 [41]. The reaction gives a mixture of hexa-alkoxy, penta-alkoxy, tetra-alkoxy, etc., derivatives, but the penta-alkoxy derivative can be isolated in low yield. Hydrolysis and acylation or alkylation of the product has led to a series of penta-alkoxymonoacyloxy and unsymmetrical hexa-alkoxy derivatives. The introduction of a single ester (particularly a branched ester) side chain increases the clearing temperatures, but more notably it surpresses crystallization such that these systems readily form glasses [41, 73].

The simplest and most straightforward of the 'rational' syntheses is the oxidative coupling of a tetra-alkoxybiphenyl and a 1,2-dialkoxybenzene derivative [74–76] using the ferric chloride/methanol work up protocol shown in Scheme 20. The reaction is easy to perform, offers a good degree of regiocontrol, and has been adapted to yield products in which both the number and nature of the substituents have been varied. Through this reaction a variety of different ring substitution patterns has also been produced [75]. The biphenyl derivatives required for these phenyl/biphenyl coupling

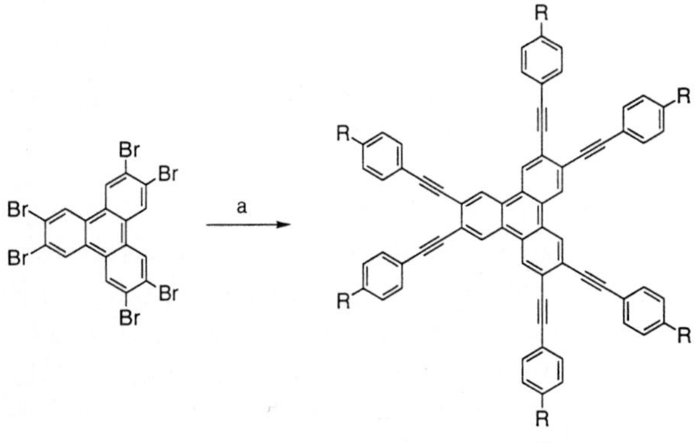

7

Scheme 17. Synthesis of multiyne derivatives of triphenylene [24]. Reagents: (a) $RC_6H_4C\equiv CH/Pd(PPh_3)_2Cl_2/PPh_3/Cu(I)/$ 100 °C/38 h, 52–57%.

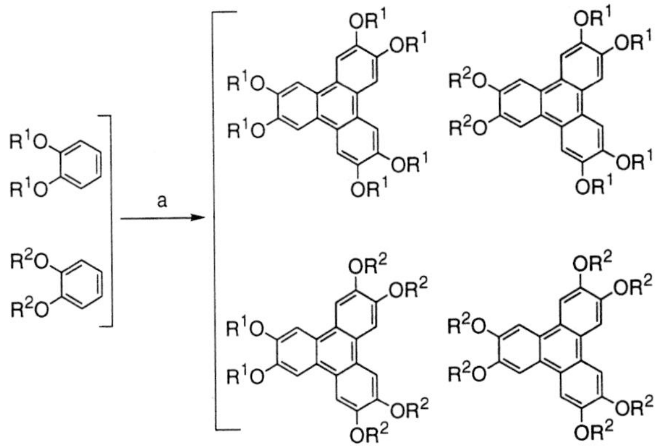

Scheme 18. 'Statistical' route to unsymmetrically substituted derivatives of triphenylene. Reagents: (a) electrochemical, FeCl₃, or chloranil oxidation.

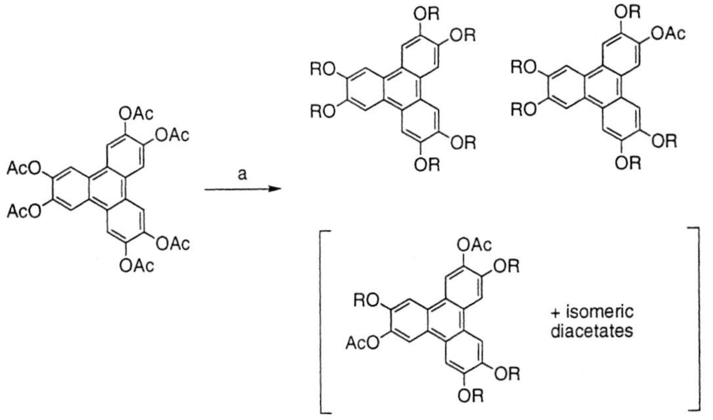

Scheme 19. 'Statistical' synthesis of unsymmetrically substituted derivatives of triphenylene [41]. Reagents: (a) RBr/K₂CO₃/ PhCOMe/116 °C/17 h, 26% monoacetate.

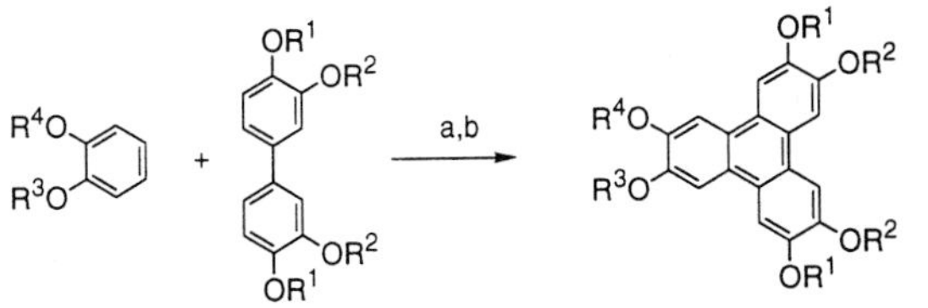

Scheme 20. 'Rational' synthesis of unsymmetrically substituted derivatives of triphenylene [74–76]. Reagents: (a) FeCl₃/CH₂Cl₂; (b) MeOH

reactions can be prepared either through Ullman or Heck coupling reactions, and the only significant limitation on this strategy arises from a lack of regiospecificity in the coupling step itself, such that, if neither the phenyl nor the biphenyl component has C_2 symmetry, a mixture of regioisomers will result. Longer and more elaborate strategies [77, 78] have been developed that allow full regiocontrol. In these, 'Heck' chemistry is

used to assemble a terphenyl, which is then subjected to oxidative cyclization to yield the triphenylene. One of these routes is illustrated in Scheme 21 [78]. A third, and even longer, route to unsymmetrically substituted triphenylenes was originally developed by Wenz in his synthesis of discotic liquid crystalline polymers [79–82], but has since been adapted for the synthesis of low molar mass discogens [83]. Scheme 22 il-

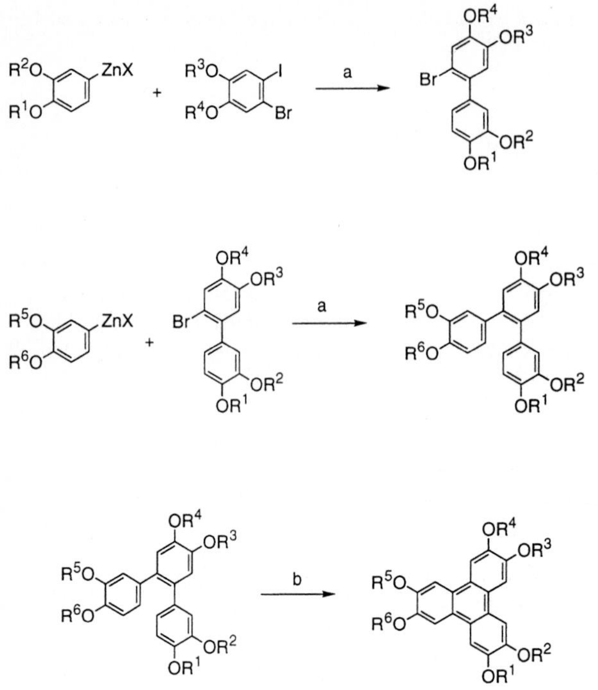

Scheme 21. 'Rational' route to unsymmetrically substituted derivatives of triphenylene [78]. Reagents: (a) $Pd_2(dba)_3/Ph_3P/THF/$ reflux/4.5 h, 45%; (b) $FeCl_3/CH_2Cl_2/1$ h, then MeOH (dba = dibenzylidine acetone, THF = tetrahydrofuran).

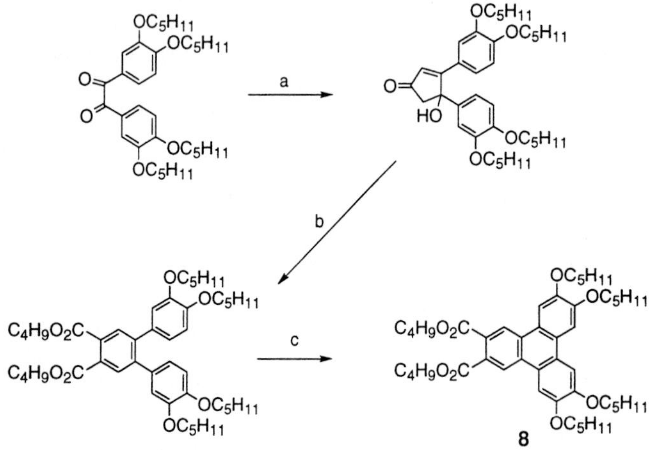

Scheme 22. 'Rational' synthesis of unsymmetrically substituted derivatives of triphenylene [79, 83]. Reagents:
(a) $CH_3COCH_3/^tBuOK/EtOH/$ reflux/4 h;
(b) $TsOH/C_6H_5Cl/C_4H_9O_2CC\equiv CCO_2C_4H_9$; (c) $I_2/h\nu$.

lustrates its use in the synthesis of the diester **8**, which gives a columnar phase (D_{hd}) between room temperature and 167 °C and shows very interesting self-organizing behaviour at the phase interfaces [83–86]. Transition data for unsymmetrically substituted systems are given in Table 21.

2.5.5 Modifications of the Number and Nature of Ring Substituents

Reactions have now been developed that allow the properties of triphenylene hexaeth-

Table 21. Transition temperatures for unsymetrically substituted derivatives of triphenylene of the type **9**.

9

R^1	R^2	R^3	R^4	Phase transitions	Refs.
C_5H_{11}	C_5H_{11}	C_5H_{11}	CH_3	Cr 72 D_h 101 I	[87]
C_5H_{11}	C_5H_{11}	C_5H_{11}	$(CH_2)_3OH$	Cr 55 D_h 106 I	[87]
C_5H_{11}	C_5H_{11}	C_5H_{11}	$COCH_3$	Cr? D_{ho} 161 I	[41]
C_5H_{11}	C_5H_{11}	C_5H_{11}	COC_4H_9	Cr 44 D_{ho} 177 I	[73]
C_5H_{11}	C_5H_{11}	C_5H_{11}	$COCHClCH(CH_3)C_2H_5$	D_{ho} 182 I	[73]
C_5H_{11}	C_5H_{11}	C_5H_{11}	$COCHClCH(CH_3)C_3H_7$	D_{ho} 182 I	[73]
C_5H_{11}	C_5H_{11}	C_5H_{11}	$COCHClC_3H_7$	D_{ho} 191 I	[73]
C_5H_{11}	C_5H_{11}	C_5H_{11}	$(CH_2)_2CH=CH_2$	Cr 56 D_h 127 I	[87]
C_5H_{11}	C_5H_{11}	C_5H_{11}	$(CH_2)_4CH=CH_2$	Cr 60 D_h 104 I	[87]
C_5H_{11}	C_5H_{11}	C_5H_{11}	$(CH_2)_6CH=CH_2$	Cr 51 D_h 100 I	[87]
C_5H_{11}	C_5H_{11}	C_5H_{11}	$(CH_2)_9CH=CH_2$	Cr 32 D_{ho} 55 I	[41]
C_5H_{11}	C_5H_{11}	C_5H_{11}	$CH(CH_3)C_4H_9$	Cr 67 I [a]	[73]
C_6H_{13}	C_6H_{13}	C_6H_{13}	CH_3	Cr 53 D 72 I	[88]
C_7H_{15}	C_7H_{15}	C_7H_{15}	$COCH_3$	Cr 50 D_{ho} 117 I	[89]
C_5H_{11}	C_5H_{11}	CH_3	CH_3	Cr 80 I	[87]
C_5H_{11}	C_5H_{11}	C_9H_{19}	C_9H_{19}	Cr 54 D_{ho} 74 I	[50, 71]
C_5H_{11}	C_5H_{11}	$C_{10}H_{21}$	$C_{10}H_{21}$	Cr 61 D_{ho} 63 I	[50, 71]
C_5H_{11}	C_5H_{11}	$C_{11}H_{23}$	$C_{11}H_{23}$	Cr 61	[50]
C_5H_{11}	C_5H_{11}	C_2H_4OH	C_2H_4OH	Cr 128 D_h 139 I	[90]
C_5H_{11}	C_5H_{11}	$(CH_2)_3OH$	$(CH_2)_6CH=CH_2$	Cr 68 I [b]	[87]
C_6H_{13}	C_6H_{13}	C_8H_{17}	C_8H_{17}	Cr 47 D_{ho} 84 I	[50, 71]
C_8H_{17}	C_8H_{17}	C_6H_{13}	C_6H_{13}	Cr 47 D_h 84 I	[50, 71, 91]
C_8H_{17}	C_8H_{17}	$C_{12}H_{25}$	$C_{12}H_{25}$	Cr 51 D 61 I	[91]
C_9H_{19}	C_9H_{19}	C_5H_{11}	C_5H_{11}	Cr 53 D_{ho} 71 I	[50, 71]
$C_{10}H_{21}$	$C_{10}H_{21}$	C_5H_{11}	C_5H_{11}	Cr 56 I	[71]
$(CH_2)_3OH$	C_5H_{11}	C_5H_{11}	C_5H_{11}	Cr 87 D_1 99 D_2 118 I	[90]
$(CH_2)_6OH$	C_5H_{11}	C_5H_{11}	C_5H_{11}	Cr 65 I [c]	[90]

[a] Gives D_{ho} phase on cooling to 51 °C.
[b] Gives D_h phase on cooling to 64 °C.
[c] Gives D phase on cooling to 55 °C.

er discogens to be modified either by introducing extra substituents into the aromatic nucleus through electrophilic aromatic substitution or by removing some of the alkoxy groups through selective reduction. The outcome of electrophilic substitution reactions on the unsubstituted hydrocarbon triphenylene seems to be determined by a conflict between electronic effects that favour α attack and steric factors (involving mainly the interaction with the *peri* hydrogen atoms) that favour β attack, but normally β attack dominates. In those mesogens in which all the β-positions bear substituents steric crowding must be severe, but the systems can be readily nitrated, chlorinated and brominated [92–94]. The introduction of the nitro substituent is a particularly valuable step since, as shown in Scheme 23, it can be readily transformed into a wide range

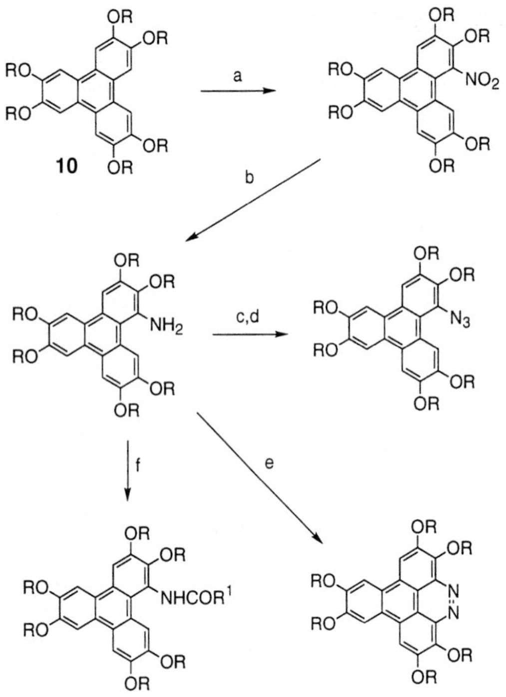

Scheme 23. Modification of the substituents through electrophilic substitution. Reagents: (a) $HNO_3/Et_2O/$ AcOH/15 min, 73% [92]; (b) Sn/AcOH/reflux/4 h, 87% [92]; (c) $NaNO_2/AcOH$ [94]; (d) KN_3, 58% [94]; (e) $NaNO_2/AcOH$, 45%, [92]; (f) $R^1COCl/K_2CO_3/$ CH_2Cl_2/18 h, 95% (R=Me) [92].

Table 26. Transition temperatures for hepta-substituted derivatives of triphenylene of the type **11**.

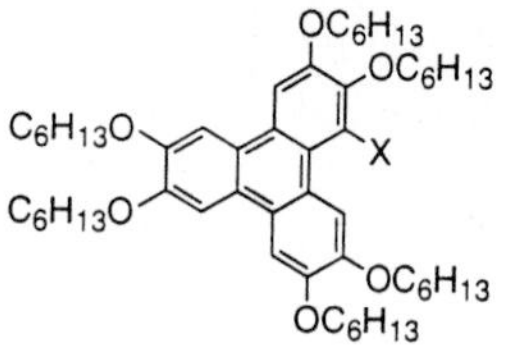

11

X	Phase transitions	Refs.
NO_2	<25 D 136 I	[92]
NH_2	Cr 35 D 77 I	[92]
N_3	Cr 70 D 118 I	[93]
$NHCOCH_3$	Cr 99 D 162 I	[92]
$NHCOC_2H_5$	Cr 104 D 167 I	[93]
$NHCOC_3H_7$	Cr 98 D 176 I	[93]
$NHCOC_4H_9$	Cr 103 D 186 I	[93]
$NHCOC_5H_{11}$	Cr 96 D 186 I	[93]
$NHCOC_6H_{13}$	Cr 90 D 191 I	[93]
$NHCOC_7H_{15}$	Cr 90 D 191 I	[93]
$NHCOC_8H_{17}$	Cr 84 D 192 I	[93]
$NHCOC_9H_{19}$	Cr 78 D 190 I	[93]
$NHCOC_{11}H_{23}$	Cr 68 D 181 I	[93]
Cl	Cr 37 D 98 I	[94]
CH_2OH	Cr 70 I	[94]

of other substituents and many of the products show enhanced mesophase properties as detailed in Table 26. Note also those compounds, shown in Scheme 24, that are not themselves mesomorphic but which give mesomorphic nitro derivatives (e.g. **12–14**, the second and third of which are readily available through variations on the 'biphenyl' route shown in Scheme 20) [93]. Hence **12** shows Cr 96 I, but its nitration product gives a columnar phase on cooling to 88 °C. Compound **13** shows Cr 83 I and its nitration product Cr 81 I, but gives a columnar phase on cooling to 64 °C. Compound **14** shows Cr 70 I and gives a columnar phase on cooling to 60 °C, but its nitra-

tion product shows Cr 45 D 89 I. Another important development has been the introduction of the 'reductive removal of substituents' by the group of Ringsdorf [90, 95]. An example of this route is shown in Scheme 25. After selective removal of the methoxy methyl group [96], tetrazenyl hydrogenolysis is used to remove the hydroxyl group [97]. The β-sites 'exposed' in this way also provide convenient entry points for the introduction of functionality into the triphenylene nucleus through electrophilic substitution reactions. In the example shown the sites are brominated, but there are other obvious possibilities. Whereas **15** is not mesomorphic (m.p. 86 °C), the bromide **16** is liquid crystalline (D_x 19 D_h 165 I). In a similar manner it has proved possible to prepare a symmetrical triphenylene with three alkoxy chains (2,6,10-trispentyloxy-triphenylene). This is also non-mesomor-

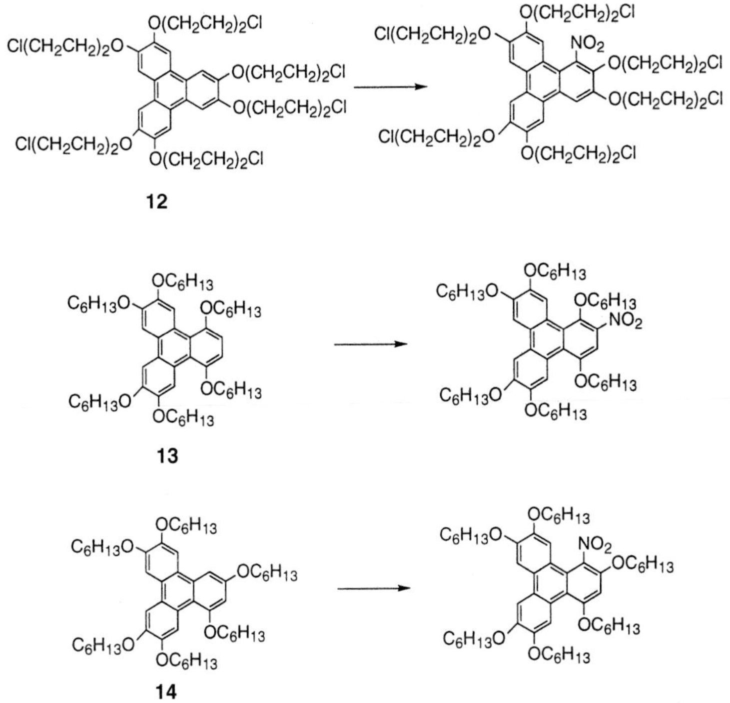

Scheme 24. Nitration reactions leading to enhanced mesophase behaviour [93].

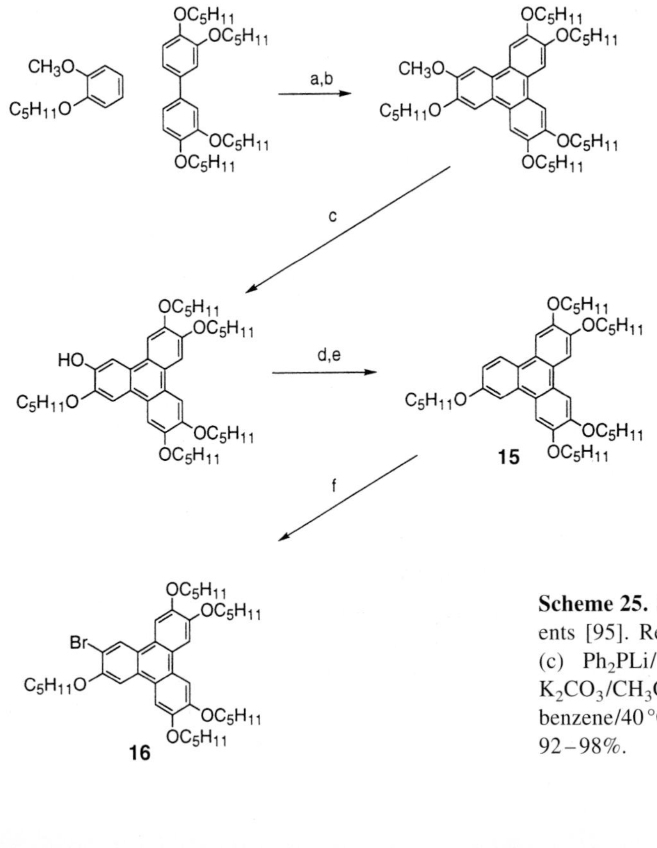

Scheme 25. Selective removal of methoxy substituents [95]. Reagents: (a) FeCl$_3$/CH$_2$Cl$_2$; (b) MeOH; (c) Ph$_2$PLi/THF; (d) 5-chloro-1-phenyltetrazine/ K$_2$CO$_3$/CH$_3$COCH$_3$/reflux/18 h; (e) Pd/C/H$_2$/THF– benzene/40 °C/24 h, 80% (two steps); (f) Br$_2$/CH$_2$Cl$_2$, 92–98%.

phic (m.p. 95 °C), but bromination gives 3,7,11-tribromo-2,6,10-trispentyloxytriphenylene which is mesogenic (Cr 189 D$_h$ 217 I) [95]. Although these substitution and dealkoxylation routes have been developed (but perhaps not yet fully exploited) for triphenylene-based liquid crystals, they are also applicable to some of the other classes of mesogen described in this chapter.

2.6   Dibenzopyrene

Derivatives of 1,2,5,6,8,9,12,13-octahydroxydibenzo-[e,1]-pyrene (17) giving discotic mesophases were discovered by Bock and Helfrich [63] in their search for ferroelectrically switchable columnar mesogens. The synthesis (Scheme 26) involved modification of a known literature procedure to produce hexamethoxydibenzopyrene-1,8-quinone [98] as the key intermediate. Reduction followed by acylation affords the chiral octaester (18), the mesophase structure of which was deduced to be tilted columnar after demonstration of its ferroelectric switching in an applied electric field [63]. The basic synthesis has also been modified to produce a discotic octapentyl derivative, which forms an hexagonal mesophase [87].

2.7   Perylene

Discotic derivatives of perylene have been prepared by Mullen and coworkers [99]. The discotic mesophase behaviour of these highly fluorescent materials is quite unusual because the molecules have low symme-

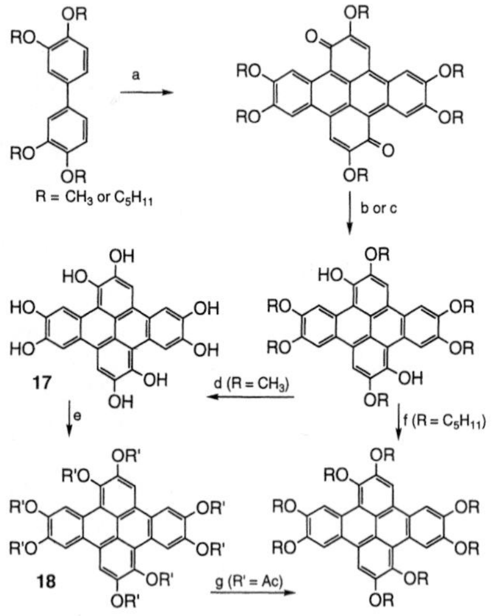

Scheme 26. Synthesis of dibenzopyrene based discogens. Reagents: (a) chloranil/H$_2$SO$_4$/CH$_3$CO$_2$H 60 °C, 24 h, 51% [87]; (b) Na$_2$S$_2$O$_4$ [98]; (c) Zn [87]; (d) BBr$_3$; (e) R'COX; (f) C$_5$H$_{11}$Br/KOH; (g) C$_5$H$_{11}$Br/K$_2$CO$_3$.

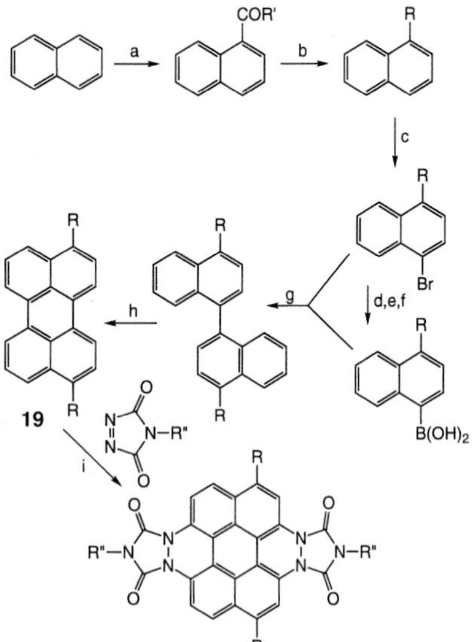

Scheme 27. Synthesis of perylene based discogens [99]. Reagents: (a) R'COCl; (b) NH$_2$NH$_2$·HCl/base; (c) Br$_2$; (d) BuLi, −78 °C; (e) B(O-^{i}Pr)$_3$; (f) H$_2$O; (g) Pd(O)/PPh$_3$/K$_2$CO$_3$/toluene, reflux; (h) K/DME; (i) xylene, reflux.

Table 30. Transition temperatures for perylene based discotic mesogens (see Scheme 27 for structures).

R	R″	Phase transitions [99]
$C_{18}H_{37}$	C_5H_{11}	Cr 69 D 261 I
$C_{18}H_{37}$	$C_{17}H_{35}$	Cr 45 D_{ho} 305 I
$C_{12}H_{25}$	$C_{11}H_{35}$	Cr −20 D_{ho} 373 I
C_5H_{11}	$C_{17}H_{35}$	Cr 85 D 200 I
C_5H_{11}	C_5H_{11}	Cr 27 D 319 I
$C_{12}H_{25}$	$C_{17}H_{35}$	Cr 15 D_{ho} 316 I

try. The synthesis is outlined in Scheme 27 with the key features being the efficient synthesis of 3,10-dialkylperylenes (**19**) [100] and their smooth double Diels–Alder reaction with 4-alkyl-3,5-dioxotriazoles.

The mesophase behaviour of the discotic perylenes is summarized in Table 30. It is apparent immediately that the columnar phase of these perylene derivatives is remarkably stable, with the largest temperature range being observed for the materials that contain perylene (R) and amide (R″) side chains of approximately equal lengths. When the molecular shape approximates to a disk the hexagonal (D_h) mesophase is produced, but deviations from this structure occur when there is a mismatch of side-chain lengths.

2.8   Truxene

The hexaesters of truxene **23** [101, 102] show a fascinating variety of phases, including nematic N_D phases, but unfortunately they are not very easy to synthesize. The synthesis used is based on the self-condensation of indanone brought about by treatment with ethyl polyphosphate [103] (Scheme 28). When the reaction is applied to the dimethoxy analogue **20** it generates a mixture of the desired trimeric truxene **22**

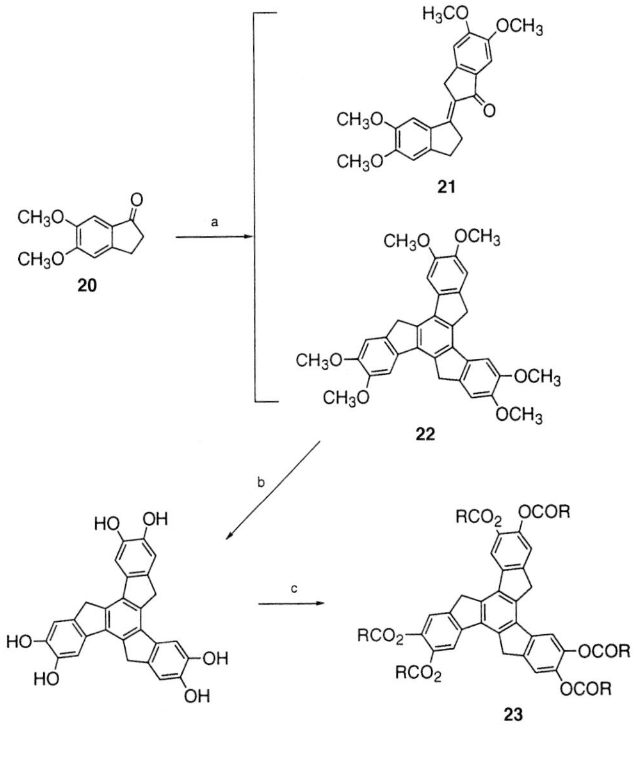

Scheme 28. Synthesis of derivatives of truxene [105]. Reagents: (a) PPE/CHCl$_3$ –EtOH/reflux, 66%; (b) BBr$_3$ · Me$_2$S/ ClCH$_2$CH$_2$Cl/reflux/72 h, 86%; (c) RCOCl/pyridine DMAP/48 h, 86% (C_{14}).

Table 31. Transition temperatures for derivatives of truxene.

Substituent	Phase transitions	Refs.
$C_6H_{13}CO_2$	Cr 112 D_r (D_{rd})[a] 138 D_h (D_{ho}) 280 I	[101 (104)]
$C_7H_{15}CO_2$	Cr 98 D_r (D_{rd})[b] 140 (158) D_h (D_{ho}) 280 I	[101 (104)]
$C_8H_{17}CO_2$	Cr 88 D_r[c] 141 D_h (D_{ho}) 280 I	[101 (104)]
$C_9H_{19}CO_2$	Cr 68 N_D 85 D_r (D_{rd}) 138 D_h (D_{ho}) 280 I	[101 (104)]
$C_{10}H_{21}CO_2$	Cr 62 N_D 89 D_r (D_{rd}) 118 D_h 250 (277) I	[101 (104)]
$C_{11}H_{23}CO_2$	Cr 64 N_D 84 D_r (D_{rd}) 130 D_h 250 (253) I	[101 (104)]
$C_{12}H_{25}CO_2$	Cr 57 N_D[d] 84 D_r (D_{rd}) 107 (112) D_h (D_{hd}) 249 (246) I	[101 (104)]
$C_{13}H_{27}CO_2$	Cr 58 (61) N_D[e] 83 D_h (D_{rd} 112 D_{ho}) 235 (241) I	[101 (104)]
	[Cr 61 N_D 84 D_{rd} 112 D_{ho} 241 I]	[106, 107]
$C_{14}H_{29}CO_2$	Cr 64 N_D[f] 82 D_{rd} 95 D_{ho} (D_{hd}) 241 (221) I	[108 (104)]
$C_{15}H_{31}CO_2$	Cr 69 N_D[g] 84 D_{rd} 95 D_{hd} 210 I	[104, 108]
$C_{11}H_{23}OC_6H_4CO_2$	Cr 90 D_{rd} 137 N_D 171 (179) D_{rd} 284 N_D 297 I	[109 (105, 106)]
$C_{12}H_{25}OC_6H_4CO_2$	Cr 90 N_D[h] 173 D_{rd} 260 N_D >297 I	[105, 106]

[a]   Gives N_D on cooling to 96 °C.
[b]   Gives N_D on cooling to 85 °C.
[c]   Gives N_D on cooling to 87 °C.
[d]   Gives D_{hd} on cooling to 53 °C.
[e]   Gives D_{hd} on cooling to 56 °C.
[f]   Gives D_{hd} on cooling to 56 °C (58 °C).
[g]   Gives D_{hd} on cooling to 62 °C.
[h]   Gives D_{rd} on cooling to 79 °C.

and the dimeric indanylindanone **21**. The original paper by Tinh et al. [104] gives no clues as to how this rather intractable mixture should be treated. However, more recently, a detailed experimental procedure has been provided by Lee et al. [105]. The trimer is so insoluble that it cannot be purified by recrystallization or chromatography, but it can be freed of the dimer and most other impurities simply by washing with solvent. Lee et al. then recommend demethylation using boron tribromide–dimethyl sulfide complex and the use of 4-dimethylaminopyridine (DMAP) to catalyse the esterification. The shorter chain-length esters (C_7 to C_9) give D_h and D_r phases and a monotropic N_D phase is produced on cooling the D_r phase. The longer chain-length esters give enantiotropic N_D, D_h and, up to C_{12}, D_r phases (Table 31). The alkoxybenzoate esters give re-entrant N_D and D_{rd} phases [105, 109], and a mixture of 13% dodecyloxybenzoate and 87% tetradecanoate shows a re-entrant isotropic phase (Cr 67 N_D 112 I 129 D_{ho} 214 I) [106], although there has been some disagreement in the literature over the nature of the D_{ho} phase [105].

2.9 Decacyclene

The decacyclene derivatives **24a** and **24b** are unusual in that they have only three side chains. They are readily synthesized by a direct Friedel–Crafts acylation of the parent hydrocarbon, a reaction claimed to proceed with very good regioselectivity (Scheme 29). Both compounds show a high tendency to aggregate, even in dilute solution, and both give two mesophases which are stable over a wide temperature range (**24a**, heating, Cr 92.8M_1 115 M_2 262 I; **24b**, heating, Cr 98 M_1 108.5 M_2 240.5 I). Hysteresis is observed on cooling. Reduction of the keto groups to methylene groups gives non-mesomorphic products [110].

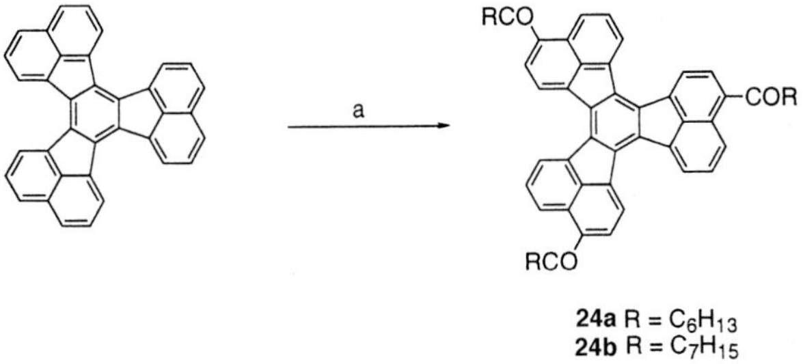

Scheme 29. Synthesis of derivatives of deca-cyclene. Reagents: (a) RCOCl/AlCl₃/-ClCH₂CH₂Cl/reflux, 38% (24).

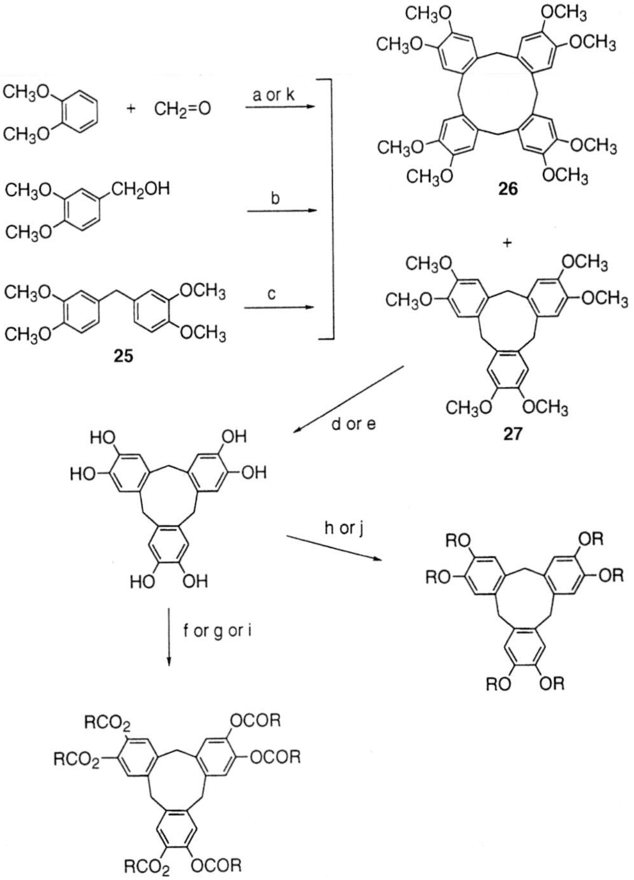

24a R = C_6H_{13}
24b R = C_7H_{15}

Scheme 30. Synthesis of tribenzocy-clononatriene based discotic mesogens. Reagents: (a) AcOH/1% H_2SO_4/ 100 °C/15 min, 87% [114, 118, 119]; (b) 70% H_2SO_4/0 °C [118, 119, 124]; (c) CH_2O/HCl [118, 119, 124]; (d) BBr₃/CHCl₃ or C_6H_6/reflux, 85% [124]; (e) pyridine hydrochloride [120]; (f) RCOCl/185 °C, 58% (C_{13}) [111]; (g) RCOCl/0 °C/24 h, 67% (C_{15}) [111]; (h) RBr/K₂CO₃/DMF/re-flux/24 h [111]; (i) RCO₂H/DCCI/ DMAP [120]; (j) RBr/K₂CO₃/i-Bu-COCH₃/reflux/6 days [114]; (k) 37% HCl/room temp. 3 h, 60 °C 5 h, 20% [120].

2.10 Tribenzocyclo-nonatriene

The tribenzocyclononatriene nucleus has three-fold symmetry and adopts a shallow cone-shaped conformation. Consequently, it has been suggested that the liquid crystal-line materials based on this nucleus should be termed 'pyramidic' rather than discotic [111]. As a result of this particular confor-mation the molecule has a net dipole along its C_3 axis, and this has the interesting con-sequence that the columnar phases in which

Table 32. Transition temperatures for derivatives of tribenzocyclononatriene.

Substituent	Phase transitions	Refs.
CH_3O	Cr 232 I	[111]
C_4H_9O	Cr 136 I[a]	[111]
$C_5H_{11}O$	Cr 104 I[b]	[111]
$C_6H_{13}O$	Cr 41 D_1 92 I	[111]
$C_7H_{15}O$	Cr 25 D_1 80 I	[111]
$C_8H_{17}O$	Cr 25 D_h 72 I	[111, 114]
$C_9H_{19}O$	Cr 19 D_h 66 I	[111, 112]
$C_{10}H_{21}O$	Cr 26 D_h 63 I	[111, 114]
$C_{11}H_{23}O$	Cr 35 D_2 44 D_1 62 I	[111]
$C_{12}H_{25}O$	Cr 48 D_1 62 I	[111]
$C_3H_7CO_2$	Cr 197 I	[120]
$C_7H_{15}CO_2$	Cr 5 D_4 153 I (Cr 150 I)	[111 (120)]
$C_8H_{17}CO_2$	Cr 24 D_4 153 I	[111]
$C_9H_{19}CO_2$	Cr 33 (50) D_4 146 (145) I	[111 (120)]
$C_{10}H_{21}CO_2$	Cr 32 D_5 39 D_4 132 D_3 141 I	[111]
$C_{11}H_{23}CO_2$	Cr 59 (55) D_4 119 (116) D_3 141 (138) I	[111 (120)]
$C_{12}H_{25}CO_2$	Cr 67 D_r 100 D_h 139 I	[111, 114]
$C_{13}H_{27}CO_2$	Cr 73 D_4 81 D_3 136 I	[111]
$C_{14}H_{29}CO_2$	Cr 81 D_3 135 I	[111]
$C_{15}H_{31}CO_2$	Cr 80 D_3 129 I	[120]
$C_7H_{15}C_6H_4CO_2$	Cr 33 D_6 57 D_7 149 I	[112]
$C_8H_{17}C_6H_4CO_2$	Cr 12 D_6 21 D_7 157 I	[112]
$C_{10}H_{21}C_6H_4CO_2$	D_6 43 D_7 162 I	[112]
$C_{10}H_{21}OC_6H_4CO_2$	Cr 51 D_6 100 D_7 190 I	[112]
$C_{12}H_{25}OC_6H_4CO_2$	Cr <25 M_1 75 M_2 188 I	[120]

[a] Gives D_1 on cooling to 127 °C.
[b] Gives D_1 on cooling to 96 °C.

these molecules are stacked one on top of the other are potentially ferroelectric or antiferroelectric [112, 113]. However, the situation is complicated by the fact that the cone can be inverted quite easily. At the temperatures at which the mesophases are formed this inversion is rapid for isolated molecules in isotropic solution, but it is somewhat slower in the mesophase itself. This is presumably because of the need for cooperative behaviour in the latter case [113, 114]. The basic reaction through which this nucleus is made (Scheme 30), the acid catalysed condensation of veratrole and formaldehyde, has a long and complicated history. Working in the 1920s, Robinson identified the product as the dimer (a dihydroanthracene) [115]. Once this structure had been disproved, a hexameric structure was proposed [116, 117]. This also proved to be false. Using more modern analytical techniques, Lindsey [118, 119] was able to show that the main product was the trimer (the tribenzocyclononatriene **27**) and the minor product the tetramer (the tetrabenzocyclododecatetraene **26**). He further claimed that, under his reaction conditions, the same approximately 5:1 mixture of **27** and **26** was obtained from the acid-catalysed condensation of veratrole and formaldehyde, from the acid-catalysed oligomerization of 3,4-dimethoxybenzyl alcohol and, most surprisingly and most significantly, from an attempted chloromethylation of the diphenylmethane derivative **25**! Of all these routes the veratrole/formaldehyde reaction is normally chosen. Complete separation of the mixture of **27** and **26** requires careful fraction-

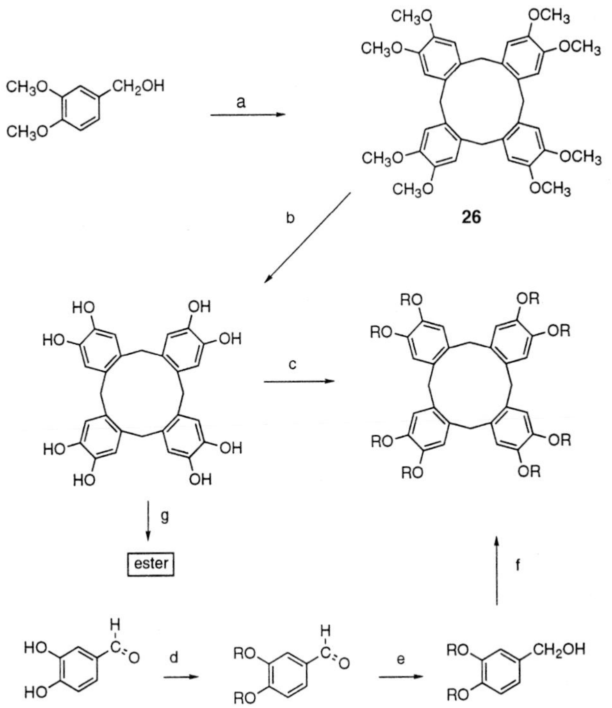

CH$_3$O CH$_2$OH
CH$_3$O

a →

CH$_3$O OCH$_3$
CH$_3$O OCH$_3$

CH$_3$O OCH$_3$
CH$_3$O OCH$_3$

b

26

HO OH
HO OH

HO OH
HO OH

c →

RO OR
RO OR

RO OR
RO OR

g ↓

ester

f ↑

HO C≡O
HO H

d →

RO C≡O
RO H

e →

RO CH$_2$OH
RO

Scheme 31. Synthesis of derivatives of tetrabenzododecotetraene. Reagents: (a) TFA/CH$_2$Cl$_2$/0 °C/4 h, 55% [125, 126]; (b) BBr$_3$/CHCl$_3$/0 °C, 78% [125, 127]; (c) RBr/K$_2$CO$_3$/EtOH–DMF/reflux, 56% (C$_8$) [125]; (d) RBr/K$_2$CO$_3$/EtOH/reflux, 50% (C$_5$) [125]; (e) NaBH$_4$, 58–96% [125]; (f) TFA/CH$_2$Cl$_2$/4 h, 19% (C$_{18}$) [126]; (g) RCOCl [127].

al crystallization. Repeated crystallization from benzene gives the pure trimer as a 1:1 benzene complex. Demethylation of **27** can be achieved using either boron tribromide [118–120] or pyridine hydrochloride [120], and the hexahydroxy product alkylated or acylated. As well as simple ether [111] and ester [111, 120] derivatives, alkoxybenzoates [121] and acyloxybenzoates [118, 119] have also been reported and simple methods have been described for deuteration of the nucleus [114]. Table 32 gives the mesophase ranges for both esters and ethers [111]. Since this nucleus is also an important building block in the synthesis of host molecules such as cryptophanes, a number of other variations have been developed on the basic synthesis. It is now possible to make a multitude of analogous systems with substituents at both the methylene and the aryl ring positions [122]. Of particular rel-

evance to the future development of this area is the recently reported development of routes to derivatives bearing aryl amino substituents – derivatives for which further elaboration is clearly possible [123].

2.11 Tetrabenzocyclo-dodecatetraene

The tetrabenzocyclododecatetraene nucleus, like the tribenzocyclononatriene nucleus, is non-planar. It is, however, much more flexible. It has, on average, four-fold symmetry, undergoing rapid interconversion between two symmetry-related 'sofa' conformations. Whereas it was believed that the reactions shown in Scheme 30 always gave the trimer as the major product and initial work on the tetramer relied on the isolation

Table 33. Transition temperatures for derivatives of tetrabenzocyclododecatetraene.

Substituent	Phase transitions[a]	Refs.
C_4H_9O	Cr 223 I	[125]
$C_5H_{11}O$	Cr 182 (177) D_1 190 (186) I	[125 (129)]
$C_6H_{13}O$	Cr 162 (160) D_1 172 (170) I	[125 (129)]
$C_7H_{15}O$	Cr 139 (136) D_1 162 (160) I	[125 (129)]
$C_8H_{17}O$	Cr 137 (137) D_1 154 (153) I	[125 (129)]
$C_9H_{19}O$	Cr 114 (109) D_1 148 (148) I	[125 (129)]
$C_{10}H_{21}O$	Cr 113 (112) D_1 (137) D_2 140 (142) I	[125 (129)]
$C_{11}H_{23}O$	Cr 101 (98) D_1 (129) D_2 137 (137) I	[125 (129)]
$C_{12}H_{25}O$	Cr 97 (94) D_1 (122) D_2 131 (133) I	[125 (129)]
$C_{13}H_{27}O$	Cr 93 (92) D_1 (113) D_2 127 (128) I	[125 (129)]
$C_{14}H_{29}O$	Cr 97 (96) D_1 (106) D_2 120 (123) I	[125 (129)]
$C_{15}H_{31}O$	Cr 96 (93) D_1 (102) D_2 118 (119) I	[125 (129)]
$C_{16}H_{33}O$	Cr 98 D_1 100 D_2 115 I	[129]
$C_{11}H_{23}CO_2$	Cr 82 D_r 246 (246) I	[129 (127)]
$C_{13}H_{25}CO_2$	Cr 90 (90) D_r 237 (236) I	[129 (127)]
$C_{14}H_{29}CO_2$	Cr 92 D_r 230 I	[129]
$C_{15}H_{31}CO_2$	Cr 95 D_r 223 I	[129]
$C_{10}H_{21}C_6H_4CO_2$	Cr 96 M 151 I	[127]
$CH_3OCH_2CH_2O$	Cr 82 M 211 I	[130]
$CH_3O(CH_2CH_2O)_2$	Cr 30 M 109 I	[130]
$CH_3O(CH_2CH_2O)_3$	Cr 22 M 63 I	[130]

[a] Figures in parenthesis are as in the reference also given in parenthesis.

of the minor product, Percec [125, 126] has shown that, by varying the reaction conditions, the tetramer can be made the major product. The key observation was that the trimer could be obtained by chloromethylation of the diphenylmethane (**25**). Clearly, the reactions are all reversible. According to Percec's analysis, the trimer is the kinetic product of the oligomerization of 3,4-dimethoxybenzyl alcohol and is formed in highest yield using very strong acids (e.g. $HClO_4$ or CF_3SO_3H) and a poor or a nonsolvent, whereas the tetramer is the thermodynamic product and it becomes the major product if the reaction is carried out in dilute dichloromethane solution using an excess of trifluoroacetic acid as the catalyst [125, 126] (up to a 6:1 ratio of tetramer to trimer, Scheme 31). The octamethoxy compound **26** can be demethylated using boron tribromide, and alkylated or acylated in the normal way [127–129]. Preparation of the octaether derivatives by alkylation of the re-

sultant octaphenol gives both higher yields and purer products than the alternative route shown in Scheme 31, in which the alkoxy substituents are introduced before the oligomerization step [125]. Initial reports suggested that these ethers gave a single enantiotropic mesophase [125, 128], but it was later shown that, while the C_5 to C_9 ethers give a single D_r mesophase, the C_{10} to C_{17} ethers from both D_r and D_h phases [129]. The transition between these two has not been detected by DSC, but is observed by polarizing microscopy. The ethyleneoxy derivatives (Table 33) show both thermotropic and lyotropic behaviour giving what is probably a D_h phase [130]. Generally the tetramers give higher transition temperatures than do their trimeric counterparts, and show less tendency to polymorphism. Transition data are summarized in Table 33.

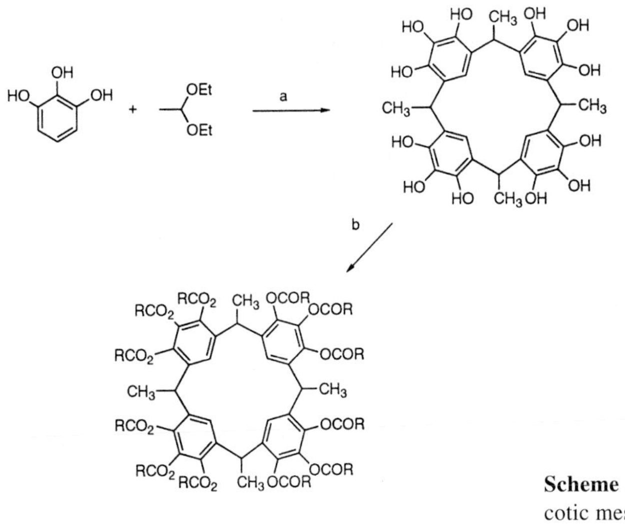

28

Scheme 32. Synthesis of metacyclophane based discotic mesogens. Reagents: (a) HCl/EtOH/0 °C, 72%; (b) RCOCl/180 °C.

2.12 Metacyclophane

Metacyclophanes of the type **28** have a rigid cone-shaped or crown-shaped conformation in which the ring methyl substituents adopt an *endo* orientation. Like the tribenzocyclononatrienes they give potentially ferroelectric columnar phases. The key step of the synthesis shown in Scheme 32 is the acid-catalysed condensation of pyrogallol and 1,1-diethoxyethane [131–133]. Although many dodeca-ether and dodeca-ester derivatives have been examined, only the C_{13} to C_{18} esters with methyl substituents on the ring proved to be mesmorphic (Table 34) [134]. They give a D_h phase. The

Table 34. Transition temperatures for metacyclophane based discotic mesogens [134].

Substitutent	Phase transitions
$C_{12}H_{25}CO_2$	Cr 46 D_{ho} 66 I
$C_{13}H_{27}CO_2$	Cr 30 D_{ho} 67 I
$C_{15}H_{31}CO_2$	Cr 48 D_{ho} 61 I
$C_{17}H_{35}CO_2$	Cr 58 D_{ho} 68 I

properties of mesogens with a random, statistical mixture of ester chains [131] and the synthesis of deuteriated derivatives [135] have also been reported.

2.13 Phenylacetylene Macrocycles

Discotic liquid crystals of phenylacetylene macrocycles **29** were first reported by Zhang and Moore in 1994 [136]. Their synthesis involves the stepwise build-up of phenylacetylene hexamers [137] followed by a macrocyclization step [138]. The key reactions of the synthesis (Scheme 33) are the palladium-catalysed coupling of aryl iodides and acetylenes, and the selective 'deprotection' steps allowing the ends of the growing oligomer to be distinguished. The synthesis shown is just one of the many possible sequences that can be used to prepare the macrocycles. The mesophase behaviour of the discotic derivatives prepared to date is shown in Table 35 [136]. In some ways

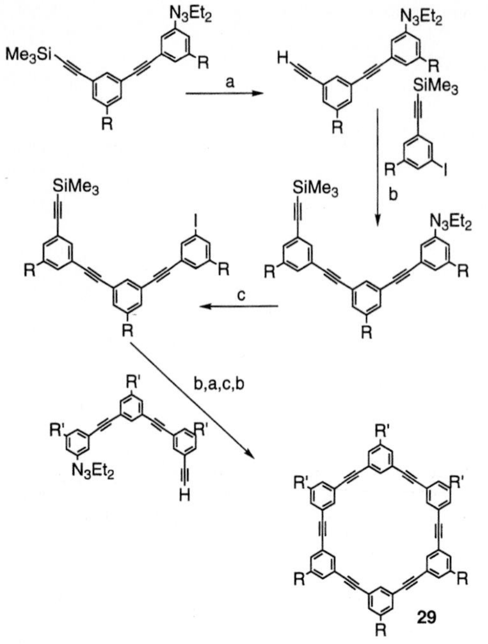

Scheme 33. Synthesis of phenylacetylene macrocycles. Reagents: (a) K_2CO_3/MeOH; (b) Pd(0); (c) MeI/110 °C.

Table 35. Transition temperatures for discotic phenylacetylene macrocycles (see Scheme 33 for structures).

R	R'	Phase transitions [136]
OC_7H_{15}	OC_7H_{15}	I 192 N_D 168 Cr
$OCOC_7H_{15}$	$OCOC_7H_{15}$	I 241 N_D 121 Cr
$COOC_6H_{13}$	OC_6H_{13}	[I 202 N_D 130 D_h 103 Cr][a]

[a] Monotropic phases.

these macrocycles can be compared with the 'tubular' azacrowns. They are, however, much more rigid and, unlike the azacrowns, tend to form nematic phases. This behaviour is similar to the 'super-disk' benzene, naphthalene and triphenylene multiynes, the core-sizes of which are of the same order.

3 Heterocyclic Cores

3.1 Pyrillium

The pyrillium salts **30** are relatively simply synthesized. As shown in Scheme 34, the key step is the acid-catalysed condensation reaction between a chalcone and an acetophenone derivative [139, 140]. The C_2 to C_{12} derivatives all give a D_{ho} columnar phase [139–142]. The dependence of phase behaviour on chain length is illustrated in Table 36. Unfortunately, these systems are not very thermally stable, so the actual transition temperatures are subject to considerable uncertainty. The really surprising aspect of the phase behaviour of these systems is that a mesophase is obtained for systems with side chains as short as C_2 and C_3 [142]. It is believed that, in these systems, the disordered nature of the BF_4^- counterions is important in helping to bring about the disordered stacking of the pyrillium ring inherent to the columnar phase [142]. Analogous systems bearing aryl substituents at the 2 and 6 positions (only) of the pyrillium ring have also been synthesized. Despite the fact that they are very non-disk-like and bear only four side-chains there are some indications that these also give D_h phases, but unfortunately thermal instability problems have, once again, prevented full characterization [141, 143].

Table 36. Transition temperatures for pyrillium based discotic mesogens.

Substituent	Phase transitions	Refs.
C_2H_5O	Cr 194/200 D_h 213/217 I	[142]
C_3H_7O	Cr 167 D_h 240/247 I	[142]
C_4H_9O	Cr 100/117 D_h 277 I	[140]
$C_5H_{11}O$	Cr <25 D_h 265/283 I	[140]
$C_8H_{17}O$	Cr <25 D_h 200/237 I	[140]
$C_{12}H_{25}O$	Cr <25 D_h 231/288 I	[140]

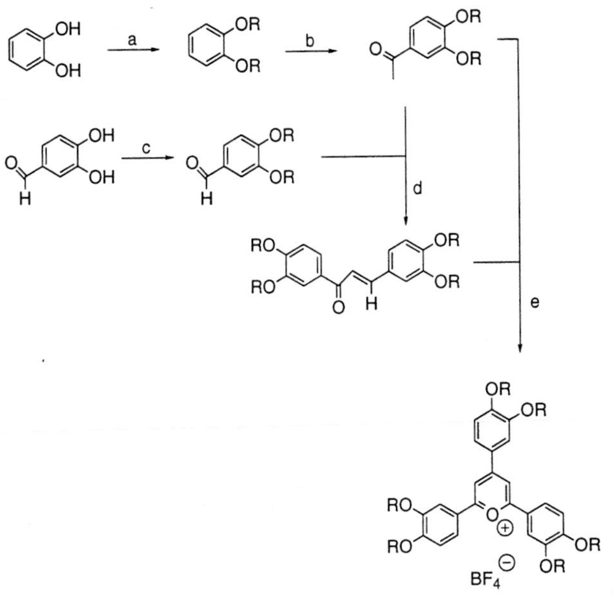

30

Scheme 34. Synthesis of pyrillium based discotic mesogens. Reagents: (a) RX/phase transfer conditions/cat. Aliquat 366 ($Oc_3NMe^+Cl^-$) [139, 144]; (b) $MeCOCl/AlCl_3/CH_2Cl_2$ [139]; (c) $RX/K_2CO_3/DMF/80\,°C$ [139]; (d) NaOH/EtOH/60 °C [139]; (e) $HBF_4/Ac_2O/100\,°C$ [139].

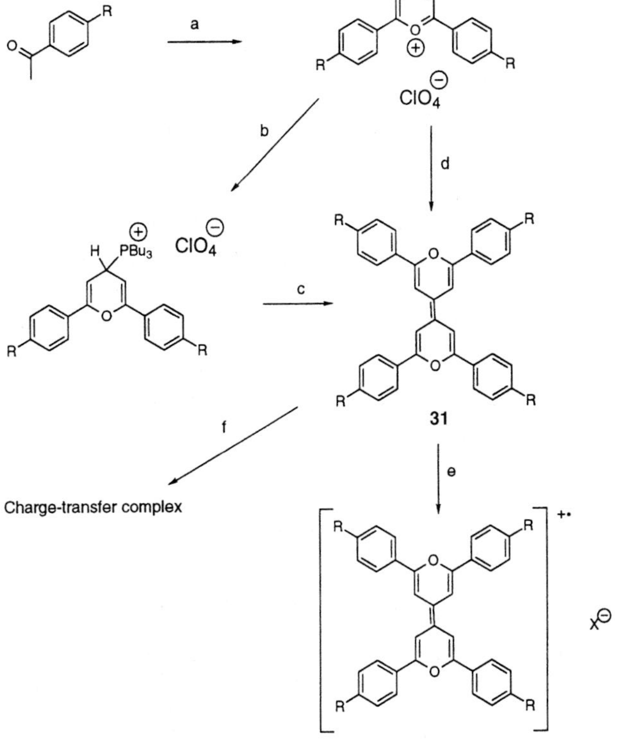

32a X = BF_4^-
32b X = ClO_4^-

Scheme 35. Synthesis of bispyran based discotic mesogens. Reagents: (a) $HC(OEt)_3$/ $HClO_4$; (b) PBu_3/MeCN; (c) Et^iPr_2N/MeCN/reflux/2 h; (d) Zn/MeCN; (e) electro-oxidation/ $R_4N^+X^-$; (f) TCNQ.

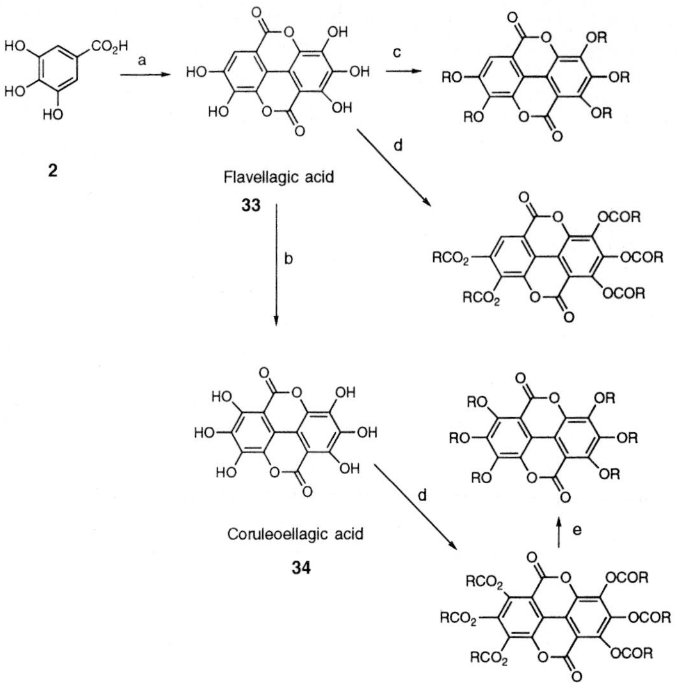

2

Flavellagic acid
33

Coruleoellagic acid
34

Scheme 36. Synthesis of discotic mesogens based on condensed benzyprone nuclei (flavellagic acid and coruoellagic acid). Reagents: (a) $K_2S_2O_8/H_2SO_4/160\,°C$; (b) As_2O_3/H_2SO_4; (c) RBr/ K_2CO_3/DMAA/90 °C/3 days or RBr/Aliquat 336 (Oc$_3$NMe$^+$Cl$^-$)/ KOH/130 °C/4 h; (d) RCOCl; (e) RBr/K_2CO_3/70 °C/ 4 days.

Table 37. Transition temperatures for bispyran based discotic mesogens.

Substituent	Phase transitions [145]
C_5H_{11}	Cr 228 I
C_9H_{19}	Cr 54 M 172 I
$C_{12}H_{25}$	Cr 96 M 147 I

3.2 Bispyran

The bispyrans shown in Scheme 35 can be generated from the 1,6-biarylpyrillium salts either directly by reaction with zinc [145], or indirectly via the trialkyl phosphine complex [146, 147]. They show a marked tendency to aggregate in organic solvents, a property that is doubtless related to their ability for form columnar phases [148].

Their propensity to form columnar phases is not destroyed by the introduction of charge into the nucleus [146]. Hence, electro-oxidation produces mesogenic salts **32 a** and **32 b**, and treatment with tetracyanoquinodimethane (TCNQ) produces a mesogenic charge transfer complex. In the C_{12} series the parent bispyran **31** has a mesophase range of 96–147 °C, the radical cation tetrafluoroborate gives two discotic phases between 142 °C and about 240 °C, the perchlorate two phases between 144 °C and about 261 °C, and the TCNQ charge-transfer salt also gives two phases between 11 °C and 242 °C [146]. The dependence of the phase behaviour of the bispyrans on chain length is illustrated in Table 37 [145].

3.3   Condensed Benzpyrones (Flavellagic and Coruleoellagic Acid)

The reaction of gallic acid **2** with potassium persulfate and concentrated sulfuric acid gives the condensed α-pyrone, flavellagic

Table 38. Transition temperatures for discotic mesogens based on condensed benzopyrone nuclei.

Substituent	Phase transitions [149]
Flavellagic acid (33)	
$C_6H_{13}O$	Cr 16 D_{ho} 198 I
$C_7H_{15}O$	Cr 37 D_{ho} 186 I
$C_8H_{17}O$	Cr 23 D_{ho} 176 I
$C_9H_{19}O$	Cr 45 D_{ho} 166 I
$C_{10}H_{21}O$	Cr 42 D_{ho} 146 I
$C_{11}H_{23}O$	Cr 45 D_{ho} 141 I
$C_{12}H_{25}O$	Cr 46 D_{ho} 134 I
$C_{13}H_{27}O$	Cr 55 D_{ho} 125 I
$C_{14}H_{29}O$	Cr 42 D_{ho} 118 I
$C_{15}H_{31}O$	Cr 45 D_{ho} 116 I
$C_{16}H_{33}O$	Cr 53 D_{ho} 109 I
$C_7H_{15}CO_2$	Cr 181 I
$C_8H_{17}CO_2$	Cr 104 D_h 183 I
$C_9H_{19}CO_2$	Cr 79 D_h 175 I
$C_{10}H_{21}CO_2$	Cr 90 D_h 174 I
$C_{11}H_{23}CO_2$	Cr 70 D_h 171 I
$C_{12}H_{25}CO_2$	Cr 83 D_h 176 I
$C_{13}H_{27}CO_2$	Cr 88 D_h 168 I
$C_{14}H_{29}CO_2$	Cr 97 D_h 165 I
$C_{15}H_{31}CO_2$	Cr 93 D_h 164 I
$C_{16}H_{33}CO_2$	Cr 102 D_h 162 I
Couleoellagic acid (34)	
C_3H_7O	Cr 143 D_h 160 I
C_4H_9O	Cr 73 D_h 170 I
$C_5H_{11}O$	Cr −4 D_h 170 I
$C_6H_{13}O$	Cr −85 D_h 159 I
$C_7H_{15}O$	Cr −77 D_h 152 I
$C_8H_{17}O$	Cr −35 D_h 138 I
$C_9H_{19}O$	Cr 4 D_h 123 I
$C_{10}H_{21}O$	Cr 6 D_h 117 I
$C_8H_{17}CO_2$	Cr 159 D_{h1} 217 I
$C_9H_{19}CO_2$	Cr 154 D_{h1} 200 D_{h2} 211 I
$C_{10}H_{21}CO_2$	Cr 146 D_{h1} 189 D_{h2} 209 I
$C_{11}H_{23}CO_2$	Cr 145 D_{h1} 175 D_{h2} 206 I
$C_{12}H_{25}CO_2$	Cr 141 D_{h1} 167 D_{h2} 202 I
$C_{13}H_{27}CO_2$	Cr 140 D_{h1} 157 D_{h2} 199 I
$C_{14}H_{29}CO_2$	Cr 138 D_{h1} 148 D_{h2} 195 I
$C_{15}H_{31}CO_2$	Cr 137 D_{h1} 141 D_{h2} 190 I
$C_{16}H_{33}CO_2$	Cr 134 D_{h2} 187 I

acid **33**, which can be oxidized to coruleoellagic acid **34** using As_2O_3 in sulfuric acid (Scheme 36). Both the ester and the ether derivatives of **33** and **34** give D_h phases. Their phase behaviour is summarized in Table 38 [149]. Esters of coruleoellagic acid may form two mesophases, the transition between which cannot be observed by DSC but is clear by polarizing microscopy, the lower temperature M_1 phase being much more birefringent. Both 'phases' are miscible with other D_h phases, and both seem to be of the D_h type. It is not clear that the transition between these two 'phases' is a true phase transition. If so, it must be second order or very weakly first order.

3.4   Benzotrisfuran

Destrade and coworkers [150] have prepared a series of benzotrisfuran derivatives that give discotic mesophases. The synthesis, outlined in Scheme 37, hinges on the condensation of phloroglucinol with p,p-dimethoxybenzoin to give hexakis(methoxyphenyl)benzotrisfuran. This intermediate is demethylated with pyridinium hydrochloride, and the resultant hexaphenol is acylated by treatment with an acid chloride

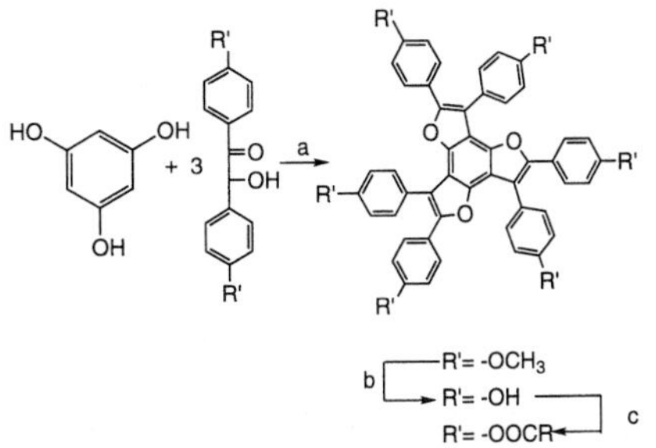

Scheme 37. Synthesis of discotic mesogens based on benzotrisfuran [150]. Reagents: (a) $H_2SO_4/159\,°C$, 5 min, 35%; (b) pyridine · HCl/220 °C, 7 h; (c) RCOCl/Mg/benzene, reflux.

Table 39. Transition temperatures for discotic benzo-trisfurans (RCOO−) (see Scheme 37 for structure).

Substituent	Phase transitions [150]
$C_5H_{11}CO_2$	Cr 186 D_{ho} 244 I
$C_6H_{13}CO_2$	Cr 134 D_{ho} 177 I
$C_7H_{15}CO_2$	(I 100 D_{ho})
$C_8H_{17}CO_2$	(I 95 D_{ho})
$C_8H_{17}O$	Cr 203 $N_{D, decomp}$

and magnesium metal in refluxing benzene. The mesophase behaviour of the discotic derivatives is summarized in Table 39. Only two members of the series (pentyl and hexyl) show enantiotropic columnar mesophases. Longer chain homologues give monotropic (heptyl and octyl) or no (nonyl or longer) mesophases. The octyloxy derivative is an exception in that it gives a nematic phase (as determined by optical microscopy), but slow decomposition of the material at these high temperatures prevented any further analysis.

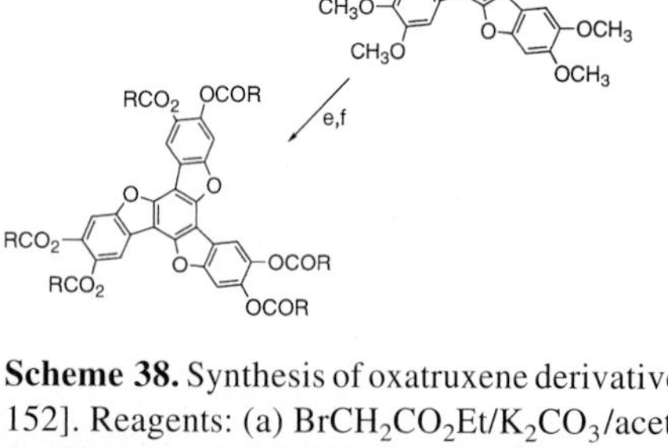

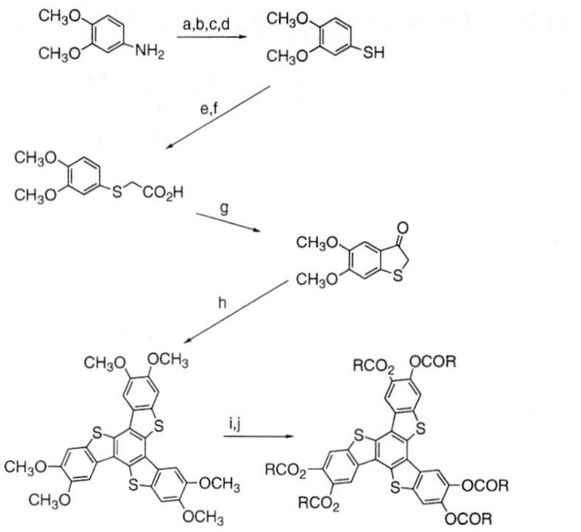

Scheme 39. Synthesis of derivatives of thiatruxene. Reagents: (a) $NaNO_2$/HCl; (b) KSCSOEt; (c) KOH/EtOH; (d) H^+; (e) $BrCH_2CO_2Et$/base; (f) KOH/EtOH; (g) H^+; (h) PPE; (i) pyridine·HCl; (j) RCOCl/pyridine.

Table 40. Transition temperatures for derivatives of oxatruxene.

Substituent	Phase transitions	Refs.
$C_7H_{15}CO_2$	Cr 95 D_{rd}[a] 194 I	[152]
$C_8H_{17}CO_2$	Cr 90 D_{rd}[b] 197 I	[152]
$C_9H_{19}CO_2$	Cr 82 D_{rd}[c] 192 I	[152]
$C_{10}H_{21}CO_2$	Cr 76 D_{rd}[d] 194 I	[152]
$C_{11}H_{23}CO_2$	Cr 78 D_{rd}[e] 184 I	[152]
$C_{12}H_{25}CO_2$	Cr 78 D_{rd}[f] 172 D_{hd} 177 I	[152]
$C_{13}H_{27}CO_2$	Cr 74 D_{rd}[g] 158 D_{hd} 166 I	[151, 152]
$C_{15}H_{31}CO_2$	Cr 80 D_{obd}[h] 84 D_{rd} 138 D_{hd} 152 I	[152]

[a] Gives N_D on cooling to 89 °C.
[b] Gives N_D on cooling to 75 °C.
[c] Gives N_D on cooling to 68 °C.
[d] Gives N_D on cooling to 62 °C.
[e] Gives D on cooling to 64 °C and N_D at 58 °C.
[f] Gives D on cooling to 64 °C and N_D at 59 °C.
[g] Gives D_{obd} on cooling to 71 °C and N_D at 58 °C.
[h] Gives N_D on cooling to 60 °C.

Scheme 38. Synthesis of oxatruxene derivatives [151, 152]. Reagents: (a) $BrCH_2CO_2Et$/K_2CO_3/acetone/reflux/90 min; (b) KOH/EtOH/reflux/150 min; total (a) and (b) 38%; (c) P_2O_5/H_3PO_4/110 °C/30 min, 55%; (d) PPE/140 °C/25 min, 46%; (e) pyridine · HCl/140 °C/30 min; (f) RCOCl/pyridine/48 h, 40% (C_{14}).

3.5 Oxatruxene and Thiatruxene

Like the truxenes themselves, esters of oxatruxene [151, 152] and thiatruxene [153,

Table 41. Transition temperatures for derivatives of thiatruxene.

Substituent	Phase transitions	Refs.
$C_6H_{13}CO_2$	Cr 133 $D_{rd(P21/a)}$ 218 D_{hd} 241 I	[153]
$C_7H_{15}CO_2$	Cr 103 $D_{rd(P21/a)}{}^a$ 212 D_{hd} 236 I	[153]
$C_8H_{17}CO_2$	Cr 90 $D_{rd(P21/a)}{}^b$ 173 $D_{rd(C2/m)}$ 191 D_{hd} 229 I	[153]
$C_9H_{19}CO_2$	Cr 87 N_D 93 $D_{rd(P21/a)}$ 185 D_{hd} 210 I	[153, 154]
$C_{10}H_{21}CO_2$	Cr 62 N_D 98 $D_{rd(P21/a)}$ 155 (151) D_{hd} 193 I	[(153) 154]
$C_{11}H_{23}CO_2$	Cr 78 (64) N_D 93 $D_{rd(P21/a)}$ 146 $D_{rd(C2/m)}$ 149 D_{hd} 180 I	[(153) 154]
$C_{12}H_{25}CO_2$	Cr 79 N_D 87 $D_{rd(P21/a)}$ 136 D_{hd} 191 I	[153]
$C_{13}H_{27}CO_2$	Cr 82 $D_{rd(P21/a)}{}^c$ 132 $D_{rd(C2/m)}$ 134 D_h 179 I	[153, 154]
$C_{14}H_{29}CO_2$	Cr 88 $D_{rd(P21/a)}{}^d$ 121 $D_{rd(C2/m)}$ 125 D_{hd} 178 I	[153]
$C_{10}H_{25}OC_6H_4CO_2$	Cr 83 N_D >300 I	[153]
$C_{12}H_{25}OC_6H_4CO_2$	Cr 81 N_D 295 I	[153]
$C_{13}H_{27}OC_6H_4CO_2$	Cr 86 N_D 280 I	[153]
$C_{16}H_{33}OC_6H_4CO_2$	Cr 96 N_D 241 I	[153]

[a] Gives N_D on cooling to 92 °C.
[b] Gives N_D on cooling to 82 °C.
[c] Gives N_D on cooling to 72 °C.
[d] Gives N_D on cooling to 81 °C.

154] show complex polymorphism. They are also made in a similar way, the only essential difference lying in the somewhat lengthy syntheses that need to be employed to generate the heteroatom analogues of the indanone precursors. These synthesis are summarized in Schemes 38 and 39, and the phase behaviour is given in Tables 40 and 41.

are 'banana-shaped' and the nucleus has only four side chains, they seem to give a D_h phase [143]. It has been suggested that some non-covalent dimer is the effective mesogen, but there seems to be no evidence to support this [141]. The phase behavior of these systems is summarized in Table 42.

Table 42. Transition temperatures for derivatives of dithiolium.

Substituent	Counterion	Phase transitions [143]
$C_5H_{11}O$	BF_4^-	Cr 160 I
$C_8H_{17}O$	BF_4^-	Cr 133 D_h 141 I
$C_{10}H_{21}O$	BF_4^-	Cr 129 D_h 154 I
$C_{12}H_{25}O$	BF_4^-	Cr 120 D_h 145 I
$C_9H_{19}O$	ClO_4^-	Cr 144 D_h 154 I
$C_{10}H_{21}O$	ClO_4^-	Cr 144 D_h 163 I
$C_{12}H_{25}O$	ClO_4^-	Cr 138 D_h 176 I

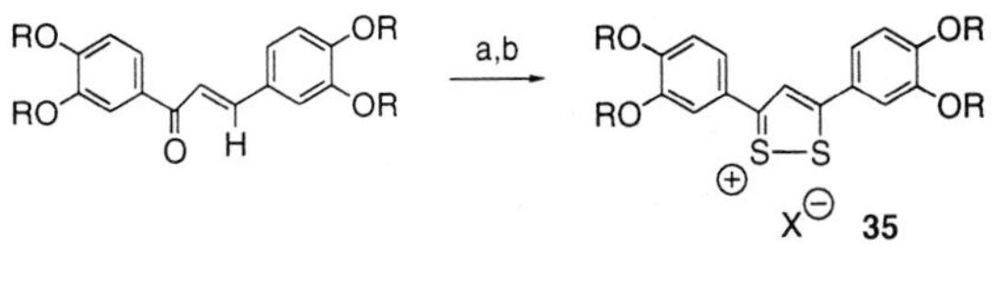

3.6 Dithiolium

The dithiolium salts **35** are prepared by the reaction of the corresponding chalcone and phosphorus pentasulfide followed by an acid work-up as shown in Scheme 40 [141, 143]. Despite the fact that these molecules

Scheme 40. Synthesis of dithiolium derivatives [143]. Reagents: (a) P_4S_{10}/150 °C; (b) HX or LiX/AcOH (X = BF_4 or ClO_4), 3–47% overall.

3.7   Tricycloquinazoline

The tricycloquinazoline nucleus possesses C_3 symmetry and is attractive for a variety of reasons. Firstly, the parent hydrocarbon possesses remarkable thermal and chemical stability. Secondly, the electron-deficient nature of the nucleus renders liquid crystal derivatives susceptible to doping by reducing agents giving n-doped semiconductors [155]. Thirdly, these derivatives are coloured. Against these advantages must be weighed the fact that the parent heterocycle, tricycloquinazoline, is a potent carcinogen. The effect on the carcinogenicity of introducing substituents into the tricycloquinazoline nucleus seems to be difficult to predict [156, 157]. The general observation that polysubstitution destroys the carcinogenic activity suggests that the risk involved in using these mesogens is small. However, until proven otherwise, they must all be treated as potential carcinogens. It has been suggested that the carcinogenic activity of tricycloquinazoline and its derivatives arises from their ability to intercalate into DNA, and that this may not be wholly unrelated to their ability to form stable 'stacked' columnar phases. Although the tricycloquinazoline nucleus can be assembled in a variety of ways, the published syntheses of the mesogens due to Keinan et al. [158, 159] have both been based on the ammonium acetate/acetic acetate mediated trimerisation of an anthranil (Schemes 41 and 42). In general this is a poor-yielding process, and in our experience it has proved difficult to match even the moderate yields claimed for these steps in the literature. Both the hexa-alkoxy

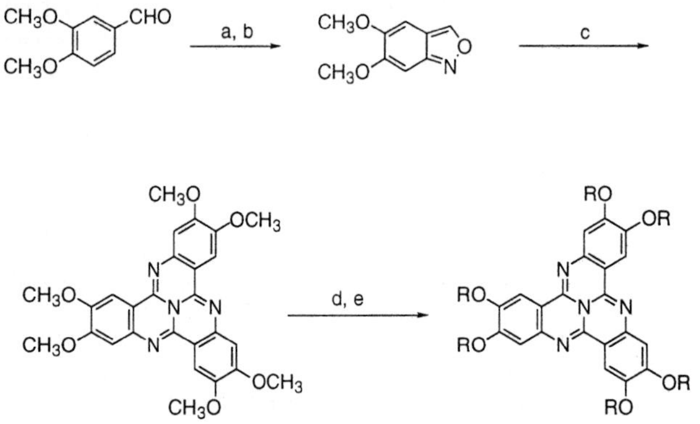

Scheme 41. Synthesis of tricycloquinazoline hexaether derivatives [158]. Reagents: (a) HNO_3, 90%; (b) Sn/AcOH, 54%; (c) NH_4OAc/AcOH/sulfolane/reflux, 27%; (d) HCl/pyridine/230 °C, 50%; (e) RBr/KOH/DMSO, 20–30%.

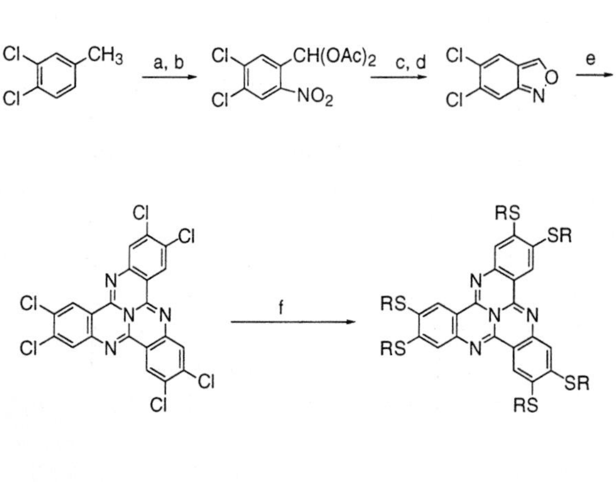

Scheme 42. Synthesis of tricycloquinazoline hexathioether derivatives [159]. Reagents: (a) $AcOH/H_2SO_4/HNO_3$/0–5 °C, 95%; (b) Ac_2O/ H_2SO_4/CrO_3/0–5 °C, 28%; (c) EtOH/ H_2O/HCl/reflux, 58%; (d) AcOH/Sn; (e) NH_4OAc/AcOH/sulfolane/150 °C, 40%; (f) RSK/N-methylpyrrolidone/100 °C, 8–50%.

Table 43. Transition temperatures for derivatives of tricycloquinazoline.

Substituent	Phase transitions	Refs.
CH_3O	Cr 350 I	[158]
C_2H_5O	Cr 315 I	[158]
C_3H_7O	Cr 286 D_{ho} 306 I	[158]
C_4H_9O	Cr 240 D_{ho} 306 I	[158]
$C_5H_{11}O$	Cr 188 D_{ho} 301 I	[158]
$C_9H_{19}O$	Cr 80 D_{ho} 237 I	[158]
$C_{11}H_{23}O$	Cr 72 D_{ho} 215 I	[158]
$C_{16}H_{33}O$	Cr 67 D_{ho} 166 I	[158]
C_3H_7S	Cr 160 D_h 274 I	[159]
C_4H_9S	Cr 156 D_h 241 I	[159]
$C_5H_{11}S$	Cr 114 D_h 223 I	[159]
$C_6H_{13}S$	Cr 95 D_h 214 I	[159]
$C_8H_{17}S$	Cr 81 D_h 207 I	[159]
$C_{12}H_{25}S$	Cr 54 D_h 170 I	[159]
$C_{16}H_{33}S$	Cr 57 D_h 152 I	[159]
$C_{18}H_{37}S$	Cr 92 D_h 184 I	[159]

[158] and hexathioalkyl [159] derivatives show good mesophase stability with a single D_{ho} phase over a wide temperature range and over a wide range of chain lengths (Table 43). Whereas, in the alkoxy series, both the melting and the clearing transitions are reversible, in the thioalkyl series only the clearing point transition shows good reversibility. The corresponding derivatives with only three symmetrically disposed side chains (C_3 to C_{18}) to not exhibit any liquid crystalline behaviour [159].

3.8   Porphyrin

3.8.1   Octa-Substituted Porphyrin

Fox and coworkers [16] have synthesized two closely related series of porphyrin discogens. Esters of porphyrin octa-acetic acid were prepared according to Scheme 43, porphyrin octa-acetic acid being prepared following the procedure of Franck [160]. Consequently, a Knorr condensation be-

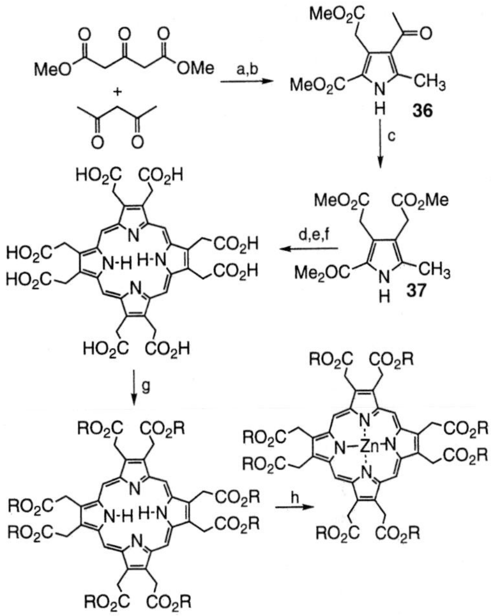

Scheme 43. Synthesis of the esters of porphyrin octa-acetic acid [161]. Reagents: (a) isopentyl nitrite, HCl; (b) Zn, AcOH, 75 °C; (c) Tl(NO$_3$)$_3$/MeOH/HClO$_4$; (d) Pb(OAc)$_4$; (e) 20% HCl, reflux; (f) O$_2$; (g) ROH, H$_2$SO$_4$, 50 °C; (h) Zn(OAc)$_2$, CHCl$_3$/MeOH, reflux.

Table 44. Transition temperatures for the esters of porphyrin octaacetic acid (see Scheme 43 for structures).

R	M	Phase transitions [161]
C_4H_9	H_2	Cr 178 D 222 I
C_4H_9	Zn	Cr 184 D 273 I
C_6H_{13}	H_2	Cr 59 D 132 D_1 220 I
C_6H_{13}	Zn	Cr 61 D 136 D_1 232 I
C_8H_{17}	H_2	Cr 96 D 99 D_1 166 I
C_8H_{17}	Zn	Cr 91 D 101 D_1 208 I

tween acetyl acetone and 3-aminodimethylpentandioate (generated in situ) gives the pyrrole 36. Treatment with thallium nitrate yields the triester, which is oxidized with lead tetraacetate to give compound 37. Boiling with 20% hydrochloride acid produces the macrocycle (and hydrolyses the esters), which is air oxidized to porphyrin octaacetic acid. Esterification proceeds smoothly, simply by warming the acid with the appro-

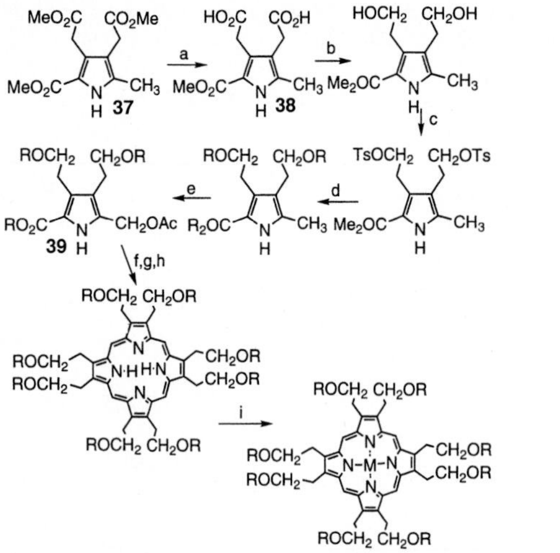

Scheme 44. Synthesis of porphyrin octaethers [162]. Reagents: (a) NaOMe/MeOH; (b) BH$_3$/THF; (c) TsCl/pyridine; (d) ROH/toluene; (e) Pb(OAc)$_4$; (f) KOH/EtOH; (g) HBr/EtOH; (h) chloranil; (i) M(OAc)$_2$, CH$_2$Cl$_2$/MeOH, reflux.

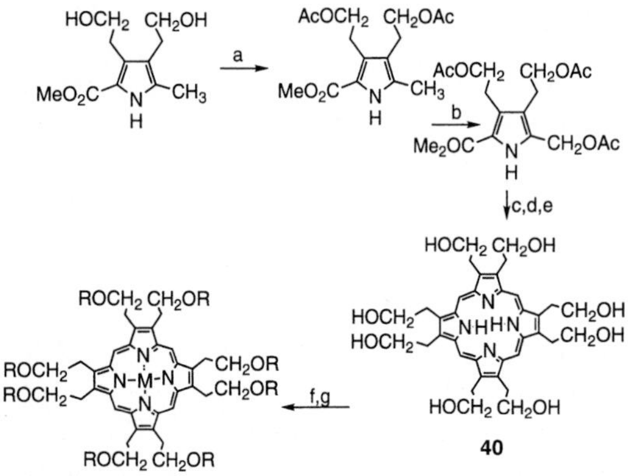

Scheme 45. Synthesis of porphyrin octaethers via porphyrin octaethanol. Reagents: (a) AcCl/NEt$_3$; (b) Pb(OAc)$_4$; (c) KOH/EtOH; (d) HBr/EtOH; (e) O$_2$; (f) NaH/DMSO, RX; (g) M(OAc)$_2$, CH$_2$Cl$_2$/MeOH, reflux.

Table 45. Transition temperatures for discotic porphyrin octaethers (ROCH$_2$CH$_2$–), (see Schemes 44 and 45 for structures).

R	M	meso	Phase transitions [162]
C$_4$H$_9$	H$_2$	H	Cr 154 I
C$_4$H$_9$	Zn	H	Cr 159 D 164 I
C$_6$H$_{13}$	H$_2$	H	Cr 111 I
C$_6$H$_{13}$	Zn	H	Cr 114 D 181 I
C$_8$H$_{17}$	H$_2$	H	Cr 84 D 89 I
C$_8$H$_{17}$	Zn	H	Cr 107 D 162 I
C$_8$H$_{17}$	Cu	H	Cr 84 D 132 I
C$_8$H$_{17}$	Pd	H	Cr 89 D 123 I
C$_8$H$_{17}$	Cd	H	Cr 103 D 136 I
C$_8$H$_{17}$	Zn	CN	Cr 85 D 140 I
C$_8$H$_{17}$	Zn	NO$_2$	Cr 93 D 118 I
C$_{10}$H$_{21}$	H$_2$	H	Cr 69 I
C$_{10}$H$_{21}$	Zn	H	Cr 86 D 142 I
COC$_5$H$_{11}$ (**41**)	Zn	H	Cr 169 I

chain lengths (hexyl), with a central zinc ion tending to raise the clearing temperature. The mesophases give fan-like textures, indicating a columnar architecture, but their precise nature has not been determined.

The synthesis has been modified to produce a series of porphyrin octaethanol derivatives (ethers) [162]. Pyrrole trimethyl ester (**37**) is selectively hydrolysed with sodium methoxide in methanol to give the diacid monoester (**38**), which is reduced (BH$_3$/THF) and tosylated. The ditosylate was heated with the appropriate alcohol to afford the diether (with concommital trans-esterification of the pyrrolic α-ester). Oxidation with lead tetra-acetate gave compound **39**, which was converted to the metal-free porphyrin in one pot by treatment with potassium hydroxide to hydrolyse the ester, hydrobromic acid to decarboxylate and cyclize, and chloranil to oxidize the macrocycle to the porphyrin (Scheme 44).

In a different, lower yielding method the derivatives were prepared from the common intermediate porphyrin octaethanol (**40**), the synthesis of which is outlined in Scheme 45. The octa-alcohol could also be esterified to produce structural isomers of the porphy-

priate alcohol using concentrated sulfuric acid as catalyst [161]. Metallated derivatives are then synthesized by heating the metal-free compounds with zinc acetate.

The mesophase behaviour of the porphyrin octaesters is summarized in Table 44 [161]. The most stable mesophase is formed from derivatives with intermediate siderin octaesters already described. Standard procedures were employed to synthesize the *meso*-nitro [163] and cyano [164] substituted derivatives.

The thermal behaviour of the octaethanol porphyrin derivatives is shown in Table 45. It is immediately apparent that the liquid crystal behaviour of these compounds is extremely sensitive to relatively small structural changes. The octaester **41** shows no liquid crystal phases at all, whereas its structural isomer (Table 44, R = C$_6$H$_{13}$, M = Zn) gives two mesophases over a wide temperature range. Within the octaether series the most striking effect appears to be that of the central metal ion. The metal-free derivatives show little tendency to form mesophases, whereas the metallated compounds give stable mesophases, typically through raising the clearing points. *meso*-Substitution tends to lower both transition temperatures, with the nitro compound also having a reduced mesophase range.

3.8.2   meso-Tetra(*p*-alkylphenyl)-porphyrin

Shimitzu et al. [165] have reported the mesophase behaviour of 5,10,15,20-tetrakis(4-n-dodecylphenyl)porphyrin (Scheme 46) [165]. The material is easily synthesized from pyrrole and 4-n-dodecylbenzaldehyde by refluxing in propionic acid [166]. The DSC trace shows the presence of two distinct mesophases. X-ray analysis of the two mesophases indicated that their structure is very similar, and is defined as discotic lamellar (D$_l$).

3.8.3   Tetraazaporphyrin

A series of 2,3,7,8,12,13,17,18-octakis-(alkylthio)-5,10,15,20-tetraazaporphyrins

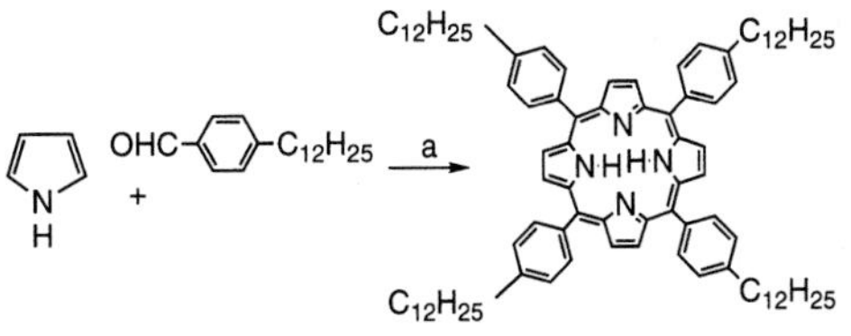

Cr 31 D$_l$ 52 D$_{L,1}$ 155 I

Scheme 46. Synthesis of 5,10,15,20-tetrakis(4-n-dodecylphenyl)porphyrin [165]. Reagents: (a) propionic acid, reflux.

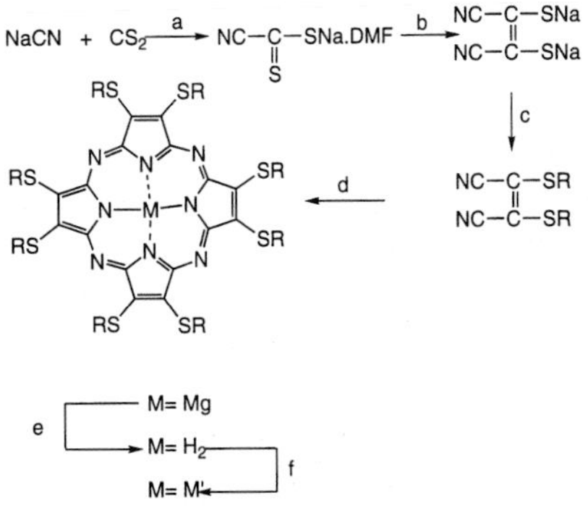

Scheme 47. Synthesis of tetra-azoporphyrin based discogens. Reagents: (a) DMF, 0 °C; (b) H$_2$O, 12 h; (c) RBr, MeOH; (d) Mg(OPr)$_2$/PrOH, 100 °C, 12 h; (e) CF$_3$CO$_2$H; (f) M'Cl$_2$.

(and their metal complexes) has been prepared and some members were found to be mesomorphic. The synthesis is shown in Scheme 47 [167]. Treatment of carbon disulfide with sodium cyanide in DMF yields the addition product. Addition of water couples the salt to give the ethylene dithiolate, which is alkylated with an alkyl bromide. Refluxing with magnesium propoxide/propanol yielded the magnesium porphyrin derivative, which could be demetallated with TFA. The metal-free compound was then metallated by treatment with the appropriate metal chloride/acetate in warm dioxane.

The transition data for these derivatives are given in Table 46 [168–170]. The most

Table 46. Transition temperatures for discotic materials based on tetra-azoporphyrin (RS−) [169] (see Scheme 47 for structures).

R	Phase transitions				
	H$_2$	Co	Ni	Cu	Zn
C$_4$H$_9$	Cr 113 D 122 I	Cr 95 D 271 I$_{decomp}$	Cr 117 D 184 I	Cr 112 D 223 I	Cr 54 D 208 I
C$_6$H$_{13}$	Cr 78 D 92 I	Cr 45 D 243 I	Cr 75 D 154 I	Cr 80 D 190 I	Cr 46 D 168 I
C$_7$H$_{15}$	Cr 75 I	–	–	Cr 70 D$_h$ 170 I	–
C$_8$H$_{17}$	Cr 82 I	Cr 55 D$_h$ 208 I	Cr 68 D$_h$ 119 I	Cr 68 D$_h$ 152 I	Cr 41 D 49 D$_1$ 120 I
C$_9$H$_{19}$	Cr 79 I	–	–	Cr 62 D$_h$ 133 I	–
C$_{10}$H$_{21}$	Cr 81 I	Cr 62 D 177 I	Cr 76 D 90 I	Cr 69 D 122 I	Cr 54 D 87 I
C$_{12}$H$_{25}$	Cr 85 I	Cr 73 D 153 I	–	Cr 83 D 102 I	–

striking feature of the data is the mesophase stabilizing effect of the central metal ions. It can be seen that the only metal-free derivatives to give mesophases are the butyl and hexyl homologues, whereas all the metal complexes (up to dodecyl) show enantiotropic mesophases. The clearing temperatures decrease linearly with increasing side-chain length, and for a given homologue the stabilizing effect of the central metal ion (higher clearing temperature) is Co > Cu > Zn, Ni > H$_2$. A non-hexagonal columnar structure is proposed for the metal-free derivatives, whereas a hexagonal columnar structure is most likely for the metal complexes.

3.9   Phthalocyanine

3.9.1   Peripherally Substituted Octa(alkoxymethyl)phthalocyanine

2,3,9,10,16,17,23,24-Octa(dodecyloxymethyl)phthalocyanine was the first discotic phthalocyanine derivative to be reported [171]. Many more discotic variants have since been synthesized by changing the side chains and/or the central metal ion, but all the syntheses follow a similar route to that

originally developed by Pawlowski and Hanack [172] (Scheme 48). o-Xylene is ring-brominated with bromine [173] and α-brominated with NBS [174, 175] to give tetrabromoxylene **42**. Nucleophilic displacement with alkoxide yields bis(alkoxymethyl)dibromobenzene, which is converted to the phthalonitrile by treatment with copper(I) cyanide in DMF. Conversion of phthalonitriles to phthalocyanines has been achieved in a number of ways. Direct routes include treatment of the phthalonitrile with

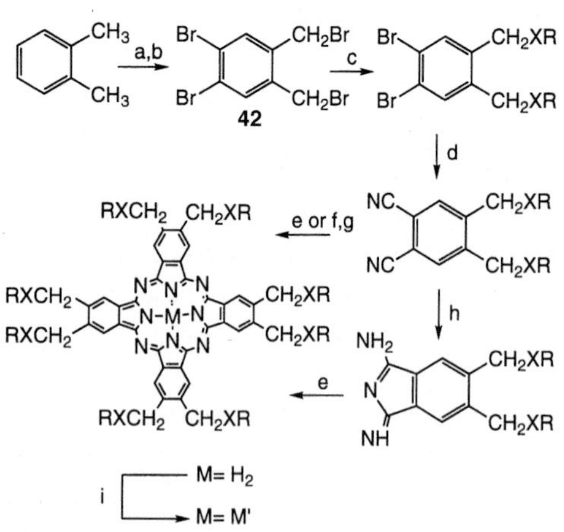

Scheme 48. Synthesis of peripherally substituted octa(alkoxymethyl)phthalocyanines (X = O) and related thio analogues (X = S). Reagents: (a) Br$_2$; (b) NBS; (c) RX$^−$; (d) CuCN/DMF; (e) N,N-dimethylethanolamine, reflux; (f) C$_5$H$_{11}$OLi/C$_5$H$_{11}$OH, reflux, (g) H$^+$; (h) NH$_3$/MeOH; (i) M' acetate.

Table 47. Transition temperatures for octa(alkoxymethyl)phthalocyanines and related thio analogues (see Scheme 48 for structures).

XR	M	Phase transitions	Refs.
OC_8H_{17}	Ni	Cr 68D I_{decomp}	[176]
OC_8H_{17}	Co	Cr 72D I_{decomp}	[176]
OC_8H_{17}	Pb	Cr $-45(53)$ D_h 158 (115) I	[177, 178 (176)]
$OC_{12}H_{25}$	H_2	Cr 78 D_h 264 I	[179]
$OC_{12}H_{25}$	Cu	Cr 53 D_h >300 I	[171]
$OC_{12}H_{25}$	Zn	Cr 78 D_h >300 I	[180]
$OC_{12}H_{25}$	Mn	Cr 44 D_h280 I	[180]
$OC_{12}H_{25}$	Pb	Cr -12 D_h 125 I	[178]
$OC_{12}H_{25}$	$Sn(OH)_2$	Cr 59 D_r 114 I	[181]
$OC_{18}H_{37}$	Pb	(Cr 46? 60 I)	[178]
$O(CH_2CH_2O)_2CH_3$	H_2	Cr <25 D_h >300 I	[180, 182]
$OCH_2CH(Me)OC_{12}H_{25}$	H_2	Cr 23 D_1 66 D_2 158 I	[183, 184]
$OCH_2CH(Me)OC_{12}H_{25}$	Cu	Cr 29 D 191 I	[184]
$SC_{12}H_{25}$	H_2	Cr 95 D 267 I	[184]
$SC_{12}H_{25}$	Cu	Cr 108 D 304 I	[184]
$OC_2H_4SC_{12}H_{25}$	H_2	Cr 52 D 247 I	[184]
$OC_2H_4SC_{12}H_{25}$	Cu	Cr 70 D 255 I	[184]
$O\text{-}p\text{-}C_6H_4C_{12}H_{25}$	H_2	<rt D_{ho} >300 I	[185]

lithium alkoxide in refluxing alcohol (e.g. n-pentanol), or simply refluxing in *N,N*-dimethylethanolamine. Alternatively, the phthalonitrile can be converted to its diiminoisoindoline, by treatment with ammonia, followed by cyclization in refluxing *N,N*-dimethylethanolamine. Metallated derivatives are then synthesized by reaction of metal-free phthalocyanines with the appropriate metal salt. It is common for these two steps to be performed together.

Examination of the mesophase behaviour within a series of derivatives (Table 47) reveals several trends. Most simple derivatives bearing straight side chains and with no central metal or a small central metal ion give hexagonal mesophases. Incorporation of small, divalent metal ions into the central phthalocyanine cavity tends to stabilize the discotic mesophase, while lengthening of the side chains has the opposite effect. Replacement of oxygen by sulfur has little effect on the clearing temperatures, but melting points are raised.

3.9.2 Peripherally Substituted Octa-alkoxyphthalocyanines

The synthesis of peripherally substituted octa-alkoxyphthalocyanines is straightforward (Scheme 49) [186]. Catechol is alkylated then brominated to give a dialkoxydibromobenzene. Reaction with copper cyanide in DMF affords the phthalonitrile, which can be converted to a phthalocyanine under standard cyclization conditions. The lithium alkoxide/alcohol method is usually avoided to prevent any possibility of trans-etherification.

Many compounds have been synthesized, and their mesophase behaviour is summarized in Table 48. Straight-chain derivatives tend to form very stable, ordered hexagonal mesophases [186–188], often decomposing before their clearing temperatures are reached (>300 °C). Within an homologous series, the general trend is for the melting points to decrease with increasing chain

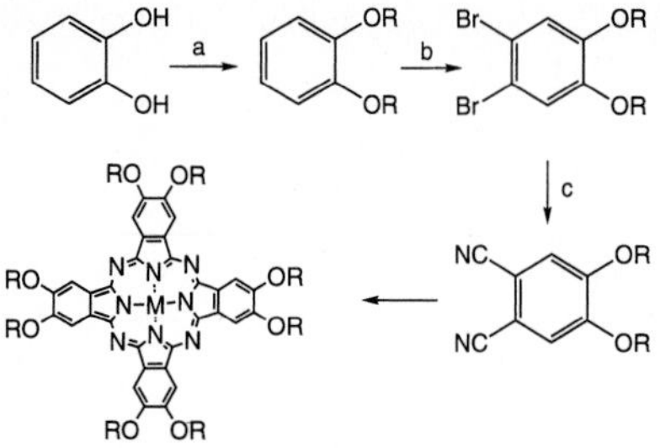

Scheme 49. Synthesis of peripherally substituted octa-alkoxyphthalocyanines. Reagents: (a) RX/base; (b) Br_2; (c) CuCN/DMF.

length, although some odd–even alternations are apparent for the lower homologues. Melting points are marginally higher for the copper derivatives but, in general, incorporation of small metal ions has a small stabilizing effect on the mesophase. An exception appears to be nickel (octadecyloxy)phthalocyanine [187], the clearing temperature of which is significantly lower than that of the parent metal-free compound. Incorporation of a silicon atom bearing two

Table 48. Transition temperatures for peripherally substituted octa-alkoxyphthalocyanines.

R	M	Transition data	Refs.
C_5H_{11}	H_2	Cr 121 D >350 I	[191]
C_6H_{13}	H_2	Cr 119 (102) D_{ho} >345 I	[188 (192)]
C_6H_{13}	Cu	Cr 120 D_{ho} >345 I	[188]
C_7H_{15}	H_2	Cr 104 D_{ho} >345 I	[188]
C_7H_{15}	Cu	Cr 110 D_{ho} >345 I	[188]
C_8H_{17}	H_2	Cr 94 D_{ho} ? I_{decomp}	[186, 188]
C_8H_{17}	Cu	Cr 112 D_{ho} ? I_{decomp}	[186, 188]
C_8H_{17}	$Si(OH)_2$	Cr 59 D_{hd} 84 I	[189]
C_9H_{19}	H_2	Cr 101 (107) D_{ho} >345 I	[188 (189)]
C_9H_{19}	Cu	Cr 106 D_{ho} >345 I	[188]
$C_{10}H_{21}$	H_2	Cr 94 D_{ho} 345 I_{decomp}	[188]
$C_{10}H_{21}$	Cu	Cr 104 D_{ho} >345 I	[188]
$C_{10}H_{21}$	$Si(OH)_2$	Cr 58 D_{hd} 87 I	[189]
$C_{11}H_{23}$	H_2	Cr 83 D_{ho} 334 I_{decomp}	[188]
$C_{11}H_{23}$	Cu	Cr 92 D_{ho} >345 I	[188]
$C_{12}H_{25}$	H_2	Cr 83 D_{ho} 309 I	[186, 188]
		Cr 95–110 D_{ho} 333 I	[187, 196, 197]
		Cr 75 D_{ho} 269 I	[184]
$C_{12}H_{25}$	Co	Cr 85–91 D 345 I	[187]
$C_{12}H_{25}$	Ni	Cr 96–98 D 254 I	[187]
$C_{12}H_{25}$	Cu	Cr 105 D_{ho} 310 I	[187]
		Cr 95 D_{ho} ? I_{decomp}	[186]
		Cr 95 D_{ho} >345 I	[188]
		Cr 78 D_{ho} 310 I	[184]
$C_{12}H_{25}$	Zn	Cr 99 D 375 I	[187]
$C_{12}H_{25}$	$Si(OH)_2$	Cr 59 D_{hd} 94 I	[189]
$C_{14}H_{29}$	$Si(OH)_2$	Cr 55 D_{hd} 107 I	[189]
$C_{16}H_{33}$	$Si(OH)_2$	Cr 56 D_{hd} 89 I	[189]
$C_{18}H_{37}$	$Si(OH)_2$	Cr 56 D_{hd} 87 I	[189]
$CH_2CH(Et)C_4H_9$	H_2	Cr 170 (158) $D_{tet.d}$ 223 N_D 270 (255) I	[190 (194)]
$CH_2CH(Et)C_4H_9$	Cu	Cr 204 D_t 242 N_D 290 I	[194, 195]
$CH_2CH(Et)C_4H_9$	Pt	Cr <−100 D_{ob} 205 I	[194]
$CH_2CH(Et)C_4H_9$	Pb	<−100 D_s 79 D ? I	[194]
$C_2H_4CH(Me)C_3H_6CH(Me)_2$	H_2	70 D_{hd} 295 I	[191, 193]
$C_2H_4CH^*(Me)C_3H_6CH(Me)_2$	H_2	$D_?$ 16 $D_h{}^*$ 111 $D_?$ 295 I	[191]
$C_2H_4CH(Me)C_3H_6CH(Me)C_3H_6CH(Me)_2$	H_2	D_h 34 $D_{h,1}$ 173 $D_{h,2}$ 185 I	[191, 193]

D_{tet}, tetragonal discotic, D_s, square discotic. *, chiral.

axial hydroxyl groups has a dramatic effect on the mesophase properties of the series [189]. As would be expected, the sterically demanding axial substituents destabilize the columnar mesophases and the derivatives have relatively low clearing temperatures and narrow mesophase ranges.

Chain-branching leads, in some cases [190–195], to more complicated phase behaviour and polymesomorphism. The octa(2-ethylhexyloxy)phthalocyanine (as a mixture of diastereomers) provides a good example [190, 192]. The crystalline solid melts to form a discotic tetragonal (D_{tet}) mesophase at 170 °C. Heating further yields a nematic mesophase (still relatively rare for discotic systems) between 233 °C and 270 °C. Even more complex phase sequences are observed when large central metal ions are introduced into the central cavity of this material [194].

3.9.3 Peripherally Substituted Octa-alkylphthalocyanine

The synthesis of peripherally substituted octa-alkylphthalocyanines was first described

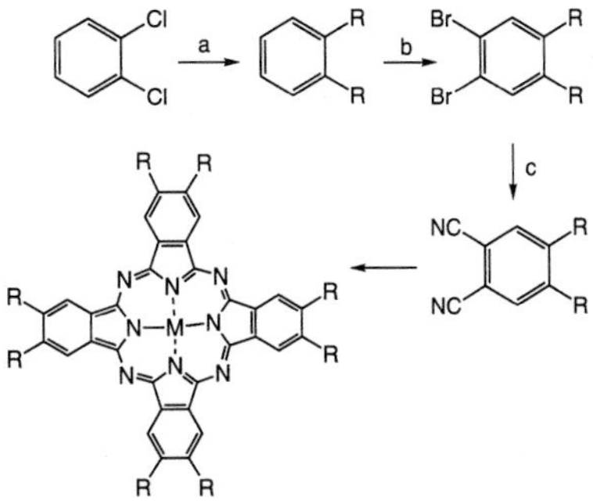

Scheme 50. Synthesis of peripherally substituted octa-alkylphthalocyanines. Reagents: (a) RMgBr, Ni catalyst; (b) Br_2; (c) CuCN/DMF.

by Cuellar and Marks [198] and has been modified by other workers to prepare liquid crystalline derivatives (Scheme 50). o-Dichlorobenzene is alkylated with a Grignard reagent in the presence of a nickel catalyst [199] to give dialkylbenzenes. Ring bromination with molecular bromine followed by reaction with copper cyanide in DMF yields the dialkylphthalonitrile. The phthalonitrile is cyclized to the phthalocyanine using standard procedures.

The mesophase behaviour of these materials (Table 49) can be compared with the

Table 49. Transition temperatures for peripherally substituted octa-alkylphthalocyanines (see Scheme 50 for structure).

R	M	Phase transitions	Refs.
C_5H_{11}	H_2	Cr 323 D 379 I_{decomp}	[200]
C_5H_{11}	Cu	Cr 342 D >450 I	[200]
C_6H_{13}	H_2	Cr 250 D 363 I_{decomp}	[200]
C_6H_{13}	Cu	Cr 259 D >450 I	[200]
C_6H_{13}	Ni	Cr 260 D 412 I	[200]
C_8H_{17}	H_2	Cr 186 D 325 I	[200]
C_8H_{17}	Cu	Cr 180 D >450 I	[200]
C_8H_{17}	Ni	Cr 190 D 373 I	[200]
$C_{10}H_{21}$	H_2	Cr 163 D 282 I	[200]
$C_{10}H_{21}$	Cu	Cr 169 D 351 I	[200]
$C_{10}H_{21}$	Ni	Cr 168 D 333 I	[200]
$C_{12}H_{25}$	H_2	Cr 120 D 252 I	[197]
$CH_2CH(Et)C_4H_9$	H_2	$D_{rd(P2m)}$ 63 $D_{rd(C2/m)}$ 269 I (Cr 267 I)	[190 (200)]
$CH_2CH(Et)C_4H_9$	Ni	Cr 293 I	[200]

closely related alkoxy derivatives. The oc-ta-alkylphthalocyanines give higher melting points, and often lower clearing temperatures, implying that the columnar arrangement is less well favoured. It should be noted also that most of the literature implies that the octa-alkylphthalocyanines tend to form rectangular mesophases. Metallation stabilizes the mesophases by raising the clearing temperatures.

3.9.4   Tetrapyrazinoporphyrazine

Ohta et al. [201] have prepared octakis-dodecyltetrapyrazinophyrazine and its copper(II) complex. It can be seen that these compounds are related to peripherally substituted octa(dodecyl)phthalocyanine. The synthesis (Scheme 51) involves the reaction of diketone 43 [202, 203] with diaminomalononitrile [204] in refluxing glacial acetic acid to give the dialkyldicyanopyrazine (44). The porphyrazine was prepared using standard methods for converting phthalonitriles to phthalocyanines. The transition

temperatures are very similar to the metal-free phthalocyanine analogue (Cr 120 D 252 I), suggesting that the extra nitrogen atoms have little effect on mesophase behaviour.

3.9.5   Peripherally Substituted Octa(alkoxycarbonyl)phthalo-cyanines

A series of liquid crystalline esters of 2,3,9,10,16,17,23,24-phthalocyanine octa-carboxylic acid have been synthesized by Dulog and Gittinger [205]. The octacarboxylic acid is prepared [206] from benzenetet-racarbonitrile (tetracyanobenzene) [205] in two steps: cyclization under standard conditions, and hydrolysis of phthalocyanine-octacarbonitrile with potassium hydroxide in triethylene glycol. The octa-acid is ester-ified with an alkyl bromide using DBU in acetonitrile (Scheme 52). The mesophase behaviour is essentially the same for all the derivatives studied (pentyl to dodecyl), with the materials exhibiting an hexagonal co-

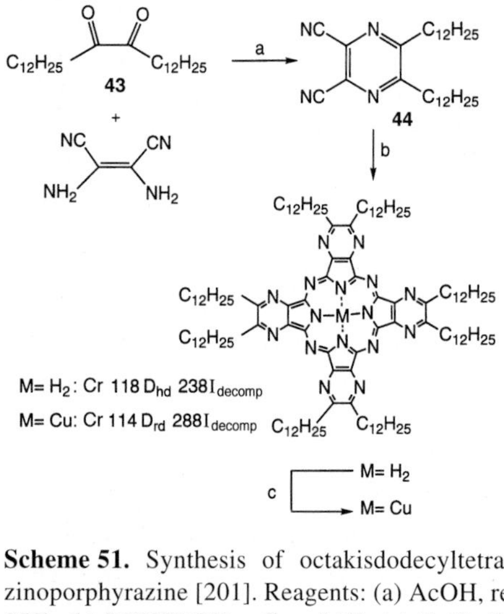

Scheme 51. Synthesis of octakisdodecyltetrapyrazinoporphyrazine [201]. Reagents: (a) AcOH, reflux, 85%; (b) DBU/EtOH, reflux, 14%; (c) CuCl$_2$/PrOH, 8% overall.

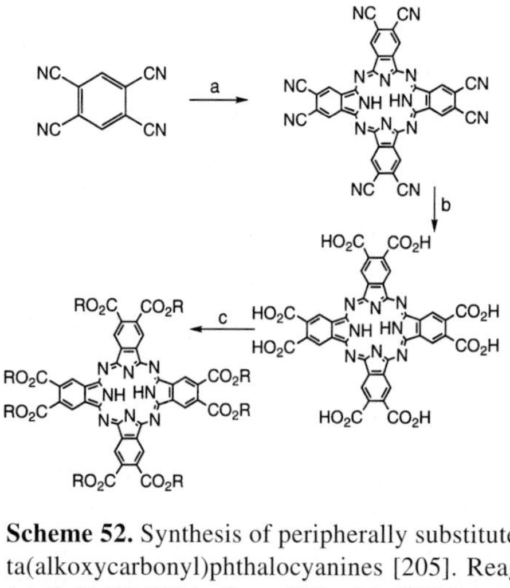

Scheme 52. Synthesis of peripherally substituted octa(alkoxycarbonyl)phthalocyanines [205]. Reagents: (a) PrOLi/PrOH, reflux, 50%; (b) KOH/triethylene glycol, 190 °C, 72%; (c) RBr/DBU/CH$_3$CN, reflux, 26–38%.

lumnar mesophase (D_{ho}) from below room temperature to above 300 °C.

3.9.6 Peripherally Substituted Octa-(p-alkoxylphenyl)phthalo-cyanine

Ohta et al. [185] have prepared a series of octa-(p-alkoxyphenyl)phthalocyanines using the synthetic route shown in Scheme 53. Intermediates **45** were prepared following the method used by Wenz [79] to prepare a series of triphenylene-based mesogens (see Sect. 2.5.4). The cyclopentadienone intermediate is trapped with dicyanoacetylene [207], which spontaneously re-aromatizes with loss of carbon monoxide to give 3,4-bis(p-alkoxyphenyl)phthalonitrile **46**. The phthalocyanines are then prepared by treating the phthalonitriles with DBU in refluxing pentanol. The dodecylphenoxymethyl derivative was also prepared (see Scheme 48 for the synthesis) to compare the effect of introducing increased flexibility in the side chains (phenyl rotation).

The mesophase behaviour of these materials is summarized in Table 50. Compounds with straight chains (decyloxy or longer) from discotic mesophases. At high temperatures a second phase is formed with concomitant decomposition. The mesophase structure is dependent on the central ions, with the metal-free derivatives giving exclusively hexagonal mesophases, whereas the copper derivatives from rectangular structures. Chain branching, which is known to lower transition temperatures, has the opposite effect in this series. The more flexible dodecyloxymethylphthalocyanine has a remarkably wide mesophase range, and Ohta et al. argue that this is because the phenyl rings are able to achieve coplanarity with the phthalocyanine core, creating a 'super-disk' structure.

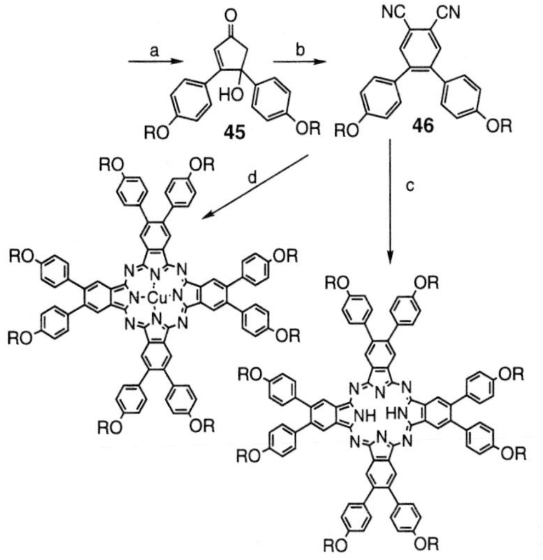

Scheme 53. Synthesis of peripherally substituted octa-(p-alkoxyphenyl)phthalocyanines [185]. Reagents: (a) see Scheme 22; (b) dicyanoacetylene/TsOH, 17–81%; (c) DBU/C_5H_{11}OH, 15–42%; (d) DBU/C_5H_{11}OH/CuCl$_2$, 70–73%.

Table 50. Transition temperatures for peripherally substituted octa-(p-alkoxyphenyl)phthalocyanines (see Scheme 53 for structures).

Substituent	M	Phase transitions [185]
$-C_6H_4OC_8H_{17}$	H_2	Cr 285 ?$_{decomp}$
$-C_6H_4OC_{10}H_{21}$	H_2	Cr 229 D 267 ?$_{decomp}$
$-C_6H_4OC_8H_{17}$	Cu	Cr 222 ?$_{decomp}$
$-C_6H_4OC_{12}H_{25}$	H_2	Cr 192 D$_{hd}$ 261 ?$_{decomp}$
$-C_6H_4OC_{12}H_{25}$	Cu	Cr 120 D$_{rd}$ 227 ?$_{decomp}$
$-C_6H_4OC_{18}H_{37}$	H_2	Cr 78 D$_{hd}$ 253 ?$_{decomp}$
$-C_6H_4OCH_2CH(Et)C_4H_9$	H_2	Cr 275 ?$_{decomp}$
$CH_2OC_6H_4C_{12}H_{25}$	H_2	<rt D$_{ho}$ >300 I

3.9.7 Peripherally Substituted Tetrabenzotriazaporphyrin

Reaction of a phthalonitrile with a Grignard reagent leads to a phthalocyanine analogue in which one of the 'meso' nitrogens is replaced by carbon – the so-called tetrabenzotriazaporphyrins [208, 209]. Leznoff and coworkers [210] have used this reaction to prepare a discotic liquid crystalline derivative (Scheme 54). Compound **47** forms a

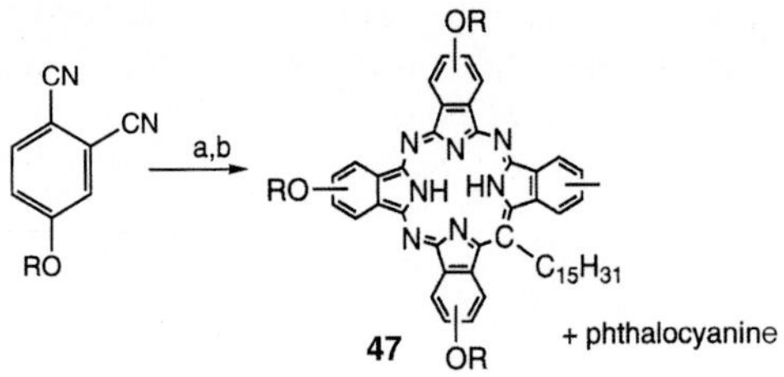

R= $CH_2C(CH_3)_3$: Cr 100–125 D > 300 I

Scheme 54. Synthesis of peripherally substituted tetrabenzotriazaporphyrins [209]. Reagents: (a) $CH_3(CH_2)_{15}MgCl$/ether; (b) H^+; 19% overall.

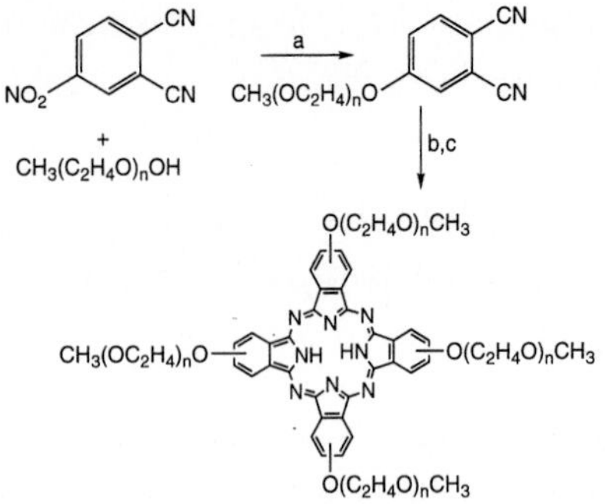

n= 3 : Cr 78 D_{ho} >300 I

n= 8 : Cr 11 D_{ho} 170-180 I

Scheme 55. Synthesis of tetrakis[oligo(ethyleneoxy)]-phthalocyanines [211]. Reagents: (a) K_2CO_3/DMF; (b) $C_5H_{11}OLi/C_5H_{11}OH$, reflux; (c) H^+.

stable mesophase (possibly with lamellar structure [210]). The synthesis presumably yields **47** as a mixture of regioisomers, which could account for the broad melting range.

3.9.8 Tetrakis[oligo(ethyleneoxy)]phthalocyanine

McKeown and Painter [211] have synthesized mesomorphic phthalocyanines bearing four oligoethyleneoxy substituents in the peripheral positions. Their synthesis is shown in Scheme 55 along with the mesophase behaviour of the final products. The ethyleneoxy substituted phthalonitrile is prepared via a base-catalysed nitro displacement of 4-nitrophthalonitrile [212] with polyethylene glycol monomethyl ether. Cyclization under normal conditions yields the phthalocyanines, presumably as a mixture of regioisomers. The compounds give a single, thermotropic, hexagonal mesophase, as well as a range of lyotropic mesophases in water.

3.9.9 Non-Peripherally Substituted Octa(alkoxymethyl)-phthalocyanines

In general, the synthesis of the non-peripherally substituted phthalocyanines is less straightforward than their peripherally substituted counterparts, requiring routes to 1,4-disubstituted phthalonitriles. For the liquid crystalline materials, this has been achieved using variations on a synthesis which uses the Diels–Alder reaction between a furan or thiophene-1,1-dioxide and fumaronitrile (Scheme 56).

The synthesis of non-peripherally substituted octa(alkoxymethyl)phthalocyanines is outlined in Scheme 57 [213, 214]. 2,5-*bis*(Alkoxymethyl)furan can be prepared by lithiation of furan followed by quenching the anion with (bromomethyl)alkyl ether to give the mono-substituted furan, and repetition of the procedure to give the required product [213]. A more simple synthesis involves chlorination of 2,5-furan dimethanol with thionyl chloride, followed by nucleophilic displacement of chloride with sodium alkoxide [214]. The furan is equilibrated with fumaronitrile for about a week and the Diels–Alder adduct aromatized by treatment with lithium *bis*(trimethylsilyl)-amide (a non-nucleophilic base), followed by an acidic work-up. The 3,6-*bis*(alkoxymethyl)phthalonitrile is cyclized under standard conditions (lithium/pentanol).

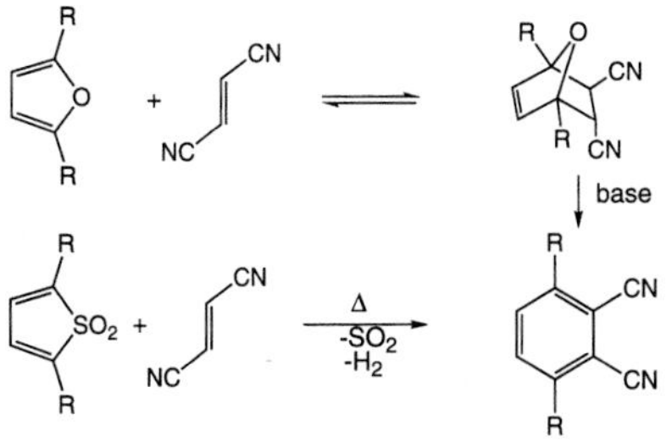

Scheme 56. The synthesis of 1,4-disubstituted phthalonitriles via a Diels–Alder reaction between a furan or thiophene-1,1-dioxide (diene) and fumaronitrile (dienophile).

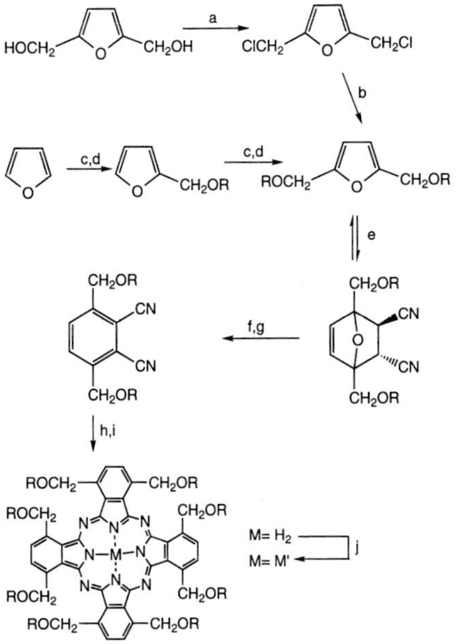

Scheme 57. Synthesis of non-peripherally substituted octa(alkoxymethyl)phthalocyanines [214]. Reagents: (a) $SOCl_2$, (b) $RONa/ROH$, 36–78% over two steps; (c) BuLi; (d) $ROCH_2Br$, 20% overall (C_8); (e) fumaronitrile; (f) $LiN(SiMe_3)_2/THF$; (g) H^+, 4–23%; (h) $ROLi/ROH$, Δ; (i) H^+, 7–34%; (j) M′ acetate, 72–87%.

Table 51. Transition temperatures for non-peripherally substituted octa(alkoxymethyl)phthalocyanines ($ROCH_2$–) (see Scheme 57 for structures).

R	M	Phase transitions [214, 215]
C_4H_9	H_2	Cr 185 D_{rd} >300 I
C_5H_{11}	H_2	Cr 123 D_{rd} >300 I
C_6H_{13}	H_2	Cr 85 D_{rd} >300 I
C_7H_{15}	H_2	Cr 79 D_{rd} ? (90–120) D_h >300 I
C_7H_{15}	Cu	Cr 84 D >300 I
C_7H_{15}	Zn	Cr 70 D >300 I
C_8H_{17}	H_2	Cr 67 D_{rd} ? (80–220) D_h >300 I
C_9H_{19}	H_2	Cr 75 D_{hd} >300 I
$C_{10}H_{21}$	H_2	Cr 64 D_{hd} >300 I
$CH_2CH(Et)C_4H_9$	H_2	D_{rd} 236 I
$CH(Me)C_6H_{13}$	H_2	D_{rd} 106 I

oxymethyl) form both, with the hexagonal phase being formed at higher temperatures. Chain branching leads to derivatives (produced as mixtures of diastereomers) that are liquid crystalline (rectangular) at room temperature. These low melting points contrast with the peripherally substituted octa(2-ethylhexyloxy)phthalocyanine, which has a significantly higher melting point than its straight-chain isomer.

3.9.10 Non-Peripherally Substituted Octa-alkylphthalocyanine

3,6-Dialkylphthalonitriles have been prepared from both 2,5-dialkylfuran [213, 216] and 2,5-dialkylthiophene [216]. Both heterocycles are synthesized in a similar manner by lithiation and alkylation of either furan or thiophene, respectively. Generally, the thiophene route is preferred because dialkylation can be performed in a single step, and the final Diels–Alder reaction (between fumaronitrile and the corresponding sulfone) goes to completion and yields the required phthalnotriles without the need for a strong base or inert conditions. The dialkylthiophenes can be oxidized to the correspond-

It can be seen from Table 51 that short-chain derivatives form rectangular mesophases and longer homologues form hexagonal mesophases. The intermediate compounds (n-heptyloxymethyl and n-octyl-

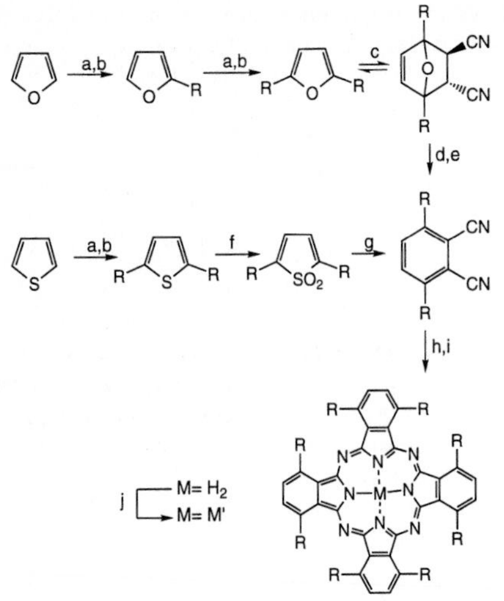

Scheme 58. Synthesis of non-peripherally substituted octaalkylphthalocyanines. Reagents: (a) BuLi; (b) RBr; (c) fumaronitrile; (d) LiN(SiMe$_3$)$_2$/THF, [213]; (e) H$^+$, 25–30%; (f) [O], 35–51% [216]; (g) fumaronitrile, Δ, 40–48% [216]; (h) C$_5$H$_{11}$OLi/ C$_5$H$_{11}$OH, reflux; (i) H$^+$, 18–25%; (j) M′ acetate, 80–94%.

ing sulfones using either *m*-CPBA, sodium perborate or dimethyldioxirane [217]. The phthalonitriles are cyclized using the lithium/pentanol method. The synthesis is shown in Scheme 58.

An homologous series of compounds, each substituted with a variety of central metal ions, has now been synthesized. Their mesophase behaviour is summarized in Table 52. A number of trends are apparent. Melting points decrease as the side-chain length increases, the effect being most marked for the lower homologous. At longer chain lengths (octyl and above) a pronounced odd–even fluctuation is observed. Clearing temperatures decrease linearly with increasing chain length. The central ion(s) also has a significant influence on mesophase behaviour. The largest mesophase stabilization is observed when zinc is incorporated in the discogen, giving derivatives with higher transition temperatures

Table 52. Transition temperatures for non-peripherally substituted octa-alkylphthalocyanines (see Scheme 58 for structures) [a].

R	H$_2$ [213, 218]	Cu [213, 218]	Ni [219]	Zn [219]	Co [217]
C$_5$H$_{11}$	Cr 218 I	Cr 261 I	Cr 220 I	Cr 279 D$_{hd}$ 292 I	
C$_6$H$_{13}$	Cr 161 D$_{hd}$ 171 I	Cr 184 D$_{hd}$ 236 D$_{hd'}$ 242 I	Cr 145 D$_{hd}$ 164 D$_{hd'}$ 169 I	Cr 209 D$_{hd}$ 280 D$_{hd'}$ 285 I	Cr 177 D 252 I
C$_7$H$_{15}$	Cr 113 D$_{hd}$ 145 D$_{hd'}$ 163 I	Cr 145 D$_{hd}$ 205 D$_{hd'}$ 236 I	Cr 118 D$_{hd}$ 169 I [D$_{hd}$ 104 D$_{rd}$]	Cr 158 D$_{hd}$ 248 D$_{hd'}$ 272 I	
C$_8$H$_{17}$	Cr 85 D$_{rd}$ 101 D$_{hd}$ 152 I [D$_{h}$ 74 D$_{rd}$]	Cr 96 D$_{hd}$ 156 D$_{hd'}$ 220 I	Cr 66 D$_{hd}$ 153 I [D$_{hd}$ 55 D$_{rd}$]	Cr 105 D$_{hd}$ 224 D$_{hd'}$ 258 I	
C$_9$H$_{19}$	Cr 103 D$_{hd}$ 142 I [D$_{hd}$ 74 D$_{rd}$]	Cr 108 D$_{hd}$ 208 I [D$_{hd}$ 99 D$_{rd}$]	Cr 92 D$_{hd}$ 152 I	Cr 114 D$_{hd}$ 177 D$_{hd'}$ 242 I	
C$_{10}$H$_{21}$	Cr 78 D$_{hd}$ 133 I	Cr 88 D$_{hd}$ 198 I [D$_{hd}$ 69 D$_{rd}$]	Cr 64 D$_{hd}$ 137 I	Cr 90 D$_{rd}$ 107 D$_{hd}$ 225 I	Cr 75 D 189 I

[a] Transitions in square brackets refer to monotropic phases.

of mesophase stabilization, the order is $Zn > Co > Cu > Ni$, H_2 [217].

This class of phthalocyanine discogens shows a rich polymesomorphism. All samples give an hexagonal phase when cooled from the isotropic liquid. In some cases, further cooling leads to a second hexagonal or a rectangular phase, and, more rarely, an additional, monotropic re-entrant hexagonal phase is observed. It is interesting to note that the closely related non-peripherally substituted octa-alkoxyphthalocyanines and non-peripherally substituted octa-alkoxyphthalocyanines are non-mesomorphic [217].

3.9.11 Unsymmetrically Substituted Phthalocyanines

A number of unsymmetrically substituted phthalocyanines have been synthesized and found to be mesogenic. Their synthesis is generally difficult, involving cyclization of a mixture of phthalonitrile derivatives, followed by tedious separation of the products (Scheme 59).

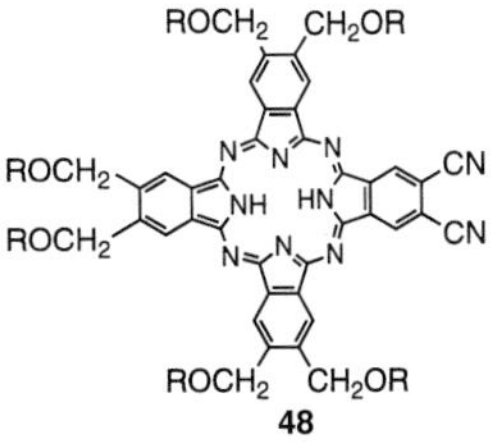

48

R = $C_{12}H_{25}$: Cr 40 Dh > 300 I

Simon and coworkers [180, 220] have synthesized peripherally substituted dicyanohexakis(dodecyloxymethyl)phthalocya-

49

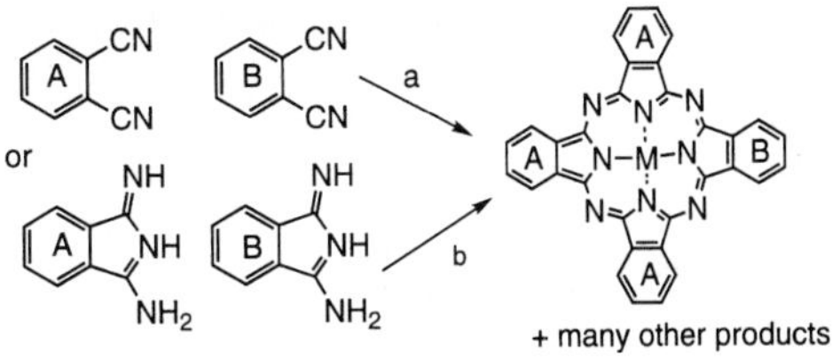

Scheme 59. Routes to unsymmetrically substituted phthalocyanines. Reagents: (a) RO^-; (b) N,N-dimethylaminoethanol, reflux.

nine 48 and found that the compound has a remarkably stable hexagonal mesophase.

Cook and coworkers [216] have synthesized a range of unsymmetrical, non-peripherally substituted phthalocyanines 49. Their mesophase behaviour (Table 53) is extremely sensitive to the degree of asymmetry (substituents) in these non-peripherally substituted systems. Mesophase behaviour is destroyed if the diester 50 is saponified to give the diacid ($R_2 = R_3 = CH_2CH_2CH_2CO_2H$). Similarly, substitution of an additional alkyl group in the penta-alkyl mono-3-hydroxypropylphthalocyanines leads to materials ($R_2 = R_3 = (CH_2)_3OH$) that give only monotropic mesophases [217, 221].

4 Saturated Cores

4.1 Cyclohexane

The synthesis of discotic liquid crystalline cyclohexanes is relatively straightforward. Both classes of derivative (the scylloinositol ethers and esters) are synthesized from commercially available scylloinositol (Scheme 60). The esters are prepared by treatment with an excess of acid chloride in trifluoroacetic acid [222], whereas the ether is synthesized using potassium hydroxide and hexyl bromide [223]. The mesophase

Table 53. Transition temperatures for unsymmetrically substituted phthalocyanines **49**.

R_1	R_2	R_3	R_4, R_5	Phase transitions	Refs.
$C_{10}H_{21}$ (**50**)	$(CH_2)_3CO_2C_5H_{11}$	R_2	H	Cr 71 D 120 I	[216]
C_6H_{13}	$(CH_2)_3OH$	R_1	H	Cr 94 D 148 I	[217, 221]
C_7H_{15}	$(CH_2)_3OH$	R_1	H	Cr 75 D$_?$ 117 D$_h$ 129 I	[221]
C_8H_{17}	$(CH_2)_3OH$	R_1	H	Cr 55 D$_?$ 74 D$_h$ 112 I	[221]
$C_{10}H_{21}$	$(CH_2)_3OH$	R_1	H	Cr 39 D 82 I	[217]
C_6H_{13}	H	H	$-OC(Me)_2O-$	Cr 244 D 259 I	[217, 221]
C_7H_{15}	H	H	$-OC(Me)_2O-$	Cr 186 D$_?$ 235 D$_h$ 237 I	[221]
C_8H_{17}	H	H	$-OC(Me)_2O-$	Cr 183 D$_?$ 208 D$_h$ 226 I	[221]
C_9H_{19}	H	H	$-OC(Me)_2O-$	Cr 155 D$_?$ 192 D$_h$ 216 I	[221]
$C_{10}H_{21}$	H	H	$-OC(Me)_2O-$	Cr 138 D$_?$ 180 D$_h$ 207 I	[217, 221]

behaviour of these materials (Table 54) has a number of interesting features. The fact that mesogenic materials are produced with side chains as short as acetyl is itself suprising. Comparing the ether and ester derivatives with similar side-chain lengths reveals a much enhanced mesophase stability for the ester. It is worth noting that neither the myoinositol ester (**51**) [224] or the mytilitol derivative (**52**) [224], each bearing an axial substituent, gives any discotic mesophases. Many related hexa (equatorial) substituted cyclohexanes have been synthesized (e.g. **53**) in which the oxygen linkage to the ring has been replaced by carbon, but to date all are non-mesogenic [225].

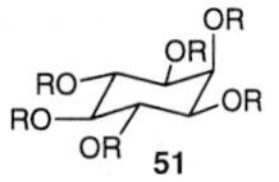

Table 54. Transition temperatures for cyclohexane (inositol) based discotic mesogens (see Scheme 60 for structures).

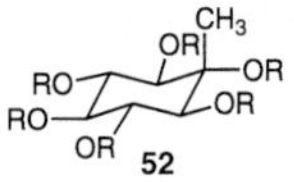

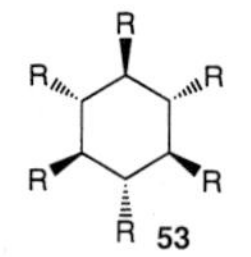

Substituent	Phase transitions	Refs.
OC_6H_{13}	Cr 18 D$_{ho}$ 91 I	[223]
$OCOCH_3$	Cr 288 D 292 I	[226]
$OCOC_2H_5$	Cr 212 D 276 I	[226]
$OCOC_3H_7$	Cr 213 D 259 I	[226]
$OCOC_4H_9$	Cr 185 D 208 I	[226]
$OCOC_5H_{11}$	Cr 69 D 200 I	[222, 226]
$OCOC_6H_{13}$	Cr 70 D 202 I	[226]
$OCOC_7H_{15}$	Cr 76 D 199 I	[222, 226]
$OCOC_8H_{17}$	Cr 81 D 196 I	[226]
$OCOC_9H_{19}$	Cr 84 D$_h$ 189 I	[222, 226]
$OCOC_{10}H_{21}$	Cr 88 D 183 I	[226]
$OCOC_{11}H_{23}$	Cr 92 D 176 I	[226]
$OCOC_6H_4C_9H_{19}$	Cr 184 H 191 I	[224]
$OCOCH(Me)C_6H_{13}$	Cr 42 D 116 I	[10]
$OCOCH_2CH(Me)C_5H_{11}$	Cr 48 D 195 I	[10]
$OCOC_2H_4CH(Me)C_4H_9$	D 181 I	[10]
$OCOC_3H_6CH(Me)C_3H_7$	D 186 I	[10]
$OCOC_4H_8CH(Me)C_2H_5$	Cr 54 D 197 I	[10]
$OCOC_5H_{10}CH(Me)CH_3$	Cr 87 D 201 I	[10]
$OCOC_4H_8CH(Et)C_2H_5$	Cr 38 D 186 I	[10]
$OCOC_4H_8Cp$	Cr 139 D 199 I	[10]
$OCOC_3H_6CH(Me)C_2H_5$	Cr 68 D 196 I	[10]
$OCOC_5H_{11}CH(Me)C_2H_5$	Cr 44 D 192 I	[10]
$OCOC_2H_4OC_4H_9$	Cr 47 D 87 I	[9]
$OCOC_5H_{10}OCH_3$	Cr 44 D 144 I	[9]
$OCOCH_2SC_5H_{11}$	Cr 57 D 106 I	[9]
$OCOC_2H_4SC_4H_9$	Cr 86 D 183 I	[9]
$OCOC_3H_6SC_3H_9$	Cr 64 D 189 I	[9]
$OCOC_5H_{10}SCH_3$	Cr 56 D 172 I	[9]
$OCO(CH_2)_4Cl$	Cr 205 D 235 I	[227]
$OCO(CH_2)_4Br$	Cr 190 D 250 I	[227]
$OCO(CH_2)_5Br$	Cr 93 D 230 I	[227]
$OCO(CH_2)_6Br$	Cr 94 D 199 I	[227]
$OCO(CH_2)_7Br$	Cr 92 D$_{ho}$ 175 I	[227]
$OCO(CH_2)_{10}Br$	Cr 90 D$_{ho}$ 117 I	[227]

R = COC$_5$H$_{11}$: Cr 48 I   R = COC$_9$H$_{19}$: Cr 75 I   R = C$_6$H$_{13}$: Cr 66 I

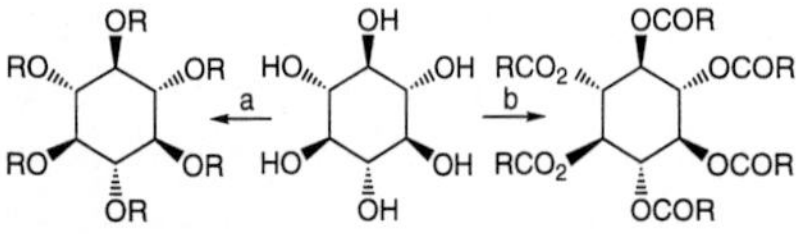

Scheme 60. Synthesis of cyclohexane (inositol) based discogens. Reagents: (a) KOH/RBr; (b) RCOCl/TFA.

Collard and Lillya have made systematic changes to the side chains of cyclohexane-based discotic mesogens by introducing branching [10] and heteroatoms [9]. Introduction of a methyl group into the side chain tends to reduce the melting point, the largest effect being observed when the branch point is in the middle of the side chain. Moving the branch point closer to the core tends to lower the clearing temperature, presumably because steric effects begin to disfavour a columnar arrangement. Introduction of a heteroatom into the chains also tends to reduce both transition temperatures, with the effects depending on its nature (oxygen or sulfur) and position. Halogen substitution at the chain terminus stabilizes the mesophase for lower homologues, but a destabilization is observed for longer chain derivatives [227].

RCO_2CH_2, RCO_2, RCO_2, RCOO, OCOR

54 α-glycopyranoside derivative

RCO_2CH_2, RCO_2, RCO_2, RCOO, OCOR, H

55 β-glycopyranoside derivative

RCO_2CH_2, RCO_2, RCO_2, X, H, O, RCO_2, RCO_2CH_2, O, X, H, OCOR

56 cellobiose derivative   X = OCOR
57 chitobiose deivative   X = NHCOR

RCO_2CH_2, RCO_2, RCO_2, X, H, O, RCO_2, RCO_2CH_2, O, X, O, RCO_2, RCO_2CH_2, X, H, OCOR

58 cellotriose derivative   X = OCOR
59 chitotriose deivative   X = NHCOR

RCO_2CH_2, RCO_2, RCO_2, RCOO, H, O, RCO_2, RCO_2CH_2, X, O, RCO_2, RCO_2CH_2, O, $OC_{10}H_{21}$

60 cellobiose derivative, X = RCO_2
61 cellobiose derivative, X = HO

4.2   Tetrahydropyran (Pyranose Sugars)

The pyranose sugars provide a ready-made source of laterally substituted cyclic nuclei and hence of potential discotic systems and the esters **54–61** are all mesogenic. All are made simply by treating the appropriate natural sugar with the required acid chloride in pyridine/chloroform for 1–3 days [228]. The C_{10} to C_{18} esters of the α- and β-glucopyranosides **54** and **55** show complex phase behaviour which is found to depend on the thermal history of the sample as well as the structure of the mesogen. There is a degree of disagreement in the literature over the phase behaviour [228–230]. In view of the fact that these systems represent fairly minor deviations from the six-fold symmetric inositol structures, the formation of D_h phases does not seem too surprising. Much

less expected and less easy to understand is that the cellobiose **56** [231, 232], chitobiose **57** [233], cellotriose **58** [231, 232] and chitotriose **59** [231] derivatives behave in a similar way. It is proposed that in the columnar phases formed by these di- and trisaccharides the pyran rings lie across the column, as in a conventional discotic phase, but that in the hexagonal columnar phases formed by higher oligomers, such as the pentamer in the cellulose series, the polysaccharide backbone lies parallel to the axis of the column [234]. The mesophase behaviour of the esters of cellubiose is summarized in Table 55 [231]. The chitin-based oligomers generally give wider mesophase ranges than do the cellulose-based oligomers [233].

Table 55. Transition temperatures of liquid crystalline sugar derivatives.

Sugar	Acyl substituent	Phase transitions	Refs.
54	$C_9H_{19}CO$	Cr 27/8 D 34 I	[230]
55	$C_9H_{19}CO$	Cr 32/3 D 38/40 I (Cr 40 I)	[230 (235)]
56	$C_6H_{13}CO$	Cr 4 D 91 I	[231]
	$C_7H_{15}CO$	Cr 25 D 62 D_{ho} 87 I	[231]
	$C_8H_{17}CO$	Cr 37 D_{ho} 86 I	[231]
	$C_9H_{19}CO$	Cr 43 D_{ho} 83 I	[231]
	$C_9H_{19}CO$	Cr 58 D_{ho} 93 I	[232, 234]
	$C_{11}H_{23}CO$	Cr 49 D_{ho} 81 I	[231]
	$C_{13}H_{27}CO$	Cr 56 D_{ho} 73 I	[231]
58	$C_9H_{19}CO$	Cr 45 D_h 93 I	[232, 234]
57	$C_9H_{19}CO$	Cr 65 M 207 I	[233]
	$C_{13}H_{27}CO$	Cr 70 M 196 I	[233]
	$C_{17}H_{33}CO$	Cr 73 M 185 I	[233]
59	$C_9H_{19}CO$	Cr 60 M 195 I	[233]
	$C_{13}H_{27}CO$	Cr 65 M 175 I	[233]
	$C_{17}H_{33}CO$	Cr 70 M 170 I	[233]
60	$C_{11}H_{23}CO$	Cr 74 D 93 I	[229]
	$C_{12}H_{25}OC_6H_4CO$	Cr ? D 120 I	[229]
61	$C_{12}H_{25}OC_6H_4CO$	Cr ? D 145 I	[229]

4.3 Hexacyclens and Azamacrocycles

Mesophase behaviour of hexa-n-acylated hexacyclens (**62**) was first observed by Lehn et al. [236] for the hexa-(p-n-dodecyloxy-benzoyl) derivative. A columnar arrangement was assumed, but the lack of high-order X-ray reflections prohibited absolute assignment. Nevertheless, the term 'tubular mesophase' (T) was introduced to describe the proposed structure, which is wholly analogous to that of the D_h mesophase. In the case of hexacyclens (and similar compounds), however, the 'cores' possess a central void region, giving rise to supramolecular 'tubes'. Many other derivatives have been prepared, but in only two cases (**63** and **64**, see Table 56) has the hexagonal columnar structure been confirmed [237]. More evidence to support the columnar structure **65** (Table 56) has come from X-ray measurements of mixtures of **65** with brominated analogues [238].

The hexacyclen mesogens are easily synthesized from commercially available hexacyclen by condensation with a benzoyl chloride in dimethylacetamide/p-dimethylaminopyridine [236] (Scheme 61). The mesophase behaviour of the discotic hexacyclens is shown in Table 56. It should be noted that an additional feature of these materials is that their thermal behaviour is dependent on the history/recrystallization conditions of the sample, and different transition temperatures are reported from different laboratories [239].

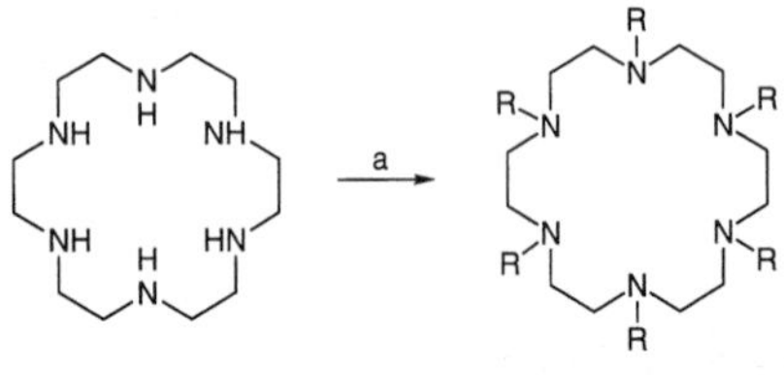

62 18aneN$_6$

63: R = (3-Br-4-C$_{12}$H$_{25}$O)C$_6$H$_4$CH:CHCO
64: R = (3,5-Cl$_2$-4-C$_{12}$-H$_{25}$O)C$_6$H$_4$CH:CHCO
65: R = C$_{12}$H$_{25}$OC$_6$H$_4$CO

Scheme 61. Acylation of hexacyclen. Reagents: (a) R'COCl/base.

Table 56. Transition temperatures for hexacyclen based discotic mesogens **62**.

R	Phase transitions[a]	Refs.
$C_8H_{17}OC_6H_4CO$	Cr 120 T 139 I	[241]
$C_{11}H_{23}OC_6H_4CO$	Cr ? T 144 I	[240]
$C_{12}H_{25}OC_6H_4CO$ (**65**)	Cr 122 (97) T 142 I	[236 (240)]
$C_{14}H_{29}OC_6H_4CO$	Cr 106 D_{hd} 136 I	[242]
(3-Br-4-$C_{12}H_{25}O)C_6H_3CH$:CHCO (**63**)	Cr ? D_h ? I	[237]
(3,5-Cl_2-4-$C_{12}H_{25}O)C_6H_2CH$:CHCO (**64**)	Cr ? D_h ? I	[237]
$C_{14}H_{29}C_6H_4CH$:CHCO	Cr 217 D_h 233 I[b]	[243]
$C_8H_{17}OC_6H_4N$:NC_6H_4CO	Cr 237 D 245 I	[242]
3,4-$(C_{10}H_{21}O)_2C_6H_3CO$	Cr 104 D 140 I	[244]

[a] T, tubular mesophase.
[b] Mesophase behaviour dependent on E/Z ratio.

66 12aneN$_3$ **67** 9aneN$_3$ **68** 14aneN$_4$

Cr 38 D 60 I[244]   D 44 D' 66 I[244]   D95 D' 132 I[245]

R =

$OC_{10}H_{21}$
$OC_{10}H_{21}$

The presence of the benzamide moiety appears essential for mesophase production. If the benzamides are reduced to the benzylamines [240], or straight-chain amides ($C_{11}H_{13}CO$) [241] are used, all mesophase behaviour disappears. Some smaller ring azamacrocycles **66–68** have been prepared and found to give discotic phases. Their synthesis is wholly similar to that of the hexacyclens. Benzamides of larger macrocycles (e.g. [30] N_{10} [241]) have also been synthesized, but these show only monotropic mesophases.

5   References

[1] S. Chandrasekhar, B. K. Sadashiva, K. A. Suresh, *Pramana*, **1977**, *9*, 471.
[2] P. G. DeGennes (1974) *The Physics of Liquid Crystals*, Oxford University Press, Oxford.
[3] J. D. Brooks, G. H. Taylor, *Carbon*, **1965**, *3*, 185.
[4] C. Destrade, N. H. Tinh, H. Gasparoux, J. Malthete, A. M. Levelut, *Mol. Cryst. Liq. Cryst.*, **1981**, *71*, 111–135.
[5] S. Chandrasekhar, B. K. Sadashiva, K. A. Suresh, N. V. Madhusudana, S. Kumar, R. Shashidhar, G. Venkatesh, *J. Phys. Colloq.*, **1979**, C120.
[6] A. J. Fatiadi, W. F. Sager, *Org. Synth.* **1973**, *V*, 595.
[7] I. Tabushi, K. Yamamura, Y. Okada, *J. Org. Chem.* **1987**, *52*, 2502–2505.
[8] A. Maliniak, S. Greenbaum, R. Poupko, H. Zimmermann, Z. Luz, *J. Phys. Chem.* **1993**, *97*, 4832–4840.
[9] D. M. Collard, C. P. Lillya, *J. Org. Chem.* **1991**, *56*, 6064–6066.
[10] D. M. Collard, C. P. Lillya, *J. Am. Chem. Soc.* **1989**, *111*, 1829–1830.
[11] C. P. Lillya, R. Thakur, *Mol. Cryst. Liq. Cryst.* **1989**, *170*, 179–183.
[12] A. J. Fatiadi, W. F. Sager, *Org. Synth.* **1973**, *V*, 1011.
[13] O. B. Akopova, G. G. Maidachenko, G. A. Tyuneva, L. S. Shabyshev, *J. Gen. Chem. U. S. S. R.*, **1984**, *54*, 1657–1659.
[14] O. B. Akopova, G. A. Tyuneva, L. S. Shabyshev, Y. G. Erykalov, *J. Gen. Chem. U. S. S. R.*, **1987**, *57*, 570–574.

[15] Y. Kobayashi, Y. Matsunaga, *Bull. Chem. Soc. Jpn,* **1987**, *60*, 3515–3518.

[16] R. Nietzki, T. Benckiser, *Berichte*, **1885**, *18*, 499–515.

[17] A. Fitz, *Berichte*, **1875**, *8*, 631–634.

[18] H. Kawada, Y. Matsunaga, *Bull. Chem. Soc. Jpn*, **1988**, *61*, 3083–3085.

[19] Y. Harada, Y. Matsunaga, *Bull. Chem. Soc. Jpn.* **1988**, *61*, 2739–2741.

[20] F. Kehrmann, N. Poehl, *Helv. Chim. Acta* **1926**, *9*, 485–491.

[21] J. Kawamata, Y. Matsunaga, *Mol. Cryst. Liq. Cryst.,* **1993**, *231*, 79–85.

[22] Y. Matsunaga, N. Miyajima, Y. Nakayasu, S. Sakai, M. Yonenaga, *Bull. Chem. Soc. Jpn.* **1988**, *61*, 207.

[23] S. Takenaka, K. Nishimura, S. Kusabayashi, *Mol. Cryst. Liq. Cryst.* **1984**, *111*, 227–236.

[24] K. Praefcke, B. Kohne, D. Singer, *Angew. Chem. Int. Ed. Engl.* **1990**, *29*, 177–179.

[25] B. Kohne, K. Praefcke, *Chimia*, **1987**, *41*, 196–198.

[26] M. Ebert, D. A. Jungbauer, R. Kleppinger, J. H. Wendorff, B. Kohne, K. Praefcke, *Liq. Cryst.* **1989**, *4*, 53.

[27] K. Praefcke, D. Singer, B. Gundogan, K. Gutbier, M. Langner, *Ber. Bunsenges. Phys. Chem.* **1993**, *97*, 1358–1361.

[28] D. M. Kok, H. Wynberg, W. H. De Jeu, *Mol. Cryst. Liq. Cryst.* **1985**, *129*, 53–60.

[29] V. W. Poules, K. Praefcke, *Chem. Z.,* **1983**, *107*, 310.

[30] K. Praefcke, W. Poules, B. Scheuble, R. Poupko, Z. Luz, *Naturforsch., Teil b,* **1984**, *39*, 950–956.

[31] M. Sarkar, N. Spielberg, K. Praefcke, H. Zimmermann, *Mol. Cryst. Liq. Cryst.,* **203**, 159–169.

[32] N. Spielberg, Z. Luz, R. Poupko, K. Praefcke, B. Kohne, J. Pickard, K. Horn, *Z. Naturforsch., Teil a,* **1986**, *41*, 855–860.

[33] K. Praefcke, B. Kohne, K. Gutbier, N. Johnen, D. Singer, *Liq. Cryst.* **1989**, *5*, 233–249.

[34] A. P. Krysin, N. V. Bodoev, V. A. Koptyug, *J. Org. Chem. U. S. S. R.,* **1977**, *13*, 1183–1186.

[35] J. Billard, J. C. Dubois, C. Vaucher, A. M. Levelut, *Mol. Cryst. Liq. Cryst.,* **1981**, *66*, 115–122.

[36] J. Grimshaw, R. D. Haworth, *J. Chem. Soc.,* **1956**, 4225–4232.

[37] G. Carfugua, A. Roviello, A. Sirigu, *Mol. Cryst. Liq. Cryst.* **1985**, *122*, 151–160.

[38] C. Carfagna, P. Iannelli, A. Roviello, A. Sirigu, *Liq. Cryst.* **1987**, *2*, 611–616.

[39] G. Scherowsky, X. H. Chen, *Liq. Cryst.* **1994**, *17*, 803–810.

[40] I. M. Matheson, O. C. Musgrave, C. J. Webster, *J. Chem. Soc., Chem. Commun.* **1965**, 278–279.

[41] W. Kreuder, H. Ringsdorf, *Makromol. Chem. Rapid Commun.* **1983**, *4*, 807–815.

[42] M. Piattelli, E. Fattorusso, R. A. Nicolaus, S. Magno, *Tetrahedron* **1965**, *21*, 3229–3236.

[43] L. Y. Chaing, C. R. Safinya, N. A. Clark, K. S. Liang, A. N. Bloch, *J. Chem. Soc., Chem. Commun.,* **1985**, 695–696.

[44] C. Destrade, M. C. Mondon, J. Malthete, *J. Phys. Colloq.,* **1979**, *C3*, 17–21.

[45] N. Boden, R. C. Borner, R. J. Bushby, A. N. Cammidge, M. V. Jesudason, *Liq. Cryst.* **1993**, *15*, 851–858.

[46] O. Karthaus, H. Ringsdorf, V. V. Tsukruk, J. H. Wendorff, *Langmuir,* **1992**, *8*, 2279–2283.

[47] A. Ronlan, B. Aalstad, V. D. Parker, *Acta Chem. Scand.,* **1982**, *36B*, 317–325.

[48] V. LeBerre, L. Angely, N. Simonet-Gueguen, J. Simonet, *J. Chem. Soc., Chem. Commun.,* **1987**, 984–986.

[49] C. Destrade, M. C. Mondon-Bernaud, N. H. Tinh, *Mol. Cryst. Liq. Cryst. Lett.* **1979**, *49*, 169.

[50] C. Destrade, N. H. Tinh, H. Gasparoux, J. Malthete, A. M. Levelut, *Mol. Cryst. Liq. Cryst.,* **1981**, *71*, 111–135.

[51] C. Vauchier, A. Zann, P. leBarny, J. C. Dubois, J. Billard, *Mol. Cryst. Liq. Cryst.,* **1981**, *66*, 103–114.

[52] R. Breslow, B. Jaun, R. Q. Kluttz, C.-Z. Xia, *Tetrahedron,* **1982**, *38*, 863–867.

[53] R. C. Borner, Ph. D. thesis, Leeds, **1992**.

[54] W. H. Laarhoven, *Rec. Trav. Chim. Pays Bas,* **1983**, *102*, 185–204 and 241–254.

[55] B. Kohne, W. Poules, K. Praefcke, *Chem. Z.* **1987**, *108*, 113.

[56] B. Kohne, K. Praefcke, T. Derz, W. Frischmuth, C. Gansau, *Chem. Z.,* **1984**, *108*, 408.

[57] C. M. Buess, D. D. Lawson, *Chem. Rev.,* **1960**, *60*, 313–330.

[58] E. F. Gramsbergen, H. J. Hoving, W. H. de Jeu, K. Praefcke, B. Kohne, *Liq. Cryst.* **1986**, *1*, 397–400.

[59] N. Boden, R. J. Bushby, J. Clements, M. V. Jesudason, P. F. Knowles, G. Williams, *Chem. Phys. Lett.,* **1988**, *152*, 94–99.

[60] E. Fontes, P. A. Heiney, W. H. de Jeu, *Phys. Rev. Lett.,* **1988**, *61*, 1202–1205.

[61] D. Adam, P. Schuhmacher, J. Simmerer, L. Haussling, K. Siemensmeyer, K. H. Etzbach, H. Ringsdorf, D. Haarer, *Nature* **1994**, *371*, 141–143.

[62] A. M. Levelut, P. Oswald, A. Ghanen, J. Malthete, *J. Phys.* **1984**, *45*, 745–754.

[63] H. Bock, W. Helfrich, *Liq. Cryst.* **1992**, *12*, 697–703.

[64] I. Tabushi, K. Yamamara, Y. Okada, *Tetrahedron Lett.,* **1987**, *28*, 2269–2272.

[65] C. P. Lillya, D. M. Collard, *Mol. Cryst. Liq. Cryst.,* **1990**, *182B*, 201–207.

[66] N. H. Tinh, H. Gasparoux, C. Destrade, *Mol. Cryst. Liq. Cryst.* **1981**, *68*, 101–111.

[67] N. H. Tinh, C. Destrade, H. Gasparoux, *Phys. Lett.* **1979**, *72A*, 251–254.

[68] T. J. Phillips, J. C. Jones, D. G. McDonnell, *Liq. Cryst.* **1993**, *15,* 203–215.

[69] P. Hindmarsh, M. Hird, P. Styring, J. W. Goodby, *J. Mater. Chem.,* **1993**, *3,* 1117–1128.

[70] D. R. Beattie, P. Hindmarsh, J. W. Goodby, S. D. Haslam, R. M. Richardson, *J. Mater. Chem.,* **1992**, *2,* 1261–1266.

[71] N. H. Tinh, M. C. Bernaud, G. Sigand, C. Destrade, *Mol. Cryst. Liq. Cryst.,* **1981**, *65,* 307–316.

[72] J. M. Chapuzet, N. Simonet-Gueguen, I. Taillepied, J. Simonet, *Tetrahedron Lett.,* **1991**, *32,* 7405–7408.

[73] M. Werth, S. U. Vallerien, H. W. Spiess, *Liq. Cryst.* **1991**, *10,* 759–770.

[74] N. Boden, R. J. Bushby, A. N. Cammidge, *J. Chem. Soc., Chem. Commun.* **1994**, 465–466.

[75] N. Boden, R. J. Bushby, A. N. Cammidge, G. Headdock, *Synthesis,* **1995**, 31–32.

[76] N. Boden, R. J. Bushby, A. N. Cammidge, *J. Am. Chem. Soc.,* **1995**, *117,* 924–927.

[77] R. J. Bushby, C. Hardy, *J. Chem. Soc., Perkin Trans. 1* **1986**, 721–723.

[78] R. C. Borner, R. F. W. Jackson, *J. Chem. Soc., Chem. Commun.,* **1994**, 845–846.

[79] G. Wenz, *Makromol. Chem. Rapid Commun.* **1985**, *6,* 577–584.

[80] H. Ringsdorf, B. Schlarb, J. Venzmer, *Angew. Chem. Int. Ed. Engl.,* **1988**, *27,* 113–158.

[81] W. Kreuder, H. Ringsdorf, P. Tschirner, *Makromol. Chem. Rapid Commun.,* **1985**, *6,* 367–373.

[82] H. Ringsdorf, R. Wusterfeld, *Phil. Trans. R. Soc., Ser. A* **1990**, *330,* 95–108.

[83] J. Y. Josefowicz, N. C. Maliszewskyj, S. H. J. Idziak, P. A. Heiney, J. P. McCauley, A. B. Smith, *Science* **1993**, *260,* 323–326.

[84] N. C. Maliszewskyj, P. A. Heiney, J. Y. Josefowicz, J. P. McCauley, A. B. Smith, *Science* **1994**, *264,* 77–79.

[85] N. C. Maliszewskyj, P. A. Heiney, J. K. Blasie, J. P. McCauley, A. B. Smith, *J. Physique (Ser. 2)* **1992**, *2,* 75–85.

[86] E. Orthmann, G. Wegner, *Angew. Chem. Int. Ed. Engl.,* **1986**, *12,* 1105.

[87] P. Henderson, H. Ringsdorf, P. Schuhmacher, *Liq. Cryst.* **1995**, *18,* 191–195.

[88] N. Boden, R. J. Bushby, A. N. Cammidge, G. Headdock, *UK Patent 9411828.8,* **1994**.

[89] W. Kranig, B. Huser, H. W. Spiess, W. Kreuder, H. Ringsdorf, H. Zimmermann, *Adv. Mater.* **1990**, *2,* 36–40.

[90] F. Closs, L. Haussling, P. Henderson, H. Ringsdorf, P. Schuhmacher, *J. Chem. Soc., Perkin Trans. 1,* **1995**, 829–837.

[91] J. W. Goodby, M. Hird, K. J. Toyne, T. Watson, *J. Chem. Soc., Chem. Commun.,* **1994**, 1701–1702.

[92] N. Boden, R. J. Bushby, A. N. Cammidge, *Liq. Cryst.* **1995**, *18,* 673–676.

[93] N. Boden, R. J. Bushby, A. N. Cammidge, G. Headdock, *J. Mater. Chem.* **1995**, *5,* 2275–2281.

[94] N. Boden, R. J. Bushby, A. N. Cammidge, G. Headdock, *UK Patent 9405793.2,* **1994**.

[95] P. Henderson, S. Kumar, J. A. Rego, H. Ringsdorf, P. Schuhmacher, *J. Chem. Soc., Chem. Commun.* **1995**, 1059.

[96] F. G. Mann, M. J. Pragnell, *J. Chem. Soc.,* **1965**, 4120–4127.

[97] W. J. Musliner, J. W. Gates, *J. Am. Chem. Soc.,* **1966**, *88,* 4271–4273.

[98] O. C. Musgrave, C. J. Webster, *J. Chem. Soc., Chem. Commun.,* **1969**, 712–713.

[99] C. Goltner, D. Pressner, K. Mullen, H. W. Spiess, *Angew. Chem. Int. Ed. Engl.* **1993**, *32,* 1660–1662.

[100] U. Anton, C. Goltner, K. Mullen, *Chem. Ber.,* **1992**, *125,* 2325–2330.

[101] C. Destrade, H. Gasparoux, A. Babeau, N. H. Tinh, *Mol. Cryst. Liq. Cryst.,* **1981**, *67,* 37–48.

[102] C. Destrade, J. Malthete, N. H. Tinh, H. Gasparoux, *Phys. Lett.,* **1980**, *78A,* 82–87.

[103] J. Chenault, C. Sainson, *Compt. Rend. C,* **1978**, *287,* 545–547.

[104] N. H. Tinh, P. Foucher, C. Destrade, A. M. Levelut, J. Malthete, *Mol. Cryst. Liq. Cryst.* **1984**, *111,* 277–292.

[105] W. K. Lee, B. A. Wintner, E. Fontes, P. A. Heiney, M. Ohba, J. N. Haseltine, A. B. Smith, *Liq. Cryst.* **1989**, *4 ,* 87–102.

[106] C. Destrade, P. Foucher, J. Malthete, N. H. Tinh, *Phys. Lett.,* **1982**, *88A,* 187–190.

[107] E. Fontes, P. A. Heiney, M. Ohba, J. N. Haseltine, A. B. Smith, *Phys. Rev. A,* **1988**, *37,* 1329–1334.

[108] N. N. Tinh, J. Malthete, C. Destrade, *Mol. Cryst. Liq. Cryst. Lett.,* **1981**, *64,* 291–298.

[109] N. H. Tinh, J. Malthete, C. Destrade, *J. Phys. Lett.,* **1981**, *42,* L417–L419.

[110] E. Keinan, S. Kumar, R. Moshenberg, R. Ghirlando, E. J. Wachtel, *Adv. Mat.,* **1991**, *3,* 251–254.

[111] H. Zimmermann, R. Poupko, Z. Luz, J. Billard, *Z. Naturforsch., Teil a,* **1995**, *40,* 149–160.

[112] H. Zimmermann, R. Poupko, Z. Luz, J. Billard, *Z. Naturforsch., Teil a,* **1986**, *41,* 1137–1140.

[113] J. Malthete, A. Collet, *J. Am. Chem. Soc.* **1987**, *109,* 7544–7545.

[114] R. Poupko, Z. Luz, N. Spielberg, H. Zimmermann, *J. Am. Chem. Soc.,* **1989**, *111,* 6094–6105.

[115] G. M. Robinson, *J. Chem. Soc.,* **1915**, 267–276.

[116] A. Oliverio, C. Casinovi, *Ann. Chim. Roma,* **1952**, *42,* 168–184; **1956**, *46,* 929–933.

[117] F. Bertinotti, V. Carelli, A. Liquori, A. M. Nardi, *Struct. Chimi. Suppl. Ricerca Sci.* **1952**, *22,* 65.

[118] A. S. Lindsey, *Chem. Ind.,* **1963**, 823–824.

[119] A. S. Lindsey, *J. Chem. Soc.,* **1965**, 1685.

[120] J. Malthete, A. Collet, *Nouv. J. Chim.*, **1985**, *9*, 151–153.

[121] A. M. Levelut, J. Malthete, A. Collet, *J. Phys. Paris,* **1986**, *47*, 351–357.

[122] A. Collet, *Tetrahedron,* **1987**, *43*, 5725–5759.

[123] C. Garcia, J. Malthete, A. Collet, *Bull. Soc. Chim. France,* **1993**, 93–95.

[124] J. D. White, B. D. Gesner, *Tetrahedron* **1974**, *30*, 2273–2277.

[125] V. Percec, C. G. Cho, C. Pugh, *J. Mater. Chem.* **1991**, *1*, 217–222.

[126] V. Percec, C. G. Cho, C. Pugh, *Macromolecules,* **1991**, *24*, 3227–3234.

[127] H. Zimmermann, R. Poupko, Z. Luz, J. Billard, *Liq. Cryst.* **1988**, *3*, 759–770.

[128] V. Percec, C. G. Cho, C. Pugh, D. Tomazos, *Macromolecules,* **1992**, *25*, 1164–1176.

[129] N. Spielberg, M. Sarkar, Z. Luz, R. Poupko, J. Billard, H. Zimmermann, *Liq. Cryst.* **1993**, *15*, 311–330.

[130] H. Zimmermann, R. Poupko, Z. Luz, J. Billard, *Liq. Cryst.* **1989**, *6*, 151–166.

[131] S. Bonsignore, A. Du Vosel, G. Guglielmetti, E. Dalcanale, F. Ugozzoli, *Liq. Cryst.* **1993**, *13*, 471–482.

[132] G. Cometti, E. Dalcanale, A. Du Vosel, A. M. Levelut, *J. Chem.* Soc., *Chem. Commun.* **1990**, 163–165.

[133] S. Bonsignore, G. Cometti, E. Dalcanale, A. Du Vosel, *Liq. Cryst.* **1990**, *8*, 639–649.

[134] E. Dalcanale, A. Du Vosel, A. M. Levelut, J. Malthete, *Liq. Cryst.* **1991**, *10*, 185–198.

[135] L. Abis, V. Arrighi, G. Cometti, E. Dalcanale, A. Du Vosel, *Liq. Cryst.* **1991**, *9*, 277–284.

[136] J. Zhang, J. S. Moore, *J. Am. Chem. Soc.,* **1994**, *116*, 2655–2656.

[137] J. Zhang, J. S. Moore, Z. Xu, R. A. Aguirre, *J. Am. Chem. Soc.,* **1992**, *114*, 2273–2274.

[138] J. S. Moore, J. Zhang, *Angew. Chem. Int. Ed. Engl.* **1992**, *31*, 922–924.

[139] P. Davidson, C. Jallabert, A. M. Levelut, H. Strzelecka, M. Verber, *Liq. Cryst.* **1988**, *3*, 133–137.

[140] H. Strzelecka, C. Jallabert, M. Verber, P. Davidson, A. M. Levelut, *Mol. Cryst. Liq. Cryst.,* **1988**, *161*, 395–401.

[141] H. Strzelecka, C. Jallabert, M. Verber, *Mol. Cryst. Liq. Cryst.* **1988**, *156*, 355–359.

[142] M. Verber, P. Sotta, P. Davidson, A. M. Levelut, C. Jallabert, H. Strzelecka, *J. Phys. France,* **1990**, *51*, 1283–1301.

[143] H. Strzelecka, C. Jallabert, M. Verber, P. Davidson, A. M. Levelut, J. Malthete, G. Sigaud, A. Skoulios, P. Weber, *Mol. Cryst. Liq. Cryst.* **1988**, *161*, 403–411.

[144] J. Barry, G. Bram, G. Decodts, A. Loupy, P. Pigeon, J. Sansoulet, *Tetrahedron Lett.,* **1982**, *23*, 5407–5408.

[145] R. Fugnitto, H. Strzelecka, A. Zann, J.-C. Dubois, J. Billard, *J. Chem. Soc., Chem. Commun.* **1980**, 271–272.

[146] F. D. Saeva, G. A. Reynolds, L. Kaszczuk, *J. Am. Chem. Soc.,* **1982**, *104*, 3524–3525.

[147] G. A. Reynolds, C. H. Chen, *J. Heterocyclic Chem.,* **1981**, *18*, 1235–1237.

[148] F. D. Saeva, G. A. Reynolds, *Mol. Cryst. Liq. Cryst.,* **1986**, *132*, 29–34.

[149] H. Zimmermann, J. Billard, H. Gutman, E. J. Wachtel, R. Poupko, Z. Luz, *Liq. Cryst.* **1992**, *12*, 245–262.

[150] C. Destrade, N. H. Tinh, H. Gasparoux, L. Mamlok, *Liq. Cryst.,* *2*, 229–233.

[151] L. Mamlock, J. Malthete, N. H. Tinh, C. Destrade, A. M. Levelut, *J. Phys. Lett.,* **1982**, *43*, L641–L647.

[152] C. Destrade, N. H. Tinh, L. Mamlock, J. Malthete, *Mol. Cryst. Liq. Cryst.,* **1984**, *114*, 139–150.

[153] R. Cayuela, H. T. Nguyen, C. Destrade, A. M. Levelut, *Mol. Cryst. Liq. Cryst.* **1989**, *177*, 81–91.

[154] N. H. Tinh, R. Cayuela, C. Destrade, *Mol. Cryst. Liq. Cryst.,* **1985**, *122*, 141–149.

[155] N. Boden, R. C. Borner, R. J. Bushby, J. Clements, *J. Am. Chem. Soc.,* **1994**, *116*, 10807–10808.

[156] G. G. Hall, W. R. Rodwell, *J. Theor. Biol.,* **1975**, *50*, 107–120.

[157] R. W. Baldwin, G. J. Cunningham, M. W. Partridge, H. J. Vipond, *Br. J. Cancer.* **1962**, *16*, 275–282.

[158] S. Kumar, E. J. Wachtel, E. Keinan, *J. Org. Chem.,* **1993**, *58*, 3821–3827.

[159] E. Kienan, S. Kumar, S. P. Singh, R. Ghirlando, E. J. Wachtel, *Liq. Cryst.* **1992**, *11*, 157–173.

[160] B. Franck, *Angew. Chem. Int. Ed. Engl.,* **1982**, *21*, 343–353.

[161] B. A. Gregg, M. A. Fox, A. J. Bard, *J. Chem. Soc. Chem. Commun.* **1987**, 1134–1135.

[162] B. A. Gregg, M. A. Fox, A. J. Bard, *J. Am. Chem. Soc.,* **1989**, *111*, 3024–3029.

[163] J. C. Fanning, F. S. Mandel, T. L. Gray, N. Datta-Gupta, *Tetrahedron,* **1979**, *35*, 1251–1255.

[164] K. M. Smith, G. H. Barnett, B. Evans, Z. Martynenko, *J. Am. Chem. Soc.,* **1979**, *101*, 5953–5961.

[165] Y. Shimitzu, M. Miya, A. Nagata, K. Ohta, A. Matsumura, I. Yamamoto, S. Kusabayashi, *Chem. Lett.,* **1991**, 25–28.

[166] A. D. Adler, F. R. Longo, J. D. Finarelli, J. Goldmacher, J. Assour, L. Korsakoff, *J. Org. Chem.,* **1967**, *32*, 476.

[167] G. Bahr, G. Schleitzer, *Chem. Ber.,* **1957**, *90*, 438–443.

[168] G. Morelli, G. Ricciardi, A. Roviello, *Chem. Phys. Lett.,* **1991**, *185*, 468–472.

[169] F. Lelj, G. Morelli, G. Ricciardi, A. Roviello, A. Sirigu, *Liq. Cryst.* **1992**, *12*, 941–960.

[170] P. Doppelt, S. Huille, *New J. Chem.,* **1990**, *14*, 607–609.

[171] C. Piechocki, J. Simon, A. Skoulios, D. Guillon, P. Weber, *J. Am. Chem. Soc.,* **1982**, *104*, 5245–5247.

[172] G. Pawlowski, M. Hanack, *Synthesis,* **1980**, 287–289.

[173] E. Klingsberg, *Synthesis,* **1972**, 29–30.

[174] W. Wenner, *J. Org. Chem.,* **1952**, *17*, 523–528.

[175] W. Offermann, F. Vogtle, *Angew. Chem. Int. Ed. Engl.,* **1980**, *19*, 464–465.

[176] M. Hanack, A. Beck, H. Lehmann, *Synthesis,* **1987**, 703–705.

[177] P. Weber, D. Guillon, A. Skoulios, *J. Phys. Chem.,* **1987**, *91*, 2242–2243.

[178] C. Piechocki, J.-C. Boulou, J. Simon, *Mol. Cryst. Liq. Cryst.,* **1987**, *149,* 115–120.

[179] D. Guillon, A. Skoulios, C. Piechocki, J. Simon, P. Weber, *Mol. Cryst. Liq. Cryst.,* **1983**, *100*, 275–284.

[180] D. Guillon, P. Weber, A. Skoulios, C. Piechocki, J. Simon, *Mol. Cryst. Liq. Cryst.* **1985**, *130*, 223–229.

[181] C. Sirlin, L. Bosio, J. Simon, *J. Chem. Soc. Chem. Commun.,* **1987**, 379–380.

[182] C. Piechocki, J. Simon, *Nouv. J. Chim.,* **1985**, *9*, 159–166.

[183] I. Cho, Y. Lim, *Chem. Lett.,* **1987**, 2107–2108.

[184] I. Cho, Y. Lim, *Mol. Cryst. Liq. Cryst.,* **1988**, *154*, 9–26.

[185] K. Ohta, T. Watanabe, S. Tanaka, T. Fujimoto, I. Yamamoto, P. Bassoul, N. Kucharczyk, J. Simon, *Liq. Cryst.* **1991**, *10*, 357–368.

[186] J. F. Van der Pol, E. Neeleman, J. W. Zwikker, R. J. M. Nolte, W. Drenth, *Rec. Trav. Chim. Pays-Bas,* **1988**, *107*, 615–620.

[187] L. M. Severs, A. E. Underhill, D. Edwards, P. Wight, D. Thetford, *Mol. Cryst. Liq. Cryst.,* **1993**, *234*, 235–240.

[188] J. F. Van der Pol, E. Neeleman, J. W. Zwikker, R. J. M. Nolte, W. Drenth, J. Aerts, R. Visser, S. J. Picken, *Liq. Cryst.* **1989**, *6*, 577–592.

[189] T. Sauer, G. Wegner, *Mol. Cryst. Liq. Cryst.,* **1988**, *162B*, 97–118.

[190] T. Komatsu, K. Ohta, T. Watanabe, H. Ikemoto, T. Fujimoto, I. Yamamoto, *J. Mater. Chem.* **1994**, *4*, 537–540.

[191] P. G. Schouten, J. M. Warman, M. P. de Haas, C. F. van Nostrum, G. H. Gelinck, R. J. M. Nolte, M. C. Copyn, J. W. Zwikker, M. E. Engel, M. Hanack, Y. H. Chang, W. T. Ford, *J. Am. Chem. Soc.,* **1994**, *116*, 6880–6894.

[192] D. Lelievre, M. A. Petit, J. Simon, *Liq. Cryst.,* **1989**, *4*, 707–710.

[193] P. G. Schouten, J. F. Van der Pol, J. W. Zwikker, W. Drenth, S. J. Picken, *Mol. Cryst. Liq. Cryst.* **1991**, *195*, 291–305.

[194] W. T. Ford, L. Sumner, W. Zhu, Y. H. Chang, P.-J. Um, K. H. Choi, P. A. Heiney, N. C. Maliszewskyj, *New J. Chem.,* **1994**, *18*, 495–505.

[195] K. Ohta, T. Watanabe, H. Hasebe, Y. Morizumi, T. Fujimoto, I. Yamamoto, D. Lelievre, J. Simon, *Mol. Cryst. Liq. Cryst.* **1991**, *196*, 13.

[196] D. Masurel, C. Sirlin, J. Simon, *New J. Chem.* **1987**, *11*, 455–456.

[197] K. Ohta, L. Jacquemin, C. Sirlin, L. Bosio, J. Simon, *New J. Chem.* **1988**, *12*, 751–754.

[198] E. A. Cuellar, T. J. Marks, *Inorg. Chem.,* **1981**, *20*, 3766–3770.

[199] K. Tamao, K. Sumitani, Y. Kiso, M. Zembayashi, A. Fujioka, S.-I. Kodama, I. Nakajima, A. Minato, M. Kumada, *Bull. Chem. Soc. Jpn.* **1976**, *49*, 1958–1969.

[200] M. K. Engel, P. Bassoul, L. Bosio, H. Lehmann, M. Hanack, J. Simon, *Liq. Cryst.* **1993**, *15*, 709–722.

[201] K. Ohta, T. Watanabe, T. Fujimoto, I. Yamamoto, *J. Chem. Soc., Chem. Commun.,* **1980**, 1611–1613.

[202] S. M. McElvain, *Org. React.,* **1948**, *4*, 256–268.

[203] D. E. Ames, D. Hall, B. T. Warren, *J. Chem. Soc. (C),* **1968**, 2617–2621.

[204] K. Kanakarajan, A. W. Czarnik, *J. Org. Chem.,* **1986**, *51*, 5241.

[205] L. Dulog, A. Gittinger, *Mol. Cryst. Liq. Cryst.,* **1992**, *213*, 31–42.

[206] D. Wohrle, U. Hundorf, *Makromol. Chem.,* **1985**, *186*, 2177–2187.

[207] A. J. Saggiomo, *J. Org. Chem.,* **1957**, *22*, 1171–1175.

[208] P. A. Barrett, R. P. Linstead, G. A. P. Tuey, J. M. Robertson, *J. Chem. Soc.* **1939**, 1809–1820.

[209] C. C. Leznoff, N. B. McKeown, *J. Org. Chem.* **1990**, *55*, 2186–2190.

[210] N. B. McKeown, C. C. Leznoff, R. M. Richardson, A. S. Cherodian, *Mol. Cryst. Liq. Cryst.,* **1992**, *213*, 91–98.

[211] N. B. McKeown, J. Painter, *J. Mater. Chem.,* **1994**, *4*, 1153–1156.

[212] T. W. Hall, S. Greenberg, C. R. McArthur, B. Khouw, C. C. Leznoff, *New J. Chem.,* **1982**, *6*, 653.

[213] M. J. Cook, M. F. Daniel, K. J. Harrison, N. B. McKeown, A. J. Thomson, *J. Chem. Soc., Chem. Commun.,* **1987**, 1086–1088.

[214] A. N. Cammidge, M. J. Cook, K. J. Harrison, N. B. McKeown, *J. Chem. Soc., Perkin Trans. 1,* **1991**, 3053–3058.

[215] A. N. Cammidge, M. J. Cook, S. D. Haslam, R. M. Richardson, K. J. Harrison, *Liq. Cryst.* **1993**, *14*, 1847–1862.

[216] N. B. McKeown, I. Chambrier, M. J. Cook, *J. Chem. Soc., Perkin Trans. 1,* **1990**, 1169–1177.

[217] M. J. Cook, *J. Mater. Sci.,* **1994**, *5*, 117–128.

[218] A. S. Cherodian, A. N. Davies, R. M. Richardson, M. J. Cook, N. B. McKeown, A. J. Thomson, J. Feijoo, G. Ungar, K. J. Harrison, *Mol. Cryst. Liq. Cryst.,* **1991**, *196*, 103–114.

[219] M. J. Cook, S. J. Cracknell, K. J. Harrison, *J. Mater. Chem.* **1991**, *1*, 703–704.

[220] C. Piechocki, J. Simon, *J. Chem. Soc., Chem. Commun.* **1985**, 259–260.

[221] I. Chambrier, M. J. Cook, S. J. Cracknell, J. McMurdo, *J. Mater. Chem.,* **1993**, *3*, 841–849.

[222] B. Kohne, K. Praefcke, *Angew. Chem. Int. Ed. Engl.,* **1984**, *23*, 82–83.

[223] K. Praefcke, B. Kohne, W. Stephan, P. Marquardt, *Chimia,* **1989**, *43*, 380–382.

[224] B. Kohne, K. Praefcke, W. Stephan, P. Nurnberg, *Z. Naturforsch., Teil b,* **1985**, *40*, 981–986.

[225] K. Praefcke, P. Psaras, B. Kohne, *Chem. Ber.,* **1991**, *124*, 2523–2529.

[226] B. Kohne, K. Praefcke, *Z. Naturforsch., Teil b,* **1986**, *41*, 1036–1044.

[227] C. P. Lillya, D. M. Collard, *Mol. Cryst. Liq. Cryst.,* **1990**, *182B*, 201–207.

[228] N. L. Morris, R. G. Zimmerman, G. B. Jameson, A. W. Dalziel, P. M. Reuss, R. G. Weiss, *J. Am. Chem. Soc.,* **1988**, *110*, 2177–2185.

[229] V. Vill, J. Thiem, *Liq. Cryst.* **1991**, *9*, 451–455.

[230] R. G. Zimmerman, G. B. Jameson, R. G. Weiss, G. Dermailly, *Mol. Cryst. Liq. Cryst. Lett.,* **1985**, *1*, 183–189.

[231] A. Takada, T. Fukuda, T. Miyamoto, Y. Yakoh, J. Watanabe, *Liq. Cryst.* **1992**, *12*, 337–345.

[232] T. Itoh, A. Takada, T. Fukuda, T. Miyamoto, Y. Yakoh, J. Watanabe, *Liq. Cryst.* **1991**, *9*, 221–228.

[233] M. Sagiura, M. Minoda, J. Watanabe, T. Fukuda, T. Miyamoto, *Bull. Chem. Soc. Jpn.,* **1992**, *65*, 1939–1943.

[234] A. Takada, K. Fujii, J. Watanabe, T. Fukuda, T. Miyamoto, *Macromolecules,* **1994**, *27*, 1651–1653.

[235] B. Kohne, K. Praefcke, *Chem. Z.,* **1985**, *109*, 121–127.

[236] J.-M. Lehn, J. Malthete, A. M. Levelut, *J. Chem. Soc., Chem. Commun.,* **1985**, 1794–1796.

[237] S. H. J. Idziak, N. C. Maliszewskyj, G. B. M. Vaughan, P. A. Heiney, C. Mertesdorf, H. Ringsdorf, J. P. McCauley, A. B. Smith, *J. Chem. Soc., Chem. Commun.,* **1992**, 98–99.

[238] J. Malthete, A. M. Levelut, J.-M. Lehn, *J. Chem. Soc., Chem. Commun.* **1992**, 1434–1436.

[239] M. Zhao, W. T. Ford, S. H. J. Idziak, N. C. Maliszewskyj, P. A. Heiney, *Liq. Cryst.,* **1994**, *16*, 583–599.

[240] S. H. J. Idziak, N. C. Maliszewskyj, P. A. Heiney, J. P. McCauley, P. A. Sprengeler, A. B. Smith, *J. Am. Chem. Soc.,* **1991**, *113*, 7666–7672.

[241] J. Malthete, D. Poupinet, R. Vilanove, J.-M. Lehn, *J. Chem. Soc., Chem. Commun.* **1989**, 1016–1019.

[242] C. Mertesdorf, H. Ringsdorf, *Liq. Cryst.,* **1989**, *5*, 1757–1772.

[243] C. Mertesdorf, H. Ringsdorf, J. Stumpe, *Liq. Cryst.* **1991**, *9*, 337–357.

[244] G. Lattermann, *Mol. Cryst. Liq. Cryst.* **1990**, *182B*, 299–311.

[245] G. Lattermann, *Liq. Cryst.,* **1989**, *6*, 619–625.

Chapter VIII
Columnar, Discotic Nematic and Lamellar Liquid Crystals: Their Structures and Physical Properties

S. Chandrasekhar

1   Introduction

We begin this chapter with a few remarks about terminology. Soon after the discovery of liquid crystals of disk-like molecules [1–3], the word 'discotic' was introduced to distinguish these mesogens from the familiar rod-like or calamitic-type mesogens. However, before long, terms such as 'discogen', 'discotic phase', etc., began to be used extensively in the literature, and the word 'discotic' and related terms came to be applied loosely to describe the molecules as well as the mesophases formed by them. Strictly speaking, it is the molecules that are discotic and not the mesophases, which may be columnar, nematic or lamellar. Moreover, as we shall see later, certain molecules, which themselves are not discotic, have been found to exhibit columnar phases [4–6]. Thus, care is necessary in the use of these terms in order to avoid any ambiguity.

The first examples of thermotropic mesomorphism in discotic compounds were observed in the hexa-alkanoyloxy benzenes (**1**, $R = C_nH_{2n+1}$) [1] and the hexa-alkoxy- and hexa-alkanoyloxytriphenylenes (**2**, $R = C_nH_{2n+1}O$ and $C_nH_{2n+1}CO \cdot O$) [2, 3], and it was established by X-ray studies [1, 7] that the mesophases of these compounds have a columnar structure. About 1500 discotic mesogens are known to date [8] and a variety of new mesophase structures has been identified. The simplest of these compounds have flat or nearly flat cores surrounded by six or eight, or sometimes four, long-chain substituents. Much more complex discotic molecules have since been synthesized; for example, the star-like triphenylene derivatives (**3**) reported recently [9]. Available experimental evidence indicates that the presence of long side chains is crucial to the formation of discotic liquid crystals. The synthetic procedure and other chemical aspects of the subject are dealt with in Chapters VII and IX in this volume, and are not therefore considered here. The discussion here is limited to the structural classification of the mesophases and a description of their physical properties, and just a few illustrative examples of discotic molecules that are relevant to the topics discussed, are presented.

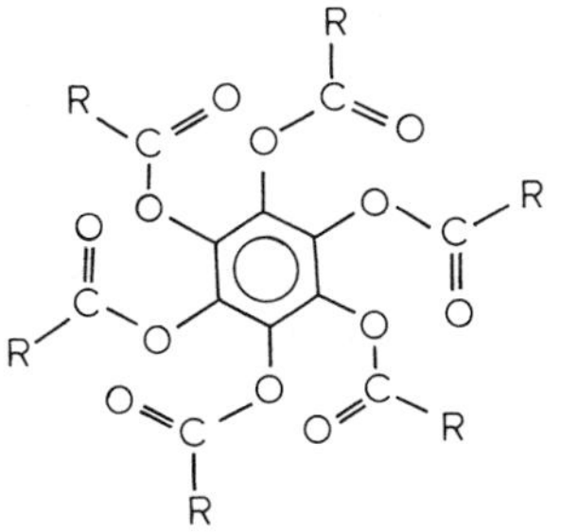

1

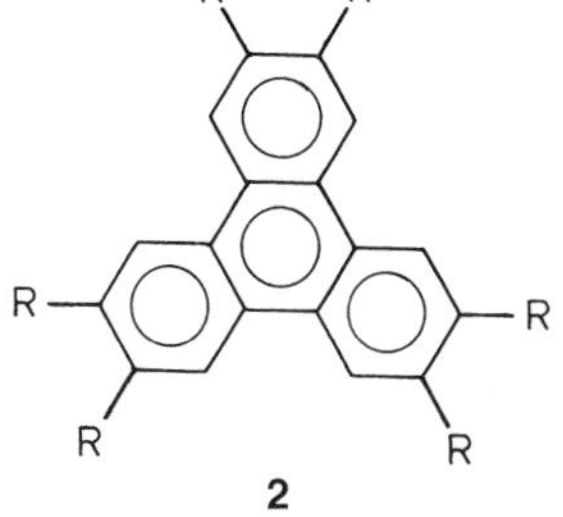

2

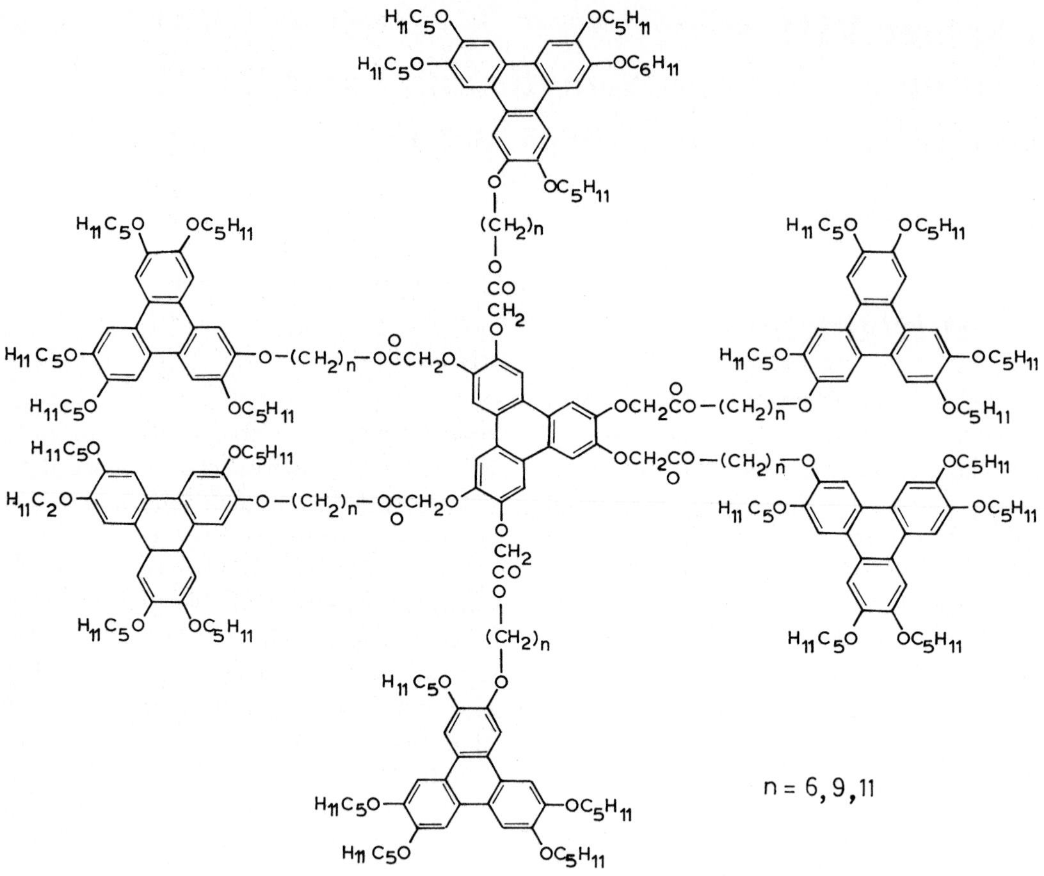

H₁₁C₅O, OC₅H₁₁, H₁₁C₅O, OC₆H₁₁, OC₅H₁₁

(CH₂)ₙ

CO CH₂ O

O—(CH₂)ₙ—OCCH₂O

OCH₂CO—(CH₂)ₙ—O

CH₂ CO O

(CH₂)ₙ

H₁₁C₅O, OC₅H₁₁, H₁₁C₅O, OC₅H₁₁, H₁₁C₂O, O—(CH₂)ₙ—OCCH₂O, H₁₁C₅O, OC₅H₁₁

H₁₁C₅O, OC₅H₁₁

n = 6, 9, 11

3

2 Description of the Liquid – Crystalline Structures

Discotic liquid crystals may be classified broadly into three types: *columnar, nematic* (including its chiral modification) and *lamellar*.

2.1 The Columnar Liquid Crystal

The basic columnar structure is illustrated in Figure 1 (a): the disks are stacked one on top of another to form columns, the different columns constituting a two-dimensional lattice. The structure is somewhat similar to the hexagonal phase of soap–water and lyotropic systems [10], but several variants of the structure have been identified – upright columns (Figure 1 a), tilted columns (Figure 1 b), hexagonal, rectangular, etc. In some cases, the columns are liquid-like, i.e. the molecular centres are arranged aperiodically within each column, as depicted in Figure 1 (a and b), while in others they are arranged in a regular, ordered fashion.

The columnar phase is denoted by the symbol D (according to new guidelines, see Chap. II of Vol. I of this Handbook, it should

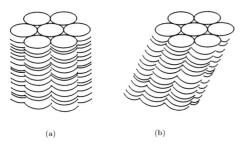

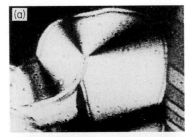

(a)   (b)

Figure 1. Schematic representation of columnar structures of discotic mesogens: (a) upright columns, (b) tilted columns.

be denoted as "Col" instead of "D" in future) with appropriate subscripts to distinguish the different modifications. For example, the subscript 'h' stands for a 'hexagonal' lattice of columns, 'r' for a 'rectangular' lattice, 'd' for a 'disordered' stacking of the molecules in each column, 'o' for an 'ordered' stacking and 't' for tilted columns. Thus:

- D_{hd} signifies columnar, hexagonal, disordered
- D_{rd} signifies columnar, rectangular, disordered
- D_{ho} signifies columnar, hexagonal, ordered
- D_t signifies columnar, tilted

A tilt of the molecular core with respect to the column axis is a common feature in these structures. In the very first report on the hexa-alkanoyloxy benzenes [1] it was noted that the X-ray photographs contained a few weak diffraction spots not quite conforming to true hexagonal symmetry, but the quality of the X-ray patterns in these early studies did not warrant a more refined analysis. Subsequently, Frank and Chandrasekhar [11] noted that, when seen under the polarizing microscope, the extinction brushes (or crosses) in the optical textures of the mesophases do not coincide with the polarizer – analyser directions, but are usually tilted

Figure 2. Circular domains with curved column axes showing oblique extinction brushes in pure benzene hexa-n-hexanoate (a b) and in its mixture with a few per cent by weight of benzene (c). The mesophase of the mixture is slightly more mobile than that of the pure material, but has the same structure. Polarizer – analyser directions are vertical and horizontal ((a) Courtesy of K. A. Suresh; (b and c), from Frank and Chandrasekhar [11], reproduced by permission of the Commission des Publications Françaises de Physique).

with respect to these directions (Figure 2). From this and other optical observations, they concluded that the plane of the rigid aromatic core (in which resides the anisotropy of molecular polarizability, and the diameter of which is only about a third of the whole molecule) is not normal to the column axis, but inclined to it at an angle of about $55-60°$, indicating that the lattice should be pseudo-hexagonal. This was confirmed by Levelut [12] who found that the columnar phase of the hexa-n-octanoate of benzene (**1**) has a rectangular lattice with an axial ratio of 1.677, which departs from the true hexagonal value of √3 by about 3%. Figure 3 shows a schematic representation of the structure of a column in which the cores are tilted with respect to the column axis and the chains are in a highly disordered state. Figure 4 presents the different types of two-dimensional lattice structures that have been identified by Levelut [12], and Table 1 gives the space groups of the columnar structures formed by some derivatives of triphenylene. (These are *planar* space groups that constitute the subset of the 230 space groups when symmetry elements relating to translations along one of the axes, in this case the column axis, are absent.)

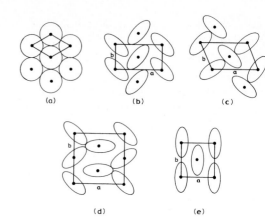

Figure 4. Plan views of two-dimensional lattices of columnar phases; ellipses denote disks that are tilted with respect to the column axis: (a) hexagonal (*P6 2/m 2/m*); (b) rectangular (*P 2₁/a*); (c) oblique (*P₁*); (d) rectangular (*P₂/a*); (e) rectangular, face-centered, tilted columns (*C2/m*). (From Levelut [12], reproduced by permission of the division des Publications Société de Chimie Physique).

2.1.1   NMR Studies

Deuterium NMR spectroscopy of the columnar phase of hexa-n-hexyloxytriphenylene has revealed the marked difference between the order parameters of the cores and the tails [13]. Spectra of two selectively deuteriated isotopic species, one in which all aromatic positions are substituted and the other in which only the α-carbon side chains

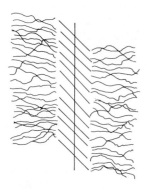

Figure 3. The structure of a column in which the molecular cores are tilted with respect to the column axis and the chains are in a highly disordered state.

Table 1. Two-dimensional lattices of columnar structures formed by some hexa-substituted triphenylenes [12].

R	Space group	Lattice parameters (nm)	Temperature (°C)
$C_5H_{11}O$	*P6 2/m 2/m*	$a=1.895$	≈ 80
$C_7H_{15}O$	*P6 2/m 2/m*	$a=2.22$	≈ 80
$C_8H_{17}O$	*P6 2/m 2/m*	$a=2.33$	≈ 80
$C_{11}H_{23}COO$	*P2₁/a*	$a=4.49, b=2.64$	117
	P6 2/m 2/m	$a=2.63$	105
$C_7H_{15}COO$	*P2₁/a*	$a=3.78, b=2.22$	100
$C_{11}H_{23}-O-C_6H_4-COO^a$	*P2₁/a*	$a=5.18, b=3.26$	165
$C_6H_{13}-O-C_6H_4-COO^a$	*C2/m*	$a=3.07, b=2.84$	185

[a] The C_6H_4 groups are *para* substituted.

are substituted, were investigated. Figure 5 gives the quadrupole splittings of the aromatic and the α-aliphatic deuterons versus temperature in the mesophase region. It is seen that the rigid core is highly ordered, the orientational order parameter S ranging from 0.95 to 0.90, where

$$S = \left\langle \frac{1}{2}(3\cos^2\theta - 1) \right\rangle$$

and θ is the angle which the molecular-symmetry axis makes with the director or column axis, whereas the α-aliphatic chains are in a disordered state. Relaxation studies have also been reported [14].

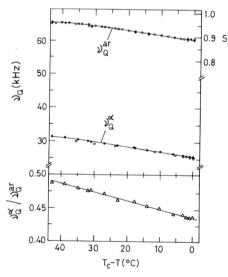

Figure 5. Quadrupole splittings for aromatic v_Q^{ar} and α-aliphatic v_Q^{α} deuterons of deuteriated hexa-n-hexyl-oxytriphenylene (THE6) as functions of temperature $(T_c - T)$ in the mesophase region, where T_c is the columnar – isotropic transition point. (○) Neat THE6-ar-d_6 amd THE6-α-d_{12} separately; (●) a 2:1 mixture of the two isotopic species. The scale on the upper right-hand side gives the orientational order parameter of the aromatic part. The curve at the bottom gives the ratio of the quadrupole splittings for the α-aliphatic and aromatic deuterons (From Goldfarb et al. [13], reproduced by permission of the Commission des Publications Française de Physique).

These results indicate that in any realistic theory of the statistical mechanics of columnar systems the molecules cannot be treated as rigid disks, but the conformational degrees of freedom of the hydrocarbon chains must also be taken into account.

2.1.2 High Resolution X-Ray Studies

High resolution X-ray studies have yielded important additional information about the columnar structures. The measurements were made on very well oriented monodomain samples obtained by preparing freely suspended liquid crystal strands (typically about $100-200\ \mu m$ in diameter and $1.5-2$ mm long) with the column axis parallel to the axis of the strand [15 – 19]. The salient results of these studies are described below.

$(C_{13}H_{27}CO \cdot O)_6$-truxene
(**4**, $R = C_{13}H_{27}CO \cdot O$)

The correlation length of the two-dimensional lattice of the hexagonal columnar phase of this compound is greater than 400 nm (or approximately 200 columns), the lower limit being set by the resolution of the instrument [19]. Within each column, on the other hand, the flat molecular cores from an oriented, one-dimensional liquid, i.e. they are orientationally ordered but translationally disordered. In contrast, the flexible hydrocarbon chains are highly disordered and produce a nearly isotropic scattering pattern. This striking difference between the ordering of the cores and the tails is also seen in the X-ray scattering from the columnar mesophases of triphenylene compounds, in conformity with the results of deuterium NMR spectroscopy, but it is not so apparent in the hexa-n-octanoate of benzene, which has a much smaller core [16, 17].

The structure of the columnar phase of the truxene compound is essentially the same as that proposed originally by Chandrasekhar et al. [1]. The possibility of such a structure i.e. a three-dimensional body in which the density is a periodic function of only two dimensions, was in fact envisaged many years ago by Landau and Lifshitz [20] (see Sec. 10.3).

$(C_{12}H_{25}CO \cdot O)_6$-triphenylene ($2$, R = $C_{12}H_{25}CO \cdot O$)

This compound exhibits a transition from a hexagonal (D_h) to a rectangular (D_r) columnar phase. The transition, which is weakly first order, is associated with a small distortion of the lattice, consistent with a herringbone arrangement in the rectangular structure, with only the core of the molecule tilted with respect to the column axis (see Figure 3). However, high resolution synchrotron X-ray studies on monodomain strands [16, 17] have established that the tilt of the molecular core persists in the D_h phase as well, except that the tilts in neighbouring columns are rotationally uncorrelated, i.e. they are free to assume different azimuthal angles. Thus the D_h–D_r transition may be looked upon as an orientational order–disorder transition. Theoretical models have been developed to describe this transition [21–23].

It may be mentioned that deuterium NMR and dielectric spectroscopy studies [24, 25] on columnar liquid crystals indicate the existence of rotational motions of the molecules or groups of molecules about the column axes.

Hexahexylthiotriphenylene (2, R = SC_6H_{13})

This compound (HHTT) shows a transition from a hexagonal ordered phase (D_{ho}) to a hexagonal disordered one (D_{hd}). As mentioned earlier, the former is a phase in which there is regularity in the stacking of the triphenylene cores in each column, and the latter one in which the column is liquid-like. X-ray studies using freely suspended strands reveal that in the ordered phase there is a helicoidal stacking of the cores within each column, the helical spacing being incommensurate with the intermolecular spacing [18]. In addition, a three-column superlattice develops as a result of the frustration caused by molecular interdigitation in a triangular symmetry (Figure 6). This helical phase of HHTT is also referred to as the H phase. Ideally, if there is no intercolumn interaction, true long-range order cannot exist within a column because of the Peierls–Landau instability. The existence of a regular periodicity in the stacking in each column therefore implies that neighbouring columns must be in register. Thus ordered columnar phases can probably be compared with the highly ordered smectic-like phases of rod-like molecules, e.g. the crystal B, E, G, H, etc. phases, which possess three-dimensional positional order. However, further high resolution studies are necessary before general conclusions can be drawn.

An unusual feature observed in the D_{hd} phase of this compound is worth mentioning. The hexagonal lattice exhibits a negative coefficient of thermal expansion (Figure 7). This has been attributed to the stiffening of the hydrocarbon tails, which therefore become longer and may provide the driving mechanism for the $D_{hd}-D_{ho}$ transition by enhanced intercolumn coupling.

2.2   Columnar Phases of 'Non-discotic' Molecules

Columnar phases are also formed when the flat core is replaced by a conical or pyramidal shaped one (**5** and **6**) [26, 27], and, rarely, even when the central core is absent altogether, as in certain macrocyclic molecules (**7**) [28]. In the latter case, the columns are in the form of tubes (Figure 8), and the mesophase has been called 'tubular'.

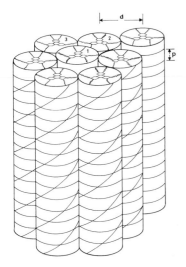

Figure 6. Structure of the D_{ho} phase of HHTT: the diagram represents a three-column superlattice (not to scale), with column 1 displaced by $p/2$ relative to columns 2 and 3, neighbouring columns having correlated helical phases. (From Fontes et al. [18], reproduced by permission of the Americal Physical Society).

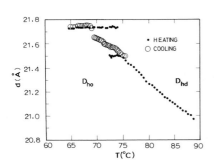

Figure 7. The intercolumn separation d in the columnar phases of HHTT plotted versus temperature, showing the negative coefficient of thermal expansion. (●) Heating mode; (○) cooling mode. (From Fontes et al. [18], reproduced by permission of the American Physical Society).

$$R = C_9 H_{19} COO-$$
$$R = C_{11} H_{23} COO-$$
$$R = C_{12} H_{25} O-\!\!\!\bigcirc\!\!\!-COO-$$

5

$$R = C_n H_{2n+1}-O$$
$$R = C_n H_{2n+1} \cdot CO \cdot O$$

6

7

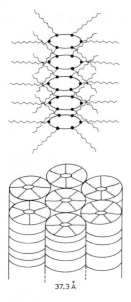

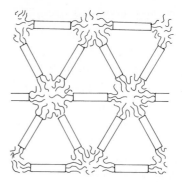

Figure 9. Structure of the hexagonal columnar meso-phase of **8**. The mesophase has been described as 'phasmidic'. (From Malthete et al. [6], reproduced by permission of Commission des Publication Françaises de Physique).

37.3 Å

Figure 8. (a) Stacking of macrocyclic molecules: for clarity of presentation the molecules are drawn with the same orientation about the tube axis. (b) The hex-agonal 'tubular' mesophase so formed. (From Lehn et al. [28], reproduced by permission of the Royal So-ciety of Chemistry).

Malthete and coworkers [6, 29] have pre-pared a series of mesogens shaped like stick insects called 'phasmids' (**8**). Some of them form columnar mesophases; the structure proposed for the hexagonal phase is shown in Figure 9.

Eaborn and Hartshorne [30] observed that the compound showed a mesophase which they were unable to classify as belonging to any of the then known liquid crystal types. Twenty-five years later, Bunning et al. [4] carried out miscibility studies of **9** with the hexa-n-heptanoate of benzene (Figure 10) and proved conclusively that the mesophase is a columnar liquid crystal. They have sug-gested that the molecule forms a dimer [4] or a polymer [5] that favours the occurrence of a columnar phase.

8

Historically, an interesting case is that of diisobutylsilane diol (**9**). In the 1950s,

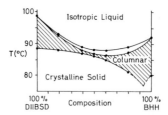

Figure 10. The miscibility diagram of diisobutylsi-lane diol (DIIBSD) with benzene hexa-n-heptanoate (BHH). (From Bunning et al. [4], reproduced by per-mission of Springer-Verlag).

9

2.3   The Nematic Phase

The nematic phase formed by disk-like molecules, designated by the symbol N_D, is a fluid phase showing schlieren textures similar to those of the classical nematic phase of rod-like molecules (Figure 11). Its structure is represented in Figure 12: it is an orientationally ordered arrangement of disks with no long-range translational order. However, unlike the usual nematic phase, N_D is optically negative, the director n now representing the preferred axis of orientation of the molecular short axis or the disk normal.

The N_D phase is exhibited by relatively few compounds: examples are hexa-n-alkyl- and hexa-n-alkoxybenzoates of triphenylene (**2**, $R = C_nH_{2n+1}-C_6H_4-CO\cdot O$, and $C_nH_{2n+1}O-C_6H_4-CO\cdot O$) [31, 32] and hexakis(4-octylphenylethynyl)benzene (**10**) [33].

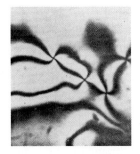

10

Figure 11. Schlieren texture of the nematic phase of hexa-n-hexyloxybenzoate of triphenylene. (From Destrade et al. [32], reproduced by permission of Heyden and Son).

Figure 12. Schematic representation of the nematic phase (N_D) of disk-like molecules.

In some compounds (usually the lower members of a homologous series), the transition to the N_D phase takes place directly from the crystal, while in others (the higher members of the series) it takes place via a columnar phase (Table 2) [31, 32]. The trend is somewhat analogous to the behaviour of the smectic A – nematic transition for a homologous series, and as will be discussed in Sec. 3 this trend can be explained by an extension of the McMillan model of the A-N transition to systems of discotic molecules. There is also X-ray evidence of skew cybotactic groups in the N_D phase prior to the transition to the D_t phase, very much like the situation seen near a nematic-smectic C transition [34].

The hexa-n-alkanoates of truxene show an unusual sequence of transitions [35, 36]. For the higher homologues the phase sequence on cooling is $I - D_h - D_r - N_D -$ Reentrant $D_h - Cr$, where I and Cr are the isotropic and crystalline phases, respectively. It has been suggested that the truxene molecules are probably associated in pairs, and that these pairs break up at higher temperatures, which might explain this extraordinary behaviour.

2.4   The Columnar Nematic Phase

The addition of electron acceptor molecules such as 2,4,7-trinitrofluorenone (TNF) to electron donating discotic molecules results in the formation of charge transfer complexes, which stabilize (or, in non-mesomorphic materials, induce) mesophases [37 – 39]. A

Table 2. Hexa-n-alkoxybenzoates of triphenylene: temperatures and heats of transition[a] [31, 32].

				Phase			
n	Cr$_3$	Cr$_2$	Cr$_1$	D$_2$	D$_t$	N$_D$	I
4			• 257			• >300	•
5			• 224 • 4.55			• 298	•
6	• 145 2.15	• 183 0.37	• 186 2.24	• 193 1.12		• 274	•
7			• 168 2.25			• 253	•
8			• 152 14.58		• 168	• 244	•
9			• 154 3.24		• 183	• 227	•
10			• 142 8.30		• 181	• 212	•
11			• 145 6.0		• 179 1.21	• 185 0.054	•
12			• 146 1.16		• 174		•

[a] The phases exhibited by a compound are indicated by points in the appropriate columns. The transition temperatures are given in degrees Celsius, and for those cases for which data are available the corresponding heats of transition are given immediately below in kilocalories per mole.

new type of induced mesophase identified in such systems is the columnar nematic (N$_c$) phase, the structure of which consists of columns (Figure 13). Transitions from

N$_c$

Figure 13. Schematic diagram of the structure of the columnar nematic phase induced by charge transfer interactions between an electron donor, pentakis(phenylethynyl)phenylalkyl ether (in black), and an electron acceptor, 2,4,7-trinitrofluorenone (hatched) [37 – 39].

N$_c$ to D$_{ho}$ have been observed in mixtures of TNF and a pentakis(phenylethynyl)phenylalkyl ether [37, 38].

2.5 The Chiral Nematic Phase

Optically active esters of benzene, triphenylene and truxene have been studied [40]. The triphenylene and truxene derivatives (structures **2** and **4**, respectively, with R = (+)- or (–)-CH$_3$-(CH$_2$)$_5$–CH(CH$_3$)–CH$_2$–CO · O) do show mesophases, but of the non-chiral columnar type. In other words, this chiral substituent does not change the nature of the phases. The benzene derivative is non-mesomorphic. However, when these compounds are mixed with the hexa-n-heptyloxybenzoate of triphenylene, which in the pure state shows the N_D phase, typical cholesteric textures are obtained. This was the first example of a twisted nematic or cholesteric phase of discotic mesogens. It is designated by the symbol N_D^* and its structure is illustrated schematically in Figure 14. However, unlike the classical nematic, a relatively high concentration ($\sim 50\%$) of the chiral additive is required to produce an appreciable twist, and, furthermore, the cholesteric texture can be seen only at higher temperatures after the initial appearance of a nematic texture, implying a very strong temperature dependence of the pitch.

A pure discotic cholesterogen, a triphenylene derivative (2) with $R = (S)\text{-}(+)\text{-}CH_3 - CH_2 - CH(CH_3)(CH_2)_3O - C_6H_4 - CO \cdot O$, has been reported [41]. The cholesteric phase occurs between 192.5 and 246.5 °C, with a pitch of about 30 μm at 200 °C. On the other hand, the optically active derivative of triphenylene with $R = (S)\text{-}(+)\text{-}CH_3 - (CH_2)_5 - CH(CH_3)(CH_2)_2O - C_6H_4 - CO \cdot O$ exhibits only a non-chiral D_{rd} phase, whereas that with $R = (S)\text{-}(+)\text{-}CH_3 - (CH_2)_3 - CH(CH_3)CH_2O - C_6H_4 - CO \cdot O$ is non-mesomorphic. Thus the occurrence of a cholesteric phase depends rather critically on the structure of the chiral chain.

2.6   The Lamellar Phase

Giroud-Godquin and Billard [42], who prepared and investigated the first discotic metallomesogen, *bis*(*p*-n-decylbenzoyl)methanato copper(II) (11, $R = C_{10}H_{21}$), suggested that its mesophase has a lamellar structure. Similar conclusions have been drawn by Ohta and coworkers [43, 44] from X-ray studies and by Ribeiro et al. [45] from NMR investigations. Sakashita et al. [44] have proposed a tilted smectic C type structure, as depicted in Figure 15.

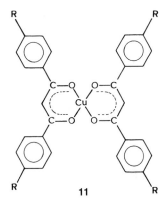

A variety of discotic metal complexes has been synthesized which exhibit columnar or lamellar phases [46]. X-ray determina-

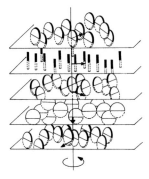

Figure 14. Schematic representation of the twisted nematic phase (N_D^*) of disk-like molecules.

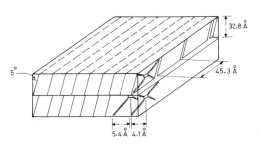

Figure 15. Schematic representation of the structure of the lamellar phase of the copper complex **11**, with $R = C_{12}H_{25}$, proposed by Sakashita et al. [44]. (Reproduced by permission of Gordon and Breach).

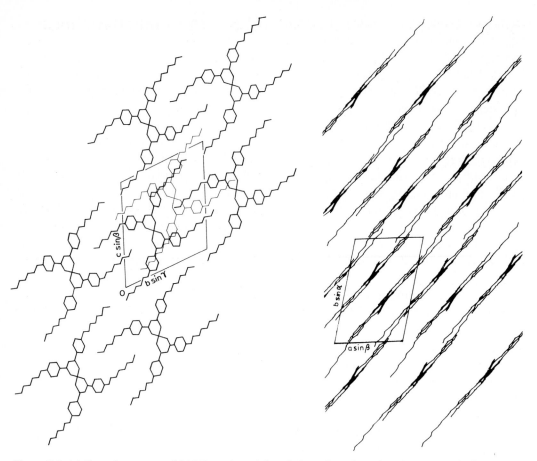

Figure 16. (a) Crystal structure of *bis*[(1-*p*-n-heptylphenyl, 3-*p*-n-heptyloxyphenyl)propane-1,3-dianato]copper(II): arrangement of the molecules perpendicular to the crystallographic *a* axis (From Usha and Vijayan [51], reproduced by permission of Taylor and Francis Ltd.). (b) Crystal structure of *bis*[(1-*p*-n-heptylphenyl, 3-*p*-n-heptyloxyphenyl)propane-1,3-dianato]copper(II): arrangement of the molecules perpendicular to the crystallographic *b* axis. (From Usha et al. [52], reproduced by permission of Taylor and Francis Ltd.).

tions of the crystal structures have been useful in elucidating the structures of the complexes and the mesophases formed by them [47–53]. Some of them have a layered arrangement in the crystalline phase (Figure 16), but it would be fair to say that the disposition of the molecules in the lamellar mesophase is not yet completely understood.

3   Extension of McMillan's Model of the Smectic A Phase to Columnar Liquid Crystals

As mentioned in Section 7.2.3, the observed trend of the columnar – nematic – isotropic sequence of transitions in some homologous series of discotic compounds is rather reminiscent of the behaviour of the smectic A – nematic – isotropic sequence in calamitic systems. This can be understood qualitatively by a simple extension of McMillan's mean field model of the smectic A phase [54], which takes account of the fact that the density wave is now periodic in two dimensions [55–58].

The hexagonal disordered columnar phase (D_{hd}) can be described by a superposition of three density waves with the wavevectors

$$A = \frac{4\pi}{\sqrt{3}\,d}\,\mathbf{j}$$

$$B = \frac{4\pi}{\sqrt{3}\,d}\left(\frac{\sqrt{3}}{2}\mathbf{i} - \frac{1}{2}\mathbf{j}\right)$$

$$C = A + B$$

where d is the lattice constant. The appropriate single particle potential, which depends on both the orientation of the short axis of the molecule and the position r of its centre of mass, may be written in the mean field approximation as

$$V_1(x, y, \cos\theta) = -V_0 P_2(\cos\theta)$$
$$\cdot \{\eta + \alpha\sigma[\cos(A \cdot r)$$
$$+ \cos(B \cdot r) + \cos(C \cdot r)]\} \quad (1)$$

retaining only the leading terms in the Fourier expansion. Here V_0 is the interaction energy which determines the nematic – isotropic transition, α is the McMillan parameter given by

$$2\exp\left[-\left(\frac{2\pi r_0}{\sqrt{3}\,d}\right)^2\right] \quad (2)$$

r_0 being the range of interaction which is of the order of the size of the aromatic core,

$$\eta = \langle P_2(\cos\theta)\rangle \quad (3)$$

is the orientational order parameter, P_2 being the Legendre polynomial of order 2, and

$$\sigma = \frac{1}{3}\langle[\cos(A \cdot r) + \cos(B \cdot r)$$
$$+ \cos(C \cdot r)] P_2(\cos\theta)\rangle \quad (4)$$

is an order parameter coupling the orientational and the translational ordering. The angle brackets represent the statistical average over the distribution function derived from the potential given by Eq. (1), the spatial integrations being carried out over a primitive cell of the hexagonal lattice. This form of the potential ensures that the energy of the molecule is a minimum when the disk is centered in the column with its plane normal to the z axis.

The free energy can then be calculated using standard arguments:

$$\frac{\Delta F}{N k_B T} = \frac{V_0}{2 k_B T}(\eta^2 + \alpha\sigma^2)$$

$$- \ln\left\{\frac{2}{\sqrt{3}\,d^2}\int dx\,dy\int_0^1 d(\cos\theta)\right.$$

$$\left.\times \exp[-V_1(x, y, \cos\theta)/k_B T]\right\} \quad (5)$$

Numerical calculations yield three possible solutions to the equation:

(1) $\eta = \sigma = 0$ (isotropic phase)
(2) $\eta \neq 0$, $\sigma = 0$ (nematic phase)
(3) $\eta \neq 0$, $\sigma \neq 0$ (hexagonal columnar phase)

For any given values of α and T, the stable phase is the one that minimizes the free energy given by Eq. (5). Interpreting α as a measure of the chain length, as in McMillan's model, the phase diagram as a function of α (Figure 17) reflects the observed trends for a homologous series of compounds. The temperature range of the nematic phase decreases with increasing α, and for $\alpha > 0.64$ the columnar phase transforms directly to

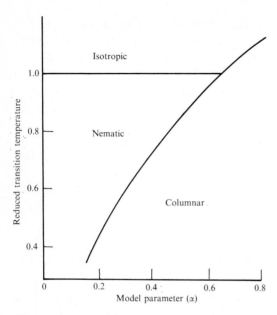

Figure 17. Theoretical plot of the reduced transition temperature against the model parameter α showing the hexagonal, nematic and isotropic phase boundaries. All the transitions are of first order [57, 58]. (Reproduced by permission of the Royal Society).

4 Pressure–Temperature Phase Diagrams

Pressure studies have been carried out on the hexa-alkanoyloxybenzenes ($\mathbf{1}$, $R=C_nH_{2n+1}$). Only three homologues ($n=6$, 7 and 8) are mesomorphic at atmospheric pressure, the last ($n=8$) being monotropic. On application of even small pressures, the $n=8$ homologue becomes enantiotropic [62]. For all three compounds, the temperature range of the mesophase decreases with increasing pressure, and finally disappears (Figure 18). The crystal–mesophase–isotropic triple points occur at 2.96 kbar and 116.5 °C, 1.43 kbar and 97.4 °C, and 1.3 kbar and 101 °C for $n=6$, 7 and 8, respectively. On the other hand, the $n=5$ homologue, which is non-mesomorphic at atmospheric pressure, shows pressure-induced mesomorphism [62]. The mesophase–isotropic tran-

the isotropic phase. It should be noted that, since the wavevectors of the hexagonal lattice satisfy the relation $A+B-C=0$, the columnar–nematic transition is always of first order; a Landau expansion of the free energy contains a non-vanishing cubic term in the order parameter σ.

Ghose et al. [59] have extended this theory using a variational principle to solve the problem with the full potential rather than the one truncated to the first Fourier component. This method, which is closely analogous to the extension of McMillan's model of the smectic A phase by Lee et al. [60], leads to some qualitative improvements in the phase diagram for a homologous series. More recently, Ghose et al. [61] have presented a simpler mean field version of their theory, which yields essentially similar results.

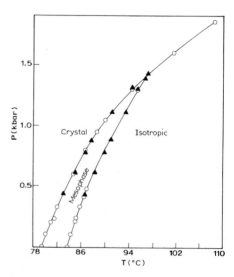

Figure 18. Pressure–temperature phase diagram of benzene hexa-n-octanoate. The circles and triangles represent two independent sets of measurements. (From Chandrasekhar et al. [62], reproduced by permission of the Commission des Publications Françaises de Physique).

(a)

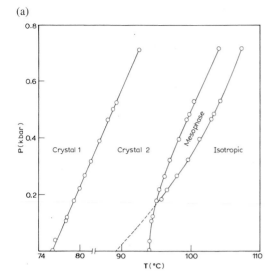

(b)

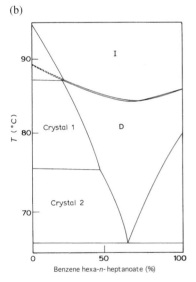

Benzene hexa-*n*-heptanoate (%)

Figure 19. (a) Pressure–temperature diagram of benzene hexa-n-hexanoate. The mesophase–isotropic transition line extrapolated to atmosphere pressure yields a virtual transition temperature of 89 °C. (From Chandrasekhar et al. [63], reproduced by permission of Academic Press). (b) Miscibility diagram of benzene hexa-n-hexanoate and the heptanoate. The virtual mesophase–isotropic transition temperature for the hexanoate is 89 °C, in agreement with the value obtained from the pressure–temperature diagram (a). From Billard and Sadashiva [64], reproduced by permission of the Indian Academy of Sciences).

sition line extrapolated to atmospheric pressure yields a virtual transition temperature of 89 °C (Figure 19 a) [63]. Remarkably, this value agrees very well with that obtained from miscibility studies carried out by Billard and Sadashiva [64] (Figure 19 b).

(a)

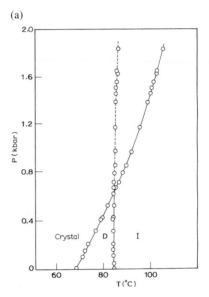

(b)

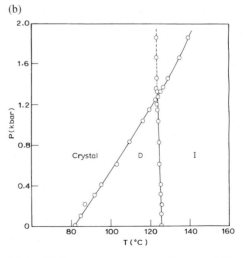

Figure 20. Pressure–temperature diagrams: (a) hexa-n-octyloxytriphenylene; (b) hexa-n-decanoyl-triphenylene. (- - - -) Monotropic transitions. (From Raja et al. [65], reproduced by permission of the Indian Academy of Sciences).

Pressure–temperature diagrams have been studied for two triphenylene compounds, hexa-n-octyltriphenylene (HOT; **2**, $R = C_8H_{17}O$) and hexa-n-decanoyloxytriphenylene (HDOT; **2**, $R = C_9H_{19}COO$), both of which exhibit the columnar phase D_{ho} at atmospheric pressure. Two interesting results, common to both compounds, have been obtained [65]. Firstly, in contrast to what is usually observed in other liquid crystals, the columnar–isotropic (D–I) transition, which is enantiotropic at atmospheric pressure, becomes monotropic at high pressures. The triple points occur at 0.64 kbar and 84.5 °C for HOT and 1.25 kbar and 123 °C for HDOT. Secondly, dT/dP is extremely small for the D–I transition, implying that, despite the drastic change in the molecular order at this transition (the heats of transition being 1.04 kcal mol^{-1} for HOT and 0.69 kcal mol^{-1} for HDOT [3, 66]), the associated volume changes are extremely small for both compounds (Figure 20).

5   Techniques of Preparing Aligned Samples

Progress in precise quantitative experimental studies on discotic materials has been hindered to some extent because of the difficulty in preparing aligned samples.

Vauchier et al. [67] investigated the effect of surface treatment on the anchoring of the director of the mesophases of certain triphenylene compounds. Glass surfaces coated with flat molecules (e.g., hexahydroxybenzene, mellitic acid, hexahydroxytriphenylene and rufigallol) orient the N_D phase, and in some cases the D phases as well, with the director perpendicular to the wall (i.e. with the disks parallel to the

surface). Similar results have been obtained by slow cooling from the isotropic phase to the columnar phase of samples contained between aluminium-coated glass or quartz plates [68, 69]. Freshly cleaved surfaces of crystals such as apophyllite and muscovite also have the same effect. Orientation of the N_D phase with the director parallel to the surface has been achieved by the standard technique of oblique deposition of SiO. Twisted N_D cells could be prepared in this manner [67].

As most D and N_D phases are diamagnetically negative, the application of a magnetic field does not result in an oriented sample as it does in the case of a calamitic system. The sample consists of domains the directors of which are distributed in a plane perpendicular to the field direction. Goldfarb et al. [13] demonstrated that a single domain can be obtained by allowing the mesogen to cool slowly from the isotropic phase in a magnetic field while spinning the sample about an axis perpendicular to the field direction. The single domain is formed with its director parallel to the spinning axis.

Another technique, which has proved to be extremely useful for high resolution X-ray studies of columnar liquid crystals (see Sec. 2.1.2), is the preparation of highly oriented freely suspended strands [15–19, 70]. Figure 21 (a) shows a drawing of the temperature-controlled oven designed by Fontes et al. [19] for growing and annealing strands of the liquid crystal material in an inert atmosphere. The material is placed in a cup and a pin inserted into it. When the pin is slowly pulled out, a fibre or strand of about 150–200 μm diameter is drawn (Figure 21 b), with the column axes parallel to the axis of the strand (Figure 21 c). The oven has cylindrical beryllium walls, which permit 360° X-ray access.

Shearing of the glass plates has been found to produce aligned films of the co-

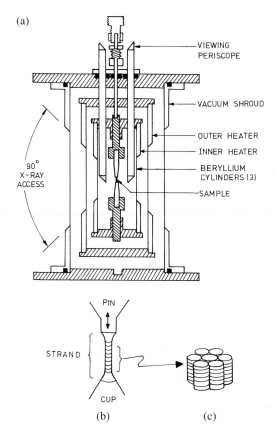

Figure 21. (a) The temperature-controlled oven designed to grow and anneal strands of the liquid crystal in an inert atmosphere. When the pin is slowly pulled out, a thin strand of the material is drawn (b), with the column axes parallel to the axis of the strand (c). (From Fontes et al. [19], reproduced by permission of the American Physical Society.).

lumnar phase [71–73]. The columns are oriented parallel to the glass surfaces and to the direction of shear.

6   Ferroelectricity in the Columnar Phase

Columnar phases can be ferroelectric if the discotic molecules are chiral and tilted with respect to the column axis [74]. From sym-

metry arguments (see Sec. 2 of Chap. VI) it follows that each column is spontaneously polarized perpendicular to its axis. Macroscopic polarization is observed when the polarizations of the different columns do not annul each other.

It is only recently that a ferroelectric columnar liquid crystal was realized experimentally. Bock and Helfrich [72, 73] demonstrated ferroelectric switching in the columnar phases of two compounds, namely 1, 2, 5, 6, 8, 9, 12, 13-octakis ((*S*)-2-heptyl-oxypropanoyloxy)dibenzo[e,1]pyrene (**12**) and a lower homologue with one less methylene unit in the hydrocarbon chain. Sheared samples, with the columns parallel to the glass surfaces and to the shear direction, exhibit the full optical tilt angle (the optical tilt angle being defined as the angle which the extinction cross makes with the polarizer – analyser directions (see Figure 2). This implies that the disk normals are inclined to the column axis, as discussed in Sec. 2.1 [11]. The polarization is then normal to the cell walls. On applying an alternating voltage, the optical tilt angle reverses with the field, going through an optically untilted state.

12

Compound **12** exhibits a field-induced transition from one ferroelectric phase to another at about $10 \ \mathrm{V} \ \mu m^{-1}$. The high field phase relaxes slowly to the low field phase when the field is switched off. The polarizations of the low and high field phases are ~ 60 and ~ 180 nC cm^{-2}, and the corresponding optical tilt angles are $\pm 24.5°$ and $\pm 37°$, respectively, and practically independent of temperature. The mechanism of reversal of the polarization is still not quite clear. It could be caused by (a) a rotation of the column as a whole, (b) the molecular tilt direction reversing through an untilted state (without rotation about the column axis), (c) the molecular tilt direction rotating around the column axis, or (d) a combination of these effects. The switching times ranged from 0.1 ms to 100 s in these experiments, depending strongly on temperature and field strength.

7   The Columnar Structure as a One-Dimensional Antiferromagnet

The copper complex 11 with $R = C_8H_{17}$ contains a paramagnetic metal ion at its centre. Single crystal X-ray analysis has established that in this homologue the disks are stacked in columns in the crystalline state [50], indicating the possibility of formation of a one-dimensional exchange coupled system. Eastman et al. [75] have carried out detailed electron spin resonance studies on this copper complex in the crystalline and liquid-crystalline phases. Using the method discussed by Bartkowski and Morosin [76] for one-dimensional systems, Eastman et al. have analysed the angular dependence and line shapes, and drawn the important conclusion that the spectra of single crystals have the features of a spin-$\frac{1}{2}$ one-dimensional Heisenberg antiferromagnet, and, furthermore, that the exchange interactions are quite significant (i.e. an appreciable degree

of antiferromagnetic long-range order persists) even in the columnar mesophase.

8   Electrical Conductivity in Columnar Phases

The columnar phases of metallomesogens have the electrical properties of molecular semiconductors [77–84]. Figure 22 presents the Arrhenius plots of the a.c. conductivity (σ) of unaligned samples of copper phthalocyanines (13) determined by Van der Pol et al. [83]. The conductivity σ is of the order of 5×10^{-8} Sm^{-1} at 175 °C, and increases with increasing temperature. The activation energy is approximately $0.5–0.6$ eV at 175 °C for the compounds 13 with $R = OC_8H_{17}$ and $OC_{12}H_{25}$.

13

Like the triphenylene compounds, the columnar phases of pure polynuclear aromatic mesogens are insulators, but they can be made to conduct by doping [84–86]. The d.c. conductivities of unaligned samples of HHTT in the pure form and after saturation doping with iodine (by exposure to iodine vapour) have been studied by Vaughan et al. [85]. The conductivities of pure HHTT and HHTT/I$_3$ are plotted in Figure 23, and as can be seen σ increases by $4–5$ orders of magnitude as a result of doping. Figure 24 gives a schematic representation of the structure

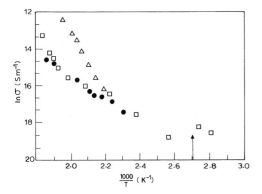

Figure 24. Proposed structure of the iodine-doped co-lumnar (D_{hd}) phase of HHTT. The molecular cores are assembled in columns with short-range intracolumnar order and long-range hexagonal intercolumnar order. The black spheres in sets of three represent I_3^- ions, which are assumed to occupy the space between the columns in a disordered fashion. For the sake of clarity, the liquid-like tails filling the space between the columns are not shown in the diagram. (From Vaughan et al. [85], reproduced by permission of the American Physical Society).

Figure 22. Arrhenius plots of the electrical conductivity (σ) of unaligned samples of copper phthalocyanines (**13**). Data points: ($\triangle$) R = H; ($\bullet$) R = OCH$_8$H$_{17}$; ($\square$) R = OC$_{12}$H$_{25}$. A slight increase in the conductivity is observed for the last compound at the transition from the mesophase to the crystalline phase (indicated by the arrow). (From Van der Pol et al. [83], reproduced by permission of Taylor and Francis Ltd.).

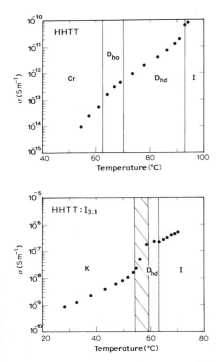

Figure 23. Temperature dependence of the d.c. conductivity (σ) of unaligned samples of (a) pure HHTT and (b) iodine doped HHTT. The shaded area in (b) indicates the Cr–D_{hd} coexistence region. (From Vaughan et al. [85], reproduced by permission of the American Physical Society).

of the iodine-doped columnar D_{hd} phase, as envisaged by Vaughan et al. [85].

Figure 25 presents the a.c. conductivity measurements carried out by Boden et al. [86] on aligned samples of hexa-n-hexyloxy-triphenylene (HAT6) (**2**, with R = C$_6$H$_{13}$O) doped with a small quantity (mole fraction $x = 0.005$) of AlCl$_3$. The conductivity $\sigma_\parallel$ measured parallel to the column axis is about 10^3 times greater than $\sigma_\perp$, the conductivity measured in the perpendicular direction. This is a striking demonstration of the quasi-one-dimensional nature of the electrical properties of these materials. Calculations indicate that the hopping probability of the electron or hole is several orders of magnitude greater along the column axis than transverse to it, as is of course to be expected [82]. Boden et al. [86] have also investigated the dependence of the conductivity as a function of frequency and interpreted the results in terms of the Scher–Lax model [87] of hopping transport. The conduction along the columns is identified with

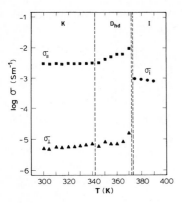

Figure 25. Low frequency (100 Hz) conductivities of aligned samples of hexa-n-hexyloxytriphenylene doped with a small quantity (mol fraction $x = 0.005$) of AlCl$_3$ in the columnar and isotropic phases plotted versus temperature. The conductivity measured parallel to the column axis ($\sigma_\parallel$) is about 10^3 times greater than that measured in the perpendicular direction ($\sigma_\perp$). (From Boden et al. [86], reproduced by permission of Chapman & Hall).

a single transport process in which the carries hop between localized states (radical cations) associated with counterions (AlCl$_4^-$) distributed alongside the columns.

The subject is, needless to say, of considerable fundamental and practical importance, so much so that rapid advances may be expected to take place in the near future.

9   Photoconduction in Columnar Phases

Photoconductivity of the discotic columnar phase is of much interest because of its potential applications, and several studies have been reported in recent years [68, 69, 88–90]. Probably the most significant of these is the discovery by Adam et al. [69] of fast photoconduction in the highly ordered columnar phase of a discotic mesogen. The experiments were performed on

HHTT, which exhibits the following sequence of transitions: Cr 62 D$_{ho}$ 70 D$_{hd}$ 93 I, where Cr is the crystalline phase, D$_{ho}$ is the highly ordered discotic phase, the structure of which is depicted in Figure 6, and I is the isotropic phase.

The photocurrent transients were studied using a time-of-flight method. HHTT was sandwiched between two glass plates coated with semi-transparent aluminium electrodes. The cell thickness was about 30 μm. The interaction of the molecules with the flat electrodes resulted in homeotropic orientation with the column axes normal (or the disks parallel) to the glass surfaces. A pulsed (10 ns) N$_2$ laser of wavelength 337 nm, which coincides with the main absorption band of HHTT, generates electron–hole pairs in the electrically biased cell. Depending on the polarity of the external field, the electrons or holes drift across the cell, causing displacement currents that can be recorded in an external circuit. Figure 26 gives the measured photocurrents for holes as a function of time t for the different phases of HHTT. The electric field applied was 2.0×10^4 V cm^{-1}. Here t_T represents the transit time. The form of the photocurrent in the D$_{hd}$ and D$_{ho}$ phases can be explained by assuming Gaussian transport. A fairly compact charge carrier 'packet' drifts across the sample at a constant velocity, broadened only by the usual thermal diffusion, leading to a broadening of the photocurrent decay when the packet reaches the counter electrode. On the other hand, in the polycrystalline Cr phase, there is dispersive behaviour; the charge carrier packet is rapidly smeared out in time and space because of trapping events, and this results in a featureless decay.

The hole mobilities are plotted versus temperature for the different phases in Figure 27. We will discuss the cooling mode first. In the isotropic phase the carrier mo-

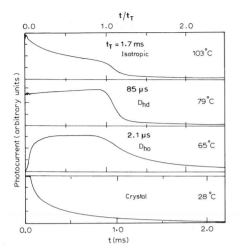

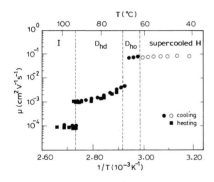

Figure 26. Photocurrent as a function of time in the different phases of HHTT. A change of slope in the profile marks the transit time (t_T) of the charge-carrier packet. The value of t_T could be determined in the I, D_{hd} and D_{ho} phases, but not in the Cr phase which gives a featureless decay profile. (Adapted from Adam et al. [69], reproduced by kind permission of MacMillan Magazines Ltd.).

Figure 27. Arrhenius plot of the hole mobility (μ) versus the reciprocal of the temperature in the different phases of HHTT. Electric field $= 2 \times 10^4$ V cm^{-1}. (From Adam et al. [69], reproduced by permission of MacMillan Magazines Ltd.).

bility is of the order of 10^{-4} cm^2 V^{-1} s^{-1}, a value which is indicative of short-range orientational order. At the transition to the liquid crystalline D_{hd} phase, μ jumps by an order of magnitude, and increases with decreasing temperature in the mesophase, suggesting a slight increase in the intracolumnar correlation.

At the transition to the highly ordered D_{ho} phase, μ increases by almost two orders of magnitude and reaches values of the order of 0.1 cm^2 V^{-1} s^{-1}. With the exception of organic single crystals, these are the highest values known to date of electronic mobility for photoinduced charge carriers in organic systems. The mobilities are independent of temperature and the externally applied field.

The high mobility and the rapid transit time in the D_{ho} phase indicate an efficient charge transport mechanism with no major perturbation by disorder or defects, and good π-orbital overlap along the columns. In the crystalline phase the formation of grain boundaries and other defects inhibits efficient charge transport, and the behaviour is dispersive.

In the heating mode, the defects in the crystalline phase persist in the D_{ho} phase, so that the photocurrents remain dispersive. However, in the D_{hd} phase, which consists of liquid-like columns, the mobilities measured in the heating and cooling modes remain the same.

As Adam et al. [69] point out, these studies will no doubt have an impact on applications such as electrophotography, and on the development of new devices such as organic transistors and active layers in light emitting diodes.

10 Continuum Theory of Columnar Liquid Crystals

10.1 The Basic Equations

The hydrodynamic equations of the hexagonal columnar phase composed of liquid-like columns are [91]:

$$-\frac{\partial \theta}{\partial t} + \partial_i V_i = 0 \qquad (\theta = -\partial \rho / \rho)$$

$$\dot{Q} = \frac{\partial \varepsilon}{\partial t} + \frac{\partial \theta}{\partial t}(\varepsilon + P)$$

$$= K_\perp \Delta_\perp T + K_\parallel \partial_{zz}^2 T + \xi(\partial_{xi}^2 \phi_i^x + \partial_{yi}^2 \phi_i^y)$$

$$\frac{\partial u_x}{\partial t} - V_x = \frac{\xi}{T} \partial_x T + \zeta \partial_i \phi_i^x$$

$$\frac{\partial u_y}{\partial t} - V_y = \frac{\xi}{T} \partial_y T + \zeta \partial_i \phi_i^y$$

$$\rho \frac{dV_x}{dt} = -\partial_x P + \partial_i \phi_i^x + \eta_{xjkl} \partial_l \partial_j V_k$$

$$\rho \frac{dV_y}{dt} = -\partial_y P + \partial_i \phi_i^y + \eta_{yjkl} \partial_l \partial_j V_k$$

$$\rho \frac{dV_z}{dt} = -\partial_z P + \eta_{zjkl} \partial_l \partial_j V_k$$

$$\phi_i^\alpha = \frac{\partial \varepsilon}{\partial(\partial_i u_\alpha)}\bigg|_{\rho,s,q_j} \qquad (\alpha = x, y)$$

where:

$$\partial_x = \frac{\partial}{\partial x} \qquad \partial_l \partial_j = \frac{\partial^2}{\partial x_l x_j}$$

$$\partial_{zz}^2 = \frac{\partial^2}{\partial z^2} \qquad \Delta_\perp = \frac{\partial^2}{\partial x^2} + \frac{\partial^2}{\partial y^2}$$

$$\partial_{xi}^2 = \frac{\partial^2}{\partial x \partial x_i} \qquad (x_i = x, y, z)$$

Q is the heat density; θ is the volume strain; ρ is the density; V_i are components of the velocity; $K_\parallel$ and $K_\perp$ are the thermal conductivities parallel and perpendicular to the column axis, respectively; u_i are components of the column displacement; P is the pressure; ε is the energy density expressible as $F + \frac{1}{2}A\theta^2 + C\theta(u_{x,x} + u_{y,y})$, where F is given by equation (6) (see Sec. 7.10.3); A is the bulk modulus; C is a coupling constant; ζ is the permeation coefficient; ξ is the thermomechanical coefficient; and η_{ijkl} are viscosity coefficients.

10.2   Acoustic Wave Propagation

In the absence of a temperature gradient, and assuming that permeation is negligible, i.e. $\dot{u}_i = V_i$, and that damping is weak, it turns out that under adiabatic conditions there are three propagating acoustic modes for any arbitrary direction not coinciding with an axis of symmetry. Of these, one is the familiar longitudinal wave arising from density fluctuations, the velocity (V_1) of which is practically independent of the direction of propagation. Another is a transverse wave, which is exactly analogous to the 'second sound' of the smectic A phase [92, 93]. The velocities V_1 and V_2 of these two waves are given by the roots of the equation

$$V_{1,2}^4 - V_{1,2}^2 \left[\frac{A}{\rho} + \sin^2 \psi \left(\frac{D + B + 2C}{\rho} \right) \right]$$

$$+ \sin^2 \psi \cos^2 \psi [(B + D)A - C^2] = 0$$

where ψ is the angle between the direction of propagation and the column axis, and B and D are defined in Eq. (6) (see Sec. 10.3). We have ignored the coupling between the lattice and the curvature deformation of the columns. The third mode is a transverse wave the polarization of which is orthogonal to that of the second sound. This wave propagates because the two-dimensional lattice can sustain a shear. Its velocity for any arbitrary direction is given by

$$V_3 = \left(\frac{D}{\rho} \right)^{1/2} \sin \psi$$

The velocities of the three waves and their angular dependence can be studied by Brillouin scattering, as has been demonstrated for the smectic A phase [94]. In the present case, there should be one, two or three pairs of Brillouin components, depending on the orientation, but no experiments have yet been reported.

10.3   Fluctuations

In his treatment of the general problem of the stability of 'low-dimensional' systems, Landau also considered the case of a three-dimensional body in which the density is a periodic function of two dimensions only. In the light of his analysis, he concluded: 'Thus, such a structure could in theory exist, but it is not known whether they do in fact exist in Nature' [20]. The structure envisaged by Landau is, of course, an accurate description of the columnar phase of discotic molecules that was proposed originally [1], i.e., liquid-like columns forming a two-dimensional lattice.

We shall now discuss the theory of thermal fluctuations in such a columnar structure. Let us suppose that the liquid-like columns are oriented along the z axis and that the two-dimensional lattice (assumed to be hexagonal) is parallel to the xy plane. The two basic deformations in such a structure are: (1) the curvature deformation (or bending) of the columns without distortion of the lattice, and (2) the lattice dilatation (or compression) without bending of the columns. There can also be coupling between the two types of distortion, but the coupling term merely rescales the bend elastic constant of the columns [95]. Considering only the vibrations of the lattice in its own plane, the free energy may be written as [91, 96]

$$F = \frac{B}{2}\left(\frac{\partial u_x}{\partial x} + \frac{\partial u_y}{\partial y}\right)^2$$

$$+ \frac{D}{2}\left[\left(\frac{\partial u_x}{\partial x} - \frac{\partial u_y}{\partial y}\right)^2 + \left(\frac{\partial u_x}{\partial y} + \frac{\partial u_y}{\partial x}\right)^2\right]$$

$$+ \frac{k_{33}}{2}\left[\left(\frac{\partial^2 u_x}{\partial z^2}\right)^2 + \left(\frac{\partial^2 u_y}{\partial z^2}\right)^2\right] \tag{6}$$

where B and D are the elastic constants for the deformation of the two-dimensional

lattice in its own plane, u_x and u_y are the displacements along x and y at any lattice point, and k_{33} is the Frank constant for the bending of the columns. In the notation of the standard crystal elasticity theory, $B = \frac{1}{2}(c_{11}+c_{12})$ and $D = \frac{1}{2}(c_{11}-c_{12})$.

We neglect here the splay and twist deformations because they give rise to a distortion of the lattice and involve considerable energy. We also neglect any contributions from the surface of the sample. Writing the displacement u in terms of its Fourier components

$$u(r) = \sum_q u(q)\exp(iq\cdot r)$$

and substituting in Eq. (6) we get, in the harmonic approximation

$$F = \frac{1}{2}\sum_q (B_0 q_\perp^2 + k_0 q_z^4)\langle u_q^2\rangle$$

and from the equipartition theorem

$$\langle u_q^2\rangle = \frac{k_B T}{B_0 q_\perp^2 + k_0 q_z^4}$$

where

$$B_0 = B + 2D, \quad k_0 = 2k_{33} \quad \text{and} \quad q_\perp = (q_x^2 + q_y^2)^{1/2}$$

The mean square displacement at any lattice point is given by

$$\langle u^2\rangle = \sum_q \langle u_q^2\rangle = \frac{1}{(2\pi)^3}\int \langle u_q^2\rangle\,dq$$

$$= \frac{k_B T}{(2\pi)^3}\int_{2\pi/L}^{2\pi/d}\int_{2\pi/L'}^{\infty}\frac{2\pi q_\perp\,dq_\perp\,dq_z}{B_0 q_\perp^2 + k_0 q_z^4}$$

where L' is the length of the columns, L is the linear dimension of the lattice in the xy plane and d is its periodicity. Assuming that $L' \gg L$,

$$\langle u^2\rangle = \left[\frac{k_B T}{4 B_0\,(\lambda d)^{1/2}}\right]\left[1 - \left(\frac{d}{L}\right)^{1/2}\right]$$

where $\lambda = (k_0/B_0)^{1/2}$ is a characteristic length [57, 96, 97]. The structure is therefore stable as $L \to \infty$. As is well known from the classical work of Peierls [98] and Landau [99], the two-dimensional lattice itself is an unstable system with $\langle u^2 \rangle$ diverging as $\ln L$ [100, 101]. Thus the curvature elasticity of the liquid-like columns stabilizes the two-dimensional order in the columnar liquid crystal.

In the two-dimensional lattice and the smectic A phase, the displacement–displacement correlation is of logarithmic form, and therefore the Bragg reflections (or the δ-function singularities in the X-ray structure factor) are 'washed out' for large samples. Instead, there results a strong thermal diffuse scattering with a power-law singularity [100–103]. This has been verified quantitatively for the smectic A phase by the fine X-ray work by Als-Nielsen et al. [104]. On the other hand, the columnar liquid crystal gives the usual Bragg reflection, as borne out by the excellent high resolution X-ray studies by Fontes et al. [19] using oriented liquid crystal strands.

The mean square amplitude of the orientational fluctuations of the director and the corresponding intensity of light scattering can be worked out in a similar way [55]. For example, when the incident and scattered beams are both polarized perpendicular to the column axis, the scattered intensity is

$$I(q) \sim T\omega^4 \left[\frac{q_\perp^2}{(B+D)q_\perp^2 + k_{33}\,q_z^4} \right]$$

where $q_\perp = (q_x^2 + q_y^2)^{1/2}$, and ω is the angular frequency of the light. In certain geometries, the scattering is similar to that from the smectic A phase.

10.4 Mechanical Instabilities

When a smectic A film is taken between glass plates with the layers parallel to the

plates, a mechanical dilatation normal to the film gives rise to an undulation instability, which has been studied in detail [105]. An equivalent type of instability may be expected to occur in the columnar liquid crystal [95]. The columns are parallel to the glass plates, and the separation d between the plates is increased (Figure 28). If $u_x = \alpha x$ (α being positive), then at a critical value α_c given by

$$\alpha_c = \frac{2\pi}{d} \left(\frac{k_{33}}{B+D} \right)^{1/2}$$

a periodic distortion should set in with a wavevector q_z given by

$$q_z^2 = \frac{\pi}{d} \left(\frac{B+D}{k_{33}} \right)^{1/2}$$

Another type of instability may be described as a buckling instability. A compressive stress is applied parallel to the columns. At a critical value of the stress given by

$$\sigma_c = \frac{\pi^2 \, k_{33}}{d^2}$$

the columns buckle. This has actually been observed [106], but there is a significant discrepancy between theory and experiment which has not been explained satisfactorily.

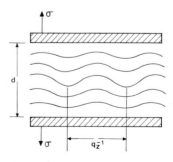

Figure 28. Undulation instability in a columnar liquid crystal subjected to a mechanical dilatation normal to the columns.

11   Defects in the Columnar Phase

The symmetry elements in the hexagonal
structure composed of liquid-like columns
are shown in Figure 29. The director n is
parallel to the column axis; a (or, equivalently, a' and a'') is the lattice vector of the
two-dimensional hexagonal lattice; L_2, T_2
and θ_2 are two-fold axes of symmetry; L_3 is
a three-fold axis and L_6 a six-fold axis. Any
point on L_2 or L_6 is a centre of symmetry.
Any plane normal to n and the planes L_6, T_2
and L_6, θ_2 are planes of symmetry.

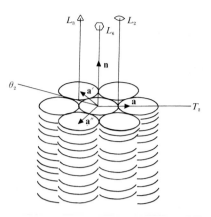

Figure 29. The hexagonal columnar phase with the
symmetry elements in the structure.

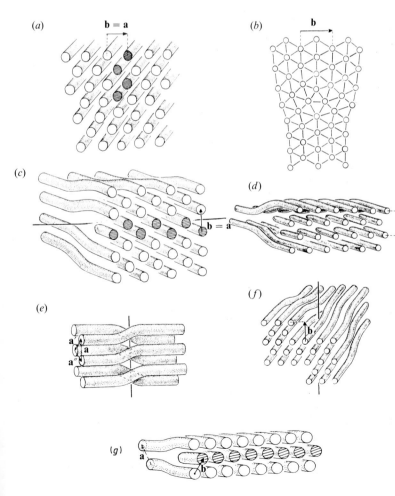

Figure 30. Dislocations in
the columnar phase: (a, b)
longitudinal edge dislocations; (c, d) transverse edge
dislocation; (e, f) screw dislocations; (g) a hybrid of
screw and edge dislocations.
It should be noted that the
Burgers vector $b=a$ for (a),
(c), (e) and (g), and $b=a-a'$
for (b), (d) and (f). (From
Bouligand [107], reproduced
by permission of the Com-
mission des Publications
Françaises de Physique).

The nature of the defects in such a structure has been investigated in detail by Bouligand [107] and by Kléman and Oswald [95, 108, 109].

11.1 Dislocations

The Burgers vector b of a dislocation in the columnar structure is normal to the columnar axis n. In general,

$$b = la + ma'$$

where l and m are positive or negative integers. For edge dislocations, b is perpendicular to the dislocation line L. Two types of edge dislocation are possible: longitudinal edge dislocations (L parallel to n, Figure 30 a and b), and transverse edge dislocations (L perpendicular to n, Figure 30 c and d). For screw dislocations b is parallel to L (Figure

30 e and f). A hybrid composed of screw and edge dislocations is shown in Figure 30 g.

11.2 Disclinations

Longitudinal wedge disclinations are the standard crystal disclinations in a hexagonal lattice. The rotation vector is L_6 or L_3 or L_2, parallel to the columns. Two examples of such defects are shown in Figure 31 (a and b). The lattice becomes compressed near a positive disclination and stretched near a negative one. In general, most of these disclinations have prohibitively large energies, and hence they occur as unlike pairs at the core of a dislocation.

In transverse wedge disclinations the rotation vector is θ_2 or T_2, normal to n. Examples of $\pm\pi$ disclinations are shown in Figure 31 (c, d and e).

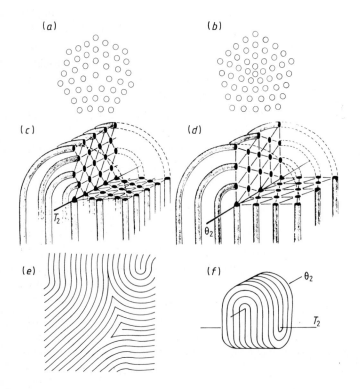

Figure 31. Disclinations in the columnar phase: (a, b) $-\pi/3$ and $\pi/3$ longitudinal wedge disclinations about the six-fold axis L_6; (c, d) π transverse wedge disclinations about the binary axes T_2 and θ_2, respectively; (e) $-\pi$ transverse wedge disclination leading to the formation of walls; (f) two π disclinations at right angles to each other, one about T_2 and the other about θ_2. (From Bouligard [107], reproduced by permission of the Commission des Publications Française de Physique).

Two transverse wedge disclinations may occur in association, as shown in Figure 31 (f). The angle between the rotation axes may be 90°, 60° or 30°. Such defects have been observed experimentally [109].

The symmetry of the columnar phase also permits the occurrence of twist disclinations in the hexagonal lattice and of hybrids consisting of a twist disclination in the hexagonal lattice and a wedge disclination in the director field. According to Bouligand, these defects are not likely to exist.

The energetics of defects in the columnar structure have been discussed in detail by Chandrasekhar and Ranganath [110].

12   The Properties of Discotic Nematic Phases

The symmetry of the discotic nematic phase N_D (Figure 12) is the same as that of the classical nematic phase of rod-like molecules. Thus identical types of defects and textures are seen in both cases.

The N_D phase is optically, and, as a rule, diamagnetically, negative. However, its dielectric anisotropy $\Delta\varepsilon\,(=\varepsilon_\parallel-\varepsilon_\perp$, where $\varepsilon_\parallel$ and $\varepsilon_\perp$ are the dielectric constants parallel and perpendicular to the director, respectively) may be positive or negative, depending on the detailed molecular structure. For example, the hexa-n-heptyloxybenzoate of triphenylene and hexa-n-dodecanoyloxy-truxene are both dielectrically positive in the nematic phase [111, 112]. This is because, while the contribution of the electronic polarizability of these disk-shaped molecules to $\Delta\varepsilon$ is negative, the permanent dipoles associated with the six ester linkage groups make a stronger positive contribution, so that $\Delta\varepsilon>0$ for the two compounds. On the other hand, hexakis(4-octylphenyl-

ethynyl)-benzene (**10**), which does not have any permanent dipoles, does in fact show a negative $\Delta\varepsilon$, as expected [13].

Very few quantitative measurements of the physical properties have been reported. The Frank constants for splay and bend have been determined using the Fréedericksz method [111–114]. Interestingly, the values are of the same order as for nematic phases of rod-like molecules. The twist constant k_{22} has not yet been measured. The fact that the diamagnetic anisotropy is negative makes it somewhat more difficult to measure these constants by the Fréedericksz technique. However, by a suitable combination of electric and magnetic fields, it is possible, in principle, to determine all three constants [112].

The hydrodynamic equations of the classical nematic phase are applicable to the N_D phase as well. There are six viscosity coefficients (or Leslie coefficients), which reduce to five if one assumes Onsager's reciprocal relations. A direct estimate of an effective average value of the viscosity of the N_D phase from a director relaxation measurement [111, 114] indicates that its magnitude is much higher than the corresponding value for the usual nematic phase.

In ordinary nematic phases of rod-like molecules, the Leslie coefficients μ_2 and μ_3 are both negative and $|\mu_2|>|\mu_3|$. Under planar shear flow, the director assumes an equilibrium orientation θ_0 given by

$$\tan^2\theta_0 = \mu_3/\mu_2$$

where θ_0 lies between 0° and 45° with respect to the flow direction (Figure 32 a). In practice, θ_0 is usually a small angle. In certain nematics, it is found that $\mu_3>0$ at temperatures close to the nematic–smectic A transient point. Under these circumstances there is no equilibrium value of θ_0, and in the absence of an orienting effect due to the walls or a strong external field the director

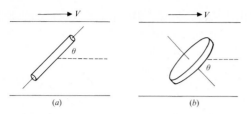

Figure 32. Flow alignment of the director in nematic liquid crystals. For rod-shaped molecules the alignment angle θ with respect to the flow direction lies between 0° and 45° (a), while for disk-shaped molecules it lies between −90° and 45° (or, equivalently, between 90° and 135°) (b).

13 Discotic Polymer Liquid Crystals

We end this chapter with a very brief account of a new class of liquid crystal polymers, i.e. discotic polymers [118–126]. The basic monomer units are discotic mesogenic moieties, which are components of the polymer main chain itself or are attached to the polymer backbone as side groups. A few examples are shown as structures **14–16**. Besides the columnar phase, some new types of mesophase have been identified. A

starts to tumble and the flow becomes unstable. Consequently, such materials have two distinct flow regimes, which have been investigated in detail both theoretically and experimentally [115].

The suggestion has been made that in the N_D phase, the disk-like shape of the molecule may have a significant effect on μ_2 and μ_3 [116, 117]. The stable orientation of the director under planar shear will now be as shown in Figure 32 (b). Thus it can be argued that both μ_2 and μ_3 should be positive, and the flow alignment angle θ_0 should

14   Annealed: glass 19 D 39 I [118]

$R = C_5H_{11}$

15 Glass 60 D 150 I [120]

lie between −45° and −90°. It then follows that when $\mu_2 < 0$ the director tumbles and the flow becomes unstable. As yet, however, no experimental studies have been carried out to verify any of these ideas.

$R_1: -\overset{O}{\overset{\|}{C}}-O-C_{12}H_{25}, R_2: -O-C_{12}H_{25}$

16 D_1 67 D_2 130 I [121]

polyester with triphenylene as the repeating unit in the main chain separated by flexible spacers [120], or with triphenylene units attached as side group to the polymer chain via flexible spacers [126], forms a hexagonal columnar mesophase (Figure 33a). On the other hand, rigid aromatic polyamides and polyesters with disk-shaped units in the main chain appear to form a nematic-like, or sanidic, structure [122], with boards stacked parallel to one another (Figure 33b). The addition of electron acceptor molecules such as 2,4,7-trinitrofluorenone, to discotic polymers results in the formation of charge transfer complexes that induce the columnar nematic phase (Figure 33c) [124, 125] similar to what is observed for low molar mass system (see Sec. 2.4).

X-ray diffraction studies of the macroscopic alignment of the columnar structures produced by stretching a film or by drawing fibres (about 1 m long and several micro-

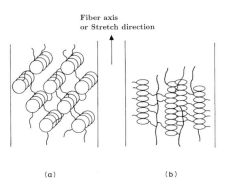

Figure 34. Orientations of the molecules in stretched samples of discotic polymers: (a) main chain discotic polymer; (b) side group discotic polymer. (From Huser et al., [126], reproduced by permission of the American Chemical Society).

metres thick) from the polymer melt have been reported [120, 126]. Huser et al. [126] have demonstrated that almost perfect alignment can be achieved by proper mechanical and thermal treatment. Interestingly, in the main chain polymer, the columns are oriented perpendicular to the stretch direction (or the chain direction), whereas in the side group polymer they are oriented parallel to it (Figure 34). A study of the mechanical and other physical properties of these oriented polymers is of much interest.

Acknowledgements

I am greatly indebted to Dr Geetha Nair and Dr S. Krishna Prasad for their valuable cooperation throughout the preparation of this chapter.

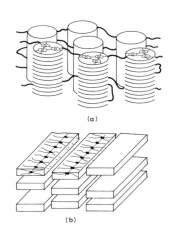

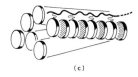

Figure 33. Mesophases of discotic polymers: (a) hexagonal columnar phase; (b) the board-like, or sanidic, nematic phase; (c) the columnar nematic phase.

14   References

[1] S. Chandrasekhar, B. K. Sadashiva, K. A. Suresh, *Pramana* **1977**, *7*, 471–480.
[2] J. Billard, J. C. Dubois, N. H. Tinh, A. Zann, *Nouv. J. Chim.,* **1978**, *2*, 535–540.
[3] C. Destrade, M. C. Mondon, J. Malthete, *J. Phys (France)* **1979**, *40*, C3–C17.
[4] J. D. Bunning, J. W. Goodby, G. W. Gray, J. E. Lydon in *Proceedings Conference on Liquid Crystals of One- and Two-Dimensional Order,*

Garmisch-Partenkirchen (Eds.: W. Helfrich, G. Heppke), Springer-Verlag, Berlin, **1980**, p. 397.

[5] J. D. Bunning, J. E. Lydon, C. Eaborn, P. M. Jackson, J. W. Goodby, G. W. Gray, *J. Chem. Soc., Faraday Trans. 1*, **1982**, *78*, 713 – 724.

[6] J. Malthete, A. M. Levelut, N. H. Tinh, *J. Phys. Lett. (France)* **1985**, *46*, L875 – L880.

[7] A. M. Levelut, *J. Phys. Lett. (France)* **1979**, *40*, L81 – L84.

[8] V. Vill, *LiqCryst – Liquid Crystal Database,* LCI, Hamburg, **1995**.

[9] T. Plesnivy, H. Ringsdorf, P. Schumacher, U. Nutz, S. Diele, *Liq. Cryst.* **1995**, *18*, 185 – 190.

[10] P. Spegt, A. Skoulios, *Acta Crystallogr.* **1966**, *21*, 892 – 897.

[11] F. C. Frank, S. Chandrasekhar, *J. Phys. (France)* **1980**, *41*, 1285 – 1288.

[12] A. M. Levelut, *J. Chim. Phys.* **1983**, *80*, 149 – 161.

[13] D. Goldfarb, Z. Luz, H. Z. Zimmermann, *J. Phys. (France)* **1981**, *42*, 1303 – 1311.

[14] S. Zumer, M. Vilfan, *Mol. Cryst. Liq. Cryst.* **1981**, *70*, 39 – 56.

[15] D. H. Van Winkle, N. A. Clark, *Phys. Rev. Lett.* **1982**, *48*, 1407 – 1410.

[16] C. R. Safinya, N. A. Clark, K. S. Liang, W. A. Varady, L. C. Chiang, *Mol. Cryst. Liq. Cryst.* **1985**, *123*, 205 – 216.

[17] C. R. Safinya, K. S. Liang, W. A. Varady, N. A. Clark, G. Anderson, *Phys. Rev. Lett.* **1984**, *53*, 1172 – 1175.

[18] E. Fontes, P. A. Heiney, W. H. de Jeu, *Phys. Rev. Lett.* **1988**, *61*, 1202 – 1205.

[19] E. Fontes, P. A. Heiney, M. Ohba, J. N. Haseltine, A. B. Smith, *Phys. Rev. Ser. A*, **1988**, *37*, 1329 – 1334.

[20] L. D. Landau, E. M. Lifshitz, *Statistical Physics, Part II,* 3rd edn., Pergamon, Oxford, **1980**, p. 435.

[21] E. I. Kats, M. I. Monastyrsky, *J. Phys. (France)* **1984**, *45*, 709 – 714.

[22] Y. F. Sun, J. Swift, *Phys. Rev., Ser. A,* **1986**, *33*, 2735 – 2739, 2740 – 2744.

[23] D. Ghose, T. R. Bose, M. K. Roy, M. Saha, C. D. Mukerjee, *Mol. Cryst. Liq. Cryst.* **1989**, *173*, 17 – 29.

[24] W. Kranig, C. Boeffel, H. W. Spiess, *Macromolecules* **1990**, *23*, 4061 – 4067.

[25] M. Moller, J. H. Wendorff, M. Werth, H. W. Spiess, H. Bengs, O. Karthaus, H. Ringsdorf, *Liq. Cryst.* **1994**, *17*, 381 – 395.

[26] A. M. Levelut, J. Malthete, A. Collect, *J. Phys. (France)* **1986**, *47*, 351 – 357.

[27] H. Zimmerman, R. Poupko, Z. Luz, J. Billard, *Z. Naturforsch., Teil a* **1985**, *40*, 149 – 160.

[28] J. M. Lehn, J. Malthete, A. M. Levelut, *J. Chem. Soc., Chem. Commun.* **1985**, 1794 – 1796.

[29] N. H. Tinh, C. Destrade, A. M. Levelut, J. Malthete, *J. Phys. (France)* **1986**, *47*, 553 – 557.

[30] C. Eaborn, N. H. Hartshorne, *J. Chem. Soc.* **1955**, 549 – 555.

[31] N. H. Tinh, H. Gasparoux, C. Destrade, *Mol. Cryst. Liq. Cryst.* **1981**, *68*, 101 – 111.

[32] C. Destrade, M. C. Bernaud, H. Gasparoux, A. M. Levelut, N. H. Tinh, in *Proceedings of the International Conference on Liquid Crystals,* (Ed.: S. Chandrasekhar), Heyden, London, **1979**, p. 29 – 32.

[33] B. Kohne, K. Praefcke, *Chimia* **1987**, *41*, 196.

[34] A. M. Levelut, F. Hardouin, H. Gasparoux, C. Destrade, N. H. Tinh in *Proceedings of the Conference on Liquid Crystal One and Two-Dimensional Order, Garmisch-Partenkirchen* (Eds.: W. Helfrich, G. Heppke), Springer-Verlag, Berlin, **1980**, p. 396.

[35] C. Destrade, H. Gasparoux, A. Babeau, N. H. Tinh, J. Malthete, *Mol. Cryst. Liq. Cryst.* **1981**, *67*, 37 – 48.

[36] N. H. Tinh, J. Malthete, C. Destrade, *J. Phys. Lett. (France)* **1981**, *42*, L417 – 419.

[37] K. Praefcke, D. Singer, B. Kohne, M. Ebert, A. Liebmann, J. H. Wendorff, *Liq. Cryst.* **1991**, *10*, 147 – 159.

[38] K. Praefcke, D. Singer, M. Langer, B. Kohne, M. Ebert, A. Liebmann, J. H. Wendorff, *Mol. Cryst. Liq. Cryst.* **1992**, *215*, 121 – 126.

[39] H. Bengs, O. Karthaus, H. Ringsdorf, C. Baehr, M. Ebert, J. H. Wendorff, *Liq. Cryst.* **1991**, *10*, 161 – 168.

[40] C. Destrade, N. H. Tinh, J. Malthete, J. Jacques, *Phys. Lett., Ser. A* **1980**, *79*, 189 – 192.

[41] J. Malthete, C. Destrade, N. H. Tinh, J. Jacques, *Mol. Cryst. Liq. Cryst. Lett.* **1981**, *64*, 233.

[42] A. M. Giroud-Godquin, J. Billard, *Mol. Cryst. Liq. Cryst.* **1981**, *66*, 147 – 150.

[43] K. Ohta, H. Muroki, A. Takagi, K. I. Hatada, H. Ema, I. Yamamoto, K. Matsuzaki, *Mol. Cryst. Liq. Cryst.* **1986**, *140*, 131 – 152.

[44] H. Sakashita, A. Nishitani, Y. Sumiya, H. Terauchi, K. Ohta, I. Yamamoto, *Mol. Cryst. Liq. Cryst.* **1988**, *163*, 211 – 219.

[45] A. C. Ribeiro, A. F. Martin, A. M. Giroud-Godquin, *Mol. Cryst. Liq. Cryst. Lett.* **1988**, *5*, 133 – 139.

[46] A. M. Giroud-Godquin, P. M. Maitlis, *Angew. Chem., Int. Ed. Engl.* **1991**, *30*, 375 – 402.

[47] K. Usha, K. Vijayan, B. K. Sadashiva, *Mol. Cryst. Liq. Cryst. Lett.* **1987**, *5*, 67 – 71.

[48] K. Usha, K. Vijayan, *Mol. Cryst. Liq. Cryst.* **1989**, *174*, 39 – 48.

[49] K. Usha, K. Vijayan, B. K. Sadashiva, P. R. Rao, *Mol. Cryst. Liq. Cryst.* **1990**, *185*, 1 – 11.

[50] K. Usha, K. Vijayan, B. K. Sadashiva, *Mol. Cryst. Liq. Cryst.* **1991**, *201*, 13 – 21.

[51] K. Usha, K. Vijayan, *Liq. Cryst.* **1992**, *12*, 137 – 145.

[52] K. Usha, K. Vijayan, S. Chandrasekhar, *Liq. Cryst.* **1993**, *15*, 575 – 589.

[53] B. Muhlberger, W. Haase, *Liq. Cryst.* **1989**, *5*, 251–263.

[54] W. L. McMillan, *Phys. Rev., Ser. A,* **1971**, *4*, 1238–1246.

[55] E. I. Kats, *Sov. Phys.-JETP* **1978**, *48*, 916–920.

[56] G. E. Feldkamp, M. A. Handschy, N. A. Clark, *Phys. Lett., Ser. A* **1981**, *85*, 359–362.

[57] S. Chandrasekhar, *Phil. Trans. R. Soc., London, Ser. A* **1983**, *309*, 93.

[58] S. Chandrasekhar, K. L. Savithramma, N. V. Madhusudana, *ACS Symp. Ser.* **1984**, *4*, 299–309.

[59] D. Ghose, T. R. Bose, C. D. Mukherjee, M. K. Roy, M. Saha, *Mol. Cryst. Liq. Cryst.* **1986**, *138*, 379–392.

[60] F. T. Lee, H. T. Tan, Y. M. Shih, C. W. Woo, *Phys. Rev. Lett.* **1973**, *31*, 1117–1120.

[61] D. Ghose, T. R. Bose, M. K. Roy, C. D. Mukherjee, M. Saha, *Mol. Cryst. Liq. Cryst.* **1988**, *154*, 119–125.

[62] S. Chandrasekhar, B. K. Sadashiva, K. A. Suresh, N. V. Madhusudana, S. Kumar, R. Shashidhar, G. Venkatesh, *J. Phys. (France)* **1979**, *40*, C3–120–124.

[63] S. Chandrasekhar, *Adv. Liq. Cryst.* **1982**, *5*, 47–78.

[64] J. Billard, B. K. Sadashiva, *Pramana* **1979**, *13*, 309–318.

[65] V. N. Raja, R. Shashidhar, S. Chandrasekhar, R. E. Boehm, D. E. Martire, *Pramana J. Phys.* **1985**, *25*, L119–L122.

[66] C. Destrade, N. H. Tinh, H. Gasparoux, J. Malthete, A. M. Levelut, *Mol. Cryst. Liq. Cryst.* **1981**, *71*, 111–135.

[67] C. Vauchier, A. Zann, P. Le Barny, J. C. Dubois, J. Billard, *Mol. Cryst. Liq. Cryst.* **1981**, *66*, 103–114.

[68] D. Adam, F. Closs, T. Frey, D. Funhoff, D. Haarer, H. Ringsdorf, P. Schuhmacher, K. Siemensmeyer, *Phys. Rev. Lett.* **1993**, *70*, 457.

[69] D. Adam, P. Schuhmacher, J. Simmerer, L. Haussling, K. Siemensmeyer, K. H. Etzbach, H. Ringsdorf, D. Haarer, *Nature* **1994**, *371*, 141–143.

[70] P. Davidson, M. Clerc, S. S. Ghosh, N. C. Maliszewskyi, P. A. Heiney, J. Hynes, A. B. Smith, *J. Phys. II (France)* **1995**, *5*, 249–262.

[71] I. Itoh, A. Tanaka, F. Fukuda, T. Miyamoto, *Liq. Cryst.* **1991**, *9*, 221–228.

[72] H. Bock, W. Helfrich, *Liq. Cryst.* **1995**, *18*, 387–399.

[73] H. Bock, W. Helfrich, *Liq. Cryst.* **1992**, *12*, 697–703.

[74] J. Prost in *Symmetries and Broken Symmetries* (Ed.: N. Boccara) IDSET, Paris **1981**, p. 159.

[75] M. P. Eastman, M. L. Horng, B. Freiha, K. W. Shen, *Liq. Cryst.* **1987**, *2*, 223–228.

[76] R. R. Bartkowski, B. Morosin, *Phys. Rev., Ser. B* **1972**, *6*, 4209–4212.

[77] J. Simon, C. Sirlin, *Pure Appl. Chem.* **1989**, *61*, 1625–1629.

[78] C. Piechocki, J. Simon, A. Skoulios, D. Guillon, P. Weber, *J. Am. Chem. Soc.* **1982**, *104*, 5245–5247.

[79] B. A. Gregg, M. A. Fox, A. J. Bard, *J. Am. Chem. Soc.* **1989**, *111*, 3024–3029.

[80] P. G. Schouten, J. M. Warman, M. P. de Haas, M. A. Fox, H. L. Pan, *Nature* **1991**, *353*, 736–737.

[81] S. Gaspard, P. Millard, J. Billard, *Mol. Cryst. Liq. Cryst.* **1985**, *123*, 369–375.

[82] Z. Belarbi, M. Maitrot, K. Ohta, J. Simon, J. J. Andre, P. Petit, *Chem. Phys. Lett.* **1988**, *143*, 400–403.

[83] J. F. Van der Pol, E. Neelemen, J. W. Zwikker, R. J. M. Nolte, W. Drenth, J. Aerts, R. Visser, S. J. Picken, *Liq. Cryst.* **1989**, *6*, 577–592.

[84] N. Boden, R. J. Bushby, J. Clements, M. V. Jesudason, P. F. Knowles, G. Williams, *Chem. Phys. Lett.* **1988**, *52*, 94–99.

[85] G. B. M. Vaughan, P. A. Heiney, J. P. McCauley Jr, A. B. Smith III, *Phys. Rev., Ser. B,* **1992**, *46*, 2787–2791.

[86] N. Boden, R. J. Bushby, J. Clements, *J. Mater. Sci. Mater. Electron.* **1994**, *5*, 83–88.

[87] H. Scher, M. Lax, *Phys. Rev., Ser. B* **1973**, *7*, 4491–4502.

[88] N. Boden, R. J. Bushby, A. N. Cammidge, J. Clements, R. Luo, K. J. Donovan, *Mol. Cryst. Liq. Cryst.* **1995**, *261*, 251–257.

[89] H. Bengs, F. Closs, T. Frey, D. Funhoff, H. Ringsdorf, K. Siemensmeyer, *Liq. Cryst.* **1993**, *15*, 565–574.

[90] F. Closs, K. Siemensmeyer, T. H. Frey, D. Funhoff, *Liq. Cryst.* **1993**, *14*, 629–634.

[91] J. Prost, N. A. Clark in *Proceedings of the International Conference on Liquid Crystals* (Ed.: S. Chandrasekhar), Heyden, London, **1979**, 53–58.

[92] P. G. deGennes, J. Prost, *Physics of Liquid Crystals*, 2nd edn., Clarendon Press, Oxford, **1993**, 421–423.

[93] S. Chandrasekhar, *Liquid. Crystals,* 2nd edn., Cambridge University Press, Cambridge, **1992**, 324–326.

[94] Y. Liao, N. A. Clark, P. S. Pershan, *Phys. Rev. Lett.* **1973**, *30*, 639–641.

[95] M. Kléman, P. Oswald, *J. Phys. (France)* **1982**, *43*, 655–662.

[96] G. S. Ranganath, S. Chandrasekhar, *Curr. Sci.* **1982**, *51*, 605–606.

[97] V. G. Kammensky, E. I. Kats, *Zh. Eksp. Teor. Fiz* **1982**, *56*, 591–594.

[98] R. E. Peierls, *Helv. Phys. Acta. Suppl.* **1934**, *7*, 81.

[99] L. D. Landau in *Collected Papers of L. D. Landau* (Ed.: D. Ter Haar), Gordon & Breach, New York, **1967**, p. 210.

[100] B. Jancovici, *Phys. Rev. Lett.* **1967**, *19*, 20–22.
[101] L. Gunther, Y. Imry, I. Lajzerowicz, *Phys. Rev., Ser. A* **1980**, *22*, 1733–1740.
[102] J. D. Litster in *Proceedings of the Conference on Liquid Crystals of One- and Two-dimensional Order, Garmisch-Partenkirchen, FRG* (Eds.: W. Helfrich, G. Heppke), Springer-Verlag, Berlin, **1980**, 65–70.
[103] J. Als-Nielsen in *Symmetries and Broken Symmetries* (Ed.: N. Boccara) IDSET **1981**, 107–122.
[104] J. Als-Nielsen, J. D. Litster, R. J. Birgeneau, M. Kaplan, C. R. Safinya, A. Lindegaard-Andersen, S. Mathiesen, *Phys. Rev., Ser. B* **1980**, *22*, 313–320.
[105] G. Durand, *Pramana*, (Suppl. 1), **1973**, 23.
[106] M. Cagnon, M. Gharbia, G. Durand, *Phys. Rev. Lett.* **1984**, *53*, 938–940.
[107] Y. Bouligand, *J. Phys. (France)* **1980**, *41*, 1297–1306, 1307–1315.
[108] M. Kléman, *J. Phys. (France)* **1980**, *41*, 737–745.
[109] P. Oswald, *J. Phys. (France)* **1981**, *42*, L-171.
[110] S. Chandrasekhar, G. S. Ranganath, *Adv. Phys.* **1986**, *35*, 507–596.
[111] B. Mourey, J. N. Perbet, M. Hareng, S. Le Berre, *Mol. Cryst. Liq. Cryst.* **1982**, *84*, 193–199.
[112] V. A. Raghunathan, N. V. Madhusudana, S. Chandrasekhar, C. Destrade, *Mol. Cryst. Liq. Cryst.* **1987**, *148*, 77–83.
[113] G. Heppke, H. Kitzerow, F. Oestreicher, S. Quentel, A. Ranft, *Mol. Cryst. Liq. Cryst. Lett.* **1988**, *6*, 71–79.
[114] T. Warmerdam, D. Frenkel, J. J. R. Zijlstra, *J. Phys. (France)* **1987**, *48*, 319–324.
[115] S. Chandrasekhar, U. D. Kini in *Polymer Liquid Crystals* (Eds.: A. Ciferri, W. R. Krigbaum, R. B. Meyer) Academic Press, New York, **1982**, 201–246.
[116] T. Carlsson, *J. Phys. (France)* **1983**, *44*, 909–911.
[117] G. E. Volovik, *Zh. Eksp. Teor. Fiz Lett.* **1980**, *31*, 273–275.
[118] W. Kreuder, H. Ringsdorf, *Makromol. Chem. Rapid Commun.* **1983**, *4*, 807–815.
[119] G. Wenz, *Makromol. Chem. Rapid Commun.* **1985**, *6*, 577–584.
[120] O. Herrmann-Schönherr, J. H. Wendorff, W. Kreuder, H. Ringsdorf, *Makromol. Chem. Rapid. Commun.* **1986**, *7*, 97–101.
[121] O. Herrmann-Schönherr, J. H. Wendorff, H. Ringsdorf, P. Tschirner, *Makromol. Chem. Rapid. Commun.* **1986**, *7*, 791–796.
[122] M. Ebert, O. Herrmann-Schonherr, J. H. Wendorff, H. Ringsdorf, P. Tschirner, *Makromol. Chem. Rapid Commun.* **1988**, *9*, 445–451.
[123] B. Kohne, K. Praefcke, H. Ringsdorf, P. Tschirner, *Liq. Cryst.* **1989**, *4*, 165–173.
[124] H. Ringsdorf, R. Wüstefeld, E. Zerta, M. Ebert, J. H. Wendorff, *Angew. Chem., Int. Ed. Engl.* **1989**, *28*, 914.
[125] H. Ringsdorf, R. Wüstefeld, *Molecular Chemistry for Electronics*, The Royal Society, London, **1990**, p. 23.
[126] B. Hüser, T. Pakula, H. W. Spiess, *Macromolecules* **1989**, *22*, 1960–1963.

Chapter IX
Applicable Properties of Columnar Discotic Liquid Crystals

Neville Boden and Bijan Movaghar

1 Introduction

We start by reminding ourselves that columnar discotic liquid crystals are comprised of disordered stacks (1-dimensional fluids) of disc-shaped molecules arranged on a two-dimensional lattice (Fig. 1) [1]. This structure imparts novel properties to these materials from which applications are likely to stem. One such property is the transport of charge along the individual molecular stacks [2–7]. The separation between the aromatic cores in, for example, the hexaalkoxytriphenylenes (HATn), the archetypal columnar discotic mesogen, is of the order of 0.35 nm, so that considerable overlap of π^* orbitals of adjacent aromatic rings is expected to lead to quasi-one-dimensional conductivity [8].

It turns out, however, that the HATn have a very low intrinsic carrier concentration, owing to the large band-gap in these materials. They are therefore good insulators (Table 1) with high breakdown fields of ≈ 4 MV cm^{-1} [9]. The mobility of an injected charge along the molecular columns (wires) is, however, fairly high [10–12].

The combination of these properties, together with other unique properties of columnar liquid crystals such as the self-organizing propensity of the molecular columns [13] which eliminates long lived grain boundaries which trap charge, and the ability readily to wet the surface of metals and

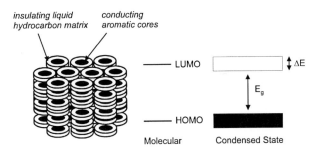

insulating liquid hydrocarbon matrix   conducting aromatic cores

LUMO $\updownarrow \Delta E$

E_g

HOMO

Molecular   Condensed State

Figure 1. Schematic view of a discotic columnar liquid crystal Col$_h$ phase, a simplified electronic structure of an isolated molecule, and the electronic structure in the gap in the condensed state showing the band gap E_g and the bandwidth ΔE.

Table 1. Comparison of some electrical properties of HAT6 with those of the organic metal TTF/TCNQ [8] and pure silicon [9].

	σ/Ω^{-1} cm^{-1}	$\mu/$cm^2 V^{-1} s^{-1}	$n/$cm^{-3}	$E_g/$eV	$\Delta E/$eV
HAT6	10^{-13}	4×10^{-4}	10^{10}	4	0.2
TTF/TCNQ	500	1	10^{21}	2	0.2–0.5
Si	10^{-3}	10^2	10^{13}	1.1	3–4

semiconductors which enables charges to be injected from electrodes into the molecular wires [14], could be harnessed in a variety of new applications.

In this short review of applicable properties, we will examine the electrical properties of columnar liquid crystals in some detail and discuss their potential application in light scanners, xerography, chemical and light sensors, molecular scale FETs, p–n junctions, ferroelectric memories and electrical contacts [8, 9]. We will also consider fluorescent and ferroelectric properties.

2   Molecular Structure-Property Relationships

We will consider some physical examples of columnar liquid crystals so as to estab-lish the relationship between molecular structure and their electrical and optical properties.

2.1   HATn Materials

Much work has been focused on the HATn materials (Scheme 1). Their electrical properties are typical of wide band-gap transparent columnar liquid crystals in general and can be considered as a model system.

The HAT-n are chemically stable, important for electro-optical applications; their chemistry is often more accessible that that of other discogens such as the phthalocyanines. These materials can now be produced in large quantities, in high purity and at an economic price. Routes for synthesizing low molar mass discotics, polymeric

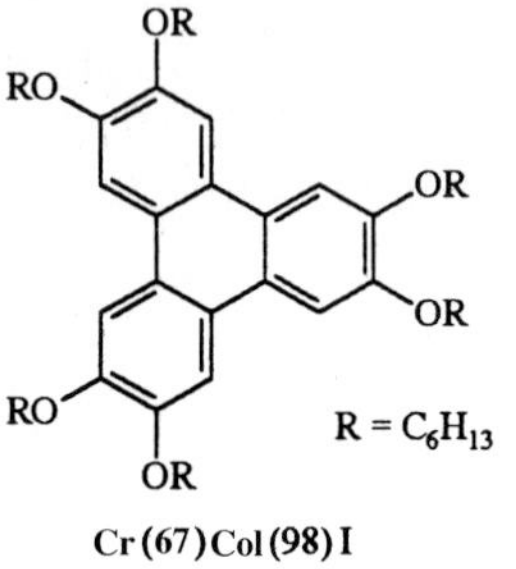

Cr (67) Col (98) I

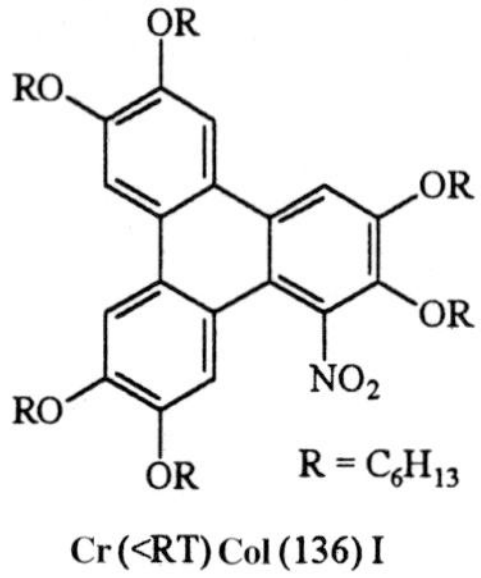

Cr (<RT) Col (136) I

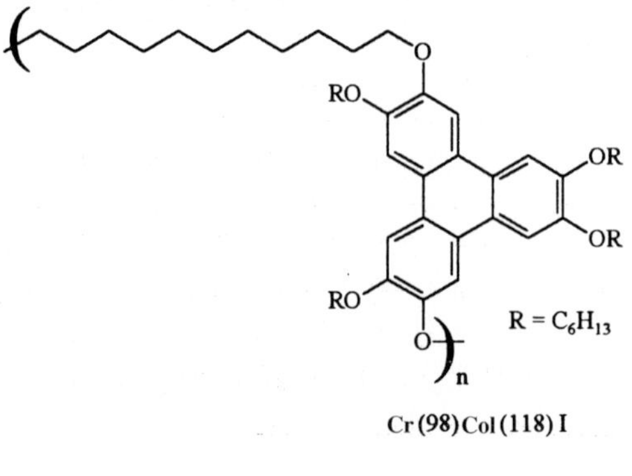

Cr (98) Col (118) I   **Scheme 1.** HAT-n materials.

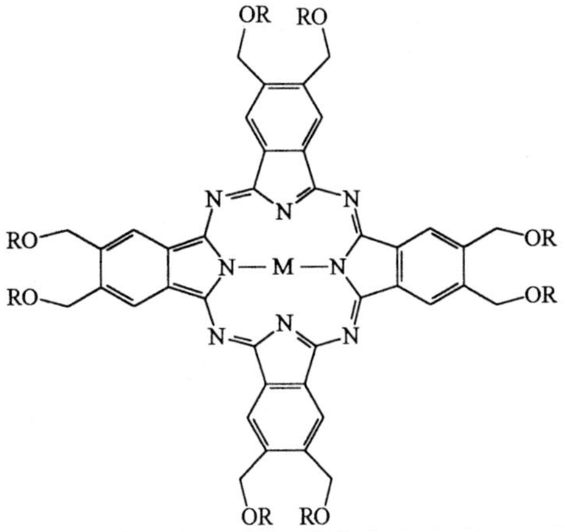

Cr (78) Col$_{ho}$ (305) I

R = C$_{12}$H$_{25}$ Pc Zn

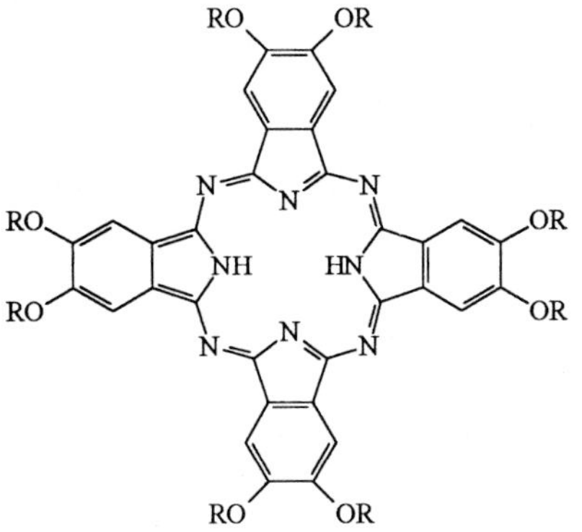

Cr (83) Col$_{ho}$ (309) I

R = C$_{12}$H$_{25}$

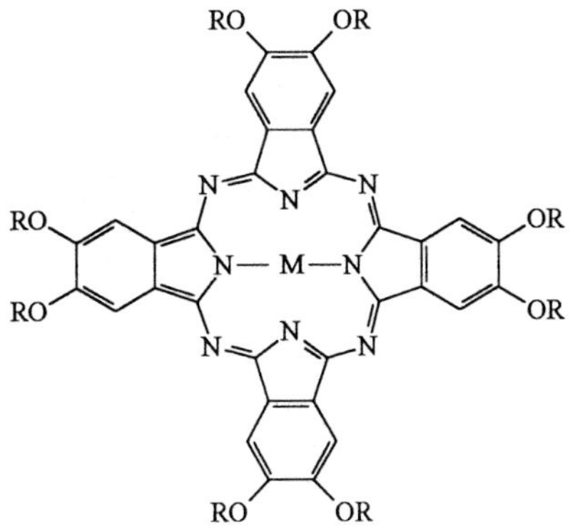

C$_{10}$CuPc : Cr (86) Col$_{rd}$ (88) Col$_{hd}$ (198) I

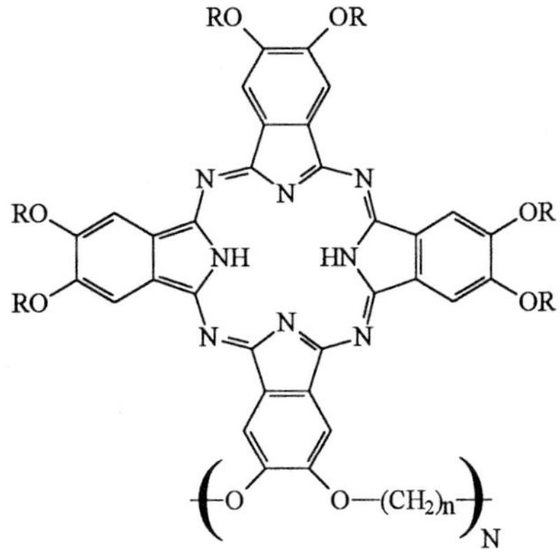

Scheme 2. The Phthalocyanines.

discotics, and for introducing substituents into the triphenylene ring are now available [15–19]. The latter is important not only for fine tuning their electrical and optical properties, but also for producing room temperature liquid crystals essential for applications.

2.2   The Phthalocyanines

The Phthalocyanine discogens (Scheme 2) have been extensively studied and their synthesis and properties are well documented in books and articles [20, 21] and references therein. These, together with the *porphyrins and metal-porphyrins* form phases which typify narrow band-gap columnar liquid crystals [21–24].

Table 2. Comparison of the main properties of wide band-gap and narrow band materials.

Key properties	HATn and related compounds	Phthalocyanines and related compounds
Band gap	4 to 5 eV	<3 eV
Electronic mobility	10^{-2} to 10^{-4} cm^2 V^{-1} s^{-1}	no transits but $\sim 10^{-3}$ cm^2 V^{-1} s^{-1}
Mobility anisotropy	$>10^3$ in undoped state $\sim 10^3$ in doped state	not measured
Undoped conductivity	$<10^{-14}$ S/cm $T=300$ K	$\sim 10^{-12}$ S/cm
Doped conductivity	$\sim 10^{-5}$ S/cm $T=300$ K	10^{-5} S/cm
Fluorescence yield	$\Phi_f \sim 0.1$, $T=300$ K	can be >0.1
O$_2$ binding	<0.5 eV	1–2 eV
Charge carrier gen. efficiency	$\sim 10^{-4}$ at $F=10^4$ V/cm and at $T=300$ K	$>10^{-4}$ same F, T
Clearing temperatures	$T_c < 100\,^{\circ}$C	$T_c > 180\,^{\circ}$C
Birefringence	$\Delta n < 0.1$	$0.5 > \Delta n > 0.1$
Ionization energy (condensed phase)	6–7 eV	5–6 eV

3   Electrical Properties

3.1   HATn Materials

The current vs. voltage (I–V) characteristics, measured along the columns, are typical of organic insulators: they exhibit (Fig. 2) a high resistance ohmic region at low voltages, and a space charge injection regime at high voltages [14].

An understanding of the conduction mechanism has been obtained from measurements of the temperature dependence of the conducticity $\sigma = n e \mu$ (Fig. 3).

The strong temperature dependence of $\sigma(T)$ could be caused either by the mobility μ or by the concentration of charge carriers (holes) $n(T)$, or both. This issue has been resolved by independently measuring the mobility μ of the carriers. This has been done in photoconductive carrier transit experiments. Figure 4 shows typical transit curves for HAT6 in the Col$_h$ phase [10, 12].

We see that the carriers (holes) are exiting the sample with a near Gaussian (weak-ly dispersive) behavior indicative of trap free transport. The mobility, μ, extracted from the transit time, is plotted against temperature in Fig. 5.

In the crystalline phase, the temperature dependence of the hole mobility, μ, was measured by charge injection by Warman et al. [25]. Phototransit measurements of mobility in HAT6 were only possible along the molecular columns in the Col$_h$ phase of HAT6 [12]. Using the temperature dependence of the conductivity of HAT6 from Fig. 3, however, enables us to deduce the change in the mobility when crossing into the isotropic phase. The value of μ drops from $\approx 5 \times 10^{-4}$ to $\approx 5 \times 10^{-6}$ cm^2 V^{-1} s^{-1} in going from Col$_h$ into the isotropic phase of HAT6. For HPAT$_2$ the value drops from 10^{-3} to 3×10^{-5} cm^2 V^{-1} s^{-1} at T$_{\text{Col}_h-I}$ as measured by Adam et al. [11] using photo time-of-flight. Similar changes have been observed in HHTT [11 b]. In general, transit times cannot be measured in the crystalline phases using photoinjection because grain boundaries cause deep traps. The pulsed radiolysis method of Warman et al. [25], permits mobility measurements in all

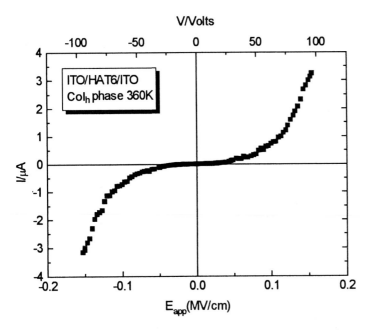

Figure 2. Applied field E_{app} dependence of the dark current in ITO/HAT6/ITO (6 μm thick) in the Col$_h$ phase at 360 K.

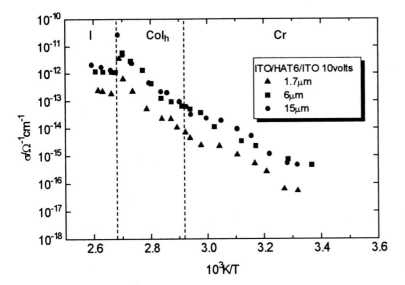

Figure 3. The temperature dependence of the conductivity for ITO/HAT6/ITO for three different sample thicknesses.

three phases and has been used to determined $\mu \approx 3 \times 10^{-4} \text{ cm}^2 \text{ V}^{-1} \text{ s}^{-1}$ in the Cr phase of the 4,5,6,11-alkoxy-triphenylenes [25, 25b]. We can conclude that by combining the three measurement techniques, we now have a reliable picture of the columnar mobility behavior in all three phases of these materials.

The invariance of μ with temperature across the entire columnar liquid crystalline phase is quite dramatic. It is a signature of nearest neighbor, random phase diffusion. The sudden change in the magnitude of the mobility at the clearing temperature observed by Adam and coworkers [11] signifies that thermal fluctuations in the positions

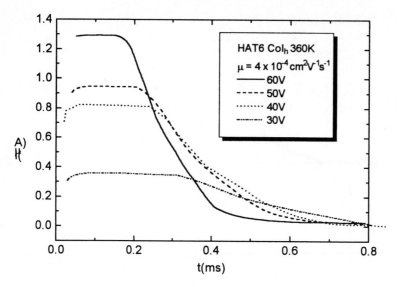

Figure 4. Transient photo-conductivity hole transits in the Col$_h$ phase of HAT6 at 360 K.

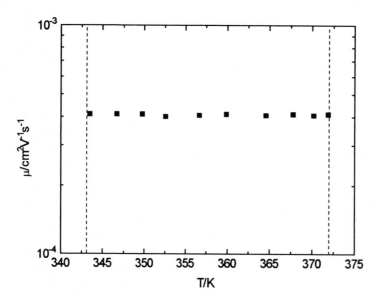

Figure 5. The temperature dependence of the mobility in HAT6 [12] measured in the Col$_h$ phase.

of the molecules are not seriously affecting the mobility of the carriers until a genuine macroscopic break occurs in the columnar organization. These breaks are no longer self-repairing on the time scale of a regular intermolecular hop and therefore the mobility drops by more than one order of magnitude. This remarkable effect has also been

observed recently in d.c. transport and by time-of-flight photoconductive measurements in the calamitic smectic liquid crystal 2-(4′-heptyloxyphenyl)-6-dodecylthio-benzothiazole by Funahashi and Hanna [26].

The fact that the conductivity drops by several orders of magnitude at the Col$_h$-to-I transition temperature reflecting the loss of columnar order also gives us a measure of the minimum expected anisotropy of the mobility. The anisotropy is difficult to measure directly in the undoped state because of the high resistance perpendicular to the columns. It is also not possible to measure it by time-of-flight. Conduction in the perpendicular direction involves crossing large insulating regions of ≈ 2 nm. This can only be done by using defects, edges, and large amplitude structural fluctuations and molecular exchange at high temperatures. Given this background on mobility [11, 12, 25], we can now deduce the origin of charge carriers in the undoped state.

The temperature dependence of σ in Fig. 3 must, according to the time-of-flight data have its origin solely in the temperature dependence of $n(T)$. The measured apparent activation energy, $E_a \approx 0.75$ eV is much smaller than 2 eV which corresponds to half the energy gap, E_g. Charge carrier generation can only be due to impurity detrapping or thermionic emission from the electrodes. Assuming $n = n_0 \exp(-E_a/k_B T)$ gives a total impurity concentration n_0 of $\approx 10^{18}$ cm^{-3}. This could arise from chemical contamination by, for example, O_2^- as is the case with the PCPS-siloxanes{phthalocyaninatopolysiloxanes} [27]. However, it has not been possible to detect any such impurities by ESR, leading us to conclude that the carriers are being injected from the electrode into the valence band of the columnar material in accordance with the Richardson equation $J = AT^2 \exp(-\phi/k_B T)$ [9], where ϕ is the effective barrier height at the interface (the difference in the work functions for the columnar phase and the ITO electrode is ≈ 1.2 eV). Unlike ordinary semiconductors, in these materials, charge exchange with metal contacts can dominate the carrier production mechanisms even at low voltages. The reason is that the high band-gap

prevents intrinsic carrier pair production, whereas the relatively low barrier to the metal Fermi energy and the high density of electronic states in the metal (10^{22} cm^{-3}), permit effective interface charge exchange. Given that the impurity generated free carrier concentration in the bulk can be as low as 10^9 cm^{-3}, charge production at the electrode surfaces dominates as soon as we are near or above room temperature.

3.2   Chemically Doped Materials

The HATn materials can be converted into p-type conductors by doping with oxidants such as $AlCl_3$ or $NOBF_4$. The dopant counterions $AlCl_4^-/BF_4^-$ are incorporated in the fluid hydrocarbon chain matrix. Up to 10 mol% of dopant can be incorporated without destroying the liquid crystalline phase. Equating the carrier concentration to the radical cation concentrations (measured by ESR) charge carrier mobilities have been calculated. The d.c. mobility measured for HAT6/NOBF$_4$ at 360 K along the columns is $\mu_\| \approx 2 \times 10^{-4}$ cm^2 V^{-1} s^{-2} ($\mu_\perp \approx 1 \times 10^{-7}$ cm^2 V^{-1} s^{-1}) and compares with the value obtained from the photoconductivity experiment. It has therefore been concluded that the carriers are holes and that the counterions do not significantly affect their migration along the interior of the columns. This is a consequence of the strong molecular fluctuations which average out the effects of the coulombic fields of the counterions. The dynamics give rise to a time dependent mobility which can be seen in the a.c. conductivity (Fig. 6).

The conductivity behaves as $\sigma(\omega) = \sigma(0) + \sigma_i \omega^s$ where $s \approx 0.8$ [3, 28]. This well known hopping-like behavior is a manifes-

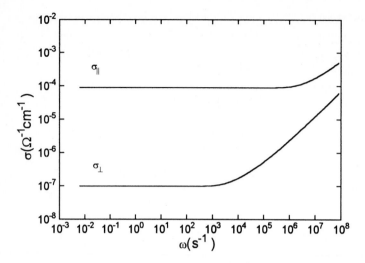

Figure 6. The frequency dependence of the conductivity of HAT6/NOBF$_4$ (1 mol%) in the Col$_h$ phase at 360 K.

tation of the fact that carriers encounter fewer major static or dynamic obstacles in the short time limit. In the short time limit, using information from band structure and triplet migration rates, an intermolecular hop rate, $\approx 10^{11}$ Hz, has been inferred. At longer times, carrier diffusion is limited by the highest barriers. The onset of the frequency dependence is, according to transport theory [28], a measure of the slowest jump frequency encountered along the transport pathway. The mobilities calculated from the limiting values of the conductivity $\mu_\parallel \approx 1 \times 10^{-4}$ cm^2 V^{-1} s^{-1} and $\mu_\perp \approx 1 \times 10^{-7}$ cm^2 V^{-1} s^{-1}) are very similar to the d.c. values [29]. The low critical a.c. frequency, and the relatively high mobility suggest long regions of relatively coherent transport with 10^{11} Hz rates. The coherent transport region shortens when the side chain is lengthened because this lowers the melting temperature [30], and the effect is to enhance fluctuations out of the columns which strongly lower the long time mobility, see [31].

4 Dielectric Properties

Measurements of the real part ε' and imaginary part ε'' of the dielectric susceptibility has given us valuable insight into the mechanisms of dipolar relaxation and thermal dipolar production of discotic liquid crystal materials [32–37].

The HATn materials have no permanent dipoles in the ideal symmetric state, but surprisingly have a relatively large low frequency dielectric constant $\varepsilon'(T)$ of the order 3 which varies only weakly with temperature [38].

The real and imaginary parts of the susceptibility of HAT6 are shown in Fig. 7 (a and b), and the temperature dependence of ε' is shown in Fig. 7(c) [38]. The frequency response can be explained by assuming a Debye law with, surprisingly, a single relaxation time ($\approx 5 \times 10^{-7}$ s). This process is most probably due to the rotation of the side chain around the oxygen bond. Numerical simulations have shown this mode to have a relatively low activation energy ($<kT$), and it is indeed a main mechanism by which the cancellation symmetry is broken and an effective overall thermal dipole can appear.

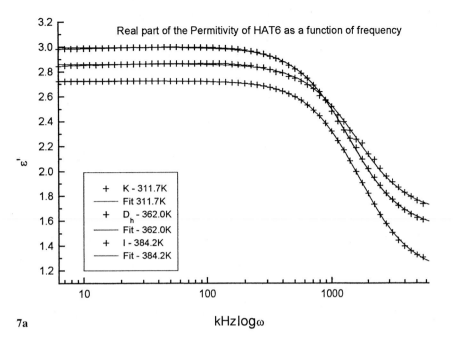

7a

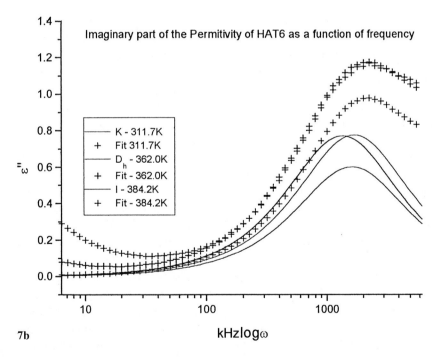

7b

The single frequency relaxation behavior and the mechanism for dipolar production is remarkably similar to that found by Phillips et al. for the compounds HET7 [34].

The HET7 molecules, however, have permanent dipoles on the side-groups which selforganize into an antiferroelectric state at low temperatures.

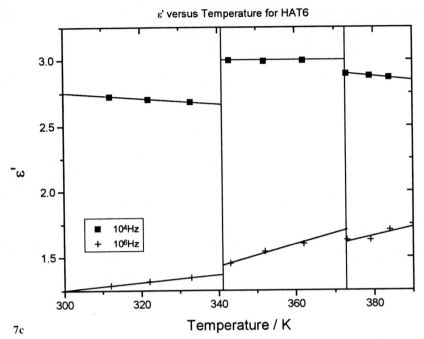

Figure 7. (a) The real part of the susceptibility ε' as a function of frequency ω for HAT6. The top curve corresponds to $T = 362$ K, the middle curve to $T = 384.2$ K, and the lower curve to $T = 311.7$ K. Solid lines are data, crosses are fits using simple Debye theory. (b) The frequency dependence of the imaginary part (bottom curve) of the dielectric constant for HAT6. Solid lines are data and crosses are theory fits using simple Debye theory with a weakly temperature dependent relaxation time of $\approx 5 \times 10^{-7}$ s [38]. Top curve $T = 384.2$ K, middle curve $T = 362$ K, lower curve $T = 311.7$ K (in (a) and (b) the old nomenclature K and D_h instead of Cr and Col_h is used). (c) The temperature dependence of the real part $\varepsilon(T)$ as a function of T at two frequencies: +, 10^6 Hz; ■, 10^4 Hz. Note that despite the typical dipolar relaxation law observed in (a) and (b), the temperature dependence only exhibits small variations which are related to structural order.

These findings allow us to conclude that with suitable synthesis, we may be able to engineer discotic materials which acquire large dipoles and long range polarization order only as a result of temperature: molecules which change from anti- to ferroelectric or to a super-paraelectric order when heated.

5   Fluorescence Properties

Fluorescence and exciton dynamics in 2,3,6,7,10,11-hexa-*n*-hexyloxytripheny-

lene, C_6HET and Phthalocyanines $\{(C_{12}OCH_2)_8PcH_2, PcZn\}$ have been extensively studied [39–43]. Markovitsi et al. have shown that the HATn are excellent transporters of excitons along the columns with $D_{singlet} \approx 2 \times 10^{-2}$ cm^2 s^{-1} at room temperature. Triplet migration of energy has also been demonstrated [43]. From the measured triplet exciton hopping rates (≈ 10 ps) we can also extract, using second order perturbation theory and taking into account exciton binding, a nearest neighbor charge hopping rate of the same order of magnitude. The singlet Foerster transfer rate between adjacent molecules is as high as 3×10^{13} s^{-1}, faster than optic phonons

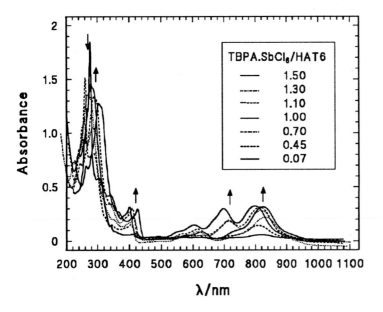

Figure 8. The absorbance of HAT6 in dichloromethane as a function of wavelength with p-doping concentration. The arrows indicate the direction of change (increase or decrease) in the absorption intensity with increasing amount of oxidizing agent TBA.SbCl$_6$.

and therefore remarkably efficient at high temperatures. This would normally imply a coherent exciton band of width ≈ 0.4 eV. The HATn absorb in the UV and emit in part in the visible range. The quantum yield Φ_F is ≈ 0.1 at room temperatures; in phthalocyanine compounds, the low T yield is ≈ 0.5 and drops to about 0.01 at room temperature [42]. The charged molecule generated by chemical doping has a quite distinct absorption spectrum [44] as shown in Fig. 8.

This spectrum (Fig. 8) implies that the charged species can be accurately counted by absorption. We can also count and therefore monitor the charge in the organic film by using the fact that absorption in the film prevents photocarrier generation in a semiconductor substrate.

6   Gaussian Transit Characteristics

The unique transport properties of columnar phases of discotic liquid crystals (well

defined transits and columnar-scale resolution) can be exploited in high-resolution light scanning or xerography (Fig. 9). A potential advantage of discotics over amorphous silicon is the prospect for wavelength sensitive scanning (employing materials with different optical absorption spectra).

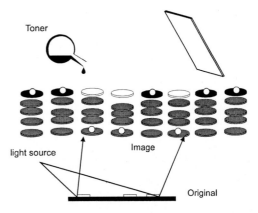

Figure 9. High resolution light scanning and xerography.

7   Anisotropy of Charge Mobility and Inertness to Oxygen

7.1   The Surface Sensor

The low clearing temperatures, excellent wetting and alignment properties of HATn give these materials a very low bulk mobility in the direction perpendicular to the molecular columns. The sensor uses the surface as a conducting channel which is formed by thermal fluctuations and chemical contamination. Injected or intrinsic charge can therefore move several orders of magnitude faster near the more 'isotropically' ordered surface (see Fig. 10). The surface relaxation and trapping times for charges on the core are sensitive to the presence of adsorbed molecules. Both shape and polarizability of the adsorbate will in this way influence the transport dynamics of the injected charge. The relaxation times can be measured by pulsed impedance spectroscopy. Current levels are at present too low for low cost commercial exploitation, but the prospects for the HATn are good because the molecules possess apart from good wetting and alignment properties, the very unique feature significant for any material namely to remain chemically stable and reversible under oxygen (air) exposure [45, 46]. As long as one has control over the signal magnitude, one can sense through the surface repeatedly and on a relatively fast time scale of less than milliseconds.

7.2   Phthalocyanine Based Gas Sensors

These have been fabricated and tested recently by Wright, Cook and coworkers [47, 48]. These sensors exploit the low band gap of the material and detect the gas by way of real charge transfer from the core to the adsorbed species. The relatively strong molecular oxygen reactivity can also be used through the indirect mechanism of oxygen displacement. The kinetics of oxygen desorption can be adjusted to some extent by changing side-chain structure and length. The response to NO_2 was shown to be as low as 1 p.p.m. at $T = 300$ K. With NO_2, the reaction is one of real charge transfer from the molecule to the gas.

8   Selforganizing Periodicity, High Resistivity, and Dielectric Tunability

The device illustrated in Fig. 11 can be used to detect incoming signals (along molecular wires) generated by light (detector),

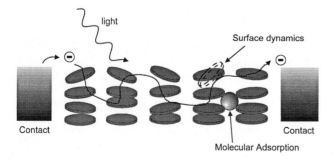

Figure 10. Schematic view of a carrier transport process in the surface conductivity of a columnar liquid crystal film.

molecules (chemical sensor) or excitons (signal processing), and to store charge (memory). Carriers in the two-dimensional gas react to changes in the interface potentials.

The two-dimensional electron gas shown in Fig. 11 can also represent carriers in an ultrathin metal film or magnet, or a superlattice specially designed to give high surface responsivity. To obtain good results and for making devices, one would need to use MBE fabrication for the substrate layer(s), and high vacuum molecular deposition technology [49, 50] for optimum contact. The adsorbed molecules pattern the surface and eventually form arrays of self-organizing molecular wires. The suitably engineered quantum dot substructures created at the interface linking molecular energy bands to 'metal' bands could be used to sense gases, light, heat, magnetic polarization and to store charge. A key innovation is to be able to write signals into the organic film and then read this 'imprint' in the plane of the substrate as a change in the electrical or photo-electrical response. The responsivity of the organic state, if it is to be usefully exploited in molecular electronic devices, must satisfy minimum criteria of durability and this implies necessarily that the material cannot be subjected to large currents, power levels and stress, and one cannot demand relaxation times on the scale achievable in normal silicon or gallium arsenide device technology [9].

On a larger than micron scale, it is now possible to store holographic information optically in columnar liquid crystals which undergo glass transitions in useful temperature ranges [51, 52]. Work by Shimizu [53] and recent unpublished work [54] have also demonstrated that the magnitude and the rectification behavior of bulk photocurrents are very much dependent on phase behavior and interface alignment at the electrodes, and that consequently the charge injection rate must be dependent on the interfacial electronic band structure. If interface bands affect carrier injection rates into the bulk, then the next step, as shown in Fig. 11, is to look for the molecular interface bands in planar conduction, and then to exploit this coupling in planar processing on ultra–thin channels.

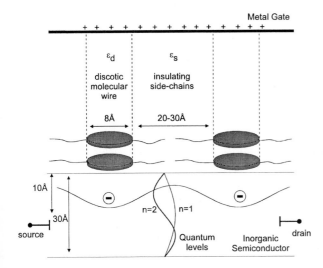

Figure 11. A two-dimensional modulated electron channel showing molecular wires addressing an electron gas. The molecular wires generate a two-dimensional band structure in the gas. The source–drain current is therefore sensitive to signals (charge/light) arriving along the columns; the integer n denotes the quantum well states in the substrate gas.

9   Ferroelectrics and Dielectric Switches

The main application of liquid crystals has been, and still is, in the area of optical switching and displays. This systems from the unique property of nematic and cholesteric phases to change orientation of structure when subjected to relatively weak perturbations. Perturbations can be electric fields (as low as 10^3 V cm^{-1}), magnetic fields (<1 T), pressure, heat or surface forces. Internal organization and pitch lengths of helical deformations can be sensitive to heat and infra-red rays. In this way one can use external and internal forces to 'structure' the refractive indices and achieve the required degree of light guiding and transmission. The optical properties and applications of nematic and cholesteric liquid crystals are reviewed elsewhere in the Handbook. We can therefore turn our attention to the possibility of realizing similar or other useful optical effects using discotic liquid crystal materials.

9.1   Nematic Phases

There are a few examples of nematic discotic liquid crystals and they have been investigated in detail [32–34]. Examples are the nematic triphenylene-hexa-4-(alkoxy-benzoates) [32, 34], truxene esters [34], and non-polar hexakisbenzenes. Phase transition temperatures, birefringence and electrical properties of pure materials are not usually in a range where immediate applications can be envisaged. Lower phase transition temperatures have been obtained using mixtures. The mixture of the nematic DB126 with the columnar HAT7, for example, gives crystal–nematic phase transitions at considerably lower temperatures [34].

Mixing calamitics with discotics leads to an increase in biaxiality and better control over the dielectric properties [34].

9.2   Ferro- and Antiferroelectric Phases

Considerable effort is, and has been, devoted in recent years to the synthesis of ferroelectric materials with disc-shaped molecules. The main applications here would be in the areas of optical information storage. When discotic molecules carry electric dipole moments, on or near the core, the energetically preferred order is in all known cases an antiferroelectric order along the columns. Perpendicular to the columns, the order in the plane should in principle be ferroelectric. This is, however, very often frustrated by rapid intercolumnar exchange of molecules which means that the relatively weak dipolar forces in the plane are frustrated by the efficient dynamics of rotation and exchange.

It was recognized as early as 1981 [54] that in order to achieve the ferroelectric state, a chiral liquid crystal is necessary. Chirality puts two molecules in the 'ideal' unit cell, and this allows new modes of fluctuations and also new forms of organisation which also give possible minima in the free energy ['structure'] function. An electric field can now take the liquid crystal into one of these alternative forms of molecular packing, some of which can have a net ferroelectric moment per unit cell and therefore an overall ferroelectric moment. Columnar pyrenes made recently by Bock and Helfrich [55] have saturation polarization of ≈ 110 nC cm^{-2} and switch from the anti- to the ferro-electric state on a time scale of $\approx 10^{-4}$ s with switching fields $E \approx 10$ V μm^{-1}. Similar behavior has been reported by Scherowsky and Chen [56] who synthesized 2,3,6,7,9,10-hexa-substituted phenanthrenes. The latter exhibit a maximum polarization of 54 nC cm^{-2} and a minimum switching field of 2 V μm^{-1}. The temperature dependent switching times range from 4 to 24 ms. Bock and Helfrich also made a dibenzopyrene columnar liquid crystal which requires at 90 °C a field strength of ≈8 × 10^4 V cm^{-1} to switch into the ferroelectric tilted columnar state [57]. The rise and decay times here are considerably shorter ≈20 μs.

The future for chiral ferroelectrically switchable columnar liquid crystals looks bright. However, applications are not likely to be in the same area as those for conventional display materials, because the switching fields and times are too large and too long, respectively, to be competitive.

One can, however, envisage new applications other than in the area of display devices. These would include new types of heat and infrared sensors, ferroelectric tapes, plastics, polymers and optical fibres. In particular, the very large number of new dipolar organizations [58–60] that can be realized in such materials could be used to form high density optical and electronic memories.

10   Novel Absorption Properties, Fluorescence, and Fast Exciton Migration

One-dimensionality of charge and exciton transport is a unique feature of HATn discotic liquid crystals. The exciton transfer anisotropy is >10^4 [40]. Given the large bandwidth along the columns, it is difficult to envisage an intercolumnar transfer unless the latter is defect or fluctuation assisted as for the 'snakelike' modes shown by Levelut [31]. The combination of highly anisotropic and well-defined charge and exciton transport can be usefully exploited in devices that may use light quenching or absorption to announce the arrival of a charge with a high degree of spatial resolution. The molecular columns can also be used as efficient light antenna to a location where the photons are needed for a chemical conversion process, in a situation where, for example, the location itself must not be in direct contact with air or sun light, or, alternatively, as light gathering antennae for triplet excitons: light travels to an exposed surface in the form of triplet excitons with hopping times as fast as [4–68 ps] such as in the octa-substituted phthalocyanines [43]. The accumulated triplets can then be used for scanning directly at the air–organic surface for example. Triplet lifetimes (≈125 μs in (C$_{12}$OCH$_2$)$_8$PcH$_2$) are sensitive to small variations in electrical and magnetic forces. This effect could be used in gas or biosensing.

11   Applications of Doped Columnar Conductors

The p- and n-type chemically doped columnar materials are fairly stable and have potential as electrical contacts (hole injectors) [3], structured electrolytes in microbatteries, p–n junctions, cheap disposable solar cells, and in charge storage devices.

12   Conclusions

There are no significant commercial applications for discotic liquid crystals to date.

Work has been focussed mainly on development of synthetic protocols, and establishing the physical properties of discotic materials. However, the unique properties of the materials suggest that there are existing prospects for applications in the not too distant future. Arguably the unique electrical properties of the discotic liquid crystals could be exploited in the development of molecular sensors. In particular the stability of the tryphenylene core to oxygen exposure gives this class of materials a unique advantage even when measured against inorganic semiconductors and metals. Here we have a class of materials which transports charge and remains apparently chemically stable at the immediate interface to air.

In some applications one can argue that durability and reproducibility are not absolutely essential; discotic liquid crystals could be used, for example, in the fabrication of conducting, rectifying (p–n junctions) or electrochromic soft solids. Undoped HATn can, however, exhibit long time stability and might therefore provide a useful emitter of light when used as a thin barrier layer for double injection. Electroluminescent applications of HAT6 are under serious consideration in several groups [61, 62]. One can also envisage making wave guides and possibly even non-linear optical fibres, which would exploit the temperature and electric field dependence of the refractive index. 'One-dimensional' exciton migration and triplet fluorescence can also be useful in chemical and bio-sensing.

Of all potential future applications, the most exciting ones are perhaps those associated with making ferroelectric memories which can be addressed by light or with two-dimensional electronic chips. The idea is to store dielectric information from the three-dimensional bulk into electronic configurations of a two-dimensional chip. The high degree of 'near structural degeneracy' (very different structures with small free energy differences) of self-assembled systems is invaluable to optical display technology, and this too can be usefully exploited. Every dipolar configuration gives, in the ideal case, a specific electronic signal in the chip. The change in configuration is achieved with small perturbations and without having to subject the 'organic state' to excessive levels of electro-optical stress. The change in configuration then gives rise to a new interfacial electronic coupling to the channel. The new coupling changes the electronic transport properties. Truly fast switching and relaxation ($<10^{-9}$ s), however, cannot be achieved; to complete with conventional electronics, speed must be replaced with parallelism, and inevitable degradation must be compensated for by self-repairing and self-organizing dynamics.

Acknowledgements

Contributors to the work described: Ruth Borner, Richard Bushby, Andrew Cammidge, Jonathan Clements, Kevin Donovan, Theo Kreouzis, Gareth Headdock, Philip Martin. Bijan Movaghar is a Leverhulme Fellow at the 'SOMS Centre' of Leeds University.

13   References

[1] S. Chandrasekhar, G. S. Raganath, *Rep. Prog. Phys.* **1990**, *53*, 57.
[2] N. Boden, R. J. Bushby, J. Clements, M. V. Jesudason, P. F. Knowles, G. Williams, *Chem. Phys. Lett.* **1988**, *152*, 94; **1989**, *154*, 613.
[3] N. Boden, R. J. Bushby, J. Clements, *J. Phys. Chem.* **1993**, *98*, 5920.
[3b] N. Boden in *Proceedings of the ONRI Workship on Advanced Materials*, Osaka, Japan 11–12 Dec. **1995**, p. 47.
[4] L. Y. Chiang, J. P. Stokes, C. R. Safinya, A. N. Bloch, *Mol. Cryst. Liq. Cryst.* **1985**, *125*, 279.
[5] J. van Keulen, T. W. Warmerdam, R. J. M. Nolte, W. Drenth, *Recl. Trav. Chim. Pays-Bas* **1987**, *106*, 534.
[6] G. B. M. Vaughan, P. A. Heiney, J. P. McCauley Jr., A. B. Smith III, *Phys. Rev. B* **1992**, *46*, 2787.

[7] N. Boden, R. J. Bushby, R. C. Borner, J. Clements, *J. Am. Chem. Soc.* **1994**, *116*, 10807.

[8] M. Pope, C. E. Swenberg, *Electronic Processes in Organic Crystals*, Oxford University Press, New York, **1982**.

[9] S. M. Sze, *Physics of Semiconductor Devices*, John Wiley, New York, **1981**.

[10] D. Adam, F. Closs, T. Frey, D. Funhoff, H. Ringsdorf, K. Siemensmeyer, *Phys. Rev. Lett.* **1993**, *70*, 457.

[11] D. Adam, P. Schumacher, J. Simmerer, L. Haussling, K. Siemensmeyer, K. H. Etzbach, H. Ringsdorf, D. Haarer, *Nature* **1994**, *371*, 141.

[11b] D. Adams, P. Schumacher, J. Simmerer, L. Haussling, W. Paulus, K. Siemensmeyer, K.-H. Etzbach, H. Ringsdorf, D. Haarer, *Adv. Mater.* **1995**, *7*, 276.

[12] N. Boden, R. J. Bushby, J. Clements, B. Movaghar, K. Donovan, T. Kreouzis, *Phys. Rev. B* **1995**, *52*, 13274.

[13] I. G. Voigt-Martin, R. W. Garbella, M. Schumacher, *Liq. Cryst.* **1994**, *17*, 775.

[14] B. Movaghar, J. Clements, N. Boden, J. Clements, *J. Appl. Phys.*, preprint **1997**.

[15] I. M. Matheson, O. C. Musgrave, C. J. Webster, *J. Chem. Soc., Chem. Commun.* **1965**.

[16] N. Boden, R. J. Bushby, R. C. Borner, A. N. Cammidge, M. V. Jesudason, *Liq. Cryst.* **1993**, *15*, 851.

[17] N. Boden, R. J. Bushby, A. N. Cammidge, *J. Am. Chem. Soc.* **1995**, *117*, 924.

[18] N. Boden, R. J. Bushby, A. N. Cammidge, *Liq. Cryst.* **1995**, *18*, 673.

[19] N. Boden, R. J. Bushby, A. N. Cammidge, G. Headdock, *J. Mater. Chem.* **1995**, *5*, 2275.

[20] *Phthalocyanines Properties and Applications*, Vol. 2 (Eds.: C. Leznoff, A. B. Lever), VCH Publishers NY **1993**.

[21] M. J. Cook in *Proc. ONRI Workshop On Advanced Materials*, Osaka, Japan **1996**, p. 76.

[22] "Porphyrins" Vol. I–VII (Ed.: D. Dolphin), Academic Press, New York **1978**.

[23] S. A. Hudson, P. M. Maitlis, *Chem. Rev.* **1993**, *93*, 861.

[24] G. C. Bryant, M. J. Cook, C. Ruggiero, T. G. Ryan, A. J. Thorne, *J. Chem. Soc., Chem. Commun.* **1995**, 467.

[25] J. Warman, P. G. Schouten, *J. Physical Chemistry* **1995**, *99*, 17181.

[25b] A. M. van de Craats, J. M. Warman, M. P. de Haas, D. J. Simmerer, D. Haarer, P. Schumacher, *Adv. Mater.* **1996**, *8*, 823.

[26] M. Funahashi, Jun-ichi Hanna, *Jpn. J. Appl. Phys.* **1996**, *35*, L703–705; and *Appl. Phys. Lett.* **1997**, *71*, 602.

[27] G. Wegner, *Ber. Bunsenges. Phys. Chem.* **1991**, *95*, 1326.

[28] B. Movaghar, B. Pohlmann, D. Wuertz, *Zeit für Physik B* **1987**, *66*, 523.

[29] N. Boden, R. J. Bushby, J. Clements, B. Movaghar, preprint **1997**.

[30] E. O. Arikainen, N. Boden, R. J. Bushby, J. Clements, B. Movaghar, A. Wood, *J. Mater. Chem.* **1995**, *5*, 2161.

[31] A. M. Levelut, *J. Phys. Lett.* **1979**, *40*, L-81.

[32] J. T. Phillips, J. C. Jones, D. G. McDonnell, *Liq. Cryst.* **1993**, *15*, 203.

[33] J. T. Phillips, J. C. Jones, *Liq. Cryst.* **1994**, *16*, 805.

[34] J. T. Phillips, Ph. D. Thesis, University of Bristol **1992**.

[35] H. Groothues, F. Kremer, P. G. Schouten, J. Warman, *Adv. Mater.* **1995**, *7*, 283.

[36] M. Werth, S. U. Vallerien, H. W. Spiess, *Liq. Cryst.* **1991**, *10*, 759.

[37] S. U. Vallerien, M. Werth, F. Kremer, H. W. Spiess, *Liq. Cryst.* **1990**, *8*, 889.

[38] N. Boden, R. J. Bushby, J. Clements, G. Headdock, P. Martin, B. Movaghar, preprint **1997**.

[39] D. Markovitsi, I. Lecuyer, P. Lianos, J. Malthete, *J. Chem. Soc., Faraday Trans.* **1991**, *87*, 1785.

[40] D. Markovitsi, F. Rigaut, M. Mouallem, J. Malthete, *Chem. Phys. Lett.* **1987**, *135*, 236.

[41] O. Braitbart, R. Sasson, A. Weinreib, *Mol. Cryst.* **1988**, *159*, 233.

[42] G. Blasse, G. J. Dirksen, A. Meijerink, J. F. van der Pol, E. Neelman, W. Drenth, *Phys. Lett. Chemical* **1989**, *154*, 420.

[43] D. Markovitsi, I. Lecuyer, J. Simon, *J. Phys. Chem.* **1991**, *95*, 3620.

[44] N. Boden, R. J. Bushby, J. Clements, Rong Luo, *J. Mater. Chem.* **1995**, *5*, 1741.

[45] N. Boden, J. Clements, B. Movaghar, UK Patent Application No. 9509729,1, *Electronic Devices Based on Discotic Liquid Crystals*, **1995**.

[46] N. Boden, J. Clements, B. Movaghar, UK Patent Application No. 96D0452,7, *Sensing Devices*, **1996**.

[47] J. Wright, P. Roisin, G. P. Rigby, R. Nolte, M. Cook, S. Thorpe, *Sensors and Actuators B*, **1993**, Vols. *13–14*, 276.

[48] A. Cole, R. McIlroy, S. C. Thorpe, M. J. Cook, J. McMurdo, A. K. Ray, *Sensors and Actuators B*, **1993**, Vols. *13–14*, 416.

[49] N. Boden, R. Bissel, J. Clements, B. Movaghar, *Liq. Cryst. Today* **1996**, *6*, No. 1, 3.

[50] R. H. Tredgold, *J. Mater. Chem.* **1995**, *5*, 1095.

[51] K. Praefke, A. Eckert, D. Blunk, *Liq. Cryst.* **1997**, in press.

[52] J. Contzen, G. Heppke, H. J. Kitzerow, D. Kruerke, H. Schmid, *Appl. Phys. B* **1996**, *63*, 605.

[53] Yo Shimizu in *Proc. of the ONRI workshop on Advanced Materials*, Osaka, Japan, 11–12 Dec. **1995**, p. 64.

[54] J. Prost, *Compt. Rend. Colloq. Pierre Curie* (Ed.: N. Boccara, IDSET) **1981**.

[55] H. Bock, W. Helfrich, *Liq. Cryst.* **1992**, *12*, 697.

[56] Xin Hua Chen, G. Scherowsky, *J. Mater. Chem.* **1995**, *5*, 417.

[57] D. Kruerke, H. S. Kitzerow, G. Heppke, *Ber. Bunsenges. Phys. Chem.* **1993**, *97*, 1371.

[58] G. Heppke, D. Kruerke, M. Mueller, H. Bock, *Ferroelectrics* **1996**, *179*, 203.

[59] M. Langner, K. Praefke, D. Kruerke, G. Heppke, *J. Mater. Chem.* **1995**, *5*, 693.

[60] A. Fukuda, Y. Takahashi, T. Isozaki, K. Ishikawa, T. Takezoe, *J. Mater. Chem.* **1994**, *4*, 997.

[61] T. Christ, J. H. Wendorff, *Makromol. Rapid Commun.* **1997**, *18*, 93–98.

[62] R. Friend, University of Cambridge, personal communication, **1997**.

Part 3:
Non-Conventional Liquid-Crystalline Materials

Chapter X
Liquid Crystal Dimers and Oligomers

Corrie T. Imrie and Geoffrey R. Luckhurst

1 Introduction

Semiflexible main chain liquid crystal polymers are composed of semirigid anisometric or mesogenic groups separated via flexible spacers, normally alkyl chains [1]. The length and parity of the flexible spacer has a dramatic influence on the transitional properties of the polymer [2]. In the early 1980s, Griffin and Britt synthesized model compounds of these polymers and showed that materials consisting of molecules having just two mesogenic units connected via an alkyl chain exhibited liquid crystalline behavior [3]. This was a most surprising result as it contravened the conventional wisdom that low molar mass liquid crystals required a semirigid anisometric core [4]. Instead these new materials possessed a highly flexible core and are now referred to as liquid crystal dimers which is preferred to other names such as twins and bimesogens. As with so many discoveries in liquid crystal chemistry, however, Vorländer actually reported the first dimers in the 1920s [5] although this important discovery appears to have been overlooked. Indeed, in the 1970s dimers were also rediscovered [6] but it has only been since the 1980s that dimers have been the focus of much research activity. Two main reasons have stimulated this interest. First, the transitional properties of dimers are markedly different to those of conventional low molar mass liquid crystals containing a single mesogenic group and thus provide a demanding challenge to our current understanding of the factors that promote self assembly in condensed phases. Secondly, the transitional properties of dimers depend critically on the length and parity of the flexible spacer in a manner strongly reminiscent of that observed for semiflexible main chain polymers and, as we shall see later this similarity has resulted in their use as model compounds for the polymeric systems [7].

2 Structure–Property Relationships in Liquid Crystal Dimers

The overwhelming majority of dimers reported in the literature consist of molecules comprising two rod-like mesogenic units linked via a flexible alkyl chain normally containing between 3 and 12 methylene groups. Some notable exceptions to this general description include dimers containing discotic moieties [8–11], oligo(ethylene oxide) [12–14] or oligo(siloxane) spacers [12, 15]. Indeed discotic dimers are of particular interest because it has been claimed that selected examples exhibit the biaxial nematic phase [8], and this approach to obtaining the biaxial nematic phase has been extended to considering dimers containing one rod-like and one disc-like moiety [16]. A further structural variant are dimers in which the mesogenic groups are linked lat-

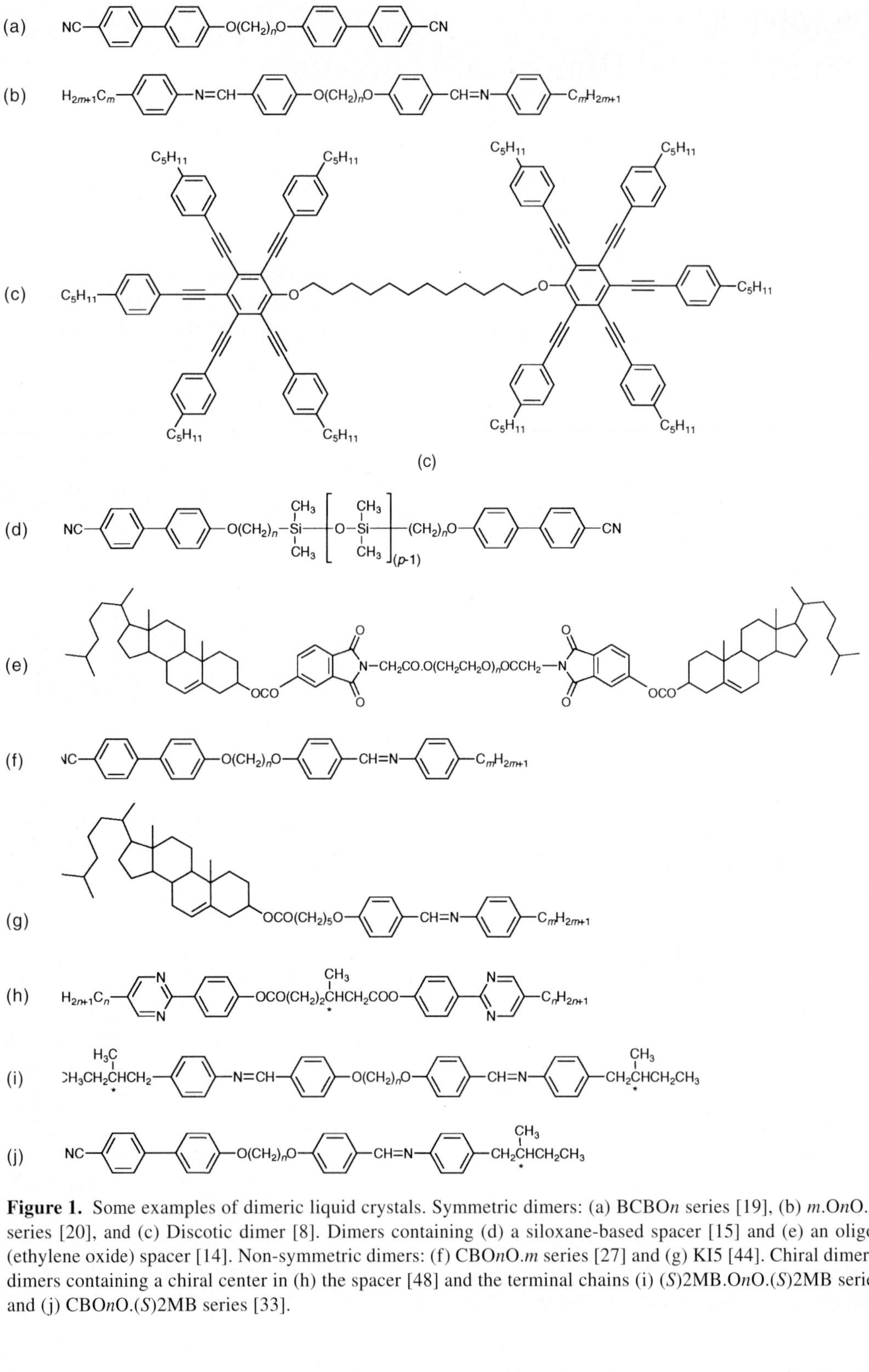

Figure 1. Some examples of dimeric liquid crystals. Symmetric dimers: (a) BCBO*n* series [19], (b) *m.On*O.*m* series [20], and (c) Discotic dimer [8]. Dimers containing (d) a siloxane-based spacer [15] and (e) an oligo-(ethylene oxide) spacer [14]. Non-symmetric dimers: (f) CBO*n*O.*m* series [27] and (g) KI5 [44]. Chiral dimers: dimers containing a chiral center in (h) the spacer [48] and the terminal chains (i) (*S*)2MB.O*n*O.(*S*)2MB series and (j) CBO*n*O.(*S*)2MB series [33].

erally rather than in terminal positions [17, 18]. Broadly speaking all dimers can be subdivided into two groups: symmetric dimers in which the two mesogenic units are identical and nonsymmetric dimers which contain two different mesogenic groups. A range of dimeric structures are shown in Fig. 1.

The two most extensively studied series of liquid crystal dimers are the α,ω-bis(4′-cyanobiphenyl-4-yloxy)alkanes [19] and the α,ω-bis[4-(4-alkylphenyliminomethyl)-phenoxy)alkanes [20] (structures (a) and (b) in Fig. 1, respectively) and we shall use these to illustrate the characteristic behavior of dimers. The acronyms used to refer to these series are BCBOn and m.OnO.m, respectively, in which n refers to the number of methylene groups in the flexible spacer and m that in the terminal alkyl chains. The particular interest in these series arises, in part, because they may be considered to be the dimeric analogues of the 4-n-alkyloxy-4′-cyanobiphenyls (nOCB) and the N-(4-n-alkyloxybenzylidene)-4′-n-alkylanilines, respectively, which are probably the most widely studied series of conventional low molar mass liquid crystals. In addition, members of the m.OnO.m series were the first dimers for which extensive smectic polymorphism was observed.

Figure 2 shows the dependence of the nematic–isotropic temperatures, T_{NI} on the number of methylene groups, n, for the BCBOn series. A very large odd-even effect is apparent in which the even members of the series exhibit the higher values although this alternation is attenuated on increasing n [19]. For comparative purposes the clearing temperatures for the analogous conventional or monomeric low molar mass series, the 4-n-alkyloxy-4′-cyanobiphenyls, are also shown in Fig. 2. The clearing temperatures of the monomeric compounds are considerably lower than those of the dimers

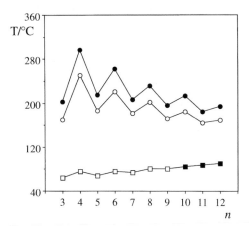

Figure 2. Dependence of the nematic–isotropic transition temperatures (○) on the number of methylene groups in the flexible spacer, n, for the BCBOn series [19]. Also shown are the nematic–isotropic (□) and smectic A–isotropic transition (■) temperatures for the nOCB series for which n now indicates the number of carbon atoms in the terminal alkyl chain, and the nematic–isotropic transition temperatures (●) for the TCBOn series.

and their dependence on the number of carbon atoms in the terminal chain is very much weaker than that seen on varying the length and parity of the spacer in the dimers.

The entropy changes associated with the N–I transition for the BCBOn series also show a pronounced alternation as the length and parity of the spacer is varied but which does not attenuate on increasing n, see Fig. 3 [19]; again, it is the even members which exhibit the higher values. Indeed the values for the N–I entropy for even members are typically several times larger than those for the odd members. The clearing entropies for the monomeric nOCB series are also shown in Fig. 3. The propyl to nonyl memers of this series exhibit a N–I transition and the associated entropy change is considerably smaller than either the odd or even membered dimers. The decyl, undecyl and dodecyl members of the monomeric series exhibit a SmA–I transition but the associated

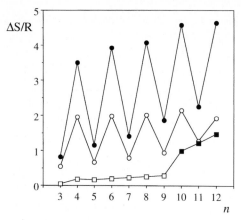

Figure 3. Dependence of the entropy change associated with the nematic–isotropic transition (○) on the number of methylene groups in the flexible spacer, n, for the BCBOn series [19]. Also shown are entropy changes associated with the nematic–isotropic (□) and smectic A–isotropic transition (■) for the nOCB series for which n now indicates the number of carbon atoms in the terminal alkyl chain, and the nematic–isotropic entropies (●) for the TCBOn series.

clearing entropies are smaller than those associated with the N–I transition for the even members of the dimers and comparable to those for the odd members.

Thus, both the N–I temperature, T_{NI}, and the associated entropy change for a series of liquid crystal dimers show very large odd–even effects as the length and parity of the spacer is varied in which the even members exhibit the higher values. This dependence of both T_{NI} and the associated entropy change on n is archetypal behavior for liquid crystal dimers. The differences between odd and even membered dimers are thought to reflect, at least in part, the difference in their average molecular shapes which are governed to a large extent by the parity of the flexible spacer. Furthermore, the N–I temperatures and entropies for even membered dimers are considerably higher than for the analogous conventional or monomeric compounds. We shall return

to a detailed discussion of the molecular significance of these observations later.

In general, the effects on the N–I temperature of a liquid crystal dimer on varying the chemical structure of the mesogenic units tend to mirror those seen for monomeric liquid crystals [4]; for example, increasing the length of the core enhances the N–I temperature [21] while lateral substituents cause this temperature to fall [22]. In comparison, the role the spacer plays in determining the liquid crystalline properties of a dimer both through controlling the shape of the molecules as well as simply by constraining the relative orientational dispositions of the mesogenic groups has no precedence in conventional low molar mass liquid crystals. In consequence it is primarily these issues which this chapter will address.

3   Smectic Polymorphism

3.1   Conventional Smectics

The full range of conventional smectic phase behavior has been observed for dimers including smectic A, B, C and F phases and the crystal B, E, G and H phases [20]. Rare phase transitions have also been observed, including SmF–SmA and crystal G–isotropic transitions [20]. Figure 4 shows the dependence of the phase behavior on the number of methylene groups in the flexible spacer, n, for a series of dimers, the 5.OnO.5 series (see Fig. 1 b), which exhibits both smectic and nematic behavior [20]. The characteristic, pronounced alternation in the clearing temperatures on varying n can be seen in Fig. 4. On increasing the length of the spacer, the tendency is for nematic behavior to emerge and for smectic phases to be extinguished (see Fig. 4). Similar behavior has been observed for many

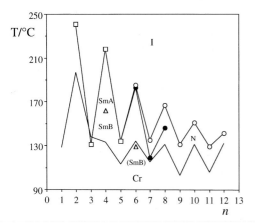

Figure 4. Dependence of the transition temperatures on the number of methylene groups, n, in the flexible alkyl spacer for the 5.OnO.5 series [20]. □ indicates smectic A–isotropic transitions, ○ the nematic–isotropic A transitions. The solid line represents the melting points. SmA, smectic A; N, nematic; SmB, smectic B; I, isotropic; Cr, crystal.

dimer series [20] but is somewhat surprising, given the very general observation that increasing the length of alkyl chains in mesogenic structures stabilizes smectic phases relative to the nematic phase. Examples of this include increasing the length of the terminal alkyl chain in a conventional low molar mass compound [23], increasing the length of a spacer in a semiflexible main chain liquid crystal polymer [24] and increasing the spacer length in a side-chain liquid crystal polymer [25].

To understand why the effect of increasing the spacer length in a dimer appears anomalous we must first consider why smectic phases form. Smectic phase formation may be considered in terms of a microphase separation in which the mesogenic units form one domain while the alkyl chains constitute another. There are two possible driving forces behind this separation: energetically if the mean of the mesogenic unit–mesogenic unit and chain–chain interactions is more favorable than the

mixed mesogenic unit-chain interaction then phase separation will occur, or entropically the interaction between a core and a chain acts to order the chain and hence, is unfavorable. Thus, for conventional low molar mass materials the very general observation is that increasing the length of a terminal alkyl chain promotes smectic phase behavior [23].

This is not, however, the case for increasing the length of the spacer in liquid crystal dimers. A study of 11 homologous series belonging to the m.OnO.m family (see Fig. 1 b) revealed a simple empirical relationship between the occurrence of smectic phase behavior and molecular structure [20]. Thus, for a smectic phase to be observed the length of the terminal chains has to exceed half the spacer length. This rule, established for the m.OnO.m compounds, is in fact obeyed by the overwhelming majority of dimers reported in the literature. The molecular interpretation of this observation considers the possible smectic structures symmetric dimers could adopt. In essence there are just two plausible molecular arrangements for symmetric dimers residing in extended conformations within a smectic phase and these are shown in Fig. 5. In a monolayer arrangement (see Fig. 5 a) smectic phase formation can be described as microphase separation giving three distinct regions containing either mesogenic cores, alkyl spacers or terminal chains. In the second arrangement, the molecules intercalate (see Fig. 5 b) and there are just two domains, one containing the mesogenic units and the other the alkyl chains. The difference between these structures is that in the intercalated arrangement the spacers and terminal chains are mixed while in the monolayer they phase separate. The empirical rule that the length of the terminal chain must exceed half the length of the spacer for smectic phases to be observed, effectively eliminates the possibility that the

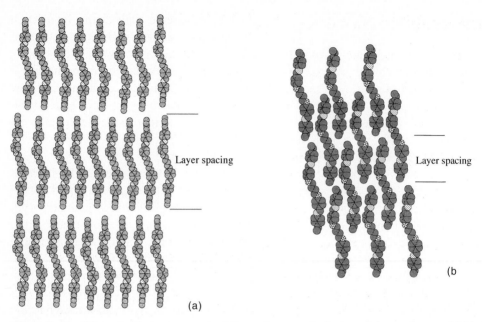

(a)

(b

Figure 5. Possible molecular arrangements of dimers in a smectic A phase: (a) monolayer and (b) intercalated.

dimers form an intercalated structure simply because the terminal chains can only be accommodated within such a structure if the total length of the two terminal chains is equal to or less than the length of the spacer (see Fig. 5b). This view is supported by X-ray diffraction studies which reveal that symmetric dimers generally give rise to smectic phases in which the layer spacing is approximately the molecular length, that is a monolayer structure [20]. By comparison, the intercalated structure would in fact give rise to a smectic phase in which the layer spacing is approximately half the molecular length and as we will see later such an arrangement is adopted by nonsymmetric dimers. The mesogenic unit–mesogenic unit interactions in both smectic structures shown in Fig. 5 are identical, strongly suggesting that the relative stabilities of the two arrangements must rest with the nature of the interaction between the spacer and the terminal alkyl chains. Furthermore, the ob-

servation of only the monolayer structure implies that the terminal chain–spacer interaction is an unfavorable one and so destabilizes the intercalated arrangement.

We have seen, therefore, that the very general observation is that for symmetric dimers increasing the length of the alkyl spacer decreases the tendency towards smectic phase formation. We have noted already that part of the interest in dimers arises from their ability to act as model compounds for semiflexible main chain liquid crystal polymers but in this respect they behave quite differently. Specifically, increasing the length of the spacer in a semiflexible polymer increases the tendency towards smectic phase formation [24]. Thus the driving force for smectic phase formation must differ between the dimers and polymers and presumably for the polymers this must be an entropic one in order to disentangle the polymer chains.

3.2 Intercalated Smectics

As we have seen, nonsymmetric dimers consist of molecules containing two differing mesogenic units linked via a flexible spacer (see Fig. 1); particular interest in these has focused on systems in which the mesogenic groups exhibit a specific molecular interaction [21, 26–33]. This was stimulated, in part, by the behavior of binary mixtures of conventional low molar mass liquid crystals in which one component consisted of molecules containing electron rich mesogenic units while the other consisted of molecules having electron deficient moieties. For many years it has been known that such mixtures exhibit clearing temperatures higher than the composition weighted average of those of the individual components and show a greater tendency to form smectic phases [34]. Both these observations are attributed to a specific interaction between the unlike cores being more favorable than the geometric mean of the interactions between the like cores. The smectic phases exhibited by these mixtures have conventional structures but the intriguing question was to what extent would the behavior of such mixtures and the structure of the phases they exhibit, be effected if the unlike mesogenic units were covalently linked in the same molecule? Thus many series of nonsymmetric dimers have now been characterized and as we shall see this resulted in the discovery of a novel family of smectic phases.

Figure 6 shows the dependence of the transition temperatures on the number of methylene units in the spacer, n, for the CBOnO.10 series (see Fig. 1f) [27]. The clearing temperatures exhibit the characteristic, pronounced alternation with the length and parity of the spacer which attenuates as the spacer length is increased. The clearing temperatures of the nonsymmetric dimers

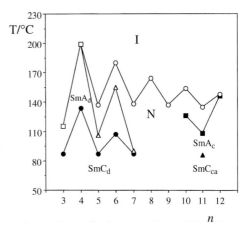

Figure 6. Dependence of the transition temperatures on the number of methylene groups, n, in the flexible alkyl spacer for the CBOnO.10 series [27]. □ indicates interdigitated smectic A–isotropic transitions, ○ nematic–isotropic transitions, ● interdigitated smectic A–interdigitated smectic C transitions, △ interdigitated smectic A–nematic transitions, ■ intercalated smectic A–nematic transitions and ▲ intercalated smectic A–intercalated smectic C transitions. The melting points have been omitted for the sake of clarity. SmA$_d$, interdigitated smectic A; SmC$_d$, interdigitated smectic C; SmA$_c$, intercalated smectic A; SmC$_{ca}$, intercalated alternating smectic C; N, nematic; I, isotropic.

are higher than the weighted averages of the corresponding symmetric dimers and such behavior is analogous to that seen for binary mixtures of the corresponding conventional low molar mass materials. For the nonsymmetric dimers, this enhancement of the clearing temperature is also interpreted in terms of a favorable specific interaction between the unlike mesogenic groups. In contrast to such anticipated behavior, however, smectic phases are observed for short and long spacer lengths but not for intermediate values contradicting the observation made for symmetric dimers that the length of the terminal chains had to exceed half the spacer length if smectic behavior was to be observed. If, instead of varying the spacer length, we vary the terminal chain length in

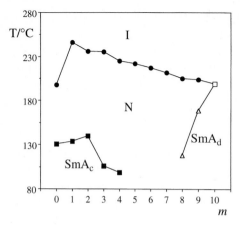

Figure 7. Dependence of the transition temperatures on the number of carbon atoms, m, in the terminal chain for the CBO4O.m series. □ indicates the interdigitated smectic A–isotropic transition, ● the nematic–isotropic transitions, △ interdigitated smectic A–nematic transitions and ■ intercalated smectic A–nematic transitions. The melting points have been omitted for the sake of clarity. SmA$_d$, interdigitated smectic A; SmA$_c$, intercalated smectic A; N, nematic; I, isotropic.

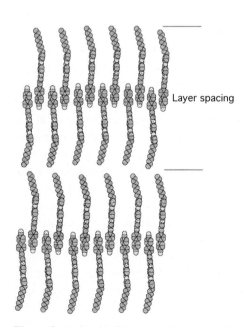

Layer spacing

Figure 8. A sketch of the molecular organization in the interdigitated smectic A phase exhibited by a nonsymmetric dimer, CBO10O.4.

a series of nonsymmetric dimers even more surprising behavior is observed. Figure 7 shows the dependence of the transition temperatures on the number of carbon atoms, m, in the terminal chain for the CBO4O.m series [27], here smectic behavior is observed for terminal chain lengths including up to four carbon atoms, not for the pentyl, hexyl or heptyl chains but re-emerges for chains containing eight carbon atoms or greater. This behavior is unique to nonsymmetric dimers and contravenes the very general observation that increasing the length of a terminal chain in a mesogenic structure promotes smectic phase formation.

To understand the unusual behavior of these two series we must consider the molecular arrangements within the observed smectic phases. For the CBOnO.10 series the ratio of the smectic periodicity to the estimated all-*trans* molecular length, d/l, for the SmA phase shows a dramatic dependence on n [27]. Thus, for $n = 3–7$ $d/l \approx 1.8$ while for the decyl, undecyl and dodecyl homologues $d/l = 0.5$. It is clear from these data that the structure of the smectic phase must somehow be governed by the length of the spacer. The d/l ratio of 1.8 was interpreted in terms of a conventional interdigitated smectic structure in which the like parts of the molecules overlap, see Fig. 8. The driving force for the formation of this phase is presumably the electrostatic interaction between the polar and polarizable cyanobiphenylyl groups while the smectic phase results from the molecular inhomogeneity arising from the long terminal alkyl chains as for the corresponding monomers. The apparent voids in the structure are presumably filled by the flexibility of the terminal chains. The higher homologues, however, must exhibit a quite different molecular arrangement in which it has been proposed it is the unlike parts of the molecule that overlap, see Fig. 9 [27]. This structure was

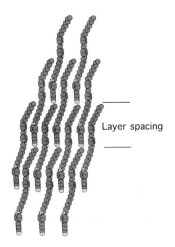

Layer spacing

Figure 9. A sketch of the molecular organization in the intercalated smectic A phase exhibited by a nonsymmetric dimer, CBO10O.10.

termed the intercalated SmA phase and the notation used to designate smectic phases was extended to include a subscript c to denote intercalated by analogy to the use of a subscript d to denote interdigitated. Thus, SmA_c represents the intercalated smectic A phase. The intercalated structure is thought to be stabilized via the specific interaction between the unlike mesogenic cores. The sketch of the phase shown in Fig. 9 implies that the structure has ferroelectric ordering but this is presumably removed by a random arrangement of such domains at the macroscopic level. This view is supported by the failure to detect any macroscopic polarization for several typical intercalated SmA phases [35]. The ability to accommodate the terminal chains in the space between the intercalated layers of mesogenic groups is determined largely by the length of the spacer. Thus, for the intercalated phases to be observed the terminal chain must be equal to or shorter in length than the spacer. If the terminal chain is considerably longer than the spacer then interdigitated phases are observed. For intermediate chain lengths

the disappearance of smectic behavior, see Fig. 6, suggests that neither smectic modification is favorable and hence, nematic behavior is observed. There is a strong similarity here to re-entrant nematic behavior which is also driven by two different length scales. It should be noted that this dependence of smectic behavior on the relative lengths of the spacer and terminal chain is not restricted to the CBOnO.m family but is observed for other nonsymmetric dimers possessing electron rich and deficient groups [32].

The precise nature of the specific interaction between the unlike mesogenic groups is unclear but often assumed to be a charge transfer interaction. This assumption has been questioned and more recently it has been suggested that it is, in fact, electrostatic quadrupolar interactions between groups with quadrupole moments which are opposite in sign that give rise to the intercalated structure [32]. It has also been noted that the increase in entropy arising from the mixing of the unlike groups provides an entropic contribution driving the formation of the intercalated arrangement. In addition, the intercalated arrangement mixes the terminal chains and flexible spacers again giving rise to an increase in entropy although this should operate also for symmetric dimers but apparently does not for the vast majority of systems.

The unique phase behavior shown by the CBO4O.m series, see Fig. 7, may also be rationalized in terms of intercalated and interdigitated smectic phases. For short terminal chain lengths the intercalated variant may be formed while for longer terminal chains the interdigitated arrangement is adopted. The crossover in structure is a result of insufficient space between the layers to accommodate the terminal chains. A surprising feature of the intercalated SmA phase was observed for the CBOnO.2 series for which intercalated behavior was observed for $n = 4-12$ and the stability of the phase tended to increase with increasing spacer length for odd and even members. This is unexpected because presumably space is filled more efficiently for CBO4O.2 than for CBO12O.2. This is thought to imply that the translational order in the phase is low but we will discuss this in more detail later.

The SmA_c phase was the first intercalated phase identified [26] but examples of intercalated SmC and I phases and intercalated crystal B and J phases have since also been observed [27]. The tilted intercalated phases are, however, only observed for homologues containing odd-membered spacers and it is thought that the difficulty experienced by these bent molecules packing efficiently into an intercalated structure is a driving force for the molecules to tilt. Of particular interest is the SmC_c phase in which the tilt direction alternates between the layers, see Fig. 10. In this respect the SmC_c phase is analogous both to the antiferroelectric SmC phase [36] and to a SmC phase exhibited by semiflexible main chain liquid crystal polymers [37]. In each of these phases, the tilt direction alternates between the layers and the notation used to refer to smectic phases has now been extended to include reference to this structural feature. Thus, a subscript a denotes an *alternating* tilt direction; for example, SmC_{ca} represents the intercalated and alternating SmC phase. It has been suggested that the transition from the SmA_c to SmC_c phase may be described as a biasing of the precessional distribution of the molecules about their long axes which gives rise to a long range correlation of the tilt angle within the layers. It is interesting to note that the global tilt angle in the SmC_c phase is zero but locally, within a layer is nonzero. In consequence, the SmC_c phase exhibits a rather unusual optical texture [38]. On cooling from the SmA phase, light

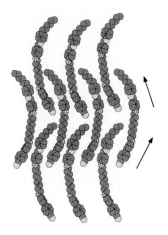

Figure 10. A sketch of the molecular organization in the intercalated smectic C phase exhibited by a non-symmetric dimer, CBO9O.6.

lines develop across the backs of the fans while in homeotropic regions a schlieren texture develops containing both types of point singularity. The latter texture is normally indicative of the nematic phase and is not observed for conventional SmC phases.

The proposed structure for the SmA_c phase shown in Fig. 9 does not fully explain all the experimental data and we have noted already that the stability of the SmA_c phase increases unexpectedly with spacer length for the CBOnO.2 series [27]. In particular it is difficult to see how for a compound, such as CBO12O.2, in which the terminal chain is very much shorter than the spacer, the molecules can pack efficiently into the intercalated structure without the presence of a considerable concentration of voids, see Fig. 11. We mentioned earlier that this problem could be resolved if the translational order within the phase is low. This would in turn suggest, however, that the overlap between the mesogenic groups was limited but it is the interactions between these groups which are normally considered to be responsible for driving the smectic

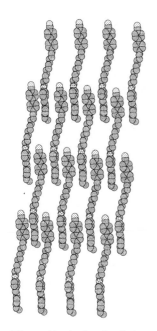

Figure 11. A sketch of the molecular organization in the intercalated smectic A phase exhibited by CBO12O.2.

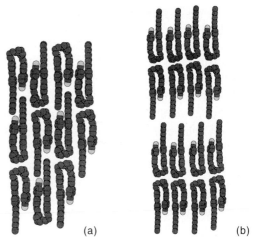

(a)   (b)

Figure 12. Horseshoe-like conformations of molecules arranged within a smectic phase for which (a) $d/l = 0.5$ and (b) $d/l = 1.8$.

arrangement. This difficulty with filling space has led to the suggestion that the dimers may adopt horseshoe-like structures which are stabilized by the favorable interaction between unlike mesogenic groups [32]. These horseshoe conformations can then be packed into a SmA phase in which the ratio of the smectic periodicity to the all-*trans* molecular length is 0.5, see Fig. 12a. Indeed it is also possible to pack these horseshoe-like molecules into a SmA phase for which $d/l = 1.8$ (Fig. 12b), that is the experimentally observed ratio when the terminal chain length exceeds that of the spacer. Further experiments are now required to establish unambiguously the local molecular arrangements within these phases. However, an indication that the intercalated smectic phases have the network structures shown in Figs. 9 and 11 and not the monolayer structure of Fig. 12a is given by the consid-

erably higher field-induced director relaxation times of the SmA_c phase in comparison with those of the interdigitated smectic phases measured at comparable temperatures [39].

The intercalated phases are not only observed for nonsymmetric dimers but are also exhibited by mixtures of the appropriate symmetric dimers, for example, a binary mixture of 2.O6O.2 and BCBO8 [27]. This observation tends to support the intercalated structures as opposed to the horseshoe conformations because it is far from clear why the symmetric dimers should adopt the horseshoe arrangements. It has also been reported that intercalated behavior can be exhibited by symmetric dimers in which the linking group of the spacer to the mesogenic units differs from that of the terminal chain to the mesogenic moieties [40] although this observation has been made for just a single set of compounds. It is also of interest to note that the nematic phase above an intercalated smectic phase has a local molecular packing which is intercalated [41].

3.3   Modulated Smectics

A second class of novel smectic phases are also exhibited by liquid crystal dimers, namely modulated smectic phases having tilted hexatic molecular ordering. The most common example of this has been termed the Sm1 phase and is exhibited by members of the $m.OnO.m$ family with $m = 10$, 12 and 14 and $n = 9$ and 11 [42]. The molecular arrangement within the Sm1 phase is shown schematically in Fig. 13 in which the molecules are assumed to have a bent molecular shape; the molecular shapes of dimers will be discussed in some detail later. In the Sm1 phase the mesogenic groups are tilted symmetrically with respect to the layer normal and the modulation is purely displacive along the b-axis. In a more realistic representation of the Sm1 phase than that shown in Fig. 13, the domain boundaries would be somewhat less well-defined. In essence, the Sm1 phase can be considered as the modulated SmF or SmI phase but the tilt direction of the director with respect to the local hexagonal lattice has yet to be determined for the Sm1 phase. The Sm1 phase has been observed in a number of liquid crystal phase sequences and has been obtained on cooling the I, SmA and SmC phases.

It is widely believed that modulated smectic phases arise as a result of a competition between different characteristic length scales; they are most commonly observed for mesogens possessing a terminal polar group, for example, the cyano group [43]. For such compounds the competing periodicities are the molecular length and the length of antiparallel molecular pairs. It is far from clear, however, where such a competition arises for the symmetric dimers which, as we have seen, give rise to monolayer smectic structures in which the periodicity is approximately the molecular length. A plausible explanation considers

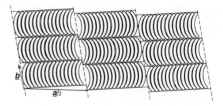

Figure 13. A sketch of the molecular organization in the Sm1 phase composed of bent molecules.

that the Sm1 phase has been observed only for odd membered dimers which strongly suggests that molecular shape plays an important role in the formation of the phase as this is the most striking difference between odd and even dimers. Thus, the bent odd dimers may interlock giving rise to the periodicity L' with $L < L' < 2L$ which competes with the monolayer periodicity L (see Fig. 13).

On cooling the Sm1 phase for 10.O9O.10 a second modulated hexatic phase is observed, the Sm2 phase [42]. This is a monotropic phase and its optical texture is indistinguishable from the preceding Sm1 phase. In addition, the phase transition could not be observed using differential scanning calorimetry. Surprisingly, the two phases appear to coexist over a relatively wide temperature range although further speculation on the nature of the Sm1 phase must await detailed structural studies.

Modulated smectic phases are just one of a number of classes of frustrated smectic phases, all of which arise from the competition between different characteristic length scales but differ in the manner by which this frustration is relieved. In the incommensurate SmA phases the competing periodicities coexist along the layer normal and the first example of such a phase was observed for a nonsymmetric dimer, KI5 (see Fig. 1 g). In this phase the larger periodicity appears to correspond to the molecular length while the shorter period may indicate a coexisting intercalated arrangement of the dimers [44, 45].

4   Chiral Dimers

Symmetric [44, 46–48] and nonsymmetric [33] chiral liquid crystal dimers have also been reported. The chiral center can be placed in either the flexible spacer [44, 48], or in the terminal chain [33, 49]. The rationale for placing the chiral center in the spacer, at least for even dimers, is that its orientational order should be higher than for a terminal chain with a subsequent enhancement of the form chirality of the phase. Examples of both these types of structure are shown in Fig. 1.

We have already seen that the transitional behavior of dimers depends strongly on the length and parity of the spacer. Blatch et al. [33] studied two series of chiral dimers, one symmetric (S)2MB.OnO.(S)2MB, and the other nonsymmetric, CBO-nO.(S)2MB, (see Fig. 1) in order to establish whether the form chirality of the chiral phases would also depend critically on the parity of the spacer. For the CBO-nO.(S)2MB series with $n = 7$ and 9 a blue phase was observed but not for $n = 6$ and 8. This was rationalized in terms of the smaller pitch for the odd relative to the even membered dimers which arises from the smaller twist elastic constant of odd dimers and is related to their lower orientational order. Surprisingly, the helical twisting powers of the dimers in a common monomeric nematic solvent appear to depend solely on the nature of the chiral group and not on its environment. Thus similar helical twisting powers are observed for both odd and even membered dimers.

5   Oligomeric Systems and Relation to Dimers

The transitional behavior of liquid crystal dimers and in particular the dramatic dependence of the clearing temperature and entropy on the length and parity of the flexible spacer is strongly reminiscent of that observed for semiflexible main chain liquid crystalline polymers. In order to investigate how these properties evolve from the dimers to the polymers a small range of higher oligomers have been characterized. The majority of these may be described as liquid crystal trimers as they are composed of molecules containing three mesogenic groups and two flexible spacers [31, 50–53], see Fig. 14, although liquid crystal tetramers containing four mesogenic groups and three spacers have also been reported [54].

The dependence of the N–I temperatures for the trimeric TCBOn series, see Fig. 14, on the length and parity of the flexible spacers is shown in Fig. 2 while Fig. 3 shows the corresponding entropy changes [55]. A pronounced alternation is evident in both these quantities as n is varied in which the even members exhibit the higher values. For any given value of n the N–I temperature is highest for the trimer and lowest for the monomer, see Fig. 1; such a trend is to be expected given the enhanced shape anisotropy on passing from the monomer to the dimer and trimer. The magnitude of the alternation exhibited by the N–I temperatures of the trimers is slightly higher than that exhibited by the N–I temperatures of the dimers on a relative scale. By comparison, the alternation exhibited by the N–I entropies of the trimers is considerably larger than that observed for the dimers, again using a relative scale. These data suggest that the mesogenic units in the trimers are correlated to

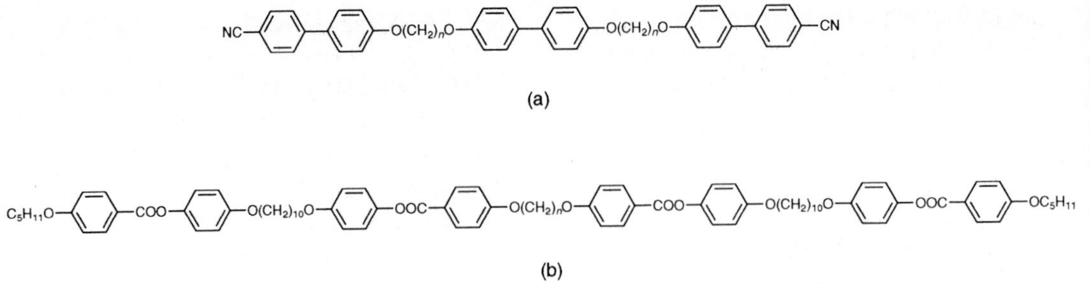

(a)

(b)

Figure 14. Examples of (a) trimeric, the TCBO*n* series [55], and (b) tetrameric liquid crystals [54].

the same extent as in the dimers and hence the similar alternation in the N–I temperatures, but the differences in the transitional entropies suggest that the orientational order is significantly higher in the nematic phases exhibited by the trimers. These observations are based, however, on a very limited data set and much research is now required to establish the generality of their behavior. Throughout this discussion we have not considered the need to normalise, in some way, the data with respect to the number of mesogenic groups. In calculating the enthalpies and hence entropies of transition for polymeric systems the molar mass of the repeat unit is used rather than the number or weight average molar mass of the polymer. The justification for this is that the intramolecular correlations decay along the polymer chain. Model calculations are clearly required to establish how the data for oligomers should also be scaled for comparative purposes.

6   Molecular Theories for Liquid Crystal Dimers

As we have seen, one of the main characteristics of liquid crystal dimers is the alternation of the transitional properties with the number of groups in the flexible spacer linking the two mesogenic units. The alternation in the change in the entropy at the N–I transition is particularly marked and essentially unattenuated with increasing length of the spacer, unlike the N–I transition temperature (see Figs. 2 and 3). Since the transitional entropy reflects the order of the system it is to be expected that the second rank orientational order parameter of the mesogenic group will also alternate with the parity of the spacer as will the order of solutes dissolved in the nematic phase of the dimer; such behavior has indeed been observed [56]. Similarly, the elastic constants which are related to the orientational order of the nematic phase are also found to be significantly larger for an even than for an odd dimer [57]. The alternation in the transitional entropy should also be reflected in the pretransitional behavior and again this is observed, for $(T_{N-I} - T^*)$ is significantly larger for dimers with even spacers in comparison for those with odd spacers [58, 59], where T^* is the divergence temperature.

The often dramatic difference in behavior between dimers with odd and even spacers is sometimes attributed to the difference in their shape when the spacer has the all-*trans* conformation (see Fig. 15). Thus, the two mesogenic groups are antiparallel for the even dimer whereas for the odd dimer

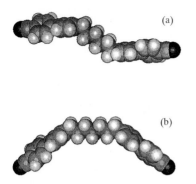

(a)

(b)

Figure 15. The structure of (a) even and (b) odd liquid crystal dimers with the spacers in their all-*trans* form.

they are inclined. This structure for the even dimer is more compatible with the molecular organization in the nematic phase than for the odd dimer; it is then argued that this greater compatibility results in, for example, the larger transitional entropy of even dimers. However, this cannot be the entire explanation for the even dimers with their more or less linear molecular structures should behave like monomers whereas the odd dimers with their bent structures should exhibit weaker transitions. In fact, essentially the reverse is the case with the even dimers having stronger N–I transitions than the monomers whereas the odd dimers behave more or less like the monomers (see Fig. 3). The missing factor in the argument is the flexibility of the spacer linking the two mesogenic groups which allows the molecules to adopt a wide range of conformations in which the groups are either antiparallel (or parallel) or inclined for both odd and even dimers. We shall find it convenient to refer to the conformers with antiparallel (or parallel) mesogenic groups as linear and to those conformers with inclined mesogenic groups as bent. The difference between the dimers is not in the geometry of the conformers but rather in the relative amounts of the two classes of conformer. Thus for even

dimers, approximately half the conformers are linear whereas for the odd dimers only about 10% are linear. There is a synergy between this conformational distribution and the orientational order of the nematic phase. For the even dimers this means that at the transition to the nematic phase many of the bent conformations are converted to the linear form which enhances the orientational order and increases the entropy over that expected for a monomer. In contrast, for the odd dimers the difference in the free energy between the linear and bent conformers is such that the orientational order of the nematic phase is insufficient to convert bent to linear conformers. As a result, there is no increase in the orientational order over that expected for a bent monomer.

Although this predicted behavior is qualitatively in accord with that found for dimers it is important to see whether a model based on the synergy between conformational and orientational order can account, at least semiquantitatively, for the unusual properties of liquid crystal dimers. In the next two sections we describe two models which have been proposed and which have been treated using the molecular field approximation. The predictions of the theories are outlined and, where possible, contact is made with experiment.

6.1   The Generic Model

The development of a theory for mesogens composed of flexible molecules is complicated by the large number of conformational states which the molecules can occupy. Even within the Flory rotational isomeric state model [60] where the conformations are restricted to just *trans* and *gauche* forms the number of discrete conformers, 3^{N-1}, adopted by a spacer containing N groups can become excessive and the extensive calculations can mask the basic physics of the problem. We shall describe such a theory in the following section but here we are concerned with the so-called generic model in which only the essential features are retained [61, 62].

The spacer in a liquid crystal dimer serves two functions, first it contributes to the molecular anisotropy and second, it controls the relative orientation of the two mesogenic groups. This latter function is the most important as far as the behavior of the dimers is concerned and so, for the moment, the molecular anisotropy of the chain is ignored. For a spacer with a tetrahedral geometry, within the rotational isomeric state model the mesogenic groups will be either antiparallel (or parallel) or inclined at the tetrahedral angle of $\cos^{-1}(-1/3)$. This reduces the problem, apparently, to one in which only three conformers need to be considered, one bent and the other two linear. However, the dominant interaction responsible for nematic behavior is apolar and so conformations in which the mesogenic groups are parallel are equivalent to those in which they are antiparallel; at least within the form of this theory where the spatial arrangement of the mesogenic groups is ignored. The generic model, therefore, replaces the many conformational states available to the dimer with just two, one linear and the other bent (see Fig. 16). The nematic behavior of this binary mixture has been studied using the molecular field approximation in which the many-body interactions are replaced by an effective or molecular field. The potential of mean torque experienced by a conformer is then a function of the orientation of the director with respect to an axis system set in the molecule. For the linear conformer the potential of mean torque takes the Maier–Saupe form [63]

$$U_1(\omega) = -X_1 P_2(\cos\beta) \tag{1}$$

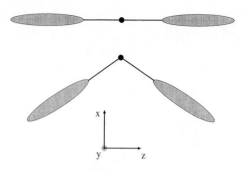

Figure 16. The linear and bent conformers of the generic model for liquid crystal dimers. The axis system associated with the bent conformer and reflecting its symmetry which is used to define the director orientation is also shown.

where $P_2(\cos\beta)$ is the second Legendre polynomial $(\equiv(3\cos^2\beta-1)/2)$ and the strength parameter, X_1, is related to the orientational order of the phase, although for the moment we can make some progress without specifying the form of this. X_1 also depends on the molecular geometry and for the segmental approximation proposed for flexible molecules by Marcelja [64] it is the tensorial sum of the interaction parameters for the two mesogenic groups, that is

$$X_1 = 2X_a \tag{2}$$

The potential of mean torque for the bent conformer depends on the polar angle, β, and the azimuthal angle, γ, which the director makes in the principal molecular axis system (see Fig. 16); these angles are denoted collectively by ω. This potential is assumed to take the form proposed by Luckhurst et al. [65], that is

$$U_b(\omega) = -\{X_b P_2(\cos\beta) \tag{3}$$
$$+ \delta X_b (3/8)^{1/2} \sin^2\beta \cos 2\gamma\}$$

where

$$X_b = X_a \tag{4}$$

and δX_b, which reflects the biaxiality of the bent conformer, is

$$\delta X_b = X_a/\sqrt{6} \tag{5}$$

The second rank orientational order parameters of the two conformers can be obtained by taking the Boltzmann average using the appropriate potential of mean torque. For example, for the linear conformer the order parameter for the molecular symmetry axis is

$$\bar{P}_2^l = Q_1^{-1} \int P_2(\cos \beta)$$
$$\cdot \exp\{-U_1(\omega)/k_B T\} \, d\omega \qquad (6)$$

where the orientational partition function Q_1, is

$$Q_1 = \int \exp\{-U_1(\omega)/k_B T\} \, d\omega \qquad (7)$$

There are analogous expressions for the major order parameter, $\bar{P}_2^b$, and the biaxial order parameter $(3/8)^{1/2} \sin^2\beta \cos 2\gamma$ for the bent conformer. From these principal order parameters it is possible to determine the order parameters for the symmetry axes of the mesogenic groups in the two conformers $\bar{P}_2^{la}$ and $\bar{P}_2^{ba}$. It is the weighted average of these which is available from NMR experiments [66],

$$\langle \bar{P}_2^a \rangle = x_l \bar{P}_2^{la} + x_b \bar{P}_2^{ba} \qquad (8)$$

where x_l and x_b are the mole fractions of the linear and bent conformers, respectively. One important feature of the model is that the composition of the system can change in response to variations in the orientational order. In the isotropic phase where there is no long range order the composition is determined by

$$x_b^0 = g_b \exp(-\Delta E/k_B T)/$$
$$[g_l + g_b \exp(-\Delta E/k_B T)] \qquad (9)$$

Here, ΔE is the energy difference between the two conformers and g_b is a degeneracy factor which allows for the fact that for real dimers about 75% of the conformers have the mesogenic groups inclined with respect to each other irrespective of the length and parity of the spacer. It is the energy differ-

ence which favors the linear form for even spacers and which enhances the bent form for the odd spacers. Within the model, therefore, ΔE for even dimers is positive so that x_b^0 is small while for odd dimers ΔE is negative and so x_b^0 is large. Over the temperature range of most liquid crystal dimers the Boltzmann factor, $\exp(-\Delta E/k_B T)$, can be taken to be constant. Then the composition in the nematic phase is related to that in the isotropic phase by [66]

$$x_b = x_b^0 Q_b/(x_l^0 Q_1 + x_b^0 Q_b) \qquad (10)$$

By taking the ratio x_l/x_b we can see that the nematic phase will favor the conformer with the largest orientational partition function which in this model is the linear conformer. The N–I transition is located by finding when the molar Helmholtz free energy in the isotropic phase

$$A_I = -RT \qquad (11)$$
$$\cdot \{\ln 4\pi + \ln[g_l + g_b \exp(-\Delta E/k_B T)]\}$$

is equal to that in the nematic phase

$$A_N = -RT \{\ln[g_l Q_1 + g_b Q_b \exp(-\Delta E/k_B T)]\}$$
$$+ (1/2) [x_l X_l \bar{P}_2^l + x_b X_b \bar{P}_2^b \qquad (12)$$
$$+ \delta X_b (3/8)^{1/2} \overline{\sin^2\beta \cos 2\gamma}]$$

Given this brief outline of the generic model, we are now able to consider certain of its predictions. The first is the dependence of the entropy of transition on the composition of the isotropic phase. In the limit that this is unity, the system corresponds to a pure system of rods and so is analogous to a rod-like monomer. At the other limit, of x_l^0 equal to zero, the system is again a single component but now of bent molecules; indeed within our parametrization of the model the particles have the maximum biaxiality. In between these two extremes, the system corresponds to a liquid crystal dimer; even dimers would have x_l^0 of about 0.5 whereas odd dimers should have x_l^0 of less than 0.1. From

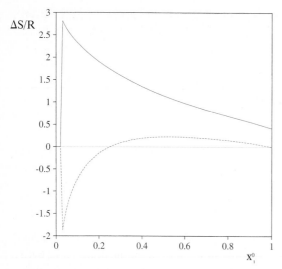

Figure 17. The dependence of the entropy change, $\Delta S/R$, at the nematic–isotropic transition on the mole fraction, x_1^0, of the linear conformer in the isotropic phase. The conformational contribution to $\Delta S/R$ is shown as the dashed line.

Fig. 17, we see that $\Delta S/R$ is equal to 0.417 when $x_1^0=1$ which is the Maier–Saupe prediction for monomers [63]. Then as the mole fraction of bent conformers increases the transitional entropy is also predicted to increase which is paradoxical because for a pure system of biaxial particles $\Delta S/R$ is predicted to decrease with increasing molecular biaxiality [65]. We shall return to this important point shortly but for the moment we note that for x_1^0 of 0.5 $\Delta S/R$ has increased to 1.3 which is over three times the value found for monomers and so is analogous to the behavior of even liquid crystal dimers. As x_1^0 continues to decrease so $\Delta S/R$ increases until for a mole fraction of the linear conformer of less than 0.03 the transitional entropy falls catastrophically to a value of less than 0.03. This behavior is reminiscent of odd dimers although $\Delta S/R$ is somewhat less than the values usually found, as is the critical composition for x_1^0. In the limit that x_1^0 vanishes so too does the transitional entropy, in

fact the isotropic phase is predicted to undergo a second order transition direct to a biaxial nematic phase [67], in keeping with the maximal molecular biaxiality of the bent conformer within this version of the generic model. The transitional entropy is made up of two contributions, one resulting from the onset of long range orientational order and the other from the change in the conformational composition. It is possible to use the generic model to assess the relative magnitudes of these contributions and in Fig. 17 we show the conformational entropy as the dashed line, and how it varies with the mole fraction for the linear conformer in the isotropic phase. We can see that for the values of x_1^0 corresponding to the even dimers the conformational entropy makes a minor contribution to the total entropy of transition. For compositions which would be associated with odd dimers, the conformational contribution is still smaller than that for the even dimers although it makes about the same relative contribution as for the even dimers.

The model can also predict the average order parameter, $\langle \bar{P}_2^a \rangle_N$, of the mesogenic groups in the nematic phase at the N–I transition, again without making any assumptions as to how the strength parameter X_a depends on the orientational order. The results of this calculation are shown as a function of the mole fraction of the linear conformer in the isotropic phase in Fig. 18. At one extreme, the order parameter is 0.429 as predicted by the Maier–Saupe theory for a monomer composed of linear molecules [63]. The other extreme corresponds to a system of pure biaxial particles and here $\langle \bar{P}_2 \rangle_N$ is necessarily zero at the second order transition to the biaxial nematic phase [67]. In between these two limits $\langle \bar{P}_2^a \rangle_N$ increases with the mole fraction of the bent conformer. When x_1^0 is 0.5 which corresponds to even dimers the order parameter has increased to just 0.55; unlike the transitional entropy the change in $\langle \bar{P}_2 \rangle_N$ relative to the monomer is not quite so dramatic. This behavior is in accord with that found for real liquid crystal dimers [56]. Again there is a catastrophic reduction in the order parameter $\langle \bar{P}_2^a \rangle_N$ to approximately 0.07 when the mole fraction of the linear conformer is about 0.03. This is analogous to the behavior of the odd liquid crystal dimers but as for $\Delta S/R$ the reduction occurs at too low a mole fraction of the linear conformer and the order parameter is reduced to too small a value.

To use the model to predict other properties of liquid crystal dimers, for example, the N–I transition temperature and the temperature dependence of the order parameter it is necessary to make an additional approximation. This is to relate the strength parameter X_a for a mesogenic group to the orientational order of the nematic mesophase. By analogy with the Maier–Saupe theory [63] and the extension of this to multicomponent mixtures [68] it is assumed that

$$X_a = \varepsilon_{aa} \langle \bar{P}_2^a \rangle \qquad (13)$$

where ε_{aa} reflects the anisotropic interactions between two mesogenic groups. Indeed this parameter is directly proportional to the N–I transition temperature for the monomeric mesogen [63]:

$$T_{NI} = 0.2203 \, \varepsilon_{aa}/k_B \qquad (14)$$

The dependence of T_{NI} on the composition of the isotropic phase predicted by the generic model is shown in Fig. 19 [69]. At the limiting composition when x_1^0 is unity the scaled transition temperature, $k_B T_{NI}/\varepsilon_{aa}$, is equal to 0.4406, because the molecules contain two mesogenic groups. At the other extreme when the system contains just the bent conformer the scaled transition temperature is found to be 0.1333 in accord with previous molecular field predictions for this particle with its maximal biaxiality [67]. In

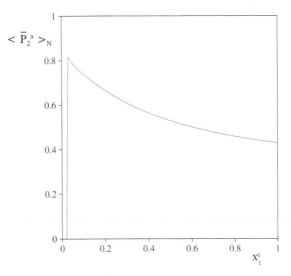

Figure 18. The variation of the average second rank orientational order parameter at T_{NI} for the mesogenic groups, $\langle \bar{P}_2^a \rangle_N$, with the mole fraction of the linear conformer in the isotropic phase.

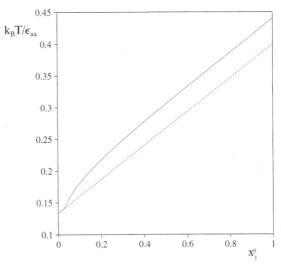

Figure 19. The dependence of the scaled nematic–isotropic transition temperature, $k_B T_{NI}/\varepsilon_{aa}$, (—), and the scaled divergence temperature, $k_B T^*/\varepsilon_{aa}$, (---) on the mole fraction of the linear conformer in the isotropic phase.

between these two limits, the introduction of the bent conformer is seen to depress the N–I transition. This is analogous to the reduction in T_{NI} found when a liquid crystal dimer is created by the introduction of a spacer between two mesogenic groups. We also see that the even dimers with their large values of x_1^0 are predicted to have higher transition temperatures than odd dimers with smaller values of x_1^0, in keeping with experiment [19]. However, it is noticeable that unlike the transitional entropy the transition temperature decreases continuously with the mole fraction of the linear conformer.

The generic model can also be used to predict the divergence temperature which is identified as the bifurcation point for the orientational order parameter. The scaled divergence temperature, $k_B T^*/\varepsilon_{aa}$, is shown in Fig. 19 as a function of x_1^0 [69]. For the pure system of rods ($x_1^0 = 1$) the divergence temperature is 0.4000 which is simply twice that for a monomeric mesogen; this is lower than the scaled T_{NI} [70]. At the other extreme ($x_1^0 = 0$) T^* is equal to the transition temperature, as required for a second order phase transition. In between these two extremes the divergence temperature varies linearly with the mole fraction of the linear conformer [70]. In fact, it is the difference $(T_{NI} - T^*)$ or the ratio T^*/T_{NI} which is of particular interest. From the predictions in Fig. 19 the ratio T^*/T_{NI}^* has been calculated as a function of the mole fraction of the linear conformer. At the limit when x_1^0 is unity the ratio T^*/T_{N-I}^* is predicted to be 0.9070 which is the Maier–Saupe result for monomers [70]. As the mole fraction of the linear conformer decreases so T^*/T_{NI}^* also decreases which is in accord with the behavior of even dimers where the difference in T^* and T_{NI} is observed to be larger than for analogous monomers [58, 59]. In addition, at the other extreme when x_1^0 is essentially zero T^*/T_{NI}^* is close to unity which is comparable to that found for odd dimers [58, 59]. Unlike the transitional entropy which exhibits a catastrophic collapse when x_1^0 falls below 0.03 no such behavior is

found for the ratio T^*/T_{NI}^*. This is found to decrease with decreasing x_1^0 but then to pass through a minimum when the mole fraction of the linear conformer in the isotropc phase is about 0.2. It then increases rapidly before levelling out at x_1^0 of approximately 0.03 and reaching the limiting value of unity when x_1^0 vanishes.

The temperature dependence of the average order parameter for the mesogenic groups, $\langle \bar{P}_2^a \rangle$, and the mole fraction of the linear conformer, x_1, are also available from the theory and some typical results are shown in Fig. 20. One is for x_1^0 of 0.5 which is expected to behave in a manner analogous to even dimers and the other has x_1^0 of 0.01 which should mimic the odd dimers. For x_1^0 of 0.5 the order parameter jumps to a large value at the N–I transition and then changes relatively slowly with decreasing temperature. In contrast, for x_1^0 of 0.01 the order parameter, $\langle \bar{P}_2^a \rangle$, undergoes a very small jump at the transition but then increases rapidly with decreasing temperature. This predicted difference in the temperature dependence of the average order parameter is in agreement with that observed for pure liquid crystal dimers [71] and for solutes dissolved in them [72]. Perhaps of greater interest is the observaton of a second phase transition for the system with a very low concentration of the linear conformer in the isotropic phase (see Fig. 20b). At a scaled temperature of about 0.119 the average order parameter undergoes a large change, corresponding to a strong first order phase transition. Associated with this is a very large increase in the mole fraction of the linear conformer from about 0.05 to 0.85 (see Fig. 20b). The two nematic phases involved in this transition differ in their orientational order and composition. This predicted transition appears to be analogous to that observed for semiflexible main chain liquid crystal polymers [73]. Indeed, it is signifi-

Figure 20

(a)   (b)

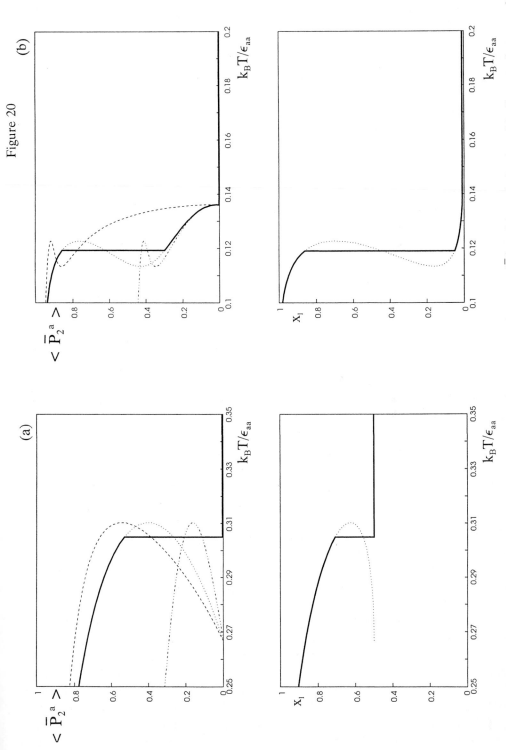

Figure 20. The temperature dependence of the average order parameter for the mesogenic groups, $\langle \bar{P}_2 \rangle$, and the mole fraction of the linear conformer, x_1, for a mole fraction x_1^0 in the isotropic phase of (a) 0.5 and (b) 0.01.

cant that this transition is only observed when the flexible spacer contains an odd number of units, as predicted by the generic model. The composition is also predicted to change at the N–I transition; this change is extremely small for the system with x_1^0 of 0.01 but is considerably larger when x_1^0 is 0.5; that is the conformational distribution is predicted to change significantly at the N–I transition for even dimers but not for odd dimers.

It is clear that the generic model provides a valuable account of most of the properties of the nematic behavior of liquid crystal dimers which depend on the parity of the spacer. In addition, the model gives a valuable insight into why the parity of the spacer is so important in influencing the nematogenic properties. Thus the addition of the bent conformer depresses the N–I transition temperature for the pure linear conformer. However, for concentrations of the bent conformer equivalent to an even dimer a point is reached where the gain in orientational free energy produced by converting bent to linear conformers more than compensates for the loss of conformational free energy. In other words within the nematic phase the orientationally ordered environment converts the bent to the linear conformer. As a result the effective N–I transition temperature increases and the effective reduced temperature necessarily decreases. This causes the orientational order to increase which also contributes to the large transitional entropy. In contrast when the concentration of bent conformers is so high, as for an odd dimer, the gain in orientational free energy at the transition is insufficient to convert bent to linear conformers. As a result the N–I transition is weak and the orientational order in the nematic phase is small. However, at some lower temperature it is possible to convert bent to linear conformers with the orientational free ener-

gy now being sufficient to account for the loss in conformational free energy, leading to a N–N transition.

In concluding this section on the generic model we should note that it has been used to explore the dependence of the predictions of the model on the bond angle for the bent conformer [62]. It is found that as this angle is increased beyond the tetrahedral value so the strength of the N–I transition decreases for compositions, x_1^0, corresponding to even dimers and increases for those associated with odd dimers; thus diminishing the odd–even effect. This behavior is entirely in keeping with that found for real dimers [7]. In addition, the N–N transition is predicted to change dramatically with the bond angle for the bent conformer. Thus, increasing the bond angle causes the N–N transition to weaken while the N–I transition strengthens, as found for the semiflexible main chain liquid crystal polymer [73]. The N–N transition is also predicted to exhibit a critical behavior with increasing bond angle; that is as the bond angle is increased so the difference in the two nematic phases decreases and vanishes for a bond angle between 112° and 115°.

6.2 A More Complete Model

Although the generic model which we have discussed in the preceding section provides a valuable route to understanding the properties of liquid crystal dimers it cannot be used to predict their behavior quantitatively. To achieve this it is necessary to include in the model all of the conformational states which the spacer can adopt. The simplest way to describe these is with Flory rotameric state model [60] which restricts the conformers to a discrete number. The conformational distribution in the isotropic phase is given by the Boltzmann average

$$p_n = \exp[-U_{int}(n)/k_BT]/$$
$$\sum \exp[-U_{int}(n)/k_BT] \quad (15)$$

where n labels a particular conformation whose internal or conformational energy is

$$U_{int}(n) = n_g E_{tg} + n_{g\pm g\mp} E_{g\pm g\mp} \quad (16)$$

Here, E_{tg} is the energy difference between a *trans* and a *gauche* link and the number of *gauche* links in the spacer is n_g. The term $n_{g\pm g\mp} E_{g\pm g\mp}$ is the additional energy resulting from the so-called pentane effect produced by a $g\pm g\mp$ sequence in the spacer. Within the nematic phase the more anisotropic conformers are favored, as we have seen, and the conformational distribution becomes [66]

$$p_n = Q_n \exp[-U_{int}(n)/k_BT]/$$
$$\sum Q_n \exp[-U_{int}(n)/k_BT] \quad (17)$$

Here the orientational partition function for the nth conformer is

$$Q_n = \int \exp\{-U_{ext}(n,\omega)/k_BT\}\, d\omega \quad (18)$$

where the external potential in the molecular field theory has the same form as that given in Eq. (3) for the bent conformer. However, the interaction parameters, X_n and δX_n, now depend on the conformation as does the orientation of the principal axes of the interaction tensor with respect to the molecular frame.

A number of theories have been proposed with which to calculate the interaction tensor for each conformer; this is clearly essential given the large number of conformers. The first approach was by Marcelja [64] who suggested that $\mathbf{X}_n$ should be a tensorial sum of contributions from the segments constituting the molecule. These segments were taken to be the mesogenic groups and the carbon–carbon bonds in the alkyl chain, both were assumed to be cylindrically sym-

metric and assigned the interaction parameters X_a and X_c, respectively. An alternative segment-like approach has been developed by Photinos et al. [74] and in its simplest form this adds terms to those from the segmental building blocks proposed by Marcelja. These additional terms are contributions from vectors joining the midpoints of adjacent bonds; these vectors are known as chords which give the model its name. They are especially important because this allows correlations between bonds to be introduced into the model.

It is often argued that it is the shape anisotropy which is largely responsible for liquid crystal formation. Two methods have been proposed to introduce this view into the calculation of the interaction tensor for each conformer. In one it is assumed that the tensor is proportional to the moment of inertia tensor which is readily calculated from a knowledge of the molecular geometry [75]. However, it is found that this parametrization results in too great a dependence of the N–I transition temperature on the molecular length [76]. This observation was partly responsible for the development of the surface tensor model [77]. In this the interaction tensor is defined in irreducible form as

$$X_n^{(2m)} = -\int_s dS C_{2m}(\theta,\phi) \quad (19)$$

where $C_{2m}(\theta,\phi)$ is a modified spherical harmonic, θ,ϕ denote the spherical polar angles which the normal to the surface makes in a molecular frame and the integration is over the surface of the molecule. This approach has the merit that the number of adjustable parameters is kept to a minimum provided the molecular geometry is known. This geometry is not, on its own, sufficient because the molecular surface must also be defined and this is usually achieved by the addition of van der Waals spheres to the nuclear coordinates. When the radii of these is small so that only spheres on bonded atoms overlap then the model becomes equivalent to the segmental approach of Marcelja. However, when the van der Waals spheres on nonbonded atoms overlap, as is usually the case, then the additional terms are analogous to the chord contributions appearing in the model proposed by Photinos et al. [74].

Our aim here is not to make a detailed comparison of the various parametrizations which have been proposed for the potential of mean torque. Instead we wish to illustrate the nature of the results which can be obtained with models which include all of the conformations for the dimer, suitably weighted [78]. The calculations proceed in an analogous manner to the generic model; for example to determine the N–I transition temperature it is necessary to determine when the molar Helmholtz free energy of the isotropic phase

$$A_I = -RT \ln Z_I \tag{20}$$

is equal to that for the nematic phase

$$A_N = N_A (X_a \langle \bar{P}_2^a \rangle + X_c \langle \bar{P}_2^c \rangle)/2 - RT \ln Z_N \tag{21}$$

In these expressions the conformational–orientational partition functions are given by

$$Z_I = 4\pi \sum \exp\{-U_{int}(n)/k_B T\} \tag{22}$$

and

$$Z_N = \sum Q_n \exp\{-U_{int}(n)/k_B T\} \tag{23}$$

where $\langle \bar{P}_2^a \rangle$ is the conformationally averaged order parameter for the mesogenic groups and $\langle \bar{P}_2^c \rangle$ is the sum of the conformationally averaged order parameters for the bonds in the spacer. As in the generic model the interaction parameters are related to the orientational order in the nematic phase; for example

$$X_a = (\varepsilon_{aa} \langle \bar{P}_2^a \rangle + \varepsilon_{ac} \langle \bar{P}_2^c \rangle)/V_m \tag{24}$$

where V_m is the molecular volume, with an analogous expression for X_c. This introduces three scaling parameters ε_{aa}, ε_{ac} and ε_{cc} into the model and to remove one of these the geometric mean approximation

$$\varepsilon_{ac} = (\varepsilon_{aa}\varepsilon_{cc})^{1/2} \tag{25}$$

is invoked [78]. This assumption has the additional merit of reducing the complexity of the calculations significantly.

Some of the parameters occurring in the model, such as E_{tg} and $E_{g\pm g\mp}$, can be taken from other studies while quantities such as ε_{aa} are obtained by fitting the predictions of the theory to experiment. As an example of the quality of the fit we show in Fig. 21 the predicted dependence of the N–I transition temperature on the number of methylene units in the spacer of the α,ω-bis(4-cyano-biphenyl-4'-yloxy)alkanes [27]; for comparison the experimental T_{NI} are also included. As we can see the theory predicts a strong alternation in T_{NI} together with an attenuation combined with an overall reduction in the transition temperature with increasing spacer length. This predicted behavior is seen to be in good but not perfect agreement with experiment. The theory has also been used to predict the dependence of

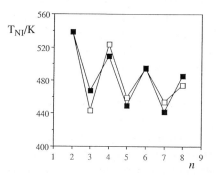

Figure 21. The observed (□) and predicted (■) dependence of the nematic–isotropic transition temperature on the number of methylene groups in the flexible spacer for the α,ω-bis(4-cyanobiphenyl-4'-yloxy)alkanes.

the transitional entropy for the cyanobiphenyl dimers on the number of methylene groups in the flexible spacer. The predicted variation was calculated with the same parameter set as that used to determine the N–I transition temperatures. The results of the calculations are shown in Fig. 22 where the experimental values are included for comparison. The theory predicts a strong alternation in $\Delta S/R$ which is not attenuated with increasing length of the spacer; indeed the change in the transitional entropy is predicted to grow with the spacer length. In addition there is a slight increase in $\Delta S/R$ for the odd as well as for the even dimers as the length of the spacer grows. Again, these predictions are found to be in rather good agreement with experiment.

Experimentally, the transitional behavior of the cyanobiphenyl dimers is observed to depend quite critically on the nature of the group linking the mesogens to the flexible spacer. For example, the N–I transition temperatures are higher for the ether than for the methylene linked dimers. Similar differences have been found for the cyanobiphenyl monomers [4] where the difference is usually related to an increase in the polarizability anisotropy resulting from the conjugation of the oxygen to the phenyl ring. Clearly the same explanation could be applied to the dimers but there is an alternative explanation as we shall now explain. The methylene and ether links also differ in their geometry, thus the $C_{ar}\hat{O}C_{al}$ bond angle is greater than that $C_{ar}\hat{C}_{al}C_{al}$ for the methylene linked dimers. In consequence, the all-*trans* conformer for an odd spacer has a greater anisotropy for the ether than for the methylene linked dimer. This greater anisotropy should result in a higher N–I transition temperature for the ether linked dimers than for their methylene linked counterparts. To see if this expectation is fulfilled the $T_{N–I}$ were calculated as a function of the

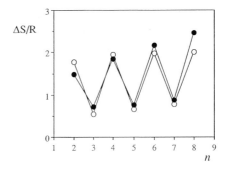

Figure 22. The observed (□) and predicted (■) dependence of the transitional entropy of the number of methylene groups in the flexible spacer for the α,ω-bis(4-cyanobiphenyl-4′-yloxy)alkanes.

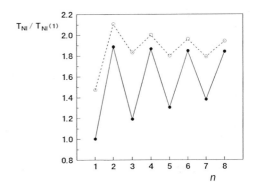

Figure 23. The dependence of the nematic–isotropic transition temperature on the number of atoms in the flexible spacer predicted for liquid crystal dimers with methylene (●) and ether (○) linkages between the spacer and the mesogenic groups.

spacer length for two series of liquid crystal dimers differing solely in the geometry of the link between the spacer and the mesogenic group [79]. The results of these calculations are shown in Fig. 23 where we can see quite clearly that the $T_{N–I}$ for the ether linked dimers are greater than those for the methylene linked dimers. The predicted increase in the N–I transition temperature is seen to be greater for the odd dimers than for the even dimers which means that the alternation in T_{NI} is greater for the methy-

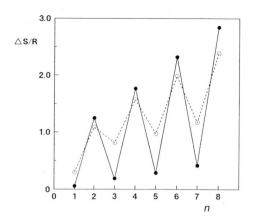

Figure 24. The dependence of the transitional entropy on the number of atoms in the flexible spacer for liquid crystal dimers with the mesogenic groups connected to the spacer by methylene (●) and ether (○) linkages.

lene than the ether linked dimers. This pronounced difference in the transitional behavior is found to be in good agreement with the behavior of real liquid crystal dimers which suggests the geometric effect alone is sufficient to account for these observed differences in T_{NI} [79]. In addition to the influence of the nature of the link on T_{NI} it also has a major effect on the entropy change at the N–I transition [7]. To see if this dependence can be explained by a geometric effect alone the variation of $\Delta S/R$ with the number of atoms in the flexible spacer has been calculated with the same parameters as those used to predict the T_{NI}. The results of these calculations are shown in Fig. 24 where both homologous series of dimers are seen to exhibit a strong and unattenuated alternation in the transitional entropy. However, what is also apparent is that the alternation is predicted to be far greater for the methylene linked dimers in comparison with those having ether links. This enhanced alternation comes in part from an increse in $\Delta S/R$ for the even membered dimers but, of greater significance, is the reduction in the transitional entropy for the odd membered dimers with methylene links. The theoretical predictions are in good agreement with experiment for ether and methylene linked cyanobiphenyl dimers [80]. In addition, the importance of the geometrical factor in determining the transitional properties is supported at least qualitatively by calculations based on the surface tensor parametrization of the potential of mean torque for the various conformers [80].

One of the novel outcomes of the generic model was the discovery that systems with high concentrations of bent conformers in the isotropic phase exhibit a N–N transition [61]. It is clearly of some interest and importance to see if this transition is also predicted by the complete model. The results of calculations, for the temperature dependence of the conformationally averaged order parameter for the mesogenic groups, $\langle \bar{P}_2^a \rangle$, for dimers with even membered spacers having a tetrahedral geometry do not show a N–N transition [81] in accord with the predictions of the generic model and with experiment. However, this is not the case for odd-membered dimers as the results in Fig. 25 clearly show. Thus, for the dimer with a propane spacer (see Fig. 25a) there is a very weak N–I transition followed by a rapid increase in $\langle \bar{P}_2^a \rangle$ at a scaled temperature of about 0.073 which is suggestive of a N–N transition. However, it may be that the dimer is close to the critical point at which the order parameter changes continuously rather than discontinuously with temperature. Our results for $\langle \bar{P}_2^a \rangle$ for the dimer with the pentane spacer clearly do not show a N–N transition (see Fig. 25b). Again the N–I transition is predicted to be weak, although not as weak as for the propane dimer. Then, however, as the scaled temperature is reduced the order parameter is seen to increase rapidly before tending towards its limiting value of unity. Although there is a

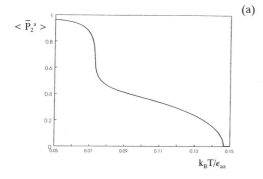

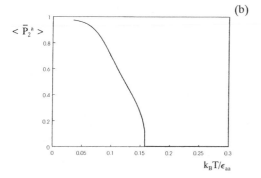

Figure 25. The scaled temperature dependence of the conformationally averaged order parameter, $\langle \bar{P}_2^a \rangle$, for the mesogenic groups, predicted for the liquid crystal dimers with (a) propane and (b) pentane spacers.

slight increase in the slope near a scaled temperature of 0.1 it is clear that the system is somewhat removed from critical behavior. It seems likely that this difference in behavior between the dimers with the propane and pentane spacers results from the higher concentration of bent conformers for the shorter spacer. This variation in the nature of the N–N transition is in accord with the predictions of the generic model as the concentration of the bent conformer in the isotropic phase decreases [62].

The prediction of the complete model that the alternation in the entropy change at the N–I transition seems to proceed unattenuated and indeed is enhanced as the spacer

length increases is intriguing. It seems likely that it is a direct consequence of the use of the Flory rotameric state model to describe the conformational states adopted by the dimers. Within this approach based on discrete conformers, for a tetrahedral geometry of the spacer the relative orientation of the two mesogenic groups can only adopt two values, irrespective of the length of the spacer. In reality, of course, the torsional potential governing rotation about, for example, carbon–carbon bonds in the flexible spacer is a continuous function and not discrete as in the Flory model. One consequence of this is that the torsional angle can adopt a range of values close to the minima in the potential. Such deviations from the minima can accumulate along the spacer and result in the two mesogenic groups adopting a range of relative orientations and not just the two predicted by the Flory model. Clearly, the longer the spacer the greater will be the range of relative orientations and so it might be expected that the alternation in the transitional entropy should also be attenuated with increasing spacer length. Such an attenuation has not been observed experimentally possibly because the longest spacer in an homologous series tends to have 14 atoms [19]. However, the question has been studied theoretically [82], again within the framework of the molecular field theory. The torsional potential was taken to have the form proposed by Ryckaert and Bellemans [83] namely

$$U_{\text{int}}(\phi) = \sum C_b \cos^b \phi \qquad (26)$$

where ϕ is the torsional angle which is defined to be zero in the *trans* conformation and b takes integral values from 0 to 5. The molecular field equations for the continuous potential can be obtained more or less from those which we have given for the discrete model (see Eqs. 15, 17, 18, 20–23) by replacing the summations over the discrete

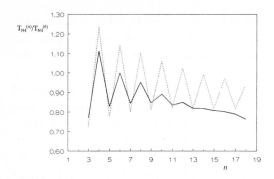

Figure 26. The dependence of the nematic–isotropic transition temperature, $T_{N-I}^{(n)}$, scaled with that for the sixth homologue on the number of atoms, n, in the spacer of the cyanobiphenyl dimers predicted with the continuous model for the torsional potential (——) and by the discrete model (....).

conformations with integrations over the torsional angles [82]. For long chains the number of such integrations is clearly large which complicates the calculations considerably and makes it impossible to include all conformational states, unlike the discrete model. This difficulty may be overcome by using a biased Monte Carlo scheme to select the torsional angles based on the internal energy, $U_{int}(\phi)$, alone and with the additional weighting resulting from the orientational partition function (see Eq. 17) being introduced specifically, as in umbrella sampling [84].

The results of these calculations for the N–I transition temperature based on the continuous torsional model are shown in Fig. 26 for the methylene linked cyanobiphenyl dimers with the spacer containing from 3 to 18 atoms. For ease of comparison, the transition temperatures have been scaled with the value for the sixth member of the homologous series. The results reveal that the alternation in T_{NI} is rapidly attenuated with increasing spacer length and that for spacers containing more than about 11 atoms, the alternation in T_{N-I} is essentially

removed. This is, of course, the behavior which we had anticipated, at least, qualitatively, for a continuous model. However, to ensure that this behavior does not result from some other feature of the molecular field calculations they were repeated for the discrete model in which the spacer adopts just the *trans* and *gauche* conformations of the Flory model. In this molecular field calculation, the energy difference between the *trans* and *gauche* conformations E_{tg} was given the value predicted by the Ryckaert–Bellemans torsional potential, all other parameters were kept the same as for the continuous model. The dependence of T_{N-I} on the number of atoms in the spacer is also shown in Fig. 26 and we can see that the alternation is larger and that the attenuation is weaker. This confirms, therefore, the major influence which the continuous nature of the torsional potential has on the N–I transition temperature for dimers containing relatively long spacers.

The entropy of transition was also calculated for the continuous and discrete torsional potential models with the results shown in Fig. 27. The results for the continuous model reveal that as the spacer length increases the larger alternation in $\Delta S/R$ is attenuated although this does not occur to any significant extent until the spacer contains 15 or more atoms. The results predicted by the discrete model are also given in Fig. 27 and these exhibit an enhanced alternation in the transitional entropy with increasing spacer length. It is of interest that for short spacers the values for $\Delta S/R$ predicted by the two models are very similar. However, as the spacer length grows so the transitional entropy predicted by the discrete model become significantly greater than those given by the continuous model for even dimers. In contrast, the results predicted by both models for the odd dimers are essentially same. This is analogous to the behavior

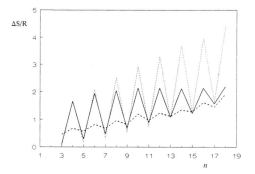

Figure 27. The variation of the transitional entropy, $\Delta S/R$, for the cyanobiphenyl dimers with the number of atoms, n, in the flexible spacer predicted with the continuous model for the torsional potential (——) and by the discrete model (....). For comparison the corresponding predictions for the alkyl cyanobiphenyl monomers based on the continuous potential are also shown (——).

found for the N–I transition temperature where the values predicted by the discrete model for even spacers are significantly larger than for the continuous model whereas the predictions for the odd dimers are essentially the same for the two models (see Fig. 26).

Another way to consider these predictions for the continuous potential model is that in the limit of long spacers the relative orientations of the two mesogenic groups will be decoupled. In this limit it would then be expected that the dimers should behave like monomers. To see if this expectation is likely to be met the transitional entropy for an alkyl cyanobiphenyl was calculated using the continuous model with the same parameters as for the dimers. The results are shown in Fig. 27 and reveal that as the length of the terminal chain is increased so $\Delta S/R$ exhibits an overall increase with a small alternation superimposed on this. In addition, the transitional entropy for the monomers is comparable to that for the odd dimers, in keeping with the fact that the conformational distribution for the dimers does

not change significantly at the N–I transition. Of greater interest is the observation that the alternation in $\Delta S/R$ for the monomers is essentially identical to that predicted for the dimers, again for the continuous torsional potential for spacers containing 19 or more atoms. It would seem, therefore, that the ability to undergo fluctuations around the torsional minima decouples the mesogenic groups and makes the dimers behave like monomers for relatively long spacers. Such dimers have yet to be made which would allow this fascinating prediction and, by implication, the validity of the Flory rotational isomeric state model to be tested.

7 Molecular Shapes of Liquid Crystal Dimers

The anisometric shape of a mesogenic molecule is thought to play an important role in determining its liquid crystal behavior. Indeed, as we have seen it is the differing shapes of odd and even dimers in their all-*trans* conformations which are often invoked to explain the strong alternation in the transitional behavior along a homologous series. However, as we have also seen, this difference in the shape of a single conformer is not the entire story and it is also necessary to allow for the many conformational states adopted by the spacer to account for the characteristic odd-even effect exhibited by liquid crystal dimers. The question then arises as to how the average shape of a nonrigid molecule which can assume different shapes can be defined and whether such average structures could be of value in understanding the behavior of liquid crystal dimers. One definition for the average shape of a nonrigid molecule which has been proposed is the following [85]. The coordinates of the atoms, r_i^n, in the nth conformer are averaged over all conformers

$$\bar{r}_i = \sum_n p_n r_i^n \qquad (27)$$

Then, in order to obtain a chemical-like structure, those atoms which are covalently bonded are linked together. Clearly in such an average structure the bond lengths and bond angles will differ from their values in rigid molecules. So as to emphasize still further that the image represents an average molecular structure and that the molecule should not be thought of as existing in this single structure each atom is decorated with an ellipsoid, rather like the thermal ellipsoids used in crystallographic structures. The dimensions of the ellipsoids are proportional to the weighted root mean square displacement of the atom away from its average position

$$\sigma_i^2 = \sum_n p_n (r_i^n r_i^n - \bar{r}_i \bar{r}_i) \qquad (28)$$

and the orientation of the ellipsoid is determined by the principal axes of σ_i^2.

This definition of an average molecular shape requires one further element to be complete and that is the axis system in which the atomic coordinates are to be expressed. One scheme is simply to locate the axis system in a rigid fragment of the molecule, which would be the same for each conformer. For example, with conventional monomers an intuitive choice would be the mesogenic group to which one or two terminal alkyl chains are attached [85]. However, this would not be appropriate for liquid crystal dimers because it would treat the two mesogenic groups differently and yield an image not in accord with the symmetry of the dimer. Indeed, even for a monomer changing the location of the axis system is known to have a profound influence on the resulting image [86]. It would seem, therefore, that attaching the coordinate system to

a rigid fragment of the dimer, which would be the same in all conformers, is not appropriate. Instead, it has been argued that the axis system should change with the conformer and would be related to some property of the molecule. For example, the axis system could be located at the center of mass of the conformer with the axes parallel to the principal axes of the moment of inertia tensor. An alternative is to select one conformer as a reference and then to overlay it with the other conformers. One way in which this can be achieved is to superimpose a conformer with the reference so that the centers of mass coincide. Then the conformer is rotated with respect to the reference so that the sum of the squares of the distances between all of the atoms in the conformer and their counterparts in the reference is a minimum. In fact, the rotation angles can be obtained analytically [87] which simplifies the construction of the average structure considerably, especially when the number of conformers is high. Once the rotation of the conformer has been made, the coordinates of the atoms in the same frame as that of the reference conformer are known. This process is repeated for all conformers and averages taken to obtain the $\bar{r}_i$ and σ_i^2 needed to construct the image of the average structure. However, there is still one problem which cannot be solved unambiguously and that is the choice of the reference conformer. Clearly this needs to be chosen to yield the correct symmetry for the resultant image of the average molecule. This criterion on its own is not sufficient because there are usually several conformers which are consistent with this constraint. The additional, intuitive criterion is to use as a reference the conformer which gives overall the smallest deviations of the atoms from their average positions.

To implement this scheme to construct the average molecular structure for a liquid

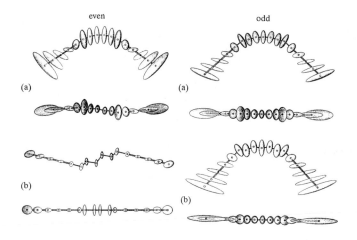

even   odd

(a)   (a)

(b)   (b)

Figure 28. The images created for the average molecular structures for liquid crystal dimers with even ($N=6$) and odd ($N=7$) dimers in (a) the isotropic and (b) the nematic phase.

crystal dimer it is necessary to know the atomic coordinates for each conformer. To simplify the calculations each mesogenic group is represented as a line of five atoms and each group in the spacer is replaced by a single atom. Next, the probability of each conformer, p_n is needed; these are taken from the Flory rotational isomeric state model [60]. In the isotropic phase the probabilities are evaluated from Eq. (15) and in the nematic phase they are calculated from Eq. (17) which includes the contribution coming from the orientational order of the mesophase [66]. Some typical images for the average molecular structures of liquid crystal dimers with even ($N=6$) and odd ($N=7$) spacers are shown in Fig. 28 for the isotropic and the nematic phase at a reduced temperature of 0.8 [81]; these are viewed within the molecular plane and orthogonal to it. In these calculations the reference conformer was ttg$^{\pm}$tt for the even dimer and ttttt for the odd dimer in the isotropic phase. However, in the nematic phase the optimum reference conformer has changed to ttttt for the even dimer but remains the same for the odd dimer. In the isotropic phase we see that the odd dimer has a bent shape which, coincidentally, has the same shape as the all-

trans conformer. Perhaps more surprisingly, the even dimer also has a bent shape in the isotropic phase. On reflection, both results should have been expected because the majority of conformations for the even and certainly for the odd dimer are bent. The average structure can change on going into the nematic phase because the conformational distribution is influenced by the orientational order. This proves to be the case for the even dimer where the average shape has changed significantly from bent in the isotropic to zig-zag in the nematic phase, which is analogous to that for the all-*trans* conformer. In addition, the fluctuational ellipsoids are considerably reduced in magnitude in the nematic in comparison to those in the isotropic phase. In contrast the average shape found for the odd dimer does not change with the phase and neither do the sizes of the fluctuational ellipsoids. The similarity of the images shows that the conformational distribution does not change to any significant extent with the nature of the phase.

The interpretation to be placed on such average structures as well as their usefulness have yet to be clearly established. At one level it can be argued that the images simply provide a pictorial representation of the numerical results of the molecular field theory, which in itself is valuable. At another level it gives the average shape of the dimers which for the odd dimers is clearly bent for both isotropic and nematic phases. In considering the molecular organization in the phases formed by the odd dimers it will clearly be important to take account of their bent shape (see, for example, Fig. 10). For even dimers, the situation is different with the bent average shape in the isotropic phase changing to a zig-zag shape in the nematic phase. Again, considerations of the molecular organization in the nematic phase formed by the even dimers will need to take account of this zig-zag shape.

8   References

[1] C. K. Ober, J.-I. Jin, R. W. Lenz, *Adv. Polym. Sci.* **1984**, *59*, 103.
[2] A. Blumstein, O. Thomas, *Macromolecules* **1982**, *15*, 1264.
[3] A. C. Griffin, T. R. Britt, *J. Am. Chem. Soc.* **1981**, *103*, 4957.
[4] G. W. Gray, Chap. I in *The Molecular Physics of Liquid Crystals* (Ed.: G. R. Luckhurst, G. W. Gray), Academic Press, London, **1979**.
[5] D. Vorländer, *Z. Phys. Chem.* **1927**, *126*, 449.
[6] J. Rault, L. Liebert, L. Strzelecki, *Bull. Soc. Chem. Fr.* **1975**, 1175.
[7] G. R. Luckhurst, *Macromol. Symp.* **1995**, *96*, 1.
[8] K. Praefcke, B. Kohne, D. Singer, D. Demus, G. Pelzl, S. Diele, *Liq. Cryst.* **1990**, *7*, 589.
[9] C. P. Lillya, Y. L. N. Murthy, *Mol. Cryst. Liq. Cryst. Lett.* **1985**, *2*, 121.
[10] B. Kohne, P. Marquardt, K. Praefcke, P. Psaras, W. Stephan, K. Turgay, *Chimia* **1986**, *40*, 360.
[11] W. Kreuder, H. Ringsdorf, O. Herrmann-Schönherr, H. Wendorff, *Angew. Chem., Int. Ed. Engl.* **1987**, *26*, 1249.
[12] D. Creed, J. R. D. Gross, S. L. Sullivan, A. C. Griffin, C. E. Hoyle, *Mol. Cryst. Liq. Cryst.* **1987**, *149*, 185.
[13] J. C. W. Chien, R. Zhou, C. P. Lillya, *Macromolecules* **1987**, *20*, 2341.
[14] I. Sledzinska, E. Bialecka-Florjanczyk, A. Orzesko, *Eur. Polym. J.* **1996**, *32*, 1345.
[15] M. Ibn-Elhaj, A. Skoulios, D. Guillon, J. Newton, P. Hodge, H. J. Coles, *Macromolecules* **1995**, *19*, 373.
[16] I. D. Fletcher, G. R. Luckhurst, *Liq. Cryst.* **1995**, *18*, 175.
[17] W. Weissflog, D. Demus, S. Diele, P. Nitschke, W. Wedler, *Liq. Cryst.* **1989**, *5*, 111.
[18] H. Ishizuka, I. Nishiyama, A. Yoshizawa, *Liq. Cryst.* **1995**, *18*, 775.
[19] J. W. Emsley, G. R. Luckhurst, G. N. Shilstone, I. Sage, *Mol. Cryst. Liq. Cryst. Lett.* **1984**, *102*, 223.
[20] R. W. Date, C. T. Imrie, G. R. Luckhurst, J. M. Seddon, *Liq. Cryst.* **1992**, *12*, 203.
[21] A. C. Griffin, S. R. Vaidya, R. S. L. Hung, S. Gorman, *Mol. Cryst. Liq. Cryst. Lett.* **1985**, *1*, 131.
[22] C. T. Imrie, *Liq. Cryst.* **1989**, *6*, 391.
[23] G. W. Gray, Chap. 12 in *The Molecular Physics of Liquid Crystals* (Eds.: G. R. Luckhurst, G. W. Gray), Academic Press, London, **1979**.
[24] H. Finkelmnn, Chap. 6 in *Thermotropic Liquid Crystals* (Ed.: G. W. Gray), John Wiley and Sons, Chichester, **1987**.
[25] C. T. Imrie, Vol. 5 of *Polymeric Materials Encyclopedia* (Ed.: J. C. Salamone), CRC Press, Boca Raton, FL, **1996**, 3770.
[26] J. L. Hogan, C. T. Imrie, G. R. Luckhurst, *Liq. Cryst.* **1988**, *3*, 645.
[27] G. S. Attard, R. W. Date, C. T. Imrie, G. R. Luckhurst, S. J. Roskilly, J. M. Seddon, L. Taylor, *Liq. Cryst.* **1994**, *16*, 529.
[28] G. S. Attard, S. Garnett, C. G. Hickman, C. T. Imrie, L. Taylor, *Liq. Cryst.* **1990**, *7*, 495.
[29] G. S. Attard, C. T. Imrie, F. E. Karasz, *Chem. Mater.* **1992**, *4*, 1246.
[30] A. C. Griffin, S. R. Vaidya, *Liq. Cryst.* **1988**, *3*, 1275.
[31] T. Ikeda, T. Miyamoto, S. Kurihara, M. Tsukada, S. Tazuke, *Mol. Cryst. Liq. Cryst.* **1990**, *182*, 357.
[32] A. E. Blatch, I. D. Fletcher, G. R. Luckhurst, *Liq. Cryst.* **1995**, *18*, 801.
[33] A. E. Blatch, I. D. Fletcher, G. R. Luckhurst, *J. Mater. Chem.* **1997**, *7*, 9.
[34] C. T. Imrie, *Trends Polym. Sci.* **1995**, *3*, 22.
[35] C. Carboni, F. J. Farrand, G. R. Luckhurst, D. de Silva, unpublished data.
[36] J. S. Goodby, *J. Mater. Chem.* **1991**, *1*, 307.
[37] J. Watanabe, M. Hayashi, *Macromolecules* **1989**, *22*, 4083.
[38] Y. Takanishi, H. Takezoe, A. Fukuda, H. Komura, J. Watanabe, *J. Mater. Chem.* **1992**, *2*, 71.
[39] P. J. Le Masurier, PhD Thesis, University of Southampton, UK, **1996**; P. J. Le Masurier and G. R. Luckhurst, *Chem. Phys. Lett.*, in submission.
[40] J. Watanabe, H. Komura, T. Niiori, *Liq. Cryst.* **1993**, *13*, 455.

[41] A. E. Blatch, G. R. Luckhurst, J. M. Seddon, un-published data.

[42] R. W. Date, G. R. Luckhurst, M. Shuman, J. M. Seddon, *J. Phys. II Fr.* **1995**, *5*, 587.

[43] F. Hardouin, A. M. Levelut, M. F. Achard, G. Sigaud, *J. Chem. Phys.* **1983**, *80*, 53.

[44] F. Hardouin, M. F. Achard, J.-I. Jin, J.-W. Shin, Y.-K. Yun, *J. Phys. II Fr.* **1994**, *4*, 627.

[45] F. Hardouin, M. F. Achard, J.-I. Jin, Y.-K. Yun, *J. Phys. II Fr.* **1995**, *5*, 927.

[46] J. Barberá, A. Omenat, J. L. Serrano, *Mol. Cryst. Liq. Cryst.* **1989**, *166*, 167.

[47] J. Barberá, A. Omenat, J. L. Serrano, T. Sierra, *Liq. Cryst.* **1989**, *5*, 1775.

[48] A. Yoshizawa, K. Matsuzawa, I. Nishiyama, *J. Mater. Chem.* **1995**, *5*, 2131.

[49] K. Shiraishi, K. Kato, K. Sugiyama, *Chem. Lett.* **1990**, 971.

[50] A. T. Marcelis, A. Koudijs, E. J. R. Sudhölterr, *Liq. Cryst.* **1996**, *21*, 87.

[51] H. Furuya, K. Asahi, A. Abe, *Polym. J.* **1986**, *18*, 779.

[52] G. S. Attard, C. T. Imrie, *Liq. Cryst.* **1989**, *6*, 387.

[53] A. T. Marcelis, A. Koudijs, E. J. R. Sudhölterr, *Liq. Cryst.* **1995**, *18*, 851.

[54] A. C. Griffin, S. L. Sullivan, W. E. Hughes, *Liq. Cryst.* **1989**, *4*, 677.

[55] A. P. J. Emerson, C. T. Imrie, G. R. Luckhurst, *J. Nat. Chem.*, submitted.

[56] J. W. Emsley, G. R. Luckhurst, G. N. Shilstone, *Mol. Phys.* **1984**, *53*, 1023.

[57] G. A. DiLisi, E. M. Terentjev, A. C. Griffin, C. Rosenblatt, *J. Phys. II* **1993**, *3*, 597.

[58] D. A. Dunmur, M. R. Wilson, *J. Chem. Soc., Faraday II* **1988**, *84*, 1109.

[59] C. Rosenblatt, A. C. Griffin, U. Hari, G. R. Luckhurst, *Liq. Cryst.* **1991**, *9*, 831.

[60] P. J. Flory, *Statistical Physics of Chain Molecules*, Wiley, New York, **1969**.

[61] A. Ferrarini, G. R. Luckhurst, P. L. Nordio, S. J. Roskilly, *Chem. Phys. Lett.* **1993**, *214*, 409.

[62] A. Ferrarini, G. R. Luckhurst, P. L. Nordio, S. J. Roskilly, *Liq. Cryst.* **1996**, *21*, 373.

[63] See, for example, G. R. Luckhurst, Chap. 4 in *The Molecular Physics of Liquid Crystals* (Eds.: G. R. Luckhurst, G. W. Gray), Academic Press, London, **1979**.

[64] S. Marcelja, *J. Chem. Phys.* **1974**, *60*, 3599.

[65] G. R. Luckhurst, C. Zannoni, P. L. Nordio, U. Segre, *Mol. Phys.* **1975**, *30*, 1345.

[66] J. W. Emsley, G. R. Luckhurst, *Mol. Phys.* **1980**, *41*, 19.

[67] M. J. Freiser, *Phys. Rev. Lett.* **1970**, *24*, 1041.

[68] R. L. Humphries, P. G. James, G. R. Luckhurst, *Symp. Faraday Soc.* **1971**, *5*, 107.

[69] G. R. Luckhurst, T. J. Payne, unpublished data.

[70] G. S. Attard, G. R. Luckhurst, *Liq. Cryst.* **1987**, *2*, 441.

[71] S. Garnett, G. R. Luckhurst, unpublished data.

[72] P. J. Barnes, A. G. Douglass, S. K. Heeks, G. R. Luckhurst, *Liq. Cryst.* **1993**, *13*, 603.

[73] G. Ungar, V. Percec, M. Zuber, *Macromolecules* **1992**, *25*, 75.

[74] D. J. Photinos, E. T. Samulski, H. Toriumi, *J. Phys. Chem.* **1990**, *94*, 4688.

[75] E. T. Samulski, R. Y. Dong, *J. Chem. Phys.* **1982**, *77*, 5090.

[76] A. P. J. Emerson, PhD Thesis, University of Southampton, UK, **1991**.

[77] A. Ferrarini, G. J. Moro, P. L. Nordio, G. R. Luckhurst, *Mol. Phys.* **1992**, *77*, 1.

[78] G. R. Luckhurst, Chap. 7 in *Recent Advances in Liquid Crystalline Polymers* (Ed.: L. L. Chapoy), Elsevier, London and New York, **1985**.

[79] A. P. J. Emerson, G. R. Luckhurst, *Liq. Cryst.* **1991**, *10*, 861.

[80] A. Ferrarini, G. R. Luckhurst, P. L. Nordio, S. J. Roskilly, *J. Chem. Phys.* **1994**, *100*, 1460.

[81] S. J. Roskilly, PhD Thesis, University of Southampton, UK, **1994**.

[82] A. Ferrarini, G. R. Luckhurst, P. L. Nordio, *Mol. Phys.* **1995**, *85*, 131.

[83] J. P. Ryckaert, A. Bellemans, *Chem. Phys. Lett.* **1975**, *30*, 123.

[84] G. M. Torrie, J. P. Valleau, *J. Comput. Phys.* **1977**, *23*, 187.

[85] A. Kloczkowski, G. R. Luckhurst, R. W. Phippen, *Liq. Cryst.* **1988**, *3*, 185.

[86] A. P. J. Emerson, G. R. Luckhurst, R. W. Phippen, *Liq. Cryst.* **1991**, *10*, 1.

[87] W. Kabsch, *Acta Cryst. A* **1976**, *32*, 922; *Acta Cryst. A* **1978**, *34*, 827.

Chapter XI
Laterally Substituted and Swallow-Tailed Liquid Crystals

Wolfgang Weissflog

1 Introduction

The relationship between the chemical constitution of molecules and their mesophase behaviour is one of the most interesting areas in the field of liquid crystal chemistry. The calamitic substances and the discotic compounds represent the two basic types of conventional liquid crystal. According to previous opinion, deviation from these structural concepts should cause the loss of mesophases. For instance, there was the general rule that clearing temperatures decrease as the volume of the lateral groups increases. Analogously, branches within the terminal alkyl chains often produce a drastic decrease in the mesophase stability, as observed in the synthesis of optically active compounds. For a long time these facts prevented further investigation into the preparation of non-conventional liquid crystalline materials. However, in the 1980s, the first results overcoming such reservations were published.

In this chapter the chemical constitution of rod-like mesogens is changed systematically by the introduction of additional long-chain substituents as well as phenyl ring containing segments. The relationship between the molecular shape and the mesophase behaviour, and some physical properties of these new materials, are summarized.

2 Laterally Alkyl Substituted Rod-Like Mesogens

2.1 Long-Chain 2-Substituted 1,4-Phenylene Bis(benzoates)

1,4-Phenylene bis(benzoates) have been widely used to check the influence of small lateral groups [1, 2] and of the t-butyl substituent [3]. In 1983, the synthesis of 2-n-alkyl-1,4-phenylene bis(4-n-alkyloxybenzoates) (**1** in Figure 1) by Weissflog and Demus [4] initiated the investigation of laterally long-chain substituted mesogens. The course of the clearing points as a function of the length of the lateral chain (Figure 1) has been proved to be typical for laterally alkyl substituted series. As the length of the lateral chain increases, the clearing points that do not alternate tend to a convergence temperature. All compounds show nematic phases, while layer structures are depressed but cybotactic groups are detected [5]. In order to understand the mesophase behaviour of laterally alkyl substituted mesogens, many experiments have been done, and the homologous series shown in Figure 1 is now one of the most well characterized homologous series in the field of liquid crystals. Enthalpies and entropies, phase transition volumina, densities, thermal expansion co-

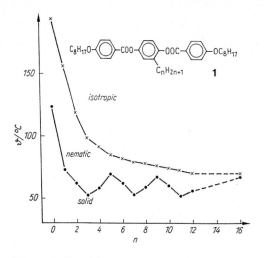

Figure 1. Transition temperatures of the 2-n-alkyl-1,4-phenylene bis(4-n-octyloxybenzoates) **1** [4].

efficients, order parameters, optical refractive index and viscosities have all been measured by Demus and coworkers [6–9]. The convergence behaviour of the clearing temperatures and of other physical properties, as well as the results of ^{13}C NMR investigations [10], X-ray studies [5] and theoretical studies all indicate that the conformations of the lateral chain deviate strongly from those of the terminal chains. Gauche conformers between the carbon atoms three to six make a great contribution to the orientation of the lateral chain in direction of the molecular length axis. Therefore, the liquid crystalline behaviour of laterally long-chain substituted meso-

Table 1. The phase transitions of some laterally long-chain substituted 1,4-phenylene bisbenzoates.

R^1—⬡—COO—⬡—OOC—⬡—R^2 **2**
（R³ below central ring）

Compound No. [a,b]	Substituents	Phase transitions	Ref.
2a	$R^3 = C_nH_{2n+1}$, $n = 1$ to 12, 16	(see Fig. 1)	[4, 10, 12]
2b	$R^3 = COC_nH_{2n+1}$, $n = 1$ to 11 $R^1 = R^2 = C_8H_{17}O$, $n = 9$	Cr 77 N 84 I	[13, 14]
2c	$R^3 = COOC_nH_{2n+1}$, $n = 1$ to 10 $R^1 = R^2 = C_8H_{17}O$, $n = 8$	Cr 52 N 93 I	[13, 15, 16]
2d	$R^3 = COSC_nH_{2n+1}$, $n = 6, 10$ $R^1 = R^2 = C_8H_{17}O$, $n = 6$	Cr 63 N 79 I	[2] [13]
2e	$R^3 = SC_nH_{2n+1}$, $n = 12$ $R^1 = R^2 = C_8H_{17}O$	Cr 55 N 55 I	[16]
2f	$R^3 = COO(CH_2CH_2O)_nC_2H_5$, $n = 1, 2$ $R^1 = R^2 = C_8H_{17}O$, $n = 2$	Cr 55 N 87 I	[2]
2g	$R^3 = COOCH_2C^*H(CH_3)C_2H_5$ $R^1 = R^2 = C_8H_{17}O$	Cr 105.5 (N* 81) I	[16]
2h	$R^3 = OCH_2C^*H(CH_3)C_2H_5$ $R^1 = R^2 = C_6H_{13}O$	Cr 78–80 I	[17]
2i	$R^3 = C_nH_{2n}OOC(Y)C=CH_2$, Y=H, Cl, CH₃, $n = 6, 11$ $R^1 = C_4H_9O$, $R^2 = CN$, Y=CH₃, $n = 11$	Cr 68 (N 40) I	[18–20] [18]
2j	$R^3 = CH_2OOCCH=CH_2$ $R^1 = R^2 = C_4H_9O$	Cr 87 I	[21]
2k	$R^3 = COOC_nH_{2n}CH=CH_2$, $n = 2, 8$ $R^1 = R^2 = C_8H_{17}O$	Cr 96 (N 95) I	[22, 23]

[a] Unless defined otherwise, R is an unbranched alkyl.
[b] Further homologues are presented in the original paper.

gens can be understood in terms of the length-to-breadth ratios of the molecules, as has been supported theoretically by the work of Demus and coworkers [5, 8]. A statistical contribution of single segments of the lateral chain does not explain the observed properties sufficiently [11].

Recently, Hoffmann et al. [12] have described the molecular and crystalline solid structure of 2-n-nonyl-1,4-phenylene bis(4-n-octyloxybenzoate), as estimated using X-ray diffraction [12]. The observed structure deviates from the model for the nematic phase in the crystalline solid state, as the lateral chains exist in all-*trans* conformation. The type of link between the lateral hydrocarbon chain and the mesogenic core influences the course of the clearing points. A comparison performed for the 2-alkyl, acyl and alkyloxycarbonyl substituted derivatives **2a–2c** in Table 1 showed a break in the clearing-point curve at **2b** and **2c** for five single units (including the –CO– or –CO–O groups) within the lateral chains. The lower the transition temperatures the more pronounced the break in the curve [13, 15].

The mobility of the laterally attached hexylcarbonyl group (**2b**, $n=6$) was investigated by Kresse et al. [14] by means of dielectric measurements. The terminal wing groups can be chosen in the usual way in liquid crystal chemistry, e.g. alkyl, alkyloxy, acyloxy and alkyloxycarbonyloxy [6], and the laterally attached chain can be varied in many ways (e.g. **2a–2k**, Table 1). The chain itself can be branched and optically active, but these changes also cause a decrease in the clearing point (see **2g** and **2h**) [16, 17, 19]. In contrast to the known effect in terminal positions, oxyethylene containing aliphatic groups in the lateral position do not disturb the liquid crystalline phases (compare **2c** and **2f**). Most of the 1,4-phenylene dibenzoates, in which the lateral long-

chain substituents bear a terminal unsaturated group (**2i–2k**), are intermediates in the synthesis of liquid crystalline side-on polymers [18–23].

2.2   Further mesogens bearing one Long-Chain Group in the Lateral Position

In principle, all rod-like mesogens can be laterally long-chain substituted if their cores themselves are long enough. The nematic–isotropic transition temperatures of two-ring compounds are strongly decreased by the addition of longer lateral alkyl groups (e.g. **3a** and **3b** in Table 2, and **4a** in Table 2) [24–26]. The lateral chain can also be situated at a connecting group, as has been proved in the substance class of oximesters (**4**) [26–29]. Other examples are given in Table 2, e.g. phenylazosalicylic acid esters (**5**), which form orthopalladated metallomesogens [31], and the alkylcyclohexanoates (**6**). On enlargement of the middle part (A) of the three-ring mesogens from naphthyl (**6d**) to anthracenyl (**6e**), the clearing points are decreased [32, 33]. In many cases (e.g. **6e** (i), **7b**, **8** and **9**) liquid crystalline behaviour can only be obtained by extending of the number of cyclic segments in the basic molecule to 4 or 5. Substances **9** were prepared instead of the laterally aryl substituted mesogens that were wanted; this is discussed in Sec. 3.1 of this Chapter.

The liquid crystalline behaviour is influenced in the same way if one of the outer rings is substituted at the 2-position, as demonstrated by the formula given in Table 3. In most cases, nematic phases and the typical decrease of the clearing point with increasing length of the lateral chain are detected. It is remarkable that in few cases

Table 2. Different basic mesogens substituted with one long-chain lateral group.

$C_8H_{17}O$—⬡—COO—⬡—OC_8H_{17} **3**
$\quad\quad\quad\quad$ |
$\quad\quad\quad\quad R^1$

Compound No.	Substituents	Phase transitions	Ref.
3a	$R^1 = O(CH_2)_{12}OOCCH=CH_2$	Cr 28 (N 17) I	[24]
3b	$R^1 = OOCCH_2C^*H(CH_3)C_2H_5$	Cr 32 I	[25]

R^1—⬡—C=N—OOC—⬡—OR^2 **4**
$\quad\quad\quad\quad$ |
$\quad\quad\quad C_nH_{2n+1}$

Compound No.	Substituents	Phase transitions	Ref.
4a	$R^1 = C_5H_{11}$, $R^2 = C_6H_{13}$, $n = 3$	Cr 54 (N 8) I	[26]
4b	(i) $R^1 = $ ⬡— , $n = 1$ to 9		[27]
	(ii) $R^2 = C_8H_{17}$, $n = 6$	Cr 85 (N 53) I	
4c	(i) $R^1 = C_mH_{2m+1}O$—⬡—COO–		
	$R^2 = CH_2C^*H(CH_3)C_2H_5$, $n = 1$ to 5		[26–29]
	(ii) $m = 8$, $n = 5$	Cr 111 (N* 103) I	

C_2H_5O—⬡—N=N—⬡—OOC—⬡—OCH_3 **5**
$\quad\quad\quad\quad\quad\quad$ |
$\quad\quad\quad\quad\quad COOC_nH_{2n+1}$

Compound No.	n	Phase transitions	Ref.
5a	(i) 1–10, 12, 14, 16		[31]
	(ii) 18	Cr 75 N 81 I	

R^1—[⬡]$_p$—COO—⟨A⟩—OOC—[⬡]$_p$—R^1 **6**
$\quad\quad\quad\quad\quad\quad\quad$ |
$\quad\quad\quad\quad\quad\quad\quad R^2$

Compound No.	Ring A	Substituents	Phase transitions	Ref.
6a	⬡R^2	$R^1 = C_5H_{11}$ $R^2 = C_nH_{2n+1}$ (i) $n = 2$ to 10, $p = 1$ (ii) $n = 9$	Cr 57 (N 47.5) I	[2, 13] [13]
6b	⬡R^2	$R^1 = C_5H_{11}$ $R^2 = COC_nH_{2n+1}$ (i) $n = 1$ to 9, $p = 1$ (ii) $n = 8$	Cr 71 (N 43) I	[13] [13]

Table 2. (continued)

Compound No.	Ring A	Substituents	Phase transitions	Ref.
6c	(benzene ring with R² lateral substituent)	$R^1 = C_5H_{11}$ $R^2 = (CH_2)_{11}OOCCH(CH_3)=CH_2$	Cr 28 (N 17) I	[18]
6d	(naphthalene ring with R² lateral substituent)	(i) $R^2 = O(CH_2)_nOOCCH(CH_3)=CH_2$, $n=10$, $p=1$ (ii) $R^1 = C_7H_{15}$	 Cr 44 (N 25) I	[32,33] [32]
6e	(anthracene ring with R² lateral substituent)	(i) $R^2 = O(CH_2)_nOOCC(CH_3)=CH_2$, $n=10$ (ii) $R^1 = C_3H_7$, $p=2$	 Cr 164 N 235 I	[32, 33]

$C^*_5H_{11}O$—(ring)—COO—(ring)—COO—[(ring)]$_m$—C_8H_{17} **7**
with $O(CH_2)_3CH=CH_2$

Compound No.	m	Phase transitions	Ref.
7a	1	Cr 68–70 I	[34]
7b	2	Cr 92 N* 142 I	[34]

NC—(ring)—(ring)—$O(CH_2)_4O$—(ring)—$O(CH_2)_4O$—(ring)—(ring)—CN **8**
with $COOC_nH_{2n+1}$ $n=1-12$

Compound No.	n	Phase transitions	Ref.
8a	3	Cr 131 (SmA 72) N 155 I	[35]
8b	10	Cr 136 (N 131) I	[35]

R^1O—(ring)—X—(ring)—$N=N$—(ring)—OOC—(ring)—OR^2 **9**
with R^3 and $C_{12}H_{25}O$

Compound No.	Substituents	Ref.
9a	$X = COO$, $R^3 = H$	[36]
9b	$X = CH=N$, $R^3 = OH$	[36]

Table 3. The phase transitions when lateral hydrocarbon chains are located in the 2-position of an outer phenyl ring.

$$R^1-X-\!\!\bigcirc\!\!-Y-\!\!\bigcirc\!\!-Z \qquad \mathbf{10}$$
(with R^2 substituent on the first ring)

Compound No.	R^1	R^2	X	Y	Z	Phase transitions	Ref.
10a	(i) $C_nH_{2n+1}O$ (ii) $n=8$	OC_nH_{2n+1}, $n=1$ to 9	CH=N	–	CN	Cr 98 (N 70) I	[153]
10b	(i) $C_nH_{2n+1}O$ (ii) $n=8$	OC_nH_{2n+1}, $n=5$ to 10	COO	COO	CN	Cr 68 (N 60) I	[37]
10c	$C_9H_{19}O$	OC_9H_{19}	COO	OOC	OC_9H_{19}	Cr 81 (N 67) I	[38]
10d	(i) $C_nH_{2n+1}O$ (ii) $n=12$	$O(CH_2)_pOOCCH=CH_2$, $p=4, 6, 12$ $p=6$	COO	COO	OC_mH_{2m+1} $m=12$	 Cr 71 (SmA 36 N 53) I	[24]
10e	(i) $C_nH_{2n+1}O$ (ii) $n=12$	$O(CH_2)_pOOCCH=CH_2$, $p=4, 6, 12$ $p=6$	COO	COO	CN	 Cr 68 (N 46) I	[24]
10f	(i) $C_nH_{2n+1}O$ $n=8$ (ii) $n=8$	$O(CH_2)_pOOCCH=CH_2$, $p=3, 6, 9$ $p=6$	COO	–	C_nH_{2n+1}, OC_nH_{2n+1}, $CH_2C^*H(CH_3)C_2H_5$ OC_9H_{19}	Cr 44.5 N 64.5 I	[30] [39] [30]
10g	(i) $C_nH_{2n+1}O$ (ii) $n=8$	$O(CH_2)_pCH=CH_2$, $p=3, 4$ $p=4$	COO	–	CN	Cr 74.5 (N 58) I	[40]
10h	$C_2H_5C^*H(CH_3)CH_2O$	$O(CH_2)_3CH=CH_2$	COO	–	C_5H_{11}	Cr ? N -2.4 I	[34]

monotropic smectic behaviour was found, as demonstrated for **10d** (ii). Dielectric investigations of the corresponding cyanophenyl derivatives **10b** and **10e** yielded relatively high values for the dielectric anisotropy. One reason for this could be the hindrance of dimerization by the lateral substituents [24, 37, 38]. Compounds **10e**–**10h**, which have an unconjugated double bond within the lateral chain, serve as precursor for side-group polymers. For synthetic reasons it should be assumed that in the heterocyclic compound **11** the chain length should be equal in the two 2,4-alkyloxy groups [41].

11   Cr 79.5 (N 76) I

If the lateral chain is attached at the 3-position, i.e. *ortho* to an other alkyloxy chain, the behaviour changes. The properties of these substances are comparable to those of the swallow-tailed compounds. However, mesogens bearing three alkyloxy chains linked directly to the outer aromatic rings are grouped as tricatenar substances, according to the definition given by Malthete et al. [42] (see Chap. XII of this volume).

In laterally alkyl substituted mesogens, each branch in a terminal or lateral substituent further decreases the mesophase stability dramatically (e.g. **2g**, **2h**, **3b**, **6c**, **6d**, **7** and **10h**). This fact is important in the synthesis of optically active compounds.

2.3   Two Long-Chain Substituents in Lateral Positions

Steric loading of rod-like compounds with two additional alkyl groups in lateral positions often leads to the loss of anisotropic properties. Some variations of rod-like molecules with two lateral chains are given in Table 4. Weissflog [2] found that the addition of two small lateral groups disturbs the mesogenity more than the addition of one large lateral group, particularly if the two substituents are located in positions 2 and 5. This relationship has been proven for isomeric C4-substituted phenylene bisbenzoates [2]. A comparison of the transition temperatures of compounds **12a**–**12d** in Table 4 supports this comment. The same dependency has been found for the oximesters **13** [27]. Both lateral chains can start from one position in the molecule, as in the ketoximesters **14**, by means of a branched functional group [16]. A diagram of the clearing points of this homologous series is given in Chapter 6 of this volume; the clearing points go through a minimum depending on the length of the lateral chains, a rare phenomenon. The azulene dicarboxylic esters **15** are mentioned here as they allow the possibility of substitution by two alkyloxycarbonyl groups [45].

Three-ring compounds bearing two lateral chains are often not liquid crystalline (e.g. [44, 46]), but corresponding four-ring mesogens, e.g. the oximesters **13b** with two laterally attached nonyl groups, can exhibit nematic phases with clearing points at about 140 °C [27].

Table 4. Mesogens containing two long-chain substituents in lateral positions.

R1—⟨O⟩—COO—⟨O⟩—OOC—⟨O⟩—R1   **12**
(with R2 top, R4 and R3 bottom)

Compound No.	R¹	R²	R³	R⁴	Phase transitions	Ref.
12a	$C_8H_{17}O$	$COOC_2H_5$	H	H	Cr 102 N 115 I	[13]
12b	$C_8H_{17}O$	H	$COOC_2H_5$	$COOC_2H_5$	Cr 96.5 N 109.5 I	[43]
12c	$C_8H_{17}O$	$COOC_2H_5$	$COOC_2H_5$	H	Cr 130 (N 52) I	[2]
12d	$C_6H_{13}O$	CH_2COOCH_3	CH_2COOCH_3	H	Cr 79 I	[44]

$C_nH_{2n+1}O$—⟨O⟩—COO—N=C—(A)—C=N—OOC—⟨O⟩—OC_nH_{2n+1} **13**
(R¹ top, R² bottom on the N=C groups)

Compound No.	Substituents	Phase transitions	Ref.
13a	(i) —(A)— = —— R¹, R²=CH₃, C₂H₅, C₃H₇, i-C₃H₇, C₅H₁₁ (ii) R¹=C₂H₅; R²=i-C₃H₇; n=8	Cr 104 (N 67) I	[27]
13b	(i) —(A)— = —⟨O⟩—⟨O⟩— n=1 to 10 (ii) R¹=R²=C₉H₁₉, n=8	Cr 88 N 148 I	[27]

$C_8H_{17}O$—⟨O⟩—COO—⟨O⟩—OOC—⟨O⟩—OC_8H_{17} **14**
n=1–11, with COO—N=C(CₙH₂ₙ₊₁)(CₙH₂ₙ₊₁)

Compound No.	n	Phase transitions	Ref.
14a	11	Cr 74 (N 43) I	[16]

(azulene) COOC_nH_{2n+1} / OOC—⟨O⟩—⟨ ⟩—R¹ / COOC_nH_{2n+1} **15**
n=1 to 3, R¹=C₃H₇, C₇H₁₅

Compound No.	n	R⁴	Phase transitions	Ref.
15a	3	C_7H_{15}	Cr 104.5 (N 89.3) I	[45]

3 Mesogens incorporating Phenyl Rings within the Lateral Segments

3.1 Mesogens with One Lateral Segment containing a Phenyl Group

In 1988, Weissflog and Demus reported a new concept that emphasized the importance of a spacer in the successful design of mesogens with lateral segments incorporating cyclic groups (Figure 2) [2, 47, 48]. According to this, the lateral aryl group should be connected to the basis mesogen by a flexible spacer that is long enough to allow good

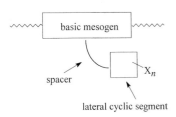

Figure 2. Schematic representation of the design of mesogens with lateral segments containing a cyclic unit.

alignment of the molecular segments and a high packing density in the anisotropic condensed phase. In the following, mesogens are discussed in the framework of increasing the number of spacer units.

Connecting a phenyl group with or without one spacer unit to the middle ring of the basis core in the compounds **16** (Table 5) produces monotropic nematic behaviour only [47–50].

Table 6. Benzoinylanilines – liquid crystalline or not?

$$R^1O-\text{C}_6H_4-\overset{|}{\underset{OH}{CH}}-\overset{||}{\underset{N}{C}}-\text{C}_6H_4-OR^1 \quad \text{17}$$

Compound No.	R^1	R^2	Phase transitions	Ref.
17a	C_4H_9	C_3H_7	Cr 119.2 (N 112.3) I	[51]
17b	$C_{10}H_{21}$	$C_{10}H_{21}$	Cr 103.2 Sm 112.7 I	[52]

$$R^1O-\text{C}_6H_4-CH=N-\text{C}_6H_4-OR^2 \quad \text{18}$$

Compound No.	R^1	R^2	Phase transitions	Ref.
18a	C_4H_9	C_3H_7	Cr 120.5 (N 114.7) I	[53]
18b	$C_{10}H_{21}$	$C_{10}H_{21}$	Cr 104 Sm 112 I	[53]

Table 5. Laterally aryl substituted three-ring compounds with or without one spacer unit Z.

$$R^1O-\text{C}_6H_4-COO-\text{C}_6H_4-OOC-\text{C}_6H_4-OR^1 \quad \text{16}$$

Compound No.	Z	R^1	R^2	Phase transitions	Ref.
16a	–	C_8H_{17}	H	Cr 81 (N 77) I	[50]
16b	(i) SO_2	C_4H_9	H	Cr 151 (N 88) I	[49]
	(ii) SO_2	C_4H_9	CH_3	Cr 137 (N 56) I	[49]
16c	CO	$-C_8H_{17}$	H	Cr 85 (N 58) I	[47]

Table 7. Connection of the lateral phenyl group, itself substituted by a two-unit spacer Z to three-ring compounds.

19

Compound No.	R^1	R^2	R^3	X	Y	Z	Phase transitions	Ref.
19a[a]	C_2H_5O	OCH_3	OC_2H_5	N=N	OOC	CH=N	Cr 160 (N 73) I	[55]
19b	$C_5H_{11}O$	OC_5H_{11}	OC_3H_7	COO	OOC	CH=N	Cr ? (N 64) I	[56]
19c	$C_8H_{17}O$	$OC_{12}H_{25}$	OC_8H_{17}	COO	N=N	OOC	Cr 80 (N 71) I	[57]
19d	$C_8H_{17}O$	OC_8H_{17}	OC_8H_{17}	OOC	OOC	OOC	Cr 70 N 75 I	[58]
19e	$C_8H_{17}O$	OC_8H_{17}	OC_8H_{17}	OOC	COO	COO	Cr 57 (N 30) I	[58]
19f	$C_8H_{17}O$	OC_8H_{17}	OC_8H_{17}	COO	OOC	OOC	Cr 51	[58–60]
19g	NC	OC_8H_{17}	OC_8H_{17}	OOC	OOC	OOC	Cr 114 (SmA 87 N 89) I	[61]
19h	$C_8H_{17}O$	OC_8H_{17}	CN	COO	OOC	COO	Cr 107 (N 50) I	[47]
19i	$C_8H_{17}O$	OC_8H_{17}	CH_2CH_2CN	COO	OOC	COO	Cr 91 (N 39) I	[47]

[a] Cr 145 (N 59) I in [54].

In the 1980s, Gallardo and Müller [51] and Bennur et al. [52] described the reaction of 4,4′-dialkyloxybenzoines with 4-substituted anilines, which they claimed yielded liquid crystalline benzoinylidene anilines **17**. However, there were indications, e.g. from the experimental details given, that the dialkyloxybenzoines were not isolated in the pure state. Therefore, by reaction of the raw intermediates with anilines simple 4,4′-disubstituted benzylidene anilines **18** were prepared (Table 6).

Table 7 gives some representative homologous of three-ring compounds **19** with connecting groups X and Y differing in position and direction, and having two units within the spacer Z. The azomethin **19a** was first prepared by Kuhrmann in 1926 [54] and was reproduced by Weissflog et al. in 1990 [55]. In general, nematic phases of **19** can exist below about 75 °C. Only the terminally cyano substituted derivatives **19g** (and not the lateral cyano analogue **19h**) exhibit higher clearing points and smectic A phases, be-

cause of the presence of association effects. As demonstrated by Takenaka et al. [58] using **19d–19f**, the direction of all the carboxylic groups X, Y and Z is of significance for the existence of mesophases.

Substances **20** could be liquid crystalline, but the transition temperatures reported by Berdague et al. [57] belong to laterally alkyloxy substituted four-ring mesogens **9**, as reported recently by the same authors [36].

20 X=COO, CH=N; Y=H, OH; R^2=$C_{12}H_{25}$

Four-ring basis mesogens can cope with a bulky lateral segment without loss of mesophases (Table 8). Derivative **21** is one of the longest known substances under discussion, being first described by Mauerhoff working in Vorländer's research group in

Table 8. Laterally aryl substituted four-ring mesogens in which the lateral phenyl group is attached by a two-unit spacer Z.

21	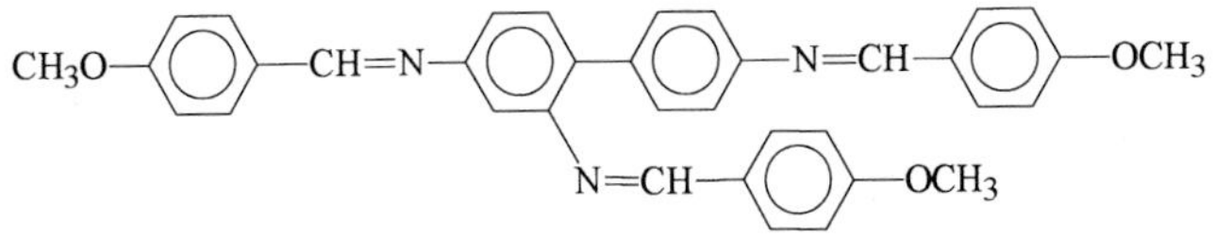 CH_3O—⬡—$CH{=}N$—⬡—⬡—$N{=}CH$—⬡—OCH_3	Cr 159 N 218 I	[62]

$N{=}CH$—⬡—OCH_3

22	⬡—$N{=}N$—⬡—$N{=}CH$—⬡—OOC—⬡—OR		

OOC—⬡—OR

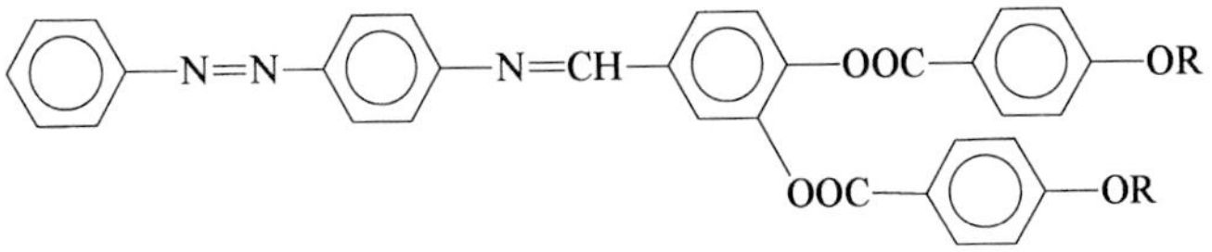

22a	$R=C_{12}H_{25}$, Cr 82 N 121 I	[63]

23	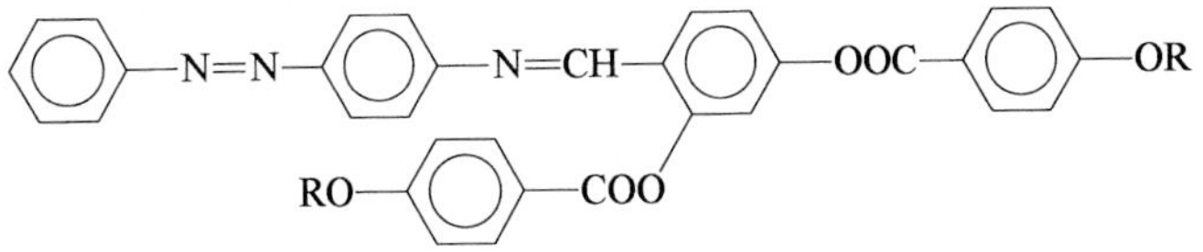 ⬡—$N{=}N$—⬡—$N{=}CH$—⬡—OOC—⬡—OR

RO—⬡—COO

23a	$R=C_{12}H_{25}$, Cr 105 (N 75) I	[63]

24	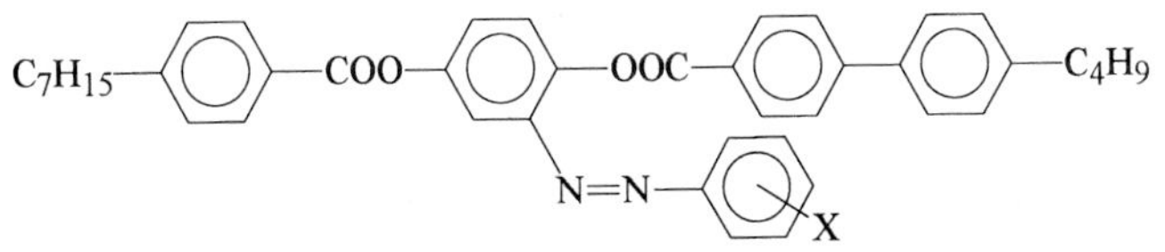 C_7H_{15}—⬡—COO—⬡—OOC—⬡—⬡—C_4H_9

$N{=}N$—⬡X

24a	$X=2\text{-}NO_2$, Cr 91 N 144 I	[64]
24b	$X=4\text{-}NO_2$, Cr 160 N 198 I	[64]

Table 9. Molecules containing 7, 10 or 13 phenyl rings.

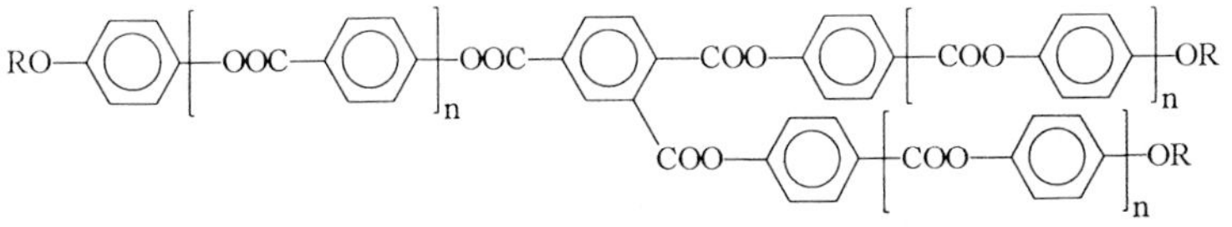

RO—⬡—[OOC—⬡]$_n$—OOC—⬡—COO—⬡—[COO—⬡]$_n$—OR **25**

COO—⬡—[COO—⬡]$_n$—OR

Compound No.	R	n	Phase transitions	Ref.
25a	C_4H_9	1	Cr 105 Sm 135 N 189 I	[65, 66]
25b	C_4H_9	2	Cr 172 Sm 279 N 326 I	[65, 66]
25c	C_4H_9	3	Cr 144 SM 286 N 306 I	[65, 66]

Halle in 1922 [62]. Interestingly, these results (**19a** and **21**) from the theses by Kuhrmann [54] and Mauerhoff [62] have never been published by Vorländer, perhaps because of their incompatibility with the rod-like molecular model.

In addition to the results given above, Braun et al. [65, 66] have reported on molecules containing 7, 10 or 13 phenyl rings **25** (Table 9). These compounds are readily soluble in some solvents.

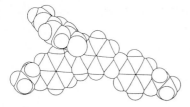

Figure 3. Space-filling plot of the crystalline substance **19a** bearing a lateral 4-ethoxyphenyl segment linked by a two-unit iminomethyl spacer. (Reproduced by permission of Baumeister et al. [54]).

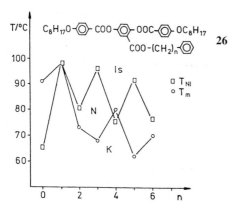

Figure 4. Transition temperatures of compounds in the series **26** versus the spacer length [47].

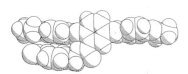

Figure 5. Space-filling model of the crystalline solid structure of 4-nitrobenzyl 2,5-bis(4-ethoxybenzoyl-oxy)benzoate having a three-unit spacer. (Reproduced by permission of Hoffmann et al. [67]).

The three-ring compound **19a** bearing a lateral 4-ethoxyphenyliminomethyl segment has been subjected to solid state X-ray investigations [54]. The results show the molecule to be Y-shaped (Figure 3). The same conclusion can be drawn for all the compounds discussed here (**19, 21–25**) that contain a relatively rigid two-unit spacer group Z.

Lengthening of the spacer Z should allow an orientation of the lateral segment in the direction of the molecular length axis and, therefore, should be of advantage in the sense of higher packing densities and clearing points. This hypothesis is supported by the transition temperatures of the compounds **26** (Figure 4) which show an alternation of the nematic–isotropic transition points according to the number of spacer units. Logically, higher mesophase stability is observed for odd-numbered spacers. Therefore, three and five single units between the three-ring basic mesogens and the lateral aryl ring are favourable conditions for high clearing temperatures, as has been verified by, for example, the synthesis of derivatives of gentisic acid benzyl esters and of phenoxyethyl esters [2, 47, 48]. As demonstrated in Table 10 for the 2,5-bis(4-n-octyloxybenzoyloxy)benzoates **27**, the lateral phenyl ring can itself be multiply substituted without loss of mesophases. The laterally attached cyclic segment can also be a cycloaliphatic, heterocyclic or polycyclic ring system (e.g. **27f–27h**). Substituents in the *para* position of the laterally attached phenyl ring cause an increase in the clearing point. This relationship emphasizes the advantages of the unconventional chemical structures under discussion. Figure 5 shows the change in shape in comparison to the structure shown in Figure 3. Increasing the length of the laterally attached segments by insertion of benzoyloxy or phenyl groups (e.g. **27i–27k**) gave a new type of laterally terminally connected unsymmetric twin molecule exhibiting high clearing temperatures [48, 68].

Unlike to the mesophase behaviour of laterally alkyl substituted mesogens, smectic phases can be stabilized depending on the substituents at the lateral phenyl ring. Figure 6 shows the polymorphism of a homologous series bearing a lateral 2-(4-n-alkan-

Table 10. Linking of cyclic segments by means of a three-unit spacer Z to 1,4-phenylene bis(4-n-octyloxyben-zoate) [47, 48].

$C_8H_{17}O$—⬡—COO—⬡—OOC—⬡—OC_8H_{17}
(lower substituent on middle ring): $COOCH_2$—⟨A⟩

27

Compound No.	—⟨A⟩	Phase transitions
27a	phenyl	Cr 98 N 98 I
27b	4-nitrophenyl (O₂N)	Cr 108 (N 80) I
27c	3-nitrophenyl (NO₂)	Cr 80 SmC 82 SmA 96 N 112 I
27d	4-nitrophenyl (—NO₂)	Cr 94 SmA 163 I
27e	3,4,5-trichlorophenyl (Cl, Cl, Cl)	Cr 120 (N 115) I
27f	anthracenyl	Cr 115 (N 102) I
27g	4-propylphenyl (—C₃H₇)	Cr 89 N 108 I
27h	furyl (O)	Cr 55 N 103 I
27i	4-octyloxyphenyl (—OC₈H₁₇)	Cr 89 (SmC 67) N 104 I
27j	—⬡—OOC—⬡—OC_8H_{17}	Cr 99 (SmC 88) N 151 I
27k	—⬡—OOC—⬡—OOC—⬡—OC_8H_{17}	Cr 130 N 197 I

oylphenoxy)ethoxycarbonyl segment [48]. Such molecules have been used as the starting point for preparing substances with chiral smectic C* phases (Table 11). However, although bulky lateral segments already exist in the molecules, the introduction of a methyl branched wing groups (e.g. (*R*)-octyl-2) depresses the phase transition temperature dramatically (compare **28a** and **28b**) [69]. Two ways of preparing the tilted layer structures wanted were elucidated. First, lengthening the basic mesogen to four aromatic rings makes possible the existence of stable smectic C* phase ranges (e.g. **28c**) [70]. Secondly, the introduction of one nitro group in a *meta* position of the basic mesogen results in glass forming low molecular mass ferroelectric liquid crystals with high spontaneous polarization (e.g. **28d**) [71].

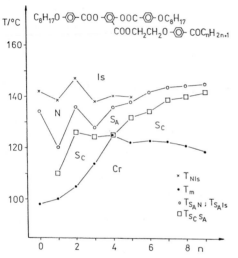

Figure 6. Polymorphism of the 4-acylphenoxyethyl esters [48].

Investigations of the helical twisting power in mixtures of laterally aryl substituted mesogens each other or with rod-like mesogens (one of the components of which is chiral) are concerned with relationships between the molecular constitution and the values of the helical twisting power [69, 72].

The unexpectedly high mesophase stability of such unconventional mesogens raises questions about the molecular shape and the packing in anisotropic phases. X-ray investigations, mostly of smectic A phases of 4-nitrobenzyl esters (**27d**, Table 10), were performed by Diele et al. [73, 74]. The layer thickness d is clearly lower than the molecular length L ($d/L = 3.42/4.20$ nm). The first proposed model of molecular packing involved intercalating layers (Figure 7(a)) [73], but the publication of X-ray measurements in the crystalline solid state by Hartung et al. [67, 75] indicated in the molecular packing shown in Figure 7(b). It is very possible that, on melting, the tilted layers existing in the crystalline solid state are lost, producing the order shown in Figure 7(a). Figure 7(a) and 7(b) show a two-dimensional layer, but within the bulk phase an antiparallel orientation should exist, as detected by dielectric measurements made by Kresse et al. [76, 77]. The packing model

Table 11. Smectic C* phases of laterally 2-(4-n-acyl-phenoxy)ethyloxycarbonyl substituted mesogens (the spacer Z contains five units).

Compound No.	R^1	R^2	R^3	n	X	Phase transitions	Ref.
28a	C_8H_{17}	C_8H_{17}	C_8H_{17}	1	H	Cr 121 SmC 140 SmA 144 I	[48]
28b	C_8H_{17}[a]	C_8H_{17}[a]	C_8H_{17}	1	H	Cr 93 I	[69]
28c	C_8H_{17}[a]	C_8H_{17}	C_8H_{17}	2	H	Cr 118 SmC 143 N 160 I	[70]
28d	C_8H_{17}[a]	C_8H_{17}	C_9H_{19}	1	NO_2	Cr 13[b] SmC 62 SmA 110 I	[71]

[a] $C_8H_{17} = (R)-C_6H_{13}*CH(CH_3)$
[b] T_G

(a)

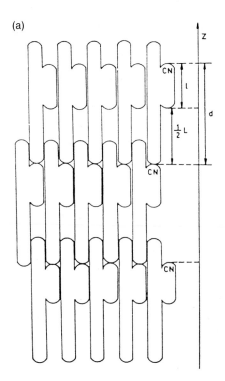

(b)

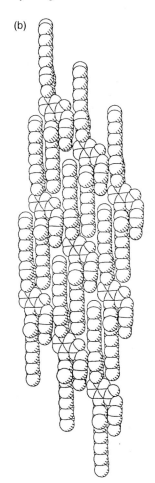

Figure 7. Model of the molecular packing of mesogens with a lateral aryl segment connected by an odd-numbered spacer. (a) Smectic state of 4-nitrobenzyl (**28c**) and 4-cyanophenoxyethyl bis(4-n-octyloxy-benzoates) (By courtesy of Diele et al. [73].) (b) Lamellar sheets in the crystal structure of 4-nitrobenzyl 2-(3-chloro-4-n-octyloxybenzoyloxy)-5-(4-n-octyl-oxy)benzoate. (By permission of Hartung et al. [75]).

of smectic A phases (Figure 7(a)) results in 'latent' holes, which is in agreement with the very unusual mixture behaviour observed (e.g. the induction of nematic re-entrant phases or smectic C phases [69, 74, 78] and the guest–host behaviour observed for small molecules. In particular, the last-mentioned effect is of interest. In binary systems, such as those composed of the derivatives **27d** and rod-like molecules, the smectic A mixed phases can be stabilized or destabilized depending on the molecular length of the rods. Short rigid molecules such as azoxyanisole, biphenyl or 4-nitroaniline can be added in properties of up to about 50% without any

change occurring in the packing parameters (e.g. the layer thickness) of the laterally aryl substituted host phase [74, 79–81]. The unusual mixing behaviour is illustrated in Figure 8; there is a wide range of stable layer thickness, at mixture compositions of up to 60% of guest molecules, in comparison to the expected d/L curve.

According to the results reported above the basic mesogens must possess a minimum of three aromatic rings if the corresponding laterally substituted derivatives are to exhibit liquid crystalline properties. Recently, Hohmuth et al. [82] provided evidence that segments incorporating a phenyl

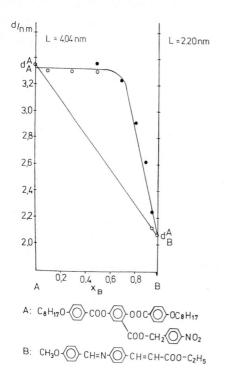

A: C₈H₁₇O⟨O⟩COO⟨O⟩OOC⟨O⟩OC₈H₁₇
          |
          COO-CH₂⟨O⟩NO₂

B: CH₃O⟨O⟩CH=N⟨O⟩CH=CH-COO-C₂H₅

Figure 8. The layer thickness *d* as a function of mixture composition *x*. (By courtesy of Diele et al. [79]).

ring can be laterally attached to two-ring compounds without loss of mesophases (**29**). In this case the terminal wing groups must be of a minimum length, and it should be noted that the nitro group attached at the lateral aryl ring gives rise to association effects.

In conclusion, comparison of the influence of the different bulky lateral substituents in the examples given (**26–29**) shows that segments containing lateral aryl groups do not only act as disruptive factor, but also make a contribution to the mesophase stability.

RO⟨O⟩COO⟨O⟩OR'
                |
                COOCH₂⟨O⟩NO₂

29

29a R=CH₃; R′=C₈H₁₇CO; Cr 86 (N 6) I [2]
29b R=C₈H₁₇; R′=C₁₂H₂₅; Cr 65 N 91 I [82]

3.2   Mesogens with Two Lateral Segments each containing a Phenyl Ring

Three-ring mesogens with two longer lateral alkyl chains (e.g. **12**) do not show mesophases. However, cross-shaped compounds (**30**) have nematic phases (uniaxial) with clearing points above 100 °C [46, 59]. That is surprising because 1,2,4-tris(4-n-alkyloxybenzoyloxy)benzenes (**19f**) are not liquid crystalline, but 1,3,5-tris(4-n-alkyloxyphenyl)benzoates exhibit nematic discotic phases [83]. Apparently, such substituted benzene derivatives are located at the boundary between calamitic and discotic compounds. However, there is no example of this type that shows both columnar and nematic or smectic phases, as occurs in polycatenar and double-swallow tailed compounds (see Sec. 5 and Chap. XII of this volume).

RO⟨O⟩COO
RO⟨O⟩COO⟨O⟩OOC⟨O⟩OR
                    |
                    OOC⟨O⟩OR

30

30a R=C₈H₁₇; Cr 109.9 (N 105.4) I [59]

4   Swallow-Tailed Mesogens

Swallow-tailed mesogens are characterized by a bulky alkyl branch connected to a terminal ring by a functional group or spacer. Thus such mesogens contain at one terminal end two long chains, and the molecular shape is therefore wedge-like and should be considered as a steric dipole. Most examples of this type are derived from three-ring basic mesogens, but shorter swallow-tailed compounds have also been described.

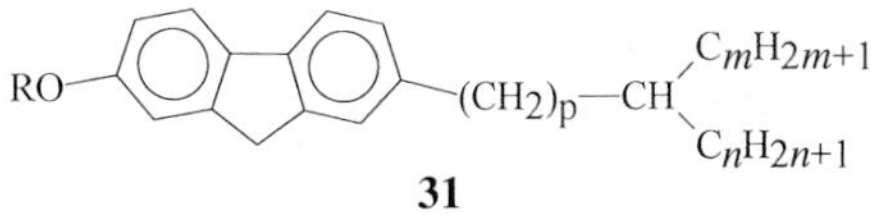

31

31a R=C_5H_{11}, p=4, m=n=6; Cr 32 N 54 I [84]

Two-ring fluorene derivatives (**31**) were the first reported terminally branched mesogens of this type [84]. Some of these substances show smectic A phases, although often on supercooling only. The two-ring cinnamic acid esters **32a** are liquid crystalline in the metastable range (Table 12) [85]. It is of interest that the clearing points in the ho-

mologous series resulting from increasing the chain length of the terminal double chain run through a minimum, which can be interpreted in terms of the length-to-breadth ratio of molecules (see Chap. VI of Volume 1). Swallow-tailed mesogens bearing a di-n-alkyl benzylidene malonate segment have been investigated in detail. The phase diagram of the series **32b** is shown in Figure 9. The nematic–isotropic transition points decrease with lengthening of the terminal double chain. However, the striking difference between this case and that of the laterally alkyl substituted mesogens is the existence of layer structures. It is very unusual that from the beginning (i.e. low chain lengths) all three transition curves T_{N-I},

Table 12. Swallow-tailed mesogens containing a di-n-alkyl benzylidenemalonate segment.

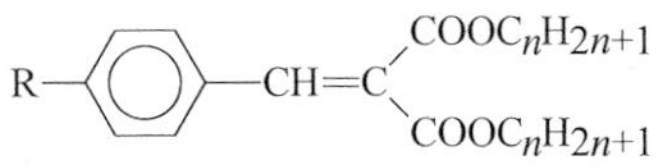

32

Compound No.	R/n	Phase transitions	Ref.
32a	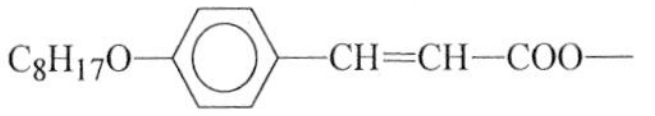 $C_8H_{17}O-$⟨○⟩$-CH=CH-COO-$ n = 10	Cr 54 (SmA 37 N 43.5) I	[85]
32b[a]	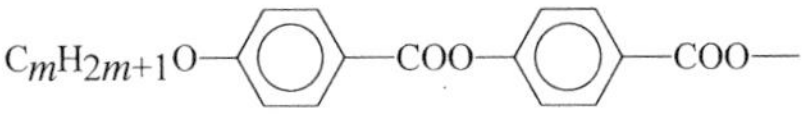 $C_mH_{2m+1}O-$⟨○⟩$-COO-$⟨○⟩$-COO-$		[85]
32c	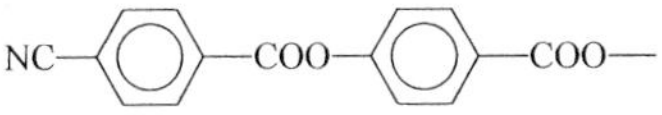 $NC-$⟨○⟩$-COO-$⟨○⟩$-COO-$ n = 10	Cr 104 N 110 I	[86]
32d	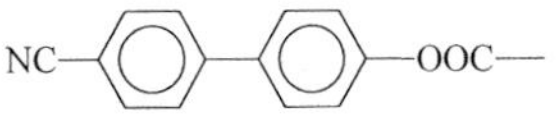 $NC-$⟨○⟩$-$⟨○⟩$-OOC-$ n = 10	Cr 89 (SmA 80) N 91 I	[87]
32e	$C_5H_{11}-$⟨A⟩$-$⟨B⟩$-COO-$ (i) A and B: phenyl, cyclohexyl (only dipropargyl malonates) (ii) A and B: phenyl	 Cr 74.7 (SmE ?) SmA 98.9 N 113.1 I	[88]
32f	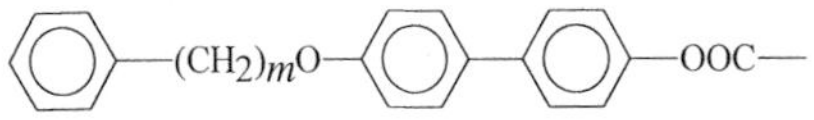 ⟨○⟩$-(CH_2)_mO-$⟨○⟩$-$⟨○⟩$-OOC-$ m = 1, n = 8	Cr 112 (SmE 105 SmB 107) N 129 I	[89]

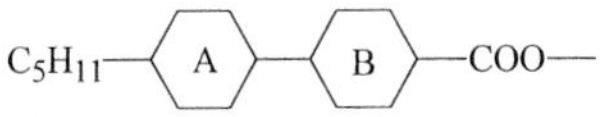

[a] See Figure 9.

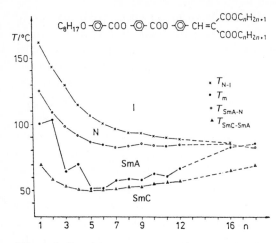

Figure 9. Transition temperatures of di-n-alkyl 4-(4-(4-n-octyloxybenzoyloxy)benzoyloxy)benzylidene-malonates **32b** versus the length of the terminal double chains [85].

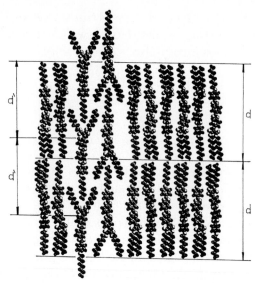

Figure 10. Proposed model of smectic A phases of swallow-tailed compounds with strings of interdigitated molecules. (By courtesy of Diele et al. [91]).

$T_{\text{SmA-N}}$ and $T_{\text{SmC-SmA}}$ run almost parallel to one another [85]. Molecular polarizability anisotropies and order parameters have been calculated by Hauser et al. [90], and X-ray studies performed by Diele et al. [85, 91] proved that the molecules are arranged antiparallel within the smectic monolayers, as would be expected for steric dipols. The layer thickness corresponds to the molecular length. In addition, diffraction patterns provide evidence for the existence of strings of interdigitated molecules (Figure 10). It can be assumed that the two different types of order coexist in such smectic A phases.

The degree of organization in the short-range structure in swallow-tailed liquid crystals of the type **32b** was investigated in detail by Kresse et al. in mixtures, using phase diagrams and dielectric methods. Antiparallel ordering is strongly distinct for geometric reasons and due to the polar interactions. Therefore, in mixtures of swallow-tailed compounds with rod-like or more wedge-like mesogens, the short-range order is retained or reduced, but the antiparallel order is destroyed by the addition of rela-

tively small concentrations of laterally alkyl substituted molecules [92–99]. The existence and the modifications of smectic phases depend on the direction (e.g. COO) and position of the polar groups (e.g. **32c** and **32d**) [86, 87]. Low-temperature smectic B and E phases of the dipropargyl malonates **32e** [88, 100] and of **32f** [89] have been detected. Unlike in **32b**, the layer thickness of the benzyl ether **32f** is clearly greater than the length of the molecules. 4-Allyloxy substituted swallow-tailed substances serve as intermediates for the synthesis of the corresponding liquid crystalline polymers [101].

Using mixtures prepared from homologous of **32b** and rod-like mesogens, Pelzl et al. [102,103] were able to detect a reentrant nematic phase in a mixture of two non-polar compounds.

Table 13 gives some further possible structures of terminally swallow-tailed branches. Compounds **33** represent dyes of low solubility [104–106]. In principle,

Table 13. Further chemical structures of terminal branches of swallow-tailed mesogens.

33 Y—⬡—CH=N—⬡—N=N—⬡—N⟨C_nH_{2n+1} / C_mH_{2m+1}

Y = CN, $n = m = 6$; Cr 119.8 (SmA 124.7) N 133.6 I   [104–106]

34 $C_{10}H_{21}O$—⬡—COO—⬡—N=CH—⬡—O—P(=O)(—OC_4H_9)(OC_4H_9)

Cr 138 Sm 176 N 198 I   [107]

35 $C_8H_{17}O$—⬡—COO—⬡—COO—⬡—COO—N=C⟨C_nH_{2n+1} / C_mH_{2m+1}

$n = 12$, $m = 15$; Cr 56 SmA 82 N 96 I   [108]

36 $C_8H_{17}O$—⬡—COO—⬡—COO—⬡—CO—N⟨C_nH_{2n+1} / C_mH_{2m+1}

$n = m = 4$; Cr 111 (N 104) I   [109]

37 RO—⬡—X—COO—⬡—⬡—COO—CH⟨C_nH_{2n+1} / C_mH_{2m+1}

X = CH=CH, CH_2–CH_2, C≡C, m, $n = 1$ to 6   [112–114]

37a R = C_9H_{19}, X = C≡C, $m = n = 3$; Cr 73 (45.9 SmC 54.8 SmA 70.3) I

38 $C_8H_{17}O$—⬡—COO—⬡—COO—⬡—CH=C⟨$COO(CH_2)_mC_nF_{2n+1}$ / $COO(CH_2)_mC_nF_{2n+1}$

$m = 2$, $n = 6$; Cr 112 SmC 156 SmA 164.5 I   [115]

39 $C_nF_{2n+1}(CH_2)_m$—O—⬡—OOC—⬡—OOC—⬡—CH=C⟨$COOC_8H_{17}$ / $COOC_8H_{17}$

$m = 2$, $n = 6$; Cr 70 SmX 72.2 SmC 81.9 SmA 155.5 I   [115]

dialkyl phosphate groups can serve as branched wing groups. However, in comparison with other three-ring swallow-tailed compounds and with results obtained by the authors research group, the mesophase stability of **34** seems to be very high [107]. Dialkylketonoxime benzoates **35** exhibit the expected trend to form smectic A phases [108]. However, the dialkylamides **36** are nematic only [109]. The effect of small terminal branches of alkylcarboxylates on the mesophase behaviour of three-ring mesogens has been tested systematically by Matsunaga et al. [110, 111]. Chiral secondary alkanols are often used to prepare smectic C* phases, the corresponding chemical structures being mentioned in many patents. Often, two long-chain substituents attached to the asymmetric carbon atom are claimed, but few examples are given with transition points. Substances **37** are of interest with regard to their antiferroelectric smectic C phases. Both the chiral and the racemic compounds ($n \neq m$), as well as the esters with two alkyl groups of the same length have been investigated intensively. The non-chiral compounds can exhibit alternating tilt smectic phases (SmC_{alt}) due to steric factors. The size of the terminally branched alkyl group has a direct influence on the type of smectic C mesophase formed [112–114].

The mesophase and dielectric behaviour of swallow-tailed substances similar 32b but containing perfluorinated alkyl chains, e.g. 38 and 39, have been studied [115]. It was found that such compounds have higher clearing points and show greater stabilization of the smectic phases than do the non-fluorinated compounds. Besides the steric and dipole interactions, the tendency for microphase separation probably also plays a role in these trends [115, 116].

In general, the mesophase behaviour of three-ring mesogens bearing two alkyloxy substituents at positions 3 and 4 of a terminal phenyl ring, as represent by the general formula 40, is often similar to that of

the swallow-tailed compounds discussed above, due to the molecular shape and the antiparallel packing in layers (e.g. [38, 117–123]). According to the definition, given by Malthete et al. [42], 3,4-dialkyloxy substituted derivatives, sometimes called 'forked' mesogens, are here classified as polycatenar mesogens (see Chap. XII of this volume).

Table 14 gives the chemical formulae of some compounds in which two alkyl chains are attached to a terminal heterocycloaliphatic ring system. These compounds are mentioned here because of their swallow-tail-like structure. They exhibit a diminished tendency to form liquid crystalline phases because of the sterically unfavourable arrangement of the double chains. Lactones 41 and 42 are not liquid crystalline and are used as dopants to induce smectic C* phases [124–128]. Only short chain lengths allow the existence of mesophases in the dioxolane 43 [129] and the dioxane derivatives 44 [2].

40

Table 14. Compounds with two long-chain substituents attached at a terminal heterocycloaliphatic ring system.

41    $C_{10}H_{21}O$—◯—◯—COO—*—C_nH_{2n+1} / C_nH_{2n+1}

$n = 1$ to 8   [124, 127]

42    $C_8H_{17}O$—[A]—◯—COO—*—C_nH_{2n+1} / C_nH_{2n+1}

$n = 1$ to 4, 6, 10, Ring A phenyl, 2,5-pyrimidyl   [125–128]

43   R—◯—$(CH_2)_3O$—◯—◯—(dioxolane)—$COOC_nH_{2n+1}$ / $COOC_nH_{2n+1}$

43a $R = C_5H_{11}$, $n = 2$; Cr 90 (SmA 77 N* 87) I   [129]
43b $R = C_5H_{11}$, $n = 4$; Cr 91 I

44    $C_8H_{17}O$—◯—COO—◯—(dioxane)—C_3H_7 / C_3H_7

Cr 88 (N 60) I   [2]

5 Double-Swallow-Tailed Mesogens

In principle, the terminal branches of swallow-tailed compounds can be attached at both ends of a rod. However, only dialkyl benzylidenemalonates have been used extensively to prepare double-swallow-tailed mesogens. The minimum size of the basic mesogen is four cyclic units, these mostly being phenyl rings [85]. The clearing points in the series **45** decrease strongly with in-

As shown by Pelzl et al. [43, 130–132], whether or not smectic A phases are induced, depends on the shape and length of the rod-like mixing components. This phenomenon can be explained by a filling of the gaps between the aromatic parts of the double-swallow-tailed molecules. Therefore, such induced smectic A phases are called 'filled smectic A phases' [130]. It is legitimate to discuss these filled phases as a new type of guest–host system. Dielectric measurements in pure substances as well as

$C_nH_{2n+1}OOC$
$C_nH_{2n+1}OOC$ $C=CH$—⬡—OOC—⬡—⬡—COO—⬡—$CH=C$ $COOC_nH_{2n+1}$ $COOC_nH_{2n+1}$

45 $n=1–12, 16$

45a $n=10$; Cr 88 (SmC 87) N 101 I

creasing alkyl chain length because of the four-fold possibility of producing *gauche* conformers. Unlike the swallow-tailed derivatives, no smectic A phases occur. The reason for this lies in the shape of the molecules, which does not allow a sufficiently high packing density without gaps of orthogonal phases. As shown in Figure 11, the package is better in the smectic C phase than in the hypothetical SmA phase. This fact also explains the behaviour of mixtures of double-swallow-tailed and rod-like mesogens.

mixtures containing mesogens of different shapes have been made by Kresse et al. [93, 133]. Evidence has been obtained to show that cyano substituted rods show higher mobility in mixtures with double-swallow-tailed compounds than in the bulk phase [134]. The range of double-swallow-tailed mesogens suitable for filled phases could be broadened to make use of five- and six-ring compounds. Pelzl et al. [135] were able to show by X-ray investigations that even two short rod-like mesogens can be incorporated into the gaps of six-ring host molecules. Depending on the binary system and the concentrations being investigated, unusual ratios of the layer spacing to the average molecular length of up to 2.13 were found. Multiple polymorphism can be induced in mixtures of the host mesogen (**46b** in Table 15) if the short guest molecules are sufficiently strong electron acceptors, e.g. trinitrofluorenone or nitrocinnamates [136].

Lengthening the aromatic part of double-swallow-tailed mesogens causes a lasting

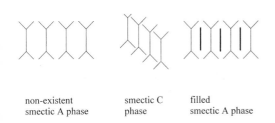

non-existent   smectic C   filled
smectic A phase   phase   smectic A phase

Figure 11. Schematic representation of the molecular packing of double-swallow-tailed mesogens in smectic and filled smectic A phases.

Table 15. Variation of the middle segment A of double-swallow-tailed compounds [137] (D=Col).

$C_nH_{2n+1}OOC$
$\quad\quad\quad\quad C=CH-$⬡$-COO-$⬡$-CH=N-A-N=CH-$⬡$-OOC-$⬡$-CH=C$
$C_nH_{2n+1}OOC$
$\quad\quad\quad\quad\quad\quad\quad\quad\quad\quad\quad\quad\quad\quad\quad\quad\quad\quad\quad\quad\quad\quad COOC_nH_{2n+1}$
$\quad\quad\quad\quad\quad\quad\quad\quad\quad\quad\quad\quad\quad\quad\quad\quad\quad\quad\quad\quad\quad\quad COOC_nH_{2n+1}$

46

Compound No.	A	n	Phase transitions
46a	⬡	12	Cr 96 D_{ob} 106 SmC 147 N 171 I
46b	⬡⬡	12	Cr 111 D_{ob} 206 I [a]
46c	⬡⬡⬡	12	Cr 124 D_{ob} 209 D_h 314 I
46d	⬡–C(=O)–⬡	12	Cr 103 I
46e	⬡–CH₂CH₂–⬡	12	Cr 95 SmC_{re} 117 D_{ob} 154 SmC 187 N 209 I
46f	⬡–C(=O)–NH–⬡	12	Cr 133 SmC 165 D_{ob} 181 D_h 188 D_X 243 I

[a] See Figure 12.

$C_nH_{2n+1}OOC$
$\quad\quad\quad\quad C=CH-$⬡$-COO-$⬡$-CH=N-$⬡⬡$-N=CH-$⬡$-OOC-$⬡$-CH=C$
$C_nH_{2n+1}OOC$
$\quad\quad\quad\quad\quad\quad\quad\quad\quad\quad\quad\quad\quad\quad\quad\quad\quad\quad\quad\quad\quad\quad COOC_nH_{2n+1}$
$\quad\quad\quad\quad\quad\quad\quad\quad\quad\quad\quad\quad\quad\quad\quad\quad\quad\quad\quad\quad\quad\quad COOC_nH_{2n+1}$

n = 5 - 14, 16

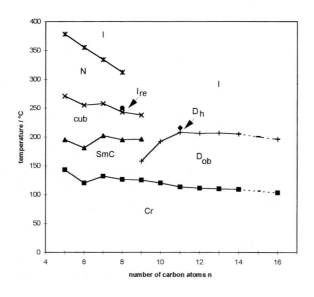

Figure 12. Phase diagram of a homologous series of six-ring double-swallow-tailed mesogens [139] (D=Col).

change in the polymorphism (Table 15). The higher the number of phenyl rings and the number of carbon atoms in the terminal alkyl chains, the greater the tendency to form columnar phases [138]. The polymorphism observed in a homologous series is demonstrated in Figure 12 [139]. Short-chain substituted derivatives exhibit nematic, cubic and smectic C phases, while the longer chain ones show oblique and hexagonal columnar phases. At the boundary between these two ranges a very interesting phase behaviour often occurs. For the octyl derivative of this series the very rare existence of a re-entrant isotropic phase has been observed [140]. That is, on heating of the optical isotropic re-entrant phase a nematic phase arises above 250 °C! This is the second known pure mesogen that exhibits this phenomenon, and the first for which all transition points have been measured calorimetrically. The first known substance, described by Warmerdam et al. [141, 142] was 2,3,7,8,12,13-hexakis(4-n-octadecanoyloxy)truxene, in which the re-entrant isotropic phase is surrounded by columnar phases.

Our knowledge about the relationship between the chemical constitution of double-swallow-tailed molecules and their liquid crystalline behaviour is increasing. For example, variation of the middle part A in molecules of the type **46** has shown that rigid and longer segments increase the clearing points and the range over which columnar phases exist (e.g. **46a**–**46c**) [137]. Surprisingly, the presence of angled units in the middle of the molecules can prevent mesophase formation in such six-ring compounds (**46d**) [137]. On the other hand, the use flexible but even-numbered spacers gives liquid crystals with a new polymorphism. For the dodecyloxy derivative **46e**, the unusual phase sequence SmC–D_{ob}–SmC–N is found with increasing temperature. At present this is the only known case of a re-entrant lamellar phase that arises from a columnar phase [137, 143]. Another example (**46f**) exhibits, in addition to a smectic C phase, two columnar phases and a high-temperature mesophase D_X which is characterized by a very weak birefringence and unusual optical textures [144]. The introduction of small lateral substituents, such as methoxy or ethoxy groups, at certain positions of the mesogens **46** can give rise to cubic phases over temperature ranges that are amenable to physical measurement [145].

The simultaneous occurrence of nematic, cubic, layer and columnar structures in pure mesogens is of great importance to bridging the gap between calamitic and discotic molecules and the corresponding mesophases. Smectic, cubic and columnar phases have also been described in polycatenar compounds, and thus double-swallow-tailed mesogens can also be considered to be a special case of tetracatenar compounds (see Chap. XII of this volume).

6   Further Aspects and Concluding Remarks

Geometrical factors, dispersion forces and dipolar interactions are at the forefront of the discussion and understanding of the relationship between chemical constitution and mesophase behaviour. Thus this field of unconventional liquid crystals can be viewed as comprising different groups.

The first group includes rod-like mesogens substituted by long-chain groups at any position, e.g. laterally alkyl substituted and swallow-tailed mesogens. In such compounds the liquid crystalline behaviour is largely determined by the repulsion forces of the surrounding molecules, which press on the additional hydrocarbon chains attached to the basic core and increase the molecule's length-to-breadth ratio.

The second group includes laterally aryl substituted molecules, in which special packing effects (e.g. antiparallel ordering) give rise to high packing densities. In such cases the bulky lateral groups do not prevent liquid crystalline properties, but add a contribution to an additional attractive interaction. The shape of the anisometric molecules should make possible an 'anisotropic molecular puzzle' type arrangement. That is, to design liquid crystals the molecules need not be calamitic or discotic. As an example of the possible structural variations, consider the combination of lateral aryl segments and swallow-tailed branches in the novel mesogens demonstrated **47** and **48** [69]. Introduction of the lateral 4-nitrobenzyloxycarbonyl segment into the 2'-position of the swallow-tailed mesogen isomer with compound **32b** (with $m = 8$ and $n = 10$; Cr 73 N 91 I) [85] decreases the nematic–isotropic transition point only a little (**47a**), while attachment at the 3'-position clearly increases the clearing point (**47b**).

ble-swallow-tailed mesogens **48** (Table 16) is influenced in such a way that the more bulky 4-nitrobenzyloxycarbonyl groups give rise to clearly higher transition temperatures than when the n-octyl chain is located in the lateral position. These results show that both theoretical and practical prediction of the mesophase behaviour of unconventional liquid crystals is virtually impossible at present. The special phase structures, particularly those in smectic A phases formed by mesogens having polar and steric asymmetry, have been reviewed by Ostrovskii [116].

In this means of interpretation, geometrical factors, including repulsion and dispersion forces and dipolar interactions, dominate. For example, the existence of thermotropic phases of ionic amphiphiles is driven by the formation of strong ion bonding lattices between the head groups, the molecular shape being a secondary factor. Additional groups, that are capable of association can influence the situation dramatically. Thus, the ability of non-ionic amphiphiles to form the anisotropic liquid state must be discussed separately [116]. In par-

47a   Cr 95 (SmA 87) I [69]

47b   Cr 111 (SmA 114) I [69]

No such mesophase stabilization (up to 23 K) by introduction of a lateral substituent has yet been observed. The mesophase behaviour of the laterally substituted dou-

ticular, the existence of groups that are capable of hydrogen bonding can change the behaviour of thermotropic liquid crystals

Table 16. The influence of a bulky lateral substituent R on the mesophase behaviour of double-swallow-tailed compounds.

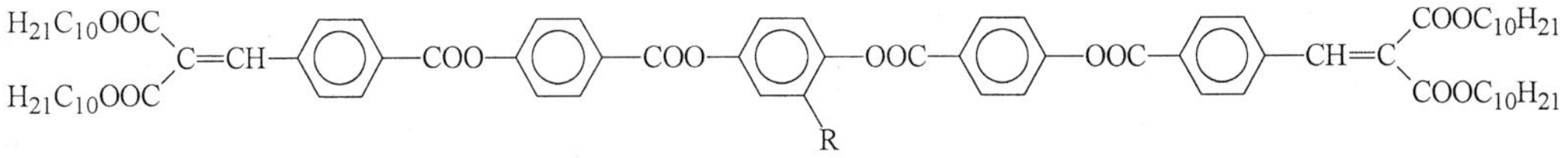

48

Compound No.	R	Phase transitions	Ref.
48a	H	Cr 97 SmC 147 N 181 I	[69]
48b	C$_8$H$_{17}$	Cr 86 SmA 118 N 121 I	[69]
48c	COOCH$_2$PhNO$_2$	Cr 103 SmC 148 SmA 161 N 163 I	[69]

Table 17. Amphiphilic mesogens by hydrophilic lateral groups [149].

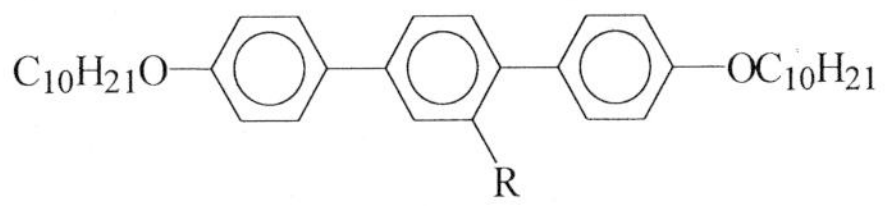

49

Compound No.	R	Phase transitions
49a	CONH–CH$_2$CH$_2$OH	Cr 115 SmA 129 I
49b	CONH–CH$_2$(OH)CH$_2$OH	Cr 115 SmA 142 I

[147, 148]. Recently, Tschierske and coworkers [149] have reported facial amphiphiles **49** (Table 17), in which the lateral groups are themselves substituted with hydroxyl groups. The mesophases are clearly stabilized by increasing the number of such groups, even though the lateral chain is lengthened, which is in contradiction to the results given for laterally alkyl substituted mesogens (see Sect. 11.2). The combination of amphiphilic segments with the molecular shape of the unconventional liquid crystals discussed here can be expected to lead to interesting results in the future.

The situation for unconventionally designed carboxylic acids, which form dimers, is clear. The existence of nematic and columnar phases of swallow-tailed-substituted benzoic acids **50** [109] can be interpreted in the same way as those of double-

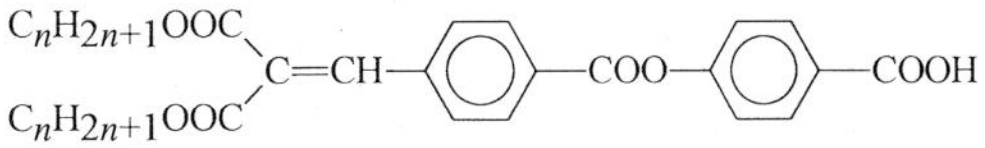

50

50a $n = 10$; Cr 107 D 119 N 123 I

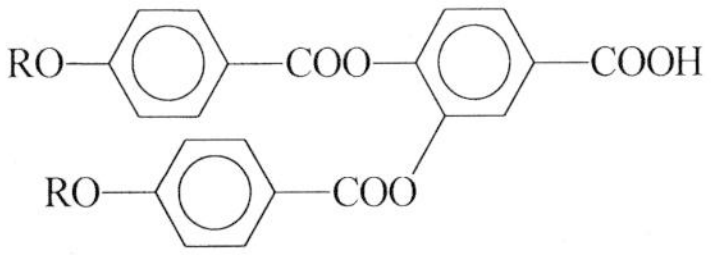

swallow-tailed compounds. The mesophase behaviour of laterally aryl substituted benzoic acids **51** [58] and **52** (Table 18) [150] is more surprising, because no covalent molecules having the shape of such dimers are known. A restricted comparison is possible with mesogens bearing aryl groups in two lateral segments **30** (see Sec. 11.3.2). The general conclusion, given for laterally aryl substituted mesogens, that odd-numbered spacers Z stabilize mesophases better can be confirmed by comparing the clearing points of derivatives **52** and **51**.

51

51a R = C$_8$H$_{17}$; Cr 117 (N 109) I

A further aspect in the design of unconventional liquid crystals is the formation of hydrogen-bonded associates produced from different components. In fact, the association of two or more single components bearing proton accepting and donating groups

Table 18. Two-ring benzoic acids incorporating a phenyl group within the lateral segment [150].

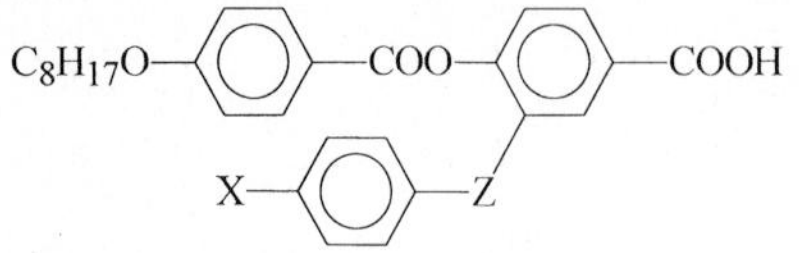

$C_8H_{17}O$—⬡—COO—⬡—COOH

                        52

X—⬡—Z

Compound No.	X	Z	Phase transitions
52a	H	$-CH_2OOC-$	Cr 151 (N 130) I
52b	NC–	$-CH_2OOC-$	Cr 147 (SmA 145) N 156 I
52c	O_2N-	$-CH_2OOC-$	Cr 172 (SmA 160 N 161) I
52d	$C_8H_{17}CO-$	$-OC_2H_4OOC-$	Cr 162 (N 131) I

can yield aggregates of defined molecular shape [147]. In principle, swallow-tailed and double-swallow-tailed as well as laterally aryl substituted mesogens should be designed in such a way. However, the hydrogen bond is more flexible than a covalent connection, and also shows dynamic behaviour. Therefore, only a few examples of this type have been reported so far [151, 152]. The associate **53** is composed of the

shape of these compounds can be used as the starting point in the search for biaxial nematic phases. All these molecules can be inserted into or attached to polymer backbones to give new types of liquid crystalline main-chain and side-on polymers and elastomers. Therefore, the unconventional liquid crystals discussed here will be of increasing interest in the future.

$C_8H_{17}O$—⬡—COO—⬡—COOH···N⬡—COO—⬡—COO—⬡—OC_7H_{15}

NC—⬡—CH_2OOC

                        53

laterally aryl substituted benzoic acid **52b** (Cr 147 (SmA 145) N 156 I) and a three-ring pyridine (Cr 130 N 140 I). The presence of a smectic A phase up to 188 °C provides evidence of the formation of thermodynamically stable associates within the 1:1 mixture.

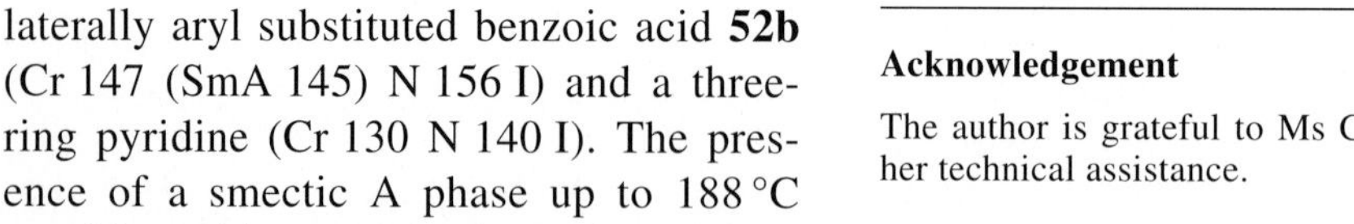

Swallow-tailed and double-swallow-tailed compounds, as well as laterally alkyl and aryl substituted mesogens, often show behaviours that lie at the boundary to another field, e.g. many of the derivatives exhibit the phenomenon of a glassy state on supercooling [123]. The special molecular

Acknowledgement

The author is grateful to Ms Christiane Lischka for her technical assistance.

7   References

[1] V. Vill in *Landolt–Börnstein, Numerical Data and Functional Relationships in Science and Technology, Group IV, Vol. 7e, Liquid Crystals,* Springer-Verlag, Berlin, **1995**.
[2] W. Weissflog, Ph. D. Thesis B, Halle, **1989**.
[3] W. Weissflog, R. Schlick, D. Demus, *Z. Chem.* **1981**, *21*, 452.

[4] W. Weissflog, D. Demus, *Cryst. Res. Technol.* **1983**, *18*, K21.

[5] S. Diele, K. Roth, D. Demus, *Cryst. Res. Technol.* **1986**, *21*, 97.

[6] W. Weissflog, D. Demus, *Cryst. Res. Technol.* **1984**, *19*, 55.

[7] D. Demus, A. Hauser, A. Isenberg, M. Pohl, Ch. Selbmann, W. Weissflog, S. Wieczorek, *Cryst. Res. Technol.* **1985**, *20*, 1413.

[8] D. Demus, S. Diele, A. Hauser, I. Latif, Ch. Selbmann, W. Weissflog, *Cryst. Res. Technol.* **1985**, *20*, 1547.

[9] C. Rein, D. Demus, *Liq. Cryst.* **1993**, *15*, 193.

[10] S. Grande, K. Singh, unpublished results. K. Singh, Ph. D. Thesis, Leipzig, **1986**.

[11] M. Ballauff, *Liq. Cryst.* **1987**, *2*, 519.

[12] F. Hoffmann, H. Hartung, W. Weissflog, P. G. Jones, A. Chrapkowski, *Mol. Cryst. Liq. Cryst.* **1995**, *258*, 61.

[13] W. Weissflog, A. Wiegeleben, S. Haddawi, D. Demus, *Mol. Cryst. Liq. Cryst.* **1996**, *281*, 15.

[14] H. Kresse, M. Keil, W. Weissflog, *Cryst. Res. Technol.* **1983**, *18*, 563.

[15] C. T. Imrie, L. Taylor, *Liq. Cryst.* **1989**, *6*, 1.

[16] W. Weissflog, D. Demus, *Mol. Cryst. Liq. Cryst.* **1985**, *129*, 235.

[17] K. Fujishiro, R. W. Lenz, *Macromolecules* **1992**, *25*, 81.

[18] F. Hessel, R.-P. Herr, H. Finkelmann, *Makromol. Chem.* **1987**, *188*, 1597.

[19] F. Hessel, H. Finkelmann, *Polym. Bull.* **1985**, *14*, 375.

[20] F. Hessel. H. Finkelmann, *Macromol. Chem.* **1988**, *189*, 2275.

[21] Qi-Feng Zhou, Hui-Min Li, Xin-De Feng, *Mol. Cryst. Liq. Cryst.* **1988**, *155*, 73.

[22] P. Keller, F. Hardouin, M. Mauzac, M. F. Achard, *Mol. Cryst. Liq. Cryst.* **1988**, *155*, 171.

[23] F. Hardouin, S. Mery, M. F. Achard, M. Mauzac, P. Davidson, P. Keller, *Liq. Cryst.* **1990**, *8*, 565.

[24] S. Takenaka, K. Yamazaki, *Mol. Cryst. Liq. Cryst.* **1994**, *241*, 119.

[25] H. Ishizuka, I. Nishiyama, A. Yoshizawa, *Liq. Cryst.* **1995**, *18*, 775.

[26] K. Mohr, W. Weissflog, H. Zaschke, *Wiss. Z. Univ. Halle* **1988**, *37*, 16–26.

[27] W. Weissflog, A. Wiegeleben, D. Demus, *Mater. Chem. Phys.* **1985**, *12*, 461.

[28] K. Mohr, H. Zaschke, W. Weissflog, *Z. Chem.* **1987**, *27*, 407.

[29] K. Mohr, H. Zaschke, W. Weissflog, *Z. Chem.* **1987**, *27*, 445.

[30] G. W. Gray, J. S. Hill, D. Lacey, *Mol. Cryst. Liq. Cryst.* **1991**, *197*, 43.

[31] N. Hoshino, H. Hasegawa, Y. Matsunaga, *Liq. Cryst.* **1991**, *9*, 267.

[32] H. F. Leube, H. Finkelmann, *Makromol. Chem.* **1990**, *191*, 2707.

[33] H. F. Leube, H. Finkelmann, *Makromol. Chem.* **1991**, *192*, 1317.

[34] R. A. Lewthwaite, G. W. Gray, K. J. Toyne, *J. Mater. Chem.* **1992**, *2*, 119.

[35] G. S. Attard, C. T. Imrie, *Liq. Cryst.* **1986**, *6*, 387.

[36] P. Berdague, J. P. Bayle, Mei-Sing Ho, B. M. Fung, *Liq. Cryst.* **1995**, *18*, 823.

[37] S. Takenaka, H. Morita, S. Kusabayashi, Y. Masuda, M. Iwano, T. Ikemoto, *Chem. Lett.* **1988**, 1559.

[38] S. Takenaka, H. Morita, M. Iwano, Y. Sakurai, T. Ikemoto, S. Kusabayashi, *Mol. Cryst. Liq. Cryst.* **1990**, *182B*, 325.

[39] G. W. Gray, J. S. Hill, D. Lacey, *Mol. Cryst. Liq. Cryst. Lett.* **1990**, *7*, 47.

[40] M. S. K. Lee, G. W. Gray, D. Lacey, K. J. Toyne, *Makromol. Chem., Rapid Commun.* **1989**, *10*, 325.

[41] M. A. Michaleva, B. M. Bolotin, E. S. Serebryakova, R. U. Safina, T. A. Kitzner, E. G. Sklyarova, *Khim. Geterosikl. Soedin.* **1992**, 377.

[42] J. Malthete, H. T. Nguyen, C. Destrade, *Liq. Cryst.* **1993**, *13*, 171.

[43] G. Pelzl, A. Humke, S. Diele, D. Demus, W. Weissflog, *Liq. Cryst.* **1990**, *7*, 115.

[44] U. Rötz, J. Lindau, W. Weissflog, G. Reinhold, W. Unseld, F. Kuschel, *Mol. Cryst. Liq. Cryst.* **1989**, *170*, 185.

[45] K. Praefcke, D. Schmidt, *Z. Naturforsch., Teil b* **1981**, *36*, 375.

[46] S. Berg, V. Krone, H. Ringsdorf, U. Quotschalla, H. Paulus, *Liq. Cryst.* **1991**, *9*, 151.

[47] W. Weissflog, D. Demus, *Liq. Cryst.* **1988**, *3*, 275.

[48] W. Weissflog, D. Demus, S. Diele, *Mol. Cryst. Liq. Cryst.* **1990**, *191*, 9.

[49] A. Furukawa, R. W. Lenz, *Macromol. Chem., Macromol. Symp.* **1986**, *2*, 3.

[50] R. Cox, W. Volksen, B. L. Dawson in *Liquid Crystals and Ordered Fluids, Vol. 4* (Eds.: A. Griffin, J. F. Johnson), Plenum, New York, **1984**, p. 33.

[51] V. Gallardo, H. J. Müller, *Mol. Cryst. Liq. Cryst.* **1984**, *102*, 13.

[52] S. C. Bennur, T. Kroin, G. R. Ouriques, T. R. Taylor, *Mol. Cryst. Liq. Cryst.* **1988**, *154*, 277.

[53] V. Vill in *Landolt–Börnstein, Numerical Data and Functional Relationships in Science and Technology, Group IV, Vol. 7b, Liquid Crystals*, Springer-Verlag, Berlin, **1992**, p. 230.

[54] C. Kuhrmann, Ph. D. Thesis, Halle, **1926**.

[55] U. Baumeister, Z. Kosturkiewicz, H. Hartung, D. Demus, W. Weissflog, *Liq. Cryst.* **1990**, *7*, 241.

[56] U. Caruso, S. Pragliola, A. Roviello, A. Sirigu, *Macromolecules* **1993**, *26*, 221.

[57] P. Berdague, J. P. Bayle, Mei-Sing Ho, B. M. Fung, *Liq. Cryst.* **1993**, *14*, 667.

[58] S. Takenaka, Y. Masuda, M. Iwano, H. Morita, S. Kusabayashi, H. Sugiura, T. Ikemoto, *Mol. Cryst. Liq. Cryst.* **1989**, *168*, 111.

[59] W. D. A. Norbert, J. W. Goodby, M. Hird, K. J. Toyne, J. C. Jones, J. S. Patel, *Mol. Cryst. Liq. Cryst.* **1995**, *260*, 339.

[60] S. Takenaka, K. Nishimura, S. Kusabayashi, *Chem. Lett.* **1986**, 751.

[61] S. Takenaka, H. Morita, M. Iwano, S. Kusabayashi, T. Ikemoto, Y. Sakurai, H. Miyake, *Mol. Cryst. Liq. Cryst.* **1989**, *166*, 157.

[62] E. Mauerhoff, Ph. D. Thesis, Halle, **1922**.

[63] T. Masuda, Y. Matsunaga, *Bull. Chem. Soc. Jpn.* **1991**, *64*, 2192.

[64] V. G. Rumyantsev, A. V. Ivashenko, V. M. Muratov, V. T. Lazareva, E. K. Prudnikova, L. M. Blinov, *Mol. Cryst. Liq. Cryst.* **1983**, *94*, 205.

[65] D. Braun, M. Reubold, M. Wegmann, J. H. Wendorff, *Makromol. Chem., Rapid Commun.* **1991**, *12*, 151.

[66] D. Braun, M. Reubold, L. Schneider, M. Wegmann, J. H. Wendorff, *Liq. Cryst.* **1994**, *3*, 429.

[67] F. Hoffmann, H. Hartung, W. Weissflog, P. G. Jones, A. Chrapkowski, *Mol. Cryst. Liq. Cryst.* **1996**, *281*, 205.

[68] W. Weissflog, D. Demus, S. Diele, P. Nitschke, W. Wedler, *Liq. Cryst.* **1989**, *5*, 111.

[69] C. Stützer, Ph. D. Thesis, Halle, **1995**.

[70] C. Stützer, W. Weissflog, *Mol. Cryst. Liq. Cryst.* **1997**, *293*, 67

[71] J. Bömelburg, G. Heppke, K. Wuthe, *SPIE: Proc. SPIE – Int. Soc. Opt. Eng.* **1994**, *2175*, 10.

[72] C. Stützer, W. Weissflog, H. Stegemeyer, *Liq. Cryst.* **1996**, *21*, 557.

[73] S. Diele, W. Weissflog, G. Pelzl, H. Manke, D. Demus, *Liq. Cryst.* **1986**, *1*, 101.

[74] S. Diele, G. Pelzl, A. Mädicke, D. Demus, W. Weissflog, *Mol. Cryst. Liq. Cryst.* **1990**, *191*, 37.

[75] F. Hartung, F. Hoffmann, C. Stützer, W. Weissflog, *Liq. Cryst.* **1995**, *19*, 839.

[76] H. Kresse, W. Weissflog, *Phys. Stat. Sol. (a)* **1988**, *106*, K89.

[77] H. Kresse, S. Heinemann, A. Hauser, W. Weissflog, *Cryst. Res. Technol.* **1990**, *25*, 603.

[78] G. Pelzl, S. Diele, W. Weissflog, D. Demus, *Cryst. Res. Technol.* **1989**, *24*, K57.

[79] S. Diele, A. Mädicke, K. Knauft, J. Neutzler, W. Weissflog, D. Demus, *Liq. Cryst.* **1991**, *10*, 47.

[80] S. Haddawi, D. Demus, S. Diele, H. Kresse, G. Pelzl, W. Weissflog, A. Wiegeleben, *Mol. Cryst. Liq. Cryst.* **1994**, *238*, 109.

[81] S. Haddawi, S. Diele, H. Kresse, G. Pelzl, W. Weissflog, *Cryst. Res. Technol.* **1994**, *29*, 745.

[82] A. Jacobi, W. Weissflog, *Liq. Cryst.* **1997**, *22*, 107.

[83] S. Takenaka, K. Nishimura, S. Kusabayashi, *Mol. Cryst. Liq. Cryst.* **1984**, *111*, 227.

[84] J. Malthete, J. Canceill, J. Gabard, J. Jacques, *Tetrahedron* **1981**, *37*, 2823.

[85] W. Weissflog, A. Wiegeleben, S. Diele, D. Demus, *Cryst. Res. Technol.* **1984**, *19*, 583.

[86] W. Weissflog, G. Pelzl, H. Kresse, D. Demus, *Cryst. Res. Technol.* **1988**, *23*, 1259.

[87] S. Heinemann, R. Paschke, H. Kresse, *Liq. Cryst.* **1993**, *13*, 373.

[88] I. Cabrera, H. Ringsdorf, M. Ebert, J. H. Wendorff, *Liq. Cryst.* **1990**, *8*, 163.

[89] H. Bernhardt, W. Weissflog, S. Diele, H. Kresse in *Proceedings of 24th Freiburger Arbeitstagung Flüssigkristalle*, 1995, p. 51.

[90] A. Hauser, M. Hettrich, D. Demus, *Mol. Cryst. Liq. Cryst.* **1990**, *191*, 339.

[91] S. Diele, S. Manke, W. Weissflog, D. Demus, *Liq. Cryst.* **1989**, *4*, 301.

[92] H. Kresse, P. Rabenstein, H. Stettin, S. Diele, D. Demus, W. Weissflog, *Cryst. Res. Technol.* **1988**, *23*, 135.

[93] H. Stettin, H. Kresse, W. Weissflog, *Mol. Cryst. Liq. Cryst.* **1988**, *162B*, 139.

[94] H. Kresse, S. Heinemann, A. Hauser, W. Weissflog, *Cryst. Res. Technol.* **1990**, *25*, 603.

[95] S. Heinemann, H. Kresse, W. Weissflog, *J. Mol. Liq.* **1992**, *53*, 93.

[96] H. Kresse, P. Rabenstein, *Phys. Stat. Sol. (a)* **1987**, *100*, K83.

[97] H. Kresse, W. Weissflog, F. Kremer, A. Schönfels, *Cryst. Res. Technol.* **1992**, *27*, K5.

[98] S. Heinemann, H. Kresse, W. Weissflog, *Cryst. Res. Technol.* **1993**, *27*, 131.

[99] H. Kresse, S. Heinemann, R. Paschke, W. Weissflog, *Ber. Bunsenges. Phys. Chem.* **1993**, *97*, 1337.

[100] S. Diele, W. Wedler, G. Pelzl, I. Cabrera, H. Ringsdorf in *Proceedings of 21st Freiburger Arbeitstagung Flüssigkristalle*, 1992, p. 27.

[101] S. Heinemann, G. Pirwitz, H. Kresse, *Mol. Cryst. Liq. Cryst.* **1993**, *237*, 277.

[102] G. Pelzl, S. Diele, I. Latif, W. Weissflog, D. Demus, *Cryst. Res. Technol.* **1982**, *17*, K78.

[103] G. Pelzl, J. Szabon, A. Wiegeleben, S. Diele, W. Weissflog, D. Demus, *Cryst. Res. Technol.* **1990**, *25*, 223.

[104] P. L. K. Hung, A. Bloom, *Mol. Cryst. Liq. Cryst.* **1980**, *59*, 1.

[105] A. Bloom, L. K. Hung, *U. S. Patent 4 105,654*, 1978.

[106] M. E. Neubert, S. J. Hummel, J. C. Bhatt, S. S. Keast, A. M. Lackner, J. D. Margerum, E. Sherman, *Mol. Cryst. Liq. Cryst.* **1995**, *260*, 287.

[107] Y. G. Galyametdinov, I. N. Alikina, G. I. Ivanova, S. V. Fridland, *Izv. Akad. Nauk SSSR, Ser. Khim.* **1988**, 2877.

[108] H. Kresse, P. Rabenstein, W. Weissflog, *Z. Chem.* **1984**, *24*, 110.

[109] W. Weissflog, S. Diele, G. Pelzl in *Proceedings of 22nd Freiburger Arbeitstagung Flüssigkristalle*, 1993, 15.

[110] Y. Matsunaga, N. Miyajima, *Mol. Cryst. Liq. Cryst.* **1990**, *178*, 157.

[111] Y. Matsunaga, H. Matsuzaki, N. Miyajima, *Bull. Chem. Soc. Jpn.* **1990**, *63*, 886.

[112] I. Nishiyama, J. W. Goodby, *J. Mater. Chem.* **1992**, *2*, 1015.

[113] I. Nishiyama, *Adv. Mater.* **1994**, *6*, 966.

[114] R. P. Tuffin, J. W. Goodby, D. Bennemann, G. Heppke, D. Lötzsch, G. Scherowsky, *Mol. Cryst. Liq. Cryst.* **1995**, *260*, 51.

[115] E. Dietzmann, W. Weissflog, S. Markscheffel, A. Jakli, D. Lose, S. Diele, *Ferroelectrics*, **1996**, *180*, 341.

[116] B. I. Ostrovskii, *Liq. Cryst.* **1993**, *4* 131.

[117] Nguyen Huu Tinh, J. Malthete, C. Destrade, *Mol. Cryst. Liq. Cryst. Lett.* **1985**, *2*, 133.

[118] Nguyen Huu Tinh, C. Destrade, *Mol. Cryst. Liq. Cryst.* **1989**, *6*, 123.

[119] M. F. Achard, Nguyen Huu Tinh, H. Richard, M. Mauzac, F. Hardouin, *Liq. Cryst.* **1990**, *8*, 533.

[120] Nguyen Huu Tinh, G. Sigaud, M. F. Achard, F. Hardouin, R. J. Twieg, K. Betterton, *Liq. Cryst.* **1991**, *10*, 389.

[121] H. Takenaka, S. Takenaka, H. Miyake, S. Kusabayashi, *Mol. Cryst. Liq. Cryst.* **1991**, *202*, 111.

[122] H. Yamasu, S. Takenaka in *18th Japanese Symposium on Liquid Crystals*, Niigata, 1992, Abstract 1B506.

[123] W. Wedler, D. Demus, H. Zaschke, K. Mohr, W. Schäfer, W. Weissflog, *J. Mater. Chem.* **1991**, *1*, 347.

[124] T. Ikemoto, K. Sakashita, Y. Kageyama, F. Terada, Y. Nakaoka, K. Ichimura, K. Mori, *Chem. Lett.* **1992**, 567.

[125] K. Sakashita, Y. Nakaoka, T. Ikemoto, F. Terada, Y. Kageyama, M. Shindo, K. Mori, *Chem. Lett.* **1991**, 1727.

[126] K. Sakashita, T. Ikemoto, Y. Nakaoka, F. Terada, Y. Sako, Y. Kageyama, K. Mori, *Liq. Cryst.* **1993**, *13*, 71.

[127] T. Ikemoto, K. Sakashita, Y. Kageyama, F. Onuma, Y. Shibuya, K. Ichimura, K. Mori, *Mol. Cryst. Liq. Cryst.* **1994**, *250*, 247.

[128] T. Ikemoto, Y. Kageyama, F. Onuma, Y. Shibuya, K. Ichimura, K. Sakashita, K. Mori, *Liq. Cryst.* **1994**, *17*, 729.

[129] S. M. Kelly, M. Schadt, H. Seiberle, *Liq. Cryst.* **1992**, *11*, 761.

[130] S. Diele, G. Pelzl, W. Weissflog, D. Demus, *Liq. Cryst.* **1988**, *3*, 1047.

[131] G. Pelzl, A. Humke, S. Diele, S. Ziebarth, W. Weissflog, D. Demus, *Cryst. Res. Technol.* **1990**, *25*, 587.

[132] G. Pelzl, A. Humke, D. Demus, W. Weissflog, *Cryst. Res. Technol.* **1990**, *25*, 597.

[133] H. Kresse, H. Stettin, W. Weissflog, *Wiss. Z. Univ. Halle* **1985**, *34*, 133.

[134] S. Haddawi, S. Diele, H. Kresse, G. Pelzl, W. Weissflog, A. Wiegeleben, *Liq. Cryst.* **1994**, *17*, 191.

[135] G. Pelzl, S. Diele, K: Ziebarth, W. Weissflog, D. Demus, *Liq. Cryst.* **1990**, *8*, 765.

[136] I. Letko, S. Diele, G. Pelzl, W. Weissflog, *Liq. Cryst.* **1995**, *19*, 643.

[137] W. Weissflog, I. Letko, S. Diele, G. Pelzl, *Adv. Mater.* **1996**, *8*, 76.

[138] S. Diele, K. Ziebarth, G. Pelzl, D. Demus, W. Weissflog, *Liq. Cryst.* **1990**, *8*, 211.

[139] W. Weissflog, G. Pelzl, I. Letko, S. Diele, *Mol. Cryst. Liq. Cryst.* **1995**, *260*, 157.

[140] W. Weissflog, I. Letko, G. Pelzl, S. Diele, *Liq. Cryst.* **1995**, *18*, 867.

[141] T. Warmerdam, D. Frenkel, R. J. J. Zijlstra, *Liq. Cryst.* **1988**, *3*, 149.

[142] T. W. Warmerdam, R. J. M. Nolte, W. Drenth, J. C. van Miltenburg, D. Frenkel, R. J. J. Zijlstra, *Liq. Cryst.* **1988**, *3*, 1087.

[143] W. Weissflog, M. Rogunova, I. Letko, S. Diele, G. Pelzl, *Liq. Cryst.* **1995**, *19*, 541.

[144] W. Weissflog, A. Saupe, I. Letko, S. Diele, G. Pelzl, *Liq. Cryst.* **1996**, *20*, 483.

[145] W. Weissflog, I. Letko, S. Diele, G. Pelzl in *Proceedings of 23rd Freiburger Arbeitstagung*, 1994, 23.

[146] C. Tschierske in *Progress in Polymer Science* (Ed.: V. Percec), Elsevier, Amsterdam, **1996**, *21*, 775.

[147] C. M. Paleos, D. Tsiourvas, *Angew. Chem.* **1995**, *107*, 1839.

[148] H. Prade, R. Miethchen, V. Vill, *J. Prakt. Chem.* **1995**, *337*, 427.

[149] F. Hildebrandt, J. A. Schröter, C. Tschierske, R. Festag, R. Kleppinger, J. H. Wendorff, *Angew. Chem.* **1995**, *107*, 1780.

[150] W. Weissflog, E. Dietzmann, C. Stützer, M. Drewello, F. Hoffmann, H. Hartung, *Mol. Cryst. Liq. Cryst.* **1996**, *275*, 75.

[151] A. Treybig, C. Dorscheid, W. Weissflog, H. Kresse, *Mol. Cryst. Liq. Cryst.* **1995**, *260*, 369.

[152] H. Kresse, A. Treybig, A. Kolbe, W. Weissflog, *SPIE: Proc. SPIE – Int. Soc. Opt. Eng.* **1995**, *2372*, 222.

[153] W. Weissflog, S. Diele, D. Demus, *Mater. Chem. Phys.* **1986**, *15*, 475.

Chapter XII
Phasmids and Polycatenar Mesogens

Huu-Tinh Nguyen, Christian Destrade, and Jacques Malthête

1 Introduction

Rod-like liquid crystals [1] have been known for more than a hundred years, the first one, cholesteryl benzoate, being discovered in 1888 by Reinitzer. In materials of this type, nematic N, cholesteric N* and different lamellar mesophases such as SmA, SmC, SmF, and SmI are obtained. Beside these classical liquid crystals, thermotropic mesophases – consisting of two-dimensional aromatic flat molecules – that exhibit various columnar phases (e.g., Col_h, Col_r, Col_{ob}) have been known since 1977 [2, 3]. In these two types of systems, the lamellar and columnar phases are observed separately. So, it was interesting to examine the mesomorphic properties of the hybrid molecules, i.e., molecules with a long rod-like rigid core ending in two half-disc moieties (Fig. 1). In fact, the phasmids [4, 5] fill the gap between rod-like and disc-like mesogenic compounds and allow the relationship between molecular architecture and mesomorphic properties to be explored. In these materials, only those columnar mesophases having the same symmetry as the disc-like mesogens are observed. The real gap between these two types of liquid crystals is filled by biforked molecules [6] which exhibit, in the same series or in a pure compound, nematic N, lamellar SmC and columnar mesophases. The phasmids and biforked mesogens have a similar molecular architecture with more than two paraffinic chains grafted on the rigid core. For this reason we name them *polycatenar mesogens*.

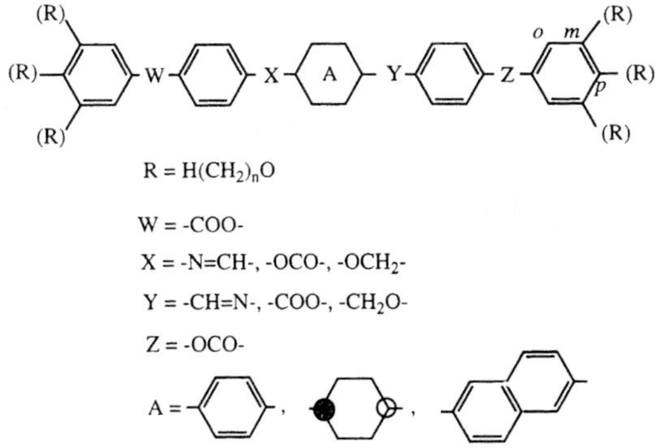

2 Nomenclature

Originally, the term "phasmid" referred to rod-like mesogens ending in two trialkyloxy phenyl groups because of the similarity of the rod-like molecules to a six-legged stick-like insect called a phasma. We propose the term phasmidic phases (denoted Φ, the initial letter of the Greek root) to distinguish this new type of columnar mesophase, which differs from disc-like mesogens in its molecular organization. On the other hand, we use the traditional nomenclature for the lamellar (S) and nematic (N) phases. Taking into account the number of aliphatic chains and the possible combination of their

$R = H(CH_2)_nO$

W = -COO-

X = -N=CH-, -OCO-, -OCH$_2$-

Y = -CH=N-, -COO-, -CH$_2$O-

Z = -OCO-

Figure 1. General molecular architecture of polycatenars.

anchoring points in ortho (*o*), meta (*m*) and para (*p*) positions with regard to the linking group of the central core, we propose a nomenclature which relates to the number and positions of the paraffinic chains (fluorinated or not).

Six-chain compounds are named hexacatenar mesogens or hexacatenars for short (from the Latin *catena* for chain). In the same way, the five-, four-, and three-chain compounds are labeled penta-, tetra-, and tricatenars, respectively. To distinguish different isomers, we label in each extremity the number and the positions of the paraffinic chains. For example, the first reported phasmids are hexacatenars *3mpm-3mpm* (Fig. 1). To avoid any confusion, it must be emphasized that the term catenar is reserved, by convention, for mesogens with a clear rod-like core, i.e., with an elongated linear polyaromatic part with rings connected in the para positions, and with normal aliphatic chains only grafted on the two terminal benzene rings, because disc-like mesogens have also more than two aliphatic chains. Only the polycatenars thus defined are described in this article. Consequently, this definition excludes disc-like mesogens, fused twins [7], bent elongated mesogens [8], and rod-like liquid crystal compounds with a lateral chain on a non-terminal benzene ring [9, 10] or with a branched chain on a terminal benzene ring [11, 12]. Nevertheless, some disc-like mesogens with a two-fold symmetry such as ruffigallol hexaalkanoates [13], tetra-alkanoates of ellagic acid [14], or cello-oligosaccharide derivatives [15] are on the boundary between disc-like mesogens and polycatenars.

3   Synthesis Principles

Symmetrical hexacatenars such as the first described phasmids were obtained as follows (Scheme 1): The three-chain derivative **2** was obtained by etherification of ester **1** with three equivalents of the required *n*-alkyl bromide, followed by saponification and hydrolysis (a). Esterification of acid **2** chloride with *p*-nitrophenol (b) afforded nitroester **3**, which was reduced to the corresponding aminoester **4** (c). Finally, phasmid **5** was obtained by reaction of two equivalents of **4** with terephthaldehyde.

Compounds **5** were generally recrystallized from absolute ethanol. For different symmetrical phasmids and biforked mesogens, the synthetic route is similar with protected and deprotected phenols or anilines. The central ring moiety is often terephthalic acid, 1,4-cyclohexanedicarboxylic acid, or terephthaldehyde. With non-symmetric polycatenar mesogens, the different intermediate compounds are protected or deprotected phenols, anilines and acids or aldehydes. The same method is repeated to add a benzoate. The last step is often an esterification reaction (Scheme 2): unlike the imine derivatives, compounds with ester linkages can be easily purified by chromatography on silica gel with dichloromethane as eluent.

4   Mesomorphic Properties

In Sec. 4.1 of this chapter, only mesogens substituted with aliphatic chains are reviewed. Then, in Sec. 4.2 of this chapter, we describe mesogenic compounds fitted with aliphatic chains and more or less polar substituents (Br, CN, NO_2, $(CF_2)_nF$).

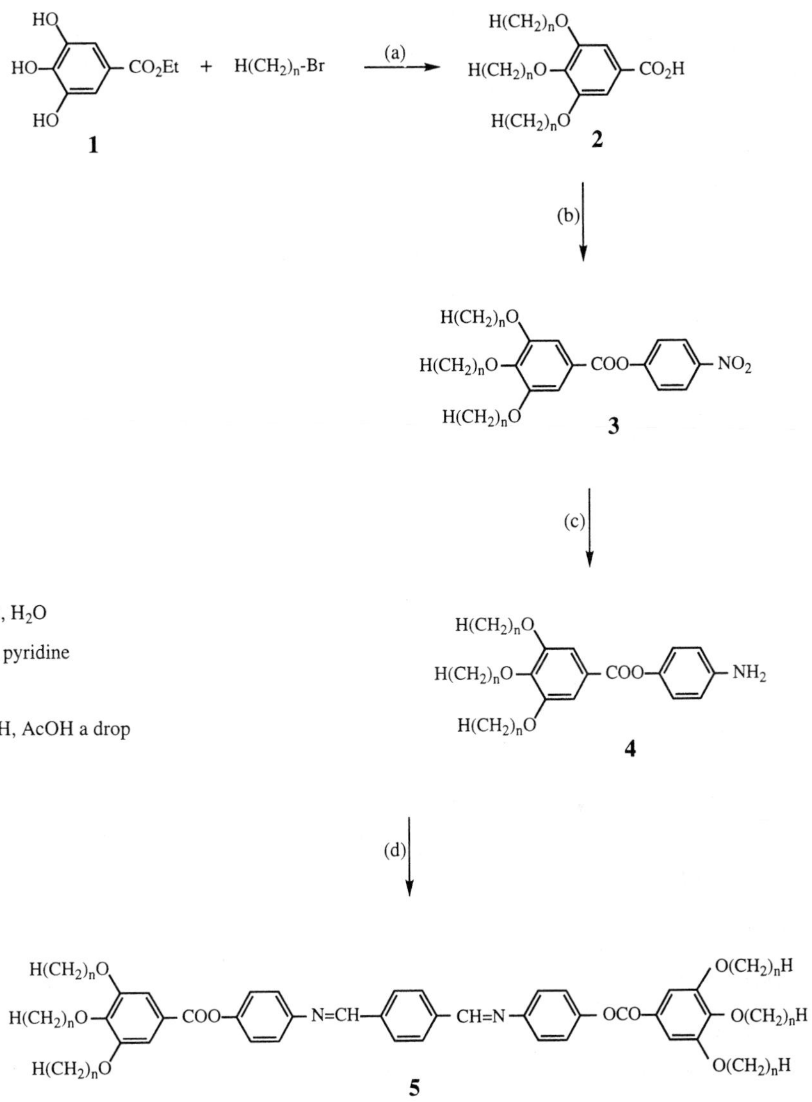

a) K$_2$CO$_3$, DMF; KOH; HCl, H$_2$O

b) SOCl$_2$, p-HO-C$_6$H$_4$-NO$_2$, pyridine

c) H$_2$, Pd/C, EtOAc, EtOH

d) p-OHC-C$_6$H$_4$-CHO, EtOH, AcOH a drop

Scheme 1

4.1   Polycatenars with Only Aliphatic Chains

4.1.1   Phasmids or Hexacatenar Mesogens 3mpm-3mpm

Beside some series of hexacatenar mesogens 3*omp*-3*omp*, in which biaxial nematic properties seem to have been observed [16], hexacatenars 3*mpm*-3*mpm* have been especially studied. In contrast to the former, up to now, a nematic N phase has not been observed in the latter.

The first studies were performed on phasmids **5** (Scheme 1) with two different chain lengths: the short chain derivative exhibits only a hexagonal columnar phase while that with longer chains provides

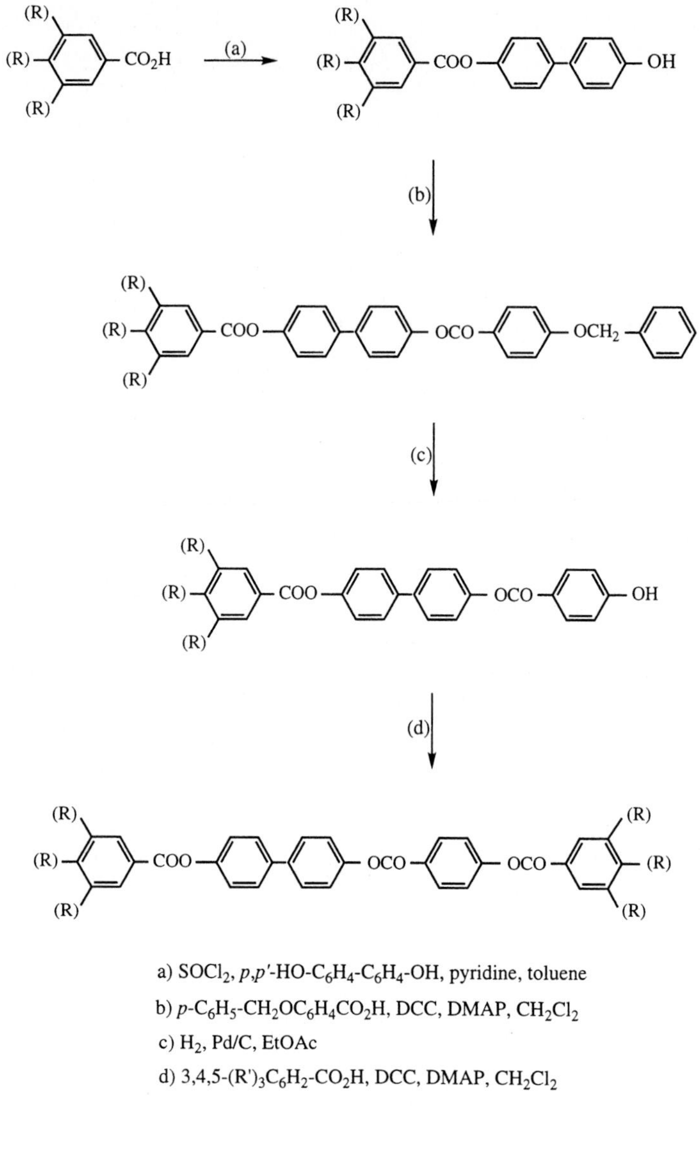

a) $SOCl_2$, p,p'-HO-C_6H_4-C_6H_4-OH, pyridine, toluene

b) p-C_6H_5-$CH_2OC_6H_4CO_2H$, DCC, DMAP, CH_2Cl_2

c) H_2, Pd/C, EtOAc

Scheme 2

d) 3,4,5-$(R')_3C_6H_2$-CO_2H, DCC, DMAP, CH_2Cl_2

the first evidence of a transition between an oblique and hexagonal mesophases (Table 1) [4, 17].

Notice the low melting temperatures – about 70–80 °C – in spite of a five-phenyl-ring core. This behavior, observed in other polycatenar series, is discussed in Sec. 5. The core may be non-symmetric; for example, the transition temperatures of triester **6** are K 67 °C Φ_h 71 °C I.

Table 1. Transition temperatures [°C] and *enthalpies* [kJ mol⁻¹] of phasmids **5** [a].

n	Cr		Φ_{ob}		Φ_h		I
7	•	80	–		–		•	82	•
12	•	70	•	81.5	•	92	•
		65.2		0.25		5.85	

[a] Cr = crystal; Φ_{ob} = phasmidic phase with an oblique 2D lattice; Φ_h = phasmidic phase with a hexagonal 2D lattice; •: the phase exists; –: the phase does not exist.

R = n-C$_{12}$H$_{25}$O-

6

The existence and stability of columnar phases depend on the flexibility or rigidity of the central groups (C$_6$H$_4$: phenyl, C$_6$H$_{10}$: cyclohexyl, C$_{10}$H$_6$: naphthyl). Table 2 shows that mesomorphic properties disappear with too flexible central groups (such as CH$_2$O) while more rigid groups (such as C$_{10}$H$_6$) enhance the mesophase stability.

It seems that the number of aromatic or alicyclic rings cannot be lower than five if mesomorphic properties are to be obtained, but beside some unsuccessful attempts with four or less rings, a particular case is worth reporting: the compound **7** [18, 19] displays a columnar phase with a four-benzene-ring core ending in six paraffinic chains grafted by means of benzyloxy groups, i.e., in two large triangular parts, nearly half-super-discs, namely 3,4,5-tris(p-n-dodecyloxy-benzyloxy)benzoyloxy groups (DOBOB in short).

R = n-C$_{12}$H$_{25}$O- CH$_2$O-

7

Table 2. Transition temperatures [°C] of phasmids (Fig. 1) according to flexibility of the central part (Y and Z correspond to inverted X and W, respectively).

n (6 chains)	A (Fig. 1)	W	X	Cr		Φ$_h$		I
10	C$_6$H$_4$	COO	COO	•	60	•	90	•
10	C$_6$H$_{10}$	COO	COO	•	79	•	(77)	•
10	C$_6$H$_4$	COO	OCH$_2$	•	84	–	–	•
10	C$_{10}$H$_6$	COO	COO	•	75	•	110	•

The main features within the hexacatenar series are:

- The clearing points are surprisingly low (about 80–90 °C) in spite of the core length. This weak mesomorphic stability may be attributed to the large number (six) of the paraffinic chains and the lateral *meta* positions for four of them. The clearing temperature increases with the number of cycles in the core. For example, when a sixth cycle is added to the core, the mesomorphic range rises to 120–150 °C [20].
- The only mesomorphic phases observed are columnar or ribbon mesophases.

4.1.2   Pentacatenars

A few pentacatenars have been reported with two isomer types at the two-chain end: *mp* and *mm*. For example, compounds **8** and **9** display the same hexagonal columnar mesophase but the melting point of **9** is at room temperature and the clearing point is significantly lower than that of the *mp* derivative [21]. This behavior is common to all compounds with *meta,meta* substitution and will be discussed in Sec. 5 of this chapter.

R = n-C$_{12}$H$_{25}$O-, R' = n-C$_{10}$H$_{21}$O- ; K 75°C Φ$_h$ 132°C I

8

R = n-C$_{12}$H$_{25}$O- ; K 25°C Φ$_h$ 86°C I

9

4.1.3   Tetracatenars

Four isomers have been studied.

Tetracatenars 3mpm-1p: Tetracatenar mesogens of this type display a cubic mesomorphism. From **10** to **14** one can observe an obvious correlation between the length of the rigid part and the stability of the cubic mesophase. Therefore the *3mpm-1p* configuration is a good source of cubic me-

R = *n*-C$_{12}$H$_{25}$O-, R' = *n*-C$_{10}$H$_{21}$O- ; K 71°C (Cub 65°C) I

10

R = *n*-C$_{12}$H$_{25}$O-, R' = *n*-C$_{10}$H$_{21}$O- ; K 95°C Cub 108°C I

11

R = *n*-C$_{12}$H$_{25}$O- ; K 75°C Cub 96°C I

12

R = *n*-C$_{12}$H$_{25}$O- ; R' = H(CH$_2$)$_n$O-

13

R = *n*-C$_{12}$H$_{25}$O- ; R' = H(CH$_2$)$_n$O-

14

sophases. The core of compound **11** is longer than that of compound **10** and thus the cubic phase becomes enantiotropic. This behavior is confirmed with the compound **12** even in the absence of a paraffinic chain. Transition temperatures of cubic mesogens **13** and **14** are given in Tables 3 and 4. In contrast, compounds such as **15a, b**, with

Table 3. Transition temperatures [°C] of tetracatenars **13**.

n	Cr		$\Phi_?$		Cub		Φ_h		I
6	•	80		–	•	142		–	•
7	•	87		–	•	126		–	•
8	•	92	•	119	•	141		–	•
9	•	85		–	•	140	(• 132)[a]		•
10	•	80	•	117	•	139	(• 132)[a]		•
11	•	80	•	118	•	138	(• 132)[a]		•
12	•	90	•	116	•	136		–	•

[a] Observed only on cooling.

Table 4. Transition temperatures [°C] of tetracatenars **14**.

n	Cr		Cub		I
6	•	84	•	151	•
7	•	62	•	146	•
8	•	85	•	146	•
9	•	71	•	144	•
10	•	79	•	146	•

only one DOBOB group, exhibit a hexagonal (**15a**) [18, 19] or a nematic phase (**15b**) [18, 22, 23].

Biforked mesogens 2mp-2mp: Biforked mesogens are the richest family and exhibit all the types of mesophases: lamellar, columnar, cubic, and nematic. They correspond to the general formula **16** [6, 24–27]. Several examples corresponding to different values of W, X, Y, Z and A (Fig. 1), as well as various chain lengths, are given in Table 5.

R = *n*-C$_{12}$H$_{25}$O-⟨⟩-CH$_2$O-

15a R' =

b R' = ⟨⟩-⟨⟩-OCO-⟨⟩-OC$_{12}$H$_{25}$

Table 5. Transition temperatures [°C] of compounds **16** (Y and Z correspond to inverted X and W, respectively).

n	W	X	A	K	S$_C$	Φ$_{ob}$	Φ$_h$	N	I
7	COO	OCO	C$_6$H$_4$	• 191	–	–	–	• 206	•
7	COO	OCO	C$_6$H$_{10}$	• 137	• 169	–	–	• 187	•
7	CH=N	OCO	C$_6$H$_4$	• 165	• 187	–	–	–	•
7	CH=N	OCO	C$_6$H$_{10}$	• 152	• 183	–	–	–	•
7	COO	N=CH	C$_6$H$_4$	• 139	• 199.5	–	–	• 224	•
11	COO	OCO	C$_6$H$_4$	• 152	• 172	–	–	–	•
11	COO	OCO	C$_6$H$_{10}$	• 114	• 158	–	–	–	•
11	CH=N	OCO	C$_6$H$_4$	• 155	–	–	• 173	–	•
11	COO	N=CH	C$_6$H$_4$	• 130	• 155	• 169	• 186	–	•
14	COO	OCO	C$_6$H$_4$	• 143	–	–	• 162	–	•
14	CH=N	OCO	C$_6$H$_4$	• 144	–	–	• 169	–	•
14	CH=N	OCO	C$_6$H$_{10}$	• 141	–	–	• 164	–	•

R = H(CH$_2$)$_n$O

16

The polymorphism within the series **17** is strongly dependent on alkyl chain length. Short-chain derivatives exhibit two mesophases unambiguously identified by their optical textures as a lamellar phase of smectic C type and a nematic phase (e.g., n = 7; Cr 138 °C SmC 199.5 °C N 223.5 °C I). Derivatives **17** with n = 9 and 10 show only a SmC phase over a large temperature range (Cr 130 °C SmC 191 °C I and Cr 125 °C SmC 185°C I, respectively). In contrast, derivatives with n = 11 and 12 exhibit an interesting polymorphism with a lamellar phase (SmC) at lower temperature; a columnar phase (Φ$_h$) at higher temperature is realized in both cases via a lamello-columnar mesophase (Φ$_{ob}$). Finally, the longest chain homologue n = 14 exhibits only one mesomorphic phase (Cr 123 °C Φ$_h$ 176 °C I).

Table 6 [24] gives another example of a series (**18**) in which a lamellar smectic C phase is observed for short chains while a

Table 6. Transition temperatures [°C] and *enthalpies [kJ mol^{-1}]* of tetracatenars **18**.

n	Cr	SmC	Cub	Φ$_h$	I
7	• 152	• 183	–	–	•
8	• 148 *81.9*	• 176 *5.73*	–	–	•
9	• 146.5 *81.2*	• 168.5 *5.49*	–	–	•
10[a]	• 144 *93.5*	• 156 *2.30*	• 165 *3.00*	–	•
11[b]	• 144 *102.8*	• 146 *2.42*	• 163 *3.02*	–	•
12[c]	• 142 *106.7*	–	• 162 *3.04*	–	•
13	• 141 *94.1*	–	–	• 163 *3.83*	•
14	• 140 *127.7*	–	–	• 163 *4.27*	•

[a] On cooling: I 157 SmC 138 Cr.
[b] On cooling: I 158 Φ$_h$ 147 Cub 140 SmC 135 Cr.
[c] On cooling: I 160 Φ$_h$ 138 Cr (the hexagonal columnar phase has a large lattice constant).

R = H(CH$_2$)$_n$O   **17**

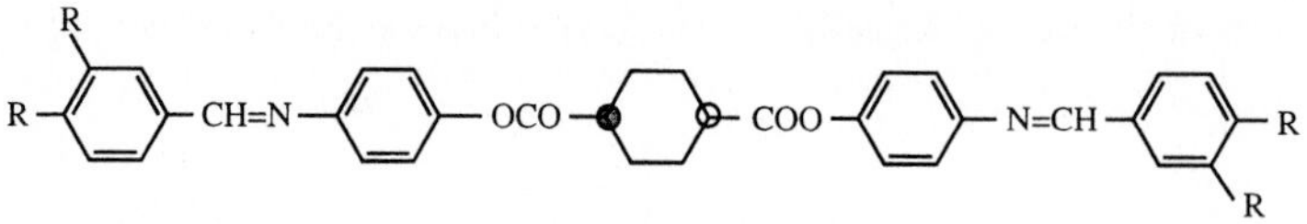

R = H(CH$_2$)$_n$O

18

columnar phase appears for long chains. But, contrary to the previous series, a cubic phase is observed instead of a Φ_{ob} phase, as is the case for lyotropic phases. Other examples of mesogens of this kind can be found in the literature [6, 27].

In another series with four ester linkages a rectangular columnar phase Φ_r is observed. For the compounds **19** [27], a comparable behavior is observed for short chains: SmC and N mesophases ($n = 6$: Cr 195 °C N 217 °C I; $n = 8$: Cr 175 °C SmC 182 °C N 205.5 °C I; $n = 10$: Cr 156 °C SmC 176 °C I) while for long chains a two-dimensional (2D) rectangular mesophase ($n = 14$: Cr 142.5 °C Φ_r 162 °C I) occurs instead of a hexagonal Φ_h one.

A similar mesomorphism with some peculiar features has been reported by Diele et al. [28] in the series **20**. Although these compounds are not strictly speaking polycatenar mesogens, the analogy with the tetracatenars is obvious. A rectangular columnar phase is claimed to be obtained at lower temperature than the smectic C phase and the authors observed a reentrant isotropic phase with the C$_8$ derivative [29].

In addition, structures with an iron complex in place of the central ring in series such as **17**, **18**, and **19** may also exhibit nematic, SmC, and columnar mesophases. These new metallomesogens are of interest because they are easily prepared in optically pure forms and therefore can lead to ferroelectric properties in the chiral SmC* form. For example, complex **21** with $n = 11$ exhibits SmC* and cholesteric properties (Cr 52.5 °C SmC* 111 °C N* 119 I), while only a hexagonal columnar phase is observed with $n = 12$ (Cr 60.5 °C Φ_h 112.5 I)

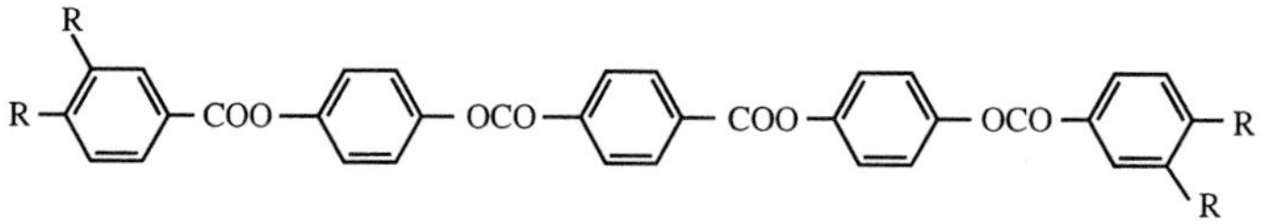

R = H(CH$_2$)$_n$O

19

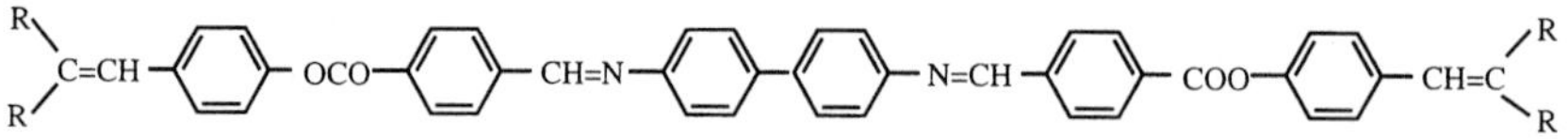

n-C$_6$H$_{13} \leq$ R $\leq n$-C$_{16}$H$_{33}$

20

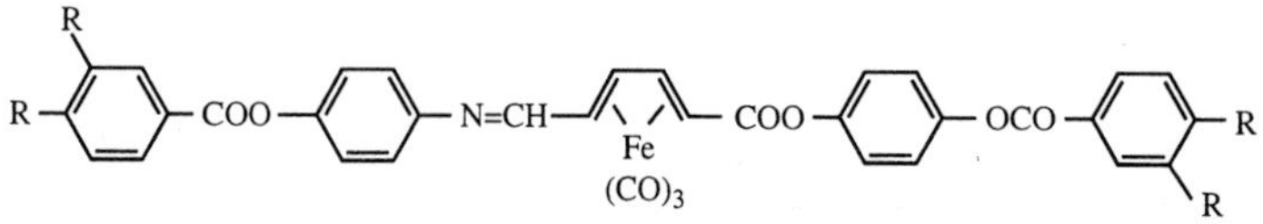

R = H(CH$_2$)$_n$O

21

Table 7. Transition temperatures [°C] and *enthalpies* [kJ mol⁻¹] of complexes **21**.

n	Cr		SmC*		Φ_h		N*		I
6 (+)	•	84 *18.1*	–		–		•	139 *0.9*	•
7 (–)	•	71.5 *38.1*	•	94.5 *2.9*	–		•	146 *0.6*	•
8 (–)	•	84.5 *61.8*	•	94 *2.9*	–		•	130.5 *0.6*	•
9 (–)	•	68.5 *45.3*	•	82 *2.2*	–		•	100.5 *0.4*	•
10 (–)	•	73.5 *18.4*	•	109 *2.3*	–		•	127 *0.6*	•
11 (–)	•	52.5 *28.7*	•	111 *2.3*	–		•	119 *0.3*	•
12 (–)	•	60.5 *33.5*	–		•	112.5 *3.1*	–		•
13 (–)	•	64.5 *59.8*	–		•	103.5 *2.8*	–		•
14 (–)	•	88.5 *78.1*	–		•	94 *3.2*	–		•
15 (–)	•	71 *69.0*	–		•	106 *2.4*	–		•
16 (–)	•	89.5 *87.1*	–		•	103.5 *3.8*	–		•

(Table 7) [30]. The biforked ligands **22** display an SmC phase with short chains and an oblique columnar one with long chains (Table 8). The Φ_{ob} is also obtained in Cu complexes **23**.

In the series **24** [25], with a long core ending in two flexible parts, a complex columnar polymorphism was found (Table 9).

Tetracatenars 2mp-2mm: Only one example (**25**) [21] of this type of tetracatenar has been studied: Cr 78 °C Φ_h 99 °C I. On the other hand, no mesomorphic properties were observed with a –N=CH– instead of –CH=N– linkage (F = 124 °C).

Table 8. Transition temperatures [°C] of phenols **22**.

m	Cr		SmC		Φ_{ob}		N		I
7	•	128	•	177	–		•	188	•
12	•	128	•	176	–		–		•
14	•	130	•	135	•	171	–		•

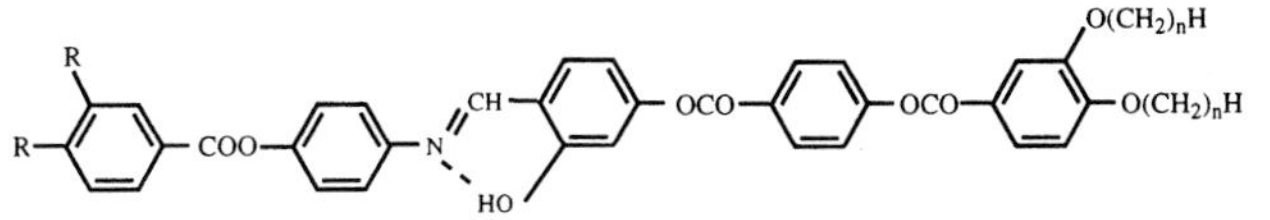

22

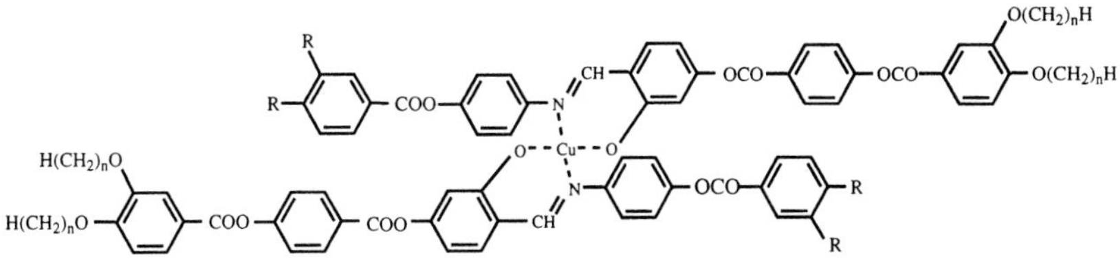

23

n = 14 ; K 198°C Φ_{ob} 215°C I

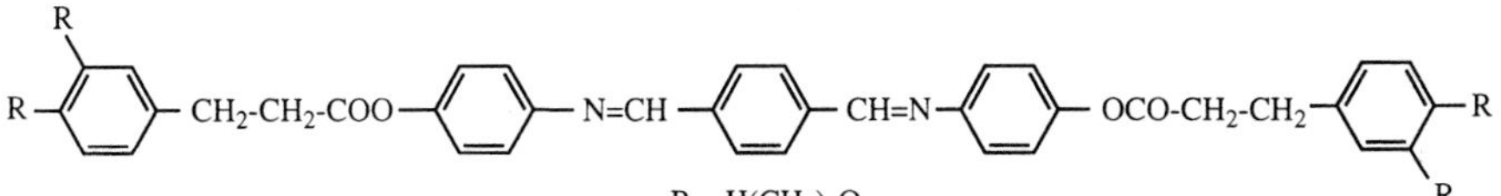

24

R = H(CH₂)ₙO-

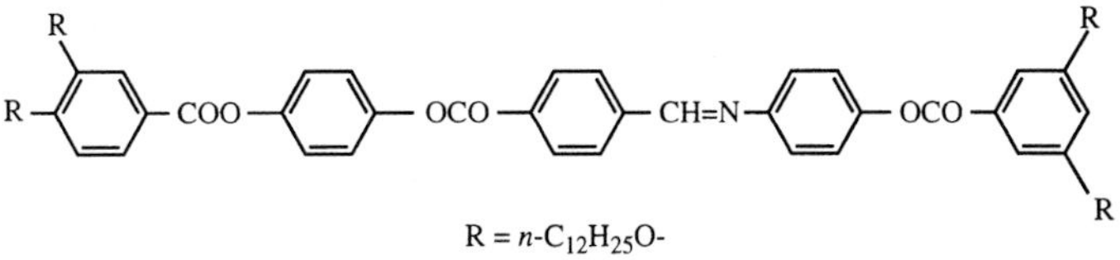

25

R = n-C₁₂H₂₅O-

Table 9. Transition temperatures [°C] and *enthalpies [kJ mol^{-1}]* of tetracatenars **24**[a].

n	Cr		Φ_{rh1}		Φ_{rh2}		Φ_{hex}		Φ_{ob}		Φ_{h}		I
9	•	109 51.8	•	126	•	152	−		•	155 5.84	−			•
10	•	110 57.1	−		−		•	150	−		•	155 6.99	•
11	•	111 44.0	−		−		−		−		•	156 5.74	•
12	•	111 67.2	−		−		−		−		•	155.5 8.39	•
13	•	112 69.6	−		−		−		−		•	155.5 8.89	•
14	•	111.5	−		−		−		−		•	154	•

[a] Φ_{rh1} and Φ_{rh2}: rhombohedral mesophases; Φ_{hex}: hexagonal 3D mesophase.

R = H(CH$_2$)$_n$O-

26

Tetracatenars 2mm-2mm: Within the homologous series with a six-benzene-ring core **26** [24], only one type of mesophase has been detected by optical observations. The textures are fan-shaped, with developable domains in every way similar to those of the Φ_h phase of tetracatenars *2mp-2mp*. As an example, for the compound with n = 10: K 76 °C Φ_h 110 °C I. Despite the unusual length of the central rigid part, the transition temperatures are surprisingly low (Table 10), a fact evidently connected with the *meta* position of all the alkyloxy chains.

Correlatively, the homologous five-benzene-ring derivatives are isotropic at room temperature and somewhat difficult to isolate.

Tricatenars: Several other mesogens of this type have been synthesized. They have a three-benzene-ring core and display the classical mesomorphic properties of rod-like mesogens. For instance, compounds **27** [5] and **28** [5, 31] are smectogenic C and nematogenic, but when the alkyloxy chain is moved from *meta* to *ortho* position, the compound becomes only nematogenic.

Table 10. Transition temperatures [°C] and *enthalpies [kJ mol^{-1}]* of tetracatenars **26**.

n	Cr		Φ_h		I
7	•	110	(•	104)	•
8	•	96	•	112	•
10	•	76	•	110	•
11	•	68 36.6	•	116 2.72	•

27 W = -CH=N- ; R = H(CH$_2$)$_n$O-, R' = *n*-C$_7$H$_{15}$O-

28 W = -COO- ; R = H(CH$_2$)$_n$O-, R' = *n*-C$_9$H$_{19}$-

The same mesomorphic sequence was observed with the ligand **29** (Table 11)

R = H(CH$_2$)$_n$O-

29

Table 11. Transition temperatures [°C] of tetracatenars **29**.

n	Cr	SmH	SmC	N	I
2	138.5	–	–	(• 132.5)	•
3	102	–	(• 74)	• 108	•
4	92	–	(• 69)	• 108.5	•
5	81	–	(• 65)	• 108	•
6	69.5	(• 49.5)	• 74.5	• 108.5	•
7	83	(• 65.5)	(• 78)	• 104.5	•
8	87.5	(• 73.5)	(• 74)	• 104.5	•
9	88.5	(• 74)	(• 83.5)	• 101	•
10	86	–	(• 82.5)	• 98.5	•
12	93	–	(• 82.5)	(• 89.5)	•
14	91	–	(• 84.5)	(• 86)	•
16	84.5	–	• 91	• 92.5	•

[32] and with other similar systems. But if the *m*-chain is grafted on the first benzene ring in *ortho* position with regard to the ester group, nematic mesomorphism is promoted. The same behavior is observed when an *o*-[4-H(CH$_2$)$_n$O-C$_6$H$_4$-COO+ group replaces the *o*-chain.

Another interesting example is the five-benzene-ring tricatenar *2mm*-1*p* **30**, which displays a cubic phase over a large temperature range, from 83 up to 172 °C [5, 33].

The mesomorphic polymorphism is, evidently, connected to the number of rings (core length) and also to the number, position, and length of the paraffinic chains.

4.2 Polycatenars with Polar Substituents

4.2.1 Polycatenars with Hydrogenated and Fluorinated Chains

Although earlier work [34–38] on rod-like mesogens has shown that smectic mesophases are favored by perfluoralkyl or perfluoralkyloxy chains, the effect of fluorination on mesomorphic properties has not been extensively explored.

Perfluoro chains have two special behaviors:
- CF$_2$–CF$_2$ bonds adopt a *trans* conformation along the fluorinated chain, which adopts a helical configuration. As a consequence of this higher order, the chain is much stiffer than the hydrocarbon homologue.
- Fluorocarbons are known to be highly incompatible with both saturated and aromatic hydrocarbons.

Therefore, using perfluorinated or semiperfluorinated chains as tails for liquid crystal materials should allow the consequences of the variation of both the flexibility and the dissymmetrization of the molecule to be studied.

Tricatenars: Several tricatenar mesogens **31a, b** with a three-benzene-ring core bearing a perfluorinated (**31a**) or semi-perfluorinated (**31b**) chain have been synthesized

R = *n*-C$_{12}$H$_{25}$O-

30

R = H(CH$_2$)$_n$O-

31a R$_F$ = F(CF$_2$)$_n$-

b R$_F$ = F(CF$_2$)$_p$(CH$_2$)$_q$O-

[38]. The mesomorphic properties were given in Table 12.

The difference between the hydrogenated and fluorinated compounds lies in the fact that the former give only N and SmC phases (see Sec. 4.1.4. of this chapter) while the latter display SmA and SmC phases with a perfluorinated (**31a**) chain and SmC or SmC̃ phases for a semi-perfluorinated chain (**31b**). First discovered in 1982 with polar rod-like molecules [39], the ribbon SmC̃ phase was then found in low molecular weight mesogens devoid of a CN or NO$_2$ group. With their perfluorinated stiff rod, the compounds **31a** are similar to polar

forked mesogens: an S$_A$ phase was obtained in spite of the forked group. In the case of a semi-perfluorinated chain, the presence of the flexible (CH$_2$)$_n$ groups favors the formation of a tilted lamellar phase to the detriment of S$_A$. Another example of this kind of fluorinated mesogens is given by perfluorinated swallow-tailed compounds **32** and **33** [40], in which the fluorinated part can be situated either in the unbranched chain (**32**) or in the branched end (**33**). They show nematic, SmA and SmC phases, and a mixture of compounds of both types could lead to an achiral ferroelectric phase as predicted by Petschek et al. [41] and probably found by Tournilhac et al. [42]

Tetracatenars 3mpm-1p: As seen in Sec. 4.1.4. of this chapter, a metastable cubic phase can be obtained with a four-benzene-ring core fitted with four paraffinic chains 1p-3mpm (compound **10**). When a perfluorinated stiff rod was grafted in place of the p-alkyloxy chain, a stable cubic phase

Table 12. Transition temperatures [°C] of compounds **31**.

	n			Cr		S$_?$		SmC̃		SmC		SmA		I
(a)	10			•	87	–		•	89.5	–		•	107	•
	12			•	98	•	(70)	•	(85)	–		•	100	•
	14			•	87	–		•	89	–		•	95	•

	n	p	q	Cr		S$_?$		SmC̃		SmC		SmA		I
(b)	9	6	4	•	110	–		–		•	138	–		•
	11	6	4	•	109	–		•	132	–		–		•
	12	6	4	•	103	–		•	130	–		–		•
	14	6	4	•	104	–		•	124	–		–		•

32

33

was obtained with a large temperature range. This behavior is explained by the fact that the perfluoro chain lengthens the rigid core (see, for instance, the compound **12**).

Special Cases: This cubic phase could be observed also with tricatenar mesogens *2mm-1p* such as compound **30**, in which the *para* chain is perfluorinated or semi-perfluorinated (**34**). It is also possible to have this phase with a mesogen bearing a bromo substituent in *ortho* position with regard to the semi-perfluorinated chain (**35**).

Compounds **36a** and **36b** show that fluorinated chains can be compatible with the

4.2.2 Polycatenars with Other Polar Substituents

Up to now, in most cases, polycatenars such as phasmids and biforked mesogens have been symmetrical and have not contained any polar substituents such as Br, CN, or NO_2. In order to study the influence of a lateral polarity on the mesomorphic properties, some new polar and non-symmetrical polycatenar mesogens have been prepared.

Bromo Derivatives: The influence of the bromine on the mesomorphic properties was more closely studied by synthesizing four series of tetracatenar (**37**, **38**) and pentacatenar (**39**, **40**) mesogens. In each group the bromo end is the same, while the other end bears one *p*-dodecyloxy chain (**a**), two *m,p*-dodecyloxy chains (**b**), *m,m*-dodecyl-

R = n-C₁₂H₂₅O- ; K 71°C Cub 108°C I

34

K 105°C Φ₍?₎ 137°C Cub 171°C I

35

36 a X = H ; K 88°C Φᵣ 148°C Φₕ 170°C I

b X = Br ; K 80°C Φᵣ 118°C Φₕ 145°C I

columnar mesomorphism [43]. The structure and the lattice parameters will be given in Sec. 6. Moreover, this case shows that the bromo-substitution seems to behave as if an aliphatic chain were on this position, as will be seen in the next section.

oxy chains (**c**), or three *m,p,m*-dodecyloxy chains (**d**).

Tables 13–16 show that everything works as expected if the bromine is assumed to occupy the same volume as the first methylene of an alkyloxy chain: thus the chain grafted close to the halogen is confined, as a chain in *ortho* position to another one would be, and forced to fill the space taken up by such a chain. So, it is not surprising

37

n-C$_{12}$H$_{25}$O— (Br) —COO— —OCO— —OCO— (R$_1$, R$_2$, R$_3$)

38

n-C$_{12}$H$_{25}$O— (Br, Br) —COO— —OCO— —OCO— (R$_1$, R$_2$, R$_3$)

39

n-C$_{12}$H$_{25}$O— (Br, n-C$_{12}$H$_{25}$O) —COO— —OCO— —OCO— (R$_1$, R$_2$, R$_3$)

40 a R$_1$ = R$_3$ = H, R$_2$ = n-C$_{12}$H$_{25}$O-

    b R$_1$ = H, R$_2$ = R$_3$ = n-C$_{12}$H$_{25}$O-

    c R$_2$ = H, R$_1$ = R$_3$ = n-C$_{12}$H$_{25}$O-

    d R$_1$ = R$_2$ = R$_3$ = n-C$_{12}$H$_{25}$O-

40

that columnar phases with only three par-affinic chains are obtained (compounds **37c**, **38b**, **38c**, **39a**, and **40a**). The last compound, with its two cubic phases separated

by an unknown ordered mesophase, is of great interest.

Nitro and Cyano Derivatives: Most of the studied compounds are polar forked mesogens with a three-benzene-ring core (**41**). Table 17 gives the data for the cyano series. The observed mesophases are classical: nematic, S$_A$, and S$_C$ phases. The lamellar thicknesses are higher than the molecular length, so the smectic phases are classified as partially bilayer phases [10, 31, 44, 45].

R— (R) —W— —OCO— —CN
                            or NO$_2$

R = H(CH$_2$)$_n$O- ; W = -COO-, -CH=N-

41

As in the case of non-polar tricatenar mesogens, a lateral chain in *ortho* position with regard to the –COO– or –CH=N– link that promotes the nematic phase is predominant. The same behavior is also observed with a lateral chain grafted onto the central ring.

Table 13. Transition temperatures [°C] of compounds **37**.

Compound	Number and positions of chains	Cr		SmC		Cub	Φ_h		N		I
a	1p	•	110	•	277	–	–		•	>300	•
b	2mp	•	126	•	211	–	–		–		•
c	2mm	•	120	–		–	•	125	–		•
d	3mpm	•	85	–		–	•	141	–		•

Table 14. Transition temperatures [°C] of compounds **38**.

Compound	Number and positions of chains	Cr		SmC		Cub		Φ_h		N		I
a	1p	•	111	•	261	–		–		•	297	•
b	2mp [a]	•	122	•	142	•	170	–		–		•
	[b]	•		•	118	•	161	•	163	–		•
c	2mm	•	89	–		–		(•	80)	–		•
d	3mpm	•	71	–		–		•	112	–		•

[a] On heating; [b] on cooling.

Table 15. Transition temperatures [°C] of compounds **39**.

Com-pound	Number and positions of chains	Cr		Cub		$\Phi_?$		Φ_h				I
a	1p	●	107	●	114	●	177	–				●
b	2mp	●	120	–		–			●	125		●
c	2mm	●	72	–		–			(●	50)		●
d	3mpm	●	81	–		–			●	84		●

Table 16. Transition temperatures [°C] of compounds **40**.

Com-pound	Number and positions of chains	Cr		Cub		$\Phi_?$		Φ_h		Cub		I
a	1p	●	74	●	99	●	121	–		●	147	●
b	2mp	●	89	–		–		●	107	–		●
c	2mm	●	80	–		–		–		–		●
d	3mpm	●	47	–		–		●	60	–		●

Table 17. Transition temperatures [°C] of compounds **41**.

n	Cr		SmC		SmA		N		I
1	●	210	–		–		●	218	●
2	●	203	–		–		(●	183)	●
3	●	160	–		–		●	161	●
4	●	154	–		(●	144)	●	159	●
5	●	141	–		●	154.5	●	156	●
6	●	128	–		●	156	–		●
7	●	120	–		●	154	–		●
8	●	116	–		●	153	–		●
9	●	118	(●	113)	●	153	–		●
10	●	118	–		●	151	–		●
11	●	119	●	136	●	147	–		●
12	●	119	●	124	●	147	–		●

ω-Cyano Chains. The ω-cyano chains associate the flexibility of the paraffins with the polarity of the cyano group. They are used in order to obtain orthogonal lamellar mesophases such as an SmA phase instead of tilted SmC phases. In fact, the biforked mesogens **42a** displayed an SmA phase owing to the *2mp* cyanohydrogenated chains. In the same way an SmA phase was also observed with the pentacatenar **42b** in which there is a second *meta* chain.

5   The Core, Paraffinic Chains, and Mesomorphic Properties

5.1   Core Length

For the reviewed compounds, at least three benzene rings are needed in order to obtain mesomorphic properties (Sec. 4.1.4 of this chapter, compounds **27–29** and Sec. 4.2.1 of this chapter, compounds **31–33**) of clas-

42 a R_1 = H, R_2 = R_3 = n-$C_{12}H_{25}O$; K 127°C S_C 158°C S_A 189°C I

  b R_1 = R_2 = R_3 = n-$C_{12}H_{25}O$; K 101°C S_A 155°C I

sical elongated molecules. It is clear that a balance must be kept between the number of chains directly grafted on the core and the length of the latter in order to obtain mesomorphic properties. For example, with four chains, a minimum of a four-ring core is necessary to obtain mesomorphic properties. With five and six chains, at least five rings are necessary for mesomorphism. Let us also recall that, owing to its stiffness and its incompatibility with hydrocarbons, a fluorinated chain acts simultaneously as a core and as a chain and thus allows the ring number to be reduced without losing the mesomorphic properties (Sec. 4.2.1.). Both core length and chains are determinant for the existence of lamellar and columnar phases.

5.2   Number and Position of Chains

This discussion only concerns hydrogenated chain compounds with five- and six-benzene cores, which present the richest polymorphism.

Table 18 shows that when $Nm > Np$, columnar phases are obtained, when $Nm < Np$, lamellar phases are obtained, and when $Nm = Np$, a competition occurs between lamellar and columnar mesomorphisms. In this situation, the chain length is crucial.

The molecular structure of polycatenars holds an intermediate position between mesogenic discs and mesogenic rods, and the observed polymesomorphism is very similar to that of lyotropic systems [46]. For example, in homologous series 2D oblique or rectangular columnar phases or 3D cubic phases are inserted between lamellar and hexagonal columnar phases. The long-chain tetracatenars $2mp$-$2mp$ still display a hexagonal columnar mesomorphism owing to their strong curvature in the core–chain interface. As the chain length decreases, a smectic C phase appears to the detriment of the hexagonal phase: the passage between columnar and lamellar phases is realized via a lamello-columnar phase and/or a cubic phase. With short chain tetracatenars, i.e., with a weak curvature, as with a lower number of chains, the classical mesomorphism occurs: smectic C and nematic phases. Anyway, the smectic C phase is the only observed lamellar mesophase in compounds of this kind. Nevertheless, a SmA phase can be observed with polar aliphatic chains (see Sec. 4.2.2 of this chapter).

For all the polycatenars, the textures of their columnar mesophases observed under the polarizing microscope are generally fan-shaped (see [4, 6]) and the birefringence

Table 18. Comparison of the influence of number and positions of paraffinic chains on mesomorphic properties.

	Number of chains in meta position (Nm)	Number of chains in para position (Np)	Lamellar phase	Cubic phase	Columnar phase
Hexacatenars	4	2	–	–	+
Pentacatenars	4	1	–	–	+
	3	2	–	–	+
Tetracatenars	4	0	–	–	+
	3	1	–	+	+
	2	2	+	+	+
Tricatenars	2	1	–	+	–
	1	2	+	–	–

decreases strongly from a smectic C to a columnar mesophase [21].

Concerning the disputed structural models of phasmidic columnar mesophases, we can emphasize the presence of more than one molecule in a slice of column (two or three molecules for a six-chain phasmid, four for a biforked mesogen) [4, 17, 19, 21]. On the other hand, X-ray measurements on the crystalline phase of a biforked mesogen that exhibits a hexagonal phase show a layered structure similar to that of a smectic C phase (molecular tilt angle of 60°). In addition, due to the bulky paraffinic chains, the interactions between cores are very weak perpendicular to the xz-plane, i.e., along the future column axis parallel to the crystalline layers. Very recently, using dilatometry and X-ray measurements on a biforked mesogen, Guillon et al. proposed a model for the SmC and Φ_h phases [48]. So, according to these data, we can assume that above the SmC-columnar transition temperature the paraffinic volume is large enough to allow the four-molecule clusters resulting from an enhanced undulation to twist slightly around the direction of weak coupling and realize a paraffinic sheath, which leads to columns (Fig. 2). Actually, that is consistent with the weak transition heats measured between lamellar and columnar mesophases in various series (about 1 kJ mol^{-1}): non-drastic structural changes occur between the different mesophases, as is suggested by the permanence of an angle near to 60° in the various mesophases as well as in the crystalline state [49].

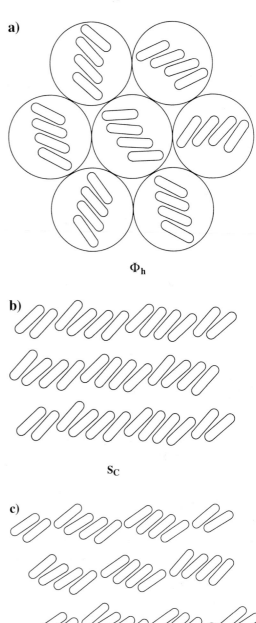

Figure 2. (a) Structure of hexagonal columnar phase. (b) Structure of smectic C phase. (c) Transition between Φ_h and S_C (= SmC).

6   Structures

6.1   Structures of Mesophases

6.1.1   Nematic

The nematic phase observed with polycatenars has the optical properties of the classical bicatenar nematic. Nevertheless, in some cases, as with biforked mesogens [6], observations on nematic free droplets reveal, on heating above the isotropic transition, a Schlieren texture with quite a weak birefringence, but it is difficult to decide whether this originates in Brownian motion or cybotactic fluctuations. Moreover, in spite of the strong geometrical anisotropy of some tetracatenars such as **15b** (Sec. 4.1.3 of this chapter), possible biaxial properties have not been confirmed.

6.1.2   Lamellar Mesophases

As a general rule, polycatenars display S_C mesomorphism, except in the case of smectogenic A tetra- and pentacatenars with polar ω-cyano chains (Sec. 4.2.2 of this chapter).

SmA Phase: In contrast to polar compounds **41** exhibiting a partially bilayer S_A phase (Sec. 4.2.2 of this chapter), the lamellar mesophase of biforked ω-cyano chain compounds **42a** (also Sec. 4.2.2 of this chapter) is a monolayer S_A.

SmC Phase: In the immense majority of cases, the lamellar mesophases displayed by these mesogens are S_C (Fig. 2b), except for non-polar penta- and hexacatenars, which do not exhibit any lamellar mesomorphism. For example, the X-ray pattern of magnetically aligned samples of biforked mesogens **16** (typically Table 5, compound with $n = 7$,

W $= -COO-$, X $= -N=CH-$ and A $= C_6H_4$) [6] is that of a usual nematic phase of elongated molecules with evidence of tilted S_C cybotactic groups; the apparent length of the molecule is 51.3 Å and the tilted angle reaches 60° near the $N-S_C$ transition. The found angles are those of the crystalline phase: about 50–60°.

6.1.3   Columnar Phases

The main area per paraffinic chain increases with temperature and chain length, corresponding to an increase of the curvature between the aromatic and the paraffinic moieties. For this reason columnar mesomorphism is always the high temperature one and is obtained directly from the isotropic liquid. Correlatively, with two dense paraffinic ends (phasmids and pentacatenars) a columnar phase occurs, while smectic and nematic phases appear as the number of paraffinic chains is decreased [50].

In addition to the influence of the core linkages – angles, dipole sense (see for example Table 5 and the literature [49]) – on the appearance of columnar mesomorphism, this arrangement is also quite sensitive to the presence of lateral groups on the core. Compounds **43** with two methyls and **44** with two chlorine atoms, for instance, do not display any columnar mesomorphic organization. The former exhibits only SmC properties and the latter is only nematogenic, whereas the respective analogues **16** (W $= -CH=N-$, Y $= -OCO-$, $n = 14$) and **17** ($n = 12$) present the sequences K 146 °C Φ_h 168 °C I [6, 27] and K 127.5 °C S_C 128.5 °C Φ_{ob} 161 °C Φ_h 186 °C I [24]. Evidently, both methyl and chlorine are too bulky as substituents to keep a good parallelism and a suitable inter-core distance for making mesomorphic columns occur at high temperature.

R = n-C$_{14}$H$_{29}$O- ; K 133°C (S$_C$ 110°C) I

43

R = n-C$_{12}$H$_{25}$O- ; K 140°C (N 126°C) I

44

Hexagonal Phases: The average number of molecules in a slice of column was found to be 2–3 for phasmids and 4 for biforked mesogens (Fig. 2a). Lattice constants lie between 40 and 50 Å for chain lengths from C$_7$ to C$_{15}$ and the main core–core distance is 4.6 Å. A hexagonal phase of different type has been observed with a large lattice constant corresponding to twice the molecular length (135 Å) on cooling from the isotropic phase before the appearance of the cubic phase (compound **18**, n = 12). In some cases, the transition between Φ$_h$ and SmC occurs via a Φ$_{ob}$ phase, the structure of which may be represented by Figure 2c.

Oblique and Rectangular Phases: The first example of an oblique columnar phase has been observed on the first phasmid described (**5**, n = 7, Table 1). A central terephthalylidene ring fitted with two benzoyloxyphenylimines seems to favor Φ$_{ob}$ phases (Tables 1, 5 and 9). Regarding the rectangular columnar mesophase, chain-rich polycatenars such as phasmids do not display this mesomorphic symmetry, which so far has appeared only in the biforked tetraester series **19**.

Other Columnar Phases: Some other ordered columnar phases (rhombohedral)

have been observed on compounds **24** for n = 9 and 10 (Table 9 and [25]).

6.1.4 Cubic Mesophases

Usually, cubic phases are observed on symmetrical polycatenars (maximum of four chains, see Table 18) with a flexible core such as compounds **18** (Table 6). But they may also be obtained with non-symmetrical rigid cores bearing three or four paraffinic chains. The cubic phases have the same space group Im3m [33].

6.2 Crystalline Structures

Up to now, only biforked mesogens have provided single crystals and they have invariably presented a lamellar crystalline structure, whatever the mesophase observed on heating. For example, this is the case for compounds **24** (n = 12) with the flexible group CH$_2$–CH$_2$–COO in the core and the unsaturated homologue (CH=CH–COO, n = 10), which do not display lamellar and columnar mesophases, respectively. The

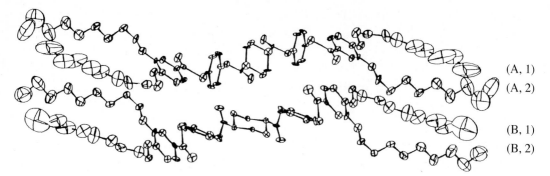

(A, 1)

(A, 2)

(B, 1)

(B, 2)

Figure 3. ORTEP drawing of the molecule **16** (R = C$_9$H$_{19}$O–, W = COS, Z = SCO, X = OCO, Y = COO, A = C$_6$H$_{10}$).

molecules adopt a zigzag form: the two ortho chains are stretched and parallel to the *xy*-plane; they make an angle close to 120° with the central core, which makes an angle of about 60° with the normal to the layers (Fig. 3).

The fact that these mesophases of various symmetries display the same crystalline structures (unchanged angle between crystalline phase and lamellar phase) and that lamellar-columnar transition enthalpies remain very weak (see the literature [27]) indicates that structural modifications between the crystalline phase and the lamellar mesophase are minor.

7   Conclusion

Phasmids and polycatenar mesogens are just ten years old. Their intermediate shape between classical rod-like and disc-like liquid crystals allows practically all the mesomorphic symmetries (Φ_{rh}, Φ_{ob}, Φ_r, Φ_h, Cub, SmA, SmC, N) to be obtained with the same molecular core by varying the temperature and/or the number and positions of grafted chains. This great versatility might be exploited to build up some novel materials using, for example, the ability to dramati-

cally change liquid crystalline symmetries and thus the physico-chemical properties by heating or cooling.

8   References

[1] D. Demus, *Mol. Cryst. Liq. Cryst.* **1988**, *165*, 45. D. Demus, *Mol. Cryst. Liq. Cryst.* **1989**, *5*, 75.
[2] S. Chandrasekhar, B. K. Shadashiva, K. A. Suresh, *Pramana* **1977**, *9*, 471.
[3] C. Destrade, P. Foucher, H. Gasparoux, H. T. Nguyen, A.-M. Levelut, J. Malthête, *Mol. Cryst. Liq. Cryst.* **1984**, *106*, 121.
[4] J. Malthête, A.-M. Levelut, H. T. Nguyen, *J. Phys. Lett. (Paris)* **1985**, *46*, L-875.
[5] J. Malthête, H. T. Nguyen, C. Destrade, *Liq. Cryst.* **1993**, *13*, 171.
[6] H. T. Nguyen, C. Destrade, A.-M. Levelut, J. Malthête, *J. Phys. (Paris)* **1986**, *47*, 553.
[7] J. Malthête, J. Billard, J. Jacques, *C. R. Acad. Sci. Paris* **1975**, *C281*, 333. J. Malthête, *C. R. Acad. Sci. Paris, Ser. II* **1983**, *296*, 435.
[8] H. Strzelecka, C. Jellabert, M. Veber, P. Davidson, A.-M. Levelut, J. Malthête, G. Sigaud, A. Skoulios, P. Weber, *Mol. Cryst. Liq. Cryst.* **1988**, *161*, 403.
[9] D. Demus, S. Diele, A. Hauser, I. Latif, C. Selbmann, W. Weissflog, *Cryst. Res. Technol.* **1985**, *20*, 1547. W. Weissflog, D. Demus, *Mol. Cryst. Liq. Cryst.* **1985**, *129*, 235. S. Diele, A. Mädicke, K. Knauft, J. Neutzler, W. Weissflog, D. Demus, *Liq. Cryst.* **1991**, *10*, 47.
[10] S. Takenaka, H. Morita, M. Iwano, S. Kusabayashi, T. Ikemoto, Y. Sakurai, H. Miyake, *Mol. Cryst. Liq. Cryst.* **1989**, *166*, 157.
[11] S. Diele, G. Pelzl, W. Weissflog, D. Demus, *Liq. Cryst.* **1988**, *3*, 1047. S. Diele, S. Manke, W.

Weissflog, D. Demus, *Liq. Cryst.* **1989**, *4*, 301. G. Pelzl, A. Humke, S. Diele, D. Demus, W. Weissflog, *Liq. Cryst.* **1990**, 7, 115.

[12] J. Malthête, J. Canceill, J. Gabard, J. Jacques, *Tetrahedron* **1981**, *37*, 2823.

[13] J. Billard, J.-C. Dubois, C. Vaucher, A.-M. Levelut, *Mol. Cryst. Liq. Cryst.* **1981**, *66*, 115.

[14] J. Billard, H. Zimmermann, R. Poupko, Z. Luz, *J. Phys. (Paris)* **1989**, *50*, 539.

[15] T. Itoh, A. Takada, T. Fukuda, T. Miyamoto, Y. Yakoh, J. Watanabe, *Liq. Cryst.* **1991**, *9*, 221. V. Vill, J. Thiem, *Liq. Cryst.* **1991**, *9*, 451.

[16] K. Praefcke, B. Kohne, B. Gündogan, *Mol. Cryst. Liq. Cryst. Lett.* **1990**, *7*, 27. K. Praefcke, B. Kohne, B. Gündogan, D. Singer, D. Demus, S. Diele, G. Pelzl, U. Bakowsky, *Mol. Cryst. Liq. Cryst.* **1991**, *198*, 393. S. M. Fan, I. D. Fletcher, B. Gündogan, N. J. Heaton, G. Kothe, G. R. Luckhurst, K. Praefcke, *Chem. Phys. Lett.* **1993**, *204*, 517.

[17] D. Guillon, A. Skoulios, J. Malthête, *Europhys. Lett.* **1987**, *3*, 67.

[18] J. Malthête, H. T. Nguyen, A.-M. Levelut, *J. Chem. Soc., Chem. Commun.* **1986**, 1548; ibid. **1987**, 40.

[19] J. Malthête, A. Collet, A.-M. Levelut, *Liq. Cryst.* **1989**, *5*, 129.

[20] I. Letko, S. Diele, G. Pelzl, W. Weissflog, *Mol. Cryst. Liq. Cryst.* **1995**, *260*, 171.

[21] A.-M. Levelut, J. Malthête, C. Destrade, H. T. Nguyen, *Liq. Cryst.* **1987**, *2*, 877.

[22] J. Malthête, L. Liébert, A.-M. Levelut, Y. Galerne, *C. R. Acad. Sci. Paris* **1986**, *303*, 1073.

[23] J. Malthête, P. Davidson, *Bull. Soc. Chim. Fr.* **1994**, *131*, 812.

[24] C. Destrade, H. T. Nguyen, A. Roubineau, A. M., Levelut, *Mol. Cryst. Liq. Cryst.* **1988**, *159*, 163.

[25] C. Destrade, H. T. Nguyen, C. Alstermark, G. Lindsten, M. Nilsson, B. Otterholm, *Mol. Cryst. Liq. Cryst.* **1990**, *180B*, 265.

[26] C. Alstermark, M. Eriksson, M. Nilsson, C. Destrade, H. T. Nguyen, *Liq. Cryst.* **1990**, *8*, 75.

[27] H. T. Nguyen, C. Destrade, J. Malthête, *Liq. Cryst.* **1990**, *8*, 797.

[28] S. Diele, K. Ziebarth, G. Pelzl, D. Demus, W. Weissflog, *Liq. Cryst.* **1990**, *8*, 211.

[29] W. Weissflog, G. Pelzl, I. Letko, S. Diele, *Mol. Cryst. Liq. Cryst.* **1995**, *260*, 157.

[30] P. Jacq, Ph. D. Thesis, Université Claude Bernard, Lyon I, **1994**, No. 202–94. P. Jacq, in *15th Int. Liquid Crystal Conf.*, Budapest, Hungary, 3–8 July, **1994**, Vol. 2, poster J-Sbp29, Book of Abstracts, p. 884. P. Jacq, J. Malthête, *Liq. Cryst.* **1996**, *21*, 291.

[31] H. T. Nguyen, C. Destrade, *Mol. Cryst. Liq. Cryst. Lett.* **1989**, *6*, 123.

[32] P. Berdaguer, F. Perez, J. Courtieu, J. P. Bayle, *Bull. Soc. Chim. Fr.* **1994**, *131*, 335.

[33] A. M. Levelut, Y. Fang, *J. Phys. (Paris), Colloq. C7,* **1990**, *51*, 229. Y. Fang, A. M. Levelut, C. Destrade, *Liq. Cryst.* **1990,** *7*, 265. Y. Fang, Ph. D. Thesis, Université de Paris-Sud, Centre d'Orsay, **1988**.

[34] A. V. Ivashenko, E. I. Koshev, V. T. Lazareva, E. K. Prudnikova, V. V. Titov, T. I. Zverkova, H. I. Barnik, L. M. Yagupolski, *Mol. Cryst. Liq. Cryst*, **1981**, *67*, 235.

[35] E. P. Janulis, J C. Norack, G. A. Papapolymerou, M., Tristani-Kendra, W. A., Huffman, *Ferroelectrics* **1988**, *85*, 375.

[36] F. Tournilhac, L. Bosio, J. F. Nicoud, J. Simon, *J. Chem. Phys. Lett.* **1988**, *145*, 452.

[37] H. T. Nguyen, G. Sigaud, M. F. Achard, F. Hardouin, R. J., Twieg, K. Betterton, *Liq. Cryst.* **1991**, *10*, 389.

[38] T. Doi, Y. Sakurai, A. Tamatani, S. Takenaka, S. Kusabayashi, Y. Nishihata, H. Teraushi, *J. Mater. Chem.* **1991**, *1*, 169.

[39] F. Hardouin, H. T. Nguyen, M. F. Achard, A. M. Levelut, *J. Phys. Lett.* **1982**, *43*, 327.

[40] E. Dietzmann, W. Weissflog, S. Markscheffel, A. Jakli, D. Lose, S. Diele, *Ferroelectrics* **1996**, *180*, 341.

[41] R. J. Petschek, K. M. Wiefling, *Phys. Rev. Lett.* **1987**, *59*, 343.

[42] F. Tournilhac, L. M. Blinov, J. Simon, S. V. Yablonsky, *Nature* **1992**, *359*, 621.

[43] L. Navailles, P. Barois, H. T. Nguyen, unpublished results. H.-T. Nguyen, R. J. Twieg, K. Betterton, L. Navailles, P. Barois, C. Destrade, in *14th Int. Liquid Crystal Conf.*, Pisa, Italy, 21–26 June, **1992**, A-P30, Book of Abstracts, p. 52.

[44] H. T. Nguyen, J. Malthête, C. Destrade, *Mol. Cryst. Liq. Cryst. Lett.* **1985,** *2*, 133. H. T. Nguyen, C. Destrade, *Mol. Cryst. Liq. Cryst. Lett*, **1989**, *4*, 123.

[45] W. Weissflog, S. Diele, D. Demus, *Mater. Chem. Phys.* **1986**, *15*, 475.

[46] V. Luzzati, *Biological Membranes* (Ed: D. Chapman), Academic, New York **1968**.

[47] J.-P. Bideau, G. Bravic, M. Cotrait, H. T. Nguyen, C. Destrade, *Liq. Cryst.* **1991**, *10*, 379. H. Allouchi, J. P. Bideau, M. Cotrait, C. Destrade, H. T. Nguyen, *Mol. Cryst. Liq. Cryst.* **1994**, *229*, 153.

[48] D. Guillon, B. Heinrich, C. Cruz, A. C. Ribeiro, H. T. Nguyen, unpublished.

[49] H. T. Nguyen, C. Destrade, H. Allouchi, J. P. Bideau, M. Cotrait, D. Guillon, P. Weber, J. Malthête, *Liq. Cryst.* **1993**, *15*, 435. H. Allouchi, Ph. D. Thesis, Université Bordeaux I, **1994**, No. 1125.

[50] Nevertheless, a high temperature phase of unknown structure (probably a 3D mesophase) has been observed above with a double-swallow-tailed compound W. Weissflog, A. Saupe, I. Letko, S. Diele, G. Pelzl, *Liq. Cryst.* **1996**, *20*, 483.

Chapter XIII
Thermotropic Cubic Phases

Siegmar Diele and Petra Göring

1 Historical Remarks

The origin of the research and discussion of the thermotropic cubic liquid crystalline phase may be dated as 1957, in which Gray et al. [1] published the results of the investigations of compounds **1a-2** and **1a-4** in Table 1 and explained: "For the hexadecyl and octadecyl ether the changes occurring on heating are solid–smectic I–smectic II–smectic III–isotropic, … . Smectic III has a marked tendency to homeotropy."

The reinvestigation of these substances by Demus et al. in 1968 [2] confirmed the observed results and gave evidence for a new type of a liquid crystalline phase by miscibility studies. The measurements of the refractive index in 1971 by Pelzl and Sackmann [3] and the X-ray studies in 1972 by Diele et al. [4] proved the existence of a cubic phase. A structural model built up by micelles has been assumed for reasons of simplicity.

In the same period, extensive investigations of lyotropic liquid crystalline phases, including cubic mesophases, were started by several groups. This fascinating subject has been developed into a separate field and will be covered in Vol. 3 of this handbook. However, some candidates exhibit cubic phases in the water-free state, these will be mentioned in this chapter. Luzzati and Spegt [5] and Spegt and Skoulios [6] have published investigations of strontium alkanoates which form (likewise) thermotropic

cubic phases (compounds **9** in Table 3). On the basis of detailed X-ray studies, they were able to ascertain the cubic space group *Ia*3*d* of the phase and to develop a complicated structural model consisting of a network of rods linked three-by-three and interwoven but not connected (see Fig. 5). These ideas have been applied in many later cases.

The repeated X-ray measurements of compound **1a-4** by Tardieu and Billard [7] allowed a more precise interpretation of the X-ray data (six reflections instead of four, as in [4]), giving a body-centered cubic phase which is compatible with the space group *Ia*3*d*. The structural model was adopted from the results obtained for the strontium alkanoates.

The cubic phase of compound **1a-2** has been observed between the SmC and SmA phases. Since it was not miscible with one of the known thermotropic liquid crystalline phases (SmA, SmB, and SmC), the new phase was designated SmD [2]. However, the results obtained later gave evidence that the phase under discussion could not belong to the group of smectic phases that are characterized by layer structures. Therefore Demus et al. [8, 9] and later Etherington et al. [10] and others proposed to avoid the term smectic. It was thought in [8] that the cubic phases would represent a new group of thermotropic liquid crystalline phases other than the smectic and discotic ones. Therefore the designation SmD is avoided and substituted by the abbreviation Cub in the following discussion.

2   Chemical Structures and Phase Sequences

Since the discovery of the cubic phase numerous substances of quite different chemical structure (Tables 1–4) have been found to be able to form this phase. Therefore the intermolecular interactions leading to the existence of the cubic phases must be of a different nature. In Tables 1–4 it was tried to group the substances according to their possible intermolecular interactions, which mainly determine the appearance of the cubic phase.

In Table 1 substances are listed that contain groups giving rise to hydrogen bonds.

Through the hydrogen bonds, dimers and other molecular assemblies are formed, which in the presence of additional lateral interactions constitute the cubic structure. It has been shown [2] that, e.g., the loss of the lateral dipole moment by substitution of the NO_2 group (compound 1a-2) by a hydrogen or chlorine atom results in the loss of the cubic phase.

As pointed out in [12, 13], the appearance of the cubic phase in the case of polycatenar substances (Table 2) is strongly determined by the balance of the rigid rod-like middle part of the molecule and the flexible terminal alkyl chains. It can be assumed that here the steric interaction based on the molecular shape dominates the structure.

Table 1. Compounds with groups able to form associations ($-OH$, $-COOH$, $-CONH-$, $-CN$, $-NO_2$).

Compound no.	Compound	Ref.
1	4'n-Alkoxy-3'-substituted-biphenyl-4-carboxylic acids	
	$C_nH_{2n+1}O$—⟨ ⟩(X)—⟨ ⟩—COOH	
1a	4'n-Alkoxy-3'-nitro-biphenyl-4-carboxylic acids $X=NO_2$	
1a-1	$n=15$ Cr 127 SmC 187 Cub 198 SmA 204 I	[41]
1a-2	$n=16$ Cr 126.8 SmC 171.0 Cub 197.2 SmA 201.9 I	[1, 2, 4, 7, 22, 39–41]
1a-3	$n=17$ Cr 135 SmC 169 Cub 199 I	[41]
1a-4	$n=18$ Cr_1 85 Cr_2 124.6 SmC 158.9 Cub 195 I	[1, 2, 4, 7, 22, 39–41]
1a-5	$n=19$ Cr 127 SmC 154 Cub 198 I	[41]
1a-6	$n=20$ Cr_1 79 Cr_2 112 SmC 142 Cub 194 I	[41]
1a-7	$n=21$ Cr 114 SmC 143 Cub 197 I	[41]
1a-8	$n=22$ Cr_1 89 Cr_2 103 SmC 126 Cub 190 I	[41]
1b	4'n-Octadecyloxy-3'-cyano-biphenyl-4-carboxylic acid $X=CN$	
1b-1	$n=16$ Cub	[11]
1b-2	$n=18$ Cr 131 SmC 156 Cub 201 I	[10, 11, 25, 42]
2	Benzoylhydrazide	
2a	1,2-Bis(4n-alkoxy-benzoyl)-hydrazines	
	$C_nH_{2n+1}O$—⟨ ⟩—CO·NH·HN·OC—⟨ ⟩—OC_nH_{2n+1}	[8, 22, 43, 44]
2a-1	$n=8$ Cr_1 132.2 Cr_2 136.7 Cub 161 SmC 165 I	
2a-2	$n=9$ Cr_1 124 Cr_2 143 Cub 157 SmC 165 I	
2a-3	$n=10$ Cr_1 124 Cr_2 143 Cr_3 145 Cub 152 SmC 165 I	

Table 1. (continued).

Compound no.	Compound	Ref.
2b	Monoacylhydrazide	

$H_2N-HN-\overset{O}{\overset{\|}{C}}$—〈ring〉 with OC_nH_{2n+1} (top), $-OC_nH_{2n+1}$ (right), OC_nH_{2n+1} (bottom)   [65, 66]

2b-1	$n=5$ Cr 62 Cub 91 I	
2b-2	$n=10$ Cr 45 Cub 106 I	
2b-3	$n=14$ Cr 66 Cub 113 I	

3	Diol derivatives	
3a	cis,cis-(3,5-dihydroxycyclohexyl)-3,4-bis(alkoxy)-benzoates	

$H_{2n+1}C_nO$ / $H_{2n+1}C_nO$—〈ring〉—COO—〈cyclohexyl〉 with OH, OH   [45, 46, 61, 63]

3a-1	$n=8$ Cr ? Cub 63.4 Col$_h$ 97.5 I	
3a-2	$n=9$ Cr ? Cub 66.0 Col$_h$ 110.0 I	

3b	Cyclic siloxane compound	

$\left[\begin{array}{c} R \\ -Si-O- \\ CH_3 \end{array}\right]_4$ $R=$ —$(CH_2)_{11}$—O—〈ring with $H_{23}C_{11}O$〉—COO—〈cyclohexyl with OH, OH〉

Cub 120.5 Col$_h$ 148.5 I   [46–48, 62]

3c	Side group siloxane polymer	

$\left[\begin{array}{c} R \\ -Si-O- \\ CH_3 \end{array}\right]_n$ $R=$ —$(CH_2)_{11}$—O—〈ring with $H_{23}C_{11}O$〉—COO—〈cyclohexyl with OH, OH〉

Cub 147 Col$_h$ 158 I   [46–48]

4	4,4'-Bis(8-hydroxyoctyloxy)-azoxybenzene	

$HO(CH_2)_8$—O—〈ring〉—$\overset{O}{\overset{\|}{N=N}}$—〈ring〉—O—$(CH_2)_8OH$

Cr 112 SmC 127 Cub 138 N 141 I   [42]

5	Glycoside	
5a	n-Dodecyl-β-isomaltosid	

Cr 120 M 96 Cub 138 SmA 154 I   [26, 46, 49]

Table 1. (continued).

Compound no.	Compound	Ref.
5b	*n*-Dodecyl-α- or -β-gentiobiosid	

		[26, 46, 49]
5b-1	$R_\alpha = OC_{12}H_{25}$, $R_\beta = H$ Cr ? Cub 158 I	
5b-2	$R_\alpha = H$, $R_\beta = OC_{12}H_{25}$ Cr ? Cub 158 Col 164 I	
5c	6-Deoxy-β-L-erythro-hexapyranosides	

		[50]
5c-1	$R_1 = H_{21}C_{10}-O-Ph-Ph-COO-$, $R_2 = -CN$ Cr 85.5 (Cub 56.3) Ch 91.5 BP 92 I	
5c-2	$R_1 = H_{17}C_8-Ph-Ph-COO-$, $R_2 = -CN$ Cr 99.8 (Cub 85) (N* 76.7 BP 76.6) I	
5d	1,1-di-(β-D-glucosyloxymethyl)pentadecane	

		[26, 46]
	Cr 111 Cub 205 I	
6 **6a**	Amide Diethylenetriamide	

		[51, 52]
6a-1	*n* = 5	Cr$_1$ 55 Cr$_2$ 86 Cub 104 I
6a-2	*n* = 13	Cr$_1$ 45 Cr$_2$ 81 Cub 100 I
6a-3	*n* = 14	Cr$_1$ 50 Cr$_2$ 86 Cub 106 I
6a-4	*n* = 15	Cr$_1$ 54 Cr$_2$ 84 Cub 107 I
6a-5	*n* = 16	Cr$_1$ 43 Cr$_2$ 85 Cub 107 I
6b	Triethylenetetramide	

		[51, 52, 64]
6b-1	*n* = 15	Cr 116 (Col$_h$ 85.5 Cub 105) I
6b-2	*n* = 16	Cr 116 (Col$_h$ 84 Cub 194.5) I

Table 2. Polycatenar compounds[a].

Compound no.	Compound	Ref.

7a

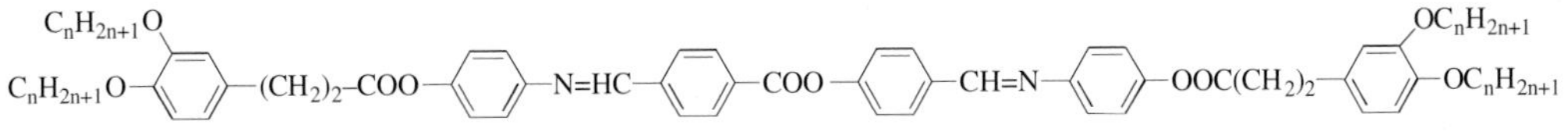

$C_nH_{2n+1}O$ ... $(CH_2)_2$–COO ... N=HC ... COO ... CH=N ... OOC$(CH_2)_2$ ... OC_nH_{2n+1}

$n = 10$ Cr 93 SmC 128 Sm? 143 Sm? 148 Cub 207 I   [13, 20]

7b

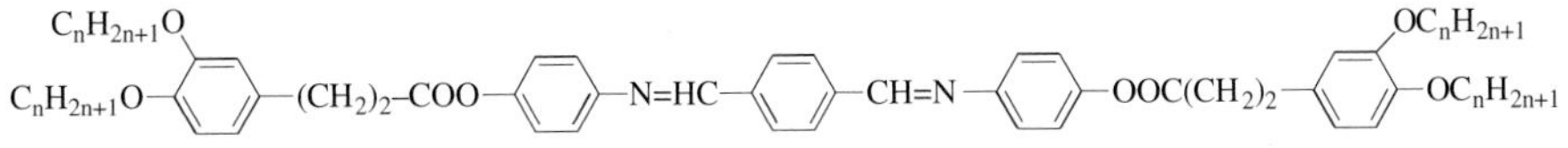

| **7b-1** | $n = 9$ | Cr 97 SmC 146 Cub 211 I | [20, 53] |
| **7b-2** | $n = 10$ | Cr 93 SmC 128 Col$_{ob}$ 143 Col$_h$ 148 Cub 207 I | |

7c

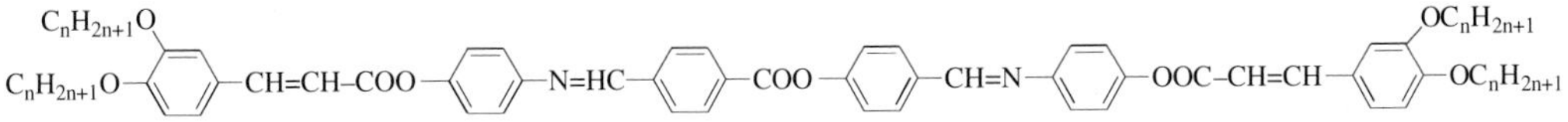

$n = 11$ Cr 131 SmC 224 Cub 239 I   [20, 53]

7d

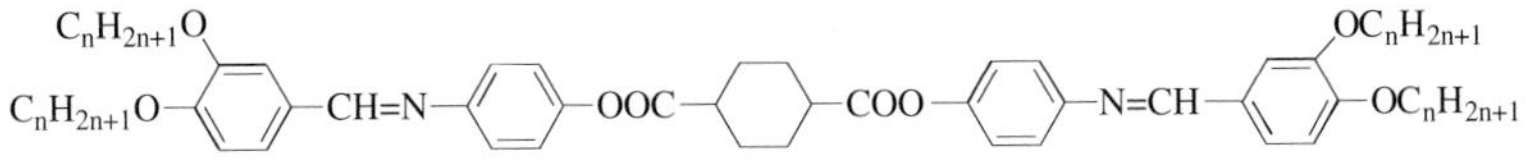

7d-1	$n = 10$	Cr 144 SmC 156 Cub 165 I	[20, 22, 54, 55]
7d-2	$n = 11$	Cr 144 SmC 146 Cub (147 Col$_h$ 158) 163 I	
7d-3	$n = 12$	Cr 142 Cub 162 I, I 160 Col$_h$ 138 Cr	

7e

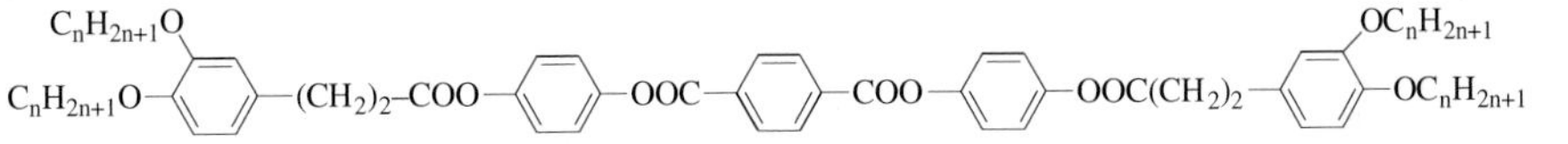

7e-1	$n = 8$	Cr 121 SmC 127.5 Cub 135 I	[20, 56]
7e-2	$n = 9$	Cr 123 Col$_h$ 132.5 Cub 137.5 I	
7e-3	$n = 10$	Cr 122 Col$_{ob}$ 134 Cub 137 I	

7f

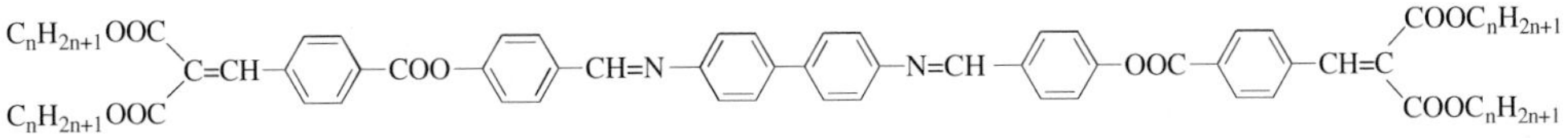

7f-1	$n = 5$	Cr 143 SmC 195 Cub 271 N 378 I	[38]
7f-2	$n = 6$	Cr 120 SmC 181 Cub 255 N 355 I	
7f-3	$n = 7$	Cr 132 SmC 202 Cub 258 N 334 I	
7f-4	$n = 8$	Cr 126 SmC 195 Cub 243 I$_{re}$ 250 N 312 I	
7f-5	$n = 9$	Cr 125 Col$_{ob}$ 158 SmC 195 Cub 238 I	
7f-6	$n = 10$	Cr 120 Col$_{ob}$ 192 Cub 220 I	

7g

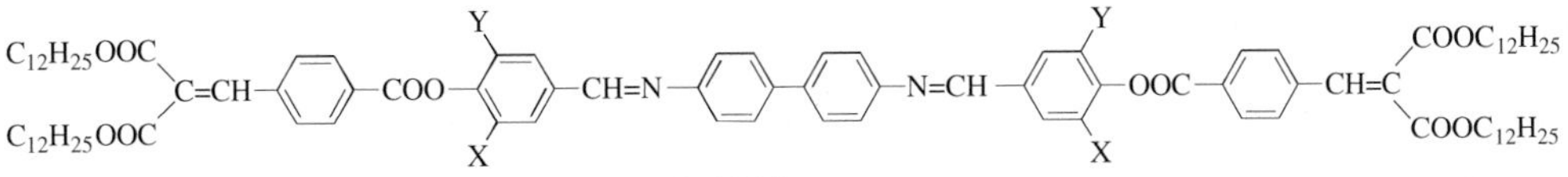

| **7g-1** | X = OCH$_3$ Cr 135 Col$_h$ 145–150 Cub 164 I Y = H | [19] |
| **7g-2** | X = OC$_2$H$_5$ Cr 69 Col$_h$ 71–79 Cub 109 I Y = H | |

892

XIII Thermotropic Cubic Phases

Table 2. (continued).

Compound no.	Compound	Ref.
7h	H25C12O, H25C12O —〇— COO —〇— OCO —〇— N=HC —〇— OCO —〇— OC12H25 Cr 83 Cub 172 I	[20, 67]
7i	H25C12O, H25C12O —〇—, H25C12O —〇— COO —〇—〇— OCO —〇— OCO —〇— OC10H21 Cr 79 Cub 146 I	[67]
7j	H25C12O, H25C12O —〇—, H25C12O —〇— COO —〇—〇— OOC —〇— OC12H24 Cr 73.4 (Cub 60) I	[68]
7k	H25C12O, H25C12O —〇—, H25C12O —〇— COO —〇—〇— OOC —〇— O—C8F17 Cr 71 Cub 108 I	[68]

[a] The term 'polycatenar' shall include all terminal branched substances irrespective of whether they are designated as swallow-tailed or biforked in other cases.

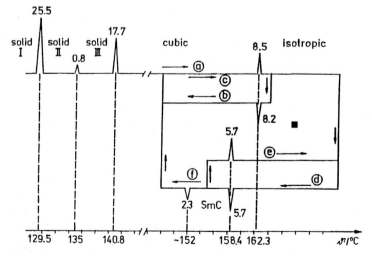

Figure 1. Transition scheme of compound **2a-1**. (Unpublished results from Wiegeleben and Demus according to [46]).

Table 3. Metal-containing substances.

Compound no.	Compound	Ref.
8	Polymorphic ionic mesogens of silver(I)	
8a	Bis(alkoxystilbazole)silver(I)dodecylsulphate complexes	

$$C_nH_{2n+1}O\!-\!\langle\rangle\!-\!HC{=}HC\!-\!\langle\rangle\!-\!N\overset{\oplus}{-}Ag\!-\!N\!-\!\langle\rangle\!-\!CH{=}CH\!-\!\langle\rangle\!-\!OC_nH_{2n+1}$$

$$\overset{\ominus}{C_{12}H_{25}OSO3}$$

[27, 57, 58, 59]

8a-1	$n=6$ Cr 138 Cub 156 SmA 185 I	
8a-2	$n=7$ Cr 132 SmC 139 Cub 152 SmA 179 I	
8a-3	$n=8$ Cr 135 SmC 139 Cub 153 SmA 178 I	
8a-4	$n=9$ Cr 133 SmC 139 Cub 153 SmA 180 I	
8a-5	$n=10$ Cr 123 SmC 135 Cub 158 SmA 171 I	
8a-6	$n=11$ Cr 116 SmC 135 Cub 156 SmA 180 I	
8a-7	$n=12$ Cr 108 SmC 123 Cub 153 SmA 168 I	

8b	Fluorinated bis(alkoxystilbazole)silver(I)-dodecylsulphate complexes	

$$C_nH_{2n+1}O\!-\!\langle{}^F\rangle\!-\!HC{=}HC\!-\!\langle\rangle\!-\!N\overset{\oplus}{-}Ag\!-\!N\!-\!\langle\rangle\!-\!CH{=}CH\!-\!\langle{}_F\rangle\!-\!OC_nH_{2n+1}$$

$$\overset{\ominus}{C_{12}H_{25}OSO3}$$

[57–59]

$n=7$ Cr 97 SmC 125 Cub 131 N 138.5 I

8c	$Ag{-}S{-}C_nH_{2n+1}$	[59]
8c-1	$n=4$ Cr 139 SmA 185 Cub 202 M 204 I	
8c-2	$n=6$ Cr 129 SmA 171 Cub 183 M 219 I M =	
8c-3	$n=8$ Cr 128 SmA 133 Cub 172 M 217 I micellar	
8c-4	$n=10$ Cr 127 SmA 129 Cub 153 M 206 I	

9	Divalent cation soaps	

$$C_nH_{2n+1}\!-\!\underset{O}{\overset{\parallel}{C}}\!-\!O\!-\!Sr\!-\!O\!-\!\underset{O}{\overset{\parallel}{C}}\!-\!C_nH_{2n+1}$$

with $n=11, 13, 15, 17, 19, 21$

[5]

With respect to the different molecular interactions, cubic phases with different molecular aggregation can be expected.

The competition between different intermolecular interactions results in a complicated thermal behavior. Mostly, the cubic phase is found in a sequence with the SmC phase, in which the latter one is the low temperature form. But the hydrazine derivatives exhibit the inverse sequence. In some cases (compounds **1a-1**, **1a-2**, and **8a-2–8a-7**), the cubic phase is observed between the SmC and SmA phase. In polycatenar substances the cubic phase is found together with columnar phases (Col$_h$ or Col$_{obl}$), where the columnar phases are either the high or the low temperature form. Also, in compound **1a-4**, a phase S_4 – not classified up to now – is discussed as a columnar one [14]. So in some cases a phase sequence smectic–cubic–columnar can be found which has a close analogy to the lyotropic systems. Here the cubic phase is considered as an intermediate state between the Lα phase (layer structure) and the Hα phase (hexagonal columnar structure) [15–17].

However, it has been pointed out [8] that the phase sequences including cubic phases

Table 4. Compounds with disc or globular molecular shapes.

Com-pound no.	Compound	Ref.
10	2,3,7,8-tetra-*n*-alkanoyloxy derivatives of ellagic acid	[22]

10-1	$n = 11$ Cr$_1$ 99 Cr$_2$ 166 M 176 (Cub) 175 I
10-2	$n = 12$ Cr$_1$ 80 Cr$_2$ 158 M 173 (Cub) 172 I
10-3	$n = 13$ Cr$_1$ 107 Cr$_2$ 156 M 171 (Cub) 169 I
10-4	$n = 14$ Cr$_1$ 96.5 Cr$_2$ 151.5 M 168 (Cub) 167 I
10-5	$n = 15$ Cr$_1$ 115 Cr$_2$ 151 M 166 (Cub) 165.5 I

11 Hexa-o-pentanoyl-scyllo-inosit

C$_4$H$_9$OCO, OCOC$_4$H$_9$, C$_4$H$_9$OCO, OCOC$_4$H$_9$, C$_4$H$_9$OCO, OCOC$_4$H$_9$, C$_4$H$_9$OCO   [21, 22]

Cr 184.5 Col 208 Cub 232 I

12 Tetrasubstituted tribenzosilatrane

R$_1$, O–Si–O, O, Ni, R$_2$, R$_2$, R$_2$ [60]

R$_1$ = C$_8$H$_{17}$
R$_2$ = C$_{12}$H$_{25}$

Cr $\xrightleftharpoons[115]{135}$ Col$_h$ $\xrightarrow{145-150}$ Cub $\xrightarrow{164}$ I

164

115

115

Figure 2. Transition scheme of compound **7g-1** [19].

undergo a complicated transition scheme dependent on the thermal history. As an example, that of compound **2a-1** will be considered. If the sample is heated only a few degrees above the clearing temperature, the

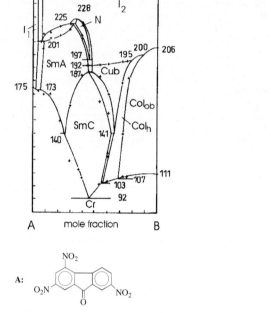

A:

NO$_2$, O$_2$N, NO$_2$, O

B: [H$_{25}$C$_{12}$OOC, C=CH—⬡—COO—⬡—HC=N—⬡]$_2$

Figure 3. Binary system according to [23].

transition Cub–I is found to be reversible (Fig. 1, traces a→b→c). Obviously, nuclei of the cubic structure exist in the isotropic phase, these cause the formation of the cubic structure during the cooling run. If the nuclei are destroyed at higher temperatures, the polymoprhism is changed at the cooling run and a reversible SmC–I transition is observed (Fig. 1, traces a→d→e). On further cooling, the SmC phase is transformed into the cubic phase (Fig. 1, trace f). The transition enthalpy ΔH (cub–I) is nearly the sum of the enthalpies ΔH (Cub–SmC)+ΔH (SmC–I). In polycatenar compounds, the cubic phase can be suppressed completely by cooling from the isotropic phase [18, 54,

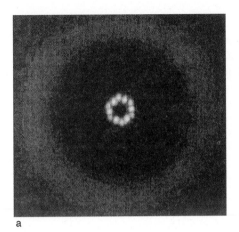
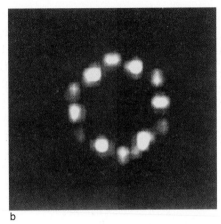

a   b

Figure 4. (a) X-ray pattern of a monodomain of compound **5d-1** [26], (b) small angle region of (a).

68]. Figure 2 shows the transition scheme of compound **7g-1** [19].

As far as miscibility studies are concerned, no example of complete miscibility between the cubic phases of substances with different chemical structures could be observed. The binary mixture of compounds **2a-1** and **1a-2** [8] shows a very restricted miscibility in the cubic phase (smaller than 5 mol% of the second component). The cubic phase of the polycatenar substance **7e-2** is able to incorporate about 80% of the rod-like component **1a-4** [20]. In the binary systems **11/1a-4** [21], **10-3/1a-4**, **10-3/2a-3**, and **10-3/11** [22] the cubic phases are separated only by a small heterogeneous region.

The lack of complete miscibility between cubic phases of substances with different chemical structures points to the peculiarities of the intermolecular interactions in these phases.

Finally, it should be mentioned that cubic phases can be induced in binary mixtures with polycatenar substances if the mixing components are pronounced electron-acceptor compounds [23, 24] (Fig. 3). Depending on the concentration, again a change from a layer structure via a cubic phase into a columnar structure takes place.

3   On the Structure of the Cubic Phases

Besides the optical and calorimetric measurements, only X-ray studies have been performed to some extent. The X-ray patterns of nonoriented samples exhibit some mostly textured reflections in the small angle region (Bragg angle: $0.8 < \Theta < 4$) with a nearly instrumentally limited line shape and a diffuse outer scattering in the wide angle region (around $\Theta \cong 10°$). Some examples of investigations of monodomains have been reported, at the pattern of which the small angle reflections are degenerated to Bragg spots, whereas the outer diffuse scattering remains as a diffuse closed ring (Fig. 4) [4, 10, 25–27].

Therefore the cubic phase must be considered as a 3D ordered lattice with a periodicity of about one [8] up to three [26] molecular lengths together with an isotropic

Table 5. Observed reflections in the cubic phase of substance **5d**.

θ_{exp} (°)	hkl	θ_{calc} (°)
1.24	211	1.24
1.44	220	1.43
1.89	321	1.89
2.02	400	2.02
2.26	420	2.26
2.36	332	2.37

Table 6. Reported X-ray data of the cubic phases.

Sub-stance	Lattice parameter a (nm)	Space group [a]	Ref.
1a-2	6.10	$Ia3d$	[4]
1b-2	8.60	$Pm3$ or $Pn3m$	[10, 25]
2a-1	4.57	$P2n3$, $Pn3$	[8]
2a-2	4.88	$P4_232$, $Pn3m$	
2a-3	5.17		
3a-1	8.67	$Ia3d$	[46]
5b-1	8.05	$Ia3d$	[26]
5d	8.74	$Ia3d$ or $I43d$	[26]
7a	18.26	$Im\bar{3}$, $I\bar{4}3m$, $Im\bar{3}m$	[13]
7c	18.26	$Im3m$	[20]
7d-2	16.20	$Im3m$	[20]
7d-3	17.18	$Im3m$	[20]
7e-1	15.57	$Im3m$	[20]
7h	16.93	$Im3m$	[20]
8a-1	8.00	$Ia\bar{3}d$	[27]
8a-2	6.60	$Ia\bar{3}d$	[27, 58]
8a-3	7.34	$Ia\bar{3}d$	[27]
9	6.97	$Ia3d$	[5]
12	2.75	RT: body-centered	[60]
	2.58	60 °C: $I4_132$	

[a] The space group symbols are presented in the form as written in the literature.

(fluid like) ordering of 0.4–0.5 nm. Etherington et al. [25] stated that "the structure is thus truly a liquid crystal, with a liquid like molecular packing modulated by the associations of cores and tails with a three-dimensional periodicity". The three-dimensional electron density function is a slowly varying wave with only a few Fourier components giving rise to an X-ray pattern with a very small number of reflections. This small number of reflections often prevents a decisive evaluation of the pattern.

The results obtained for substance **5d** [26] will be discussed as an example. The Bragg angles of the observed reflections match the proportions $\sqrt{3}$, $\sqrt{4}$, $\sqrt{7}$, $\sqrt{8}$, $\sqrt{10}$, and $\sqrt{11}$. Since in a cubic lattice $\sin\Theta \sim \sqrt{(h^2+k^2+l^2)}$, the $\sqrt{7}$ is not compatible with a cubic lattice. As already pointed out in [6], the sequence of numbers must be doubled, which leads to the Miller indices cited in Table 5.

The given indexing with the systematic absence of reflections of lower Miller indices can only be explained with the space group $Ia3d$ (No. 230) or $I43d$ (No. 220); it is not possible to distinguish between them on the basis of the performed measurements. The obtained lattice constant ($a = 8.74$ nm) is larger than the length of the molecule ($L = 2.5$ nm) estimated by CPK (space filling molecular) models. It should be emphasized that the measured intensities of the first two reflections behave as $I_{(211)} : I_{(220)} \cong 100 : 34$. On the same scale, the intensities of the other reflections are below 1. The results are the same as reported in [7] for compound **1a-4**.

Table 6 summarizes the observed lattice parameters and the space groups. The space groups $Ia3d$ and $Im3m$ are the most frequently observed ones. With the exception of compounds **2a-1** to **2a-3**, the lattice parameter is found to be greater than the length of the molecules. The unit cells contain several hundreds of molecules, suggesting the existence of molecular aggregations.

To answer the question as to the kind of molecular assemblies, two different structures of the cubic phase have in principle been discussed. For the most simple form of a molecular assembly, spherical arrangements have been considered [4]. In the case of the compounds **2a-1** to **2a-3**, the lattice parameter corresponds to the length of the molecule. Therefore bundles of mole-

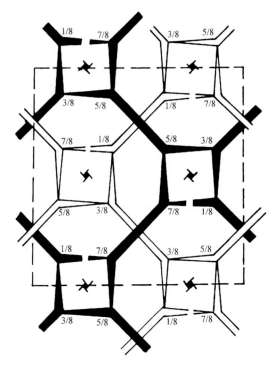

Figure 5. Representation of the *Ia3d* structure of compound **9** according to [5] and [12]. The dashed lines represent the base plane. The ratios are the *z* coordinates of the intersections of the rods.

cules with a nonfixed preferred direction have been assumed [8, 9], similar to the structural units of plastic crystals [28].

A quantitative analysis of the X-ray intensity scattered by a monodomain of compound **1b-2** [10, 25] led to four possible models of the primitive cubic lattice built up by micellar assemblies. Two of them are described by linked spherical micelles clustering around the body-center and face-center positions. In the two other models, the electron density is concentrated in spherical regions at the corner and in the body-center of the cell, as well as in the faces along nonintersecting rods in the [100], [010], and [001] directions.

In contradiction to the micellar model, the cubic structure can also be described by a network of rods. On the basis of the intensity data of 13 observed X-ray reflections, the structure of the cubic phase of compound **9** could be established to consist of rods linked three-by-three, forming two networks interwoven but not connected (Fig. 5) [5, 6]. The rods are assumed to be built up by the polar heads (cations) while the aliphatic chains fill out the free volume between the rods.

This model is the general accepted explanation of the cubic structure with the space group *Ia3d* in lyotropic systems. Its application for the thermotropic phases demands an explanation as to the formation of rods by the molecules of the pure compounds. A problem arises due to the lack of a second free component which is able to fill out the gaps between the rods. As a first proposal Lydon [14] used dimers of compound **1a**, which are arranged star-like to form flat discs. These discs act as the constituting units of rods with finite length. The cross section of the rods consists of an inner part with aliphatic chains, followed by a shell with aromatic moieties, which are again surrounded by an aliphatic wrap. The problem lies in the high density of the aliphatic chains in the inner part and in the too large diameter of the rods compared with the lattice parameter. Therefore a different packing has been proposed by Guillon and Skoulios [12] in which three molecules of the polycatenar compound (or three dimers of the acid derivative) are arranged parallel to each other with the outer molecules bent to form a disc with a nearly circular cross section. The surrounding disordered aliphatic chains again fill out the gaps of the structure.

Besides the description of the structure on a molecular level, a discussion is given on the basis of an infinite periodic minimal surface (IPMS). In lyotropic systems it is well accepted that the interface between the hy-

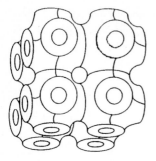

Figure 6. The Schwarz P surface.

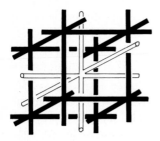

Figure 7. The labyrinths of the P surface.

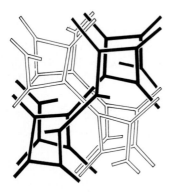

Figure 8. The labyrinths of the G surface.

drophilic part of the amphiphilic molecules and the water molecules builds an infinite bicontinuous periodic minimal surface (IPMS), which divides the space into two parts [15, 16, 29, 30].

The Schwarz P surface (Figs. 6 and 7) [31, 32] contains the symmetry elements of the space group *Im3m* (No. 229). This sur-

face forms two systems of rod-like labyrinths in which the rods of each labyrinth cross each other at the corner of the cubic cell. The origin or the second labyrinth is located at the body-center of the first one.

The space group *Ia3d* (No. 230) is related to the gyroid G [33, 34] in which the IPMS forms two interwoven labyrinths, each of which is constituted by rods linked three-by-three (Fig. 8). These two systems of rods lead to the structure at first derived for compound **9** (Fig. 5) if the cations are located around the rods and the aliphatic tails approach the IPMS. Helfrich [35] interpreted the positions of the rods as passages that are dependent on the specific values of the curvature elasticity and boundary energies. In the case of compound **1**, it is assumed that these passages are the positions of hydrogen bonds between two acid molecules. It has been pointed out by Levelut and co-workers [13, 17, 20] that the balance between the aromatic part and the aliphatic chains plays a very important role in the formation of the cubic phase in polycatenar systems. The assumption that this balance can be discussed in analogy to the hydrophobic/hydrophilic interplay in lyotropic systems offers the possibility of applying the model of the Schwarz P surface for the description of the cubic phase of polycatenar compounds, too. Comparison of the value of the P surface at a measured lattice parameter *a* with the mean area of the chains leads to a molecular packing in which an inner cylinder of the branched molecules is surrounded by an additional layer of molecules which fill out the gaps between the rods in Fig. 7.

In the case of metal complexes, Bruce et al. [27] proved the existence of space group *Ia3d* on the basis of well developed cubic monodomains. The structural model is discussed on the basis of the gyroid surface G, where the silver atoms are located on the rods (Fig. 8) and the terminal methylene groups approach the minimal surface.

Clerc and Dubois-Violette calculated the structure factor of the G surface [36], which was done earlier by Anderson et al. [37] for the P surface. The IPMS of the G type is surrounded by a thin film of constant thickness. The density of the film has been approximated by a spherical density distribution around each point of the surface. The intensities of the X-ray reflections calculated according to this model agree well with the measured values reported in the literature. These agreements can be considered as essential support of the model.

Finally it should be emphasized, that Luzzati et al. have pointed out [69], that "the topological properties of the micellar phases are profoundly different from those of the bicontinuous phases." Introducing the term "chaotic zones" for the regions of highest disorder it could be shown, that these zones occupy special crystallographic positions, which belong either to symmetry elements or to the IPMS.

4   Summary

The cubic phases of thermotropic liquid crystals display an independent class of liquid crystalline phases besides the smectic and columnar ones. The great variety of chemical structure leads to different cubic phases which can be expected to be constituted of molecular aggregates of different character.

5   References

[1] G. W. Gray, B. Jones, F. Marson, *J. Chem. Soc.* **1957**, 393–400.

[2] D. Demus, G. Kunicke, J. Neelsen, H. Sackmann, *Z. Naturforsch.* **1968**, *23a*, 84–90.

[3] G. Pelzl, H. Sackmann, *Symp. Chem. Soc. Faraday* **1971**, *5*, 68–88.

[4] S. Diele, P. Brand, H. Sackmann, *Mol. Cryst. Liq. Cryst.* **1972**, *17*, 163–169.

[5] V. Luzatti, P. A. Spegt, *Nature* **1967**, *215*, 701–704.

[6] P. A. Spegt, A. E. Skoulios, *Acta Cryst.* **1966**, *21*, 892.

[7] A. Tardieu, J. Billard, *J. Phys. Colloq. France* **1976**, *37*, C3-79–C3-81.

[8] D. Demus, A. Gloza, H. Hartung, A. Hauser, I. Raphtel, A. Wiegeleben, *Krist. Tech.* **1981**, *16*, 1445–1451.

[9] D. Demus, S. Diele, S. Grande, H. Sackmann in *Advances in Liquid Crystals*, Vol. 6 (Ed.: Glenn H. Brown), Academic Press, London **1983**, pp. 1–107.

[10] G. Etherington, A. J. Leadbetter, X. J. Wang, G. W. Gray, A. Tajbakhsh, *Liq. Cryst.* **1986**, *1*, 209–214.

[11] J. W. Goodby, G. W. Gray, *Smectic Liquid Crystals – Textures and Structures*, Leonard Hill, Glasgow **1984**.

[12] D. Guillon, A. Skoulios, *Europhys. Lett.* **1987**, *3*, 79–85.

[13] Y. Fang, A. M. Levelut, C. Destrade, *Liq. Cryst.* **1990**, *7*, 265–278.

[14] J. E. Lydon, *Mol. Cryst. Liq. Cryst. Lett.* **1981**, *72*, 79–87.

[15] J. Charvolin, J. F. Sadoc, *Colloid Polym. Sci.* **1990**, *268*, 190–195.

[16] J. M. Seddon, J. L. Hogan, N. A. Warrender, E. Pebay-Peyroula, *Progr. Colloid Polym. Sci.* **1990**, *81*, 189–197.

[17] Y. Hendrikx, A. M. Levelut, *Mol. Cryst. Liq. Cryst.* **1988**, *165*, 233–263.

[18] H. T. Nguyen, C. Destrade, A. M. Levelut, J. Malthete, *J. Physique* **1986**, *47*, 553–557.

[19] W. Weissflog, I. Letko, S. Diele, G. Pelzl, *23. Freiburger Arbeitstagung Flüssigkristalle*, **1994**, p. 23.

[20] A. M. Levelut, Y. Fang, *J. Phys. Colloq. France* **1990**, *51*, C7-229–C7-236.

[21] B. Kohne, K. Praefcke, J. Billard, *Z. Naturforsch.* **1986**, *41b*, 1036–1044.

[22] J. Billard, H. Zimmermann, R. Poupko, Z. Luz, *J. Phys. France* **1989**, *50*, 539–547.

[23] I. Letko, S. Diele, G. Pelzl, W. Weissflog, *Liq. Cryst.* **1995**, *19*, 643–646.

[24] I. Letko, S. Diele, G. Pelzl, W. Weissflog, *Mol. Cryst. Liq. Cryst.* **1995**, *260*, 171–183.

[25] G. Etherington, A. J. Langley, A. J. Leadbetter, X. J. Wang, *Liq. Cryst.* **1988**, *3*, 155–168.

[26] S. Fischer, H. Fischer, S. Diele, G. Pelzl, K. Jankowski, R. R. Schmidt, V. Vill, *Liq. Cryst.* **1994**, *17*, 855–861.

[27] D. W. Bruce, B. Donnio, S. A. Hudson, A. M. Levelut, St. Megtert, D. Petermann, M. Veber, *J. Phys. II France* **1995**, *5*, 289–302.

[28] G. W. Smith in *Advances in Liquid Crystals*, Vol. 1 (Ed.: G. H. Brown), Academic Press, London **1975**, p. 189.

[29] J. Chavrolin, J. F. Sadoc, *J. Physique* **1987**, *48*, 1559–1569.

[30] J. Seddon, R. H. Templer, *Phil. Trans. R. Soc. London* **1993**, *344*, 377–401.

[31] H. A. Schwarz, *Gesammelte Math. Abhandlungen*, Bd. I, Springer, Berlin **1890**.

[32] L. E. Scriven, *Nature* **1976**, *276*, 123–125.

[33] A. H. Schoen, *Not. Am. Math. Soc.* **1969**, *16*, 519.

[34] S. T. Hyde, S. Anderson, B. Ericsson, K. Larson, *Z. Kristallogr.* **1984**, *168*, 213–219.

[35] W. Helfrich, *J. Physique* **1987**, *48*, 291–295.

[36] M. Clerc, E. Dubois-Violette, *J. Physique II* **1994**, *4*, 275–286.

[37] D. M. Anderson, H. T. Davis, J. C. Nitsche, L. E. Scriven, *Adv. Chem. Phys.* **1990**, *77*, 337.

[38] W. Weissflog, G. Pelzl, I. Letko, S. Diele, *Mol. Cryst. Liq. Cryst.* **1995**, *260*, 157–170.

[39] M. Gongoda, B. M. Fung, *Chem. Phys. Lett.* **1985**, *120*, 527–530.

[40] P. Uklela, R. E. Siatkowski, M. E. Neubert, *Phys. Rev. A* **1988**, *38*, 4815–4821.

[41] S. Kutsumizu, M. Yamada, S. Yano, *Liq. Cryst.* **1994**, *16*, 1109–1113.

[42] S. Yano, Y. Mori, S. Kutsumizu, *Liq. Cryst.* **1991**, *9*, 907–911.

[43] H. Schubert, J. Hausschild, D. Demus, S. Hoffmann, *Z. Chem.* **1978**, *18*, 256.

[44] E. I. Demikhov, V. K. Dolganov, V. V. Korshunov, D. Demus, *Liq. Cryst.* **1988**, *3*, 1161–1164.

[45] G. Lattermann, G. Staufer, *Mol. Cryst. Liq. Cryst.* **1990**, *191*, 199–203.

[46] S. Fischer, *Diss. A, Halle* **1993**.

[47] M. Schellhorn, G. Lattermann, *Macromol. Chem. Phys.* **1995**, *196*, 211–224.

[48] M. Schellhorn, G. Lattermann, *Liq. Cryst.* **1994**, *17*, 529–542.

[49] V. Vill, *Diss. Münster* **1990**.

[50] V. Vill, F. Bachmann, J. Thiem, *Mol. Cryst. Liq. Cryst.* **1992**, *213*, 57–65.

[51] U. Stebani, G. Lattermann, R. Festag, M. Wittenberg, J. H. Wendorff, *J. Mater. Chem.* **1995**, *5*, 2243–2451.

[52] U. Stebani, G. Lattermann, M. Wittenberg, R. Festag, J. H. Wendorff, *Adv. Mater.* **1994**, *6*, 572–574.

[53] C. Destrade, H. T. Nguyen, C. Alstermark, G. Lindsten, M. Nilsson, B. Otterholm, *Mol. Cryst. Liq. Cryst.* **1990**, *180B*, 265–280.

[54] H. T. Nguyen, C. Destrade, J. Malthete, *Liq. Cryst.* **1990**, *8*, 797–811.

[55] A. M. Levelut, J. Malthete, C. Destrade, H. T. Nguyen, *Liq. Cryst.* **1987**, *2*, 877.

[56] C. Destrade, H. T. Nguyen, C. Alstermark, M. Erikson, M. Nilson, *Liq. Cryst.* **1990**, *8*, 75–80.

[57] D. W. Bruce, D. A. Dunmur, S. A. Hudson, E. Lalinde, P. M. Maitlis, M. P. McDonald, R. Orr, P. Stryring, A. S. Cherodian, R. M. Richarson, J. L. Feijoo, G. Ungar, *Mol. Cryst. Liq. Cryst.* **1991**, *206*, 76–92.

[58] D. W. Bruce, S. C. Davis, D. A. Dunmar, S. A. Hudson, P. M. Maitlis, P. Styring, *Mol. Cryst. Liq. Cryst.* **1992**, *215*, 1–11.

[59] M. J. Baena, P. Espinet. M. C. Lequerica, A. M. Levelut, *J. Am. Chem. Soc.* **1992**, *114*, 4182–4185.

[60] C. Soulie, P. Bassoul, J. Simon, *J. Chem. Soc., Chem. Commun.* **1993**, 115–116.

[61] G. Staufer, *Diss. Bayreuth* **1990**.

[62] M. Schellhorn, *Diss. Bayreuth* **1990**.

[63] G. Staufer, M. Schellhorn, G. Lattermann, *Liq. Cryst.* **1995**, *18*, 519–527.

[64] U. Stebani, G. Lattermann, *15th Int. LC Conference* **1994**, Abstracts Vol. I, 18.

[65] U. Beginn, *Diss. München* **1994**.

[66] U. Beginn, G. Lattermann, *Makromol. Rep. A* **1995**, *32*, 985–997.

[67] J. Malthete, H. T. Nguyen, C. Destrade, *Liq. Cryst.* **1993**, *13*, 171–187.

[68] H. T. Nguyen, G. Sigaud, M. F. Achard, F. Hardouin, R. J. Twieg, K. Betterton, *Liq. Cryst.* **1991**, *10*, 389–396.

[69] V. Luzzati, R. Vargas, P. Mariani, A. Gulik, H. Delacroix, *J. Mol. Biol.* **1993**, *229*, 540–551.

Added in proof:
Some recent papers published after the deadline:

(a) V. S. K. Balagurusamy, G. Ungar, V. Percec, G. Johansson, *J. Am. Soc.* **1997**, *119*, 1539–1555. Rational design of the first spherical dendrimers self-organised in a novel thermotropic cubic liquid-crystalline phase and the determination of their shape by X-ray analysis.

(b) K. Borisch, S. Diele, P. Göring, C. Tschierske, *Liq. Cryst.* **1997**, *22*, 427–443. Amphiphilic *N*-benzoyl-1-amino-1-deoxy-D-glucitol derivatives forming lamellar, columnar and different types of cubic mesophases.

(c) B. Donnio, D. W. Bruce, H. Delacroix, T. Gulik-Krzywicki, *Liq. Cryst.* **1997**, *23*, 147–153. Freeze-fracture electron microscopy of thermotropic cubic and columnar mesophases.

(d) A. M. Levelut, B. Donnio, D. W. Bruce, *Liq. Cryst.* **1997**, *22*, 753–756. Characterisation by X-ray diffraction of the S4-phase of some silver(I)complexes of alkoxystilbazoles.

Chapter XIV
Metal-containing Liquid Crystals

Anne Marie Giroud-Godquin

1 Introduction

Metal-containing liquid crystals, also known as *Metallomesogens*, combine the physical properties exhibited by liquid crystals with the variety and range of metal-based coordination chemistry due to the presence of one or more metals. Thermotropic metallomesogens have been made incorporating many metals representative of s, p, d and even f block elements.

Both rod-like and disc-like metallomesogens are known, and examples of all the main mesophase types have been observed. Many different varieties of ligand can be used: monodentate, bidentate, and polydentate. As with organic mesogens, molecular shape and intermolecular forces play an important role. By introducing a metal atom into organic liquid crystals the molecular shape can be altered. The presence of one or more metals opens many exciting possibilities: geometries and hence shapes not easily found in organic chemistry can result from coordinating a metal; many of the d- and f-block transition metal complexes are in an oxidation state which gives colored compounds, and many of them have unpaired electrons and exhibit paramagnetism [1]. Profound effects arise from the large and polarizable concentration of electron density that every metal atom possesses, since the molecular polarizability is a key factor in determining whether a molecule will form a liquid crystal. As a result of this, the physical properties (e.g. birefringence) of the compound will be enhanced and many interesting possibilities regarding optical, magnetic, and electrical properties may result. This chapter will give just an overview of the area of metallomesogens with selected references up to early 1995, the deadline asked by the editors. For more details the readers are referred to six reviews [2–7].

2 Early Work

The first thermotropic liquid crystals containing a metal were the alkali metal salts of carboxylic acids reported by Vorländer in 1910 [8a]. These alkali–metal carboxylates, $R(CH)_n COONa$, formed on heating classical lamellar phases. In 1923 Vorländer also reported some diarylmercury complexes, **1**, showing smectic phases [8b].

1 Vorländer's diarylmercury complexes

Between 1959 and 1961 Skoulios et al. [9] synthesized alkali and alkaline earth salts of carboxylic acids and showed them to possess a large variety of organized mesophases: lamellar, ribbonlike and cylindrical.

In 1976 Malthête and Billard [10] reported ferrocene derivatives, **2**, which were the first well characterized organotransition metallomesogens.

2   Monosubstituted ferrocene derivatives

In 1977, Giroud and Mueller-Westerhoff [11] reported some mesomorphic nickel and platinum dithiolenes, **3**. This work is widely accepted as being the beginning of systematic research into metallomesogens. Since then, many laboratories have entered the field and many new types of metallomesogen have been synthesized.

3   Dithiolene complexes with two chains

3   Metallomesogens with Monodentate Ligands

3.1   Organonitrile Ligands

The alkyl and alkoxy cyanobiphenyls (*n*-CB and *n*-OCB) derivatives synthesized by Gray and coworkers and the bicyclohexyl derivatives (PCH-*n* and CCH-*n*), **4**, bind rapidly to a variety of metals. The Sheffield Liquid Crystal group [12] used these compounds for the synthesis of metallomesogens with a linear geometry: M = Pd, Pt, Rh.

In each series the predominant phase is nematic, although with longer chain length SmA and SmC phases were found. The trans geometry of the complexes [MCl$_2$(*n*-CB)$_2$],

4   Nitrile ligands

5, was established by single crystal X-ray diffraction.

5   Cyanobiphenyl complexes

In addition Pd(II) complexes of related ligands PCH and CCH were made. All complexes with PCH ligands show monotropic nematic phase while with the CCH ligands, CCH-3 (*n* = 3) is not liquid crystal, CCH-5 (*n* = 5) is monotropic nematic and CCH-7 (*n* = 7) is enantiotropic nematic. Orientational order of the dimer [PtCl$_2$(5-CB)$_2$] formed by linking of the two cyanobiphenyl via platinum dichloride bridge, dissolved in a nematic solvent was studied using deuterium NMR spectroscopy [13].

The main problem of these complexes is the lability of the metal–ligand bond for palladium compounds.

3.2   *n*-Alkoxystilbazole Ligands

The 4′-alkyloxy-4 stilbazoles (*n*-OST) **6**, are 4-substituted pyridines and have been widely used by the Sheffield liquid crystal group as ligands for metallomesogen formation.

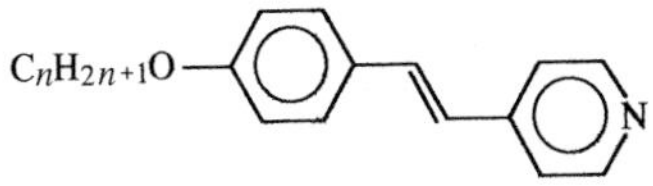

6 4-Aloxystilbazole

3.2.1 Distilbazole Ligands

3.2.1.1 Palladium and Platinum

Complexes of Pd(II) and Pt(II), **7**, analogous to those described for the cyanobiphenyls were synthesized, and it was found that few of these complexes showed mesomorphic behavior except for the longest chain which show at elevated temperature ($>200\,^{\circ}$C) a SmC phase. Replacing the two chloride ligands in the Pd(II) complexes by aliphatic carboxylic acids lowers melting and clearing temperatures [14a]. Trisubstituted ligands lead to complexes showing columnar organisation [14b].

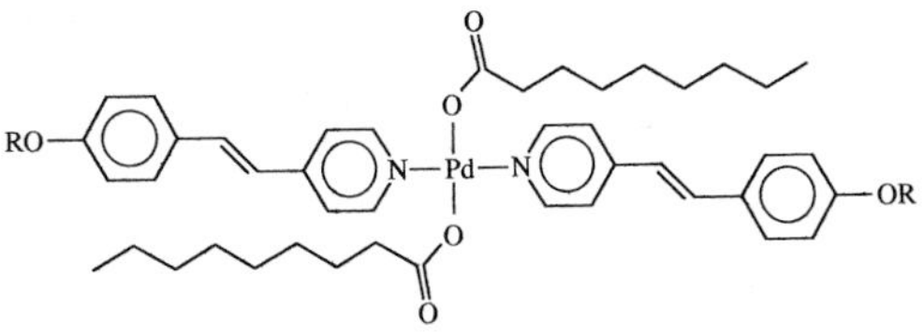

7 Palladium carboxylate complexes

3.2.1.2 Silver

Reaction of the stilbazoles with silver salts, AgY, led to a quite unexpected series of linear mesogenic materials: the ionic bis-ligand silver salts, **8**, [Ag(n-OST)$_2$]Y with Y = BF$_4^{\ominus}$, NO$_3^{\ominus}$, C$_{12}$H$_{25}^{\ominus}$, SO$_4^{\ominus}$, CF$_3$OSO$_3^{\ominus}$. The long chain homologues of each series

show SmA and SmC mesophases, while for shorter chain length the nitrate salts show only a SmA phase and the triflate salts only a nematic phase [15].

The dodecyl sulfate (Y = C$_{12}$H$_{25}$SO$_4^{\ominus}$) salts formed mesophases and showed unusual behavior. A rich polymorphism is found: N, SmA, SmC and cubic phases [20]. The single crystal X-ray structure of the methoxy homologue has been determined [16]. This rich and varied polymorphism in these ionic systems may have some electrochemical applications. This mesomorphism is disturbed by lateral fluorination of the ligand: 3-fluorinated isomers promote smectic phases and suppress the cubic phase while the 2-fluorinated isomers promote both the nematic and the cubic mesophase [17].

3.2.1.3 Iridium

Iridium complexes of dimeric stilbazole, **9**, have been synthesized recently. These materials show a very large odd–even effect in their melting behavior, and are called metal-containing 'Siamese Twin' liquid crystals [18].

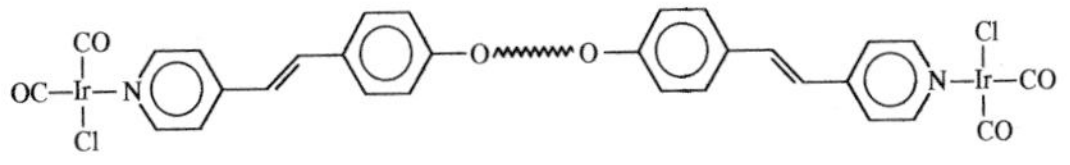

9 'Siamese Twin' complexes

3.2.2 Monostilbazole Ligands

A way to lower transition temperatures is to incorporate some asymmetry by replacement of one of the stilbazoles with an alkene [19].

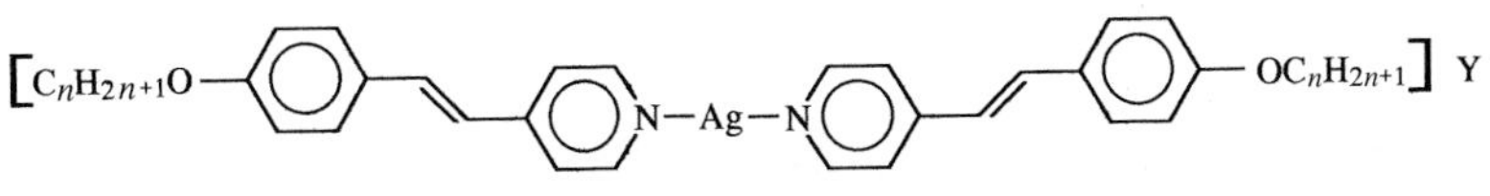

8 Silver complexes

3.2.2.1   Platinum

In the Pt complexes, **10**, the chain lengths n and m were systematically varied ($n = 3 - 12$; $m = 0 - 8$).

SmA phases were seen for most derivatives. It was also noted that complexes with $n + m < 8$ were not mesomorphic; those with $8 < n + m < 13$ showed monotropic mesomorphism, while those with $n + m > 13$ showed enantiotropic SmA phase [14].

10   Platinum alkene complexes

3.2.2.2   Rhodium and Iridium

A series of unsymmetric rhodium and iridium stilbazoles, **11**, was prepared.

11   Stilbazole complexes of rhodium and iridium

The rhodium complexes were yellow/orange in the solid state characteristic of mononuclear Rh(I), while the burgundy iridium complexes became yellow both in solution and on melting. Both series of materials showed very similar mesomorphism: nematic phases were found at short chain lengths giving way to the SmA phase for higher homologues for the two metals [20].

Similar 2- and 3-fluorinated derivatives have been synthesized to lower the transition temperatures. The 2-fluorinated iridium complexes show monotropic N and SmA phases while the 3-fluorinated compounds are not mesomorphic [21].

3.2.2.3   Tungsten

Monosubstituted and disubstituted tungsten carbonyl complexes, **12**, are intensely col-

ored and the disubstituted one shows an unusual behavior on cooling giving an unidentified liquid crystal mesophase [22].

12   Tungsten carbonyl complexes

3.3   Other Pyridine Ligands

3.3.1   Rhodium and Iridium

Serrano et al. [23] have used other substituted pyridines as ligands for metallomesogen formation. Thus the cis-iridium complexes, **13**, with short chains formed monotropic nematic phases, while those with longer chains led to SmA phases. The related rhodium complexes are also mesogenic. This is an example of a mesogenic complex formed with a nonmesogenic ligand.

13   Alkoxypyridylbenzylidene aniline complexes

3.3.2   Silver

The silver complexes, **14**, with the same iminopyridine ligand (*n*-OIP) and with

$$\left[C_nH_{2n+1}O - \bigcirc - N \diagdown \diagdown \bigcirc \diagup N - Ag - N \bigcirc \diagdown \diagup N - \bigcirc - OC_nH_{2n+1}\right] Y$$

$$\left[C_nH_{2n+1}O - \bigcirc - O \diagdown \bigcirc \diagup N - Ag - N \bigcirc \diagup O - \bigcirc - OC_nH_{2n+1}\right] Y$$

14   Silver iminopyridine and carboxylatepyridine complexes

the pyridine carboxylate ligand (*n*-OCP) showed mesogenic behavior: $Y = BF_4, CF_3, NO_3, PF_6$.

Only the salts ($Y = BF_4$, $n = 2$) are nematic: all the others show SmA and SmC mesophases. The ester salts show wider mesophase range and greater smectic polymorphism than the imino salts [24].

sessed a negative diamagnetic anisotropy and can be aligned in a magnetic field [26].

This work with polymeric acetylides was extended to the study of related low molar mass systems and the platinum(II) complexes, **16**, were found to exhibit thermotropic N and SmA phases [27].

16   Acetylide platinum complexes

3.4   Acetylide Ligands

The very unusual organometallic liquid crystals, **15**, [M, M′ = Pd, Pt or Ni; M = M′ or M ≠ M′] were made by Takahashi et al. in 1978 [25], by a copper chloride-triethylamine coupling of the appropriate metal halide with an alkyne. These materials do not show thermotropic properties, however they are lyotropic nematic in trichloroethylene solution.

Bruce et al. [28] looked at the effect of changing the acetylide ligand to make it longer and more structurally anisotropic, **17**. They found that platinum complexes show enantiotropic nematic phase, and that increasing the size of the phosphine decreases the temperatures of transition.

17   Polymeric acetylide complexes

15   Takahashi's acetylide complexes

The physical properties of these 1,4-butadiyne metal polymers were studied using both homometallic polymers (M = M′) and heterometallic polymers (M = Pt, M′ = Pd or Ni), which showed that all of them pos-

Changing the direction of the ester groups strongly influences the thermal stability of the mesophases of the complexes [29]. The mono substituted complexes were not mesomorphic.

3.5   Isonitrile Ligands

Gold–isonitrile complexes, **18**, as a new family of metallomesogens, are prepared and even nonmesomorphic isonitrile ligands have been found to form N and SmA liquid crystals when coordinated to an Au(I)Cl moiety [30 a].

$C_nH_{2n+1}O$—⬡—O—(C=O)—⬡—$NC \rightarrow AuCl$, OC_mH_{2m+1}

18   Gold–isonitrile complexes

The same group [30 b] prepared platinum and palladium complexes. The coordination of the isonitrile to a metal increases the thermal stability of the ligand. Both Pt and Pd complexes show mesomorphic properties. The liquid crystalline complex $[PtI_2(CNC_6H_4C_6H_4OC_7H_{15})_2]$ has been crystallographically characterized.

Compounds where the gold atom links an alkynyl and an isonitrile, **19**, give rise consistently to SmA mesophases [31].

X—⬡—NC—Au—≡—⬡—C_mH_{2m+1}

$X = H, OC_nH_{2n+1}$

19   Gold–isonitrile–alkynyl complexes

4   Metallomesogens with Bidentate Ligands

4.1   Carboxylate Ligands

Metal soaps have long been known to have mesogenic properties. Thus, Vorländer [8 a] in 1910 reported the existence of lamellar phases in the anhydrous salts of alkali carb-

oxylates, and in 1938 Lawrence [32] noted that copper stearate melted to a plastic fluid before clearing.

4.1.1   Alkali and Alkaline Earth Carboxylates

Structural data, indicating the existence of columnar and even disc-like mesophases for the anhydrous alkali, alkaline earth, and cadmium salts of long-chain fatty acids were first reported by Skoulios et al. [9]. For example the potassium salts can be lamellar, ribbon-like or disc-like in a three dimensional lattice [33], see Fig. 1.

4.1.2   Lead, Thallium, and Mercury Carboxylates

SmC phases have been reported for the lead(II) carboxylates, $[Pb(C_nH_{2n+1}COO)_2]$ for $n = 7, 9, 11, 15, 17$ [34]. The compounds were investigated further by various methods, particularly X-ray diffraction, and were found to transform from a lamellar crystal-

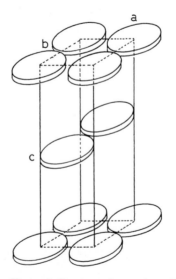

Figure 1. Structure of potassium alkanoate.

line to another lamellar highly viscous phase and then to melt to a SmC phase [35]. Later, Attard et al. [36] describes the synthesis, thermal properties, and response to UV radiation of the lead(II) salts of alkadiyonic acids $[(CH_3(CH_2)_m-C\equiv C-C\equiv C-(CH_2)_nCOO)_2Pb]$. These compounds were found to exhibit mesomorphism over a wide temperature range. They also appear to form glassy mesophases at lower temperature. All the materials could be polymerized by low intensity UV radiation.

The mercury carboxylates $[Hg(C_nH_{2n+1}COO)_2]$ for $n = 7-17$, are not mesogenic, while thallium carboxylates $[Tl(C_nH_{2n+1}COO)_2]$ gave stable lamellar phases with branched alkyl chains [37].

4.1.3 Dinuclear Copper Carboxylates

The liquid crystalline phase of dinuclear copper stearate has been known since 1938, and was reinvestigated in 1964 [38]; but it was not until 1984 [39] that this mesophase was definitely characterized by Giroud-Godquin et al. as columnar discotic by X-ray diffraction. This was the first example of a thermotropic hexagonal discotic meso-phase bearing only four alkyl chains. All the compounds $[Cu_2(\mu-O_2CC_nH_{2n+1})_4]$ for $n = 4-24$, show the same structure, see Fig. 2. The X-ray shows that the complexes crystallize in a lamellar structure in the solid state and above $120\,°C$ exist as a columnar two dimensional hexagonal structure in which polar groups are surrounded by disordered alkyl chains [40].

The phase transition in these materials has been investigated by a variety of techniques: X-ray single crystal diffraction shows them to be dinuclear with each copper surrounded by five oxygen atoms, one belonging to the neighboring dimer [41].

Cu-K EXAFS (extended X-ray absorption fine structure) confirmed this and also showed that the dinuclear arrangement of the core was preserved on passing from the crystal into the mesophase [42].

Magnetic susceptibility measurements show a small, sharp decrease in the magnetic moment at the transition to the mesophase, implying that there was some change which was probably attributable to a modification of the relative disposition of the dimers within the columns [43].

Isotopic labeling has facilitated band assignment in the infrared spectra [44a] and dynamic behavior of the alkyl chains studied by quasielastic neutron scattering [44b].

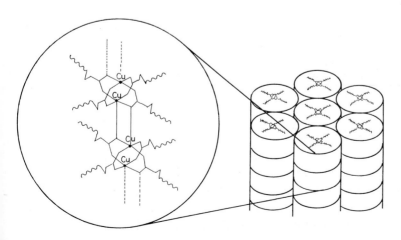

Figure 2. Structure of copper alkanoate in the discotic hexagonal mesophase.

Copper laurate can be processed into oriented fibers by melt spinning of its thermotropic columnar mesophase, X-ray diffraction and electron microscopy revealed a high degree of orientation of the spun fibers both in the crystalline and in the columnar states [45a]. Neutron experiments have been performed on these fibers to study the motion of the hydrogen atoms of the chain [45b].

The first example of a fully characterized metallomesogen giving an authentic columnar mesophase near room temperature in a pure state, was described by Maldivi and co-workers. These mesogens are obtained by simple insertion of carbon–carbon double bonds in the carboxylate ligands [46].

In contrast to the case of straight chain salts, complexation of the copper(II) branched chain carboxylate with a difunctional ligand as pyrazine or 4,4'-dipyridyl result in a liquid crystalline material whose mesophase structure has not been established [47].

4.1.4 Dinuclear Rhodium, Ruthenium, and Molybdenum Carboxylates

The rhodium carboxylates $[Rh_2(\mu\text{-}O_2CC_nH_{2n+1})_4]$ for $n = 4 - 24$ behave very similarly to the copper complexes and form, on heating, columnar mesophases. Rh-K-edge EXAFS measurements showed no change in the core structure passing from the solid state to the mesophase [48].

The complexes $Rh_2[(\mu\text{-}O_2CC_6H_4\text{-}4\text{-}OC_nH_{2n+1})_4]$ for $n = 8 - 14$, were investigated by optical microscopy, DSC and X-ray diffraction. The complexes display a columnar mesophase of rectangular symmetry. The eight chain complex shows two columnar phases, one rectangular at room temperature and a hexagonal one above 138 °C [49].

Mixtures of rhodium eicosanoate or copper dodecanoate with solvents such as toluene, decahydronaphthalene, and (+) camphene have been studied. The mesophases found at high temperature turns from columnar to nematic when the weight fraction of the solvent is increased beyond a value of about 50% [50].

A wide series of diruthenium (II, II) tetracarboxylates of general formula $[Ru_2(\mu\text{-}O_2CC_nH_{2n+1})_4]$ has been made by the Grenoble group; the carboxylate substituents include linear alkyl chains ($n = 4 - 19$), unsaturated chains, and fluorinated chains [51].

The mesomorphic behavior of these compounds was investigated by DSC, optical microscopy and X-ray diffraction. A thermotropic columnar mesophase was observed for all complexes except the perfluoro-compounds. Unsaturation in the peripheral chains depresses transition near room temperature. The magnetic susceptibility was measured. A sharp change in the magnetic moment at the solid–mesophase transition is ascribed to a core distortion similar to Cu(II) and Rh(II) complexes.

Important work was carried out by Cukiernik et al. of the Grenoble group on mixed valence ruthenium carboxylates of general formula $[Ru_2(O_2CC_nH_{2n+1})_4]X^\ominus$, with X = anion [52]. The anion has a strong influence on the mesomorphic behavior: the chloro-complexes (X = Cl$^\ominus$) are not mesomorphic. With a carboxylate (X = R'COO$^\ominus$) or an alkylsulfate (X = C$_{12}$H$_{25}$OSO$_3^\ominus$) as the anion, the compounds show a hexagonal columnar mesophase on heating.

To favor the interdimer interactions an adduct pyrazine has been intercalated between two dimers. The compound, **20**, is supposed to have a polymeric structure with the pyrazine axially coordinated to the two dimers [53].

20 Diruthenium tetracarboxylate complexes with an adduct pyrazine

The quadruply bonded molybdenum carboxylates $[Mo_2(O_2C_nH_{2n+1})_4]$ present a similar behavior to that observed for $M = Cu$, Rh, Ru. When a side chain is branched on the complex the transition temperature is lowered to about 37 °C. Substitution of fluorine increases the temperatures but an enantiotropic hexagonal discotic mesophase is still found [54].

4.2 β-Diketonate Ligands

β-Diketonates form flat bis-ligand complexes with a variety of metals (including Ni, Pd and Cu) that can take up square planar geometries. When appropriate substituents are present, these complexes can give rise to nematic, discotic and occasionally to lamellar phases.

The first β-diketonate, **21**, examined for mesophase behavior was the palladium complex of Bulkin et al. [55]. They suggested that it may be mesomorphic but have been unable to confirm this.

21 Bulkin's β-diketonate complexes

4.2.1 Copper Complexes with Two or Four Peripheral Chains

The mesogenic copper (II) β-diketonate, **22**, first described by Giroud-Godquin et al. [56] in 1981 has a square planar geometry about the metal and a d^9 configuration (one unpaired electron). Both the symmetrically and the unsymmetrically substituted complexes ($m=n$ or $m\neq n$) exhibit very organized discotic mesophase. Later work and particularly NMR investigation and X-ray diffraction suggested that the discotic phase was, in fact, a disordered lamellar crystal [57].

Ohta et al. [58] reported later the related complexes ($R = R' = C_8H_{17}$, $M = Cu$; $R = R' = OC_8H_{17}$, $M = Cu$) comiscible with Giroud-Godquin's compound and therefore of the same type.

22 Giroud-Godquin and Ohta's β-diketonate complexes

$$R = C_mH_{2m+1}, \quad R' = C_nH_{2n+1}$$
$$R = C_mH_{2m+1}O, \quad R' = C_nH_{2n+1}O$$
$$R = C_mH_{2m+1}, \quad R' = C_nH_{2n+1}O$$

The structures of the single crystal of two complexes of this series ($M = Cu$, Pd) have been established. According to X-ray diffraction data, the copper atom has a square planar configuration and the packing of the molecules in the crystal is of the layered-columnar type [59].

An interesting development in this field was made by Chandrasekhar et al. [60] in

1986 with the synthesis of copper complexes, **23**, which were monotropic nematic and paramagnetic. These complexes incorporate features of both rod-like and disc-like molecules and are claimed to be the first example of biaxial nematic metallomesogen [61]. Unfortunately no other workers have been able to reproduce this observation of biaxiality. The effect of different terminal substituents ($R = C_nH_{2n+1}$ or OC_nH_{2n+1}; $R' = CN, OCH_3, OC_2H_5, C_2H_5, C_3H_7, C_4H_9$; $M = Cu$ or Pd) on the mesomorphic properties were studied later: mesophases of the SmC, SmA and N types were observed [62].

23   Chandrasekhar's β-diketonate complexes

Mühlberger and Haase [63] synthesized the saturated derivatives, **24**, which also show a monotropic nematic phase. X-ray studies showed that the diketonate ring was planar and that identical substituents were trans to each other.

24   Mühlberger and Haase's β-diketonate complexes

Ohta [64] synthesized the biphenyl equivalent ($n = 16$, $m = 1$, $M = Cu$) and found that

this complex curiously shows a dimer discotic rectangular ordered columnar mesophase.

A more systematic investigation of such structures was performed by Thomson et al. [65] who found that complexes, **25**, with the rod-like structure **A** showed *monotropic* nematic mesophase of discotic type while those with the more disc-like molecule **B** show *enantiotropic* phases of calamitic type.

25   Thomson et al.'s β-diketonate complexes
A   Ring X = phenyl; R = $C_{10}H_{21}$, R' = CH_3, H
Ring X = cyclohexyl; R' = CH_3, C_7H_{15}

B   Ring Y = phenyl; R'' = $C_{10}H_{21}CH_3$, C_2H_5, F
Ring Y = cyclohexyl; R'' = OCH_3

4.2.2   Copper Complexes with Eight Peripheral Chains

Accounting for the empirical fact that at least six substituents are needed to get columnar mesophases, Giroud-Godquin et al. [66] reported synthesis and properties of new copper diketonate complexes, **26**, substituted on each phenyl, in 3,4 positions, by two alkoxy chains giving a resultant complex with eight chains. These complexes were shown to have a hexagonal columnar mesophase, D_h, by miscibility studies and X-ray diffraction.

The corresponding nickel complex is not mesogenic neither is the complex in which the phenyl is substituted in the 3,5 positions. Similar complexes were studied by Ohta [67] who found discotic mesophases for alkoxy ($R = OC_8H_{17}$) chains and no mesophase for alkyl tails ($R = C_8H_{17}$).

26 Giroud-Godquin's eight chained β-diketonate complexes

4.2.3 Malondialdehyde Complexes

Other interesting compounds are copper and nickel complexes of (p-alkoxyphenyl)-malondialdehydes, **27**. The synthesis, characterization, thermal behavior, X-ray diffraction studies and magnetic properties of a number of these complexes were performed by Haase et al. [68]. It was found that nickel complexes show liquid crystalline SmA phases over a broad temperature range with low melting points (85 °C), while copper complexes show nonidentified mesomorphic behavior. This behavior is different from that observed for $n = 5$ copper complex reported by Blake et al. [69] who observed a nematic phase. Models for the molecular

arrangement in the crystalline and liquid crystalline phases are discussed based on magnetic data [70].

27 Malondialdehyde complexes

4.2.4 Dicopper Complexes

Dicopper complexes, **28**, have been synthesized and characterized by Lai et al. [71]. These complexes melt to give birefringent fluid phases with columnar superstructures.

28 Dicopper β-diketonate complexes

4.2.5 Polymers

Polymers containing β-diketonate metal complexes of copper(II), nickel(II), and cobalt(III) were synthesized and examined by DSC, polarizing microscopy and X-ray diffraction. The copper complex exhibits a SmA phase while both the nickel and the cobalt complexes do not work as mesogens [72].

4.2.6 Other Metal Complexes

Synthesis, mesomorphic properties and crystal structures of copper(II), palladium(II), and nickel(II) complexes have been performed. Despite the unambiguous structural isomorphism only the palladium and the copper complexes are mesogenic. The chains are fully extended in an all trans conformation [73].

The first octahedral metallomesogen was an iron(III) tris-diketonate, **29**, which showed a birefringent structure and endothermic transitions but the mesophase has not been defined [74].

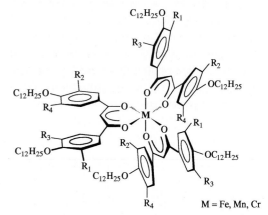

31   Iron, manganese and chromium
tris-β-diketonate complexes

Iron(III), manganese(III) and chromium(III), **31**, have been also investigated and the bargraph shown in Fig. 3 illustrates the phase behavior of these octahedral metallomesogens [77].

29   Iron tris-β-diketonate complexes

It has been shown that a β-diketonate liquid crystal containing palladium(II) and oxovanadium(IV), analogous to known copper complexes (which display discotic lamellar and columnar mesophases), has been prepared and characterized; palladium complexes are mesomorphic but their identification is difficult due to their decomposition in the isotropic phase; the complex $n=10$ seems to present a N_d phase on the basis of optical microscopy. The vanadium complexes are enantiotropic discotic [75], while the vanadium complex, **30**, presented by Styring et al. [76] ($n=10$), exhibited a short monotropic columnar discotic phase.

30   Vanadium β-diketonate complexes

The effect of the number of side chains on the mesophase stability has been investigated by the same authors. The results indicate that complexes with four side chains are not mesomorphic (in disagreement with-

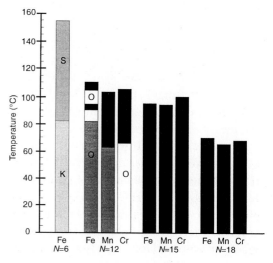

Figure 3. Bar graph showing the phase behavior of octahedral metallomesogens.

in Styring's work), while those with eight chains can display limited mesomorphism and complexes with ten side chains exhibit a discotic hexagonal disordered mesophase over a wide range of temperature [78].

Several diketonate thallium(I) complexes with a different number of alkoxy groups on the aromatic ring have been synthesized [79]. The mesogenic behavior of the thallium(I) and of the copper(II) derivatives are qualitatively similar. The authors propose a dimer structure.

4.3 Glycoximate Ligands

Ohta et al. [80] reported the discotic mesomorphism, thermochromism and solvatochromism of bis[1,2-bis(3,4-dialkoxyphenyl) ethanedione dioximato]metal complexes, **32** (abbreviated to [M{(C$_n$O)$_4$dpg}$_2$]), and of bis[1,2-bis(3,4-dialkylphenyl) ethanedione dioximato] metal complexes (abbreviated to [M{(C$_n$)$_4$dpg}$_2$] with M = Pd, Ni and Pt.

32   Glyoxamate platinum complexes

All the complexes showed a discotic hexagonal disordered columnar mesophase (D$_{hd}$). The mesophases were identified by X-ray diffraction. Platinum complexes exhibit a very clear thermochromism: the color of the liquid crystalline discotic mesophase turns from green to red with increasing temperature from room temperature.

4.4 Sulfur Containing Ligands

4.4.1 Dithiolene

The first systematic study of d-block metallomesogens were carried out by Giroud and Mueller-Westerhoff [11] and marked, in 1977, the beginning of interest in metal-containing systems. They were dithiolene complexes of nickel, palladium and platinum with two p-alkyl phenyl substituents (see structure **3**). Nickel and platinum complexes showed nematic and smectic phases, depending on the chain length. Palladium analogs do not possess any mesomorphic properties [81].

Analogous di- and tetrasubstituted nickel dithiolenes prepared by Strzelecka et al. [82] were originally described as discotic mesogens. Later, X-ray data showed that these compounds were not mesomorphic [83]. However, the related nickel dithiolenes, **33**, with four 3,4-bis(dodecyloxy)-phenyl substituents was shown by X-ray diffraction to give rise to a hexagonal disordered columnar mesophase D$_{hd}$ between 84° and 112°C for R = C$_{12}$H$_{25}$ by Ohta et al. [84].

33   Dithiolene complexes with four chains

4.4.2 Dithiobenzoate

Most of the work on dithiocarboxylates have been centered on metal complexes of 4-alkoxydithiobenzoic acids (abbreviated to n-odtbH) **34**.

$$C_nH_{2n+1}O—\langle\bigcirc\rangle—C(=S)SH$$

34   4-alkoxydithiobenzoic acid

Many complexes [M(n-odtb)2] have been synthesized for M = Ni, Pd, Zn, Hg, Ag, and some planar gold(III) complexes showing SmA phases [AuX2(n-odtb)2] with X = Cl, Br, Me [85].

4.4.2.1   Nickel and Palladium

Studies by the Sheffield group on the bis-(4-alkyloxy-dithiobenzoato)complexes, **35**, showed that the Ni(II) and the Pd(II) complexes were mesogenic. Palladium complexes were shown by X-ray studies to be composed of monomeric species and to have a square planar geometry about the metal. These complexes show a SmC phase for long chains and a N phase for shorter chains [85]. Complexes with branched chains have a melting point lower than those with straight chains [86].

$$C_nH_{2n+1}O—\langle\bigcirc\rangle—C(S)(S)M(S)(S)C—\langle\bigcirc\rangle—OC_nH_{2n+1}$$

35   Alkoxydithiobenzoate complexes

The palladium complexes were further studied both in the solid state and in the mesophase using EXAFS technique [87]. The data confirmed the known crystal structure. The nickel complexes have the same mesogenic properties than the related palladium. The major difference is that at high temperature, the blue bis-(dithiobenzoates) rearranged to form the red trithiobenzoate complexes, **36**, which are also mesogenic [88].

$$C_nH_{2n+1}O—\langle\bigcirc\rangle—C(S)(S)M(S)(S)C(S)—\langle\bigcirc\rangle—OC_nH_{2n+1}$$

36   Trithiobenzoate complexes

Lateral fluorination on these compounds promotes the SmC phase [89].

4.2.2.2   Zinc

X-ray diffraction studies on alkoxy-dithiobenzoato zinc complexes [Zn(n-odtb)$_2$] showed that in the solid state the material is a dimer containing an eight-membered Zn–S–C–S–Zn–S–C–S ring; the dimer held together in the nematic phase and disappeared in solution where is found a monomeric structure [85]. This work is confirmed by EXAFS studies [90].

Later, 4-substituted-piperazine-1-dithiocarboxylate complexes, **37**, have been synthesized by Hoshino-Miyajima [91] with different metals (M = Ni, Pd, Cu, Zn) and different chain lengths. They all show N and SmC mesophases at high temperature.

$$Y—N(\ )N—\langle C(S)(S)M(S)(S)C\rangle—N(\ )N—Y$$

37   Piperazine dithiobenzoate complexes

M = Pd, Ni, Cu, Zn

Y = Phenyl, C$_6$H$_4$OC$_n$H$_{2n+1}$, C$_n$H$_{2n+1}$

4.4.2.3   Silver

Primary silver thiolate compounds [AgSC$_n$H$_{2n+1}$] (n = 4, 6, 8, 10, 12, 16, 18), behave as thermotropic liquid crystals. On heating they display successively lamellar (SmA), cubic, and micellar mesophases. X-ray studies show that the micellar phase is an hexagonal columnar mesophase [92].

4.4.3   Other Dithio Ligands

Other dithio systems are the alkyldithioacetate and alkylxanthates of nickel described by Ohta [93]. Dimeric tetrakis (n-alkyldithiocarboxylato) dinickel complexes (R = C$_5$H$_{11}$ to C$_{12}$H$_{25}$), **38**, show monotropic lamellar mesophases, while the related monomeric xanthate [Ni(ROCS$_2$)$_2$] (R=CH$_3$ to C$_{12}$H$_{25}$), **39**, exhibit complex double- and triple-melting behavior.

38   Dimeric dithiocarboxylate complexes

39   Alkylxanthate complexes

4.5   N–O Donor Sets: Salicylaldimine Ligands

Schiff-bases derived from substituted salicylaldehydes, **40**, are very versatile ligands which form (N–O) chelates with many metals: copper, nickel, palladium, vanadium and iron.

40   Salicylaldehyde

4.5.1   Copper, Nickel, and Palladium

Most of the paramagnetic salicylaldimine copper(II) complexes, first reported by Ovchinnikov et al. in 1984, showed smectic mesophases [94]. However, nematic behavior was observed by Galyametdinov et al. for **41** [95]. They were the first examples of materials with a paramagnetic nematic phase.

41   Ovchinnikov's salicylaldimate complexes

R = C$_m$H$_{2m+1}$, F, O$_2$CC$_6$H$_4$OC$_{12}$H$_{25}$
R' = O$_2$CC$_6$H$_4$OC$_{12}$H$_{25}$

All the complexes show both smectic and nematic phases. This was confirmed by X-ray diffraction studies. Following this work, the study of mesomorphic salicylaldimate complexes became very popular [96].

Marcos et al. [97] and Hoshino et al. [98] examined at the same time several series of copper and nickel complexes, **42**. Both Cu(II) and Ni(II) complexes derived from the n-alkylamine (R=C$_n$H$_{2n+1}$) show mainly nematic mesophase, while those derived from arylamine (R=4-C$_6$H$_4$OC$_n$H$_{2n+1}$) show largely smectic mesophases.

42   Marcos and Hoshino's salicylaldimate complexes

Then, Sirigu [99] and later Courtieu [100] synthesized a series of copper, nickel and palladium complexes, **43**, with high-temperature smectic and nematic behavior.

43   Sirigu and Courtieu's salicylaldimate complexes

Courtieu [101] lowered the melting and the clearing temperatures by halogenation of the complexes (X = Cl, Br), **44**.

44   Halogenated salicylaldimate complexes

Sadashiva and coworkers [102] investigated a series of salicylaldimate complexes of nickel, copper and palladium with R = RO-PH-CH$_2$O, **45**, showing SmA and SmC phases.

45   Sadashiva's salicylaldimate complexes

The copper and the vanadium complexes have been studied by ESR [103].

Polar Schiff-base complexes, **46**, have been synthesized to study the relation between mesogenic behavior and the molecular structure. This family showed SmA and N phases [104]. The lateral polar groups X, Y, Z can be F, CF$_3$, CN, CH$_2$CN.

46   Polar salicylaldimate complexes

Chiral ferroelectric complexes, **47** and **48**, of palladium, copper and vanadium have

also been made by Marcos et al. [105, 106] and Ghedini et al. [107].

47   Marcos' chiral salicylaldimate complexes
M = Cu, Pd, VO
Z = –CH=CH–COO

48   Ghedini's chiral salicylaldimate complexes
(RCH$_n$)* = (–)-myrtalyl
= (–)-menthyl
= S(–)-β-citronellyl
= R(–)-2-octyl

Some of these materials display the potentially chiral ferroelectric SmC phase in addition to a SmA or cholesteric mesophase. Pyzuk [108] found for copper and nickel complexes some blue phase or novel type of amorphous phase between a tightly twisted chiral nematic phase and the isotropic liquid. These ferroelectric metallomesogens are interesting as they can be aligned parallel or perpendicular to a magnetic field, depending on the anisotropy of the magnetic susceptibility of the complex.

Other N-substituted salicylaldimines have been used as complexing ligands. All copper, nickel and vanadium complexes studied, **49**, show N and/or SmC phases depending on the chain length [109], and characterized by X-ray diffraction studies [110].

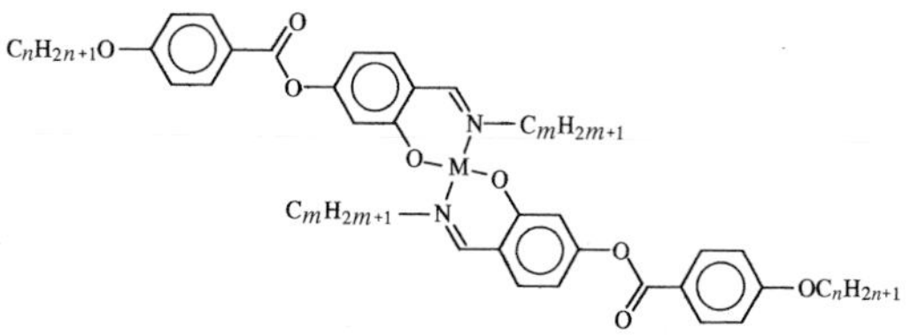

49   Marcos and Hoshino's N-substituted salicylaldimate complexes

4.5.2   Platinum, Vanadyl, and Iron

Of the complexes, **43** (M = Zn, Co, Ni, Cu, VO, Pd, Pt), only the copper, the vanadyl, the platinum and the palladium complexes showed smectic mesophases. Both platinum(II) and oxovanadium(IV) complexes containing only a chain on one biphenyl moiety (R′ = H) exhibit nematic phase [111–113].

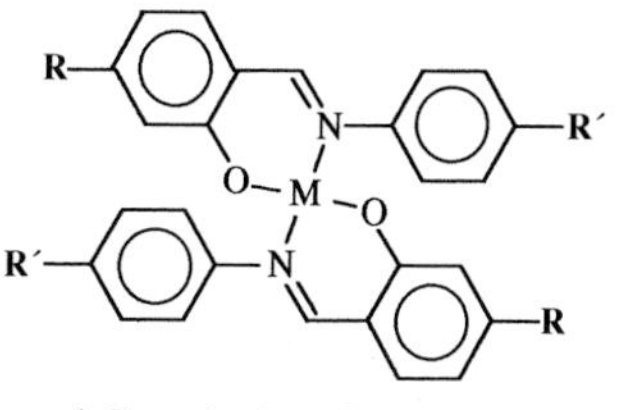

43   Sirigu and Courtieu's salicylaldimate complexes

$$R = C_nH_{2n+1}$$
$$R′ = Ph-C_mH_{2m+1}$$
$$R = OC_nH_{2n+1}$$
$$R′ = C_mH_{2m+1}$$

The mesogenic iron(III) complex reported by Galyametdinov is paramagnetic and formed SmA phases. A μ-oxo-bridged complex of Fe(III), **50**, has also been synthesized by the same authors and the magnetic behavior investigated. It is the first complex of Fe(III) exhibiting a nematic mesophase [114].

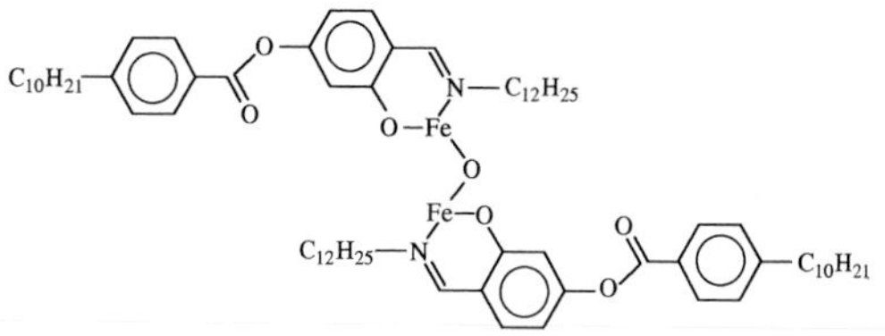

50   Galyametdinov's iron complexes

Note that many magnetic studies have been performed as well as ESR and X-ray diffraction studies [115, 116] on all the paramagnetic complexes.

4.5.3   Rhodium and Iridium

Mesomorphic Schiff-bases can also complex two square planar cis-[M(CO₂)]-moities with M = rhodium or iridium [117]. The complexes, **51**, show SmA behavior.

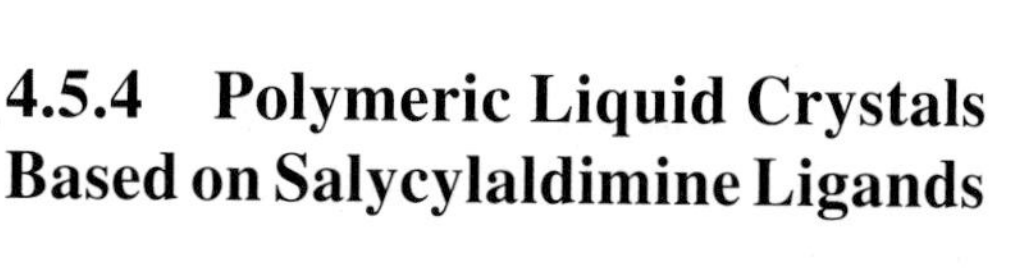

51   Rhodium and iridium salicylaldimate-carbonyl complexes

4.5.4   Polymeric Liquid Crystals Based on Salycylaldimine Ligands

Carfagna [118] reported the first polymeric metallomesogen with a Schiff-base ligand. The polymer showed a monotropic phase presumably smectic. Later, Caruso et al. [119] showed that a copper complex could give a polymer with nematic behavior. Different series of homopolymers containing paramagnetic units of copper(II) salicylaldiminates have been synthesized and studied by ESR [120]. Later, magnetic susceptibilities of copper and iron containing metallopolymers have been investigated [121].

There were two reports of the same bis(alkoxylsalicylidene) ethylenediamine complexes of Ni(II) and Cu(II), **52**. The material was reported as giving a SmA phase at elevated temperature [122] and a nonidentified mesophase [123]. Ohta later [124] found that the nickel complexes showed two smectic liquid crystalline phases, SmE and SmA. The SmE–SmA phase transition in the nickel complex is accompanied by a reversible stretching of the interlayer distances. This phenomenon can be ascribed to a dimer SmE–monomer SmA phase transition.

52 Bis(alkoxysalicylidene)ethylenediamine complexes

4.6   Cyclometalated Complexes

4.6.1   Azobenzene

Up to now, most of cyclometalated mesogens are palladium-containing. Only one mercury complex is known.

Ghedini et al. [125] in Calabria prepared the first cyclopalladated liquid crystal, **53**.

These dinuclear chlorobridged complexes formed nematic phases at about 200 °C at much higher temperature and in a narrower range than the free ligand. The role

53   Ghedini's ortho-palladated complexes
R = EtO or Et; R′ = C$_n$H$_{2n+1}$

of the bridging halogens was also investigated: the chloride showed a nematic phase while the bromide and the iodide showed enantiotropic SmA and N phases. X-ray diffraction [126], quasielastic neutron scattering [127], and electron energy loss spectroscopy [128], have been performed on these compounds. Chiral dinuclear cyclopalladated complexes have also been synthesized [129]. SmC* mesophases have been found. NMR and circular dichroism have been performed on these chiral molecules. The investigation of the viscoelastic properties has been studied and compared with the uncomplexed ligand [130].

The monomeric complexes, **54** (L = PPh$_3$), obtained by cleaving **53** with PPh$_3$ are not mesogenic while the amine adducts (L = pyridine or quinoline) form smectic and nematic mesophases.

54   Mononuclear palladated complexes

Mononuclear cyclopalladated liquid crystalline materials whose molecular structure consists of two different thermotropic ligands, connected by a palladium atom, **55**, have been also prepared. The mesomorphic properties depend on the length of the chain. Monotropic or enantiotropic nematic and/or smectic phases can be found [131].

Other orthometallated azobenzenes were reported by Hoshino [132]. The complexes presented the same nematic phase as the ligand, at higher temperature.

55 Azpac compounds

Azoxymercury complexes, **56**, have also been synthesized [133]. The ligand bears in 4 and 4′ position, an alkoxy chain and the chiral S(+)-2-octyloxy, respectively. Both the ligand and their HgCl derivatives are liquid crystals. The mercury complexes show an enantiotropic SmC* mesophase.

56 Azoxymercury complexes

4.6.2 Arylimines

Related complexes based on imines were reported by Espinet [134]. The complex **57** showed smectic phases (usually SmA) when X = Cl, Br and SCN. The acetato complexes were generally nonmesogenic. It was suggested that this was due to the nonplanarity of the molecules. Again for this series, the

melting and especially the clearing temperatures of the complexes are very significantly higher than those of the free ligand. The substitution of the alkoxy substituents by polar groups (Cl, CN, NO_2) modified the behavior only slightly [135]. By substitution of the halogeno bridges by chiral chains they prepared compounds giving cholesteric mesophases at about 145–150 °C. Recently, Espinet and coworkers reported a comparative study of metal-containing liquid crystals for which the ferroelectric behavior has been fully characterized [136].

57 Espinet's orthopalladated imine complexes

Baena et al. [137] introduced the idea that perturbing the high symmetry of the molecular shape of this family of compounds might lead to lower melting points and less ordered mesophases. Azomethines were orthopalladated to give the binuclear complex and these were treated with Tl(acac) (acac = acetylacetonate) to give a mononuclear species. These mononuclear complexes gave rise to mesophases below 90 °C. N, SmA and SmC phases were found.

New compounds containing two or even four palladium atoms and eight or twelve flexible side chains have been synthesized by Praefcke et al. [138]. The compound, **58**, shows an enantiotropic discotic hexagonal columnar mesophase over a wide range of temperature and a lyotropic nematic phase in various organic solvents. X-ray diffraction studies reveal an oblique arrangement of disordered columns.

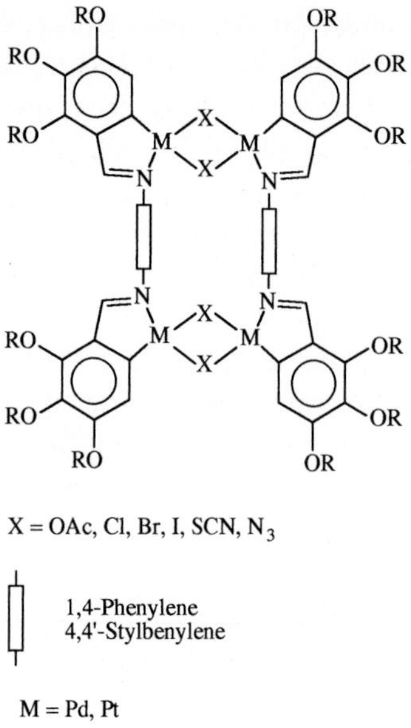

RO
OR
RO
OR
RO
OR
M
X
X
M
N
N
N
N
M
X
X
M
RO
OR
RO
OR
RO
OR
X = OAc, Cl, Br, I, SCN, N_3
1,4-Phenylene
4,4'-Stylbenylene
M = Pd, Pt

58 Tetranuclear metal complexes

The dinuclear pallado compounds, **59**, are the first (monotropic) cases of metallomesogens exhibiting the nematic discotic phase. On doping with a strong electron acceptor, the stabilization and/or induction of mesophases was observed [139].

OR
OR
OR
O
Pd
O
N
N
Pd
O
O
RO
RO
RO

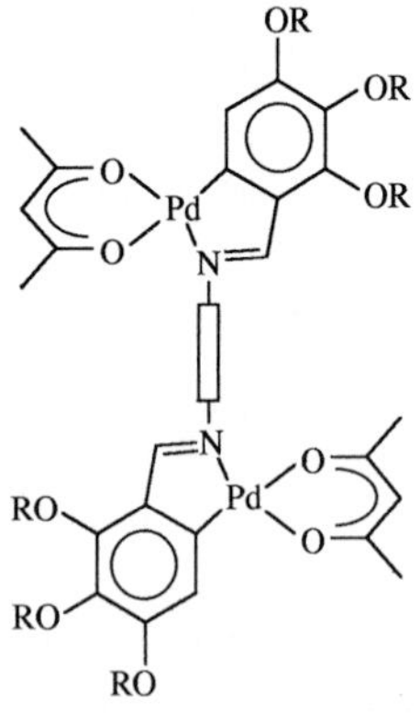

59 Dinuclear pallado complexes

Dinuclear platinum and tetraplatinum compounds have also been studied, **58** (M=Pt). Synthesis and lyotropic-crystalline behavior (nematic), are discussed for this platinum compound, disc-shape in structure due to its six alkoxy groups [140]. Compared to the analogous palladomesogens, the mesophases observed are of the same type, nematic-discotic and hexagonal columnar, but are more stable in the case of the platinum liquid crystals. X-ray diffraction has been performed on the dinuclear complex [141].

Manganese(I) and rhenium(I) complexes, **60**, have been synthesized by Bruce et al. [142]. They are the first examples of calamitic thermotropic liquid crystals containing a metal with a simple octahedral coordination environment. Manganese and rhenium complexes show nematic phases between about 120 and 190 °C.

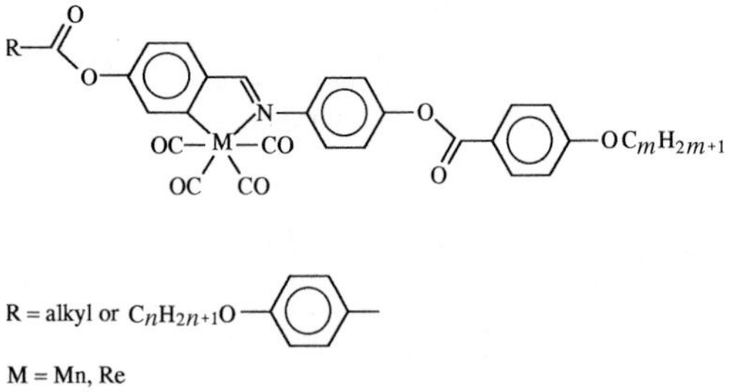

R
O
O
N
OC—M—CO
OC
CO
O
O
OC_mH_{2m+1}
R = alkyl or $C_nH_{2n+1}O$
M = Mn, Re

60 Bruce's manganese and rhenium complexes

4.6.3 Orthopalladated Diarylazines

These studies were then extended to the synthesis of palladium complexes of symmetric azines. In contrast to the above-mentioned compounds the di-μ-acetato complexes, **61**, showed SmC, and in two cases, also nematic, mesophases. The μ-acetato ligands constrain the complexes to be nonplanar, and a novel type of structure based upon open-book-shaped molecules [143] has been proposed. With R′ optically active, a mixture of *cis*- and *trans*-isomers was observed.

61 Open-book shape structure of the chiral trans isomer of the μ-carboxylato palladium azines

4.6.4 Aroylhydrazine

A series of aroylhydrazinato-nickel and -copper complexes, **62**, was synthesized in high yield and shown to form metallomesogens with SmC and N phases around 150 °C. The nickel complexes were found to be highly stable while the copper complexes decompose soon after entering the isotropic phase [144].

$$R = C_{10}H_{21} \text{ or } C_{12}H_{25}$$

62 Aroylhydrazine complexes

4.6.5 Orthopalladated Pyrimidine Complexes

Other work in this area has been done by Ghedini and coworkers [145] with pyrimi-

dine ligands, **63**. When X–Y=acac ($n=0$) a material with a monotropic SmA phase resulted. When X–Y=2–2′ bipyridine ($n=1$ and the counterion = BF_4) a material with an enantiotropic nematic phase is produced.

63 Mononuclear 2-phenylpyrimidine complexes

Further studies [146] of the same system looked at the dimeric precursors to such monomeric species. The molecular structure consists of two ligands in a *transoid* geometry with respect to the Pd–Pd axis, connected by the Pd_2X_2 bridges, **64**. On the basis of X-ray diffraction, the dinuclear derivatives show on heating, first a solid phase, then a SmA phase.

$$X = Cl, Br, I, OAc$$

64 Dinuclear 2-phenylpyrimidine complexes

4.7 Metallocen Ligands

The largest group of materials in this family are derived from the ferrocene unit [147].

4.7.1 Ferrocene

4.7.1.1 Monosubstituted Ferrocene

The first examples of mesogenic ferrocenes were reported by Malthête and Billard in

1976 [10] and showed SmC mesophases. Another ferrocenyl Schiff-base derivative, **65**, described by Galyametdinov gave rise to enantiotropic nematic phases [182].

65   Monosubstituted ferrocene

Interestingly these compounds can complex a copper anion to lead to trimetallic compounds, **66**, with a high temperature nematic phase. In these series the ferrocene moiety is at the end of the molecule rather than being incorporated into the core.

66   Copper complexes of 1,1′ monosubstituted ferrocene

4.7.1.2   1,1′ Disubstituted Ferrocene

In 1988 Bhatt et al. [148] reported a series of diesters in which the ferrocene moiety is

The crystal structure reveals that these molecules adopt the extended S geometry, with the two alkyl chains extending in the opposite direction. Those with $X = Y = p$-C_6H_4OR were shown to give monotropic SmA or SmC phases. With an additional benzene ring link by an ester, the mesomorphic properties depend of the direction of the ester: with $X = Y = p$-$OCOC_6H_4OR$, none of the ferrocenes showed liquid crystal behavior on heating; with $X = Y = p$-$COOC_6H_4OR$ $(R = C_nH_{2n+1})$ $(n = 2-18)$, the derivatives with $n = 2-4$ showed monotropic nematic phases, for $n = 6$ N and SmA phases. Increasing the alkyl chain length caused the disappearance of the nematic mesophase [149]. The unsymmetrical derivatives $X \neq Y$ led to compounds giving SmA and SmC mesophases [150]. Replacing the ester group of Y by a Schiff-base $X = Y = p$-$CH = N$ $C_6H_4OC_nH_{2n+1}$ led to nematic mesophase for short chains and smectic for longer chain lengths [151].

4.7.1.3   1,3 Disubstituted Ferrocene

Compounds of 1,3 disubstituted ferrocene have been reported by Deschenaux et al., **68**. The symmetrical compound [152] led to SmA and N phases, the unsymmetrical

68   1,3 disubstituted ferrocene

at the center of the molecule, **67**.

67   1,1′-ferrocene diesters

[153] to SmC and SmA as well as N phases.

They showed the strong influence of structural isomerism on the liquid crystal behavior by comparison between 1,1′ and 1,3 disubstituted ferrocenes. X-ray diffraction studies on the SmA phase suggested a monolayer molecular organization [154].

Synthesis and mesomorphic properties of the first ferrocene-containing side-chain liquid crystalline polymers, **69**, obtained by grafting either a 1,1'- or a 1,3-disubstituted ferrocene derivative functionalized by a vinyl group onto a polysiloxane have been reported [155]. These polymers showed enantiotropic SmC and/or SmA mesophases.

69 Ferrocene containing liquid crystalline polymers

4.7.2 Ruthenocene

Synthesis and mesomorphic behavior of ruthenocene substituted in the 1,1' positions have been reported by Deschenaux et al. [156]. With long substituted alcoxy chains ($n = 11 - 14$), compounds give rise to enantiotropic SmA phases. Comparison of the mesogenic properties with those of the analogous ferrocene shows that both series, on heating, give SmA mesophases. The ruthenocene derivatives melt at higher temperature and in a shorter anisotropic domain than the corresponding ferrocene.

4.7.3 Iron Tricarbonyl Derivatives

Another group were the butadiene iron tricarbonyl complexes, **70**, reported by Ziminsky and Malthête [157]. All the terminally-substituted complexes showed nematic phases over a wide range. The disubstituted complexes showed a nematic phase for short chains and a SmA phase for longer chains.

70 Iron tricarbonyl complexes

5 Metallomesogens with Polydentate Ligands

5.1 Phthalocyanine Ligands

5.1.1 Copper Complexes

Most work on metallophthalocyanines has been carried by Simon and coworkers [158] who synthesized in 1982 the first octasubstituted phthalocyanines and their copper complexes, **71**. These compounds were reported to show a stable mesophase over a very wide range of temperature and which was characterized by X-ray diffraction as discotic columnar mesophases.

71   Octasubstituted metallophthalocyanines

Copper phthalocyanines bearing four benzo(15)crown-5 substituents, **72**, have also been made: they stack to give channels through which ions may be transported [159].

72   Crown-ether metallophthalocyanines

5.1.2   Manganese, Copper, Nickel, and Zinc Complexes

Complexes of Co, Ni and Pb with the phthalocyanine ($R = CH_2OC_8H_{17}$) exhibited mesophases below 100 °C. Cobalt compound reacts with NaCN to give a polymeric material [160]. Discotic zinc and manganese

complexes have also been reported [161]. The fact that these compounds are brightly colored has been considered for laser addressed devices.

5.1.3   Lutetium Complexes

Simon et al. [162] found that bis(phthalocyanines) lutetium, **73**, formed on heating discotic mesophases which were molecular semiconductors. X-ray data for the mesophase indicated an intercolumnar distance twice the thickness of the phthalocyanine ring.

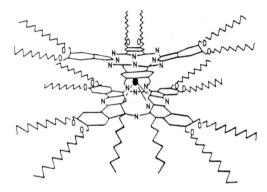

73   Luthetium phthalocyanines

5.1.4   Silicon, Tin, and Lead Complexes

Columnar phthalocyanines linked by O–Si–O or O–Sn–O spacers have also been described. The tin complex shows a columnar mesophase between 54 and 114 °C with a rectangular packing of the columns.

Silicon forms a dihydroxy silylphthalocyanine, **74**, which shows a hexagonal columnar mesophase. On heating, loss of water occurred and led to polymeric materials with a polysiloxane spine [163].

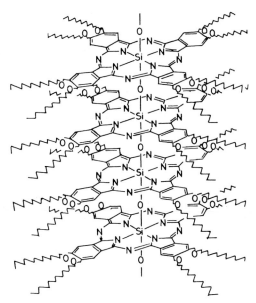

74   Silylphthalocyanines

Lead forms a columnar discotic liquid crystal with the metal sitting out of the plane of the molecule, **75**. X-ray data led to the suggestion that the molecules were stacked antiferroelectrically in a tilted stack [164].

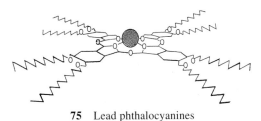

75   Lead phthalocyanines

Phthalocyanines with a different substitution pattern [165] were also described and again stable discotic mesophases were obtained.

5.2   Porphyrin Ligands

The first most systematic work in this area were the porphyrin complexes, **76**, with Cu(II), Cd(II), Zn(II), and Pd(II) reported and carefully characterized by Gregg and coworkers [166]. The results indicate a clear influence of the metal and of the length of the alkyl chains on the mesomorphic behavior.

76   Octasubstituted metalloporphyrins

The synthesis and mesomorphic properties of Co, Cu, Ni, and Zn complexes of octakis(alkylthio)tetraazaporphyrins has been reported by Doppelt and Huille [167] and by Morelli and coworkers [168] showing the hexagonal columnar nature of the mesophases. Some 5, 10, 15, 20-tetrasubstituted tetraphenyl porphyrins and their metal complexes (Co, Ni, Zn, and Pd) with square planar geometry were reported to exhibit discotic lamellar phases in contrast to the non-mesomorphic alkoxy derivatives [169]. In the D_1 phase, molecular planes align in a parallel manner to form a layered structure. However, Kujimiya et al. [170] reported that some alkoxy compounds are mesomorphic [171]. This is to compare with the zinc and copper complexes of tetrakis(carboxyphenyl) porphyrins which exhibit thermotropic phase transition with no evidence of liquid cristallinity [172]. More recently, Bruce et al. [173] reported zinc(II) complexes of 5,15-disubstituted porphyrins, **77**. By extending the porphyrin along one axis via substitution at the 5 and 15 positions, the disc-like porphyrin can behave as a rod, showing calamitic N and SmA phases at elevated temperatures. Recently the same authors, to reduce these temperatures, introduced lateral chains that folded over the faces of the porphyrin. These new compounds melt, for example with $n=7$, at 140°C to a nematic phase, which clears at 177°C, whereas the previous one melts into a nematic phase at 305°C and clears with decomposition at 433°C [173b].

77   Disubstituted metalloporphyrins

5.3   Other Amine Ligands

The annelides form mesogenic copper complexes, **78**. The lamellar phases, intermediate between crystalline solids and liquid crystals, have been termed *tegma crystals* [183].

78   Annelide complexes

Aza and aza-thia macrocyclic complexes have been developed. The induction of new mesomorphic phases in saturated macrocyclic ligands was introduced by Lehn et al. [184]. Macrocyclic [18]ane N_2O_4 and [18]ane N_2S_4 complexed with palladium give cationic complexes, **79**, showing smectic mesophases [174].

Copper(I) complexes show a lamellar structure not conclusively assigned [175]. Liquid crystalline complexes with chromium, molybdenum and tungsten as metallic centers have also been reported. The 1,4,7-trisubstituted 1,4,7-triazacyclononane and three carbonyl groups, **80**, are coordinated in an octahedral geometry.

80   Triazacarbonyl complexes

The observed mesophases are characterized as disordered rectangular columnar of a pyramidic type of X-ray measurements [176].

6   Lyotropic Metal-Containing Liquid Crystals

While thermotropic metallomesogens are widely studied, there are very few examples of lyotropic systems. Some amphiphilic Fe(II) complexes, **81**, have been synthe-

79   Macrocyclic complexes

sized and are shown to give a hexagonal (H_1) mesophase in water [177].

81 Amphiphilic iron complexes

Surfactant complexes of Cr(III) and Co(III), **82**, with long chain and bidentate ligands have been reported and lyotropic liquid crystal properties have been determined for a few of them [178]. They have to be compared with Simon et al. annelide complexes.

82 Surfactant Co(II) complexes

It has also been seen that disc-shape tetrapalladium compounds [179] form charge transfer complexes exhibiting lyotropic cholesteric properties.

Interesting surfactant tris(bipyridine) derivatives of ruthenium(II), **83**, containing one alkyl chain show lyotropic mesomorphism in water [180].

83 Lyotropic ruthenium complexes

New surfactants incorporating paramagnetic oxo-vanadium centers, **84**, were studied and some of the complexes were found to display both hexagonal and lamellar phases [181]. A review on transition metal based ionic mesogens has been published in 1996 by F. Neve [185].

84 Surfactant oxo-vanadium complexes

7 Conclusions

Metallomesogens now form an interdisciplinary topic which is properly established. The introduction of a metal into a liquid crystal may reasonably be expected to introduce a substantial number of novel features, such as color, paramagnetism, or conductivity. Other interesting and useful properties, arising from the large polarizable electron density associated with the metal center, may also be expected. A very interesting new development is the strong evidence that a continuum may exist between the rod- and the disc-like shapes and that rodlike shapes can give rise to discotic phases. A topic of increasing interest, so far not really explored, is the field of lyotropic metal-containing liquid crystals. Although many metals have already been incorporated into metallomesogens many others have not yet been used. Designs of new ligands have to be imagined. New applications will require new types of complexes. The field is young and there is still so much to do before finding applications for these materials.

Acknowledgements

I should like to thank all our colleagues who have been working in the field of metallomesogens and with whom I always have valuable discussions. I am particularly grateful to P. Maitlis (Sheffield) for the first review we did together, D. Bruce (Sheffield) and J. C. Marchon for reading the manuscript, W. Moneta for his kind assistance with the drawings and all my coworkers for fruitful collaborations. I also thank the CNRS and the CENG for their financial support.

8 References

[1] M. Marcos, J. L. Serrano, *Adv. Mater.* **1991**, *30*(5), 256–257.

[2] A. M. Giroud-Godquin, P. M. Maitlis, *Angew. Chem. Int. Ed. Engl.* **1991**, *30*, 375–402.

[3] P. Espinet, M. A. Esteruelas, L. A. Oro, J. L. Serrano, E. Sola, *Coord. Chem. Reviews* **1992**, *117*, 215–274.

[4] S. A. Hudson, P. Maitlis, *Chem. Rev.* **1993**, *93*, 861–885.

[5] D. W. Bruce in *Inorganic Materials* (Eds.: D. W. Bruce, D. O'Hare), **1992**, Wiley, Chichester, pp. 405–490.

[6] A. P. Polishchuk, T. V. Timofeeva, *Russian Chem. Rev.* **1993**, *62*(4), 291–321.

[7] D. W. Bruce, *J. Chem. Soc. Dalton Trans.* **1993**, 2983–2989; D. W. Bruce, *Adv. Mater.* **1994**, *6*(9), 699–701.

[8] (a) D. Vorländer, *Ber. Dtsch. Chem. Ges.* **1910**, *43*, 3120; (b) D. Vorländer, *Z. Phys. Chem. Stoechiom. Verwandtschaftsl.* **1923**, *105*, 211.

[9] A. Skoulios, *Ann. Phys. Paris* **1978**, *3*, 421–450.

[10] J. Malthête, J. Billard, *Mol. Cryst. Liq. Cryst.* **1976**, *34*, 117.

[11] A. M. Giroud, U. T. Mueller-Westerhoff, *Mol. Cryst. Liq. Cryst. (Lett.)* **1977**, *41*, 11.

[12] H. Adams, N. A. Bailey, D. W. Bruce, D. A. Dunmur, E. Lalinde, M. Marcos, C. Ridgway, A. J. Smith, P. Styring, P. M. Maitlis, *Liq. Cryst.* **1987**, *2*, 381.

[13] D. W. Bruce, S. M. Fan, G. R. Luckhurst, *Liq. Cryst.* **1994**, *16*(6), 1093–1099.

[14a] J. P. Rourke, F. P. Fanizzi, N. J. Salt, D. W. Bruce, D. A. Dunmur, P. M. Maitlis, *J. Chem. Soc. Chem. Commun.* **1990**, 229. *J. Chem. Soc. Dalton Trans.* **1992**, 3009–3014.

[14b] B. Donnio, D. W. Bruce, *J. Chem. Soc. Dalton Trans* **1997**, 2745–2755.

[15] D. W. Bruce, D. A. Dunmur, S. A. Hudson, P. M. Maitlis, P. Styring, *Adv. Mater. for Optics and Electronics* **1992**, *1*, 37–42.

[16] H. Adams, N. A. Bailey, D. W. Bruce, S. C. Davis, D. A. Dunmur, P. D. Hempstead, S. A. Hudson, S. Thorpe, *J. Mater. Chem.* **1992**, *2*(4), 395–400; D. W. Bruce, B. Donnio, S. A. Hudson, A. M. Levellut, S. Megtert, D. Petermann, M. Veber, *J. Phys. (France)* **1995**, *5*(2), 289.

[17] D. W. Bruce, S. Hudson, *J. Mater. Chem.* **1994**, *4*(3), 479–486.

[18] D. W. Bruce, M. D. Hall, *Mol. Cryst. Liq. Cryst.* **1994**, *250*, 373–375.

[19] J. P. Rourke, F. P. Fanizzi, D. W. Bruce, D. A. Dunmur, P. M. Maitlis, *Dalton J. Chem. Soc. Trans* **1992**, 3009–3014.

[20] D. W. Bruce, S. C. Davis, D. A. Dunmur, S. A. Hudson, P. M. Maitlis, P. Styring, *Mol. Cryst. Liq. Cryst.* **1992**, *215*, 1–11.

[21] H. A. Adams, N. A. Bailey, D. W. Bruce, S. A. Hudson, J. R. Marsden, *Liq. Cryst.* **1994**, *16*(4), 643–653.

[22] D. W. Bruce, J. P. Rourke, *Polyhedron* **1995**, *14*, 1915–1922.

[23] M. A. Esteruelas, E. Sola, L. A. Ora, M. B. Ros, J. L. Serrano, *J. Organomet. Chem.* **1990**, *387*, 103.

[24] M. Marcos, M. B. Ros, J. L. Serrano, M. A. Esteruelas, E. Sola, L. A. Oro, J. Barbera, *Chem. Mater.* **1990**, *2*, 748.

[25] S. Takahashi, M. Kariya, T. Yakate, K. Sonogashira, N. Hagihara, *Macromolecules* **1978**, *11*, 1063.

[26] S. Takahashi, Y. Takai, H. Morimoto, K. Sonogashira, N. Hagihara, *Mol. Cryst. Liq. Cryst.* **1982**, *82*, 139–143.

[27] T. Kaharu, H. Matsubara, S. Takahashi, *J. Mater. Chem.* **1991**, *1*, 145–146; *ibid.* **1992**, *2*(1), 43–47.

[28] D. W. Bruce, M. S. Lea, J. R. Marsden, *Mol. Cryst. Liq. Cryst.* **1996**, *275*, 183.

[29] T. Kaharu, H. Matsubara, S. Takahashi, *Mol. Cryst. Liq. Cryst.* **1992**, *220*, 191–199.

[30] (a) T. Kaharu, R. Ishii, S. Takahashi, *J. Chem. Soc. Chem. Commun.* **1994**, 1349–1350; (b) T. Kaharu, T. Tanaka, M. Sawada, S. Takahashi, *J. Mater. Chem.* **1994**, *4*(6), 859–865.

[31] P. Alejos, S. Coco, P. Espinet, *New J. of Chem.* **1995**, *19*(7), 799.

[32] A. S. C. Lawrence, *Trans. Faraday Soc.* **1938**, *34*, 660.

[33] B. Gallot, A. Skoulios, *Kolloid Z. Z. Polym.* **1966**, *209*(2), 164–169.

[34] H. A. Ellis, *Mol. Cryst. Liq. Cryst.* **1986**, *139*, 281–290.

[35] C. G. Bazuin, D. Guillon, A. Skoulios, A. M. Amorim da Costa, H. D. Burrows, C. F. G. C. Geraldes, J. J. Teixeira-Dias, E. Blackmore, G. J. T. Tiddy, *Liq. Cryst.* **1988**, *3*(12), 1655–1670.

[36] G. S. Attard, Y. D. West, *Liq. Cryst.* **1990**, *7*(4), 487–494.

[37] J. Lindau, W. Hillman, H. D. Doerfler, H. Sachmann, *Mol. Cryst. Liq. Cryst.* **1986**, *133*, 259.

[38] R. F. Grant, *Can. J. Chem.* **1964**, *42*, 951–953.

[39] A. M. Giroud-Godquin, J. C. Marchon, D. Guillon, A. Skoulios, *J. Phys. Lett. (Orsay, Fr.)* **1984**, *45*, L681–L684.

[40] H. Abied, D. Guillon, A. Skoulios, P. Weber, A. M. Giroud-Godquin, J. C. Marchon, *Liq. Cryst.* **1987**, *2*(3), 269–279; M. Ibn-Elhaj, D. Guillon, A. Skoulios, A. M. Giroud-Godquin, P. Maldivi, *Liq. Cryst.* **1992**, *11*(5), 731–744.

[41] J. R. Lomer, K. Perrera, *Acta Crystallographica* **1974**, *B30*, 2912–2915.

[42] P. Maldivi, D. Guillon, A. M. Giroud-Godquin, H. Abied, H. Dexpert, A. Skoulios, *J. Chim. Phys.* **1989**, *86*, 1651–1664.

[43] A. M. Giroud-Godquin, J. M. Latour, J. C. Marchon, *Inorg. Chem.* **1985**, *24*, 4452.

[44] (a) O. Poisat, D. P. Strommen, P. Maldivi, A. M. Giroud-Godquin, J. C. Marchon, *Inorg. Chem.* **1990**, *29*, 4853; (b) A. M. Giroud-Godquin, P. Maldivi, J. C. Marchon, M. Bée, L. Carpentier, *Mol. Phys.* **1989**, *68*, 1353.

[45] (a) A. M. Giroud-Godquin, P. Maldivi, J. C. Marchon, P. Aldebert, A. Peguy, D. Guillon, A. Skoulios, *J. Phys. (Orsay, Fr.)* **1989**, *50*, 513–519; (b) M. Bée, A. M. Giroud-Godquin, P. Maldivi, J. Williams, *Mol. Phys.* **1994**, *81*(1), 57–68.

[46] P. Maldivi, L. Bonnet, A. M. Giroud-Godquin, M. Ibn-Elhaj, D. Guillon, A. Skoulios, *Adv. Mater.* **1993**, *5*(12), 909–912.

[47] G. S. Attard, P. R. Cullum, *Liq. Cyst.* **1990**, *8*(3), 299–309.

[48] J. C. Marchon, P. Maldivi, A. M. Giroud-Godquin, D. Guillon, A. Skoulios, D. P. Strommen, *Phil. Trans. Roy Soc. Lond. A* **1990**, *330*, 109–116.

[49] J. Barbera, M. A. Esteruelas, A. M. Levelut, L. A. Oro, J. L. Serrano, E. Sola, *Inorg. Chem.* **1992**, *31*(5), 732–737.

[50] M. Ibn-Elhaj, D. Guillon, A. Skoulios, A. M. Giroud-Godquin, J. C. Marchon, *J. Phys. France* **1992**, *2*, 2197–2206.

[51] F. D. Cukiernik, P. Maldivi, A. M. Giroud-Godquin, J. C. Marchon, M. Ibn-Elhaj, D. Guillon, A. Skoulios, *Liq. Cryst.* **1991**, *9*(6), 903–906.

[52] L. Bonnet, F. D. Cukiernik, P. Maldivi, A. M. Giroud-Godquin, J. C. Marchon, M. Ibn-Elhaj, D. Guillon, A. Skoulios, *Chem. Mater.* **1994**, *6*, 31–38.

[53] F. D. Cukiernik, A. M. Giroud-Godquin, P. Maldivi, J. C. Marchon, *Inorg. Chim. Acta* **1994**, *215*, 203–207.

[54] D. V. Baxter, R. H. Cayton, M. H. Chisholm, J. C. Huffman, E. F. Putilina, S. L. Tagg, J. L. Wesemann, J. W. Zwanziger, F. D. Darrington, *J. Am. Chem. Soc.* **1994**, *116*, 4551–4566.

[55] B. J. Bulkin, R. K. Rose, A. Santoro, *Mol. Cryst. Liq. Cryst.* **1977**, *43*, 53.

[56] A. M. Giroud-Godquin, J. Billard, *Mol. Cryst. Liq. Cryst.* **1981**, *66*, 147.

[57] A. C. Ribeiro, A. F. Martins, A. M. Giroud-Godquin, *Mol. Cryst. Liq. Cryst. (Lett)* **1988**, *5*, 133.

[58] K. Ohta, A. Ishii, H. Muroki, I. Yamamoto, K. Matsuzaki, *Mol. Cryst. Liq. Cryst.* **1985**, *116*, 299; K. Ohta, A. Ishii, I. Yamamoto, K. Matsuzaki, *J. Chem. Soc. Chem.* **1984**, 1099.

[59] K. Usha, B. K. Sadashiva, K. Vijayan, *Mol. Cryst. Liq. Cryst.* **1994**, *241*, 91–102.

[60] S. Chandrasekhar, B. K. Sadashiva, B. S. Srikanta, *Mol. Cryst. Liq. Cryst.* **1987**, *151*, 93.

[61] S. Chandrasekhar, B. R. Ratna, B. K. Sadashiva, V. N. Raja, *Mol. Cryst. Liq. Cryst.* **1988**, *165*, 123.

[62] A. Ghode, U. Shivkumar, B. K. Sadashiva, *Bull. Mater. Sci.* **1994**, *17*(3), 283–297.

[63] B. Mülberger, W. Haase, *Liq. Cryst.* **1989**, *5*, 251.

[64] K. Ohta, H. Akimoto, T. Fujimoto, I. Yamamoto, *J. Mat. Chem.* **1994**, *4*(1), 61–69; K. Ohta, Y. Morizumi, H. Akimoto, O. Takenaka, T. Fujimoto, I. Yamamoto, *Mol. Cryst. Liq. Cryst.* **1992**, *214*, 143–149.

[65] N. J. Thomson, G. W. Gray, J. W. Goodby, K. J. Toyne, *Mol. Cryst. Liq. Cryst.* **1991**, *200*, 109.

[66] A. M. Giroud-Godquin, G. Sigaud, M. F. Achard, F. Hardouin, *J. Phys. Lett. (Orsay, France)* **1984**, *45*, L 387; A. M. Giroud-Godquin, M. M. Gauthier, G. Sigaud, F. Hardouin, M. F. Achard, *Mol. Cryst. Liq. Cryst.* **1986**, *132*, 35.

[67] K. Ohta, H. Ema, H. Muroki, I. Yamamoto, K. Matsuzaki, *Mol. Cryst. Liq. Cryst.* **1987**, *147*, 61.

[68] W. Haase, V. S. Reyes, M. Athanassopoulou, G. S. Garrido, K. Griesar, E. Schuhmacher, *Mol. Eng.* **1995**, in press.

[69] A. B. Blake, J. R. Chipperfield, S. Clark, P. G. Nelson, *J. Chem. Soc. Dalton Trans* **1991**, 1159.

[70] V. S. Reyes, G. S. Garrido, C. Aguilera, K. Griesar, M. Athanassopoulou, H. Finkmann, W. Haase, *Mol. Mater.* **1995**, in press.

[71] C. K. Lai, A. G. Serette, T. M. Swager, *J. Am. Chem. Soc.* **1992**, *114*, 7948–7949.

[72a] K. Hanabusa, T. Suzuki, T. Koyama, H. Shirai, *Makromol. Chem.* **1992**, *193*, 2149–2161.

[72b] L. Oriel, J. L. Serrano, *Adv. Mater.* **1995**, *7*, 348–369.

[73] K. Usha, B. K. Sadashiva, *Mol. Cryst. Liq. Cryst.* **1994**, *241*, 91–102.

[74] A. M. Giroud-Godquin, A. Rassat, *C. R. Acad. Sci. Ser. 2* **1982**, *294*, 241.

[75] S. N. Poelsma, A. H. Servante, F. P. Fanizzi, P. M. Maitlis, *Liq. Cryst.* **1994**, *16*(4), 675–685.

[76] P. Styring, S. Tantrawong, D. R. Beattie, J. W. Goodby, *Liq. Cryst.* **1991**, *10*(4), 581–584.

[77] H. Zheng, T. M. Swager, *J. Am. Chem. Soc.* **1994**, *116*, 761–762

[78] H. Zheng, P. J. Carroll, T. M. Swager, *Liq. Cryst.* **1993**, *14*(5), 1421–1429.

[79] J. Barbera, C. Cativiela, J. L. Serrano, M. M. Zurbano, *Adv. Mater.* **1991**, *3*(12), 602–605.

[80] K. Ohta, M. Moriya, M. Ikejima, H. Hasebe, T. Fujimoto, I. Yamamoto, *Bull. Chem. Soc. Jpn.* **1993**, *66*, 3553–3564.

[81] U. T. Mueller-Westerhoff, A. Nazzal, R. J. Cox, A. M. Giroud, *Mol. Cryst. Liq. Cryst. (Lett.)* **1980**, 249–255.

[82] M. Veber, R. Fugnitto, H. Strzelecka, *Mol. Cryst. Liq. Cryst.* **1983**, 221.

[83] M. Veber, P. Davidson, C. Jallabert, A. M. Levelut, H. Strzelecka, *Mol. Cryst. Liq. Cryst.* **1987**, *1*.

[84] K. Ohta, H. Hasebe, H. Ema, T. Fujimoto, I. Yamamoto, *J. Chem. Soc. Chem. Commun.* **1989**, 1610.

[85] H. Adams, A. C. Albeniz, N. A. Bailey, D. W. Bruce, A. S. Cherodion, R. Dhillon, D. A. Dunmur, P. Espinet, J. L. Feijoo, E. Lalinde, P. M. Maitlis, R. M. Richardson, G. Ungar, *J. Mater. Chem.* **1991**, *1*, 843.

[86] K. Ohta, Y. Morizumi, H. Ema, T. Fujimoto, I. Yamamoto, *Mol. Cryst. Liq. Cryst.* **1992**, *214*, 151–159.

[87] D. Guillon, D. W. Bruce, P. Maldivi, M. Ibn-El-haj, R. Dhillon, *Chem. Mater.* **1994**, *6*, 182–189.

[88] K. Ohta, H. Ema, Y. Morizumi, T. Watanabe, T. Fujimoto, I. Yamamoto, *Liq. Cryst.* **1990**, *8*, 311–330.

[89] D. W. Bruce, R. Dhillon, D. A. Dunmur, P. M. Maitlis, *Mater. Chem.* **1992**, *2*(1), 65–69.

[90] J. P. Fackler Jr, J. A. Fetchin, D. C. Fries, *J. Am. Chem. Soc.* **1972**, *94*, 7323.

[91] N. Hoshino-Miyajima, *J. Chem. Soc. Chem. Commun.* **1993**, 1442.

[92] M. J. Baena, P. Espinet, M. C. Lequerica, A. M. Levelut, *J. Amer. Chem. Soc.* **1992**, *114*, 4182–4185.

[93] K. Ohta, H. Ema, I. Yamamoto, K. Matsuzaki, *Liq. Cryst.* **1988**, *3*, 1671.

[94] I. V. Ovchinnikov, Yu. G. Galyametdinov, G. I. Ivanova, L. M. Yagfarova, *Dokl. Akad. Nauk SSSR, Ser. Khim* **1984**, *276*, 126.

[95] Yu G. Galyametdinov, I. V. Ovchinnikov, B. M. Bolotin, N. B. Etingen, G. I. Ivanova, L. M. Yagfarova, *Izv. Akad. Nauk SSSR, Ser. Khim.* **1984**, 2379.

[96] U. Caruso, A. Roviello, A. Sirigu, *Liq. Cryst.* **1991**, *10*(1), 85–93.

[97] M. Marcos, P. Romero, J. L. Serrano, C. Bueno, J. A. Cabeza, L. A. Oro, *Mol. Cryst. Liq. Cryst.* **1989**, *167*, 123; M. Marcos, P. Romero, J. L. Serrano, *Chem. Mater.* **1990**, *2*, 495.

[98] N. Hoshino, H. Murakami, Y. Matsunaga, T. Inabe, Y. Maruyama, *Inorg. Chem.* **1990**, *29*, 1177.

[99] U. Caruso, A. Roviello, A. Sirigu, *Liq. Cryst.* **1988**, *3*, 1515.

[100] J. P. Bayle, E. Bui, F. Perez, J. Courtieu, *Bull. Soc. Chim. Fr.* **1989**, *4*, 513.

[101] E. Bui, J. P. Bayle, F. Perez, L. Liebert, J. Courtieu, *Liq. Cryst.* **1990**, *4*, 513.

[102] V. Prasad, B. K. Sadashiva, *Mol. Cryst. Liq. Cryst.* **1993**, *225*, 303–312; V. Prasad, B. K. Sadashiva, *Mol. Cryst. Liq. Cryst.* **1994**, *241*, 167–174.

[103] E. Campillos, M. Marcos, J. L. Serrano, J. Barbera, P. J. Alonso, J. I. Martinez, *Chem. Mater.* **1993**, *5*, 1518–1525.

[104] E. Camppillos, M. Marcos, L. T. Oriol, J. L. Serrano, *Mol. Cryst. Liq. Cryst.* **1992**, *215*, 127–135.

[105] M. Marcos, J. L. Serrano, T. Sierra, M. J. Giménez, *Chem. Mater.* **1993**, *5*, 1332–1337; M. Marcos, J. L. Serrano, T. Sierra, M. J. Gimenez, *Angew. Chem. Int. Ed. Engl.* **1992**, *31*(11), 1471–1472.

[106] M. J. Baena, P. Espinet, M. B. Ros, J. L. Serrano, A. Ezcurra, *Angew. Chem. Int. Ed. Engl.* **1993**, *32*(8), 1203–1205.

[107] M. Ghedini, D. Pucci, E. Cesarotti, O. Francescangeli, R. Bartolino, *Liq. Cryst.* **1993**, *15*(3), 331–344.

[108] W. Pyzuk, Yu Galyametdinov, *Liq. Cryst.* **1993**, *15*(2), 265–268.

[109] N. Hoshino, R. Hayakawa, T. Shibuya, Y. Matsunaga, *Inorg. Chem.* **1990**, *29*, 5129–5131; N. Hoshino, A. Kodama, T. Shibuya, Y. Matsunaga, S. Miyajima, *Inorg. Chem.* **1991**, *30*, 3091–3096; E. Campillos, M. Marcos, A. Omenat, J. L. Serrano, *J. Mat. Chem.* **1996**, *6*(3), 349.

[110] E. Campillos, M. Marcos, J. L. Serrano, *J. Mater. Chem.* **1993**, *3*(10), 1049–1052; S. Armentano, G. De Munno, M. Ghedini, S. Morrone, *Inorg. Chim. Acta* **1993**, *210*, 125–127.

[111] J. Barbera, A. M. Levelut, M. Marcos, P. Romero, J. L. Serrano, *Liq. Cryst.* **1991**, *10*(1), 119–126.

[112] M. Ghedini, S. Morrone, R. Bartolino, V. Formoso, O. Francescangeli, B. Yang, D. Gatteschi, C. Zanchini, *Chem. Mater.* **1993**, *5*, 876–882; M. Ghedini, S. Morrone, O. Francescangeli, R. Bartolino, *Mol. Cryst. Liq. Cryst.* **1995**, *250*, 323.

[113] B. K. Sadashiva, A. Ghode, *Liq. Cryst.* **1994**, *16*(1), 33–42.

[114] Y. Galyametdinov, G. Ivanova, K. Griesar, A. Prosvirin, I. Ovchinnikov, W. Haase, *Adv. Mater.* **1992**, *4*(11), 739–741.

[115] K. Griesar, Y. Galyametdinov, M. Athanassopoulou, I. Ovchinnikov, W. Haase, *Adv. Mater.* **1994**, *6*(5), 381–384.

[116] P. J. Alonso, M. Marcos, J. L. Martinez, V. M. Orera, M. L. Sanjuan, J. L. Serrano, *Liq. Cryst.* **1993**, *13*(4), 585–596; P. J. Alonso, M. Marcos, J. I. Martinez, J. L. Serrano, T. Sierra, *Adv. Mater.* **1994**, *6*(9), 667–670; P. J. Alonso, J. A. Puertolas, P. Davidson, B. Martinez, J. I. Martinez, L. Oriol, J. L. Serrano, *Macromolecules* **1993**, *26*, 4304–4309.

[117] P. Berdagué, J. Courtieu, P. M. Maitlis, *J. Chem. Soc.: Chem. Commun.* **1994**, 1313–1314.

[118] C. Carfagna, U. Caruso, A. Roviello, A. Sirigu, *Makromol. Chem. Rapid. Commun.* **1987**, 8, 345.

[119] U. Caruso, A. Roviello, A. Sirigu, *Macromolecules* **1991**, 24, 2606–2609.

[120] M. Marcos, L. Oriol, J. L. Serrano, *Macromolecules* **1992**, 25, 5362–5368; L. Oriol, P. J. Alonso, J. I. Martinez, M. Pinol, J. L. Serrano, *Macromolecules* **1994**, 27(7), 1869–1874.

[121] W. Haase, K. Griesar, M. F. Iskander, Y. Galyametdinov, *Mol. Mat.* **1993**, 3, 115–130.

[122] R. Paschke, H. Zaschke, A. Mädicke, J. R. Chipperfield, A. B. Blake, P. G. Nelson, G. Gray, *Mol. Cryst. Liq. Cryst. (Lett.)* **1988**, 6, 81. See also: A. B. Blake, J. R. Chipperfield, W. Hussain, R. Paschke, E. Sinn, *Inorg. Chem.* **1995**, 34, 1125–1129.

[123] T. D. Shaffer, K. A. Sheth, *Mol. Cryst. Liq. Cryst.* **1989**, 172, 27.

[124] K. Ohta, Y. Morizumi, T. Fujimoto, *Mol. Liq. Cryst.* **1992**, 214, 161–169.

[125] M. Ghedini, F. Neve, D. Pucci, E. Cesarotti, M. Grassi, *J. Organomet. Chem.* **1992**, 438, 343–351.

[126] V. Formoso, M. C. Pagnotta, P. Mariani, M. Ghedini, F. Neve, R. Bartolino, M. More, G. Pepy, *Liq. Cryst.* **1992**, 11(5), 639–654.

[127] R. Bartolino, G. Coddens, F. Rustichelli, M. C. Pagnotta, C. Versace, M. Ghedini, F. Neve, *Mol. Cryst. Liq. Cryst.* **1992**, 221, 101–108.

[128] A. Amoddeo, R. Bartolino, L. S. Caputo, E. Colavita, V. Formoso, M. Ghedini, A. Oliva, D. Pucci, C. Versace, *Mol. Cryst. Liq. Cryst.* **1992**, 221, 93–99.

[129] M. Ghedini, D. Pucci, E. Cesarotti, O. Francescangeli, R. Bartolino, *Liq. Cryst.* **1994**, 16(3), 373–380.

[130] C. Versace, V. Formoso, D. Lucchetta, D. Pucci, C. Ferrero, M. Ghedini, R. Bartolino, *J. Chem. Phys.* **1993**, 98(11), 8507–8513.

[131] M. Ghedini, S. Morrone, O. Francescangeli, R. Bartolino, *Chem. Mater.* **1992**, 4(5), 1119–1123.

[132] N. Hoshino, H. Hasegawa, Y. Matsunaga, *Liq. Cryst.* **1991**, 9(2), 267–276.

[133] A. Omenat, M. Ghedini, *J. Chem. Soc.: Chem. Commun.* **1994**, 1309–1310.

[134] J. Barbera, P. Espinet, E. Lalinde, M. Marcos, J. L. Serrano, *Liq. Cryst.* **1987**, 2, 833; M. B. Ros, N. Ruiz, J. L. Serrano, P. Espinet, *Liq. Cryst.* **1991**, 9(1), 77–86.

[135] M. J. Baena, J. Buey, P. Espinet, H. S. Kitzerow, G. Heppke, *Angew. Chem. Int. Ed. Engl.* **1993**, 32(8), 1201–1203.

[136] M. J. Baena, J. Barbera, P. Espinet, A. Ezcurra, M. B. Ros, J. L. Serrano, *J. Am. Chem. Soc.* **1994**, 116(5), 1899–1906.

[137] M. J. Baena, P. Espinet, M. B. Ros, J. L. Serrano, *Angew. Chem. Int. Ed. Engl.* **1991**, 6, 711–712.

[138] K. Praefcke, D. Singer, B. Gündogan, *Mol. Cryst. Liq. Cryst.* **1992**, 223, 191–195; B. Gündogan, K. Praefcke, *Chem. Ber.* **1993**, 126, 1253–1255; K. Praefcke, S. Diele, J. Pickardt, B. Gündogan, U. Nütz, D. Singer, *Liq. Cryst.* **1995**, 18, 857.

[139] D. Singer, A. Liebmann, K. Praefcke, J. H. Wendorff, *Liq. Cryst.* **1993**, 14(3), 785–794; K. Praefcke, D. Singer, B. Gündogan, K. Gutbier, M. Langner, *Ber. Bunsenges. Phys. Chem.* **1994**, 98, 118–122.

[140] K. Praefcke, B. Bilgin, J. Pickardt, M. Borowski, *Chem. Ber.* **1994**, 127, 1543–1545; K. Praefcke, B. Bilgin, N. Usol'tseva, B. Heinrich, D. Guillon, *J. Mater. Chem.* **1995**, 5, 2257.

[141] K. Praefcke, B. Bilgin, J. Pickardt, M. Borowski, *Chem. Ber.* **1994**, 127, 1543–1545.

[142] D. Bruce, X. H. Liu, *J. Chem. Soc. Chem. Commun.* **1994**, 729–730; D. Bruce, X. H. Liu, *Liq. Cryst.* **1995**, 18, 165.

[143] P. Espinet, E. Lalinde, M. Marcos, J. Perez, A. Remon, J. L. Serrano, *Organometallics* **1990**, 9, 555.

[144] M. N. Abser, M. Bellwood, M. C. Holmes, R. W. McCabe, *J. Chem. Soc.: Chem. Commun.* **1993**, 1062.

[145] M. Ghedini, D. Pucci, G. De Munno, D. Viterbo, F. Neve, S. Armantano, *Chem. Mater.* **1991**, 3, 65.

[146] M. Ghedini, D. Pucci, R. Bartolino, O. Francescangeli, *Liq. Cryst.* **1993**, 13(2), 255–263.

[147] A. Togni, T. Hayashi (eds.), *Ferrocenes*, VCH, Weinheim, **1994**.

[148] J. Bhatt, B. M. Fung, K. M. Nicholas, C. D. Poon, *J. Chem. Soc.: Chem. Commun.* **1988**, 1439; J. Bhatt, B. M. Fung, K. M. Nicholas, *Liq. Cryst.* **1992**, 12, 263.

[149] R. Deschenaux, J. L. Marendaz, J. Santiago, *Helv. Chim. Acta* **1993**, 76, 865–876.

[150] R. Deschenaux, M. Rama, J. Santiago, *Tetrahedron Lett.* **1993**, 34(20), 3293–3296.

[151] K. P. Reddy, T. L. Brown, *Liq. Cryst.* **1992**, 12(3), 369–376.

[152] R. Deschenaux, J. Santiago, D. Guillon, B. Heinrich, *J. Mater. Chem.* **1994**, 4(5), 679–682.

[153] R. Deschenaux, J. Santiago, *Tetrahedron Lett.* **1994**, 35(14), 2169–2172.

[154] R. Deschenaux, J. L. Marendaz, *J. Chem. Soc.: Chem. Commun.* **1991**, 909–910; R. Deschenaux, I. Kosztics, J. L. Marendaz, H. Stoeckli-Evans, *Chimia* **1993**, 47, 206–210.

[155] R. Deschenaux, I. Kosztics, U. Scholten, D. Guillon, M. Ibn-Elhaj, *J. Mater. Chem.* **1994**, 4(8), 1351–1352.

[156] R. Deschenaux, J. Santiago, *J. Mater. Chem.* **1993**, 3(2), 219–220.

[157] L. Ziminski, J. Malthête, *J. Chem. Soc.: Chem. Commun.* **1990**, 1495.

[158] C. Piechoki, J. Simon, A. Skoulios, D. Guillon, P. Weber, *J. Am. Chem. Soc.* **1982**, *104*, 5245.

[159] C. Sirlin, L. Bosio, J. Simon, V. Ahsen, E. Yilmazer, O. Bekâoglu, *Chem. Phys. (Lett.)* **1987**, *139*, 362.

[160] M. Hannack, A. Beck, H. Lehmann, *Synthesis* **1987**, 703.

[161] D. Guillon, P. Weber, A. Skoulios, C. Piechoki, J. Simon, *Mol. Cryst. Liq. Cryst.* **1985**, *130*, 223.

[162] C. Piechocki, J. Simon, J. J. André, D. Guillon, P. Petit, A. Skoulios, P. Weber, *Chem. Phys. (Lett.)* **1985**, *122*, 124; Z. Belarbi, C. Sirlin, J. Simon, J. J. André, *J. Phys. (Lett)* **1989**, *93*, 8105.

[163] C. Sirlin, L. Bosio, J. Simon, *J. Chem. Soc. Chem. Communn.* **1988**, 236.

[164] P. Weber, D. Guillon, A. Skoulios, *J. Phys. Chem.* **1987**, *91*, 2242.

[165a] M. J. Cook, M. F. Daniel, K. J. Harrison, N. B. McKeown, A. J. Thompson, *J. Chem. Soc.: Chem. Communn.* **1987**, *1086*.

[165b] A. Sirigu, *Liq. Cryst.* **1993**, 14(1), 15–36.

[166] B. A. Gregg, M. A. Fox, A. J. Bard, *J. Am. Chem. Soc.* **1989**, *111*, 3024–3029.

[167] P. Doppelt, S. Huille, *New J. Chem.* **1991**, *14*, 607.

[168] F. Lelj, G. Morelli, G. Ricciardi, A. Roviello, *Chem. Phys. (Lett.)* **1991**, *185*, 468; F. Lelj, G. Morelli, G. Ricciardi, A. Roviello, A. Sirigu, *Liq. Cryst.* **1992**, 12(6), 941–960.

[169] Y. Shimizu, M. Miya, A. Nagata, K. Ohta, A. Matsumara, I. Yamamoto, S. Kusabayashi, *Chem. (Lett.)* **1991**, *25*; Y. Shimizu, T. Matsuno, M. Miyia, A. Nagata, *J. Chem. Soc. Chem. Commun.* **1994**, 2411.

[170] S. I. Kugimiya, M. Takemura, *Tetrahedron (Lett.)* **1990**, 31(22), 3157–3160.

[171] Y. Shimizu, M. Miya, A. Nagata, K. Ohta, I. Yamamoto, S. Kusabayashi, *Liq. Cryst.* **1993**, 14(3), 795–805.

[172] R. Ramasseul, P. Maldivi, J. C. Marchon, M. Taylor, D. Guillon, *Liq. Cryst.* **1993**, 13(5), 729–733.

[173a] D. Bruce, D. A. Dunmur, L. S. Santa, M. A. Wali, *J. Mater. Chem.* **1992**, 2(3), 363–364; D. Bruce, M. A. Wali, Q. Min Wang, *J. Chem. Soc.: Chem. Commun.* **1994**, 2089–2090.

[173b] Q. M. Wang, D. W. Bruce, *Angew. Chem. Int. Ed. Engl.* **1997**, *36*, 150.

[174] F. Neve, M. Ghedini, *J. Inclusion Phenomena and Molecular Recognition in Chemistry* **1993**, *15*, 259–272.

[175] F. Neve, M. Ghedini, A. M. Levelut, O. Francescangeli, *Chem. Mater.* **1994**, 6(1), 70–76.

[176] S. Schmidt, G. Lattermann, R. Kleppinger, J. H. Wendorff, *Liq. Cryst.* **1994**, 16(4), 693–702.

[177] D. W. Bruce, D. A. Dunmur, P. M. Maitlis, J. M. Watkins, G. J. T. Tiddy, *Liq. Cryst.* **1992**, 11(1), 127–133.

[178] D. W. Bruce, I. R. Denby, G. J. T. Tiddy, J. M. Watkins, *J. Mater. Chem.* **1993**, 3(9), 911–916.

[179] N. Usol'tseva, K. Praefcke, D. Singer, B. Gündogan, *Liq. Cryst.* **1994**, 16(4), 617–623.

[180] D. W. Bruce, J. D. Holbrey, A. R. Tajbakhsh, G. J. T. Tiddy, *J. Mater. Chem.* **1993**, 3(8), 905–906.

[181] S. S. Zhu, T. M. Swager, *Adv. Mater.* **1995**, 7, 280.

[182] Yu. G. Galyametdinov, O. N. Kadkin, I. V. Ovchinnikov, *Izv. Akad. Nauk., Ser. Khim. (Russia)* **1990**, 10, 2462; Yu. G. Galyametdinov, O. N. Kadkin, I. V. Ovchinnikov, *Izv. Akad. Nauk., Ser. Khim. (Russia)* **1992**, 2, 402.

[183] D. Markovitsi, J. J. André, A. Mathis, J. Simon, P. Spegt, G. Weill, M. Ziliox, *Chem. Phys. Lett.* **1984**, *104*, 48.

[184] J. M. Lehn, J. Malthête, A. M. Levelut, *J. Chem. Soc. Chem. Commun.* **1985**, 1794.

[185] F. Neve, *Adv. Mater.* **1996**, 8(4), 277–289.

Chapter XV
Biaxial Nematic Liquid Crystals

B. K. Sadashiva

1 Introduction

The nematic mesophase is exhibited by a large number of compounds composed of rigid rod-like molecules. This mesophase possesses a high degree of long range orientational order, but no long range positional order. In other words, it is somewhat similar to the ordinary liquid except that the molecules are arranged approximately parallel to one another. Because of the lack of positional order of the molecules, the nematic phase is quite fluid and the molecules can freely slide past one another. However, a well aligned sample of a nematic phase is optically uniaxial [1, 2].

It is well known that a smectic C liquid crystal which has a monoclinic symmetry is optically biaxial [3]. A biaxial nematic (N_b) phase in which the molecules are oriented along the three directions in space, i.e., they have three directors which are perpendicular to one another, is possible. Thus the shape of molecules exhibiting the N_b phase should be different from those exhibiting the uniaxial nematic phase.

2 Theoretical Prediction of the Biaxial Nematic Phase

Based on the interaction employed in the Maier–Saupe theory of the nematic state, Freiser [4] was the first to predict the possible existence of an N_b phase as an intermediate between two uniaxial nematic phases. Later, a number of other theoretical investigations were carried out [5–7] using various models to predict the possibility of obtaining an N_b liquid crystal. In all these models, a system consisting of hard rectangular plates was considered. These approaches gave the same result, an N_b phase should be obtained between two uniaxial nematics of opposite sign, i.e., those made up of rod-like and plate-like molecules. They also predicted that a transition from a uniaxial nematic to a biaxial nematic would be second order.

There was not much progress in the design and synthesis of rectangular plate-like molecules. However, the theoreticians [8–12] continued their efforts by considering systems in which mixtures of rod-like (N_c) and rectangular plate-like (N_d) molecules were used in varying proportions. These investigations led to the conclusion that mixtures containing rod-like and plate-like molecules should produce an N_b phase, and a typical phase diagram is shown in Fig. 1.

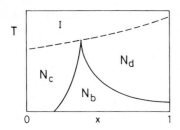

Figure 1. Phase diagram of the uniaxial and biaxial nematic phases as predicted by the microscopical theories (N$_c$ calamitic nematic; N$_b$ biaxial nematic; and N$_d$ discotic nematic). In the case of systems of hard rectangular plates, the parameter x is the shape anisotropy of the elementary units (i.e., the width to length ratio of the rectangles). In the case of mixtures of rod-like and disk-like particles, x is the relative concentration of the disk-like particles. The first order transition to the isotropic phase is marked as a dashed line. The second order N$_u$–N$_b$ phase transitions are represented with solid lines (from [8, 13]).

3   Structural Features

The N$_b$ phase was first observed experimentally by Yu and Saupe [14] in an amphiphilic system consisting of potassium laurate + 1-decanol + D$_2$O. It was suggested [15] that a convenient way of obtaining the N$_b$ phase in low molecular weight thermotropic systems is by bridging the gap between rod-like and disk-like molecules i.e., by preparing a mesogen that combines the features of the rod and the disk. It may also be possible to obtain an N$_b$ phase from discotic systems in which the constituent molecules are elliptical in shape rather than circular. By some chemical ingenuity it may be possible to synthesize compounds comprised of elliptically shaped molecules that exhibit an N$_b$ phase. The compounds designed in this manner are likely to exhibit the N$_b$ phase as a result of hindrance in the rotation of the constituent molecules along their long axis.

It is perhaps appropriate to mention here that both nematic liquid crystalline side chain polymers with laterally attached mesogenic groups [16] and sanidic aromatic polyamides [17] have been shown to exhibit N$_b$ phases.

Praefcke et al. [18] have discussed a few structural features or molecular shapes of compounds that have been synthesized with a view to obtaining an N$_b$ phase, and these are shown in Fig. 2. They are (a) elliptically shaped stretched disk, (b) two-dimensionally flat *bones* or disk–rod–disk, (c) *spoons* or disk–rod, and (d) a *pipet* or rod–disk–rod. It should be emphasized here that these are only a few examples and various other shapes are also possible.

Malthete et al. [19] were the first to make use of theoretical ideas and they synthesized 4-[3′,4′,5′-tri(p-n-dodecyloxybenzyloxy)]-benzoyloxy-4″-dodecyloxybenzoyloxy-biphenyl (compound **1**), which combined

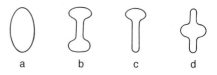

1

the features of the rod and the disk. Though they reported an N$_b$ phase for this compound, it has now been established that the characterization of the mesophase is uncon-

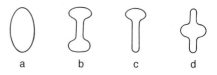

Figure 2. Shapes of low molecular weight compounds that have been synthesized with a view to obtaining an N$_b$ phase: (a) *ellipses*, (b) *bones*, (c) *spoons*, and (d) *pipets* (from [18]).

vincing and that there is a smectic C–nematic phase transition [20]. However, the possibility exists that the nematic phase could have an undetectable weak biaxiality.

4   Synthesis

Several low molecular weight compounds have been synthesized incorporating the features described above in an attempt to observe biaxiality in the nematic phase. Many such copper(II) complexes were synthesized [21] exhibiting the nematic phase in a metastable state. The overall shape of the molecules of these complexes resemble that of a *pipet*. One of these complexes, bis[1-(*p-n*-decylbiphenyl)-3-(*p*-ethoxyphenyl)propane-1,3-dionato] copper(II) (compound **2**) showed evidence of biaxiality in

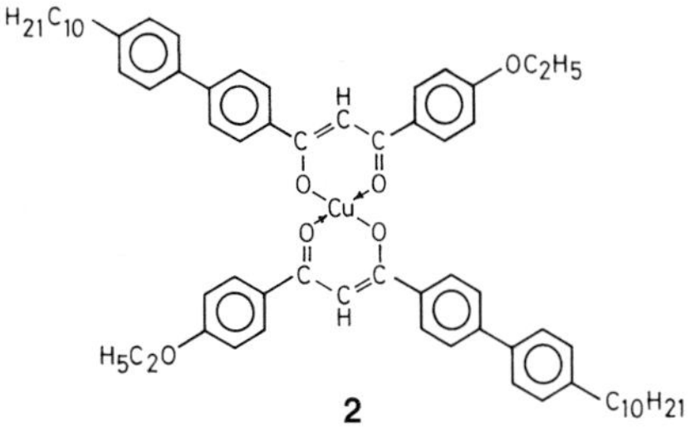

2

its nematic phase [22] and the transition temperatures of this complex are as shown

$$Cr \ (168.5 \ N) \ 186.6 \ I \qquad\qquad (1 \ a)$$

The experimental evidence put forward for the existence of the N_b phase in this complex will be described later.

Praefcke et al. [18, 23] synthesized a novel series of 2,3,4-trialkoxycinnamic acids (compound **3**) and, as expected, these car-

3

boxylic acids exist as dimers. These dimers are in *trans* conformation and are assumed to be more or less planar as a result of the conjugation between their phenyl and carboxylic groups. They also represent a new structural type of *bone*-shaped mesogens. The transition temperatures of these cinnamic acids are given in Table 1.

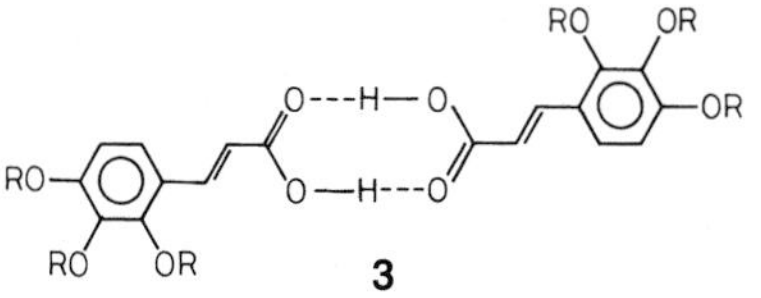

Praefcke et al. [18] have examined a number of other *bone*-shaped hexaalkoxy compounds (**4–7**), and their transition temperatures are given in Table 2. The nematic phases exhibited by these compounds have not yet been investigated.

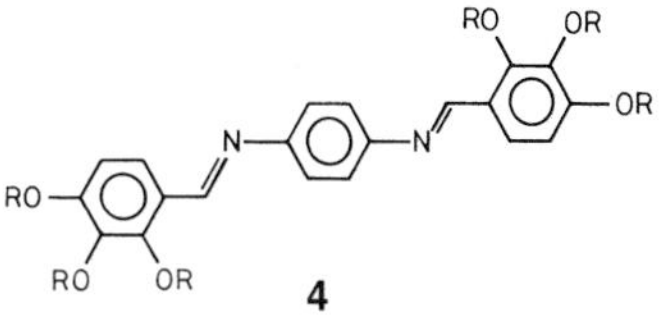

4

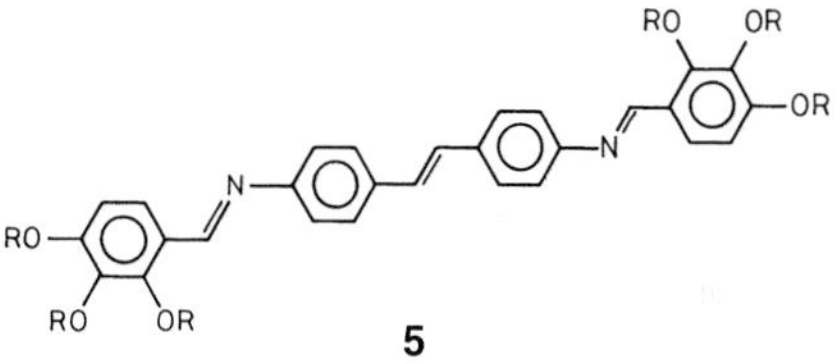

5

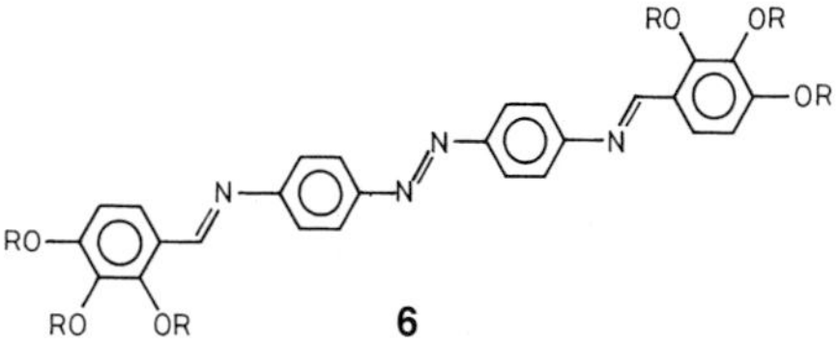

6

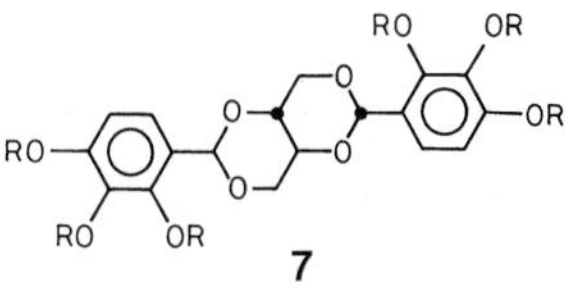

7

They also synthesized and examined [24] a number of multiethyne discotic nematogens with the hope of obtaining an N_b phase. The pentakis [(4-pentylphenyl)ethynyl]phenyl ethers (compound **8**), which are *spoon*-shaped, do exhibit nematic phases and the transition temperatures together

Table 1. Transition temperatures [a] and enthalpies [b] of a series of *trans*-2,3,4-trialkoxycinnamic acids (compound **3**) [c] (from [18]).

R	C	N_b	I
CH_3	173.2 32.2		
C_4H_9	75.6 18.6	(67.9) 0.8	
C_6H_{13}	51.2 14.1	59.6 0.9	
C_8H_{17}	54.2 48.7	(50.9) 0.5	

[a] Transition temperature in °C (top figure).
[b] Enthalpy in kJ mol^{-1} (bottom figure).
[c] Heating rate 1 K/min; C: crystalline phase, N_b: biaxial nematic phase, I: isotropic phase, the temperature in parentheses are for a monotropic transition.

with their enthalpies are summarized in Table 3.

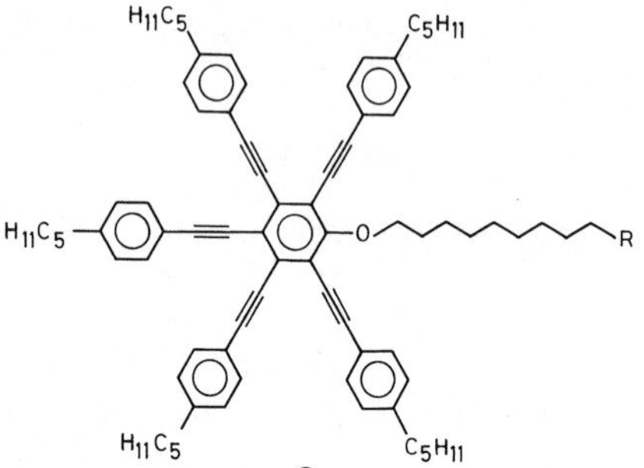

8

The α,ω-bis[penta(4-pentylphenylethynyl)phenoxy]alkanes (compound **9**), which are *bone*-shaped were obtained in reasonably good yields from palladium-catalyzed

Table 2. Transition temperatures [a] and enthalpies [b] of hexaethers (compounds **4–7**) [c] (from [18]).

Compound number	R	C	N	I
4a	CH_3	203.7 52.2	(103.4) [d]	
4b	C_4H_9	95.8 43.8	(49.1) [d]	
4c	C_6H_{13}	62.7 39.2	(33.6) [d]	
5a	CH_3	232.0 50.7	292.1 1.2	
5b	C_4H_9	145.4 39.3	154.4 1.3	
6a	CH_3	225.2 57.3	268.1 1.4	
6b	C_4H_9	150.0 41.1	(144.2) [d]	
6c	C_6H_{13}	123.8 41.9	(116.3) [d]	
7a	CH_3	204.1 55.4		
7b	C_6H_{13}	74.3 63.9		

[a] Transition temperature in °C (top figure).
[b] Enthalpy in kJ mol^{-1} (bottom figure).
[c] C: crystalline phase, N: nematic phase, and I: isotropic phase.
[d] Monotropic transitions determined by hot stage polarizing microscopy; enthalpies could not be measured.

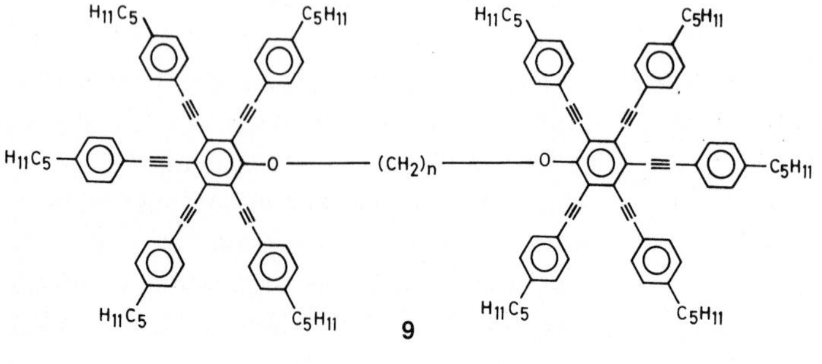

coupling reactions [18]. The transition temperatures of these twin ethers are summarized in Table 4. A plot of the transition tem-

9

Table 3. Transition temperatures[a] and enthalpies[b] of pentakis [(4-pentylphenyl)ethynyl]phenyl ethers (compound **8**)[c] (from [18]).

R	C	$N_{D,b}$	I
H	86.4	109.9	
	42.9	0.3	
CH=CH$_2$	76.7	101.4	
	36.2	0.2	

[a] Transition temperature in °C (top figure).
[b] Enthalpy in kJ mol^{-1} (bottom figure).
[c] Heating rate 5 K/min, C: crystalline phase, $N_{D,b}$: discotic nematic biaxial phase, I: isotropic phase.

Table 4. Transition temperatures[a] and enthalpies[b] of a series of α,ω-bis[penta(4-pentylphenylethynyl) phenoxy]alkanes (compound **9**)[c] (from [18]).

n	C	$N_{D,b}$	I
8	127.0	127.8	
	61.8	(...)[d]	
9	131.0	(112.5)	
	67.5	(...)[e]	
10	129.1	153.5	
	57.0	0.3	
11	118.2	140.8	
	62.2	0.3	
12	121.4	155.0	
	67.9	0.4	

[a] Transition temperatures in °C (top figure).
[b] Enthalpy in kJ mol^{-1} (bottom figure).
[c] C: crystalline phase, $N_{D,b}$: discotic nematic biaxial phase, I: isotropic phase.
[d] The enthalpy value for this transition could not be determined from the melting peak.
[e] This monotropic transition could only be determined by hot stage polarizing microscopy (heating rate 10 K/min).

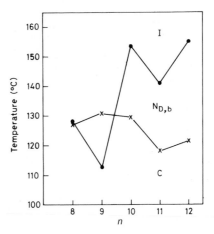

Figure 3. Plot of the transition temperatures (y-axis) versus the number of methylene units (x-axis) in the alkyl bridge of the twin ethers (compound **9**). C: crystal; $N_{D,b}$: discotic nematic biaxial; I: isotropic; x: melting points; ●: clearing points (from [18]).

5   Characterization Methods

5.1   Optical Studies

The uniaxial nematic (N_u) phase usually exhibits a schlieren texture with two and four point singularities (two and four brushes). However, the texture exhibited by complex **2**, shown in Fig. 4, often consists of [S]=1/2 (two-brush) disclinations similar to that exhibited by a smectic C phase. Also, this complex showed zig-zag disclinations [24] occasionally, and since this type of disclination has been observed in N_u phases [25] it cannot be regarded as conclusive evidence for biaxiality. It is also possible that an entanglement of disclination lines may lead to topological rigidity, as discussed by Toulouse [26] and others [27–29]. These possibilities have not been examined experimentally. Hence optical textures that

peratures of these twin ethers is shown in Fig. 3, which depicts the interesting odd–even effect of the clearing temperatures versus the number of methylene units between the disk-shaped pentayne head units.

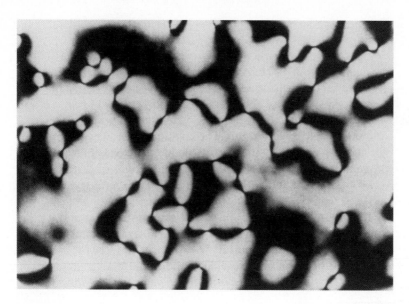

Figure 4. Schlieren texture of the nematic phase of copper(II) complex **2** (from [22b]).

have been reported for nematic phases so far do not seem to give direct evidence of biaxiality.

One of the most important and useful techniques for confirming the existence of an N_b phase is conoscopy. From a good interference pattern, it is possible to establish whether or not a system is biaxial. The evidence put forward using conoscopic observations for the compounds discussed in the previous section is given below.

A thick sample (~100 μm) of complex **2** aligned homeotropically using the combined effect of a silane coating and a 3 kHz AC (alternating current) electric field was used for conoscopic observations. Figure 5 a shows the conoscopic pattern for the pure material about 1.5 °C below the clearing point. An addition of a very small quantity of the uniaxial nematogen 4″-n-pentyl-4-cyano-p-terphenyl (5CT) to this complex seems to produce an optically positive N_u phase between the isotropic and N_b phases. Figures 5 b and c show the interference patterns for a mixture at two different temperatures. Figure 6 shows the phase diagram for the binary mixture for a concentration

range of 0–1% of 5CT. As can be seen, the temperature range of the N_u phase increases with increasing concentration of 5CT. These delicate experiments suggest a weak biaxiality of the nematic phase of complex **2**.

In the case of 2,3,4-trialkoxycinnamic acids (compound **3**) [18, 23], the conoscopic observations were made on much thinner samples (~20 μm). These acids, sandwiched between two glass plates, the inside surfaces of which were coated with transparent tin oxide, were aligned homeotropically using an electric field. The interference pattern obtained under these conditions is shown in Fig. 7. The position of the dark brushes is independent of the temperature. Rotation of the sample around 90° makes the brushes coalesce and reappear in the perpendicular direction, as shown.

Perhaps, among all the observations made so far, those conoscopic investigations carried out on (1) nonyl pentakis[(4-pentylphenyl)ethynyl]phenyl ether (compound **8**) R=H and (2) 1,12-bis(pentakis[(4-pentylphenyl)ethynyl]phenyloxy)dodecane (compound **9**), $n = 12$ are the most

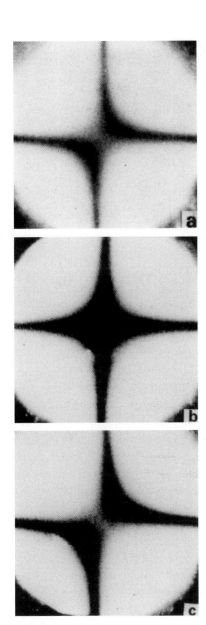

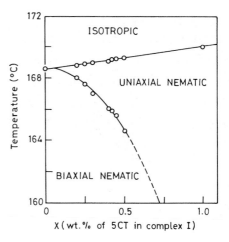

Figure 6. Partial phase diagram of the binary mixture of complex **2** and 5CT in the concentration range 0–1% 5CT, showing the I–N_u and N_u–N_b phase boundaries. The dashed portion of the curve represents an extrapolation (from [22b]).

interesting cases [24]. In these two cases, a homogeneously aligned sample (23 μm thick) between two glass plates was used. The interference pattern obtained in the nematic phase for the bis ether (compound **9**), $n=12$, is shown in Fig. 8. The isogyres form a cross when the trace of the optic axial plane lies parallel to the extinction position, as shown in (a). Rotating the sample by ±45° breaks the cross into hyperbolic brushes which are illustrated in (b) and (c). The clear splitting into two arcs could be observed only when the glass plates were rubbed in a uniform direction. The sign of biaxiality has been shown to be negative for both the compounds using sensitive red plates. It has been pointed out that the separation of the two arcs can be enhanced by a flow of the nematic fluid in the direction of rubbing, and the pattern becomes diffuse when the schlieren texture is formed from the homogeneous texture. These two compounds represent the first examples of thermotropic optically negative N_b phases [24].

Figure 5. Conoscopic figures: (a) biaxial nematic phase of pure complex **2** at $T_{N-I}-T=1.5\,°C$; (b) uniaxial nematic phase of a mixture of 99.8% complex **2** +0.2% 5CT at $T_{N-I}-T=0.7\,°C$; (c) biaxial nematic phase of the same mixture at $T_{N-I}-T=5.5\,°C$. Sample thickness ~125 μm (from [22a]).

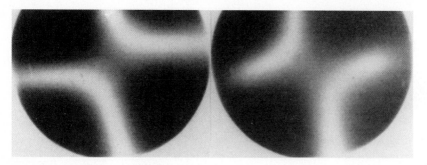

Figure 7. Conoscopic figures of an aligned sample of 2,3,4-trihexyloxycinnamic acid dimer (compound **3**, R = n-C$_6$H$_{13}$) in its nematic phase: on the left rotated 0°; on the right 90° (from [23]).

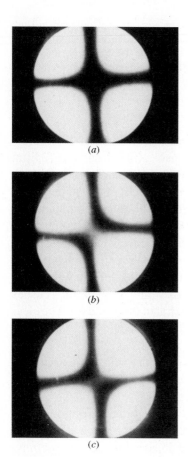

(a)

(b)

(c)

Figure 8. Interference patterns of the biaxial nematic phase of compound **9** (n = 12) at 137 °C. (a) The trace of the optic axial plane lies parallel to the polarizer or analyzer (extinction position); (b) and (c) represent the trace of the optic axial plane enclosing an angle of +45° or −45° with the polarizer position (from [24]).

5.2   X-Ray Diffraction Studies

The wide angle X-ray diffraction method was employed in the case of a sanidic aromatic polyamide [17] which exhibited a nematic phase. The characteristic feature of the X-ray pattern was the observation of two amorphous halos, which indicated the absence of rotational symmetry about the long chain axis. Based on this observation as well as the conoscopic interference pattern, it was concluded that the polyamides exhibit the N$_b$ phase.

Assuming the N$_b$ phase to be an orthorhombic fluid, X-ray diffraction methods have been used to distinguish it from the N$_u$ phase. The X-ray diffraction patterns of the uniaxial nematic phase of 4'-n-octyloxy-4-cyanobiphenyl (8OCB) and that of the nematic phase exhibited by complex **2** are shown in Figs. 9 and 10, respectively.

The additional pair of peaks obtained in the equatorial scan have been attributed to an orthorhombic structure and hence to the biaxiality of the nematic phase [22c].

Figure 11 shows the space filling molecular model of 2,3,4-trihexyloxycinnamic acid in its dimeric form and Fig. 12 shows the X-ray diffraction pattern of a magnetically aligned sample of this acid. In Fig. 12,

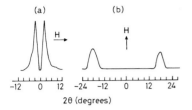

Figure 9. Raw microdensitometer scans of the X-ray intensity versus the diffracting angle (2θ) for the uniaxial nematic phase of 80CB at 77 °C: (a) meridional scan (parallel to H); (b) equational scan (perpendicular to H) (from [22c]).

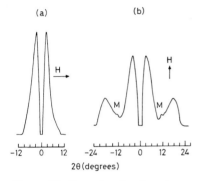

Figure 10. Raw microdensitometer scan of the X-ray intensity versus the diffracting angle (2θ) for the nematic phase of complex **2** at 166.5 °C: (a) meridional scan (parallel to H); (b) equatorial scan (perpendicular to H). M represents the diffraction peak from the mylar film (from [22c]).

the maxima (a) and (c) perpendicular to each other reveal the nematic character of the mesophase (also seen from the optical texture), whereas the third maximum (b) at small angles and perpendicular to the magnetic field points to peculiarities of the structure. The X-ray data, however, clearly indicates that the 2,3,4-trihexyloxycinnamic acid is dimeric in nature and has a flat board-like (sanidic) structure, which in principle should exhibit the N_b phase [18, 23].

6 Concluding Remarks

The controversy concerning the existence of the N_b phase in low molecular weight thermotropic liquid crystalline compounds is not yet resolved. However, the occurrence of the N_b phase in amphiphilic and polymeric systems has been clearly established. Many compounds possessing both rod-like and disk-like structural characteristics have been synthesized exhibiting the nematic phase. All these compounds have a direct transition from the nematic to the isotropic phase, in contrast to the theoretical predic-

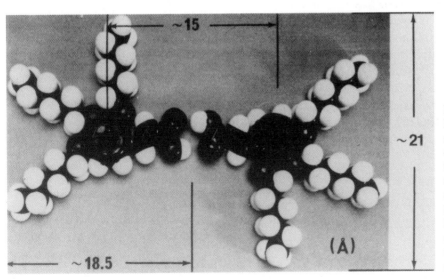

Figure 11. Photograph of a space filled model of the 2,3,4-trihexyloxycinnamic acid dimer (compound **3**, R = n-C_6H_{13}) including three estimated molecular parameters (from [23]).

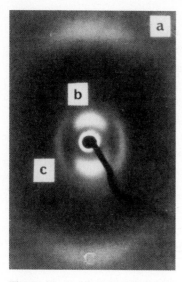

Figure 12. X-ray diffraction pattern of an oriented sample of the 2,3,4-trihexyloxycinnamic acid dimer (compound **3**, R = n-C_6H_{13}) in its nematic phase. According to the scattering maxima, the periods of the density waves in angstroms are as follows: (a) 4.4, (b) 19.5, and (c) 14.0 (from [23]).

tion that the N_b phase should occur between two N_u phases. Moreover, the thermodynamic data reported for these transitions is also first order, though weak in some cases. Though some evidence has been shown that these compounds exhibit the N_b phase, the same is yet to proved unequivocally. As mentioned in Sec. 2, there have been a number of theoretical approaches to the N_b phase and some aspects of the physical properties have also been studied. For those who are interested in these aspects of the N_b phase, a list of references [31–53] is included.

7   References

[1] P. G. de Gennes, *The Physics of Liquid Crystals*, Clarendon, Oxford **1974**.

[2] S. Chandrasekhar, *Liquid Crystals*, 2nd ed., Cambridge University Press, Cambridge **1992**.

[3] T. R. Taylor, J. L. Fergason, S. L. Arora, *Phys. Rev. Lett.* **1970**, *24*, 359.

[4] M. J. Freiser, *Phys. Rev. Lett.* **1970**, *24*, 1041.

[5] C. S. Shih, R. Alben, *J. Chem. Phys.* **1972**, *57*, 3055.

[6] R. Alben, *Phys. Rev. Lett.* **1973**, *30*, 778.

[7] J. P. Straley, *Phys. Rev. A* **1974**, *10*, 1881.

[8] R. Alben, *J. Chem. Phys.* **1973**, *59*, 4299.

[9] R. G. Caflitsch, Z. Y. Chen, A. N. Berker, J. M. Deutch, *Phys. Rev. A* **1984**, *30*, 2562.

[10] Z. Y. Chen, J. M. Deutch, *J. Chem. Phys.* **1984**, *80*, 2151.

[11] Y. Rabin, W. E. McMullen, W. M. Gelbart, *Mol. Cryst. Liq. Cryst.* **1982**, *89*, 67.

[12] A. Stroobants, H. N. W. Lekkerkerker, *J. Phys. Chem.* **1984**, *88*, 3669.

[13] Y. Galerne, *Mol. Cryst. Liq. Cryst.* **1988**, *165*, 131.

[14] L. J. Yu, A. Saupe, *Phys. Rev. Lett.* **1980**, *45*, 1000.

[15] S. Chandrasekhar, *Mol. Cryst. Liq. Cryst.* **1985**, *124*, 1.

[16] F. Hessel, H. Finkelmann, *Polym. Bull.* **1986**, *15*, 349; F. Hessel, R. P. Herr, H. Finkelmann, *Makromol. Chem.* **1987**, *188*, 1597.

[17] M. Ebert, O. H. Schonherr, J. H. Wendorff, H. Ringsdorf, P. Tschirner, *Makromol. Chem., Rapid Commun.* **1988**, *9*, 445.

[18] K. Praefcke, B. Kohne, B. Gundogan, D. Singer, D. Demus, S. Diele, G. Pelzl, U. Bakowsky, *Mol. Cryst. Liq. Cryst.* **1991**, *198*, 393.

[19] J. Malthete, L. Liebert, A. M. Levelut, *C. R. Acad. Sci. Paris* **1986**, *303*, 1073.

[20] J. C. Bhatt, S. S. Keast, M. E. Neubert, R. G. Petschek, *Liq. Cryst.* **1995**, *18*, 367.

[21] S. Chandrasekhar, B. K. Sadashiva, S. Ramesha, B. S. Srikanta, *Pramana J. Phys.* **1986**, *27*, L713; S. Chandrasekhar, B. K. Sadashiva, B. S. Srikanta, *Mol. Cryst. Liq. Cryst.* **1987**, *151*, 93.

[22] (a) S. Chandrasekhar, B. K. Sadashiva, B. R. Ratna, V. N. Raja, *Pramana J. Phys.* **1988**, *30*, L491; (b) S. Chandrasekhar, B. R. Ratna, B. K. Sadashiva, V. N. Raja, *Mol. Cryst. Liq. Cryst.* **1988**, *165*, 123; (c) S. Chandrasekhar, V. N. Raja, B. K. Sadashiva, *Mol. Cryst. Liq. Cryst. Lett.* **1990**, *7*, 65.

[23] K. Praefcke, B. Kohne, B. Gundogan, D. Demus, S. Diele, G. Pelzl, *Mol. Cryst. Liq. Cryst. Lett.* **1990**, *7*, 27.

[24] K. Praefcke, B. Kohne, D. Singer, D. Demus, G. Pelzl, S. Diele, *Liq. Cryst.* **1990**, *7*, 589.

[25] Y. Galerne, L. Liebert, *Phys. Rev. Lett.* **1985**, *55*, 2449.

[26] Y. Galerne, J. Itoua, L. Liebert, *J. Phys.* **1988**, *49*, 681.

[27] G. Toulouse, *J. Phys. Lett. (Paris)* **1977**, *38*, L-67; V. Poenaru, G. Toulouse, *J. Phys. (Paris)* **1977**, *8*, 887.

[28] N. D. Mermin, *Rev. Mod. Phys.* **1977**, *51*, 591.

[29] H. R. Trebin, *Adv. Phys.* **1982**, *31*, 195.

[30] M. Kleman, *Points, Lines and Walls*, Wiley, New York **1983**.

[31] P. B. Vigman, A. I. Larkin, V. M. Filev, *Sov, Phys. JETP* **1976**, *41*, 944.

[32] A. Saupe, P. Boonbrahm, L. J. Yu, *J. Chim. Phys.* **1983**, *80*, 7.

[33] D. L. Johnson, D. Allender, R. DeHoff, C. Maze, E. Oppenheim, R. Reynolds, *Phys. Rev.* **1977**, *B16*, 470.

[34] P. H. Keyes in *Proc. Eighth Symp. on Thermo Physical Properties – Thermophysical Properties of Fluids*, Vol. 1 (Ed.: V. Sengers), The American Soc. of Mech. Engineers, New York **1981**, p. 419.

[35] E. A. Jacobsen, J. Swift, *Mol. Cryst. Liq. Cryst.* **1982**, *87*, 29.

[36] J. W. Doane in *NMR of Liquid Crystals* (Ed.: J. W. Emsley), Riedel, Dordrecht **1985**, p. 441.

[37] S. R. Sharma, P. Palffy-Muhoray, B. Bergersen, D. A. Dunmur, *Phys. Rev.* **1985**, *A23*, 3752.

[38] D. W. Allender, M. A. Lee, *Mol. Cryst. Liq. Cryst.* **1984**, *110*, 331; D. W. Allender, M. A. Lee, N. Hafiz, *Mol. Cryst. Liq. Cryst.* **1985**, *124*, 45.

[39] T. C. Lubensky, *Mol. Cryst. Liq. Cryst.* **1987**, *146*, 55.

[40] A. Saupe, *J. Chem. Phys.* **1981**, *75*, 5118.

[41] M. Liu, *Phys. Rev.* **1981**, *A24*, 2720.

[42] H. Brand, H. Pleiner, *Phys. Rev.* **1981**, *A24*, 2777.

[43] E. A. Jacobsen, J. Swift, *Mol. Cryst. Liq. Cryst.* **1981**, *78*, 311.

[44] G. E. Volovik, E. I. Kats, *Zh. Eksp. Teor. Fiz.* **1981**, *81*, 240.

[45] U. D. Kini, *Mol. Cryst. Liq. Cryst.* **1984**, *108*, 71; U. D. Kini, *Mol. Cryst. Liq. Cryst.* **1984**, *112*, 265.

[46] E. Govers, G. Vertogen, *Phys. Rev.* **1984**, *A30*, 1998.

[47] A. Chaure, *J. Eng. Sci.* **1985**, *23*, 797.

[48] U. D. Kini, S. Chandrasekhar, *Physica* **1989**, *156A*, 364; U. D. Kini, S. Chandrasekhar, *Mol. Cryst. Liq. Cryst.* **1990**, *179*, 27.

[49] A. V. Kaznacheev, A. S. Sonin, *Sov. Phys. Crystallogr.* **1988**, *33*, 149.

[50] L. G. Fel, *Sov. Phys. Crystallogr.* **1989**, *34*, 737; **1990**, *35*, 148.

[51] D. Monselesan, H. R. Trebin, *Phys. Status Solidi* **1989**, *155*, 349.

[52] S. Chandrasekhar, G. S. Ranganath, *Adv. Phys.* **1986**, *35*, 507.

[53] G. S. Ranganath, *Curr. Sci.* **1988**, *57*, 1.

Chapter XVI
Charge-Transfer Systems

Klaus Praefcke and Dirk Singer

1 Introduction

In principle, a combination of short range repulsive forces and anisotropic dispersive interactions between mesomorphic molecules is sufficient to understand their self-organization in mesophases [1]. This basic approach is used for theoretical descriptions of mesomorphic states, especially for their simulations by molecular dynamics or Monte Carlo calculations of systems composed of idealized particles. Even on this simplified level, realistic pair potentials constitute one of the crucial problems.

For real systems, the occurrence of liquid-crystalline properties is determined by the shape of the constituent mesogenic units and, as realized to an decreasing degree, also by specific interactions of various types between them. Thus, for example, (1) electrostatic ionic forces, (2) dipole/dipole interactions, (3) hydrogen bonding, and (4) charge-transfer situations are often important in supporting the formation and stability of supramolecular structures in organized systems. In this context, in recent times, charge-transfer interactions have attracted increasing interest in the field of liquid crystal research; on the one hand, due to the fascinating possibilities concerning the variation or induction of mesomorphic self-organization, and on the other hand, connected to strategies successfully applied in materials science related to synthetic metals, organic semiconductors, molecular electronics, and nonlinear optics. Attractive interactions between π-systems are well known and have significant impact on such diverse phenomena as the vertical base–base interaction in DNA (deoxyribonucleic acid), drug-intercalation into DNA, the packing of aromatic molecules in crystals, the tertiary structures of proteins, conformational and binding properties of polyaromatic macrocycles, complexation in some types of guest-host systems, and porphyrin aggregation [2, 3]. In addition, strong π–π interactions are crucial for the formation of sufficiently large bands of electronic states required in charge-transfer complex based organic conductors [4], and might also control the molecular self-organization of such systems. However, the basic contributions to such interactions, usually denoted as π–π, charge-transfer (CT), or electron donor–acceptor (EDA) interactions, are not very well understood, despite the fact that they are often used to describe and explain the aggregation of molecules with π-electron systems in solution or even in higher condensed states [2, 3]. The self-organization of molecules in such ordered states is in general controlled by several types of interaction; some of the basic contributions are (1) van-der-Waals forces, (2) electrostatic interactions due to static distributions in the molecules, and (3) charge-transfer interactions [5]; the latter ones might be described as a stabilization due to mixing the ground state of two interacting molecules A and B with a charge-separated excited state A^+B^-.

In systems, composed of good electron donors and good electron acceptors, the formation of charge-transfer complexes is characterized either by charge-transfer transitions in the UV/VIS absorption spectra or at least by a broadening of the absorption. However, relatively little is known about the relative significance of electrostatic versus charge-transfer interactions for the molecular self-organization in condensed phases [2, 3e]. Especially for systems without spectroscopic evidence for the presence of strong CT-interactions, an approach based on electrostatic forces involving the out-of-plane π-electron density seems to be sufficient to explain geometrical requirements for interactions between aromatic molecules [2].

In the following, we will present and discuss examples where the liquid-crystalline self-organization is significantly governed by $\pi-\pi$ interactions. However, the relative strength of the charge-transfer versus electrostatic dipole–dipole or quadrupole–quadrupole interactions is generally not known. Thus we use the synonyms charge-transfer or electron donor acceptor interactions or complexes here to denote and classify those systems that show (1) spectroscopic evidence, such as, for example, a CT-band or a broadening in the UV/VIS absorption spectrum or (2) at least a significantly deeper color than the neat components as an indication of strong $\pi-\pi$ interactions in a kind of "chemists approach".

In addition, CT interactions are especially relevant for binary or multi-component systems in the context of mesophase induction or variation [6]. These terms denominate a significant change in the mesomorphic self-aggregation in the mixture compared to those of its constituting components, for example, the occurrence of a new mesophase; we also include here the stabilization of a mesophase of one of the compounds. However, these expressions do not cover the observation of a "latent" mesophase due to eutetic melting point depression.

It should also be mentioned here that the terms "liquid crystalline mixed phases" [6, 7a] of "flüssigkristalline Mischphasen" [7b], sometimes used to designate mesomorphism in multi-component systems, are misleading as they imply a mixture of phases, but in fact we are dealing with phases of mixtures, often with specific molecular aggregation.

2   Calamitic Systems

Charge-transfer interactions in liquid-crystalline systems first gained interest more than twenty years ago in connection with the report of enhanced thermal stability of the nematic phase in mixtures of N-(4-methoxybenzylidene)-4-butylaniline (MBBA) with 4-cyano-4'-pentylbiphenyl [CPB (more often referred to as 5CB)] by Park et al. [8a]. The phase diagram of this binary system shows two eutectics and a stabilized solid state, later found to be a smectic B phase (see [8c]). The maxima of the "melting" as well as of the clearing temperature are observed near to the equimolar composition. This particular phase behavior has been explained in terms of the formation of a molecular 1:1 electron donor (MBBA)/acceptor (CPB) complex due to charge-transfer interactions, evident by a broad absorption band around 550 nm. The molecular complexing is also thought to be responsible for the increased dielectric anisotropy observed for the equimolar MBBA/CPB mixture [8b]. A similar stabilization of the nematic phase of CPB has been found by doping this nematogen with 4-aminobiphenyl, a non-mesogenic electron donor compound [9] (see Table 1).

Table 1. Examples of electron donor pairs with rod-like mesogenic units exhibiting stabilized or induced mesophases.

Electron donor	Electron acceptor	Stabilized and/or induced mesophases	Ref.
1) CH$_3$O—◯—⟍N—◯—C$_4$H$_9$ **MBBA**	H$_{11}$C$_5$—◯—◯—CN **CPB**	N, SmB	[8, 13]
2) ◯—◯—NH$_2$	H$_{11}$C$_5$—◯—◯—CN **CPB**	N	[9]
3) R—◯—X ⟍Y—◯—R' X = CH or N, Y = CH, N, or NO R, R' or R', R = N(CH$_3$)$_2$ and alkyl or alkyloxy	R—◯—X ⟍Y—◯—R' R, R' or R', R = NO$_2$ and alkyl or alkyloxy or R = R' = NO$_2$	N, SmA	[10]
4) (CH$_3$)$_2$N—◯—◯—N(CH$_3$)$_2$ R—◯—⟍N—◯—R' R, R' or R', R = N(CH$_3$)$_2$ and alkyloxy	R—◯—⟍N—◯—R' R, R' or R', R = NO$_2$ and alkyloxy O$_2$N—◯—◯—NO$_2$ NC—◯—◯—CN	N, SmA	[11]
5) RHN—◯—◯—NHR R = alkyl	O$_2$N—◯—⟍C O—◯—OR	N, SmA	[14 a, e]
	RO—◯—COO—◯—NO$_2$	N, SmA	[14 a]
	RO—◯—C—O⟍ N=C—◯—CN CH$_3$	N, SmA	[14 a, b]
	RO—◯—COO—◯⟍ CN CN	N, SmA	[14 b]
	H$_{17}$C$_8$X—◯—◯—CN X = O or NH	N, SmB, SmE	[14 c]

Matsunaga et al. have studied the phase diagrams of a wide variety of electron donor acceptor systems in which both components are more or less rod-shaped and represent at least "potential" mesogens (see Table 1 and [10]). Most of these compounds under evaluation were composed of two phenyl rings connected by a rigid, two-membered double bond spacer; the components designed as electron acceptors usually carried a nitro or cyano group, while the electron donors had dimethylamino substituents (see Table 1). The phase diagrams of these binary EDA systems are characterized by congruent melting points of a molecular complex, usually near to the 1:1 composition. Furthermore, most systems show stabilized/induced nematic and/or smectic A phases. The molecular complexing in these binary mixtures is accompanied by charge-transfer bands in the visible region, or at least a broadening of the spectrum, supporting the model of strong $\pi-\pi$ interactions of the charge-transfer type.

In another study of the phase behavior of Schiff base donor compounds, with 4,4'-dinitro- or 4,4'-dicyanobiphenyl as electron acceptors, or of N,N,N',N'-tetramethylbenzidine as an electron donor with suitable Schiff base acceptors, similar mesophase inductions/stabilizations are reported [11] (see Table 1). Even p-substituted derivatives of nitrobenzene and benzonitrile, respectively, are able to induce or stabilize smectic mesophases (SmA and/or SmB) in mixtures with, for example, the mesogen N-(4-propyloxybenzylidene)-4-hexylaniline [12]. However, in these systems a dipolar mechanism for the enhanced molecular aggregation seems more likely.

Iida examined and explained the induced/stabilized nematic phase in the MBBA/CPB system and those of donor/acceptor mixtures based on N-(4-X-benzylidene)-4-Y-aniline derivatives (see Table 1) by a modified solid solution model [13]; the charge-transfer interaction energies between donor and acceptor in the nematic state were estimated as 0.16 kJ/mol for the MBBA/CPB complex [13a] and as 0.38–0.52 kJ/mol for the second type of EDA mixtures [13b].

Clear evidence of charge-transfer interactions in binary mesomorphic mixtures was provided by the homologous series of 4,4'-bis(alkylamino)-biphenyls studied by the former Halle group (Table 1). These mesogens are excellent electron donors and form colored EDA complexes with a number of liquid-crystalline compounds as electron acceptors, usually accompanied by the stabilization or induction of SmA phases [14a–c, e], but for certain combinations even induced SmE and B phases are observed [14c]. Sharma et al. have shown that the charge-transfer bands in the visible region of the absorption spectrum show negative dichroism, thus the electron transition moments responsible for the absorption are oriented perpendicular to the main axis of the molecules [14a]. Demus et al. estimated the charge-transfer complex stability constant and the formation enthalpy of 4,4'-bis(butylamino)-biphenyl/4-(2,2-dicyanoethenyl)phenyl-4'-nonyloxy-benzoate via the temperature dependence of the CT band, and of the strongly negative excess volumina in this system [15]. The complex formation enthalpies were found to range from -58 to -219 kJ/mol depending on the phase type (SmA, N, or Iso). These values are of the same order of magnitude as those of nonmesogenic systems in the isotropic or solid states, respectively [4]. In addition, complex stability constants of liquid crystalline electron donor acceptor pairs of dialkylazoxybenzene/PCB, determined by a dielectric method in cyclohexane solution [16a], are also comparable to those reported by Demus et al. [15]. Density measurements of these binary systems confirm the formation of molecular 1:1 complexes [16b]. It is noteworthy that the EDA interaction energies here are significantly higher than those reported by Iida [13], but he assumed a complete dissociation of the complex at the phase transition N–Iso. A model for induced smectic phases in binary mixtures of liquid crystals due to molecular complexing, especially due to charge-transfer interactions, has been proposed some time ago [17]. Within the mean field approximation, the influence of complex formation on the N–SmA transition temperature and the types of phase diagrams are discussed.

In this context it seems noteworthy that in mixtures of a nematogen carrying a lateral dicyanoethenyl substituent with electron donor nematogens, no phase induction or stablization occurs, despite the color change on mixing the components, indicating the formation of a CT complex [14d]. Thus the relative orientations of the donor and the acceptor moieties have to fulfill geometric requirements supporting a parallel arrangement of the molecular main axes for the formation of induced mesophases.

For a mixture of 4,4'-dinonanoylbiphenyl with N,N'-didecyl-4,4'-diaminobiphenyl, showing an induced smectic A phase and a slightly positive deviation of the clearing temperature from ideal linear behavior, an interesting synergistic combination of EDA interactions and hydrogen bridging has been proposed [18].

It should be emphasized here that electron donor–acceptor combinations are not the only ones giving rise to mesophase induction or variation. Similar phase behavior is also known for binary systems composed of suitable proton donor–acceptor pairs due to hydrogen bridging [19, 20], or for combining components of complementary molecular shapes as, for example, in the so-called "filled smectics" [21]. In mixtures of usually unpolar smectic mesogens, nematic phases are often induced if the aggregation to higher ordered layered structures is hindered by, for example, a significant difference in the molecular length [22].

Another common type of phase induction/variation is observed in mixtures of a terminal polar compound, usually a nematogen carrying a nitro or a cyano group, and a nonpolar one [23]. In such systems induced SmA-phases are often formed, but they might also exhibit higher ordered smectic phases, for example SmB, SmC, SmE, and SmI or form nematic ones, sometimes even reentrant. The phase behavior of these binary systems is often explained in terms of dipole–dipole or dipole–induced dipole interactions, partially in analogy to the specific mesophase structures observed for highly polar neat mesogens [6, 24]. According to Prost, the mesomorphic properties of such dipolar mesogens are a result of the competition between the incommensurable length of the molecules and their dimers [24e].

The electron acceptors discussed so far resemble these terminal polar compounds in their structure, or are even identical. Cladis, for example, has studied the phase behavior of mixtures of butyloxybenzylidene octylaniline with cyanooctyloxybiphenyl showing stabilized smectic phases of the A- and B-type, as well as an induced SmE phase; the results are summarized in a Landau description [23g]. However, the phase behavior of this particular system, strongly related to that studied by Park et al. [8], is discussed in terms of dipolar pair formation. Furthermore, the phase behavior and macroscopic properties, e.g., densities and rotational viscosities, in mixtures of polar 4-cyano derivatives of biphenyl with apolar azoxy compounds, were found to differ significantly from those comprising the related cyanophenylcyclohexane derivatives, but in both type of systems the formation of mixed associates occur [23i].

These examples raise the question of whether the π–π interactions of the charge-transfer type are the main cause for the enhanced aggregation in mesomorphic EDA systems or if dipole–dipole or dipole–induced dipole interactions are more important. If the latter is true, charge-transfer contributions, evident by a related absorption band, would just constitute a "side effect" of a suitable molecular arrangement as a result of dipolar forces. In this context it should be noted that up to now neither the mesomorphic molecular organization nor the dynamic behavior of the mesogens in calamitic EDA-systems have been studied systematically. Thus it is not clear if these systems show any peculiarities compared to those of binary mixtures of compounds polar–polar or polar–nonpolar in nature. Finally, an anisotropic, two-dimensional solvation model developed by Szabon for the classification of induced smectic phases, based only on the requirement of a quasi-hexagonal close-packing and a sufficient difference of the rigid and the flexible aliphatic segments of the components, should be mentioned because no specific intermolecular interactions are required within its framework [25].

Based on the scarce experimental data available, a final decision concerning the relative significance of charge-transfer versus dipolar interactions for the self-organization in these EDA-systems is not possible yet. However, recently it could be shown that 2,4,7-trinitrofluorene (TNF), an electron acceptor with only a small dipole moment and well known for inducing/stabilizing columnar aggregates of disc-shaped molecules due to charge-transfer interactions (see next section), can also induce or stabilize smectic mesophases. In contrast to the electron acceptors discussed so far, this one is much stronger and its molecular shape is far from that of even a "potential" calamitic mesogen. However, with bromo-bridged dinuclear palladium organyls, the formation of deeply colored CT complexes and a significant stabilization of the SmA phase of these centrally fused twin-like mesogens has been observed [26]. Furthermore, the bis(2,3,4-trisdodecyloxyphenyl)-substituted Schiff base of 4,4′-diaminostilbene, a nematogen, exhibits an induced SmA phase with TNF, showing the maximum clearing temperature for the 1:2 donor–acceptor ratio (see Fig. 1) [20].

With 2-(2,4,5,7-tetranitro-9-fluorenylideneaminooxy)-propionic acid (TAPA) as electron acceptor, the induction of a cholesteric nematic phase is observed instead, with a maximum clearing temperature around 45 mol% (–)-TAPA [20]. Further examples of a TNF-induced mesophase is provided by the SmA phase of various polyether macrocycles, a dimesogenic twin, and of numerous calamitic heterocycles [27]. MBBA, on the other hand, forms a red CT complex with TNF, but shows only a strong depression of the N–Iso transition temperature [26]. A similar behavior is observed for chloranil EDA complexes of 1,4-bis(phenylethinyl)benzene mesogens [28a] and for tetracyanoethylene (TCNE) in 4-cyano-4′-pentylbiphenyl (PCB) [28b]. In the latter system, TCNE was found to lower the isotropization temperature much more strikingly than, for instance, 2,3-dimethylbutane which shows no evidence of complex formation with PCB, suggesting its peculiar effect [28b]. Furthermore, despite the electron acceptor character of PCB itself, the charge-transfer interaction is clearly evident from related absorption bands; the complex formation enthalpy was determined as 14.98 kJ/mol. The unusually high order parameters found for some anthraqui-

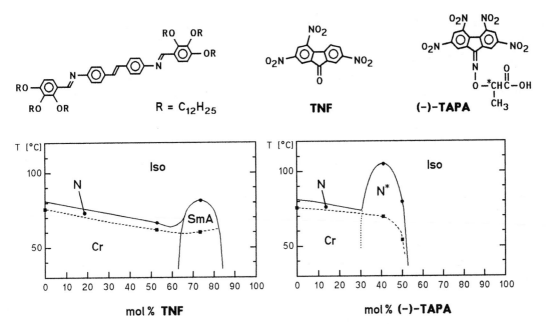

Figure 1. Phase behavior of a nematic bisimine donor with TNF and (–)-TAPA, respectively, as electron acceptors [20]; Cr: crystalline, SmA: smectic A, N and N*: nematic and cholesteric nematic phase, respectively, and Iso: isotropic liquid.

none dyes in the nematic phase of cyano-phenylcyclohexane derivatives have also been attributed to CT interactions [29].

The examples discussed above indicate that the formation of induced or stabilized mesophases in binary systems composed of calamitic compounds is usually the result of a complex combination of intermolecular interactions and packing effects. Therefore separation and analysis of the different contributions is difficult as long as none of them is clearly dominating the mesomorphic self-organization of the molecules. Especially generalizations concerning the relative significance of electron donor–acceptor properties versus permanent dipole moments of components in such binary systems, for the formation of induced/stabilized mesophases, should be viewed carefully. However, taking into account the (so far) small number of calamitic compounds investigated for charge-transfer complexation and meso-phase induction/variation, with strong non-

calamitic electron acceptors like TNF, TAPA, or 7,7,8,8-tetracyano-1,4-quinodi-methane (TCNQ), a great multitude of similar binary systems can be expected to exhibit charge-transfer induced mesophases.

The induction or stabilization of liquid-crystalline phases in binary CT systems, where the donor and the acceptor compounds are rod-shaped, has been utilized in the design of polymers to enhance their mesomorphic properties; the combination of mesomorphic properties and those of polymeric systems, e.g., glassy behavior, has attracted interest due to possible applications, for example, in the field of nonlinear optics. Side-chain copolymers with a polyester or polyacrylate backbone containing 4-me-thoxyazobenzene and 4-cyanoazobenzene units, for example, show the formation of an SmA phase with an enhanced mesomorphic temperature range [30]. Similar behavior is observed for copolystyrene derivatives carrying electron-rich donor units (4-methoxy-azobenzene), as well as for electron acceptor moieties (4-nitro- or 4-cyanoazobenzene) via flexible spacers [31]. A series of similar copolymethacrylates with ethylene spacers and carbonate linkage only shows a negative deviation from the clearing temperature of the copolymers compared to the two homopolymers [32a]. Co-methacrylates containing both an electron-donating carbazol group and an electron-accepting nitrophenyl or cyanophenyl unit show stabilized or induced SmA phases [32b], most probably due to interactions between these functions, to their longer spacer and the ether linkage to the mesogenic side groups; this tendency on thermal behavior was not seen in a related copolymer having carbazol and (4'-methoxybenzylidene)anilin groups both of which are of electron-donating nature. The mesomorphic self-organization in such polymeric systems is not only governed by the interaction profiles of the mesogenic groups, but also by backbone-related structural features leading, for example, to a variety of variations of the SmA type of mesophase [33].

Asymmetric dimeric liquid crystals with charge-transfer groups constitute good models for the mesomorphic self-organization of the related polymers discussed above. Their mesophase structures have been studied in some detail [34]. In this context, the formation of intercalated smectic phases of the A, C, and I type is especially noteworthy [34c]. However, the nature of the specific interactions between the unlike mesogenic groups, as well as the conformation of the spacer, are still to be explained [34d].

A nematic, main-chain polycarbonate polymer comprising azobenzene mesogenic units shows an induced SmA phase in mixture with a nematic nitroazobenzene dimeric liquid crystal [35]; a behavior resembling that of related monomeric mesogenic mixtures. A more specific example of charge-transfer contributions to the mesomorphic self-aggregation of a main chain polymer has been observed for rigid polyesters based on 1,4-dialkylesters of pyromellitic acid (1,2,4,5-benzenetetracarboxylic acid) copolymerized with 4,4'-biphenol [36a]. These rigid-rod polyesters are di-mesomorph; the layered, low temperature mesophase shows a high degree of ordering attributed to CT interactions between the biphenyl donor units and the pyromellitic acid ester acceptor moiety [36b].

Finally, doping surface-stabilized ferroelectric liquid crystals with charge-transfer complexes, for example, tetramethyltetrathiafulvalene/octadecyltetracyano-1,4-quinodimethane, has been utilized to improve their bistability [37]: Ions from the CT complex form an internal electric field in reverse to the applied pulse. By applying this phenomenon, a ferroelectric liquid crystal cell with perfect bistability and inverted memory characteristics was designed [37]. Ono and Nakanowatari developed a method for the determination of the internal electric field and studied a TCNQ doped ferroelectric liquid crystal [38].

In another approach towards enhanced switching properties of ferroelectric liquid crystals, a conductive overlayer of the tetrathiafulvalene/TCNQ CT complex on the SiO layers was used to shorten the response time [39].

3   Noncalamitic Systems

In the area of noncalamitic liquid crystals, especially those with a planar, sheet-like or discoid structure, the impact of charge-transfer interactions on the mesomorphic self-organization is much higher due to the favoritism of a face-to-face type, intercalated stacking in columnar supramolecular structures. However, surprisingly, the importance of CT interactions for the design of such mesomorphic systems has only been recognized in very recent years.

In this context, it is interesting that the first examples of mesomorphic CT complexes of discoid or sheet-like donor compounds and a strong electron acceptor [7,7,8,8-tetracyano-1,4-quinodimethane (TCNQ)] constitute an exception from the usually observed behavior. These complexes of bi-4-H pyran or bi-4-H thiopyran derivatives carrying four alkyl- or alkyloxy-phenyl substituents with TCNQ (see Fig. 2), have been prepared as organic conductors [40]. The neat alkylphenyl derivatives in the bi-4-H pyran series and alkyl- or alkyloxy-phenyl-substituted bi-4-H thiopyran donors are mesomorphic [40], but the structures of their mesophases are still not clear.

The equimolar charge-transfer complex of the dodecyloxyphenyl substituted bi-4-H thiopyran with tetracyano-1,4-quinodime-

thane (TCNQ) received special interest due to combining of conductivity with meso-morphic properties (Fig. 2). In the crystalline state the conductivity was found to be $\sigma = 0.7 \ \Omega^{-1} \ cm^{-1}$ for a single crystal and $10^{-3} \ \Omega^{-1} \ cm^{-1}$ for a powder sample [40d]. The latter order of magnitude is also found for the conductivity in the mesophase range [40e, f]. Based on X-ray diffraction studies, the mesophase of this CT complex has been identified as lamello-columnar, combining features of a smectic mesophase with those of a columnar one [40c]. Up to now only a few examples [41] of lamellar mesophases, first postulated by Billard [42], have been observed. Two variations are known so far: The L_D phase, without any intralamellar order of the sheet-like molecules, and the lamello-columnar L_{Col} phase with columnar stacking within each layer. A sketch of the donor self-organization of the dodecyloxy-phenyl substituted bi-4-H thiopyran with TCNQ is depicted in Fig. 2. The TCNQ acceptor molecules probably fill the space

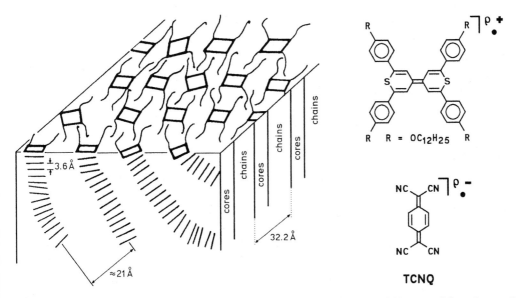

Figure 2. Schematic representation [40c] of the tetra(4-dodecyloxyphenyl) dithiapyranylidene donor self-organization in the lamello-columnar (L_{Col}) mesophase of an equimolar CT complex with TCNQ; the TCNQ molecules are not shown in this drawing.

between two stacks within a layer and may be in segregated stacks of their own. However, also an intercalated columnar structure common in mesophases of charge-transfer complexes of disc-like electron donors with strong electron acceptors cannot be ruled out, taking into account the observation of such an alternate donor–acceptor stacking in the crystalline state of the semiconductor 2,2′,6,6′-tetramethyl-$\Delta^{4,4'}$-bisthiapyran/TCNQ [43] and the X-ray diffraction pattern of charge-transfer based hexagonal columnar phases of TNF complexes of different discoid donor compounds discussed below.

2,2′,6,6′-Tetra(4-dodecylphenyl)-$\Delta^{4,4'}$-bipyran also forms a charge-transfer complex with TCNQ as well as cation–radical charge-transfer salts, the first ones of disc-like shape showing columnar mesomorphic behavior over a broad temperature range [44]. This shows liquid-crystalline properties, but no evaluation of the mesophase structure has been reported up to now. The TCNQ charge-transfer salt shows only a very small electric conductivity due to the complete charge transfer between donor and acceptor.

More recent is the idea that induced columnar stacking interactions could be beneficial for the design of mesomorphic systems composed of disc- or sheet-shaped molecules. This approach was first realized by Ringsdorf et al. in (1) charge-transfer induced nematic columnar (N_{Col})[1] phases exhibited by mixtures of amorphous side-chain polymers carrying 2,3,6,7,10,11-hexasubstituted triphenylene moieties with TNF and (2) TNF-induced columnar hexagonal (Col_{ho})[1] phases of related main-chain polymers [46]. This type of doping also makes possible the formation of one-phase blends of otherwise incompatible polymers [46] and opens new ways in molecular engineering of liquid-crystalline polymers [47].

More significant with respect to the impact of charge-transfer interactions on the mesomorphic self-organization of discoid moieties, however, is a look at electron–donor acceptor systems composed of low-molecular-weight compounds. 2,3,6,7,10,11-Hexaethers of triphenylene, for example, forming columnar hexagonal mesophases with intracolumnar stacking order (Col_{ho}) [48] show a stabilization of the existing columnar structures and an extension of the mesomorphic range on addition of TNF due to CT interactions [49]. The clearing temperature steadily increases with the TNF concentration and maximum stabilization is observed for the 6:4 electron donor acceptor ratio. The TNF–CT complexes of the triphenylene hexaethers with this composition behave like a single compound, showing a sharp transition from the Col_{ho} to the isotropic phase; higher or lower acceptor concentrations in such mixtures lead to biphasic ranges. The same type of mesophase stabilization is also observed in mixtures of 2,3,6,7,10,11-hexaester derivatives of triphenylene with TNF, while analogous nonliquid crystalline hexaethers with polyethy-

---

[1] The nomenclature of columnar mesophases formed by disc-like molecules, up to now (falsly) called discotic, and their short notation with D and indices describing the intercolumnar arrangement and the intracolumnar stacking clearly give more and more rise to misconceptions. Since the utilization of this "D-nomenclature", a wide variety of molecular shapes, often supported by specific types of intermolecular interactions, e.g., hydrogen bonding, have been found to result in the formation of columnar mesophases. Therefore, a more systematic nomenclature based on the type of mesomorphic self-organization, e.g., lamellar or columnar, was strongly desirable and has also been introduced recently [45] which quickly seems to gain growing acceptance. Although questions in this context are still under discussion we use here the novel type of notation for columnar (Col) situations in mesophases: N_{Col} (nematic columnar), N_D (nematic discotic), Col_x (columnar with, e.g., x=ob for -oblique, -tilted etc.), L_{Col} (lamello-columnar), L_D (lamellar discotic).

leneoxy chains show the induction of a Col_{ho} phase instead, even with TNF contents of only about 15 mol% [49].

X-ray studies prove the formation of mixed stacks of donor and acceptor molecules with a regular, alternating packing in the induced or stabilized columnar mesophases [49]. Very strong support for the nonstatistical, alternating donor–acceptor stacking results from the observation of X-ray diffraction maxima corresponding to a d-spacing of 6.92 Å (0.692 nm) in the mesophase of a 1:1 2,3,6,7,10,11-hexakis-pentyloxytriphenylene/TNF complex, besides the usual intracolumnar reflex at 3.46 Å (0.346 nm) [49a]. The fact that this diffraction is observed emphasizes the dissimilarity in the electron density of the donor and acceptor molecules.

Not only 2,4,7-trinitro-9-fluorenone (TNF) but also other electron acceptors, for example, TCNQ, 1,2,4,5-tetracyanobenzene, the enantiomers of 2-(2,4,5,7-tetranitro-9-fluorenylideneaminooxy)-propionic acid (TAPA) or 2,4,5,7-tetranitro-9-fluorenone, and derivatives of the latter one, are effective in this type of mesophase induction [50]. In particular, a TNF derivative carrying two hexadecanoyl side chains (2,4,7-trinitrofluorenylidene-9-malonic acid bis(hexadecyl) ester) is remarkable because of the induction of a nematic columnar type (N_{Col}) mesophase with hexakispen-

tyloxytriphenylene [51]. The formation of an electron donor–acceptor complex here also stabilizes the columnar aggregates, but at the same time the self-organization of these columns in a two-dimensional lattice is prevented by the long chains of the acceptor molecules. With 2,3,6,7,10,11-hexa-kishexadecyloxytriphenylene, however, a Col_{ho} phase is induced by this electron acceptor because its two flexible chains are well within the aliphatic periphery of the CT-based columnar aggregates. A schematic representation of the columnar electron donor–acceptor aggregates and their self-organization in the two-dimensional lattice of a Col_{ho} phase or their nematic arrangement (N_{Col} phase) is sketched in Fig. 3.

While the hexagonal arrangement of such charge-transfer based columns in the absence of intercolumnar restraints is the most common one, there is also the TNF complex of a triphenylene main chain polymer, for example, forming a rectangular two-dimensional lattice (Col_{ho} phase) [49b].

The structural features of charge-transfer based mesophases, especially those of polymeric as well as low molecular weight triphenylene derivatives, as a function of the donor–acceptor ratio, have been studied by X-ray diffraction in some detail [49].

Moreover, using the model proposed in the eighties [17] to describe the induction of smectic phases as a result of complex for-

N_{Col}

Col_{ho}

Figure 3. Schematic representation of thermotropic mesophases based on charge-transfer induced formation of intercalated donor–acceptor stacks; N_{Col}: nematic columnar and Col_{ho}: columnar hexagonal ordered mesophase (see footnote 1 in Sec. 3 commenting on the nomenclature of such columnar mesophases). ◯: electron donor, ◯: electron acceptor.

mation, Wendorff et al. could account qualitatively for the experimental findings concerning charge-transfer induced mesophases of discoid compounds [49b].

An interesting type of mesomorphic self-organization based on CT interactions is observed for charge-transfer twins consisting of triphenylene 2,3,6,7,10,11-hexaether [52a] or radial multiyne [52b] units (as discoid electron donors) each covalently linked via flexible spacers to trinitrofluorenone acceptor moieties. The structural model proposed for their mesophase assumes a columnar arrangement of the molecules in such a way that mixed stacks of donor and acceptor units are formed. The columns containing triphenylene parts are arranged in an orthorhombic two-dimensional lattice; neighboring columns are connected along specific directions via the flexible spacers, giving rise to highly anisotropic properties [52a]. In addition, these columnar aggregates and their self-organization in ordered monolayer films have been studied by atomic force microscopy [52c]. A possible borderline case with respect to these two latter examples of twin CT complexes [52a,b] could be a series of sevenfold substituted triphenylene derivatives, very recently synthesized and studied for the first time [52d], carrying one acceptor function (e.g., NO_2) in a so-called α-position and six ethergroups at the six outer (β-)positions of that aromatic ring system. These new molecules are sterically heavily congested and are, therefore, helically deformed, but show strong stabilization of their columnar phase behavior in comparison to their parent hexaether without an acceptor attached to it by only one covalent bond [52d].

Main-chain polyesters based on 2,4,7-trinitrofluorenylidene malonate are also effective electron acceptors able to stabilize the columnar hexagonal mesophase of 2,3,6,7,10,11-hexakispentyloxytriphenylene [53a]. As an interesting note, a polysiloxane with 2,4,7-trinitrofluorenylidenemalonate side groups forms a nematic mesophase on its own, most likely due to microphase separation between the polysiloxane backbone and the side groups [53b].

The molecular motion in the charge-transfer induced/stabilized columnar hexagonal mesophases of 2,3,6,7,10,11-hexakisheptyloxytriphenylene, as well as of a related dimer and a polymer, each doped with TNF, have been studied by deuterium NMR [54a]. For the monomer, fast rotation around the column axis is observed, this motion is quenched for the dimer and the polymer. The TNF molecules, however, show free rotation in all three CT complexes. Further studies by ^{2}H-NMR combined with dielectric spectroscopy have been performed (1) on mixtures of 2,3,6,7,10,11-hexakispentyloxytriphenylene with a TNF derivative with two long aliphatic chains, (2) on mixtures of this hexakispentyloxytriphenylene with a TNF-derived acceptor main-chain polymer, and (3) on two donor–acceptor twin compounds with a different spacer length between the triphenylene donor and the acceptor moiety, giving some insight into the molecular dynamics of such charge-transfer systems [54b].

The optical properties of charge-transfer complexes of the above hexakispentyloxytriphenylene and related discotic electron donor compounds with TNF-type acceptors have been studied in solution and in thin films [55]. CT complexes of that hexaalkoxytriphenylene derivative with TNF, for example, show in both cases a charge-transfer absorption corresponding to 2.51 eV; the polarized absorption spectra obtained from oriented thin films prove that the charge-transfer interaction occurs along the alternate donor–acceptor stacking [55a]. In addition, picosecond time-resolved absorption spectroscopy based on Kerr ellipsometry revealed information about spectral and temporal characteristics of the oxidized donor/reduced acceptor radical ion pairs resulting from laser excitation, and confirmed that the electron transfer occurs in a direction perpendicular to the molecular planes of the donor and acceptor molecules [55b]. The photophysical properties of the TNF complexes of 2,3,6,7,10,11-hexakisalkyloxy derivatives of triphenylene differ significantly from those of the neat triphenylene hexaalkoxy mesogen; in particular the photoconductivity is lost [55c]. The chirooptical properties of CT complexes of hexaalkoxytriphenylene main-chain polyesters carrying four chiral or achiral alkyloxy side chains, with TNF, as well as with the enantiomers of 2-(2,4,5,7-tetranitro-9-fluorenylideneaminooxy)-propionic acid propyl ester [(+)- and (−)-TAPAE], have been studied by UV and CD (circular dichroic) spectroscopy in thin films [56]. All chiral CT mixtures show high optical activity arising from helical superstructures of the columns induced by the chiral acceptor in the case of achiral polymers or as a result of the chirality of the polymer itself.

TNF–CT complexes of 2,3,6,7,10,11-hexakispentyloxytriphenylene or related amphiphilic derivatives are also able to form Langmuir–Blodgett (LB) films [57]. Polarized UV/VIS and infrared spectra of those of TNF complexes of hexakisalkyloxytriphenylenes discussed here, for example, indicate a preferred orientation of the columnar axes parallel to the surface and to the dipping direction [57a]. An amphiphilic triphenylene derivative with a pair of ω-OH chains and four pentyloxy groups, forming a lamellar mesophase in the neat state, a nematic discotic main chain polymer derived from the aforementioned monomer both exhibit induced columnar hexagonal (Col_{ho}) mesophases with TNF in the bulk state [57b]. The X-ray diffraction pattern of the LB films of these CT complexes indicate a dense packing of the columnar aggregates lying parallel to the surface in a multilayer superstructure virtually identical to that in the bulk Col_{ho} phase, quite different to the double layer structure observed for LB films of the neat monomeric and polymeric triphenylene centered mesogens [41f].

Radial multiynes constitute another class of large discoid electron donors intensively studied under the aspect of charge-transfer induced mesophases [58–60]. The CT complex formation in all the examples described here is indicated by a deep red to brown color on mixing the donor and acceptor compounds. Hexaynes with benzene (**1**) [58a], naphthalene (**2**) [58b], or triphenylene (**3**) [58c] (see Fig. 4) as the "anchors" of "super-disc-cores" with six 4-alkylphenylethinyl "spokes" each form nematic discotic (N_D) phases; in the case of the naphthalene hexaynes with peripherical alkyl chains up to $n = 8$, these are accompanied by columnar rectangular and hexagonal phases (Col_{ro} and Col_{ho}, respectively) at higher temperatures. The addition of an electron acceptor, for example, TNF, leads to the formation of charge transfer complexes and the induction of columnar hexagonal phases with intracolumnar stacking order (Col_{ho}) [26, 49a]. The phase behavior of some members of these three radial hexayne series (**1**, **2**, and **3**) is shown in Fig. 4.

The maximum clearing temperature of the CT-induced Col_{ho} phases is found for the equimolar composition. X-ray diffraction studies of the charge-transfer induced mesophases of the TNF complexes of **1** with $n = 6$ proves the hexagonal arrangement of the columnar aggregates and the high intracolumnar stacking order [49a]. Interestingly, the naphthalene and triphenylene-centered radial hexaynes (**2**) and (**3**) also show induced columnar mesophases in binary mixtures with a variety of dipolar carbocy-

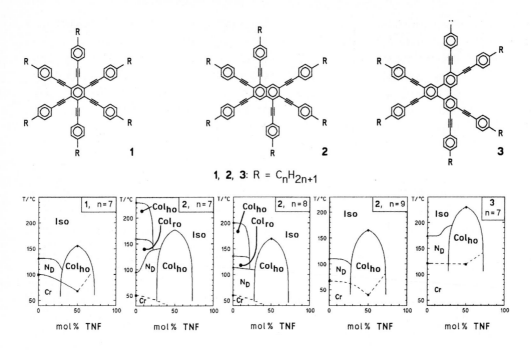

Figure 4. Schematic phase diagrams of members of the radial hexayne series **1**, **2**, and **3** as electron donors with TNF as the electron acceptor [26]; Cr: crystalline, N_D: nematic discotic, Col_r: columnar rectangular, and Col_h: columnar hexagonal mesophase (o: intracolumnar ordered), Iso: isotropic liquid (see footnote 1 in Sec. 3 commenting on the nomenclature of columnar mesophases).

clic compounds [61]. However, no indications for charge-transfer interactions are observed in these systems; the induced/stabilized columnar phases here are more likely the result of improved space filling in combination with dipolar interactions. Even more interesting is the phase behavior of radial pentayne derivatives as, for example, that of the pentayne ethers of type **4** shown in Fig. 5. These pentaynes with unsubstituted peripheral phenyl rings are nonmesomorphic in their pure states, but form liquid crystalline charge-transfer complexes with TNF [26, 59a, d]. The type of charge-transfer induced mesophase depends strongly on the length of the one alkyloxy chain present in the electron donor; an element of steric disturbance in the aggregation of the CT-based columns leads to a two-dimensional lattice (see Fig. 5).

In principle, similar behavior is observed for pentayne ethers carrying methyl groups in the para-positions of their peripheral phenyl rings with TNF, but here the neat pentayne donors are mesomorphic, exhibiting the nematic discotic (N_D) type of mesophase [26, 59d]. The members of this series with a pentyloxy or a nonyloxy chain show a slightly decreased clearing temperature of their N_D phases and induced Col_{ho} phases below the nematic state. The tridecyloxy and hexadecyloxy derivatives only exhibit the nematic state in mixtures with TNF of acceptor concentrations of more than 60 mol%. No phase boundaries are observed, but X-ray studies clearly show the presence of columnar aggregates in the nematic phases of mixtures of, for example, hexadecyl pentayne ether with more than 27 mol% TNF [26, 59d]. Several other radial

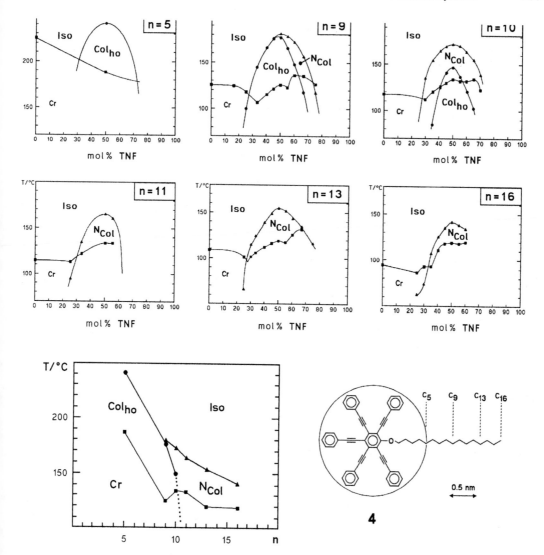

Figure 5. Schematic phase diagrams of members of the radial pentayne ether series **4** as electron donors with TNF as the electron acceptor and a plot of the transition temperatures of the equimolar CT complexes with TNF as a function of the number of carbon atoms n in the alkyloxy chain of **4**; Cr: crystalline, Col_{ho}: columnar hexagonal (o: intracolumnar ordered), N_{Col}: nematic columnar mesophase, and Iso: isotropic liquid (see footnote 1 in Sec. 3 commenting on the nomenclature of columnar mesophases).

pentakis(phenylethinyl)phenyl derivatives without side chains in the periphery also show induced nematic and hexagonal columnar mesophases with TNF as the electron acceptor [59b]. Amphiphilic pentayne ethers carrying a hydroxyl group, an ester, or a carboxylic acid function at the end of the ether chain behave similarly [59c]. These neat pentaynes form nematic discotic (N_D) phases when carrying peripheral alkyl side chains, and are nonmesomorphic without such substituents. The TNF–CT complexes of the first group show induced Col_{ho} phases, while those of the latter one display induced nematic phases, most probably of the N_{Col} type.

A nematic discotic twin-pentayne ether [62] also exhibits a charge-transfer induced Col_{ho} phase with TNF (see Fig. 6). For other examples of multiyne electron donors see [63].

By using the chiral electron acceptor (–)-TAPA, even the cholesteric versions of the nematic discotic phase N_D^* or of the nematic columnar one N_{Col}^* can be obtained, for example, by doping a nematic discotic pentayne ether with up to 30 mol% (–)-TAPA or by inducing the N_{Col}^* phase in a ternary mixture composed of pentayne ether (**4**) ($n = 16$) with TNF and (–)-TAPA; the latter electron acceptor does not itself lead to the induction of a columnar mesophase with such pentayne donors [26, 59 d].

The dielectric and elastic properties of the nematic columnar (N_{Col}) phase of charge-transfer mixtures of pentayne ether (**4**) ($n = 13$) with TNF have been studied and compared with those of the N_D type of mesophase [60]. The elastic constants K_1 and K_3 are both one order of magnitude higher in the N_{Col} phase; this dependence on the acceptor concentration indicates that the interaction is strongest for the equimolar composition.

Dinuclear palladium organyls with a discoid molecular shape constitute a class of electron donor compounds displaying an unusual and complex phase behavior with TNF as a strong electron acceptor [64] (see Fig. 7). The mesomorphism of these mesogens, exhibiting monotropic nematic discotic phases in their pure forms, in TNF–CT

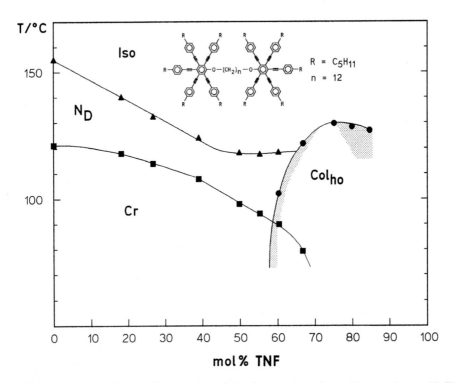

Figure 6. Schematic phase diagram of a radial twin-pentayne ether as electron donor with TNF as the electron acceptor [26]; Cr: crystalline, N_D: nematic discotic, Col_{ho}: columnar hexagonal mesophase (o: intracolumnar ordered), Iso: isotropic liquid (see footnote 1 in Sec. 3 commenting on the nomenclature of columnar mesophases).

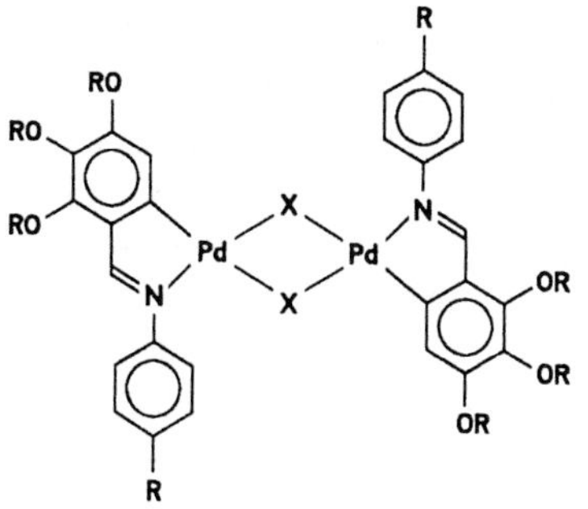

5a: X = Cl
  b: X = Br
  c: X = I
  d: X = SCN

R = hexyl

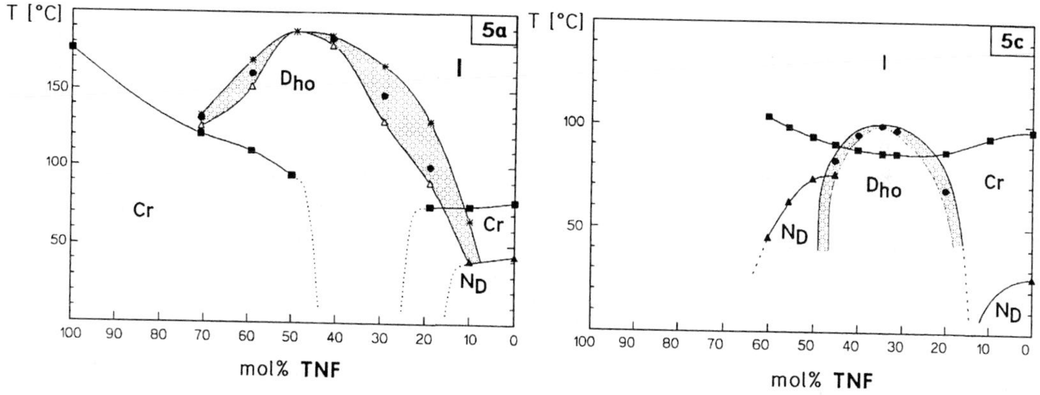

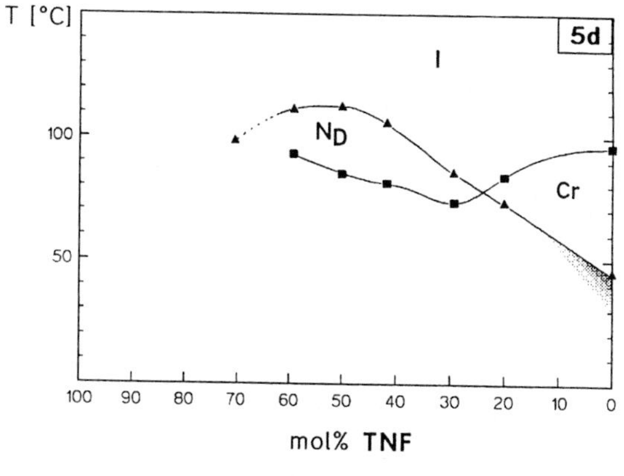

Figure 7. Schematic phase diagrams of the dinuclear palladium organyls 5a, c, and d with TNF as the electron acceptor [64]; Cr: crystalline, N_D: nematic discotic, Col_{ho}: columnar hexagonal mesophase (o: intramolecular ordered), Iso: isotropic liquid (see footnote 1 in Sec. 3 commenting on the nomenclature of columnar mesophases). Shaded areas mark biphasic regions.

complexes strongly depends on the bridging group X (Fig. 7). For the members a and b of series 5 with X=Cl or Br, for example, the induction of columnar hexagonal (Col_{ho}) mesophases is observed in mixtures with TNF, while the phase diagram of the binary system iodo-bridged compound 5c/TNF contains an additional nematic discotic area. The thiocyanato-bridged member 5d (X=SCN) shows a strong stabiliza-

tion of the nematic discotic (N_D) phase with TNF. Based on X-ray diffraction experiments, here the acceptor and donor molecules most probably aggregate only to charge-transfer pairs, not to bigger columnar aggregates. However, the N_D-phase of such an equimolar CT complex of 5d/TNF is completely miscible with the induced N_{Col} phase of a 1:1 complex of a pentayne ether (4) with TNF. Thus taking into account the absence of observable phase transitions or phase boundaries within the nematic regions in the phase diagrams of type **4** pentayne ethers/TNF systems (Fig. 5), a continuous development between these two nematic types of mesomorphic associations seems likely.

Phase diagrams similar to those of type **5** pallado-nematogens with TNF are also observed for a related series of dinuclear palladium organyls with hexyloxy instead of the hexyl chains [20] and for similar dinuclear platinum organyls [65].

Members of the 1,2,5,6,8,9,12,13-octa-(alkyloxy)dibenzopyrene series, each exhibiting a columnar hexagonal (Col_{hd}) mesophase in the pure state, show strong increases of the stability range of their mesophases on the addition of TNF, with a maximum stability for the equimolar donor/acceptor mixtures [66]. Besides being thermally more stable, the mesophases of the CT complexes are also more highly ordered and exhibit well defined intramolecular stacking order, classifying them as Col_{ho}. Deuterium NMR studies revealed planar diffusion around the columnar axis as the main molecular reorientation process in the mesophase. An X-ray diffraction peak at a position corresponding to twice the normal intracolumnar distance, as observed for the Col_{ho} phase of TNF–CT complexes of triphenylene hexaethers, indicates that the donor and acceptor molecules are stacked alternately in the columns. Moreover, when alkanes are added to these 1:1 charge transfer complexes, a lyotropic nematic phase of the N_{Col} type with columnar structures as the mesogenic units is induced [66]. Such nematic columnar lyomesophases with apolar organic solvents are only formed by compounds with especially strong intracolumnar interactions, such as, for example, long chain hexaamides of hexaaminobenzene [67a], dinuclear copper(II)- and rhodi-

um(II)carboxylates [67b, c], tetranuclear palladium [67d, e], and platinum organyls [45] or amides of azacrown ethers [67f]. However, charge-transfer interactions between disc- or sheet-shaped electron donor compounds and strong electron acceptors might lead to sufficient stability of columnar aggregates. Thus equimolar TNF–CT complexes of the hexaynes **1** or **3** with heptyl chains and benzene or triphenylene as the "anchor" display nematic lyomesophases with alkane solvents [26]. Even charge-transfer complexes of triphenylene hexaethers with TNF form nematic columnar lyomesophases with alkanes [68].

Tetranuclear palladium organyls exhibiting thermotropic columnar mesophases with an oblique two-dimensional lattice and without intracolumnar stacking order ($Col_{ob,d}$) [69] also form lyotropic mesophases with alkane solvents; in particular, some members exhibit two different lyotropic nematic phases side by side [67d, e]. Adding TNF results in a significantly increased stacking order in, for example, a 1:2 donor–acceptor complex. This CT complex forms only one nematic lyomesophase with, for example, pentadecane over an extended temperature range; by using the chiral electron acceptors (+)- or (–)-TAPA, a cholesteric nematic N* phase is observed [70a]. A dinuclear bisimine palladiumorganyl with two acetylacetonato ligands represents another example of charge-transfer induced thermotropic and lyotropic mesomorphism [70b]. This nonmesomorphic compound exhibits a charge-transfer induced columnar mesophase with TNF; the two-dimensional arrangement of the columns in this mesophase is still not clear. However, in ternary mixtures with alkanes such as, for example, pentadecane, a lyotropic nematic phase is observed [70b]. A tricyclyne, i.e., three phenyl rings connected via three ethinyl links forming an annulene type macrocycle and carrying six hexyloxy chains in its outer phenyl positions, nonmesomorphic in its pure state, also forms a highly viscous, most probably columnar mesophase with TNF [71].

Based on the observations described in this section, it seems that for a qualitative discussion of the interactions leading to mesomorphic self-organization of disc- or sheet-like compounds, separation into an intracolumnar potential and an intercolumnar one is usually sufficient. Neat discoid mesogens show only one type of potential along their column axis; the ordered or disordered nature of a columnar phase depends on the width of the potential minimum in this description, while the thermal stability reflects the depth of the minimum. For charge transfer induced or stabilized columnar phases, three types of intermolecular interactions – donor/donor, acceptor/acceptor, and donor/acceptor – must be considered. In the case of the intercalation of electron acceptors, e.g., TNF, the mixed interaction is the dominant one, with a deeper and narrower potential minimum than for the donor/donor combination of the neat mesogen, leading to alternating stacking and increased stability and intracolumnar positional order. A dominance of the donor/donor and/or acceptor/acceptor interactions might lead to the formation of separate columnar aggregates of the donor and acceptor compounds, as discussed for the mesomorphic tetracyanoquinodimethane (TCNQ) complexes of dithiapyranylidene derivatives (see Fig. 2 and [40]).

The intercolumnar interactions in a mesomorphic system built of columns of sufficient stability then can be reduced by the addition of an alkane solvent or by attaching sterical restraints. The potential minimum normal to the column axis becomes flattened and broadened, resulting in the formation of a nematic columnar mesophase and finally an isotropic solution of columnar aggregates. This behavior is well known for the so-called chromonic lyomesogens in aqueous systems [72]. Finally, dinitrophenylhydrazones of trialkyloxy derivatives of benzaldehyde should be mentioned here as mesogens with charge-transfer contributions to their mesomorphic self-organization in the pure state [20, 73]. With a 3,4,5 substitution pattern of the alkyloxy chains, a columnar hexagonal mesophase with very high intracolumnar stacking order (D_{ho} phase) is formed as a result of a regular head-to-tail packing of subsequent molecules [73]. This packing mode has been attributed to a planar conformation of the molecules supported by hydrogen bridge formation and charge-transfer interactions between the donor and acceptor moieties of subsequent molecules. The same type of intracolumnar organization is observed for a 2,3,4-trialkyloxy derivative, but here the columns form a rectangular two-dimensional lattice (Col_{ro} phase) (see Fig. 8 and [20]).

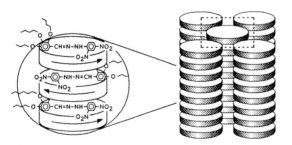 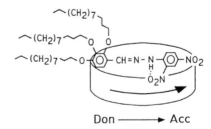

Figure 8. Model of the mesomorphic self-organization of 2,3,4-trisdodecyloxy-benzaldehyde-2',4'-dinitrophenylhydrazone in its columnar rectangular mesophase [20]; Don: donor part, Acc: acceptor part.

It is especially noteworthy that this dinitrophenylhydrazone can act as an electron acceptor in mixtures with the hexayne nematodiscogens **1** or **2** (*n* = 9, see Fig. 4), but also constitutes an electron donor in mixtures with TNF. Both types of binary systems show an induced mesophase with textures typical for columnar hexagonal mesophases [20].

Partially related to the topics discussed here is the doping of, for example, triphenylene 2,3,6,7,10,11-hexaethers with small amounts of bromine or the Lewis acid aluminum trichloride to induce electrical conductivity [74]; however, no change of the mesomorphic self-organization is caused by these dopants. In another very recent work [75] trifluoroacetic acid, a representative of a new family of dopants, i.e. that of Brønsted acids, was found to stabilize or induce columnar mesophases in disc-like alkoxy compounds, such as aforementioned hexaalkyloxytriphenylene and also hexa-alkyloxytribenzocyclononene. This acid forms 1:1 molar complexes with these two types of arene hexaethers. The mesophases exhibited by these complexes show enhanced properties for the mesogenic homologues and induced mesomorphism for the lower ones, which in the neat form are not mesogenic. It is suggested that the enhancement of the mesomorphic properties is due to the formation of oxonium complexes.

On the other hand, it is well known that perfluoro arenes, for example, hexafluorobenzene or octafluoronaphthalene, form some type of "complex" with their hydrocarbon parent compounds, having melting points higher than those of the pure components [76]. Very recently, the deuterium NMR spectra of the octafluoronaphthalene/naphthalene system have been explained in terms of a highly ordered columnar mesophase [76d]. However, the face-to-face stacking in such solids and also in liquids is better described as a result of interactions based on different molecular electric quadrupole moments [2, 76c].

4   References

[1] See, for example: a) G. R. Luckhurst, *Ber. Bunsenges. Phys. Chem.* **1993**, *97*, 1169; b) A. P. J. Emerson, G. R. Luckhurst, S. G. Whatling, *Mol. Phys.* **1994**, *82*, 113.

[2] C. A. Hunter, J. K. M. Sanders, *J. Am. Chem. Soc.* **1990**, *112*, 5525 and references cited therein.

[3] For related recent reports, as an introduction see, for example, a) M. Harmata, C. L. Barnes, *J. Am. Chem. Soc.* **1990**, *112*, 5655; b) S. B. Ferguson, E. M. Sanford, E. M. Seward, F. Diederich, *J. Am. Chem. Soc.* **1991**, *113*, 5410; c) F. Cozzi, M. Cinquini, R. Annuziata, J. S. Siegel, *J. Am. Chem. Soc.* **1993**, *115*, 5330; d) A. W. Schwabacher, S. Zhang, W. Davy, *J. Am. Chem. Soc.* **1993**, *115*, 6995; e) C. A. Hunter, *Angew. Chem.* **1993**, *105*, 1653; *Angew. Chem., Int. Ed. Engl.* **1993**, *32*, 1584; f) F. Cozzi, F. Ponzini, R. Annuziata, M. Cinquini, J. S. Siegel, *Angew. Chem.* **1995**, *107*, 1092; *Angew. Chem., Int. Ed. Engl.* **1995**, *34*, 1019.

[4] See for example a) Z. G. Soos, *Ann. Rev. Phys. Chem.* **1974**, *25*, 121; b) P. Delhaes, *Mol. Cryst. Liq. Cryst.* **1983**, *96*, 229.

[5] M. Rigby, E. B. Smith, W. A. Wakeham, G. C. Maitland, *The Forces between Molecules*, Clarendon, Oxford **1986**.

[6] For a short introduction, see: S. Diele, *Ber. Bunsenges. Phys. Chem.* **1993**, *97*, 1326.

[7] a) G. Pelzl, D. Demus, H. Sackmann, *Z. Phys. Chem.* **1968**, *238*, 22; b) H. Stegemeyer, *Nachr. Chem. Tech. Lab.* **1993**, *41*, 966, see p. 970.

[8] a) J. W. Park, C. S. Bak, M. M. Labes, *J. Am. Chem. Soc.* **1975**, *97*, 4398; b) J. W. Park, M. M. Labes, *J. Appl. Phys.* **1977**, *48*, 22; c) see related comments in [23d] and [23f].

[9] J. W. Park, M. M. Labes, *Mol. Cryst. Liq. Cryst.* **1977**, *34*, 147.

[10] a) K. Araya, Y. Matsunaga, *Bull. Chem. Soc. Jpn.* **1980**, *53*, 3079; b) K. Araya, Y. Matsunaga, *Mol. Cryst. Liq. Cryst.* **1981**, *67*, 153; c) K. Araya, Y. Matsunaga, *Bull. Chem. Soc. Jpn.* **1981**, *54*, 2430; d) K. Araya, Y. Matsunaga, *Bull. Chem. Soc. Jpn.* **1982**, *55*, 1710; e) K. Araya, N. Homura, Y. Matsunaga, *Bull. Chem. Soc. Jpn.* **1982**, *55*, 1953; f) Y. Matsunaga, I. Suzuki, *Bull. Chem. Soc. Jpn.* **1984**, *57*, 1411; g) Y. Matsunaga, *Mol. Cryst. Liq. Cryst.* **1985**, *125*, 269; h) N. Homura, Y. Matsunaga, M. Suzuki, *Mol.*

Cryst. Liq. Cryst. **1985**, *131*, 273; i) Y. Matsunaga, N. Kamiyama, Y. Nakayasu, *Mol. Cryst. Liq. Cryst.* **1987**, *147*, 85.

[11] a) M. Fukui, Y. Matsunaga, *Bull. Chem. Soc. Jpn.* **1982**, *55*, 3707.

[12] a) E. Chino, Y. Matsunaga, *Bull. Chem. Soc. Jpn.* **1983**, *56*, 3230; b) E. Chino, Y. Matsunaga, M. Suzuki, *Bull. Chem. Soc. Jpn.* **1984**, *57*, 2371.

[13] a) Y. Iida, *Bull. Chem. Soc. Jpn.* **1982**, *55*, 313; b) Y. Iida, *Bull. Chem. Soc. Jpn.* **1982**, *55*, 2661.

[14] a) N. K. Sharma, G. Pelzl, D. Demus, W. Weissflog, *Z. Phys. Chem.* **1980**, *261*, 579; b) D. Demus, G. Pelzl, N. K. Sharma, W. Weissflog, *Mol. Cryst. Liq. Cryst.* **1981**, *76*, 241; c) D. Demus, A. Hauser, G. Pelzl, U. Böttger, S. Schönburg, *Cryst. Res. Technol.* **1985**, *20*, 381; d) W. Weissflog, G. Pelzl, D. Demus, *Cryst. Res. Technol.* **1986**, *21*, 117; e) G. Pelzl, U. Böttger, S. Diele, D. Demus, *Cryst. Res. Technol.* **1987**, *22*, 1321.

[15] D. Demus, A. Hauser, G. Pelzl, U. Böttger, S. Schönburg, *Cryst. Res. Technol.* **1985**, *20*, 381.

[16] a) W. Waclawek, R. Dabrowski, A. Domagala, *Mol. Cryst. Liq. Cryst.* **1982**, *84*, 255; b) K. W. Sadowska, A. Zywocinski, J. Stecki, R. Dabrowski, *J. Physique* **1982**, *43*, 1673.

[17] W. H. de Jeu, L. Longa, D. Demus, *J. Chem. Phys.* **1986**, *84*, 6410.

[18] A. Kolbe, G. Pelzl, W. Weissflog, *Adv. Mol. Relax. Interact. Proc.* **1982**, *24*, 251.

[19] a) T. Kato, J. M. J. Frechet, *J. Am. Chem. Soc.* **1989**, *111*, 8533; b) T. Kato, J. M. J. Frechet, *Macromolecules* **1989**, *22*, 3818; c) M.-J. Brienne, J. Gabard, J.-M. Lehn, I. Stibor, *J. Chem. Soc., Chem. Commun.* **1989**, 1868; d) T. Kato, A. Fujishima, J. M. J. Frechet, *Chem. Lett.* **1990**, 919; e) T. Kato, P. G. Wilson, A. Fujishima, J. M. J. Frechet, *Chem. Lett.* **1990**, 2003; f) L. Y. Yu, J. M. Wu, S. L. Wu, *Mol. Cryst. Liq. Cryst.* **1991**, *198*, 401; g) C. Fouquey, J.-M. Lehn, A.-M. Levelut, *Adv. Mater.* **1990**, *2*, 254; h) C. M. Paleos, J. Michas, *Liq. Cryst.* **1992**, *11*, 773; i) H. Kresse. C. Dorscheid, I. Szulzewski, R. Frank, R. Paschke, *Ber. Bunsenges. Phys. Chem.* **1993**, *97*, 1345; j) C. M. Paleos, D. Tsiourvas, *Angew. Chem.* **1995**, *107*, 1839; *Angew. Chem., Int. Ed. Engl.* **1995**, *34*, 1696.

[20] K. Praefcke, B. Gündogan, D. Singer, unpublished results; Bilgi Gündogan, *Synthesen und Eigenschaften neuer flächiger Flüssigkristalle*, Köster Verlag, Berlin **1996**, ISBN 3-89574-156-6, Ph. D. Thesis, Technische Universität Berlin, Germany.

[21] a) S. Diele, G. Pelzl, W. Weissflog, D. Demus, *Liq. Cryst.* **1988**, *3*, 1047; b) G. Pelzl, A. Humke, S. Diele, D. Demus, W. Weissflog, *Liq. Cryst.* **1990**, *7*, 115; c) G. Pelzl, A. Humke, S. Diele, S. Ziebarth, W. Weissflog, D. Demus, *Cryst. Res. Technol.* **1990**, *25*, 587.

[22] See for example: a) M. E. Neubert, K. Leung, W. A. Saupe, *Mol. Cryst. Liq. Cryst.* **1986**, *135*, 283; b) B. Ziemnicka, A. de Vries, J. W. Doane, S. L. Arora, *Mol. Cryst. Liq. Cryst.* **1986**, *132*, 289; c) S. Diele, G. Pelzl, A. Humke, S. Wünsch, W. Schaefer, H. Zaschke, D. Demus, *Mol. Cryst. Liq. Cryst.* **1989**, *173*, 113.

[23] a) J. P. Schroeder, D. C. Schroeder, *J. Org. Chem.* **1968**, *33*, 591; b) J. S. Dave, P. R. Patel, K. L. Vasanth, *Mol. Cryst. Liq. Cryst.* **1969**, *8*, 93; c) C. S. Oh, *Mol. Cryst. Liq. Cryst.* **1977**, *42*, 1; d) G. Heppke, E.-J. Richter, *Z. Naturforsch.* **1978**, *33a*, 185; e) A. C. Griffin, R. F. Fisher, S. J. Havens, *J. Am. Chem. Soc.* **1978**, *100*, 6329; f) M. Domon, J. Billard, *J. Physique Colloque C3* **1979**, *40*, C3-413; g) P. E. Cladis, *Mol. Cryst. Liq. Cryst.* **1981**, *67*, 177; h) L. Songsheng, G. Jezhen, W. Liangyu, *Mol. Cryst. Liq. Cryst.* **1983**, *100*, 285; i) V. V. Belyaev, T. P. Antonyan, L. N. Lisetski, M. F. Grebyonkin, G. G. Slashchova, V. V. Petrov, *Mol. Cryst. Liq. Cryst.* **1985**, *129*, 221; j) L. K. M. Chan, G. W. Gray, D. Lacey, T. Srithanratana, K. J. Toyne, *Mol. Cryst. Liq. Cryst.* **1987**, *150b*, 335; k) K. Czuprynski, R. Dabrowski, B. Sosnowska, J. Baran, *Liq. Cryst.* **1989**, *5*, 505.

[24] a) F. Hardouin, A.-M. Levelut, M. F. Archad, G. Sigaud, *J. Chim. Phys.* **1983**, *80*, 53; b) J. Prost, P. Barois, *J. Chim. Phys.* **1983**, *80*, 65; c) H. T. Nguyen, *J. Chim. Phys.* **1983**, *80*, 83; d) R. Shashidhar, B. R. Ratna, *Liq. Cryst.* **1989**, *5*, 421; e) J. Prost, *Adv. Phys.* **1984**, *33*, 1.

[25] J. Szabon, *Mol. Cryst. Liq. Cryst.* **1985**, *124*, 343 and references cited therein.

[26] K. Praefcke, D. Singer, unpublished results; D. Singer, *Neue makrodiscotische Flüssigkristalle – mesomorphe Aggregationsformen und ihre Beeinflussung*, Köster Verlag, Berlin **1993**, ISBN 3-929937-79-4, Ph. D. Thesis, Technische Universität Berlin.

[27] B. Neumann, D. Joachimi, C. Tschierske, *Adv. Mater.* **1997**, *9*, 241, and *Liq. Cryst.* **1997**, *22*, 509.

[28] a) C. Pugh, V. Percec, *Polym. Bull.* **1990**, *23*, 177; b) K. Kato, S. Kobinata, S. Maeda, *Liq. Cryst.* **1989**, *5*, 595.

[29] S. Imazeki, A. Mukoh, N. Tanaka, M. Kinoshita, *Mol. Cryst. Liq. Cryst.* **1993**, *225*, 197.

[30] a) M. Portugall, H. Ringsdorf, R. Zentel, *Makromol. Chem.* **1982**, *183*, 2311; b) A. C. Griffin, A. M. Bhatti, R. S. L. Hung in *Nonlinear and Electroactive Polymers* (Ed.: P. N. Prasad, D. R. Ulrich, Plenum, New York **1988**; c) B. Beck, H. Ringsdorf, *Liq. Cryst.* **1990**, *8*, 247.

[31] a) C. T. Imrie, F. E. Karasz, G. S. Attard, *Liq. Cryst.* **1991**, *9*, 47; b) T. Schleeh, C. T. Imrie, D. M. Rice, F. E. Karasz, G. S. Attard, *J. Polym. Sci.* **1993**, *31A*, 1859.

[32] a) M. Sato, T. Nakano, K.-I. Mukaida, *Liq. Cryst.* **1995**, *18*, 645; b) Y. Kosaka, T. Kato, T. Uryu, *Macromolecules* **1994**, *27*, 2658.

[33] See for example C. T. Imrie, T. Schleeh, F. E. Karasz, G. S. Attard, *Macromolecules* **1993**, *26*, 539.

[34] a) J. L. Hogan, C. T. Imrie, G. R. Luckhurst, *Liq. Cryst.* **1988**, *3*, 645; b) G. S. Attard, S. Garnett, C. G. Hickmann, C. T. Imrie, L. Taylor, *Liq. Cryst.* **1990**, *7*, 495; c) G. S. Attard, R. W. Date, C. T. Imrie, G. R. Luckhurst, S. J. Roskilly, J. M. Seddon, L. Taylor, *Liq. Cryst.* **1994**, *16*, 529; d) A. E. Blatch, I. D. Fletcher, G. R. Luckhurst, *Liq. Cryst.* **1995**, *18*, 801.

[35] S. Ujiie, H. Uchino, K. Iimura, *Chem. Lett.* **1995**, 195.

[36] a) M. Sone, B. R. Harkness, J. Watanabe, T. Yamashita, T. Torii, K. Horie, *Polym. J.* **1993**, *25*, 997; b) B. R. Harkness, J. Watanabe, *Macromolecules* **1991**, *24*, 6759.

[37] a) M. Nitta, N. Ozaki, H. Suenaga, K. Nakaya, S. Kobayashi, *Jpn. J. Appl. Phys.* **1988**, *27*, L477; b) K. Nakaya, B. Y. Zhang, M. Yoshida, I. Isa, S. Shindoh, S. Kobayashi, *Jpn. J. Appl. Phys.* **1989**, *28*, L116; c) B. Y. Zhang, M. Yoshida, H. Maeda, M. Kimura, H. Sekine, S. Kobayashi, *Mol. Cryst. Liq. Cryst.* **1991**, *202*, 149.

[38] K. Ono, J. Nakanowatari, *Jpn. J. Appl. Phys.* **1991**, *30*, 2832.

[39] E. Matsui, K. Nito, A. Yasuda, *Ferroelectrics* **1993**, *149*, 97.

[40] a) V. Gionis, R. Fugnitto, G. Meyer, H. Strzelecka, J. C. Dubois, *Mol. Cryst. Liq. Cryst.* **1982**, *90*, 153; b) V. Gionis, R. Fugnitto, H. Strzelecka, *Mol. Cryst. Liq. Cryst.* **1983**, *96*, 215; c) P. Davidson, A.-M. Levelut, H. Strzelecka, V. Gionis, *J. Physique* **1983**, *44*, L-823; d) H. Strzelecka, V. Gionis, J. Rivory, S. Flandrois, *J. Physique, Colloque C3* **1983**, *44*, C3-1201; e) R. Kormann, L. Zuppiroli, V. Gionis, H. Strzelecka, *Mol. Cryst. Liq. Cryst.* **1986**, *133*, 283; f) V. Gionis, H. Strzelecka, M. Veber, R. Kormann, L. Zuppiroli, *Mol. Cryst. Liq. Cryst.* **1986**, *137*, 365.

[41] a) K. Ohta, H. Muroki, A. Takagi, K.-I. Hatada, H. Ema, I. Yamamoto, K. Matsuzaki, *Mol. Cryst. Liq. Cryst.* **1986**, *140*, 131; b) H. Sakashita, A. Nishitani, Y. Sumiya, H. Terauchi, K. Ohta, I. Yamamoto, *Mol. Cryst. Liq. Cryst.* **1988**, *163*, 211; c) A. C. Ribeiro, A. F. Martins, A.-M. Giroud-Godquin, *Mol. Cryst. Liq. Cryst. Lett.* **1988**, *5*, 133 and references therein; d) G. Latterman, S. Schmidt, R. Kleppinger, J. H. Wendorff, *Adv. Mater.* **1992**, *4*, 30; e) J. Malthete, A. Collet, A.-M. Levelut, L. Liebert, *Adv. Mater.* **1992**, *4*, 37; f) O. Karthaus, H. Ringsdorf, V. V. Tsukruk, J. H. Wendorff, *Langmuir* **1992**, *8*, 2279.

[42] J. Billard, *C. R. Acad. Sci. Paris* **1984**, *II-299*, 905.

[43] B. F. Darocha, D. D. Titus, D. J. Sandman, *Acta Cryst.* **1979**, *B35*, 2445.

[44] F. D. Saeva, G. A. Reynolds, L. Kaszczuk, *J. Am. Chem. Soc.* **1982**, *104*, 3524.

[45] K. Praefcke, B. Bilgin, N. Usol'tseva, B. Heinrich, D. Guillon, *J. Mater. Chem.* **1995**, *5*, 2257; K. Praefcke, J. D. Holbrey, N. Usol'tseva, *Mol. Cryst. Liq. Cryst.* **1996**, *288*, 189, see p. 194. – K. Praefcke, J. D. Holbrey, *J. Incl. Phen. Mol. Recogn. Chem.* **1996**, *24*, 19, see p. 25. – K. Praefcke, J. D. Holbrey, N. Usol'tseva, D. Blunk, *Mol. Cryst. Liq. Cryst.* **1997**, *292*, 123, see p. 129–130.

[46] a) H. Ringsdorf, R. Wüstefeld, E. Zerta, M. Ebert, J. H. Wendorff, *Angew. Chem.* **1989**, *101*, 934; *Angew. Chem., Int. Ed. Engl.* **1989**, *28*, 914; b) H. Ringsdorf, R. Wüstefeld, M. Ebert, J. H. Wendorff, *ACS Polym. Prep.* **1989**, *30*, 479.

[47] H. Ringsdorf, R. Wüstefeld, *Phil. Trans. R. Soc. Lond.* **1990**, *A330*, 95.

[48] See C. Destrade, P. Foucher, H. Gasparoux, H. T. Nguyen, A.-M. Levelut, J. Malthete, *Mol. Cryst. Liq. Cryst.* **1984**, *106*, 121 and references cited therein.

[49] a) H. Bengs, M. Ebert, O. Karthaus, B. Kohne, K. Praefcke, H. Ringsdorf, J. H. Wendorff, R. Wüstefeld, *Adv. Mater.* **1990**, *2*, 141; b) M. Ebert, G. Frick, C. Baehr, J. H. Wendorff, R. Wüstefeld, H. Ringsdorf, *Liq. Cryst.* **1992**, *11*, 293.

[50] M. Ebert, Ph. D. Thesis, TH Darmstadt **1990**.

[51] H. Bengs, O. Karthaus, H. Ringsdorf, C. Baehr, M. Ebert, J. H. Wendorff, *Liq. Cryst.* **1991**, *10*, 161.

[52] a) M. Möller, V. Tsukruk, J. H. Wendorff, H. Bengs, H. Ringsdorf, *Liq. Cryst.* **1992**, *12*, 17; b) D. Janietz, *Chem. Commun.* **1996**, 713; c) V. V. Tsukruk, D. H. Reneker, H. Bengs, H. Ringsdorf, *Langmuir* **1993**, *9*, 2141; d) K. Praefcke, A. Eckert, D. Blunk, *Liq. Cryst.* **1997**, *22*, 113, and *SPIE-Proceedings*, in press. – N. Boden, R. J. Bushby, A. N. Cammidge, S. Duckworth, G. Headdock, *J. Mater. Chem.* **1997**, *7*, 601.

[53] a) H. Bengs, R. Renkel, H. Ringsdorf, C. Baehr, M. Ebert, J. H. Wendorff, *Makromol. Chem., Rapid Commun.* **1991**, *12*, 439; b) M. Möller, V. V. Tsukruk, J. Wendling, J. H. Wendorff, H. Bengs, H. Ringsdorf, *Makromol. Chem.* **1992**, *193*, 2659.

[54] a) W. Kranig, C. Boeffel, H. W. Spiess, O. Karthaus, H. Ringsdorf, R. Wüstefeld, *Liq. Cryst.* **1990**, *8*, 375; b) M. Möller, J. H. Wendorff, M. Werth, H. W. Spiess, H. Bengs, O. Karthaus, H. Ringsdorf, *Liq. Cryst.* **1994**, *17*, 381.

[55] a) D. Markovitsi, H. Bengs, H. Ringsdorf, *J. Chem. Soc., Faraday Trans.* **1992**, *88*, 1275; b) D. Markovitsi, N. Pfeffer, F. Charra, J.-M.

Nunzi, H. Bengs, H. Ringsdorf, *J. Chem. Soc., Faraday Trans.* **1993**, *89*, 37; c) See for example D. Markovitsi, A. Germain, P. Millie, P. Lecuyer, L. K. Gallos, P. Argyrakis, H. Bengs, H. Ringsdorf, *J. Phys. Chem.* **1995**, *99*, 1005.

[56] M. M. Green, H. Ringsdorf, J. Wagner, R. Wüstefeld, *Angew. Chem.* **1990**, *102*, 1525; *Angew. Chem., Int. Ed. Engl.* **1990**, *29*, 1478.

[57] a) K. Ogawa, H. Yonehara, C. Pac, E. Maekawa, *Bull. Chem. Soc. Jpn.* **1993**, *66*, 1378; b) V. V. Tsukruk, J. H. Wendorff, O. Karthaus, H. Ringsdorf, *Langmuir* **1993**, *9*, 614.

[58] a) B. Kohne, K. Praefcke, *Chimia* **1987**, *41*, 196; b) K. Praefcke, B. Kohne, K. Gutbier, N. Johnnen, D. Singer, *Liq. Cryst.* **1989**, *5*, 233; c) K. Praefcke, B. Kohne, D. Singer, *Angew. Chem.* **1990**, *102*, 200; *Angew. Chem., Int. Ed. Engl.* **1990**, *29*, 177.

[59] a) K. Praefcke, D. Singer, B. Kohne, M. Ebert, A. Liebmann, J. H. Wendorff, *Liq. Cryst.* **1991**, *10*, 147; b) K. Praefcke, D. Singer, M. Langner, B. Kohne, M. Ebert, A. Liebmann, J. H. Wendorff, *Mol. Cryst. Liq. Cryst.* **1992**, *215*, 121; c) D. Janietz, K. Praefcke, D. Singer, *Liq. Cryst.* **1993**, *13*, 247; d) K. Praefcke, D. Singer, A. Eckert, *Liq. Cryst.* **1994**, *16*, 53.

[60] B. Sabaschus, D. Singer, G. Heppke, K. Praefcke, *Liq. Cryst.* **1992**, *12*, 863.

[61] K. Praefcke, D. Singer, B. Kohne, *Liq. Cryst.* **1993**, *13*, 445.

[62] K. Praefcke, B. Kohne, B. Gündogan, D. Singer, D. Demus, S. Diele, G. Pelzl, U. Bakowsky, *Mol. Cryst. Liq. Cryst.* **1991**, *198*, 393.

[63] K. Praefcke, D. Singer, B. Gündogan, K. Gutbier, M. Langner, *Ber. Bunsenges. Phys. Chem.* **1993**, *97*, 1358; Corrigendum **1994**, *98*, 118.

[64] D. Singer, A. Liebmann, K. Praefcke, J. H. Wendorff, *Liq. Cryst.* **1993**, *13*, 785.

[65] a) K. Praefcke, B. Bilgin, unpublished results; Belkiz Bilgin, *Neue flächige Mesogene des Palladiums and Platins*, Köster Verlag, Berlin **1996**, ISBN 3-89574-155-8, Ph. D. Thesis, Technische Universität Berlin, Germany; b) K. Praefcke, B. Bilgin, J. Pickardt, M. Borowski, *Chem. Ber.* **1994**, *127*, 1543.

[66] S. Zamir, D. Singer, N. Spielberg, E. J. Wachtel, H. Zimmermann, R. Poupko, Z. Luz, *Liq. Cryst.* **1996**, *21*, 39.

[67] a) B. Kohne, K. Praefcke, T. Derz, H. Hoffmann, B. Schwandner, *Chimia* **1986**, *40*, 171;

b) M. Ibn-Elhaj, D. Guillon, A. Skoulios, A.-M. Giroud-Godquin, J.-C. Marchon, *J. Phys. II France* **1992**, *2*, 2197; c) R. Seghrouchni, A. Skoulios, *J. Phys. II France* **1995**, *5*, 1385; d) N. V. Usol'tseva, K. Praefcke, D. Singer, B. Gündogan, *Liq. Cryst.* **1994**, *16*, 601; e) N. Usol'tseva, G. Hauck, H. D. Koswig, K. Praefcke, B. Heinrich, *Liq. Cryst.* **1996**, *20*, 731; f) H. Ringsdorf, K. Schimossek, M. Wittenberg, J. H. Wendorff, H. Fischer, 24. Freiburger Arbeitstagung Flüssigkristalle, 05.–07. April **1995**, Freiburg (Germany), Poster P42.

[68] K. Praefcke, D. Singer, Institute of Organic Chemistry, Technische Universität Berlin, unpublished results.

[69] K. Praefcke, S. Diele, J. Pickardt, B. Gündogan, U. Nütz, D. Singer, *Liq. Cryst.* **1995**, *18*, 857.

[70] a) N. V. Usol'tseva, K. Praefcke, D. Singer, B. Gündogan, *Liq. Cryst.* **1994**, *16*, 617; b) N. V. Usol'tseva, K. Praefcke, D. Singer, B. Gündogan, *Mol. Mater.* **1994**, *4*, 253.

[71] S. Pollack, Howard University, Washington DC, U. S. A., private communication to D. Singer.

[72] See for example T. K. Attwood, J. E. Lydon, C. Hall, G. J. T. Tiddy, *Liq. Cryst.* **1990**, *7*, 657.

[73] a) W. Paulus, H. Ringsdorf, S. Diele, G. Pelzl, *Liq. Cryst.* **1991**, *9*, 807; b) W. G. Scaife, G. McMullin, H. Groothues, *Mol. Cryst. Liq. Cryst.* **1995**, *261*, 51.

[74] See for example: a) L. Y. Chiang, J. P. Stokes, C. R. Safinya, A. N. Bloch, *Mol. Cryst. Liq. Cryst.* **1985**, *125*, 279; b) N. Boden, R. J. Bushby, J. Clements, M. V. Jesudason, P. F. Knowles, G. Williams, *Chem. Phys. Lett.* **1988**, *152*, 94; c) N. Boden, R. Borner, D. R. Brown, R. J. Bushby, J. Clements, *Liq. Cryst.* **1992**, *11*, 325; d) N. Boden, R. J. Bushby, J. Clements, *J. Chem. Phys.* **1993**, *98*, 5920; e) N. Boden, R. J. Bushby, J. Clements, J. Movaghar, K. J. Donovan, T. Kreouzis, *Phys. Rev.* **1995**, *B52*, 13274.

[75] L. Calucci, H. Zimmermann, E. J. Wachtel, R. Poupko, Z. Luz, *Liq. Cryst.* **1997**, *22*, 621.

[76] a) C. R. Patrick, G. S. Prosser, *Nature* **1960**, *187*, 1021; b) T. Dahl, *Acta Chem. Scand.* **1973**, *27*, 995; c) J. H. Williams, *Acc. Chem. Res.* **1993**, *26*, 593; d) G. R. Luckhurst, W. M. Tearle, Abstracts: Poster NM-20, European Conference on Liquid Crystals Science and Technology, ECLC 95, 5.–10. March, **1995**, Bovec, Slovenia.

Chapter XVII
Hydrogen-Bonded Systems

Takashi Kato

1 Introduction

Hydrogen bonding is one of the key interactions for chemical and biological processes in nature due to its stability, directionality, and dynamics. The history of liquid crystals and hydrogen bonding has started in the early time of 20th century when mesomorphic properties were observed for the derivatives of cinnamic acids [1] and benzoic acids [2], and argaric acid [3]. For monosaccharides with long alkyl chains, Fischer observed double melting behavior in 1911 [4]. This behavior was recognized later as thermotropic liquid crystallinity [5]. This behavior was recently established as an effect of hydrogen bonding [6]. Some compounds containing hydrogen bonding moieties, such as polyols [7], silanediols [8], amides [9], and hydrazines [10] have been shown to exhibit thermotropic mesophases. These compounds show thermotropic liquid-crystalline phases due to the intermolecular hydrogen bonding between identical molecules.

A new type of thermotropic "supramolecular" liquid crystal obtained by molecular recognition processes through hetero-intermolecular hydrogen bonds, which function between different and independent molecules species, were reported by Kato and Fréchet [11–13] and Lehn and co-workers [14, 15] in 1989 and 1993. After these findings, a wide variety of structures of molecular complexes has been prepared by supramolecular self-assembly.

This chapter focuses on supramolecular hydrogen-bonded liquid crystals built by the formation of hetero-intermolecular hydrogen bonds. The H-bonded liquid crystals obtained by the association of identical molecules are described in other chapters.

2 Pyridine/Carboxylic Acid System

2.1 Self-Assembly of Low Molecular Weight Complexes

2.1.1 Structures and Thermal Properties

Liquid-crystalline supramolecular complexes with well-defined structures can be obtained by self-assembly through hetero-intermolecular hydrogen bonding between carboxylic acid and pyridine moieties [11]. For example, the complexation of 4-methoxybenzoic acid (**1**) and *trans*-1,2-bis(4-pyridyl)ethylene (**2**) results in the formation of mesogenic complex **3**, which exhibits a nematic phase from 166 to 185 °C (Scheme 1) [16]. In this H-bonded mesogenic structure, nonmesogenic compound **2** functions as the core unit of the mesogen through the hydrogen bond, which is directional and stable. Figure 1 shows the plot of phase behavior against the length of the alkyl chain for the complex of 4-alkoxybenzoic acid

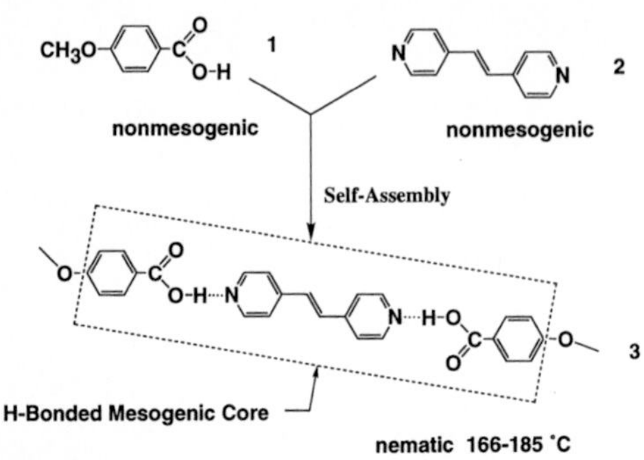

Scheme 1

nonmesogenic ... **nonmesogenic**

Self-Assembly

H-Bonded Mesogenic Core

nematic 166–185 °C

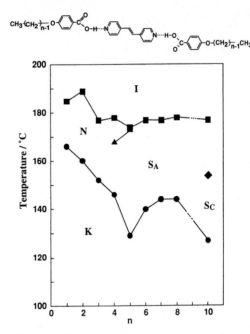

Figure 1. Plot of transition temperature against the carbon number (*n*) in the alkyl chain for the series of hydrogen-bonded complexes obtained from 4-alkoxybenzoic acid and *trans*-1,2-bis(4-pyridyl)ethylene. (S$_A$: SmA; S$_C$: SmC; K: Cr.) ■: I$_i$; ●: I$_m$.

and *trans*-1,2-bis(4-pyridyl)ethylene. These complexes show stable smectic and nematic behavior over 100 °C [16].

The relationship between the core structure and the transition temperature for the complexes of 4-hexyloxybenzoic acid (**4**) and bipyridines is shown in Table 1 [16]. The highest clearing temperature is achieved for the complex based on **2**, which has the most coplanar structure.

Table 2 compares the mesomorphic temperature ranges of hydrogen-bonded complexes (**8**, **9** and **10**) with different core lengths [16–19]. The thermal stability of the mesophase increases when the aromatic ring number increases, as is observed for normal liquid crystals. Significant mesophase stabilization is seen for complex **8** consisting of four aromatic rings, which was first reported as the dissimilar hydrogen-bonded mesogenic complex [12]. In contrast, the mesophase near room temperature was achieved by the complexation of 4-hexyloxybenzoic acid (**4**) and 4-octylpyridine [19].

The transition temperatures of the hydrogen-bonded mesogens **10** and **11**, consisting of two rings, are compared with those of the structurally related phenyl benzoates

Table 1. Liquid crystalline behavior of supramolecular 2:1 hydrogen-bonded complexes derived from 4-hexyloxybenzoic acid and bipyridines (**4**). [a]

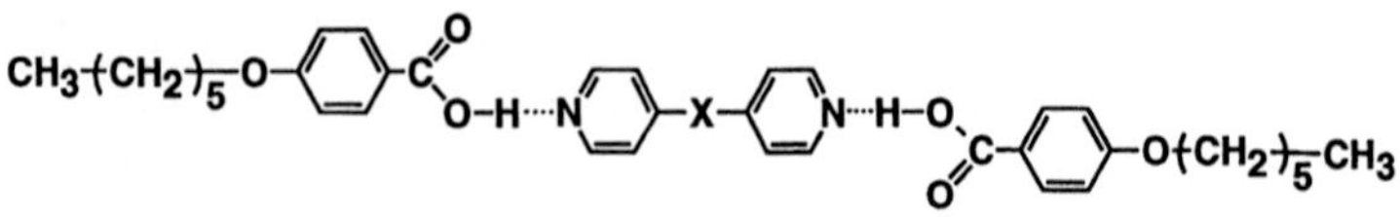

Complex	Cr		SmC		SmA		I
5 (X=none)	●	105	●	130	●	159	●
6 (X=−CH=CH−)	●			140	●	177	●
7 (X=−CH₂−CH₂−)	●			123	●	152	●

[a] Transition temperatures in °C.

Table 2. Liquid-crystalline behavior of supramolecular 1:1 hydrogen-bonded complexes formed through intermolecular hydrogen bonding.[a]

Complex	Cr		SmA		N		I
8	•	136	•	160	•	238	•
9	•	102	•	130	•	155	•
10	•			38	•	48	•

[a] Transition temperatures in °C.

Table 3. Transition temperatures of liquid-crystalline complexes **10** and **11** based on 4-hexyloxybenzoic acid and 4-octylpyridine, and of 4-octylphenyl 4-hexyloxybenzoates (**OPHOB** and **OPHOFB**).

Complex			Transition temperature [a] (°C)							
	Cr		SmC		SmA		N		I	
10	•	38	–		–		•	48	•	
OPHOB	•	43	–		–		•	61	•	
11	•	28	•	31	•	33	•	39	•	
OPHOFB	•	57	–		–		(•	45)	•	

[a] Transition temperatures of monotropic phases are in parentheses.

OPHOB and OPHOFB, which are normal liquid crystals built only from covalent bonds, in Table 3 [19]. The melting temperatures for the H-bonded liquid crystals are lower than those of the normal liquid crystals, and the wider temperature range of the mesophase can be achieved for the H-bonded complex. This difference may be caused by "softness" of the complex structure.

Ferroelectricity is induced by the self-assembly of a chiral benzoic acid which is nonmesogenic and 4-alkoxystilbazoles [20].

Complex **12** exhibits a chiral smectic C phase from 109 to 123 °C. The value of spontaneous polarization for **12** at 115 °C is 33.0 nC/cm^2. A similar structure was reported to show an SmC* phase [21].

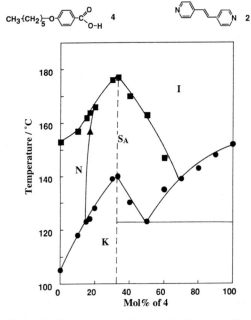

12

Molecular structures such as twins **13** [22] and angular structures **14** [23] have been prepared by the self-assembly of appropriate H-bonding components.

tion through the noncovalent interaction is observed in the diagram. The smectic A phase, which does not appear for either of the individual components, appears in the range of about 20–70 mol% of *trans*-1,2-bis(4-pyridyl)ethylene. The highest clearing temperature is observed at the point where equimolar amounts of donor and acceptor moieties is present. This hydrogen-bonded complex system can be obtained using a nonstoichiometric donor and acceptor mixure [11, 13, 16, 24]. This concept can be used efficiently to control specific properties of H-bonded mesogenic materials.

13

14

2.1.2 Phase Diagrams

The formation of supramolecular 1:1 complexes with well-defined structures was described in the preceding chapter. However, mixtures containing any ratios of donor and acceptor moieties can be prepared because each component is independent. Figure 2 presents a binary phase diagram for 4-hexyloxybenzoic acid (**4**) and *trans*-1,2-bis(4-pyridyl)ethylene (**2**) [16]. The significant effect of the mixing on mesophase stabiliza-

Figure 2. Binary phase diagram for **2** and **4**. (S$_A$: SmA; K: Cr.)

2.1.3 Stability of Hydrogen Bonds

It is noteworthy that the hydrogen-bonded complexes of carboxylic acids ($pK_a \sim 4$) and pyridines ($pK_b = 9$) exhibit thermally stable mesophases though the structure consists of noncovalent bonding [11–13, 16–23]. In the infrared specta for the complex, the O–H band at 2500 cm^{-1} is indicative of the strong hydrogen bond [16–18, 25, 26], while for benzoic acid dimers the O–H band appears at ca. 3000 cm^{-1} [18, 27].

Variable temperature infrared spectra for hydrogen-bonded liquid-crystalline complexes have shown that the structure of the hydrogen-bonded mesogen is stabilized by molecular ordering and packing in the liquid-crystalline state [16, 18, 27]. The stability of the hydrogen bond suddenly decreases once the complex becomes isotropic disordered [16, 18, 27].

2.1.4 Electrooptic Effects

Room-temperature hydrogen-bonded liquid crystals consisting of 4-alkoxybenzoic acids and simple alkylpyridines align in twisted nematic cells and show electrooptic effects [19]. This observation suggests that the hydrogen bonding is stable in electric fields. It is expected that the dynamic behavior of the complex is different from that of normal liquid crystals, because the core structure consists of noncovalent bonding, which is "soft" bonding between interacting molecular species [28].

2.2 Self-Assembly of Polymeric Complexes

2.2.1 Side-Chain Polymers

Biomacromolecules, such as nucleic acid, polypeptide, and cellulose, all have hydrogen bonding functional groups that play key roles for molecular recognition, molecular self-organization, and supramolecular self-assembly. Therefore hydrogen bonding would be useful for the molecular design of functional synthetic polymeric materials. The first approach to "supramolecular" liquid-crystalline polymers involved polyacrylates with a benzoic acid moiety through flexible spacers in the side chain [11, 13]. Polysiloxanes have also been used to build side-chain polymeric structures [29, 30]. Table 4 summarizes the thermal properties of hydrogen-bonded side-chain polyacrylates [11, 13, 17, 31], which exhibit stable mesomorphic behavior. These complexes show thermally stable mesophases. In particular, for complex **15**, the mesophase is stable up to 252 °C [13]. The ferroelectric

15: R=
16: R= -OCH₃
17: R= -CN

Table 4. Liquid-crystalline behavior of supramolecular side-chain polymers formed through intermolecular hydrogen bonding.[a]

Polymeric complex	G		SmX		SmA		N		I
15	●						140	●	252	●
16	●	38	●	74	●				194	●
17	●			38	●				200	●

[a] Transition temperatures in °C.

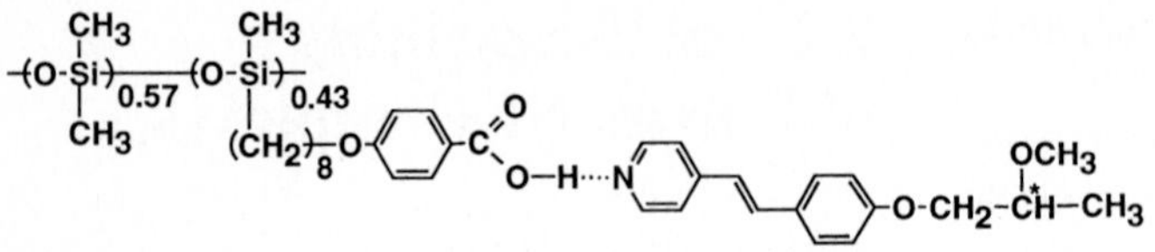

18

side-chain polymer **18** has been formed by self-assembly of the H-bonding polysiloxanes and chiral stilbazoles [30].

Another example of a side-chain supramolecular polymer is prepared from poly(4-vinylpyridine) and H-bonding side chains [32–35]. In this case, mesogenic side-chain groups are directly attached to the polymer backbone and liquid-crystalline polymers such as **19** are formed by the noncovalent interaction [33]. This molecular design has been used to incorporate functional molecules into liquid-crystalline, host–guest polymeric systems [32].

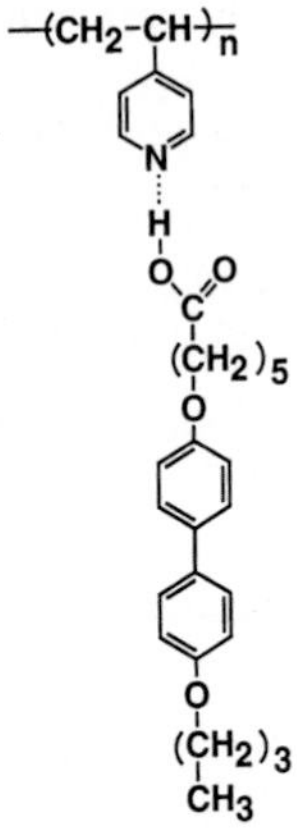

19

2.2.2 Main-Chain Polymers

The concept of the hydrogen-bonded complexation has been extended to the preparation of main-chain type polymeric complexes by Griffin and co-workers [36, 37]. Bifunctional H-bonding donor and acceptor moieties were used for the formation of a polymeric chain structure. Complex **20** exhibits a nematic phase from 174 °C to

20

183 °C. The behavior of such main-chain complexes was theoretically studied [37].

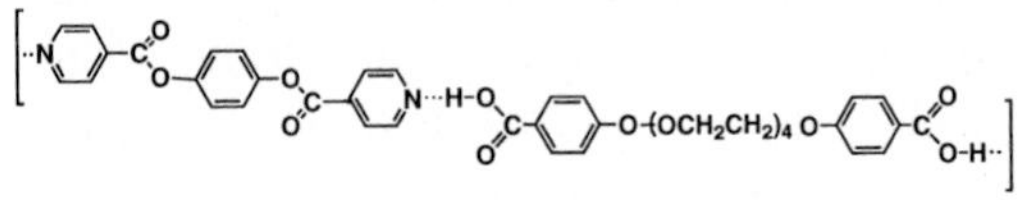

2.2.3 Networks

A liquid-crystalline network (**21**) has been constructed by the self-assembly of polyacrylate and 4,4′-bipyridine [38]. Although complex **21** has a fully cross-linked structure through the noncovalent interaction, it shows a glass transition at 95 °C and a subsequent smectic A phase up to 205 °C. The formation of a hydrogen-bonded intermo-

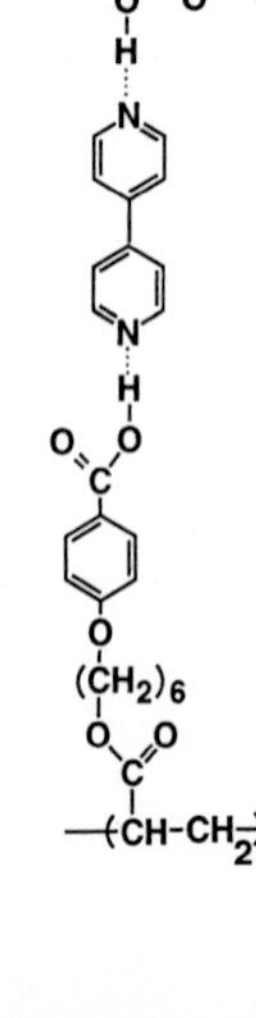

21

lecular mesogen should contribute to the induction of the stable mesophase. The smectic phase is viscous but shows shear flow. It should be noted that reversible phase transitions between the mesophase and the isotropic phase are observed for the network complexes due to the reversibility of the noncovalent interaction. The dynamic nature of the hydrogen bonding contributes to such behavior. For covalently bonded polymeric networks with a high density of crosslinks, the molecules are fixed in the solid state and no phase transitions are observed.

Multifunctional low molecular weight compounds also form a liquid-crystalline network that is dynamic and thermally reversible [39, 40]. Self-hydrogen bonding and dimerization of the benzoic acid side chain of polysiloxanes also induce network mesogenic structures [29].

3   Uracil/Diamino-pyridine System

3.1   Low Molecular Weight Complexes

Triply hydrogen-bonded mesogenic complexes can be obtained through the molecular recognition of complementary compo-

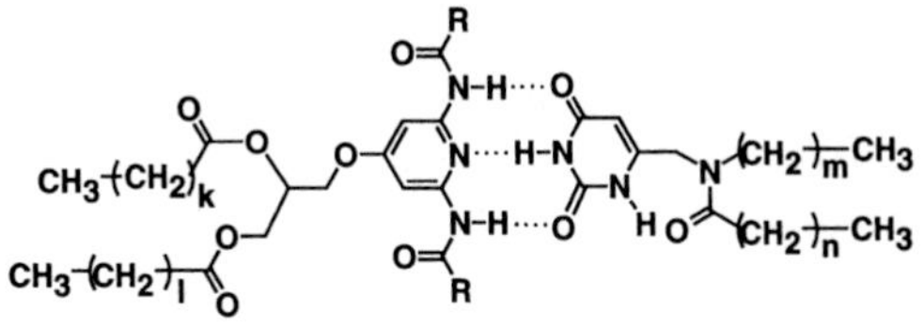

22

nents of uracil and diacylaminopyridine moieties [14, 15]. Complex **22** with R = Me, $k = l = 10$, $m = 11$, and $n = 16$ exhibits a monotropic mesophase below 72 °C. For this complex structure, long alkyl chains are needed to induce mesomorphism. An X-ray diffraction study has shown that complex **22** forms a columnar hexagonal phase with a diameter of 40.4 Å (4.04 nm). Each column contains two complexes side by side.

3.2   Polymeric Complexes

Liquid-crystalline polymeric supramolecular species **23** has been prepared by the association of bifunctional H-bonding components [41, 42]. These bifunctional individual components were derived from L(+), D(−), and meso (M) tartaric acids, respectively. It is interesting that the liquid-crystalline phases of the polymeric complexes consisting of chiral species are more stable than those of polymers consisting of meso components. The complex based on L-(+)-tartaric acids shows a mesophase from room temperature to 254 °C, while the mesophase

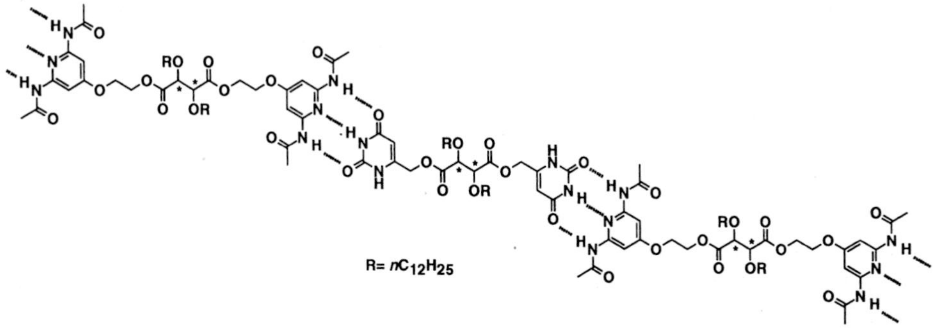

R = $nC_{12}H_{25}$

23

is observed from room temperature to 222 °C for the complex consisting of meso-tartaric acids. The X-ray patterns show that a triple helix structure is organized by the complex ased on L-(+)-tartaric acid, while the complex consisting of meso-tartaric acid forms a twofold structure.

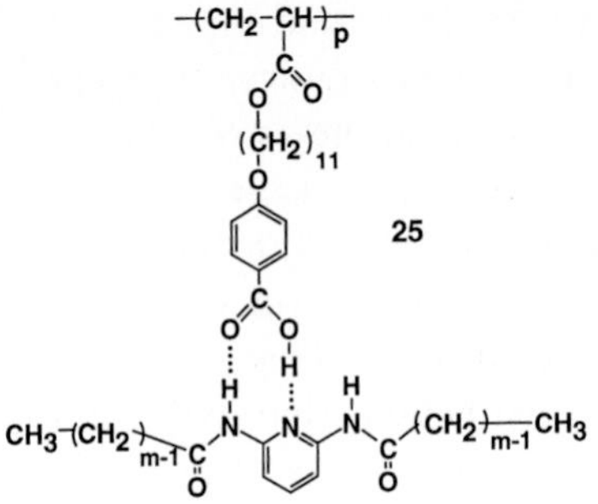

4   Miscellaneous Thermotropic H-Bonded Compounds by Inter-molecular Interaction

Doubly hydrogen-bonded supramolecular mesogenic complexes have been derived by the formation of the hydrogen bonds between 2-aminopyridyl and carboxylic acid moieties [43, 44]. Complex 24 sharply melts

24

at 90 °C and exhibits a monotropic smectic B phase from 83 to 67 °C [43]. The structure of the complex, which does not have an overall rod shape, is unique for calamitic liquid crystals. This supramolecular mesogen is attached to the polyacrylate backbone [44]. For polymeric complex 25, a monotropic smectic phase was observed between 98 and 80 °C. It should be noted that in this polymeric structure the mesogenic group is expected to align perpendicular to the polymer chain.

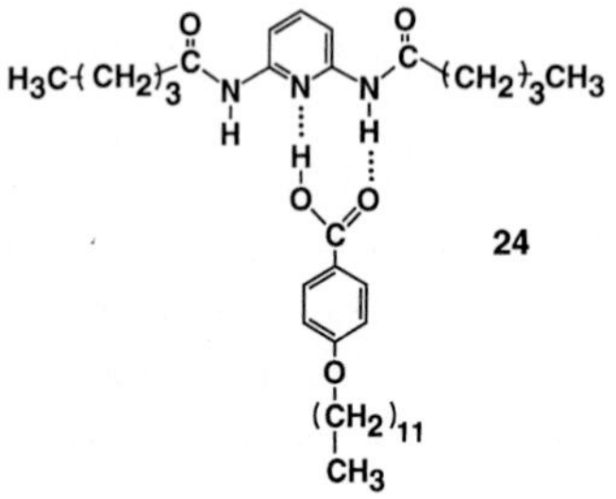

25

The interaction between phenol and pyridine, which is weaker than the carboxylic acid/pyridine interaction, has been used to effect mesomorphism [45–49]. The 1:1 mixture (26) of 4-alkoxy-4-stilbazoles and 4-cyanophenol shows a monotropic nematic phase, while the stilbazoles exhibit a very narrow temperature range of ordered smectic phases and 4-cyanophenol is nonmesogenic [45]. This is due to the interaction between the phenol and pyridine moieties. The complex based on 3-cyanophenol (27)

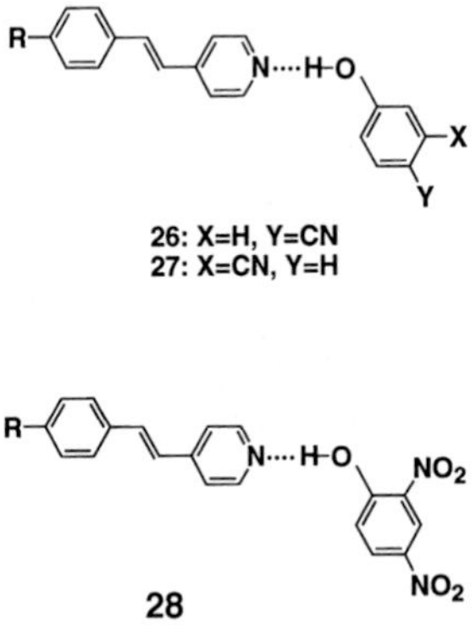

26: X=H, Y=CN
27: X=CN, Y=H

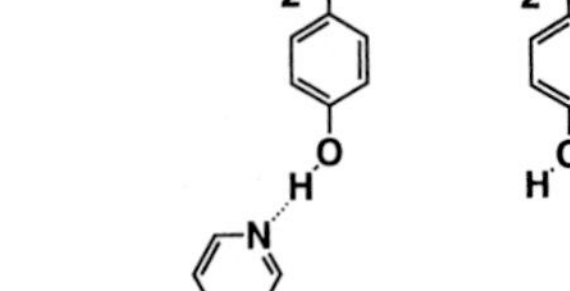

28

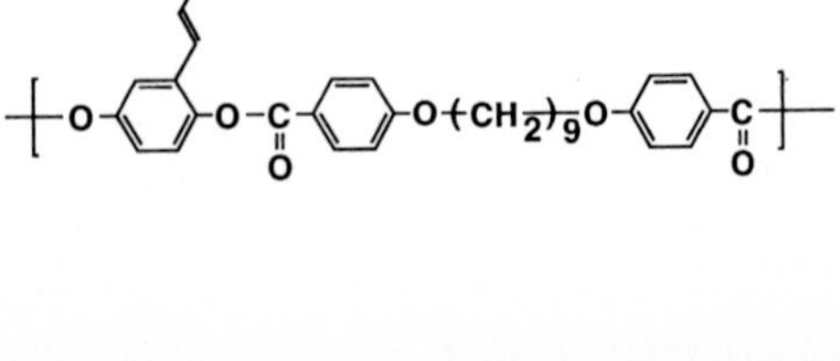

29

shows more thermally stable mesophase than that based on 4-cyanophenol because molecular linearity is well kept for the metasubstituted system [46]. A phenol/stilbazole complex (28) based on 2,4-dinitrophenol exhibits a smectic A phase between 97 and 117 °C [47]. Variable-temperature electronic spectroscopy shows that ionic hydrogen-bonded state by proton transfer is stabilized in the mesophase. The induction of smectic phases has also been observed for the mixture of 4,4′-bipyridine and 4-alkoxyphenol [44]. Liquid-crystalline polymer blends (29) involving phenol/pyridine interactions have been prepared from a polyester containing a pyridyl lateral group and poly(4-vinylphenol) [45].

The formation of mesogenic side-chain polymer complexes has been reported for pyridine-N-oxide/carboxylic acid [46] and trialkylamine/phenol [47]. Complex 30 exhibits smectic phases between 29 and 123 °C [46].

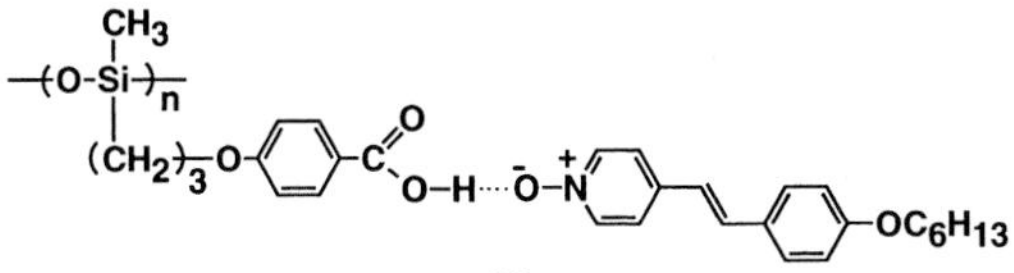

30

Stabilization of a columnar phase for a calix[4]arene by a hydrogen bonding inter-action with a low molecular weight component has been described for bowlic liquid crystals [48]. Self-assembly of these host and guest molecules leads to mesomorphism.

5   Lyotropic Hydrogen-Bonded Complexes

Self-assembly and molecular recognition of some biomolecules result in the formation of lyotropic liquid-crystalline phases with water [49]. For example, nucleic acids such as DNA and RNA, which form supramolecular complexes, show lyotropic mesophases [49].

An interesting example of a biomolecule that shows lyotropic liquid-crystallinity by association of identical molecules is a guanosine residue that exhibits columnar mesophases due to the formation of hydrogen-bonded tetramers [50]. Foric acid salts exhibit columnar liquid-crystalline phases in the presence of water [51].

Synthetic supramolecular lyotropic polymer 31 has been built through triple H bonds. The polymer shows liquid-crystalline behavior with an organic solvent, whereas it melts with decomposition at high temperatures due to its rigidity [52].

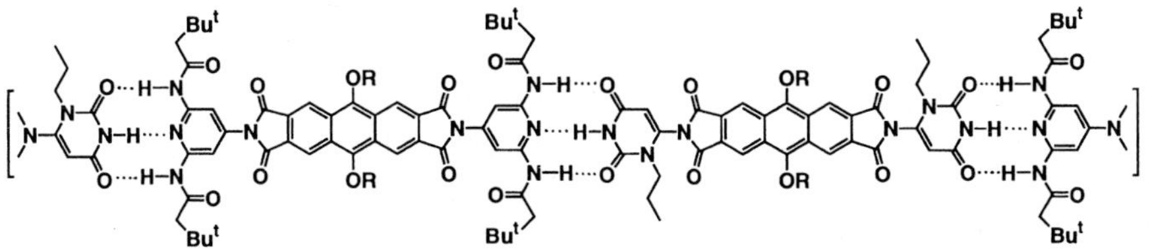

31

6   References

[1] A. C. de Kock, *Z. Phys. Chem.* **1904**, *48*, 129. D. Vorlaender, *Ber. Dtsch. Chem. Ges.* **1908**, *41*, 2033–2052. [V. Vill, LiqCryst – Database of Liquid Crystal Compounds for Personal Computers, 1994].

[2] A. E. Bradfield, B. Jones, *J. Chem. Soc.* **1929**, 2660–2661. G. M. Bennett, B. Jones, *J. Chem. Soc.* **1939**, 420–425.

[3] M. Gaubert, *C. R. Acad. Sci.* **1919**, *168*, 277–279.

[4] E. Fischer, B. Helferich, *Justus Liebigs Ann. Chem.* **1911**, *383*, 68–91.

[5] C. R. Noller, W. C. Rockwell, *J. Am. Chem. Soc.* **1938**, *60*, 2076–2077.

[6] G. A. Jeffrey, *Acc. Chem. Res.* **1986**, *19*, 168–173. J. W. Goodby, *Mol. Cryst. Liq. Cryst.* **1984**, *110*, 205–219.

[7] S. Diele, A. Madicke, E. Geissler, D. Meinel, D. Demus, H. Sackmann, *Mol. Cryst. Liq. Cryst.* **1989**, *166*, 131–142. K. Praefcke, P. Marquardt, B. Kohne, *Mol. Cryst. Liq. Cryst.* **1991**, *203*, 149–158.

[8] J. D. Bunning, J. E. Lydon, C. Eaborn, P. H. Jackson, J. W. Goodby, G. W. Gray, *J. Chem. Soc., Faraday Trans. 1* **1982**, 713–724.

[9] Y. Matsunaga, M. Terada, *Mol. Cryst. Liq. Cryst.* **1986**, *141*, 321–326. J. Malthéte, A.-M. Levelut, L. Liébert, *Adv. Mater.* **1992**, *4*, 37–41.

[10] H. Schubert, S. Hoffmann, J. Hauschild, I. Marx, *Z. Chem.* **1977**, *17*, 414–415.

[11] T. Kato, J. M. J. Fréchet, *Macromol. Symp.* **1995**, *98*, 311–326.

[12] T. Kato, J. M. J. Fréchet, *J. Am. Chem. Soc.* **1989**, *111*, 8533–8534.

[13] T. Kato, J. M. J. Fréchet, *Macromolecules* **1989**, *22*, 3818–3819; **1990**, *23*, 360.

[14] J.-M. Lehn, *Makromol. Chem., Macromol. Symp.* **1993**, *69*, 1–17.

[15] M.-J. Brienne, J. Gabard, J.-M. Lehn, I. Stibor, *J. Chem. Soc., Chem. Commun.* **1989**, 1868–1970.

[16] T. Kato, J. M. J. Fréchet, P. G. Wilson, T. Saito, T. Uryu, A. Fujishima, C. Jin, F. Kaneuchi, *Chem. Mater.* **1993**, *5*, 1094–1100; T. Kato, P. G. Wilson, A. Fujishima, J. M. J. Fréchet, *Chem. Lett.* **1990**, 2003–2006.

[17] T. Kato, H. Kihara, T. Uryu, A. Fujishima, J. M. J. Fréchet, *Macromolecules* **1992**, *25*, 6836–6841.

[18] T. Kato, T. Uryu, F. Kaneuchi, C. Jin, J. M. J. Fréchet, *Liq. Cryst.* **1993**, *14*, 1311–1317.

[19] T. Kato, M. Fukumasa, J. M. J. Fréchet, *Chem. Mater.* **1995**, *7*, 368–372; T. Kato, M. Fukumasa, T. Uryu, J. M. J. Fréchet, *Chem. Lett.* **1993**, 65–68.

[20] T. Kato, H. Kihara, T. Uryu, S. Ujiie, K. Iimura, J. M. J. Fréchet, U. Kumar, *Ferroelectrics* **1993**, *148*, 161–167.

[21] L. J. Yu, *Liq. Cryst.* **1993**, *14*, 1303–1309.

[22] T. Kato, A. Fujishima, J. M. J. Fréchet, *Chem. Lett.* **1990**, 919–923.

[23] T. Kato, H. Adachi, A. Fujishima, J. M. J. Fréchet, *Chem. Lett.* **1992**, 265–268.

[24] H. Kresse, I. Szulzewsky, S. Diele, R. Paschke, *Mol. Cryst. Liq. Cryst.* **1994**, *238*, 13–19.

[25] S. L. Johnson, K. A. Rumon, *J. Phys. Chem.* **1965**, *69*, 74–89.

[26] S. E. Odinokov, A. A. Mashkovsky, V. P. Glazunov, A. V. Iogansen, B. V. Rassadin, *Spectrochim. Acta* **1976**, *32A*, 1355–1363.

[27] T. Kato, C. Jin, F. Kaneuchi, T. Uryu, *Bull. Chem. Soc. Jpn.* **1993**, *66*, 3581–3584.

[28] S. Machida, T.-I. Urano, K. Sano, T. Kato, J. M. J. Fréchet, *Langmuir* **1997**, *13*, 576–580.

[29] U. Kumar, T. Kato, J. M. J. Fréchet, *J. Am. Chem. Soc.* **1992**, *114*, 6630–6639.

[30] U. Kumar, J. M. J. Fréchet, T. Kato, S. Ujiie, K. Iimura, *Angew. Chem.* **1992**, *104*, 1545–1547; *Angew. Chem., Int. Ed. Engl.* **1992**, *31*, 1531–1533.

[31] T. Kato, H. Kihara, S. Ujiie, T. Uryu, J. M. J. Fréchet, *Macromolecules* **1996**, *29*, 8734–8739.

[32] T. Kato, N. Hirota, A. Fujishima, J. M. J. Fréchet, *J. Polym. Sci., Part A: Polym. Chem.* **1996**, *34*, 57–62.

[33] C. G. Bazuin, F. A. Brandys, *Chem. Mater.* **1992**, *4*, 970–972.

[34] C. G. Bazuin, F. A. Brandys, T. M. Eve, M. Plante, *Macromol. Symp.* **1994**, *84*, 183–196.

[35] D. Stewart, C. T. Imrie, *J. Mater. Chem.* **1995**, *5*, 223–228.

[36] C. Alexander, C. P. Jariwala, C. M. Lee, A. C. Griffin, *Macromol. Symp.* **1994**, *77*, 283–294.

[37] P. Bladon, A. C. Griffin, *Macromolecules* **1993**, *26*, 6604–6610.

[38] T. Kato, H. Kihara, U. Kumar, T. Uryu, J. M. J. Fréchet, *Angew. Chem.* **1994**, *106*, 1728–1730; *Angew. Chem., Int. Ed. Engl.* **1994**, *33*, 1644–1645.

[39] H. Kihara, T. Kato, T. Uryu, J. M. J. Fréchet, *Chem. Mater.* **1996**, *8*, 961–968.

[40] L. M. Wilson, *Macromolecules* **1994**, *27*, 6683–6686.

[41] C. Fouquey, J.-M. Lehn, A.-M. Levelut, *Adv. Mater.* **1990**, *2*, 254–257.

[42] T. Gulik-Krzywicki, C. Fouquey, J.-M. Lehn, *Proc. Natl. Acad. Sci. USA* **1993**, *90*, 163–167.

[42a] T. Kato, Y. Kubota, M. Nakano, T. Uryu, *Chem. Lett.* **1995**, 1127–1128.

[42b] T. Kato, M. Nakano, T. Moteki, T. Uryu, S. Ujiie, *Macromolecules* **1995**, *28*, 8875–8876.

[43] D. W. Bruce, D. J. Price, *Adv. Mater. Opt. Electron.* **1994**, *4*, 273–276.

[43a] K. Willis, D. J. Price, H. Adams, G. Ungar, D. W. Bruce, *J. Mater. Chem.* **1995**, *5*, 2195–2199.

[43b] D. J. Price, T. Richardson, D. W. Bruce, *J. Chem. Soc., Chem. Commun.* **1994**, 1911–1912.

[44] T. Koga, H. Ohba, A. Takase, S. Sakagami, *Chem. Lett.* **1994**, 2071–2074.

[45] A. Sato, T. Kato, T. Uryu, *J. Polym. Sci., Part A, Polym. Chem.* **1996**, *34*, 503–505.

[46] U. Kumar, J. M. J. Fréchet, *Adv. Mater.* **1992**, *4*, 665–667.

[47] S. Malik, P. K. Dhal, R. A. Mashelkar, *Macromolecules* **1995**, *28*, 2159–2164.

[48] B. Xu, T. M. Swager, *J. Am. Chem. Soc.* **1995**, *117*, 5011–5012.

[49] S. Hoffmann in *Polymeric Liquid Crystals* (Ed.: A. Blumstein), Plenum, New York, U.S.A. **1985**, p. 423.

[50] P. Mariani, C. Mazabard, A. Garbesi, G. P. Spada, *J. Am. Chem. Soc.* **1989**, *111*, 6369–6373.

[51] F. Ciuchi, G. D. Nicola, H. Franz, G. Gottarelli, P. Mariani, M. G. P. Bossi, G. P. Spada, *J. Am. Chem. Soc.* **1994**, *116*, 7064–7071.

[52] M. Kotera, J.-M. Lehn, J.-P. Vigneron, *J. Chem. Soc., Chem. Commun.* **1994**, 197–199.

Chapter XVIII
Chromonics

John Lydon

1 Introduction

1.1 A Well-Defined Family Distinct from Conventional Amphiphiles

Over the last twenty years it has become increasingly clear that there is a well-defined family of lyotropic mesogens embracing a range of drugs, dyes, nucleic acids, antibiotics, carcinogens, and anti-cancer agents. In contrast to conventional amphiphile mesogens, such as soaps, detergents, and biological lipids, these *chromonic* materials do not generally have significant surfactant properties. Their molecules are disc-like or plank-like as opposed to rod-like, they are aromatic rather than aliphatic, and the hydrophilic ionic or hydrogen-bonding solubilizing groups are arranged around the peripheries of the molecules, not at the ends. The molecules can be regarded as being insoluble in one dimension, and the basic structural unit of all chromonic phases is a molecular stack of some kind, rather than a micelle. Chromonic systems have characteristic phase structures, phase diagrams, optical textures, X-ray diffraction patterns, and miscibility properties.

The tendency of chromonic molecules (Fig. 1) to aggregate into columns is present even in dilute solution (just as for amphiphile systems, where micelle formation occurs before a mesophase is formed). How-

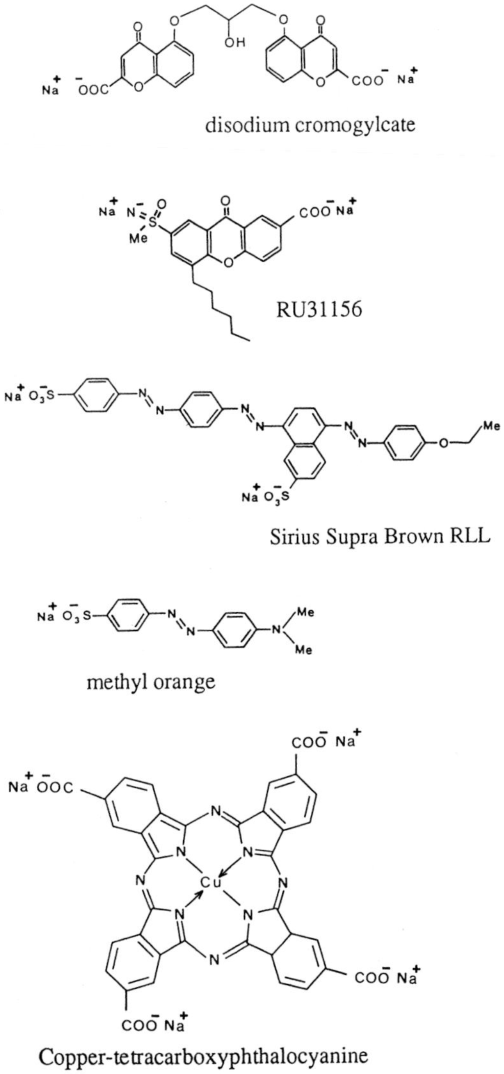

disodium cromogylcate

RU31156

Sirius Supra Brown RLL

methyl orange

Copper-tetracarboxyphthalocyanine

Figure 1. A selection of chromonic molecules: The antiasthmatic drug, disodium cromoglycate (DSCG), the antiallergic drug RU 31 156, the dye, Sirius Supra Brown RLL, the dye, methyl orange, and copper tetracarboxyphthocyanine.

ever, although there may be a threshold concentration before aggregation begins to occur, there is no optimum column length and hence no critical concentration directly analogous to a cmc. A further distinction is the absence of a Krafft temperature. Since the process of mesophase formation does not depend on the presence of flexible alkyl chains, there is no threshold temperature below which mesophases cannot form because the vital flexibility of the molecules has been frozen out.

1.2 The Chromonic N and M Phases

There are two principal chromonic mesophases, which have become known as the N and M phases (Fig. 2) (although there are some rarer additional phases). The designations N and M date back to the early polarizing microscopy studies of the mesophases of the drug disodium cromoglycate [1] (DSCG) [1, 2]. The N phase was so-named because it forms a schlieren texture, similar to that shown by the thermotropic nematics. X-ray diffraction studies indicate that it is a nematic array of columns and that the only regular repeat distance is the stacking repeat of 0.34 nm along each column [3]. The M phase was so called because it can form an optical texture similar to that of the hexagonal H_1 (middle) soap phase. X-ray diffraction investigations show that in this phase the columns are arranged on a hexagonal lattice.

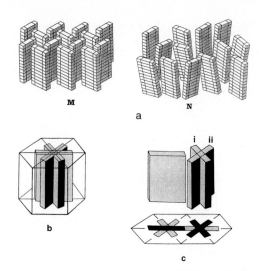

Figure 2. The structure of the chromonic N and M phases: The basic structural unit of both phases is the untilted stack of molecules. The N phase is a nematic array in which these stacks lie in a more or less parallel pattern, but where there is no positional ordering. The M phase is a hexagonal array of these columns. The six-fold symmetry is a result of orientational (but not positional) disorder. A schematic diagram of a localized region, as shown in (a) has only orthorhombic symmetry, but, averaged over the whole structure, each column actually lies in a site with six-fold symmetry (b). The restrictions to the possible orientations of the columns are shown in (c). Because of packing considerations, for any particular orientation of a column, as shown on the left, an adjacent column (right) can take up only two of the three possible orientations (i) and (ii). A representation of the orientationally disordered state of the M phase is given in Fig. 9. Note that the molecular columns are shown here in a highly stylized way. They are not necessarily such simple one-molecule-wide stacks.

1.3 Drug and Dye Systems

A general correspondence between the optical textures, the X-ray diffraction patterns, and the forms of the phase diagrams has been observed for chromonic drugs and many dye/water systems [4–7]. The only significant difference between these two families of mesogens appears to be in some optical textures and rheological properties

---

[1] Note that the widely-used name "cromoglycate" was invented for patent reasons and does not conform to IUPAC nomenclature.

of the M phase (and I suggest that these distinctions are essentially quantitative rather than qualitative). Whilst the drug M phases usually form a well-defined herringbone texture and flow in a homogeneous manner when mechanically disturbed, some dye M phases tend to form a poorly developed grainy texture and flow as distinct, fairly rigid domains. Such behavior suggests a phase tending towards a more gel-like state. However, this need not be regarded as a fundamental distinction and may be due to the difference in dimensions of the two sets of mesogens – the dye molecules being, in general, two or three times as long as those of the chromonic drugs.

1.4   Molecular Structure of Chromonic Species

At one time it was thought that DSCG was a unique mesogen and that the V-shaped structure of this molecule was vital for its mesogenic properties [3]. The subsequent discovery of other mesogenic drugs with simpler structures made it clear that this view was wrong [5]. It appears that the requirements for chromonic behavior are straightforward, i.e., the molecules must have sufficiently large aromatic regions to have a tendency to stack in columns and they must carry solubilizing groups to enable the columns to disperse in water. Just as for amphiphile systems, the solubilizing groups can utilize the ability of water to dissolve either polar or hydrogen-bonding groups (i.e., they can be ionic or nonionic). The majority of chromonic drugs and dyes studied to date are ionic, hence the major emphasis in this account will lie with such systems, but this should not be taken to imply that nonionic chromonic systems (such as aromatic nuclei with peripheral polyethylene-

oxide chains [8]) are likely to prove less important.

A survey of chromic species, especially dyes, makes it clear that there is no need for the molecules to even approximate to circular discs, as might be expected if chromonic systems are regarded merely as lyotropic discotic phases; extended blade-like molecules are capable of forming chromonic mesophases with apparently perfect hexagonal order.

2   The History of Chromonic Systems

Up until the last twenty years, the family of chromonic mesophases lay hidden like an uncharted continent. Occasionally explorers returned to give some partial account, but their reports usually fell on deaf ears, and they were largely unaware of each other's discoveries. A coherent overall picture was remarkably late coming. It is significant that in the 500 pages of the two-volume earlier attempt at a comprehensive survey of liquid crystals by Gray and Winsor, published in 1974, there is no mention whatsoever of the systems we now call chromonic [9].

2.1   The Early History

The story of the three-way connection between drugs, dyes, and nucleic acids begins in the early decades of this century when the optical microscope was still the foremost research tool in mineralogy, biology, and medicine. The use of specific stains for highlighting and identifying biological tissue had become a highly developed art form. It was apparent that stains could be highly selective for particular tissues and for distinguishing between different micro-organisms [10]. Chromosomes were so-named, not because they were naturally colored, but because they could be readily stained. Gram-positive and gram-negative bacteria were distinguished by the way in which they accepted or rejected a particular stain. Slowly, the idea grew that, since some reagents could specifically stain harmful bacteria, perhaps they could specifically kill them also. The concept of "Dr Ehrlich's magic bullet" was born.

Ehrlich and co-workers began a search through the range of artificial dyes then available, hoping to find some with medically useful antibacterial properties [11]. They achieved some success with the dye, trypan red which cured trypanosomiasis in mice and rats. (Tyrpanosomes are parasitic protozoa causing such diseases as sleeping sickness.) This started a large-scale search for other medically effective dyes, leading directly to the first generation of wonder drugs, the sulphanilamides. Note that the British synthetic dye industry was accidentally born in an almost mirror-image situation in 1856 when the young Perkin was attempting to synthesize the antimalarial drug quinine and instead produced the dye aniline purple, better known as mauve [12].

The first report of the mesogenic properties of dyes appears to have come from Sandquist in 1915. He was investigating the properties of 10 chloro- and 10 bromophenanthrene-3 (or 6)-sulphonic acid and he noticed 'nematic threads' in a solution of this material when viewed under polarized light [13]. The Bayer company then produced the antitrypanosomiasis drug Bayer 205, which was examined by Fourneau et al. [14]. Following this, Balaban and King investigated the properties of Bayer 205 and its derivatives in an attempt to discover the mode of action. In their subsequent paper, a short dis-

cussion was included describing the meso-genic properties of the drug and associated derivatives [15]. All the mesogenic compounds were salts of disulphonic acids. Commenting on their optical textures they said: "The intermediate, anisotropic fluid appears to posses a certain amount of rigidity, but always flows when under pressure of a coverslip", and noted "alternations of dark and light patches".

Between 1933 and 1935, a succession of liquid crystalline dye solutions were reported by Gaubert [16]. This was immediately followed by studies carried out independently by Jelley [17] and Schiebe et al. [18] on the aggregation of pseudo-isocyanine dyes in solution. Jelley describes the appearance of solutions of 1:1'diethyl-p-cyanine chloride in terms of: "– a fine thread-like structure of a brilliant greenish yellow colour showing streaming birefringence". He concluded that the mesophase structure had one-dimensional order and that the molecules had either "rotational freedom or random orientation" and that the mesophase was "analogous to the (thermotropic) nematic phase described by other workers". Although his paper does not include a micrograph of this texture, from the description given it is clear that the threads are actual structures (probably M ribbons) rather than lines of disclination within the single phase nematic region.

2.2 Disodium Cromoglycate and Later Studies

After the work of Jelley, no significant work appears to have been done in the field of chromonics for thirty years. In 1967, following the epic (and courageous) investigations of Althounyan [19], Fisons began to market the new antiasthmatic drug disodium cromoglycate (DSCG) under the trade name INTAL. At the instigation of Cox, the McCrone Research institute in London began studies aimed at maximizing the fluidity of the inhalant and improving the dosage formulations. This involved an optical and X-ray diffraction study of the mesophases and the hydrated crystals. Cox et al. noted the great affinity that DSCG had for water and they described two mesophases, the N and M phases observed using optical microscopy, eventually producing the peritectic phase diagram now regarded as characteristic of the simplest type of chromonic system [1]. As a postscript to their paper they remarked upon the 3.4 Å (0.34 nm) spacing found in the nematic phase (but did not identify this with a stacking repeat distance) and noted that the inner reflections in the diffraction pattern of the M phase correspond to spacings in ratios characteristic of a hexagonal phase. In the following year a more detailed optical and X-ray diffraction study by Hartshorne and Woodard was published, proposing models for the N and M phases [2], but there was no immediate follow-up. The DSCG mesophases were quietly relegated to the oddity pile.

The DSCG/water system was reconsidered in 1980 by Lydon [3] and reinvestigated experimentally in 1984 by Attwood and Lydon. New models for the N and M phases were proposed and, when the study was extended, it was found that a similar pattern of mesophase formation occurred for a number of other antiasthmatic and antiallergic drugs [5]. The 7,7′ analog of DSCG was extensively studied by Perahia et al. and found to have closely similar mesogenic properties [20].

There have been more recent reports of other mesogenic drugs, such as nafoxidine chloride (by Mlodozeniec [21]), nafcillin sodium (Bogardus [22]), diethylammonium flufenamate (Eckert and Fischer [23]), and fenprofen sodium (Rades and Muller-Goymann [24]), where the mesophases are apparently lamellar. For all of these compounds, the molecules look distinctly chromonic and some form of lamellar brick-wall structure would appear likely. The distinction between these systems and those of the DSCG type is reinforced by the observation of Kustanovitch et al. [25] who found that diethylammonium flufenate forms a mesophase with positive diamagnetic susceptibility, implying that the molecular planes lie parallel to the director. (This is in contrast to the negative diamagnetic susceptibility found by Goldfarb et al. [26] for DSCG mesophases where the director is parallel to the column axes, i.e., perpendicular to the molecular planes.)

More recently, following a preliminary survey using polarizing microscopy, each of the two azo dyes, Sirius Supra brown and "acid red", was shown to form a single lyotropic mesophase at room temperature [4]. The latter system in particular was studied in some detail and, on the basis of its optical texture, the mesophase formed was characterized as nematic. Moreover, on heating and cooling, a rich variety of optical textures was exhibited and the overall form of the phase diagram appeared to be complex. It was the similarities between the mesogenic properties of these dyes and DSCG that gave rise to the concept of the family of chromonic phases [6].

2.3 The 3-Way Link Between Drugs, Dyes, and Nucleic Acids

In view of the number of studies of dye aggregation and mesophase formation, and the presence in the literature of reports of liquid-crystalline drug solutions, the idea that there was a common pattern of behavior came surprisingly late. Once the point had been made, however, it was natural that studies of mixed drug/dye systems would follow [27–29]. Similarly, the "formal" inclusion of nucleic acids into the chromonic family occurred later than might have been expected considering that there was a growing amount of literature describing nucleic acid mesophase formation in liquid crystal journals [30–33]. The first general review of chromonic mesophases (by Vasilevskaya et al.) was published in 1989 [34].

3   The Forces that Stabilize Chromonic Systems

3.1   Hydrophobic Interactions or Specific Stacking Forces?

A simple view of the state of a chromonic system is that the molecules are effectively insoluble in one dimension. The energy involved in their aggregation is more or less comparable to kT for each aromatic ring and only at extremely low dilution are there individual, unaggregated molecules. For ionic systems, the columns attract counterions, forming electrical double layers. This results in column–column repulsion, which causes the columns to behave as if they were separated by compressed springs. This effect presumably stabilizes the hexagonal structure of the M phase and causes the columns to move apart equally when the mesophase is diluted.

It was natural that early workers on chromonic systems would attempt to explain their properties in terms of the then famil-

iar picture of molecular association and amphiphile mesophases, and there were repeated references to "hydrophobic interactions" and "micelle formation". However, there has been a growing realization that the forces that cause chromonic molecules to stack, although in a sense "hydrophobic" (because water is excluded from the space between stacked molecules), must be radically different.

The concept of hydrophobic "bonding" is a convenient myth (especially useful for physical chemists when talking to biologists). It pictures attractive forces between "oily" groups such as alkyl chains, which cause them to cluster together in the interiors of micelles and membranes. In reality, however, these are not the dominant interactions. The major interactions are those between water molecules. It is so thermodynamically unfavorable for a water molecule not to be in contact with other water molecules that hydrophobic species are excluded from the aqueous part of the phase. They are squeezed out and the process of micelle formation is said to be solvent-driven. Hydrophobicity is the absence of hydrophilicity [35].

Taken to (or perhaps beyond) its logical conclusion, this picture would suggest that there can only be one kind of hydrophobicity and that in mixed systems all hydrophobic molecules should be forced to aggregate together. This does not appear to be the case. Hydrophobic aromatic and aliphatic species do not in general mix or aggregate together in the presence of water. It appears that there are effectively two distinct types of hydrophobicity. Although one case study is hardly definitive, an investigation of a mixed amphiphile and chromonic system gave no evidence that there is any tendency for mixed micelles to form [36]. In addition, it appears that the association of chromonic molecules in solution is exothermic (i.e., enthalpically driven). It must be concluded therefore, that the interactions between parallel π-systems set aromatic systems apart from aliphatic ones [37].

3.2   The Aggregation of Chromonic Molecules in Dilute Solution and on Substrate Surfaces

The formation of conventional amphiphile mesophases is preceded by the aggregation of molecules in solution. Groups of two or three amphiphile molecules associate with reluctance. There is still an unfavorably large fraction of the hydrophobic parts of the molecules exposed to water. This situation continues until the formation of aggregates approaching the size of a micelle (Fig. 3).

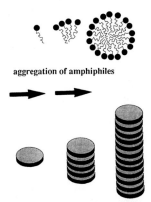

aggregation of amphiphiles

aggregation of chromonic molecules

Figure 3. The contrast between the pattern of aggregation of amphiphiles and chromonic molecules: For amphiphiles, the energetically unfavorable contact between the hydrophobic alkyl chains and water molecules is reduced to more or less zero when a closed micelle is formed. In contrast, for chromonic systems, the fraction of the total molecular surface exposed to the solvent decreases as the columns lengthen, but there is no optimum aggregate size directly comparable to a micelle.

There is then a sudden drop in the free energy as virtually all of the amphiphile molecules are incorporated into micelles, enabling their hydrophobic alkyl chains to be more or less completely shielded from the aqueous part of the phase. This leads to the familiar abrupt change in physical properties at the cmc. In chromonic systems there is also the aggregation of molecules in dilute solution before mesophase formation, but the pattern of association is different. The hydrophobic surfaces of the molecules cause them to aggregate in stacks like packs of cards. As these stacks grow, the fraction of the total hydrophobic surface area exposed to the aqueous part of the phase steadily falls, but there is no minimum free energy state, no cmc, and there is no structure directly comparable to the micelle [38].

There is widespread evidence of the association of dye molecules in solution (and of the tendency of many dyes to be adsorbed on substrates as aligned, structured aggregates rather than as individual molecules). Although the bulk phase may be isotropic, the increasing asymmetry of the aggregates causes flow alignment and there is often an easily detectable flow birefringence. Spectroscopic studies such as those of Stegemeyer and Stökel (on pseudoisocyanine dyes systems) confirm the presence of aggregates in solution above the N–I transition [39].

The changes in the absorption spectra of dye systems with concentration are of major interest to the dye industry (and to the photographic industry, because of the use of adsorbed dyes as photographic sensitizers). The situation is complex, varying from system to system, with changes occurring in both the intensity and the wavelength of maximum absorption. A range of different aggregate structures have been postulated to explain these effects. Three distinct spectroscopic effects have been identified, and repeated attempts have been made to explain these in terms of different patterns of molecular aggregation [40, 41]:

1) A change to shorter wavelength accompanying the formation of dimers (the D band),
2) a bathochromic shift (towards the red), thought to accompany the formation of tilted ("staircases") termed "J (Jelley) [17] aggregates", and
3) a hypsochromic shift (towards the blue) thought to accompany the formation of untilted or less tilted stacks ("ladders" or "H aggregates").

In completely separate investigations, early osmometry studies of the association of the nucleic acid bases and nucleosides in solution indicated that the molecules were also aggregating in columns [42]. Sedimentation equilibrium studies showed that this process is reversible and that there is a near constant free energy increment when each additional molecule is added, irrespective of the existing length of the column. Such behavior was termed "isodesmic". For purine and pyrimidine nucleosides, the association constants indicate weak associations where both ΔH and ΔS are negative (i.e., the aggregation process is enthalpically rather than entropically driven). The standard free

energy change is of the order of the thermal energy, kT (~2.5 kJ per mole).

Compared with the complicated mass of literature concerning the aggregation of dyes, that for chromonic drugs is sparse, but there is a study by Champion and Meeten [43] on the aggregation of DSCG in solution, dating back to the time of the Cox, Woodard, McCrone and Hartshorne investigations and an infrared study of the same system by Hui and Labes [44]. Both of these confirm the growth of chromonic aggregates in solution.

Simple isodesmic behavior (Fig. 4) is expected to be characteristic of chromonic systems, and this appears to be generally true. However, the situation can be more complex. Extensive studies of dye/water systems have shown that there is sometimes a distinct threshold concentration with some resemblance to the amphiphile cmc, where dimers first appear [41].

4   Phase Diagrams

The classic chromonic phase diagram is that of the DSCG/water system. This has a double peritectic form in contrast to the standard multi-eutectic form of conventional

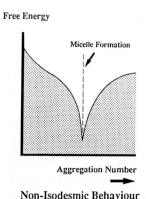

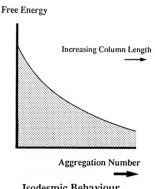

Figure 4. Isodesmic behavior: For isodesmic behavior, the addition of each molecular unit to a stack is accompanied by a constant free energy increment. There is therefore a gradual fall in free energy with column length. This is in contrast with the (nonisodesmic) behavior of amphiphiles, where micelle formation occurs and a particular aggregation state represents a pronounced free energy minimum.

amphiphile system phase diagrams. The distinction between these two patterns of behavior can be explained in terms of the different patterns of molecular aggregation. For conventional amphiphiles (at a particular pressure and temperature), each mesophase has an optimum composition corresponding more or less to the peak in the phase diagram. In the hexagonal H_1 phase, for example, the cylinders have an optimum diameter and there will be an optimum inter-cylinder separation. Higher or lower mesogen content strains the system and lowers the free energy and hence the H_1–isotropic clearing temperature. The picture for chromonic systems is different: as the phase diagram is traversed from left to right, it is not a matter of micelle-building, destruction, and rebuilding in a different manner, there is a progressive increase of order: first the aggregation into columns and then increasing levels of ordering of these columns (as indicated by the progressive rise in the clearing temperatures).

As is usually found for mesophase systems, the boundaries on the phase diagram are not all equally easy to detect. All of the known chromonic mesophases are birefringent and the clearing points can therefore be readily seen by optical microscopy. The crystalline solid–M phase boundary is often harder to detect optically, but it does show up readily in X-ray diffraction studies. Note also that some early chromonic phase diagrams determined by ^{2}H NMR did not show the $N+I \rightarrow M+I$ boundary [45].

At first sight there appears to be a paradoxical feature in the standard chromonic phase diagram (Fig. 5). This concerns the $N+I/M+I$ boundary, where on heating the N phase regions become M, i.e., a less ordered phase changes to a more ordered phase. But note that the two mesophases do not have the same composition, i.e., the M phase is more concentrated than the N. This

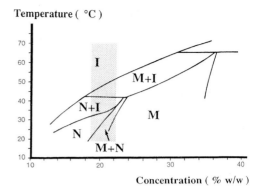

Figure 5. The phase diagram of the DSCG/water system (as determined by hot stage polarizing microscopy). The double peritectic form of this diagram has come to be regarded as the classic chromonic pattern. Note, in particular, the characteristic sequence of phases formed by samples within the composition range indicated by the shaded band.

process is an inevitable consequence of the way in which, as the temperature rises, the phase in equilibrium with the isotropic liquid becomes progressively less hydrated. At low temperatures it is the N phase, at higher temperatures the M, and at still higher temperatures, the crystalline solid.

The double peritectic form of the chromonic phase diagram often produces an interesting feature not found for conventional amphiphile systems. Samples of DSCG for example, with a particularly narrow composition range, pass on heating through the phase sequence $M \rightarrow M+N \rightarrow N \rightarrow N+I \rightarrow M+I \rightarrow I$. Thus one single sample can display all of the one-phase and two-phase liquid crystalline regions of the system.

5 Optical Textures

In optical examinations of chromonic systems, the textures of both the one-phase and the two-phase regions tend to be highly characteristic [46]. The N phase usually assumes a nematic schlieren texture, but homeotropic or homogeneous samples can be prepared by surface alignment. Occasionally an N phase sample will form the striated 'tiger skin' texture (Fig. 6). We interpret this texture in terms of a twisted rope-like arrangement of the columns. It appears that such an arrangement represents the way in which any initial misalignment in the system can be accommodated, whilst at the same time keeping the splay distortion to a minimum [47]. The characteristic M texture is much more grainy in appearance than the N schlieren.

One of the sources of confusion to early workers in the field was the way in which chromonic textures depend on the past history of the sample. A one-phase or two-phase region may have dramatically different textures depending on whether it has been reached by heating or cooling, by con-

centration or by dilution. When the N phase is heated and passes into the two-phase N + I region, more or less circular islands of isotropic phase appear, giving the reticulated texture. For compositions where the N + I becomes M + I on heating, the continuous network develops a grainy texture. Cooling produces different textures. The I–N transition produces isolated birefringent droplets which grow and coalesce to produce the bulk N phase. In more concentrated samples, which undergo the transition from I to M + I, the M phase appears as birefringent ribbons.

M phases grown by allowing an N phase sample to become more concentrated by evaporation can develop the detailed herringbone texture[2] (Fig. 7). This is very similar to the texture adopted by the middle (H_1 and H_2) phases of conventional amphiphiles, and a similar explanation of its origin is proposed [48]. As the M phase develops, the columns are packing closer and growing in length. It is energetically more favorable to add molecules to an existing column than to create another column. This causes mechanical strain to build up and eventually this is relieved by the structure buckling and fracturing, creating a series of "geological" fault lines.

Both N and M phases are uniaxial and optically negative. The N phase is uniaxial because of the nematic ordering (i.e., the random orientations of the columns about their long axes and the random column-to-column separations). The M phase is uniaxial as a consequence of the transverse hexagonal ordering. The negative sign indicates that in both mesophases the planes of the

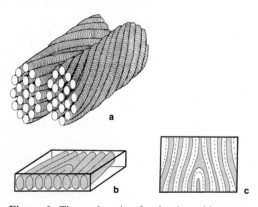

Figure 6. The explanation for the tiger skin texture sometimes adopted by chromonic N (and P) phases: The banded pattern seen with crossed polars is thought to result from a twisted rope-like arrangement of the columns. Such a pattern could be a way of accommodating initial misalignment in the sample with a minimum of splay distortion. (a) The postulated arrangement of molecular stacks in two adjacent twisted rope-like assemblies. (b) The gross structure of the sample showing the large-scale organization of the assemblies represented in (a). (c) The banded appearance of a sample when viewed vertically between crossed polars.

----

[2] Note that the term "herringbone" is used in two different contexts in chromonic systems: at the molecular level, to describe the transverse arrangement of columns in the M phase, and on a much larger scale, to describe the diagonally striped optical texture of the M phase grown by concentration of the N.

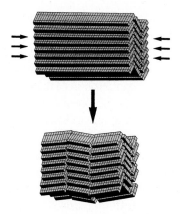

Figure 7. The origin of the herringbone texture of the M phase: This texture develops when the M phase is formed by the concentration of the N phase (by allowing water to evaporate from the edge of the sample). The explanation is the same as that proposed to explain the similar texture of the amphiphile H_1 phase. It is easier to extend a column than to create a new one. This leads to strain in the specimen which is relieved by the creation of multiple fault lines, giving the herringbone appearance.

molecules lie more or less perpendicular to the director (i.e., the column axis). The birefringence due to the molecular orientation would be expected to be reduced to some extent by the form birefringence of the columns.

The way in which an M + I sample formed by heating an M phase has a continuum of N, whereas the N + I phase formed by cooling the N has a continuum of I, appears to indicate that the surface tension at the N/I interface is relatively low.

An early paper, which could have been most influential, was apparently overlooked by most of the liquid crystal fraternity (possibly because of the journal in which it was published) [49]. This described the work of Bernal and Fankuchen on the aqueous mesophases formed by suspensions of tobacco mosaic virus. These virus particles are long cylindrical assemblies [180 Å (18 nm) in diameter and ~3000 Å (300 nm) long] and

there is a striking correspondence between the range of optical textures exhibited by this system and those of chromonic N and M phases (N-schlieren, reticulated N + I, N-droplets in I, herringbone-M, etc.). In retrospect, this appears very reasonable, since the mesogenic units in both cases are long semi-rigid columns, but note that 40 years later, some workers were still talking in terms of smectic patterns of aggregation to explain M-phase optical textures.

6 X-Ray Diffraction Studies

Chromonic phases can in general be oriented by surface alignment or magnetic fields, but when samples are introduced into untreated glass tubes for an X-ray study, there is usually sufficient spontaneous alignment for axial and equatorial reflections to be easily distinguished. (In rare cases a sample in the M phase will form a large scale zig-zag texture like that observed by Bernal and Fankuchen for TMV solutions [49].)

The diffraction patterns of both the N and M phases show fairly sharp 3.4 Å (0.34 nm) axial arcs. This spacing corresponds to the thickness of an aromatic ring and indicates that the molecules lie in untilted columns in both mesophases. (Accordingly, the position of this reflection does not vary with dilution of the sample and tends to be constant from one mesophase to another.) The 3.4 Å reflection is not as sharp as a typical reflection from a crystalline solid (but it is appreciably sharper than the broad stacking repeat reflections of the D_{hD} phase). From the width of this peak, Goldfarb et al. inferred that that the column lengths in the N phase of DSCG range from 80–200 nm [50]. However, it must be admitted that since the peak profile is a function of both the column length and any local irregularities of spacing, these values must reflect both of these parameters. In addition, there are indications from NMR studies that, for the DSCG/water system at least, the assembly of molecules into columns is "complete" in the N phase (i.e., there is no sudden increase in column length at the N → M transition) [44].

The N phase gives a diffuse inner axial reflection corresponding to the continuous range of column-to-column distances. The M phase, in contrast, gives a series of sharp axial reflections corresponding to spacings in the ratio $1:1/\sqrt{3}:1/\sqrt{4}:1/\sqrt{7}:1/\sqrt{9}$, which is typical of a hexagonal lattice (Fig. 8). The hexagonal spacing varies from mesogen to mesogen and increases linearly with the square root of the dilution (indicating that the structure is swelling in two dimensions only and that the column length is effectively infinite).

Note that in the current model for the M phase, although the centers of the columns lie on a hexagonal lattice, the herringbone

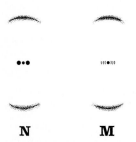

N   **M**

Figure 8. Drawings of the X-ray diffraction patterns of N and M phases: In both diffraction patterns, the axial (vertical) arcs correspond to the 3.4 Å (0.34 nm) stacking repeat distance along the columns. In the N phase patterns, the diffuse low angle equatorial reflections correspond to the range of column–column separations. The sharp low angle equatorial reflections of the M phase correspond to spacings in the ratio $1:1/\sqrt{3}:1/\sqrt{4}:1/\sqrt{7}$ (characteristic of a hexagonal lattice). The inner reflections change with dilution. Detailed studies confirm that the M phase structure swells in two dimensions as more water is added.

array itself has only orthorhombic symmetry. To explain this discrepancy, an analogy had been drawn with the arrangement of molecules within a layer of the smectic B structure. In this case comparison of the molecular dimensions with the observed hexagonal spacing appears to leave little doubt that there is a local herringbone arrangement, and that the higher hexagonal symmetry must therefore result from a pattern of rotational disorder. It is the space-averaged (and to a lesser extent, time-averaged) picture rather than the short range snapshot view, which has hexagonal symmetry. (As a corollary to this picture, there appears to be evidence of an orthorhombic chromonic O phase of certain dyes, well-formed, homeotropic, hexagonally shaped monodomain plates of the M phase grow into butterfly shapes with two-fold symmetry at the M–O transition [46].)

In addition to the outer axial 3.4 Å reflection and the inner axial equatorial hexagonal reflections, M phase diffraction patterns often show diffuse axial and off-axial reflections in the 0.4–2 nm range [51]. These are difficult to interpret. They must relate to some structural feature of the stacks, but they do not generally lie on the layer lines corresponding to multiples of 3.4 Å and hence can not be explained in terms of simple regular rational repeats along the stack.

The history of the structural studies of the DSCG/water mesophase system is unusually complex. In retrospect it can be seen that the interpretation of the early X-ray diffraction results was complicated by three factors:

1) The misleading picture given by the crystal structure [52]. In the crystalline solid the molecules are not flat. They are butterfly-shaped with an angle of 52° between the two chromone rings. It was uncertain whether the molecules had the same conformation in the mesophases or

whether the repeat distance along the stack of molecules in the M phase would be the same as that in the crystalline solid.

2) The apparatus used was a specially constructed low angle powder camera. This gave one-dimensional information only and did not indicate the geometrical relationship between the high angle and low angle reflections.

3) The 3.4 Å reflection was not at first identified with the column repeat distance and the original picture suggested for the N phase was of a nematic discotic array of individual free molecules.

7   The Extended Range of Chromonic Phase Structures

7.1   The P Phase

In some dye/water systems the N–M transition is difficult to detect by optical micros-copy. The textures are very similar and in the two-phase regions, fine details within the textures often extend across the phase boundaries from one phase into the other. When the coverslip is mechanically disturbed, there is no obvious discontinuity in viscosity to distinguish the two phases. This gives the impression that the N/M transition has become so weakly first order that it is virtually second order, and the N and M phases have effectively amalgamated into one. This combined phase has been termed the P phase [47, 53, 54].

A plausible (but as yet unsubstantiated) explanation for the distinction between N/M and P type behavior lies in terms of the effective cross sections of the columns in the two cases (Fig. 9). For conventional N/M behavior, the columns in the M phase are pictured as being blade-like and lying in a herringbone pattern. The transition from N to M therefore involves the loss of both lateral positional order and orientational order. In contrast, the P phase columns are pictured as being more or less circular and the N–M transition need only involve the loss of positional order.

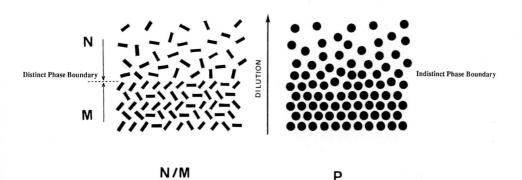

Figure 9. The postulated distinction between N/M and P-type chromonic behavior: This figure shows a schematic view of N/M and P-type phases viewed down the column axes. The distinction is explained in terms of the effective cross-sectional shapes of the columns. In the P system it is thought that the columns are effectively circular (and are able to rotate more or less independently), whereas in N/M systems they are more blade-like (and can only rotate co-operatively).

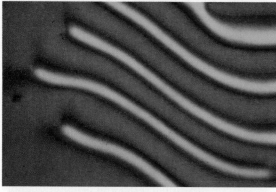

I   The schlieren texture of the N phase. (DSCG/water, ×300, crossed polars + 1 λ red plate.)

II   The tigerskin texture of the N phase. (DSCG/water, ×300, crossed polars + 1 λ red plate.)

III   The typical grainy texture of the M phase. (DSCG/water, ×300, crossed polars + 1 λ red plate.)

IV   The herringbone texture of the M phase obtained by allowing a sample in the N phase to become more concentrated by peripheral evaporation. (DSCG/water, ×300, crossed polars + 1 λ red plate.)

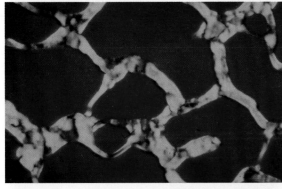

V   The reticulated texture of a two-phase
     N + I sample obtained by heating the N
     phase.
     (DSCG/water, ×300, crossed polars + 1 λ
     red plate.)

VI   The grainy reticulated texture of a two-
     phase M + I sample obtained by heating a
     sample with the reticulated N + I texture.
     (DSCG/water, ×300, crossed polars + 1 λ
     red plate.)

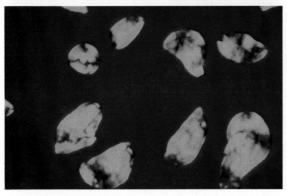

VII The two-phase N + I droplet texture
     formed on cooling an I phase, with drop-
     lets of N in an isotropic continuum.
     (DSCG/water, ×300, crossed polars + 1 λ
     red plate.)

VIII The grainy M droplet texture formed on
     cooling the N + I droplet texture.
     (DSCG/water, ×300, crossed polars + 1 λ
     red plate.)

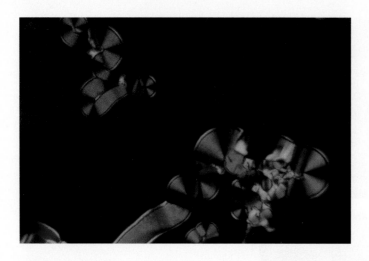

IX   M ribbons formed on cooling, when the isotropic solution passes directly into the M+I two-phase region. (DSCG/water, ×300, crossed polars.)

a

b

X   The N/M boundary and P-phase behavior: (a) An example of a very distinct N/M boundary, (b) an example of P-type behavior, where the N/M boundary has become indistinct and where features within the optical texture of one phase are continued across the phase boundary. (×300, crossed polars (λ red plate))

a

b

c

XI The alignment of intercalated dye molecules in a host N phase. These three plates show a dilute solution of the dye methylene blue in a host N phase of the drug RU 31 156. Because of the dichroism of the dye, the alignment of guest and host species can be observed independently. (a) Viewed between crossed polars – the pattern of light and dark areas results mainly from the orientation of the host phase. (b) Viewed with the polarizer alone – the colors seen result from the dichroism of the guest dye molecules. They appear blue when the electric vector lies along the major axis of the molecules and pale yellow when it is perpendicular. (c) The sample as for (b) but rotated through 90°, showing the alignment of the dichroism of the guest dye molecules.
(×300)

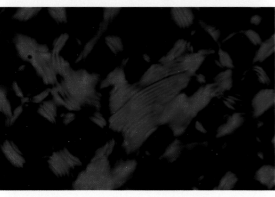

XII The fingerprint texture of a chirally doped chromonic drug N phase. This sample shows the N+I phase of a 5% solution of arginine in a 22% sample of DSCG at room temperature. The separation of the striations corresponds to 1/2 the pitch of the helicoidal structure.
(×300, crossed polars.)

7.2   Chromonic M Gels

The conditions for the formation of gels may not be very different to those required for the formation of chromonic M phases, and there may well be a smooth gradation of phase characteristics between these two states. Two quite separate observations support this notion: firstly, an appreciable number of drugs and dyes with molecular structures very closely related to those of chromonogenic species form viscous gels rather than mesophases, and secondly, it is found that the addition of electrolytes gradually increases the viscosity of the M phase and causes it to become more gel-like. The addition of 1% sodium chloride to the DSCG/water system, for example, whilst not appreciably affecting the schlieren texture of the N phase, inhibits the formation of the M-phase herringbone texture, giving a grainy texture virtually indistinguishable from that found with dye M phases [6] (see also Sec. 8).

7.3   Chiral Chromonic Phases

There appear to be no reports of small molecule chiral chromonic mesogens, but it is easy to convert the nonchiral N phase into a cholesteric phase by adding a small water-soluble chiral solute such as an amino acid. Lee and Labes carried out an extensive study of the relationship between the pitch of the induced chiral phase and the concentration of various dopants [55]. They found that the twist varies linearly with concentration over large concentration ranges, and the "helical twisting power" is characteristic of the particular dopant. This approach offers a novel form of assay for small quantities of water-soluble chiral compounds.

Low pitch chiral phases have the same characteristic optical properties as thermotropic cholesteric phases, i.e., high optical rotatory powers and selective wavelength reflection. High pitch chiral N phases (where the pitch >1 μm) have the characteristic fingerprint texture, where the spacing of the bands corresponds to half of the pitch. Preliminary reports of a blue phase [56] have been retracted [57], but there appears to be no reason why the chromonic analogs of all three thermotropic blue phases should not exist, especially since lyotropic blue phases have been reported for DNA solutions (Leforestier and Livolant [58]) and for a chiral micellar system (Radley [59]).

The effect of adding chiral dopants to dye/water systems has been examined for both small chiral molecules (e.g., tartrates) and helical polymers (poly L-glutamic acid) [37]. The effect is to create a pronounced circular dichroism in the J absorption band. This is interpreted in terms of the assembly of corkscrew columns of molecules (a recurrent theme in the disjointed history of chromonic phases). Both simple helices and more complicated intertwined assemblies have been proposed. The biologically occurring nucleic acids, DNA and RNA, can be regarded as polymerized chromic columns and, because of the chirality of the ribose sugar units, the entire assemblies are chiral and in a particular sample (under defined conditions) there are helices of only one handedness present. At low concentrations, these systems form chiral N (i.e., cholesteric) phases, but at higher concentrations, an apparently untwisted hexagonal M phase is formed [31].

7.4   More Ordered Chromonic Phases

If the analogy between the herringbone structure of the M phase and the molecular arrangement within the layer of a smectic B phase is valid, then more ordered chromonic modifications would be expected analogous to the E and higher smectic phases. There is a growing mass of evidence that this is the case and that other chromonic structures exist at lower temperatures and higher concentrations than the M phase. Goldfarb et al. have reported an additional low temperature smectic-like phase of DSCG (which they provisionally labeled phase III). This phase is difficult to study. It is metastable and it requires the addition of glycerol as an antifreeze to prevent the formation of ice [60]. A subsequent study by Lee and Labes suggested that there is a further high concentration phase to the lower right of the M phase in the phase diagram [61].

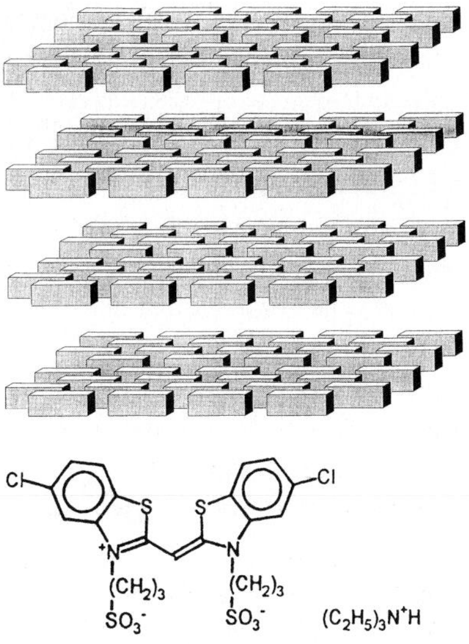

Cyanine Dye.

Figure 10. The brickwork structure proposed for the layered mesophase of a cyanine dye.

7.5   Chromonic Layered Structures

Throughout the history of chromonic mesophase studies there have been sporadic reports of smectic phases. Some of these have usually been modified later and the phase concerned later re-identified as M. However, others, including nafoxidine and the salts of fluhenamic acid mentioned earlier (Sec. 2.2), appear to be genuinely lamellar. A recent study of a cyanine dye/water system by Tiddy et al. produced unequivocal evidence of a layered chromonic phase [53]. Samples with concentrations >1.5% give X-ray diffraction patterns containing a series of inner reflections corresponding to spacings in the ratio $1:1/2:1/3:1/4$ characteristic of a simple one-dimensional lamellar spacing. A layer structure with an overlapping pattern of molecules resembling open lattice brickwork has been proposed (Fig. 10).

7.6   Corkscrew and Hollow Column Structures

A further recurrent theme in the history of chromonic structures is the proposal of complex helical and hollow column structures. An early model for the DSCG columns with a hollow square arrangement was proposed as a way of enabling V-shaped molecules to form stacks with a more or less circular cross section (a feature that at that time was thought to be necessary for the formation of a hexagonal array) [3]. This model was subsequently retracted [5] for a combination of reasons: It was realized that the hexagonal symmetry could result from a disordered herringbone packing of noncircular columns; the X-ray evidence did not support such large columns and the salt bridges proposed did not seem to be a viable way of holding the molecules together in a heavily hydrated environment.

There have been a number of proposals of complex dye columns, including a recent study of the azo dye C. I. acid red 266, where the X-ray diffraction evidence strongly indicates columns with a cross sectional area six times larger than an individual molecule [53, 54]. Furthermore, there is a sharp 0.37 nm axial reflection. This is sufficiently far removed from the aromatic ring thickness of 0.34 nm to rule out a simple untilted columnar structure. A hollow chimney made up from a number of intertwined helical stacks appears to be the most likely structural unit (Fig. 11). Note that since the molecules in this case are non-chiral, a racemic mixture of right- and left-handed columns is probable.

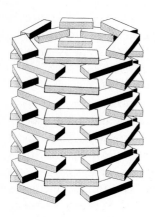

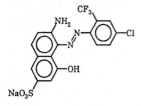

CI Acid Red 266

Figure 11. The hollow brickwork chimney model proposed for the mesophase of the azo dye, acid red 266.

8 The Effect of Additives on Chromonic Systems: Miscibility and Intercalation

From the point of view of their interactions with chromonic host phases, guest dopant molecules can be grouped into three distinct categories:

1) Electrolytes: The state of aggregation and hence the mesogenic properties of chromonic systems are sensitive to the addition of electrolytes. The effects clearly involve more than simply a modification of the ionic strength of the aqueous part of the phase. Both the column structure and the intercolumn forces seem to be affected. In most (but not all) cases, it appears that chromonic stacks are lengthened and stiffened. Systems with strongly associating chromonic molecules, which already form mesophases, tend to be converted to gels or pseudo-crystalline solids [62] and weakly associating species, which would otherwise not be mesogenic, can be induced to aggregate sufficiently to form liquid-crystalline phases.

The addition of NaCl to the DSCG/ water phase diagram has been described by Yu and Saupe [45]. The general effect is to move the phase boundaries upwards and to the left, i.e., a sample that would otherwise have been in the N phase will tend towards the M or M+I. In the dye industry, the final products are usually salted out and therefore tend to contain large quantities of salt. This reduces the solubility of the dye and can turn a respectable mesophase into a curdled gel. As might be expected, if the chromonic species is anionic like DSCG, the cation of the added electrolyte appears to be more critical than the anion. (For example, changing the added anion from Cl^- to Br^- causes only a small movement of the phase boundaries, but changing the cation from Na^+ to K^+ makes an appreciable difference.) Note that some aqueous nucleic acid systems appear to require added electrolyte to form mesophases.

2) Nonionic solutes: Nonionic water-soluble species, such as small molecules like urea and glycerol or hydrophilic polymers such as PEG have an effect different to that of electrolytes. The addition of such compounds tends to leave the qualitative features of the phase diagram unaltered, but causes the boundaries to move in the opposite direction to electrolytes. In the dye industry, the removal of salt is generally not a commercially viable option and urea has traditionally been added as a cheap way of countering its effect.

Nonionic solutes also tend to suppress the freezing point and have been used to make metastable low temperature chromonic phases accessible. For example, Goldfarb et al. used glycerol as an antifreeze in order to obtain the low temperature chromonic phase, III, of the DSCG/ water system [60].

3) Intercalating species: By analogy with the general mutual solubility of lipids and conventional alkyl chain amphiphiles or with the co-miscibility of smectic phases of the same type, it might be expected that the co-miscibility of chromonic species would be widespread. Complete miscibility has been found in only a few cases, but the intercalation [3] of sol-

uble aromatic species in chromonic systems over limited solubility ranges appears to be widespread.

An intercalating soluble aromatic molecule could be regarded as showing potential chromonic character. This view appears to be justified by the way in which intercalation occurs. In general, the guest species do not aggregate together in groups. They appear to be distributed as individual molecules within the columns of the host system. As is usual in liquid crystal systems, there may be stringent constraints for miscibility, but there do not appear to be stringent *geometrical* constraints. And for soluble aromatic molecules to be intercalated is not required an exact match with the dimensions of the host column.

The intercalation of colored dyes into colorless drug mesophases has proved to be a rewarding study [27, 28]. There is generally a metachromic color change (similar to that which occurs when large aggregates of dye molecules are formed). The absorption spectrum of the dye depends on the liquid crystalline state of the host phase, with the dye molecules acting in a loose sense like spin probes and giving information about their surroundings. The mixed system of methylene blue and DSCG is a good example: If solutions containing the same concentration of dye, but increasing concentrations of DSCG are prepared, there is a marked color change from green to blue as the absorption spectrum changes. In addition, the appearance of the sample depends on the optical conditions used and this system provides an interesting example of the use of polarizing optics. Three different situations can be explored. With no polarizer or analyzer, the sample appears homogeneously colored, showing that the dye is distributed evenly throughout the sample. With a polarizer but no analyzer, the blue/yellow

---

[3] The derivation of the word "intercalation" is interesting. It has the same root as the word calendar. Originally there was no geometrical connotation, it referred to the process of realigning an errant calendar by inserting an additional day (Oxford English Dictionary).

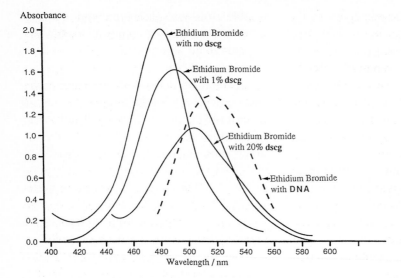

Figure 12. The metachromic color change occurring for an intercalated dye in a colorless host N phase: Absorption spectra for samples containing the same concentration (0.002%) of the DNA marker, ethidium bromide, in different concentrations of a DSCG/water host phase.

dichroism of the dye indicates the local alignment of the guest molecules, whereas with both analyzer and polarizer (in the crossed position) the orientation of the host species can be seen.

One system recently investigated is that of the biological stain, ethidium bromide, dissolved in the DSCG mesophases [63]. This reagent is widely used in biochemistry as a marker for nucleic acids. Stained material (where the stain is intercalated between the nucleic acid bases) shows a strong fluorescence, whereas the reagent on its own does not. In the mixed dye/drug system there is a metachromic color shift and the fluorescence appears (Fig. 12).

A particularly dramatic example of intercalation occurs when DNA is mixed with the square-planar platinum complex, bipyridylethylenediamine platinum(II) nitrate. Molecules of the metal complex intercalate regularly at every third position, giving the regular L (ladder) form of DNA [64].

9   The Biological Roles of Chromonic Phases

Although the mesogenic behavior of some of the most widely used antiasthmatic and antiallergic drugs has been widely studied, the mode of action of these drugs is still unclear and there is little evidence that the mesogenic properties are the direct cause of the remedial properties. (Although it may well be that the molecular properties that give rise to the formation of liquid-crystalline phases are also those that give these drugs their effectiveness.) I suggest that in the future the major biological role of the chromonic systems will be seen with relevance to the properties of nucleic acids.

The liquid crystalline state of the biological membrane (as exemplified in the Singer and Nicolson fluid mosaic model [65]) appears to be vital for its functioning. The membrane is more than just a passive hydrophobic barrier enclosing cell organelles, it must act as a substrate for membrane proteins, be flexible, self-ordering, and self-healing. These are all features of a spontaneously self-ordering system, i.e., a liquid-crystalline phase. An analogous list of the physical properties of DNA and RNA could be drawn up. It appears that the biological functioning of the nucleic acids requires them to be liquid crystalline also. They must be flexible enough to wind and unwind, to separate during DNA replication, and to pass though the ribosome reading head when coding for protein synthesis. (A crystalline solid structure would not have the flexibility and a liquid phase would lack the essential tendency towards ordering; only the liquid crystalline state combines both of these essential properties.)

In this context, DNA can be regarded as a mesogenic side chain polymer, i.e., essentially a chromonic stack held together by the two sugar-phosphate chains (Fig. 13), rather than as simply a helical polymer rod. There is at least some measure of justification for this view. As the early workers attempting to prepare aligned-fiber samples for X-ray diffraction study found, the molecule is structurally labile; the alignment of the paired bases changes with concentration, temperature, and ionic strength. The

familiar, sturdy molecular models of DNA in biology teaching laboratories (invariably of the B form) give a false impression of the rigidity of the actual structure. A more realistic picture of the physical state of the molecule model would be given by a stack of greasy plates tied together with two loose, helically wound ropes.

Just as for membrane lipids, there are two distinct experimental approaches. Firstly, the study of the physical properties of the separated components and secondly, the study of the properties of the biological assembly in the living organism.

The crucial studies justifying the view that the nucleic acids are chromonic can be listed as:

1) Investigations of the aggregation of (unpolymerized) purine and pyrimide bases in solution which show that there is a strong tendency for these molecules to stack into columns. (Note that the degree of aggregation of separate (unpolymerized) purine and pyrimidine bases appears to be, in general, just too weak to give rise to mesophases or gels. The exception is for guanosine (and many of its derivatives, but not guanine itself). The difference arises because of the formation of square, planar rings of hydrogen-bonded tetramers. The structural unit can therefore be regarded as containing four bases rather than one. When viewed in this light, the hydrogen-bonded dimer base pair units produced by Watson-Crick hydrogen bonding at the center of the DNA double helix are just on the edge of mesogenic behavior.)

2) The discovery of the liquid-crystalline phases of RNA and DNA solutions. Whilst preparing materials for X-ray diffraction studies (which eventually led to the Watson–Crick double helix model), Spencer et al. noted that liquid crystal

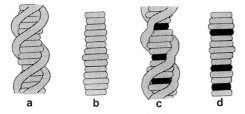

a   b   c   d

Figure 13. The chromonic nature of DNA: (a) A stylized sketch of the B form of DNA showing the central stack of base pairs and the two helical sugar-phosphate chains. (b) A chromonic column of unpolymerized molecules, to be compared with the central stack of bases in DNA. (c) The intercalation of a guest molecule in DNA. Note the untwisting of the sugar-phosphate chain to accommodate the host molecule. (d) The intercalation of a guest molecule in an unpolymerized chromonic column.

phases of t-RNA show banded optical textures [30], implying a low pitch cholesteric phase. Subsequent investigations by Robinson showed that the optical properties were comparable to those of cholesteric solutions of synthetic α-helical polypeptides [31]. Later studies notably by Livolant showed that there are at least two types of nucleic acid mesophase, i.e., a relatively low concentration cholesteric phase and a high concentration hexagonal phase [32]. In these systems, the hexagonal lattice spacing is large enough for the structure to be seen directly by electron microscopy. It is significant that the so-called peptide nucleic acids (where the central stack of bases is the same, but where the usual sugar-phosphate chain has been replaced by a modified protein chain) form a similar range of columnar structures to the conventional nucleic acids [66].

3) The ease of intercalation of other chromonic and potentially chromonic materials. A number of biochemical reagents, such as acridines and ethidium bromide, intercalate readily between the stacked bases in DNA and RNA. Anticancer drugs, such as the square planar platinum complexes, intercalate avidly (Fig. 14) and naturally occurring antibiotics, such as actinomycin, similarly act by intercalation into the stack of bases [67]. They act as tailor-made "spanners in the works" and prevent the reading and replication of DNA.

4) The liquid-crystalline nature of nucleic acids in vivo. In lower organisms, the nucleic acids are naked and exist for much of the time as concentrated droplets of cholesteric phase. In electron micrographs of the nuclei of some dinoflagellates, for example, the characteristic bands of nested arcs, "Bouligand Pat-

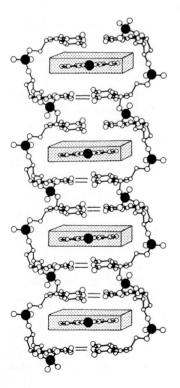

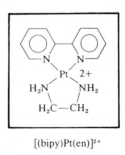

[(bipy)Pt(en)]²⁺

Figure 14. The structure of L-DNA: The planar platinum complex [(bipy)Pt(en)] [21] readily intercalates in DNA. The saturated structure has the L (ladder) form in which the DNA helix is completely unwound and the intercalation appears to occur regularly in every third position (redrawn from [64]).

terns" of a helicoidal structure are visible [68]. For higher organisms, the situation is more complex. Coiled coils of DNA are wound round histone bobbins and there is no bulk mesophase formed.

10   Technological and Commercial Potential of Chromonic Systems

Many compounds of commercial interest, e.g., dyes, drugs, antibiotics, and anticancer agents, are either chromonic or have sufficiently chromonic character to be able to intercalate into chromonic systems. However, there is a distinction between commercially useful compounds, which incidentally happen to form liquid-crystalline phases, and compounds that have technological value specifically because of their mesogenic properties.

We have hardly scratched the surface of the technological potential of chromonic systems. An occasional dye or food coloring is marketed in the form of a convenient N phase, and the drug companies know that concentrated solutions of many drugs form highly viscous M phases, but these cases are trivial compared with the potential applications. To recognize that the dye/water system can be liquid-crystalline opens the door to a wide range of possible forms of exploitation. All of the necessary ingredients appear to be present for the production of chromonic information display and storage devices, i.e., surface alignment, guest/host alignment, orientation by electric or magnetic fields, birefringence, and dichroism. Use could be made of the conflict between the surface alignment of aggregates and the alignment caused by externally applied electric or magnetic fields. There is, for ex-

ample, high security printing, where it may be possible to control not only the position of the printed image but also the orientation and state of aggregation of the dyes. Although it does not seem likely that lyotropic display devices will replace those of mainstream thermotropic technology, it would be surprising if there were not situations where chromonics had something unique to offer. As printing techniques become more sophisticated, novel processes perhaps incorporating features of both jet printers and photocopiers could utilize the mesogenic nature of dye solutions. Finally, there are indications that viable light-harvesting devices can be constructed from mixed chromonic systems involving dyes and transition metal complexes [69].

11   Conclusions

When the century of diverse literature dealing with the solution chemistry of drugs, dyes, and nucleic acids is surveyed, the way in which the same concepts recur independently, time after time is amazing – the same problems reconciling molecular structure to mesogenic properties, the same pictures of molecular aggregation, i.e., stacks of cards, piles or coins, helical aggregates, the same puzzling optical textures, "limpid fluids possessed of a sheen", "worm-like growths", and the same observations of flow birefringence, i.e., the same observations about the effects of added electrolyte.

The gathering together of drug, dye, and nucleic acid systems under the heading of chromonics has been more than just a taxonomic exercise. It has enabled observations from one area of study to throw light on another. The term "isodesmic", for example, is not generally used in the dye industry, but it neatly describes a pattern of behavior that has no other name and will, I am sure, become more widely used in future. (There was, of course, no need for such a term in the days when conventional aliphatic amphiphiles were the only recognised type of lyotropic mesogen and where only one pattern of aggregation was known.)

The identification of chromonic systems has led to a reappraisal of the meaning of the term hydrophobic, and it appears that there are two distinct forms of hydrophobic behavior, i.e., that shown by conventional amphiphiles with largely saturated alkyl chains and that shown by aromatic species where $\pi-\pi$ interactions occur.

Concerning the term chromonic, this word was coined to replace phrases such as "lyotropic mesophases of the types formed by soluble aromatic mesogens". It was intended to echo the bis-chromone structure of DSCG and to carry connotations of dyes and chromosomes (and hence nucleic acids). It was also intended to mark Hartshorne's studies of DSCG (which although by no means the first such mesophase system to be explored, was one of the most extensively examined), and it is to Norman Hartshorne, who taught me and many others a love of optical microscopy, that I respectfully dedicate this chapter.

Acknowledgements

I am grateful to Zeneca (formerly ICI Dyestuffs, Blackley, U. K.) and to Fisons Ltd., Holmes Chapel, U. K.) for their continued interest and support, and to my friend Gordon Tiddy for his encouragement. I gratefully acknowledge advice from Dr. Richard Bushby and Dr. John Griffiths (Department of Chemistry and Colour Chemistry respectively, University of Leeds) with the preparation of this chapter.

I have been privileged to have had a succession of very able and enthusiastic co-workers. I am indebted to Teresa Attwood (who was there at the beginning) to Jane Turner and Paul Alder, to Julie Cox, and finally to Kevin Mundy.

12   References

[1] J. S. G. Cox, G. D. Woodard, W. C. McCrone, *J. Pharm. Sci.* **1971**, *60*, 1458–1465.
[2] N. H. Hartshorne, G. D. Woodard, *Mol. Cryst. Liq. Cryst.* **1973**, *23*, 343–368.
[3] J. E. Lydon, *Mol. Cryst. Liq. Cryst. Lett.* **1980**, *64*, 19–24.
[4] F. Jones, D. R. Kent, *Dyes Pigm.* **1980**, *1*, 39–48.
[5] T. K. Attwood, J. E. Lydon, *Mol. Cryst. Liq. Cryst.* **1984**, *108*, 349–357.
[6] T. K. Attwood, J. E. Lydon, F. Jones, *Liq. Cryst.* **1986**, *1*, 499–507.
[7] J. E. Turner, J. E. Lydon, *Mol. Cryst. Liq. Cryst. Lett.* **1988**, *5*, 93–99.
[8] N. Boden, R. J. Bushby, C. Hardy, *J. Phys. Lett.* **1985**, *46*, L325–328; N. Boden, R. J. Bushby, C. Hardy, F. Sixl, *Chem. Phys. Lett.* **1986**, *123* (5), 359–364; N. Boden, R. J. Bushby, L. Ferris, C. Hardy, F. Sixl, *Liq. Cryst.* **1986**, *1 (2)*, 109–125.
[9] G. W. Gray, P. A. Winsor, *Liquid Crystals and Plastic Crystals*, Ellis Horwood, Chichester **1974**.
[10] See, for example, H. J. Conn, *Biological Stains*, Biotech. Publications, Geneva, N. Y., U. S. A. **1953**, or E. Gurr, *Encyclopaedia of Microscopic Stains*, Leonard H ill (Books), London **1960**.
[11] For a short review of the work of Ehrlich with detailed references see: A. S. Travis, *The Biochemist* **1991**, *13 (5)*, 9.
[12] A. S. Travis, *The Rainbow Makers – The Origin of the Synthetic Dyestuffs Industry in Western Europe*, Lehigh University Press, Bethlehem **1993**; A. S. Travis, *Text. Chem. Color.* **1988**, *20 (8)*, 13–18; A. S. Travis, *Text. Chem. Color.* **1990**, *22 (12)*, 18–20; O. Meth-Cohn, A. S. Travis, *Chem. Britain* **1995**, *31*, 547–549.
[13] H. Sandquist, *Berichte* **1915**, *48*, 2054–2055.
[14] F. Fourneau, J. Tréfouel, (Mme) J. Tréfouel, I. Valléc, *Ann. Inst. Pasteur* **1924**, *38*, 81.
[15] I. F. Balaban, H. King, *J. Chem. Soc.* **1927**, *127*, 3068.
[16] P. Gaubert, *C. R.* **1933**, *197*, 1436–1438; **1934**, *198*, 951–953; **1935**, *200*, 679–680.
[17] E. E. Jelley, *Nature* **1936**, *138*, 1009–1010; E. E. Jelley, *Nature* **1937**, *139*, 631–632.
[18] G. Scheibe, L. Kandler, H. Ecker, *Naturwissenschaften* **1937**, *25*, 75; G. Scheibe, *Koll. Z.* **1938**, *82*, 1; G. Scheibe, *Angew. Chem.* **1938**, *52*, 631–637.
[19] R. E. C. Altounyan, *Clin. Allergy* **1980**, *10*, 481; J. Pepys, A. W. Frankland, *Disodium Cromglycate in Allergic Airways Disease*, Butterworths, London **1970**.
[20] D. Perahia, Z. Luz, E. J. Wachtel, H. Zimmermann, *Liq. Cryst.* **1987**, *2*, 473–489.
[21] A. R. Mlodozeniec, *J. Soc. Cosm. Chem.* **1978**, *29*, 659–683.

[22] J. B. Bogardus, *J. Pharm. Sci.* **1982**, *71*, 105–109.

[23] T. Eckert, W. Fischer, *Colloid Polym. Sci.* **1991**, *259*, 553 and *260*, 880–887.

[24] T. Rades, C. C. Muller-Goymann, *Pharm. Pharmacol. Lett.* **1992**, *2*, 131–134.

[25] I. Kustanovitch, R. Poupko, H. Zimmermann, Z. Luz, M. M. Labes, *J. Am. Chem. Soc.* **1984**, *107*, 3494–3501.

[26] D. Goldfarb, M. E. Mosley, M. M. Labes, Z. Luz, *Mol. Cryst. Liq. Cryst.* **1982**, *89*, 119–135.

[27] J. A. Cox, Ph. D. Thesis, University of Leeds **1992**.

[28] M. Kobayashi, A. Sasagawa, *Mol. Cryst. Liq. Cryst.* **1993**, *225*, 293–301; M. Kobayashi, H. Matsumoto, T. Hoshi, I. Ono, J. Okubo, *Mol. Cryst. Liq. Cryst.* **1990**, *180B*, 253–263.

[29] K. Mundy, J. C. Sleep, J. E. Lydon, *Liq. Cryst.* **1995**, *19*, 107–112.

[30] M. Spencer, W. Fuller, M. H. F. Wilkins, G. L. Brown, *Nature* **1962**, *194*, 1014–1020.

[31] C. Robinson, *Tetrahedron* **1961**, *13*, 219–234.

[32] F. Livolant, *J. Mol. Biol.* **1991**, *218*, 165–181; F. Livolant, A. M. Levelut, J. Doucet, J. O. P. Benoit, *Nature* **1989**, *339*, 724–726.

[33] S. Bonazzi, M. De Morais, A. Garbesi, G. Gottarelli, P. Mariani, G. P. Spada, *Liq. Cryst.* **1991**, *10*, 495–506; A. Garbesi, G. Gottarelli, P. Mariani, G. P. Spada, *Pure Appl. Chem.* **1993**, *65*, 641–646; G. Gottarelli, G. P. Spada, P. Mariani, M. M. Demorais, *Chirality* **1991**, *3*, 227–232; P. Mariani, M. M. Demorais, G. Gottarelli, G. P. Spada, H. Delacroix, L. Tondelli, *Liq. Cryst.* **1993**, *15*, 757–778. See also: Y. M. Yevdokimov, S. G. Skuridin, V. I. Salanov, *Liq. Cryst.* **1988**, *11*, 1443–1459, and references therein.

[34] A. S. Vasilevskaya, E. V. Generalova, A. S. Sonin, *Russian Chemical Reviews* **1989**, *58 (9)*, 904–916; translated from *Usp. Khim.* **1989**, *57*, 1575–1596.

[35] See, for example, C. Tanford, *The Hydrophobic Effect*, Wiley, New York **1973**.

[36] T. K. Attwood, J. E. Lydon, C. Hall, G. J. Tiddy, *Liq. Cryst.* **1990**, *7*, 657–668.

[37] C. A. Hunter, J. K. M. Sanders, *J. Am. Chem. Soc.* **1990**, *112*, 5525–5534.

[38] D. F. Evans, H. Wennerstrom, *The Colloidal Domain*, VCH, New York **1994**, Chap. 4.

[39] H. Stegemeyer, F. Stökel, *Ber. Bunsenges. Phys. Chem.* **1966**, *100*, 9–14.

[40] E. G. McRae, M. Kasha, *J. Chem. Phys.* **1958**, *28*, 721–722.

[41] See, for example, N. Tyutyulkov, J. Fabian, A. Melhorn, F. Dietz, A. Tadjer, *Polymethine Dyes: Structure and Properties*, St. Kliment Ohridski University Press, Sofia **1996**.

[42] W. Saenger, *Principles of Nucleic Acid Structure*, Springer, Berlin **1984**, Sec. 6.6 and references therein, including: P. O. P. Ts'o in *Basic Principles in Nucleic Acid Chemistry* (Ed.: P. O. P. Ts'o), **1974**, *1*, 453; P. O. P. Ts'o, I. S. Melvin, A. C. Olson, *J. Am. Chem. Soc.* **1963**, *85*, 1289–1296.

[43] J. V. Champion, G. H. Meeten, *J. Pharm. Sci.* **1973**, *62 (10)*, 1589–1595.

[44] Y. W. Hui, M. M. Labes, *J. Phys. Chem.* **1986**, *90*, 4064–4067.

[45] L. J. Yu, A. Saupe, *Mol. Cryst. Liq. Cryst.* **1982**, *91*, 53–55.

[46] N. H. Hartshorne, *The Microscopy of Liquid Crystals*, Microscope Publications, London **1974**.

[47] J. E. Turner, Ph. D. Thesis, University of Leeds **1988**.

[48] J. Rogers, P. A. Winsor, *J. Coll. Interf. Sci.* **1969**, *30*, 500–510.

[49] J. D. Bernal, L. Fankuchen, *J. Gen Physiol.* **1941**, *25*, 111–146.

[50] D. Goldfarb, Z. Luz, N. Spielberg, H. Zimmermann, *Mol. Cryst. Liq. Cryst.* **1984**, *126*, 225–246.

[51] T. K. Attwood, Ph. D. Thesis, Universit of Leeds **1984**.

[52] S. Hamodrakas, A. J. Geddes, B. Sheldrick, *J. Pharm. Pharmacol.* **1974**, *26*, 54–56.

[53] G. J. T. Tiddy, D. L. Mateer, A. P. Ormerod, W. J. Harrison, D. J. Edwards, *Langmuir* **1995**, *11 (2)*, 390–393.

[54] W. J. Harrison, D. L. Mateer, G. J. T. Tiddy, *J. Phys. Chem.* **1996**, *100*, 2310–2321.

[55] H. Lee, M. M. Labes, *Mol. Cryst. Liq. Cryst.* **1982**, *84*, 137–157.

[56] H. Lee, M. M. Labes, *Mol. Cryst. Liq. Cryst. Lett.* **1983**, *82*, 355–359.

[57] M. Kuszma, H. Lee, M. M. Labes, *Mol. Cryst. Liq. Cryst. Lett.* **1983**, *92*, 81–85.

[58] A. Leforestier, F. Livolant, *Liq. Cryst.* **1994**, *17*, 651–658.

[59] K. Radley, *Liq. Cryst.* **1995**, *18*, 151–155.

[60] D. Goldfarb, M. M. Labes, Z. Luz, R. Poupko, *Mol. Cryst. Liq. Cryst.* **1982**, *87*, 259–279.

[61] H. Lee, M. M. Labes, *Mol. Cryst. Liq. Cryst.* **1983**, *91*, 53–58.

[62] D. E. Sadler, M. D. Sannon, P. Tollin, D. W. Young, M. Edmonson, P. Rainsford, *Liq. Cryst.* **1986**, *1*, 509–520.

[63] K. Mundy, J. C. Sleep, J. E. Lydon, *Liq. Cryst.* **1995**, *19*, 107–112.

[64] S. Arnott, P. J. Bond, R. Chandrasekaran, *Nature* **1980**, *287*, 561–563.

[65] S. J. Singer, G. L. Nicolson, *Science* **1972**, *175*, 720–731.

[66] P. Wittung, P. E. Nielsen, O. Buchardt, M. Egholm, B. Norden, *Nature* **1994**, *368*, 561–563.

[67] See, for example: W. Saenger, *Principles of Nucleic Acid Structure*, Springer, Berlin **1984**, Chaps. 8 and 16 and the references therein.

[68] Y. Bouligand, M. O. Soyer, S. Puiseux-Dao, *Chromosoma* **1968**, *24*, 251–287.

[69] D. Gustand, T. A. Moore, *Science* **1989**, *244*, 35–41.

Index